8th International Conference on Wind Turbine Noise (WTN 2019)

Lisbon, Portugal
12-14 June 2019

ISBN: 978-1-5108-8970-5

Printed from e-media with permission by:

Curran Associates, Inc.
57 Morehouse Lane
Red Hook, NY 12571

CURRAN ASSOCIATES INC.
proceedings.com

Some format issues inherent in the e-media version may also appear in this print version.

Printed by Curran Associates, Inc. (2019)

Institute of Noise Control Engineering/Europe (INCE/Europe)
Riverside House
4 Oakland Vale
New Brighton, Merseyside CH45 1LQ
United Kingdom

Phone: +44 (0)151 638 0181
Fax: +44 (0)151 639 5212

Mail@inceeurope.org

Additional copies of this publication are available from:

Curran Associates, Inc.
57 Morehouse Lane
Red Hook, NY 12571 USA
Phone: 845-758-0400
Fax: 845-758-2633
Email: curran@proceedings.com
Web: www.proceedings.com

8th International Conference on Wind Turbine Noise

Day	Start	Session	Subject	Chair/Moderator
Tuesday	17:00		Reception and Registration to 20:00	
Wednesday EUROPA	09:00		Opening of Conference - Jorge Patricio, President of European Acoustics Association: President of Portuguese Acoustical Society	
	09:10	A	A - Tones and Vibration	Jorge Patricio
	11:20	B	B - Low Frequency and Amplitude Modulation	Payam Ashtiani
	14:00	C	C - Propagation - 1	Kristina Conrady
	16:00	D	D - Propagation 2	Sylvia Broneske
	17:20	E	E - Forum on Propagation	David Ecotiere
Thursday EUROPA	08:30	F	F - Assessment and Compliance	Matthew Cand
	11:00	G	G - Background Noise and Measurement	Pierre Dutilleux
	14:00	H	H - Impact on People - 1	Cecilia Rocha
	16:00	I	I - Impact on People - 2	TBC
	17:00	J	J - FORUM on Impact on People	David Colby
Thursday BRUXELAS	14:00	K	K - Source Prediction - 1	Thomas Carolus
	16:00	L	L - Source Prediction - 2	Wouter van der Velden
	17:40	M	M - FORUM on Source Prediction	Franck Bertagnolio
Thursday	20:00		Conference Dinner - Restaurant Monte Mar	
Friday EUROPA	08:30	N	N - Aerofoil Design	Kathrin Stahl
	10:30	O	O - Trailing Edge Noise Mitigation	Cordula Hornung
	12:10	P	P - FORUM on "Future Design of Low Noise Wind Turbines"	Michaela Herr
	13:50	Q	Q - Regulations	Frits van den Berg
	15:35	R	R - FORUM - "WHO Guidelines for Europe" - Principal Speaker David Michaud	Andy McKenzie
	16:25	S	S - FORUM - "Where Next"	Sabine von Hunerbein
	17:00		Conference Ends	

8th International Conference on Wind Turbine Noise

8th International Conference on Wind Turbine Noise

Thursday 13th June 2019 - EUROPA			
08:30	**F - Assessment and Compliance**	**Matthew Cand**	
08:30	Measurement of wind turbine noise characteristics in receptor position - A new IEC Technical Specification	Broneske	188
08:50	Danish experiences with measuring wind turbine noise at neighbor dwellings.	Egedal	200
09:10	Atmospheric absorption in sound power measurements of wind turbines	Hansen	215
09:30	France - Germany: A Comparison of the Acoustic Assessment Procedures	Dutilleux	231
09:50	Noise Sorce Location in Wind Turbines	Alloza	244
10:10	Discussion		
10:30	Break		
11:00	**G - Background Noise and Measurement**	**Pierre Dutilleux**	
11:00	A low wind noise microphone for wind turbine noise	von Hunerbein	254
11:20	Measuring infrasound from wind turbine: the benefits of a wind shielding dome	D'Amico	259
11:40	Is it possible to predict background noise levels from measured meteorological data with machine learning techniques?	Bigot	265
12:00	Does Background Noise Vary with Seasons?	PETIT	273
12:20	A study of background noise levels measured during far-field receptor testing of wind turbine facilities	Halstead	286
12:40	Discussion		
13:00	Lunch		
14:00	**H - Impact on People - 1**	**Cecilia Rocha**	
14:00	Wind Turbines and Groundwater Contamination: An Analysis	Colby	298
14:20	Human Health Hazard - The Shirley Wind Story	Old	303
14:40	Self-reported health in the vicinity of five wind power areas in Finland	Turunen	322
15:00	Testing the human response to wind turbine emissions	Nelson (Lueker)	N/A
15:20	Discussion		
15:30	Break		
16:00	**I - Impact on People - 2**	**TBC**	
16:00	PIBE : a new French project for predicting the impact of wind turbine noise	Ecotiere	341
16:20	I can still hear it and it's making me ill	Leventhall	347
16:40	Discussion		
16:50	Short Break		
17:00	**J - FORUM on Impact on People**	**David Colby**	
18:00	Close		

Thursday 13th June 2019 - BRUXELAS			
	Plenary in EUROPA until Lunch		
14:00	**K - Source Prediction - 1**	**Thomas Carolus**	
14:00	A Brief Study on Noise Propagation of Airfoils from Wind Turbines Using the Lattice Boltzmann Method	Rodriguez	365
14:20	Towards digital wind turbine noise certification	van der Velden	386
14:40	The Quasi-3D TE Rotor Noise Prediction Tool of the PNoise Code	Saab	399
15:00	Analyses of a high fidelity aero-servo-elastic process chain to assess low-frequency emissions from wind turbines	Wenz	416
15:20	Discussion		
15:30	Break		
16:00	**L - Source Prediction - 2**	**Wouter van der Velden**	
16:00	Simulation of sound radiation of wind turbines using large-scale finite element models	Graetsch	429
16:20	Airfoil LE Noise prediction supplement for PNoise Code	Martuscelli Faria	442
16:40	Numerical study of aerodynamic radiated noise of a Coflow-Jet Vertical Axis Wind Turbine	Radulescu	455
17:00	A Wind Turbine Noise Code Benchmark - Round 1	Bertagnolio	466
17:20	Discussion		
17:30	Short Break		
17:40	**M - FORUM on Source Prediction**	**Franck Bertagnolio**	
18:40	Close		

20:00	Conference Dinner - Restaurant Monte Mar	

8th International Conference on Wind Turbine Noise

Friday 5th May 2017 - EUROPA			
08:30	**N - Aerofoil Design**	**Kathrin Stahl**	
08:30	Aeroacoustic Optimization of a Small Wind Turbine - Methodology and Experimental Validation	Volkmer	480
08:50	Scales of turbulence on a wind turbine leading edge	Serre	503
09:10	Aeroacoustic Assessment of Wind Turbine Blade Tips	Herr	519
09:30	Development of design guidelines for low noise but high yield wind turbines	Hornung	540
09:50	Discussion		
10:00	Break		
10:30	**O - Trailing Edge Noise Mitigation**	**Cordula Hornung**	
10:30	Numerical Investigation of Porous Trailing Edge Noise Reduction Mechanism using lattice-Boltzmann Method	Manegar	552
10:50	Experimental Investigation of Self-Aligning Trailing Edge Serrations for Airfoil Noise Reduction	Stahl	570
11:10	Measurement of sound efficiency of trailing edge serrations (TES) on wind turbines in the Jura mountains.	Falourd	583
11:30	Hydrodynamic Analysis of Trailing Edge Serrations with Blunt and Rounded Edges	Celik	598
11:50	Discussion		
12:00	Short Break		
12:10	**P - FORUM on "Future Design of Low Noise Wind Turbines"**	**Michaela Herr**	
13:00	Lunch		
13:50	**Q - Regulations**	**Frits van den Berg**	
13:50	Current regulations for the protection against noise from wind turbines in Germany	Bauerdorff	616
14:10	Applying a general noise regulation on Wind Turbines: an administrative interpretation to a specific noise source	Parra	624
14:30	WHO Environmental Noise Guidelines for the European Region conditional recommendation for wind turbine noise in the context of Australian regulations	Delaire	633
14:50	Discussion		
15:05	Break		
15:35	**R - FORUM - "WHO Guidelines for Europe" - Principal Speaker David Michaud**	**Andy McKenzie**	
16:25	**S - FORUM - "Where Next"**	**Sabine von Hunerbein**	
17:00	Conference Ends		

INCE*EUROPE*

8th International Conference
on
Wind Turbine Noise
Lisbon – 12th to 14th June 2019

Session A

Tones and Vibration

Organised by INCE-Europe
mail@inceeurope.org
www.inceeurope.org

Supported by Portuguese Acoustical Society

Sociedade
Portuguesa
de Acústica

INCE EUROPE

8th International Conference on Wind Turbine Noise Lisbon – 12th to 14th June 2019

A comparison of standardized methods for prominence analysis of tonal components

Tomohiro Kobayashi, Kobayasi Institute Physical Research: kobayashi@kobayasi-riken.or.jp
Sakae Yokoyama, Kobayasi Institute Physical Research: sakae@kobayasi-riken.or.jp

Summary

Tonal components contained in wind turbine noise are apt to increase psycho-acoustical annoyance. Reduction techniques of tonal noise emission caused by mechanical system of wind turbine generators are well developed in recent years, however, not small amount of wind turbines which emitting tonal noise are still working through the country and its tonal components are often observed around residential areas. Therefore, the method of objectively assessing the psycho-acoustical effect caused by tonal components should be established for the wind turbine noise problem. There are some international standards to determine the prominence of tonal components contained in various type of noise, such as IEC 61400-11 (2012), ISO 1996-2 Annex C (2007), ISO 1996-2 Annex K (2017), ISO/PAS 20065 (2016), DIN 45681 (2005), ECMA-74 15th Edition (2018) and so on. In this paper, principles of these analysis methods are compared and problems to be solved are discussed in order to apply to penalty assessment in immission area. Relationships among metrics determined by these analysis methods are also investigated by using artificially synthesized noise and wind turbine noise actually measured.

1. Introduction

Since tonal components contained in wind turbine noise are apt to increase psycho-acoustical annoyance, authors have been investigated the actual conditions of tonal components contained in wind turbine noise by filed measurements and psycho-acoustical annoyance to tonal components by subjective experiments [1, 2]. Tonal noise resulting from mechanical excitation could be reduced by damping or isolating the vibration at the source [3], however, a certain amount of wind turbine which emitting tonal noise is already exists and the impact of annoyance caused by its tonal noise to people living around the wind turbine cannot be neglected. As an international movement, IEC 61400-11-2 entitled "Measurement of wind turbine noise characteristics in receptor position" is currently under development and is expected to include an analysis method of tonal components. Meanwhile, in Japan, tonal components generated by not only wind power generating facilities but also water heating system based on electric heat pump also have been become a problem, some researches are currently being carried out [4].

Although there are some international standards specifying an analysis method of tonal components contained in environmental noises, there are few papers comparing among these analysis methods despite such a circumstance. This paper organizes concepts of analysis principles for various analysis methods of tonal components specified in international standards, and advantages and disadvantages of each method are described. And then, relationships of calculated results obtained by applying these methods to artificially synthesized noise and actually measured wind turbine noise were investigated.

2. Standardized methods for prominence analysis

There are several international standards as a method of determining tonality or prominence of tonal components contained in noise. Among them, following methods may be applicable to environmental noises and it could be categorized into three kinds of frequency analysis methods.

- FFT based method
  - ✓ ISO 1996-2 Annex C (2007) (hereinafter referred to as ISO method)
  - ✓ IEC 61400-11 (2012) (hereinafter referred to as IEC method)
  - ✓ DIN 45681 (2005) (hereinafter referred to as DIN method)
  - ✓ ISO/PAS 20065 (2016) (hereinafter referred to as PAS method)
- 1/3 octave band based method
  - ✓ ISO 1996-2 Annex K (2017) (hereinafter referred to as ISO survey method)
  - ✓ NZS 6808 (2010) (hereinafter referred to as NZS survey method)
- Loudness based method
  - ✓ ECMA-74 15th Edition (2018) (hereinafter referred to as Psychoacoustic tonality)

Algorithms were implemented into computer program by authors in order to investigate the advantages and disadvantages of those analysis methods. The results are described below.

2.1 FFT based method

Each FFT based method is based on the concept of the masking index described in reference [5]. The masking index was obtained from a psychoacoustic experiment on level difference between a UEN (uniform excitation noise) and a sinusoidal tone when a sinusoidal tone become just-audible, in which UEN is a broadband noise with the same intensity in each critical band. The masking index is shown in Fig. 1.

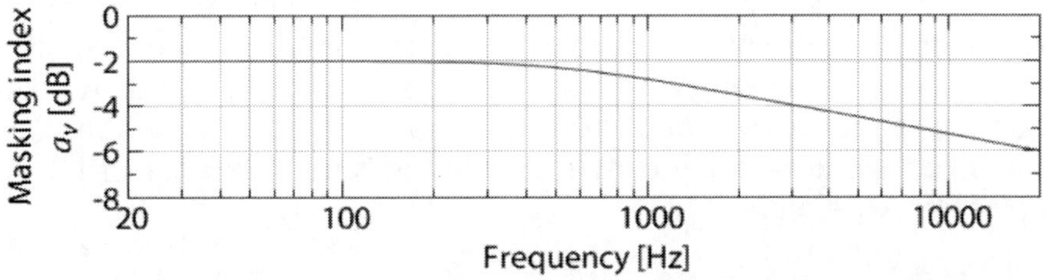

Fig.1 Masking index as a function of a frequency of tonal components

In each standard, the masking index is used to estimate how much the tonal components contained in a noise protrude from just-audible sound pressure level. That is the Tonal Audibility (hereafter referred to as TA) and is calculated by the following formula in any of the standards.

$$\Delta L = L_{\mathrm{T}} - L_{\mathrm{G}} - a_v \tag{1}$$

where, ΔL is the tonal audibility [dB], L_T is the total power of tonal components contained in a critical band [dB], L_G is the total power of masking noise contained in a critical band centred on a frequency of a tonal component [dB] and a_v is the masking index at a frequency of a tonal component [dB].

As apparent from formula (1), the tonal audibility is calculated from the relative level difference between the tonal component and the masking noise. That is, the absolute sound pressure level does not affect the calculated value.

In order to determine the tonal audibility, it is needed to calculate right-hand members of formula (1) according to the following procedure:

- ✓ Identify possible tonal components contained in noise
- ✓ Calculate critical band width centred on the frequency of tonal component
- ✓ Extract tonal components and masking noise by classifying FFT lines in a critical band
- ✓ Calculate the power ratio of tonal components to masking noise within a critical band

Differences between each analysis method are summarized in Table 1. As for the FFT based method, it can be seen from Table 1 that technical specifications such as usable frequency range, evaluation time duration, classification method of the masking noise and expression of the critical band width are slightly different. These technical differences among analysis methods lead to differences in the calculation results of the tonal audibility, which depends on identification accuracy of the tonal component and estimation accuracy of the masking noise level. Since the difference between PAS method and DIN method is substantially only the lower frequency of the usable frequency range, DIN method will not be dealt with individually in this paper.

Table 2 shows overviews of determination methods for the masking noise components within a critical band. L_G is derived using a first order linear regression for lines classified as noise in the ISO method, while that is derived from the results of energy summation for lines classified as masking noise using the 70th percentile value of the lines within a critical band in the IEC method. As for the PAS method, L_G is derived using an iterative procedure. The procedure commences with the energy averaging of lines with in a critical band with the exception of the line centred on the critical band. In the subsequent steps, energy averaging is performed for lines excluding lines with levels exceed the energy mean value determined beforehand by more than 6 dB. If energy mean value approximately equals to that of previous iteration step, the iterative procedure is discontinued. Then L_G is derived by multiplying energy mean value and the critical band width.

For each determination method of masking noise, good classification result is obtained if shape of masking noise components is approximately flat. However, errors contained in classification result may be large when shape of masking noise components has a steep slope especially in the frequency range below about 100 Hz. The factor for the shape of masking noise component with a steep slope is caused by not only the characteristics of the noise itself but also the influence of the A-weighting filter.

Table 1 Comparisons between analysis methods of the tonal audibility

Standards	ISO 1996-2 Annex C	IEC 61400-11	DIN 45681	ISO/PAS 20065
A-weighted spectrum	Yes	Yes	Yes	Yes
Frequency range [Hz]	from 50	from 20	from 90	from 50
Frequency resolution [Hz]	$\Delta f_{\mathrm{eff}} < 0.05 \times \Delta f_{\mathrm{T}}$	1 to 2	1.9 to 4.0	1.9 to 4.0
Critical band width Δf_{C} [Hz]	$0.2 f_{\mathrm{T}}$ $(f_{\mathrm{T}} > 500)$ 100 $(f_{\mathrm{T}} \leq 500)$	$25 + 75 \cdot (1 + 1.4 \cdot [f_{\mathrm{T}}/1000]^2)^{0.69}$		
Cut-off freq. of CB	$f_{\mathrm{T}} = 0.5 \times (f_{\mathrm{H}} + f_{\mathrm{L}})$		$f_{\mathrm{T}} = \sqrt{f_{\mathrm{H}} \times f_{\mathrm{L}}}$	
Estimation of masking noise	Linear regression	Statistical method	Iterative method	Iterative method
Evaluation time duration	60 s (or about 10 s)	at least 10 s × 10 spectra	at least 3 s × 12 spectra	
Penalty K_{T}	Specified	Not specified	Specified	Specified in ISO 1996-2:2017 Annex J
Sample source code	Not provided	Not provided	Provided	Provided

f_{eff}: Effective analysis bandwidth [Hz]
f_{T}: Frequency of a tonal component [Hz]
f_{L}: Lower cut-off frequency of the critical band [Hz]
f_{H}: Higher cut-off frequency of the critical band [Hz]

Table 2 Overview of determination methods of masking noise components

Linear regression (ISO method)	Statistical method (IEC method)	Iterative method (DIN/PAS method)
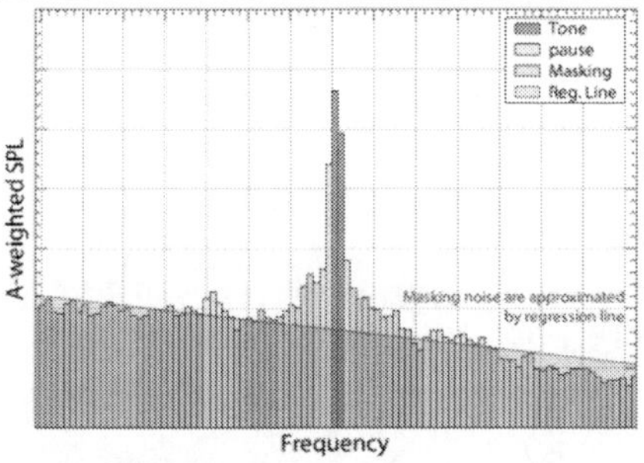	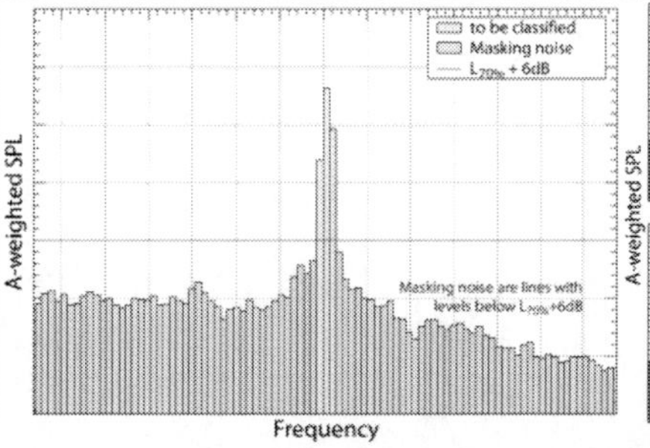	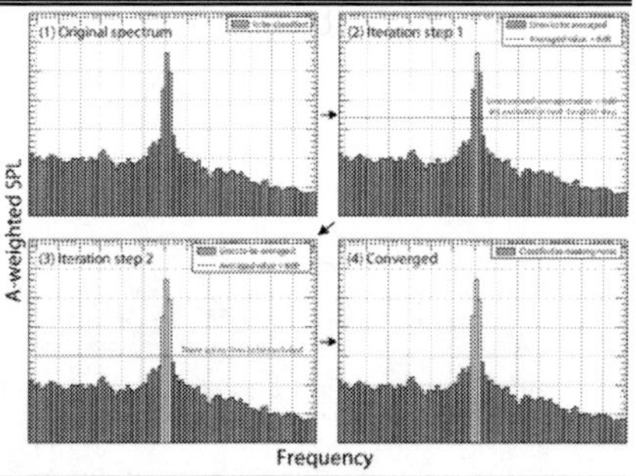

2.2 1/3 octave band based method

As mentioned before, the tonal audibility is determined by classifying lines within a critical band into tonal components or masking noise components in the FFT based method. Since the critical band width and the 1/3 octave band width are approximately same as shown in Fig. 2, classification of components within a critical band cannot be performed using the 1/3 octave band analysis. Therefore, what can be done using the analysis results of 1/3 octave band analysis is to compare the sound pressure level of a band which have a possibility of containing tonal components with the sound pressure level of both sides. That is the method used in the ISO survey method and the NZS survey method, and it is very useful when surveying the presence or absence of tonal components in residential areas. The difference between the two methods is that the ISO survey method requires that both level differences between a band of interest and its neighbouring bands exceed the threshold value, while the NZS survey method requires that arithmetic mean of level differences between a band of interest and its neighbouring bands exceeds the threshold value. The threshold value is described in table 3. Since wind turbine noise has a spectral slope of -4dB/octave [7], this fact yields level difference between a band and its neighbouring bands whether tonal components exists or not exists. Examining an arithmetic mean of level differences between a band of interest and its adjacent lines used in NZS survey method seems to be more suitable for wind turbine noise problems.

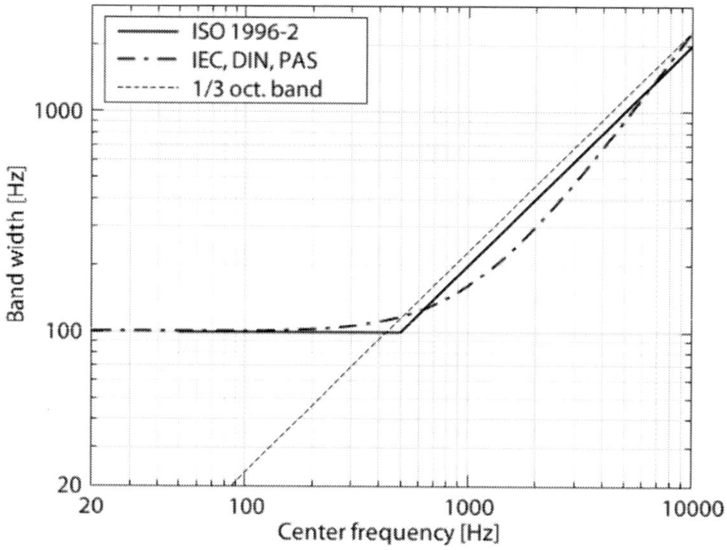

Fig. 2 Comparison between critical band width and 1/3 octave band width

Table 3 Level difference specified in ISO 1996-2:2017 or NZS 6808:2010

Frequency	Level difference
Low frequency (25 Hz to 125 Hz)	15 dB
Middle frequency (160 Hz to 400 Hz)	8 dB
High frequency (500 Hz to 10000 Hz)	5 dB

2.3 Loudness based method

Psychoacoustic tonality calculation method specified in the ECMA-74 is "a perception-model-based procedure for determining whether or not noise emissions contain prominent tonalities, and if present" [6]. Although the standard containing this method targets tonal components emitted by information technology and telecommunications equipment, it is seemed to be applicable to tonal components contained in environmental noise by reason of analysis method described below. To make a long story short, it is a method to analyse by paying attention that tonality perception and the partial loudness of tonal sound components have a strong correlation. As show in Fig. 3, this method uses waveforms divided into 53 critical bands by using auditory filter bank. From this filtered waveforms, loudness contributions of tonal components and noise components to the specific loudness for each critical band are determined. Separation of each contribution to the specific loudness is estimated by using auto-correlation function. The unit of calculated results is given in tu_{HMS}, but it seems close to sone in terms of dimension of a quantity. This psychoacoustic tonality takes account of psycho-acoustical effects such as threshold in quiet, loudness-level-dependent masking and frequency-dependent, therefore, it may have a better relationship between the calculated value and subjective response to the tonal component comparing to both the FFT based method and the 1/3 octave band based method.

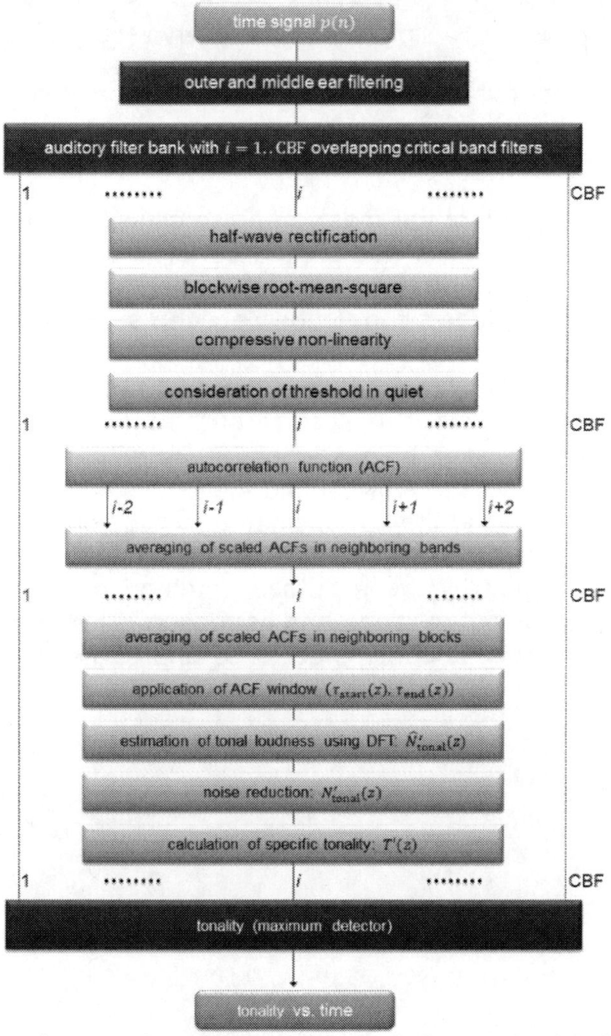

Fig. 3 Analysis scheme of psychoacoustic tonality method (adopted from [6])

3. Relationship among each method for synthesized noise

Although each method is originated from different frequency analysis method as described before, it is also important to examine the relationships among these methods and applicable range of each method. In this chapter, various analysis methods are compared by using artificially synthesized noise. This artificial noise is a broadband noise having a frequency characteristic of -4dB/octave, which is based on the filed measurements of wind turbine noise [7] and is referred to as modelled noise hereinafter. Test signals are generated by superposing a sinusoidal tone on the modelled noise such that TA is in the range from -3 dB to 15 dB in 3 dB steps. A-weighted sound pressure levels of modelled noise are set to 3 conditions of 35 dB, 45 dB and 55 dB. Sound pressure level of a sinusoidal tone to be superposed was determined by using formula (1), in which L_{G} is determined from pre-synthesized modelled noise before sinusoidal tone is superposed. This method enables to generate a test sound with exact TA conditions which does not depend on the classification method of the masking noise component. Frequency of sinusoidal tone is in range from 25 Hz to 1250 Hz in terms of the center frequency of the 1/3 octave band. Conditions of test sound are summarised in table 4.

Table 4 Conditions of test sound used for case study

Noise model	L_{Aeq} of noise	Tone frequency	Tonal audibility
-4dB/octave	35 dB 45 dB 55 dB	25 to 1.25 kHz in 1/3 oct. band interval (18 frequencies in total)	-6, -3, 0, 3, 6, 9, 12, 15

3.1 Comparing among FFT method

In order to compare the property of the masking noise classification method, TA specified at test signal generation and analysed TA was compared for each analysis method. Results are show in Fig. 4. As a result, IEC method has the best correlation between specified TA and analysed TA. ISO method also has a good correlation between specified TA and analysed TA but has a certain amount of variation comparing to the IEC method. This is due to the estimation error of L_{G} especially in the frequency range below about 100 Hz. According to the ISO method, L_{G} is estimated using a first order linear regression of the masking noise component within a critical band even in a frequency range below about 100 Hz where the shape of masking noise components tends to be curved by the A-weighting filter. This causes a deviation between the actual masking noise component and the approximated regression line. As for the PAS method, analysed TAs for tone frequency below 50 Hz are significantly smaller than the specified TAs. This phenomenon is also caused by the estimation method of masking noise components. Frequency range below 50 Hz is out of range and its reason is not described in the PAS method, however, such estimation error is considered to be one of the reasons for the limitation of frequency below 50 Hz.

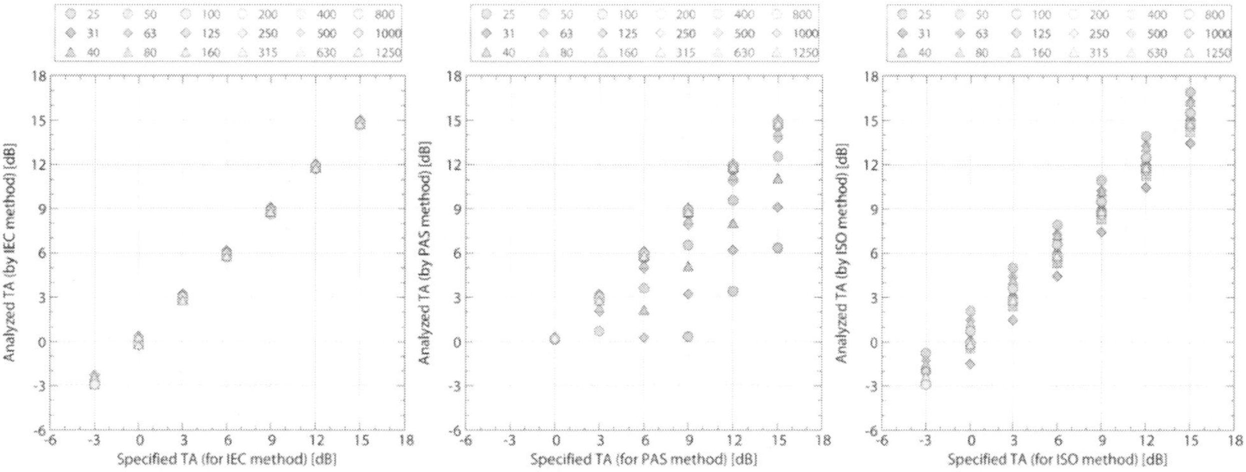

(a) IEC method (b) PAS method (c) ISO method

Fig. 4 Relationship between specified TA and analysed TA

3.2 Comparing between FFT method and 1/3 octave band method

Figure 5 shows examination results of relationships between the level difference based on 1/3 octave band analysis and the TA. Figure 5 (a) and (b) show the level difference between the frequency band containing the tonal component and the adjacent band of the lower frequency band and the higher frequency band, respectively. In the ISO survey method, level differences of both adjacent bands need to exceed the threshold described in table 3, therefore, smaller level difference is used for examination. In this case, judgement is performed using the results of Fig. 5 (a), test signal was judged as prominent tonal component is contained, when TA is 15 dB or more with the tone frequency from 25 Hz to 125 Hz and TA is 12 dB or more with the tone frequency from 160 Hz to 1250 Hz, respectively. As for the Fig. 5 (c),which is based on the arithmetic mean of level differences between the frequency band containing the tonal component and the adjacent bands, tonal component is identified when TA is 12 dB or more with the tone frequency from 25 Hz to 125 Hz and TA is 9 dB or more with the tone frequency from 160 Hz to 1250 Hz, respectively. That is, the NZS survey method is considered to have a better detection ability of the tonal components comparing to the ISO survey method. For details of this discussion, please refer to the literature [8].

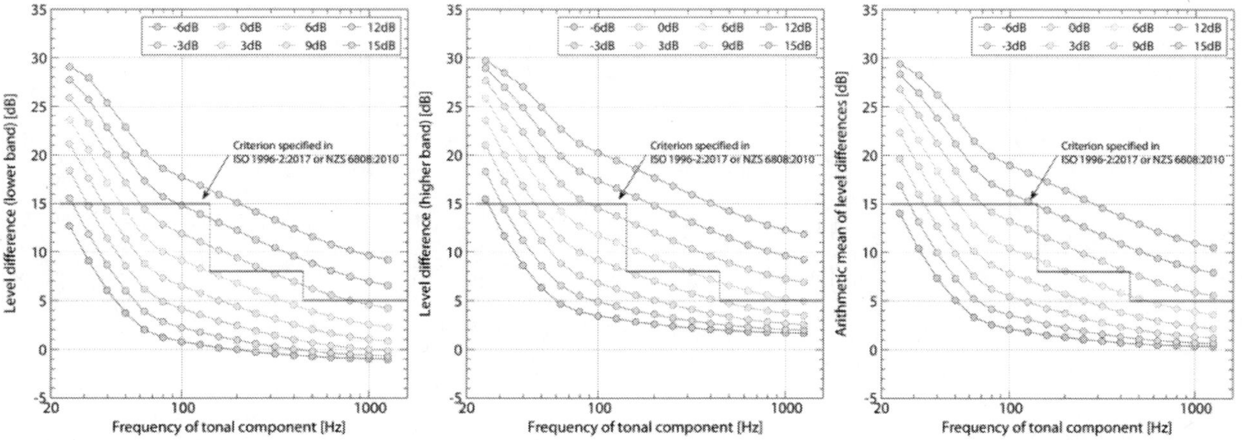

(a) Diff. with lower band (b) Diff. with higher band (c) Average of both differences

Fig. 5 Relationship between TA and level difference for -4dB/octave noise

3.3 Comparing between FFT method and psychoacoustic tonality method

Figure 6 shows results of comparison between the TA calculated by the IEC method and the psychoacoustic tonality for the modelled noise conditions of 35 dB, 45 dB and 55 dB in terms of A-weighted sound pressure level. If the calculation algorithm of the psychoacoustic tonality has a simple dependency on the sound pressure level of interest, calculated results of psychoacoustic tonality is expected to be simply shifted with increase of sound pressure level. However, results shown in Fig. 5 are not simple dependency. This result suggests that the psychoacoustic tonality include the effects of not only sound pressure level but also other factors.

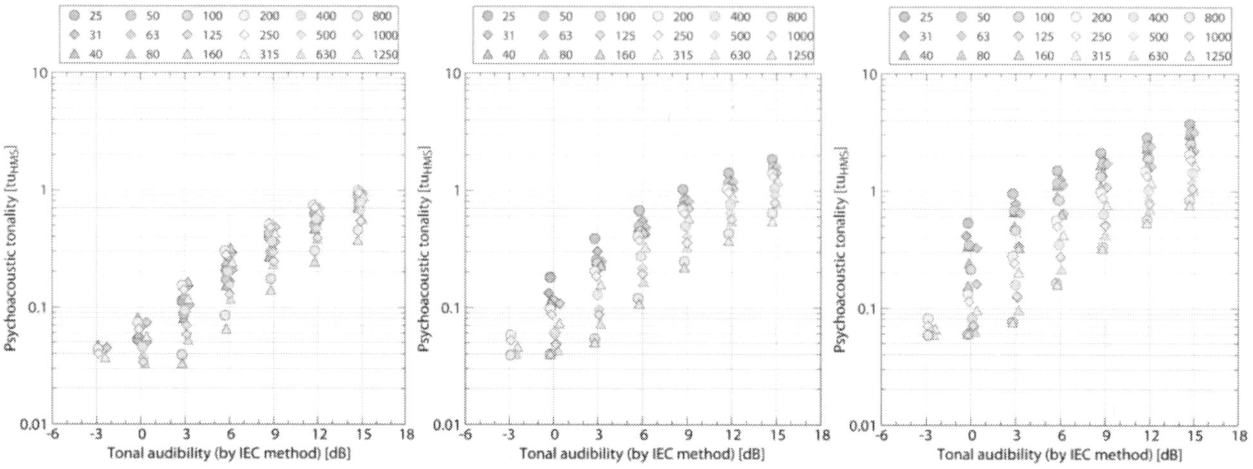

(a) 35 dB (b) 45 dB (c) 55 dB

Fig. 6 Relationship between TA and Psychoacoustic tonality for -4dB/octave noise

Figure 7 show the relationship between the psychoacoustic tonality and the frequency of the tonal component by re-arranging the data shown in Fig. 6. In this result, the shapes of the lines significantly change with increase of sound pressure level of masking noise especially in the frequency range below 200 Hz. Assuming that acoustic model of the psychoacoustic tonality is correct, it has been suggested that it is necessary to carry out a research on the influence on humans by subjective experiment using sound recordings of actual noises or test sound simulating actual sound pressure level and frequency characteristics, when investigating the value of the penalty for environmental impact assessment. In other words, blind use of the penalty value which is not limited the types of noise as specified in ISO 1996-2:2017 may cause evaluation result not reflecting the actual situation.

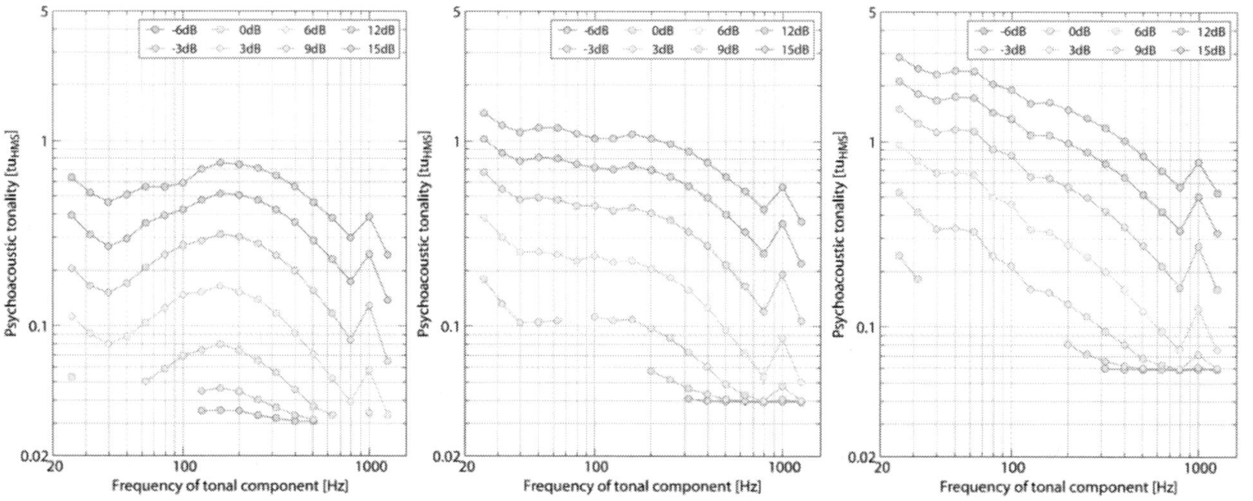

(a) 35 dB (b) 45 dB (c) 55 dB

Fig. 7 Frequency dependence of Psychoacoustic tonality for -4dB/octave noise

4. Relationship among each method for wind turbine noise

In order to examine the before mentioned analysis methods, an analysis was performed for a wind turbine noise containing tonal components actually measured. Figure 8 shows frequency characteristics of sound recording used for analysis. Analysis time duration is 10 s in this case. Tonal components are clearly seen for all of the data used in analysis.

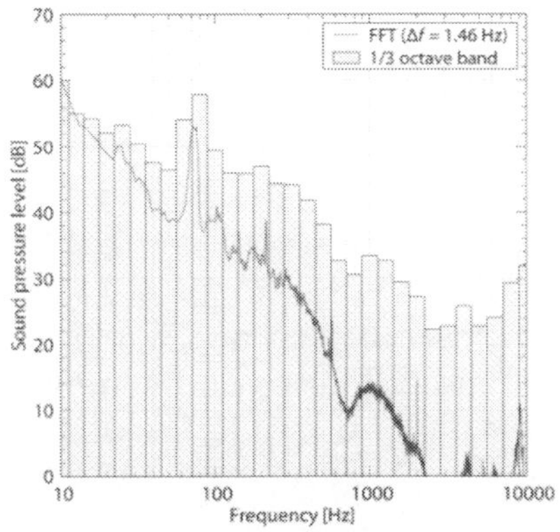

(a) W22M01

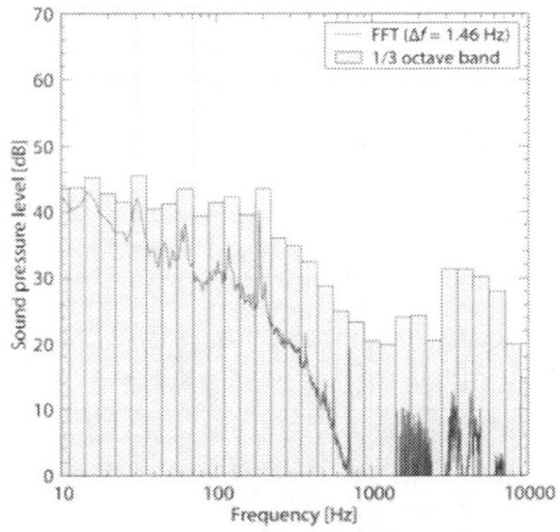

(b) W25M04

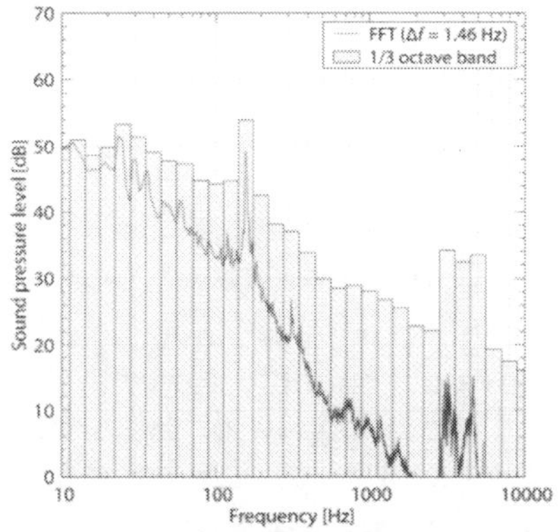

(c) W29M04

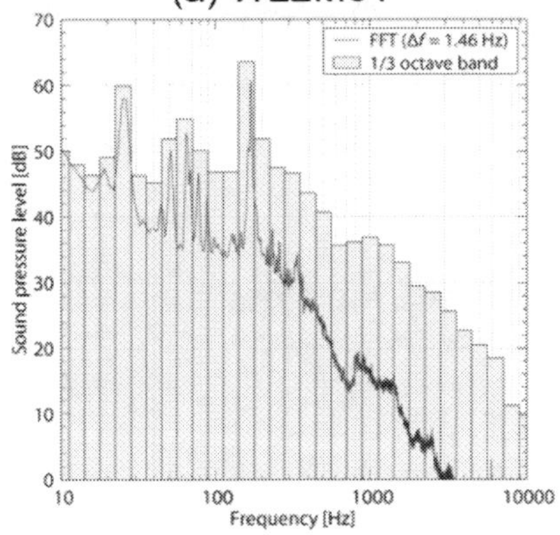

(d) W28M06

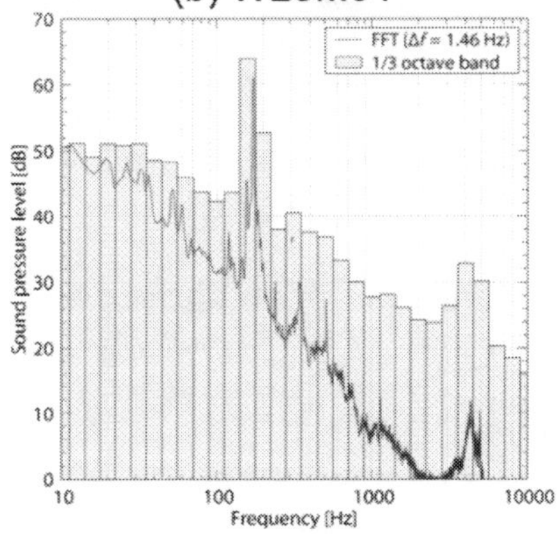

(e) W29M01

Fig. 8 Frequency characteristics of wind turbine noise for analysis

Table 5 Calculated results for wind turbine noise containing tonal components

Name	W22M01	W25M04		W29M04	W28M06	W29M01
$L_{Aeq,30s}$	45.9 dB	40.3 dB		44.5 dB	52.7 dB	52.2 dB
Frequency	76.2 Hz	189.0 Hz	720.7 Hz	158.2 Hz	169.9 Hz	171.4 Hz
TA by IEC	5.3 dB	0.9 dB	4.3 dB	6.2 dB	12.2 dB	17.1 dB
TA by PAS	2.5 dB	1.2 dB	4.5 dB	6.3 dB	12.1 dB	19.5 dB
TA by ISO	6.4 dB	0.8 dB	3.8 dB	7.1 dB	11.5 dB	18.7 dB
Frequency band	80 Hz	200 Hz	800 Hz	160 Hz	160 Hz	160 Hz
Average of level diff.	6.1 dB	5.7 dB	0.7 dB	10.2 dB	14.2 dB	15.8 dB
Prominent?	No	No	No	Yes	Yes	Yes
Frequency	(96.9 Hz)	(211.4 Hz)		(159.6 Hz)	(169.6 Hz)	(171.0 Hz)
Psychoacoustic tonality	0.255 tu_{HMS}	0.149 tu_{HMS}		0.564 tu_{HMS}	1.361 tu_{HMS}	1.763 tu_{HMS}

Table 5 shows the calculated results of the tonal audibility and the psychoacoustic tonality for the recording shown in Fig. 7. Since the detected frequency of the psychoacoustic tonality is calculated by arithmetic mean of the frequency with a maximum tu_{HMS} for instantaneous specific loudness calculated at every 0.1 ms interval, it is shown in parentheses for reference in table 5. To see the result, there is no significant difference in the detection ability of the tonal components in the present data among the methods for calculating the tonal audibility, while it is seemed to be a little varied in the TA for tonal component of 76 Hz. As for the 1/3 octave band based method, tonal components of about 160 Hz, in which TA exceeds about 6 dB, were judges as prominent. This is consistent with the theoretical value show in Fig. 4(c). The tendency was seen that calculated results of the psychoacoustic tonality increased with increase of analysed TA.

An interesting thing in this result is found at the data of W25M04. Tonal component of 720 Hz will be adopted for evaluation by the tonal audibility because TA of 720 Hz is larger than that of 189 Hz. On the other hand, tonal component of 189 Hz was predominant over that of 720 Hz for the psychoacoustic tonality. Tonal component of 189 Hz is 'audible' also in the auditory impression, while that of 720 Hz was hard to hear as if masked. In such a case, validity of simply adopting a maximum TA for the evaluation needs further investigation.

5. Conclusion

In this paper, we investigated various international standards considered to be applicable as an evaluation indicator of the tonal component contained in environmental noise, and summarized the overview of the analysis principles. Each method can be arranged by the frequency analysis method, and the each feature are as follows.

- FFT based method
  - Advantages
    - ✓ Moderately simple algorithm
  - Disadvantages
    - ✓ Does not consider absolute sound pressure levels
    - ✓ Does not consider the hearing threshold
  - Remarks
    - ✓ Need for scientific evidences or open discussions to validate the algorithm
- 1/3 octave band based method
  - Advantages
    - ✓ Extremely simple method
    - ✓ Only need for 1/3 octave band analyser
  - Disadvantages
    - ✓ Cannot detect tonal components of around 'just audible' sound pressure levels
  - Remarks
    - ✓ NZS survey method is better than ISO survey method for environmental noises
- Loudness based method
  - Advantages
    - ✓ Adoption of human auditory mechanisms
    - ✓ Reflection of absolute sound pressure levels
    - ✓ Taking into account of hearing threshold
  - Disadvantages
    - ✓ Complex algorithm to implement into computer programs
    - ✓ High computational load
  - Remarks
    - ✓ Useful for understanding results of subjective experiments
    - ✓ Need for more evidences of relationship to subjective evaluation

Based on the concept of the psychoacoustic tonality, we confirmed that the following points are important. In order to use TA for determining a penalty, it is necessary to investigate relationship to human response by subjective experiment using test sounds in consideration of both actual frequency characteristic and actual sound pressure level. In addition, it is indicated that tonal component with a maximum TA not always has maximum contribution to the auditory impression.

Although the analysis methods of tonal components to be surveyed are publish as international standard, scientific evidences or open discussion results for validity of analysis principle are seemed to be less information. If these analysis methods will continue to be used as part of the wind turbine noise measurement method, more scientific evidences or open discussion is desired.

References

1. T. Kobayashi, S. Yokoyama, A. Fukushima, T. Ohshima, S. Sakamoto, H. Tachibana, "Assessment of tonal components contained in wind turbine noise in immission areas", 6th International Meeting on Wind Turbine Noise, Glasgow, 2015.
2. S. Yokoyama, T. Kobayashi, H. Tachibana, "Subjective experiments on the perception of tonal component(s) contained in wind turbine noise", 7th International Meeting on Wind Turbine Noise, Rotterdam, 2017.
3. Colin H. Hansen, Con J. Doolan, Kristy L. Hansen, "Wind Farm Noise: Measurement, Assessment, and Control", John Wiley & Sons Inc., 2017.
4. M. Yonemura, H. Lee, S. Sakamoto, "Study on subjective evaluation of noise containing low frequency tonal components using rating scale method" (in Japanese), 2019 Spring Meeting of the Acoustical Society of Japan, 2019.
5. E. Zwicker, H. Fastl, "Psychoacoustics: facts and models", Springer, 1999.
6. ECMA-74, "Measurement of Airborne Noise emitted by Information Technology and Telecommunications Equipment (15th edition)", Ecma International, 2018.
7. H. Tachibana, H. Yano, A. Fukushima, S. Sakamoto, "Nationwide field measurements of wind turbine noise in Japan", Noise Control Engineering Journal, 62(2), pp.19-101, 2014.
8. S. Yokoyama, T. Kobayashi, H. Tachibana, "A Practical Method for Estimating a Presence of a Prominent Tonal Component in Wind Turbine Noise", Inter-noise 2018, Chicago, 2018.

INCE EUROPE

8th International Conference
on
Wind Turbine Noise
Lisbon – 12th to 14th June 2019

Advancements in continuous learning for tonality free turbine design

Mranal Gupta, Product Functional Lead Tonality: mrgup@vestas.com
Kaj Dam Madsen, Senior Product Functional Lead Noise: kadma@vestas.com

Summary

Wind turbine noise tonality is a phenomenon where a frequency component sticks out from the overall frequency spectrum shape of the noise emitted from the turbine. This is a concern for turbine manufacturers as such tonal noise could be annoying for neighbors. The various international standards and guidelines define methods for assessing tonality and based on the meta-data regarding turbine components and operation it is possible to identify the source of such tonalities.

Turbine design for good tonal noise performance requires thorough understanding of the involved phenomena and system integration capability of the highest order as it is a complicated subject and the realm of possibilities is huge. Therefore, it is important to have a focused and simplified communicable approach to find solutions where most optimal.

This paper describes the framework developed by Vestas Wind Systems to ensure integration of all the complex systems from multiple suppliers that enables our turbines to operate compliant with tonality requirements. Multiple simulation models, measurements and consistency in performance are evaluated for the complete (value) chain, starting from component suppliers to a full turbine assembly under operational conditions.

1. Introduction

Wind turbines keep expanding their role in a world with rapidly increasing demand for clean energy and therefore the individual turbines are optimized for providing more energy production to lower the cost and utilize the wind resources to a maximum - also in more densely populated areas. This together drives bigger turbines, larger rotors to capture the wind and potentially more noise emission from the individual turbine.

As overall wind turbine noise is for 90 % emitted as aerodynamic noise from the rotor and to a wide extend governed by tip speed the larger rotors challenges the entire load carrying system by the need of operation of the large rotors at very low rotational speeds and at high torques through the drive system. In addition to lowering the rotor speed overall noise improvement is gained by advanced noise optimized airfoil/blade shapes and application of noise reduction addons to the blades (serrated trailing edges). Bringing it all together and finding the optimal balance in design it has been possible to even lower the overall noise emission while improving energy production as illustrated in Figure 1.

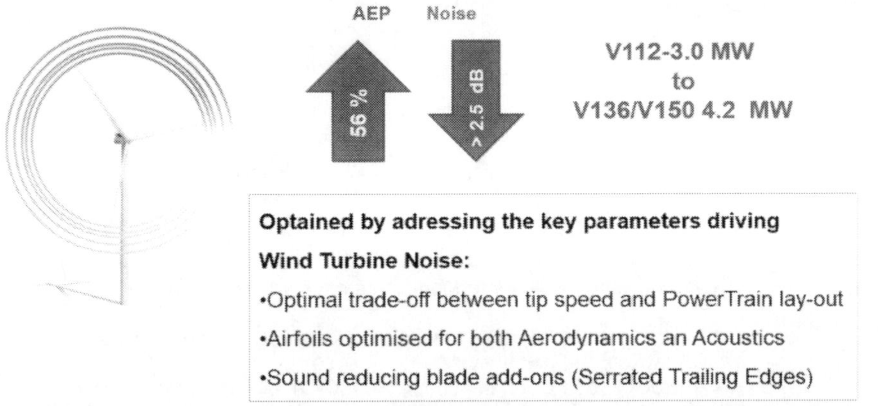

Figure 1: Balancing annual energy production (AEP) and noise emission for Vestas 4 MW platform

Although the large design effort in reducing overall noise emission from the turbines do pay of, the side effect is that less rotor noise emission at lower rotational speeds has reduced the masking energy for tones and potentially making them more audible as illustrated in Figure 2 as evaluated according to IEC 61400-11 ed.3 [1].

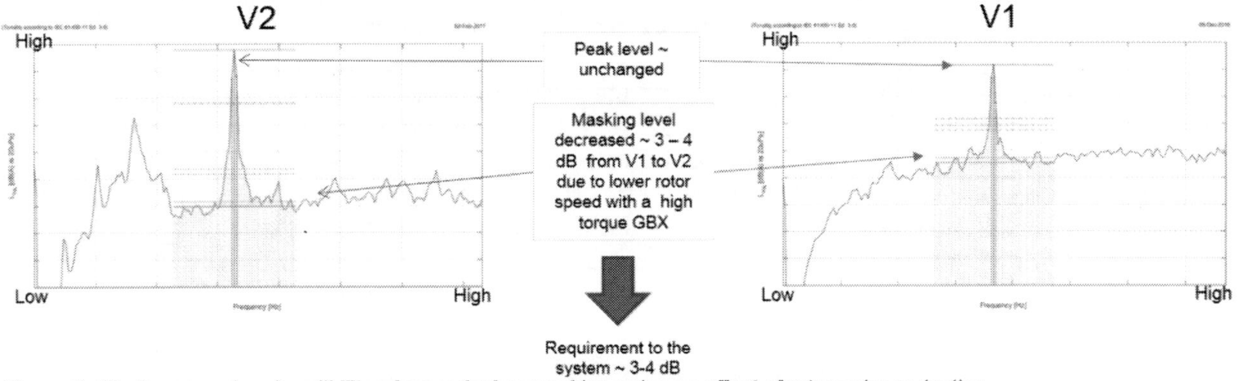

Figure 2: Challenge on tonal audibility when reducing masking noise as effect of rotor noise reduction

The optimization of noise performance for turbines therefore links heavily across both rotor noise optimization as well as optimization of the so-called mechanical noise sources normally responsible for any tonal noise emission and calls for a thorough and systematic approach.

This paper will describe the framework developed by Vestas Wind Systems focussing on tonal noise optimization with the link to aerodynamic noise models.

2. Background

When considering tonal noise emission from wind turbines a quick separation into airborne noise sources as fans, pumps etc. and structure borne (vibration) noise sources as gearbox or generator can normally be made.

For airborne tonal noise sources, it is rather straight forward to set up requirements on individual components and have that verified already on component level before assembling into the turbine.

For structure borne tonal noise sources it is more complex and calls for a good understanding of the source, the transmission path and the noise radiator as illustrated in Figure 3 to the left.

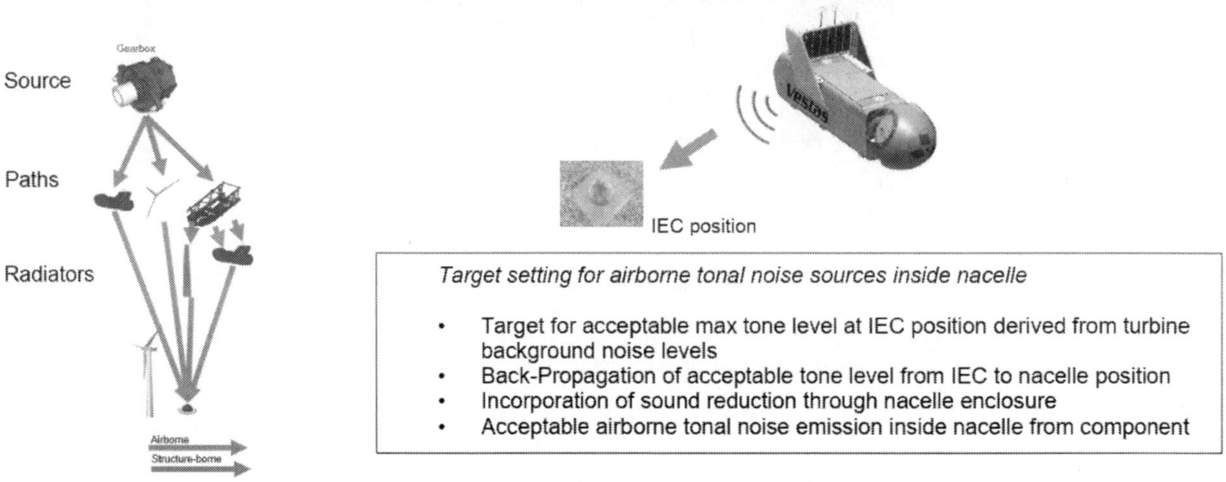

Figure 3: Structure borne noise split into source, path and radiator and target setting example for airborne sources

Also, in Figure 3 (to the right) the systematic approach for setting tonality target on component level is listed with the airborne tonal noise emission from the Gearbox taken as an example. Based on predicted masking noise levels a "back-propagated" tonality target can be derived for the different noise sources depending on frequency and turbine operation.

In Figure 4 an example of masking noise predicted for the actual turbine rotor is shown for a given gearbox order for different turbines. The masking noise is calculated based on the aerodynamic rotor noise model according to the relevant critical bands for a given gearbox order and used as described for target setting. As a conservative approach, only rotor noise masking is considered but different levels of ambient noise from surrounding vegetations etc. could be considered as well. This masking energy prediction approach is validated, and an example shown in Figure 5 plotting the simulated and measured masking energy. As can be observed, the trends and levels are quite comparable.

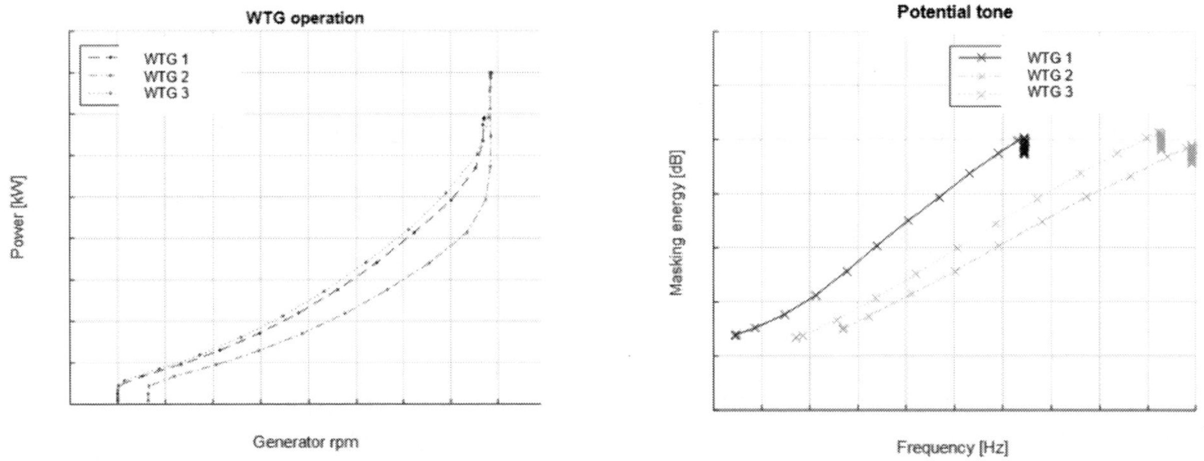

Figure 4: Predicted masking noise levels for different turbines for a given tone source order with reference to IEC 61400-11 ed.3

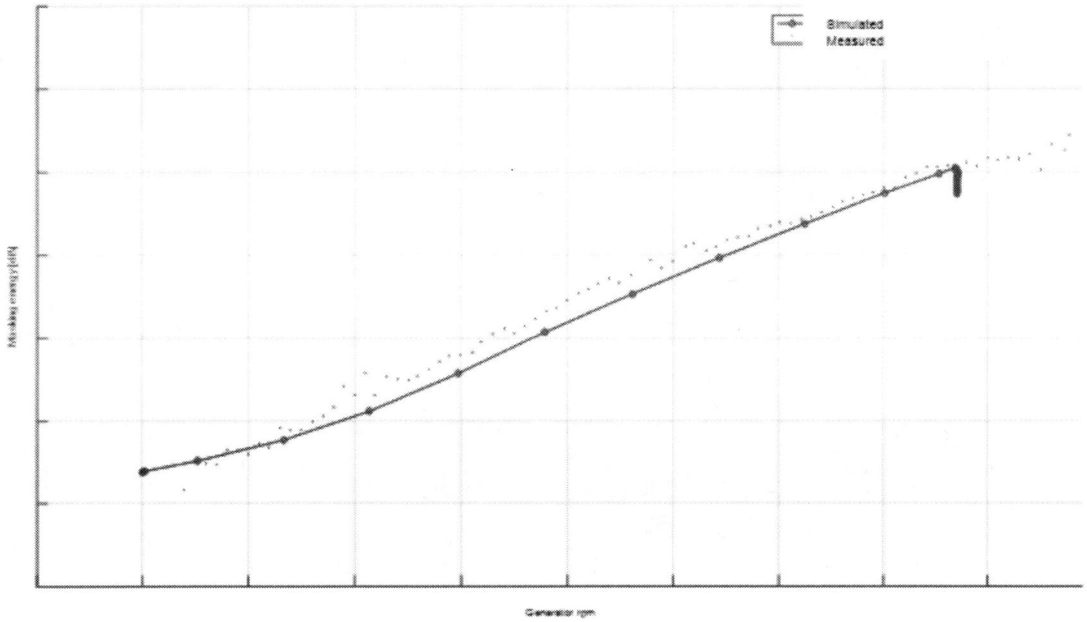

Figure 5: Validation of masking energy prediction

For structure borne tonality noise sources the approach is the same as for airborne noise sources although quite more complex and includes, as the remainder of this paper will describe, the approach taken by Vestas to support this.

3. Process for structure borne tonality free design

Turbine design for good tonal performance requires significant insight to module integration and co-ordination in a turbine development process. The phenomena that result in tonality are complex and thus without thorough understanding it is easy to oversee some potential pitfalls. However, if everything is considered at once then the process ends up being too complicated and impedes progress. While presenting a lot of challenges, tonality also manifests lots of opportunities once properly considered. There is also a need to communicate with the modules and components building the turbine in a manner that is understandable since their primary functions cannot be compromised.

3.1 Design for tonality

Keeping the above in mind, a process has been designed which paves the way for successful design. The process relies heavily on the experience and past developments. The turbine is broken down to the needed level through a combination of full FEM models and super elements. On this *reduced* model excitations from potential tonality sources are applied. The location of the excitations is chosen to fit with the actual design. As an example, for potential powertrain tones, excitation from multibody simulations are applied at the interfaces to the turbine. Based on the input loads, the transfer characteristics of the components and radiation from the blades and tower, the tonal noise radiation is simulated, as shown in Figure 6.

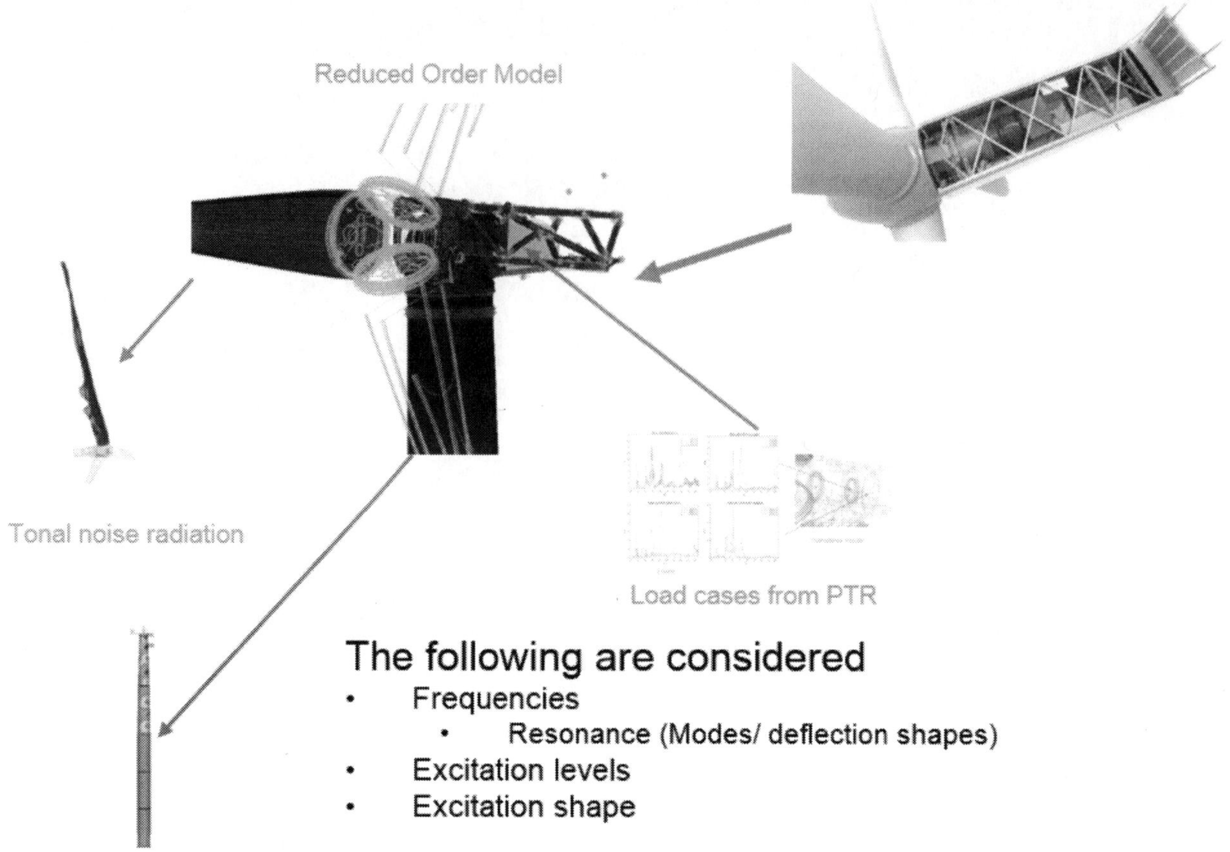

Figure 6: Workflow for tonal noise radiation simulations

This is then used to back propagate requirements to the modules building the turbine, keeping in mind the expected masking. The requirement breakdown and balancing with regards to both compliance and cost is done for the excitation source, transfer path and the radiators, with the knowledge of past experiences within the realm of possibilities. The approach is a significant development over past simulation approaches, in the sense that together with resonance frequencies and excitation levels, excitation shape is also considered. The accuracy of the prediction depends on the combination of excitation inputs, transfer path component models and the radiation simulations. The excitation shape inputs are achieved via loading the model at the interfaces.

To get to the requirements for the modules building the turbine, inputs are also taken from the masking simulations. This sets up the criteria for allowable tonal radiation for the different operational regimes for the turbine before physical turbine set-up is measured.

The tonal noise radiation simulations and masking energy simulations are performed during different stages of any project. The need and extent for simulations for a turbine type is evaluated on case by case basis. Tonal noise radiation is simulated for each load case, which might arise out of different tonal sources or changes in the source or turbine design. The simulations are also performed for sensitivity analysis of components to tonal radiation.
The masking energy simulations are needed for the excitation source only if the frequency changes as a function of speed. As an example, for a generator if the number of pole pairs is changed the slot passage frequency at the same generator speed would change and this would require a re-calculation for the expected masking.

Figure 7 presents a simplified overview of the simulation process to ensure design for tonality from kick-off. Tonal radiation simulations are performed on the turbine concept design utilizing concept loads for potential sources. This supports selection of the right concept. Once the concept is chosen more matured loads are applied on a further detailed turbine model to identify potential risks. For the risks identified, design iterations are performed to evaluate the effectiveness of mitigations. The pro-active approach is essential to ensure preparedness for identified risks, should they materialize.

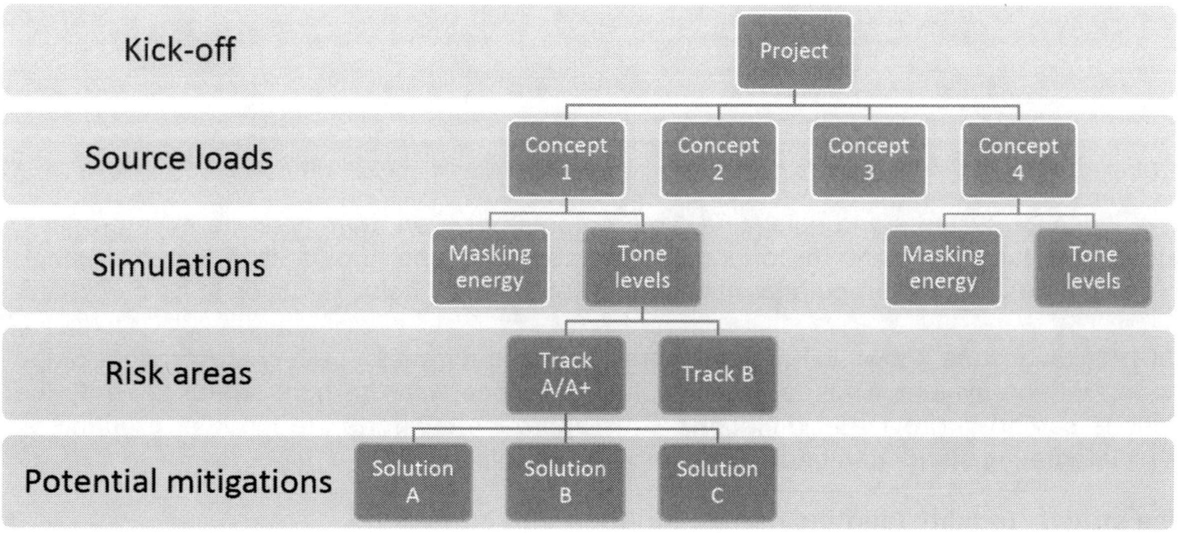

Figure 7: Schematic of the approach for tonality simulation during different stages

In Figure 8 an example is given where loads from a potential tonal source are applied on the turbine model and tonal noise radiation is estimated and compared to the relevant target. The estimate includes breakdown of the contribution from potential radiators.

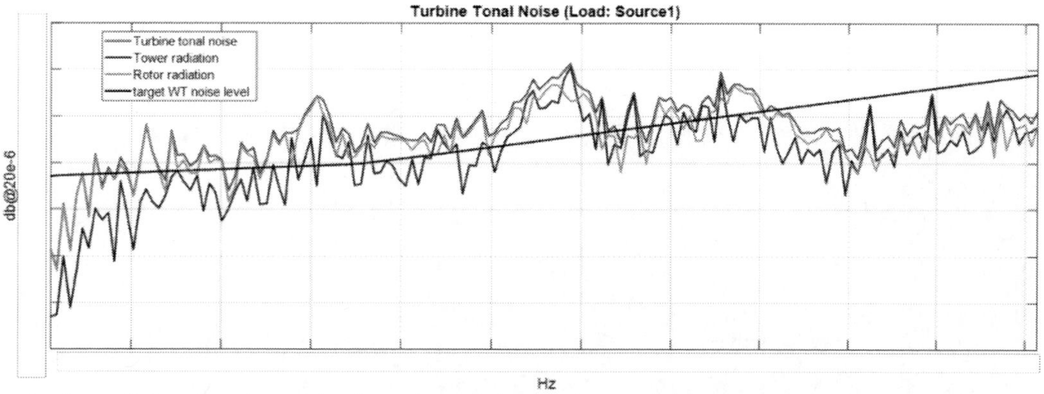

Figure 8: Tonal noise radiation from the turbine, including contribution from radiators and compared to target

The overall turbine requirements are broken down for modules from source to transfer path and radiators. The tonality models are used to set guidance in the form of needed reduction as scaling factors. This is done to preserve the excitation shape since the loading is known to influence the tonal noise radiation significantly.

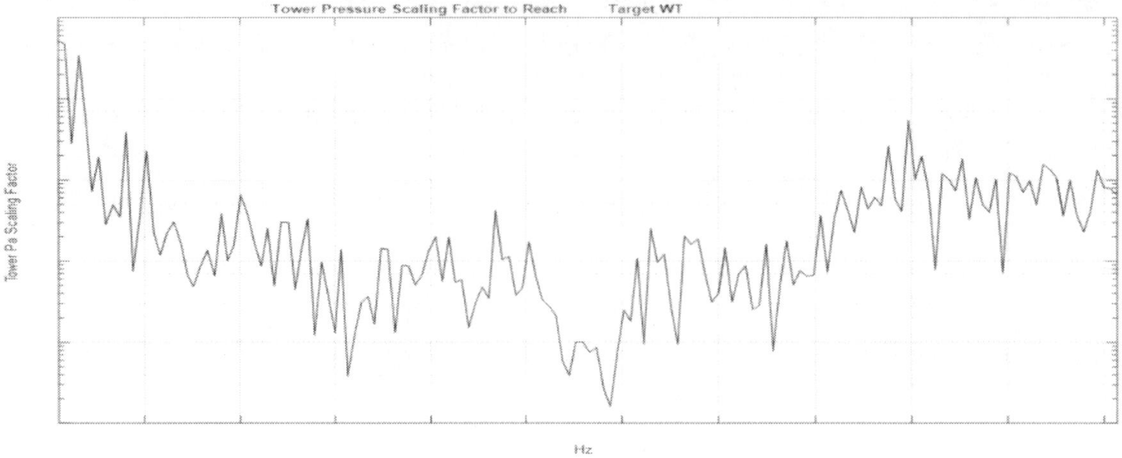

Figure 9: Tower noise radiation scaling factor to meet the target turbine tonal radiation for Source1

Setting of limits for systems and balancing requirements between them is an iterative process, even though the target limit line for tonal noise radiation from the turbine itself doesn't change much. Figure 9 illustrates the requirements that are set on tower noise radiation to be compliant.

The input excitations are known to have a dependence on the transfer path characteristics of the neighboring turbine system. Thus, a final simulation is done with the right set-up to verify the assumptions carried in previous simulations. This approach supports the modular products at Vestas by integrating standard modules into turbine variants.

3.2 Case study – tonality identification to solution and control

In addition to the simulation process, a structure borne noise model, SBN, is also employed which uses measured data to establish critical to tonality criteria, CTT. The two methods complement each other. After first IEC noise measurements are performed on a turbine, the data is thoroughly analysed using an inhouse tool with reference to the IEC analysis.

This is used to identify tonal or sound power issues. Once a tonal issue is identified the source can be easily pinpointed by means of an FFT waterfall, linking the turbine operation with the expected frequency from the source. Image digitalization software could also be used for this purpose, although this would introduce some uncertainty. The max tone levels and cause(s) for the issue are precisely extracted. Figure 10 below shows typical FFT waterfall from an internal noise measurement

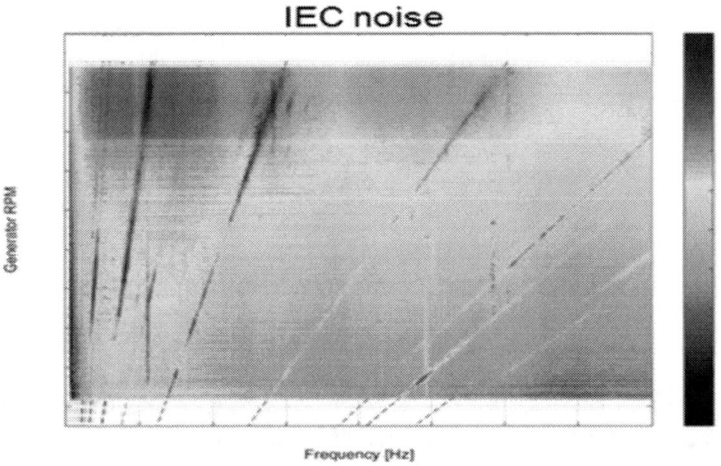

Figure 10: FFT waterfall from a typical internal noise measurement

Utilizing the advanced analysis methods available, a problem statement is made towards the source. It is expected that the potential issue was already identified during the simulations. Figure 11 is provided as guidance for the module responsible for tonality with an indication for the needed reduction and margins available

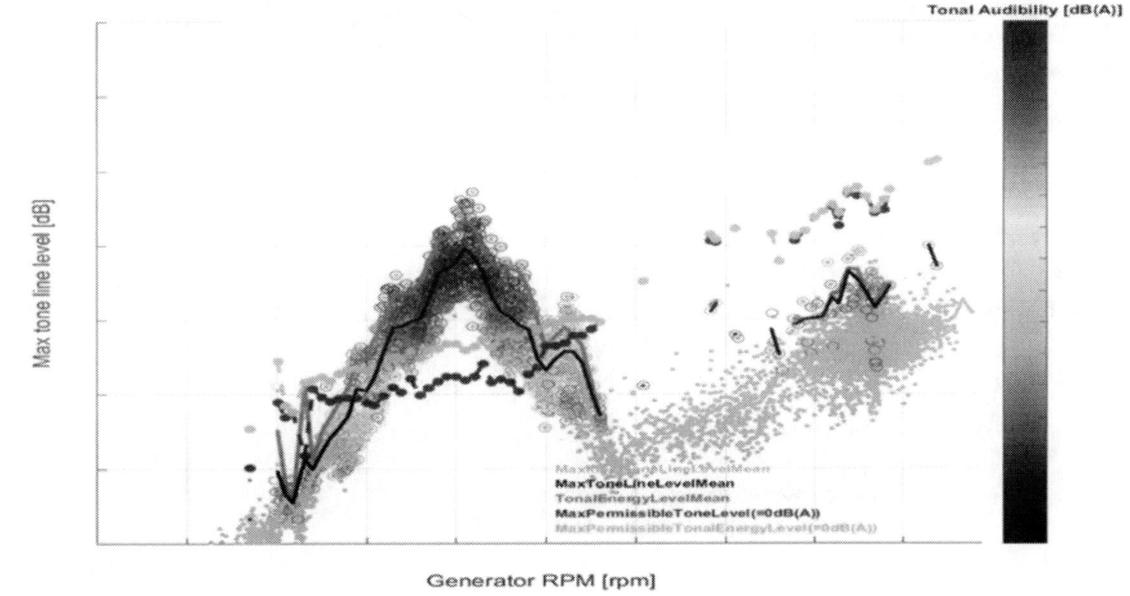

Figure 11: Problem statement towards source including exceedance and margins

This is one of the many figures that are generated to clearly pin point the source. Past measurement experience is extensively explored to set first guidance on turbine level limits. This is achieved by benchmarking against the previous measurements available in the database.

Supplementary vibration measurements are performed to identify drivers for tonality as shown in Figure 12.

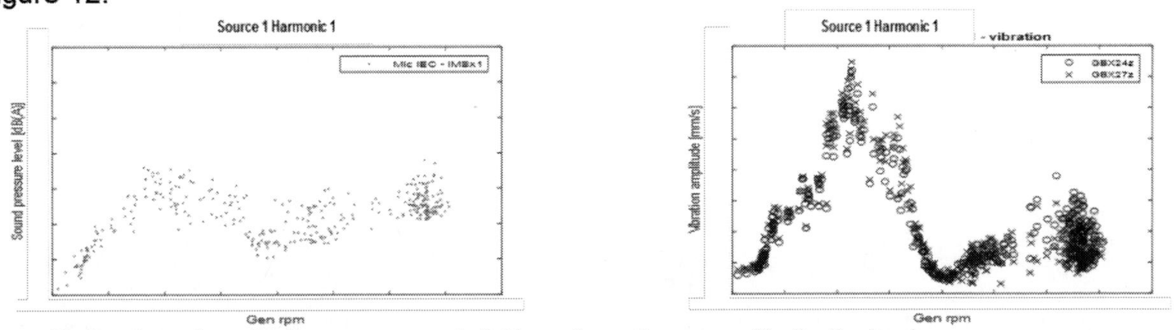

Figure 12: Supplementary vibration measurements linking noise performance with vibration levels

The operation deflection shapes around the hot-spots are animated and, if needed, modal analysis is also performed. This is done to provide inputs to the module causing the issue for mitigation design. If the issue has already been simulated, then it is used to confirm the hypothesis and validate models.

Figure 13 is obtained utilizing another in house tool. Sensors of interest for tonality are identified and a critical to tonality criteria, CTT-criteria, is established or updated. Through this, source sensors with high vibrations are identified. The same exercise is also done for transfer path and radiators and requirements are accordingly distributed. Experience regarding vibration performance is kept in mind and serves as input guidance for the systems for assessing feasibility towards the requirements.

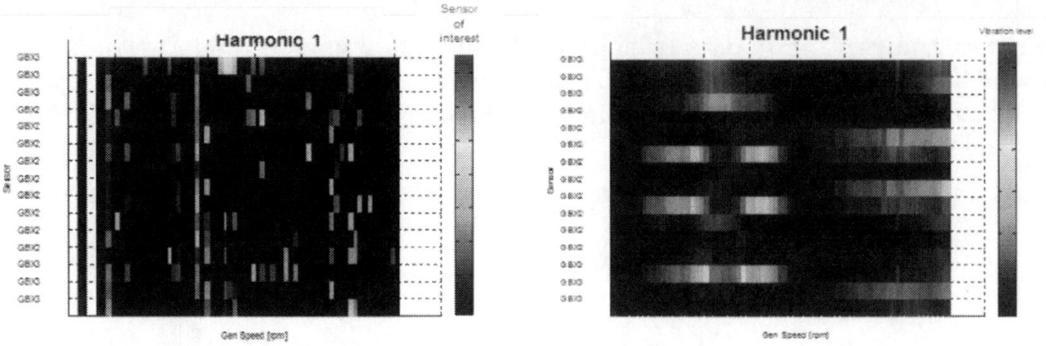

Figure 13: Sensors of interest identification

The requirements for needed reduction are even more precisely formulated and communicated to the responsible module owner or supplier. Design iterations are then measured on the test-rig for expected compliance as represented in Figure 14. Once an option provides acceptable results it is taken to the turbine and its performance verified.

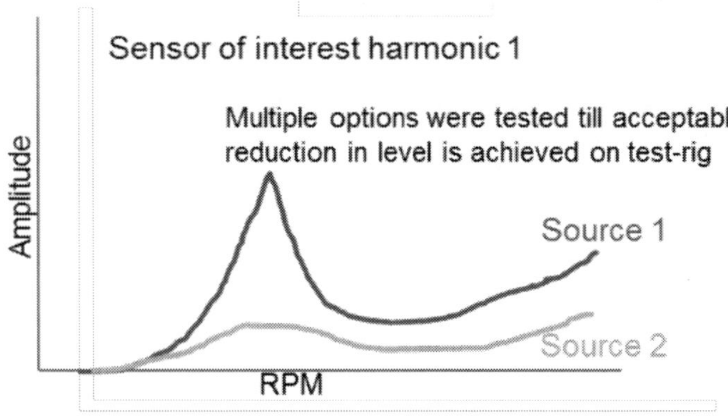

Figure 14: Design solution verification on the test-rig

Final verification is performed on the turbine. Sometimes intermediate add-on solutions representing the final design are also verified on the turbine before the final design, as indicated in Figure 15. Similar iterations are performed for the 2nd supplier, next Mk release, etc.

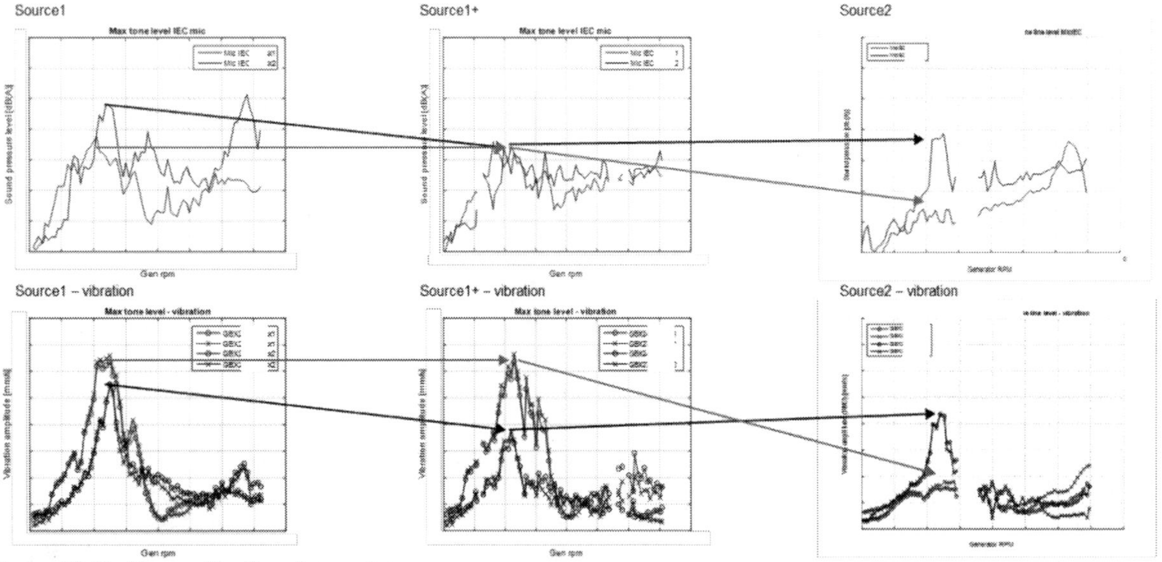

Figure 15: Turbine verification of compliance

The tool is designed and validated for turbine vibrations. A wind turbine structure borne noise model through measured data is created and critical to tonality criteria established. Although the criterion is defined on turbine level it can be used to define the sensors and load cases of interest that should be monitored during factory acceptance testing of the components' serial production.

3.2.1 Other investigations

Depending on the issue, other investigations are performed as needed, with clearly defined objectives for their usage. These include acoustic camera imaging, doppler shift analysis, etc. as shown in Figure 16. Their application is governed, or limited, by the need for identification of the radiator, quantification of contribution and the frequency of interest.

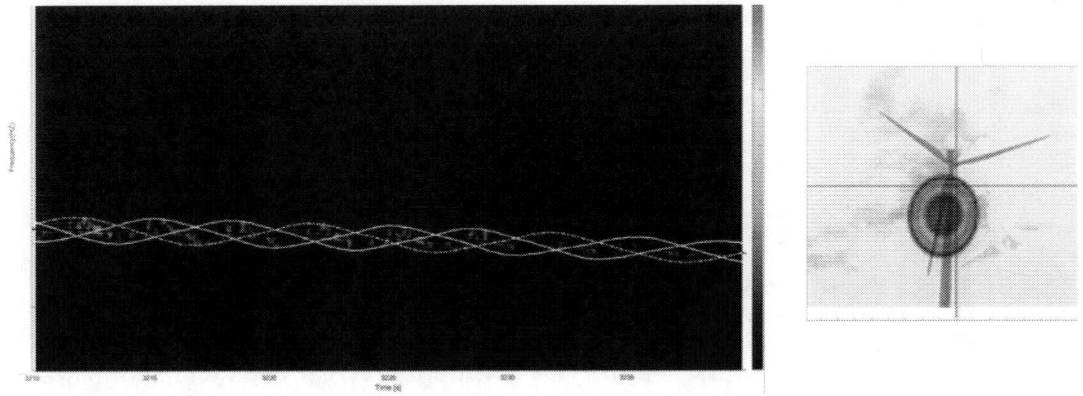

Figure 16: Doppler shift analysis and acoustic image to confirm radiating components

The investigations like in the figure above are performed on case by case basis with a line of sight towards a solution. The turbine blades and tower surfaces are known to be efficient radiators and could potentially be tonal if excited.

A parallel approach to identify the drivers for tonality from turbine operation, is usually maintained. This is done to investigate the driving factor for vibration levels leading to tonality and to understand if there could be potential intermediate control solutions that can be leveraged to reduce annoyance.

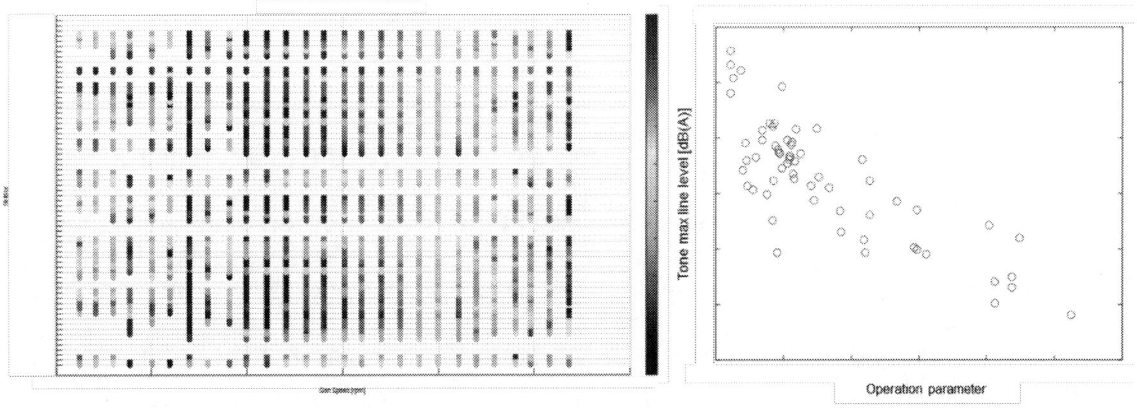

Figure 17: WTG operation parameters that could be drivers for tonality

A potential solution in Figure 17 is controlled operation parameter variation

3.3 Results and discussion

Achieving the objective of tonality free turbine design is not trivial. It requires understanding of noise and vibrations together with good module integration. Coordination and cooperation through simplified communicable targets distributed to modules facilitates design progress in the right direction. It also paves the way for needed developments.

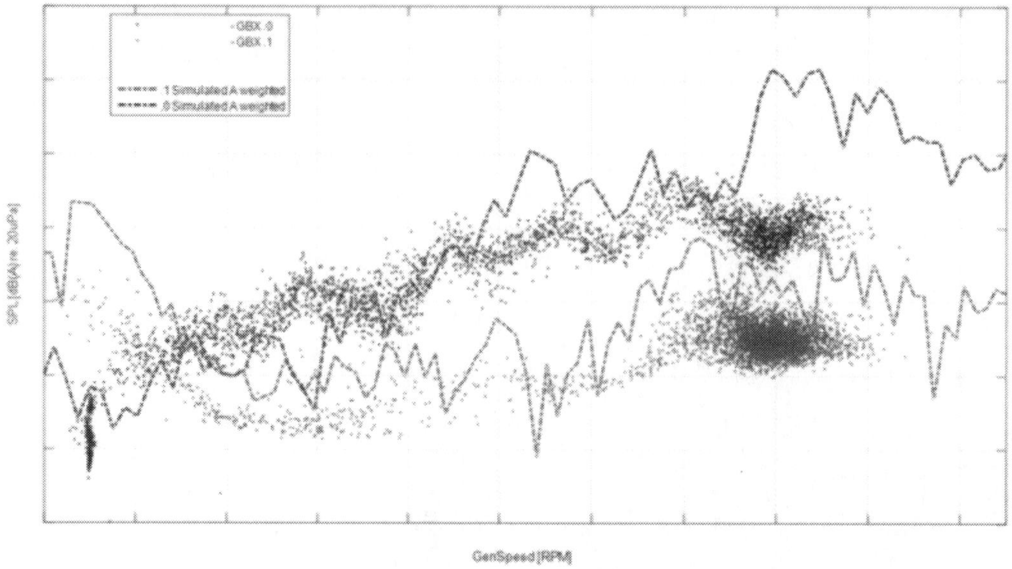

Figure 18: Simulated and measured tonal noise, arising from GBX tooth mesh frequency, for a full wind turbine

Figure 18 plots the expected turbine tonal noise from simulations of two gearbox variants with different excitation patterns and compares it with measurements. While the simulations were executed as constant torque speed run-up, the operation data was acquired when the turbine was following its power curve. The current model can capture the trends to an acceptable level. However, areas for improvements have been identified and are being worked upon. Some of these are further validation of the noise radiation model, input excitation loads validation, input loads discretization, input operational loads extraction, component damping, global model modal damping, radiation efficiency validation, etc.

Based on the simulations, proactive mitigation design and testing for potential tonal sources is executed to enable shorter time to compliance. As this approach develops, a database is maintained for regular benchmarking and evaluation against past experiences, keeping in mind the complicated nature of the challenge, which implies the learnings may not always be directly transferable.

3.3.1 Validation

Validation follows the typical V-model. Simulations are used to set the target requirements. The requirements are cascaded down from turbine to system to component for verification on a test-rig. The validation of model follows. This is executed in the reverse order, since this is the order of testing. Component is verified on a test-rig first, hence that is the first model validation. This is followed by system testing and validation. Finally, extensive turbine testing and validation is performed.
The final step also includes establishment of a structure borne noise critical to tonality criteria, only in case a system is tonal.

The validation path necessitates updates for the different models and re-evaluation of the potential mitigation designs in scope for the various phases of the project. The results should be continually benchmarked.

To obtain the maximum value from testing, it is important that the prototype components undergo extensive verification in the test rig at component and system level before the turbine. This is used to establish a proper specification for testing along the project lifecycle.

3.3.2 Control workflow

The control process is to ensure compliance to requirements. This is needed for all potentially tonal systems, for example, gearbox, generator, fans, etc.

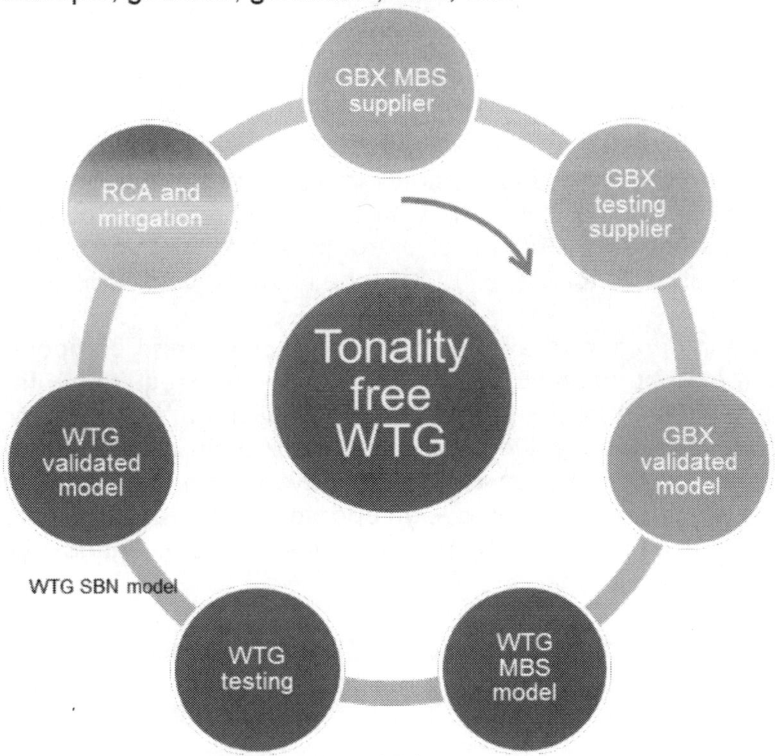

Figure 19: Workflow to be in control of SBN performance for a GBX

Figure 19 is a control workflow for root cause analysis, mitigation design and process control. After one complete cycle, from second test at supplier test rigs, we can come up with expected tone levels and evaluate for compliance. Once the expected levels are defined for the test-rig, quality follow-up keeps the serial production under control.

Ambition is to improve this even further with validated WTG acoustic simulation models using accurate excitation inputs from potential tone sources.

The explanation and understanding of reasons for tonality has evolved. The transition of focus to excitation levels from just resonance frequencies has been validated. The next step on the journey is the shape of excitations.

4. Conclusions

The advancements in continuous learning for tonality free design have been presented. The future is even more digital, with more simulations of potentially even bigger models and much more data from simulations and testing. Greater number of simulations will be performed for sensitivity studies to come up with ideal design and enable readiness for pro-active mitigation of potential issues. The simulations for probabilistic design will cover the intended component tolerances and their influence on tonality.

Noise tonality is a concern as we care about neighbours, customers, regulations, etc. We have good understanding of our turbines and tonality, as we take it to be part of our core competence. Vestas has extensive experience from field noise measurements and simulations. Thus we are well positioned to leverage the potential out of the data we acquire from measurements using advanced analysis.

The uncertainty on tonality results is a combination of the entire chain on production tolerances, equipment assembly and measurements conditions, including but not limited to wind conditions, season, vegetation, etc. A good understanding of these at Vestas enables us to make the right conclusions that support decision making, by providing directions to have the right focus. Our ambition is for more accurate excitation inputs and better understanding of source excitations (frequencies, levels and shapes).

To support continued improvement of prediction accuracy early in the design phase further work is also needed to link the turbine vibrations with test-rig vibrations. Collaboration with module owners and suppliers is ongoing. Getting in control of part-to-part variance and its influence is relevant. Setting up of critical to tonality (CTT) criteria is taking us closer to our objective.

At Vestas, we ensure integration of all the complex modules (and components) from multiple suppliers to have turbine operation compliant with our tonality requirements, in addition to the components primary function. We also leverage the operation of the turbine to avoid tonality. We achieve this through a validated masking noise tool, turbine structure borne noise model based on measurements and are working on even more advanced turbine acoustic radiation model based on simulations which would enable design updates for source, transfer path and radiators. The approach would be used for setting up permissible excitations / loads, vibration transfer and tonal noise radiation for future products.

Tonality is a complicated subject. The realm of possibilities is huge. We aim to work in a focused and simplified communicable way to achieve desired objective of tonality free turbines without unnecessarily complicating or avoiding the issue but finding solutions where most optimal.

References

[1] IEC 61400-11. Edition 3.0 2012-11. INTERNATIONAL STANDARD. *Wind turbines – Part 11: Acoustic noise measurement techniques*

8th International Conference
on
Wind Turbine Noise
Lisbon – 12th to 14th June 2019

Prominent tones in wind turbine noise - Round-robin test of the IEC 61400-11 and ISO/PAS 20065 methods for analysing tonality content

Lars Sommer SØNDERGAARD[1], Carsten THOMSEN[2], Torben Holm PEDERSEN[3]
DELTA – a part of FORCE Technology, Denmark
[1] lss@force.dk, [2] ctho@force.dk, [3] thp@force.dk

Summary

Presently, IEC 61400-11[1] is the widely used method for analysing the tonality content in wind turbine noise for sound power level measurements (made close to the turbine). However, many different methods are used worldwide for analysing the tonality content in wind turbine noise at the receptor position. One of the most used methods at the receptor position is the ISO 1996-2 Annex C method[2]. However, in 2016 the new ISO/PAS 20065 method[3] was published, replacing the ISO 1996-2 Annex C method.

In 2018 a new standardization working group, PT-61400-11-2, was formed: "Wind energy generation systems - Part 11-2: Measurement of wind turbine noise characteristics in receptor position" [8]. The PT-61400-11-2 WG is interested in comparing the IEC 61400-11:2012 and ISO/PAS 20065:2016 for analysing tones in wind turbine noise. DELTA – a part of FORCE Technology, suggested and volunteered to set up a round-robin test distributed among parties working with wind turbine noise, hereafter referred to as "laboratories".

Based on a prior project[4] 31 samples were constructed based on real wind turbine noise recordings and distributed these to participating laboratories. The tonality of the 31 samples have also been subjectively analysed in formal listening test by a panel.

This paper gives examples of analysis results of the round-robin tests and shows the uncertainties when a number of laboratories with different experience and different types of equipment analyse the same samples and identifies areas for additional research.

1. Introduction

It is critical to understand the differences between the two tone methods and understand any cases where there might be significant differences.

2. Working group TC 88/PT 61400-11-2

In 2018 the working group TC 88/PT 61400-11-2 was established, with 29 members from Europe, North America and Asia.

2.1 Scope

The scope of the working group is as follows:
"To establish standardized techniques and methods applicable for measurement of noise immission. The technical specification focuses on compiling methods for quantification of wind turbine noise in one or a set of documents in order to ensure a common reference taking the characteristics of wind turbine operational parameters into account.
Quantification of the acoustics phenomena at the receiver position includes the following topics:
- *amplitude modulation;*
- *low frequency noise;*
- *impulsivity;*
- *tonality;*
- *sound pressure levels;*
- *rating level, which includes adjustments for the above-mentioned phenomena, should be suggested if possible."*

2.2 Tonality

Presently IEC 61400-11 is the widely used method for analysing tones in wind turbine noise for sound power level measurements (i.e. measurements close to the turbine) whereas one version of the ISO tonality method (the old ISO 1996-2 or the new ISO/PAS 20065) is typically used for analysing tones in wind turbine noise at receptor position. Other methods are used as well.

As part of the work the group would like to compare the following two methods for analysing tones in wind turbine noise:

1. The tonality method described in IEC 61400-11:2012
2. ISO/PAS 20065:2016

Both methods can be described as a two-step method, where the first step evaluates the tonal audibility for each critical band of each spectra. The second step combines the tonal audibility to an overall tonal audibility.

One of the members of PT 61400-11-2 is performing an in-depth study of the two methods. In addition, the group has desired a round-robin comparison of the first step of the two methods, to ensure that differences between the two methods are found for more than one participant.

3. Round-robin comparison

3.1 Scope of the round-robin

Factors tested in the round-robin:

- Computation of the audibility of tones in each critical band.
- The correctness of the implementation of the standard
- The correct understanding of the algorithms used by each participant *
- The correct understanding of the use of the tone analysis algorithm.

*: An example of the correct understanding of the algorithms used by each participant could be the placement of the Critical Band centred around the tone. The ISO standard specifies this with a formula based square root of the product of the upper and lower frequencies of the critical band. This is different from the method implied by the IEC standard where the lower and upper frequencies of the critical band are placed symmetrically around the tone.

Factors not tested in the round-robin:

- Differences between the two standards due to
  - Data Recording
  - FFT Line Spacing: For IEC: must be from 1 to 2 Hz. ISO: from 1.9 to 4 Hz.
  - Averaging time for FFT: For IEC: 10 seconds, for ISO: about 3 seconds.
  - Impact of FFT overlap: IEC: at least 50%. ISO: Not specified
  - Differences due to frequency shifts of the wind turbine tones as related to the above parameters
- Uncertainty calculations.
- Combinations of multiple tones in a critical band.

3.2 Invitation

It was initially assumed that not many have implemented both methods, so additional participants outside the working group have been invited to ensure enough participants.
The invitation has been sent to:

- Members of IEC/TC88/MT 11
- Members of ISO/TC43/SC1/WG45
- Persons working with wind turbine noise and tonality known from for example the Wind Turbine noise conference in Rotterdam in 2017

Additionally, some of the invitees has been so kind to extend the invitation even further.
In total more than 100 persons/laboratories have been invited.

3.3 Stimuli

From a number of original recordings of wind turbines (recorded with Class 1 equipment) 24 samples of 20 second periods were selected. If several equally prominent tones in different critical bands were present in a sample, the least prominent were attenuated to facilitate an unambiguous and comparable assessment in the listening test and tone analysis. In additional, seven samples of stationary industrial noise with tones were added.
The samples - each mono recording and of duration of 20 seconds - were level aligned so that the A-weighted levels were the same for all samples. The intended presentation level after the level calibration was 50 dB(A).

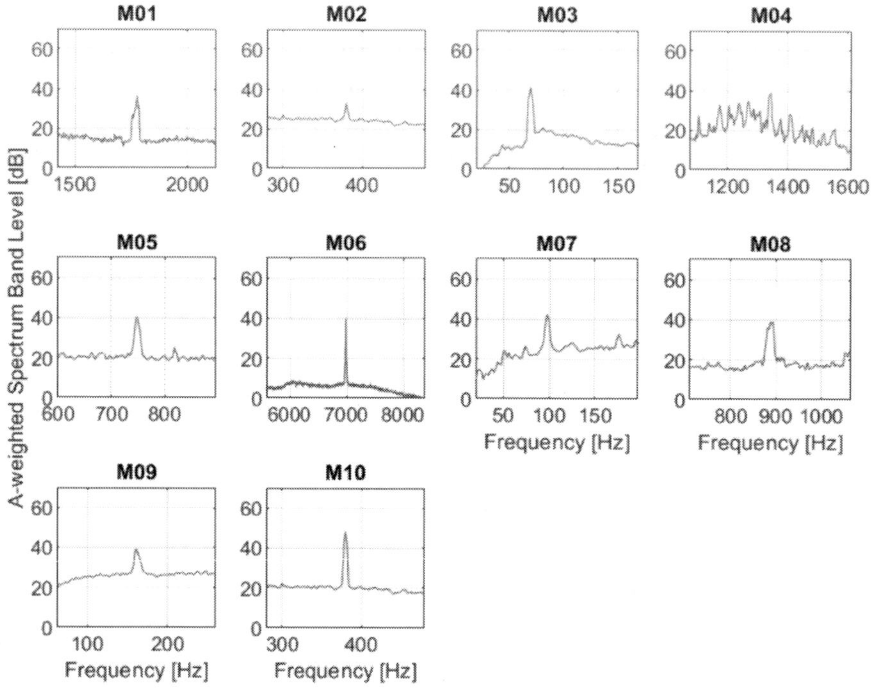

Figure 1. The 10 mandatory A-weighted spectra, M01-M10. The X axis is zoomed to only show the frequencies of interest.

The 31 spectra were randomized, numbered and grouped in ten mandatory (M01-M10) and 21 optional spectra (O01-O21). The A-weighted spectra are shown in the following three figures, Figure 1 - Figure 3 (where only 2 times the critical bandwidth surrounding the "true" tone is shown).

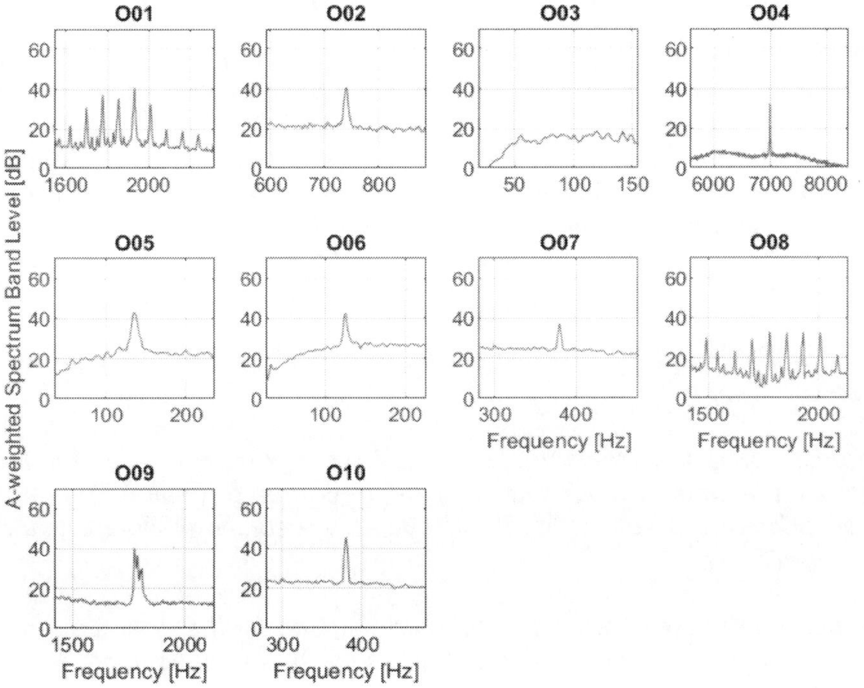

Figure 2. The first 10 optional A-weighted spectra, O01-O10. The X axis is zoomed to only show the frequencies of interest.

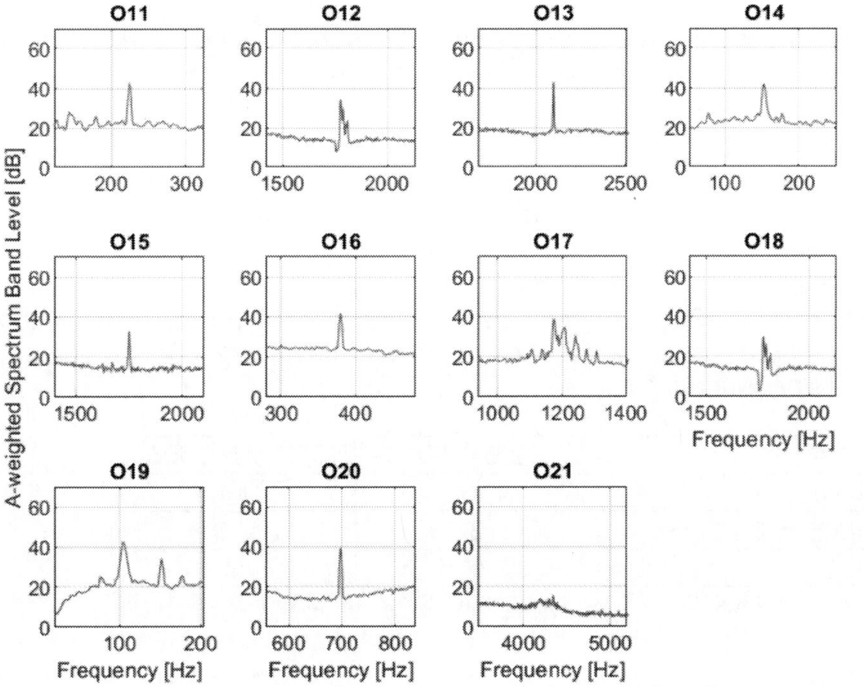

Figure 3. The last 11 optional A-weighted spectra, O11-O21. The X axis is zoomed to only show the frequencies of interest.

3.4 Instructions

The participants were given the following instructions (only relevant sections are shown):

"It is MANDATORY as a minimum to analyse the 10 spectra marked with an M with at least one of the above-mentioned methods. Preferable all 31 spectra should be analysed with both methods."

"To prevent too many variations, I hope everyone or as many as possible will be able to do the analysis based on the spectra. I will also provide a link to the wav files. You're of course very welcome to provide the results with both the spectra and the sound files. Please indicate whether you used the sound files or the enclosed spectra."

"You might find more than one tone in the spectra. You should report the most prominent/audible of the tones."

"The frequency resolution of the provided spectra is 2 Hz. If you analyse on the recordings, it is recommended to use the same frequency resolution, to compare results. You're of course very welcome to supply additional results with other frequency resolutions, where you document which settings you have used."

"The provided spectra are A-weighted, which should be used for the analysis. I however also attach the linear spectra in case your software applies the A-weighting itself. Please mark which spectra you have used."

"The sound files are 20 seconds long. If you use the sound files for analysis, please perform the analysis on the full 20 second data in order to compare it to the analysis with the spectra."

A pre-formatted xls sheet was supplied to be used when submitting the results. The submitted results will be anonymised, so no result can be tracked back to a participant.

3.5 Participation

29 persons/laboratories accepted the invitation, where 23 of those submitted results. 20 of the participants analysed using the IEC method and 15 of the participants used the ISO method. 12 participants submitted results with both methods. Two participants also submitted results using the old ISO standard, ISO 1996-2 Annex C, the results are not included in this analysis. The table below shows whether the analysis was performed based on A-weighted spectra, unweighted spectra or sound files.

	IEC 61400 -11	ISO/PAS 20065
A-weighted spectra	16	14
Un-weighted spectra Z weight)	3	1
Sound files	5	2

Table 1. Overview of number of results which each method and stimuli

4. Methods and parameters

4.1 IEC 61400-11

The analysis should be performed according to IEC 61400-11:2013/A1:2018 however notice that in section 9.5.5 the sentence "This corresponds to subtracting 10 log (1/1.5) or 1.8 dB

(1.76 dB) from the tone level." should have been "This corresponds to subtracting -10 log (1/1.5) or 1.8 dB (1,76 dB) from the tone level."
Earlier versions of the method can be followed as well, it is however then important to state which version is used. The method used, if for large turbines, and not the method in Annex F for small turbines.

For the analysis according to IEC 61400-11 the following parameters shall be stated:
- Frequency of the tone, F_c
- The sound pressure level of the tone $L_{pt,j,k}$
- The sound pressure level of the masking noise $L_{pn,j,k}$ in the critical band around the tone
- The tonality $\Delta L_{tn,j,k}$, the difference between the sound pressure level of the tone and the masking noise level
- The tonal audibility $\Delta L_{a,j,k}$, the difference between the tonality and the audibility criterion of the tone

4.2 ISO PAS 20065

For the analysis according to PAS 20065 the following parameters shall be stated:
- f_T, is the frequency of the maximum narrow-band level, in Hertz (Hz)
- L_T, is the tone level, in decibels (dB)
- L_G, is the masking noise, in decibels (dB)
- a_v, is the masking index, in decibels (dB)
- ΔL, the audibility, in decibels (dB)

5. Results

Each participant reported results in the pre-formatted xls sheet. Numerous good comments were also received and will be distributed to the respective maintenance standardization group. Unless marked otherwise, the comments will be anonymized before the distribution.

5.1 Tone frequency

For most of the spectra there was good agreement between the participants of the result of the tone frequency. Figure 4 and Figure 5 show the detection rate in % of the "true" tone frequency, where the "true" tone frequency will be referred to as the tone frequency the majority of the participants report. See also Table 2.

There are three examples of significant differences in the ability to find the "true" frequency.
- M02 (380 Hz): The ISO method only finds these in 2 of 15 instances and also identifies it as "no tone" in 11 out of 15 instances. Which method is "correct" for this type of barely audible tone may require more research. However, for this spectrum it will have no consequence, since the average tonal audibility is low (about -5 dB).
- O07 (380 Hz): This spectrum is the same as M02, but with a 5 dB higher tone level. However, the differences between the two methods in the ability to detect the tone, is virtually the same for M02 and O07. Further research may be required.
- O08 (1776 Hz): This is a complex, multi-tone spectrum. The ISO method finds the "true frequency" 67% of the time, while the IEC method finds it 25% of the time.

The above examples of the most extreme differences should be the basis for additional research to determine if the differences are reproducible, if they are "correct", and perhaps most important, if they have any significant impact on the final tonality. This research may also have high relevance to document the credibility of the respective standards.

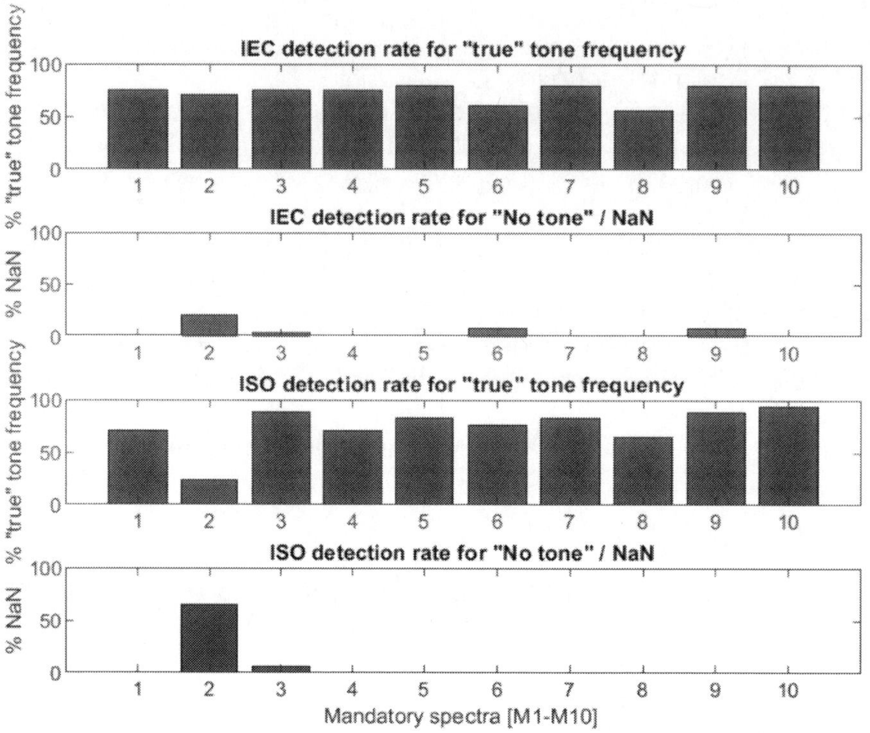

Figure 4. Detection rate for "true" tone frequency and "no tone" for mandatory spectra

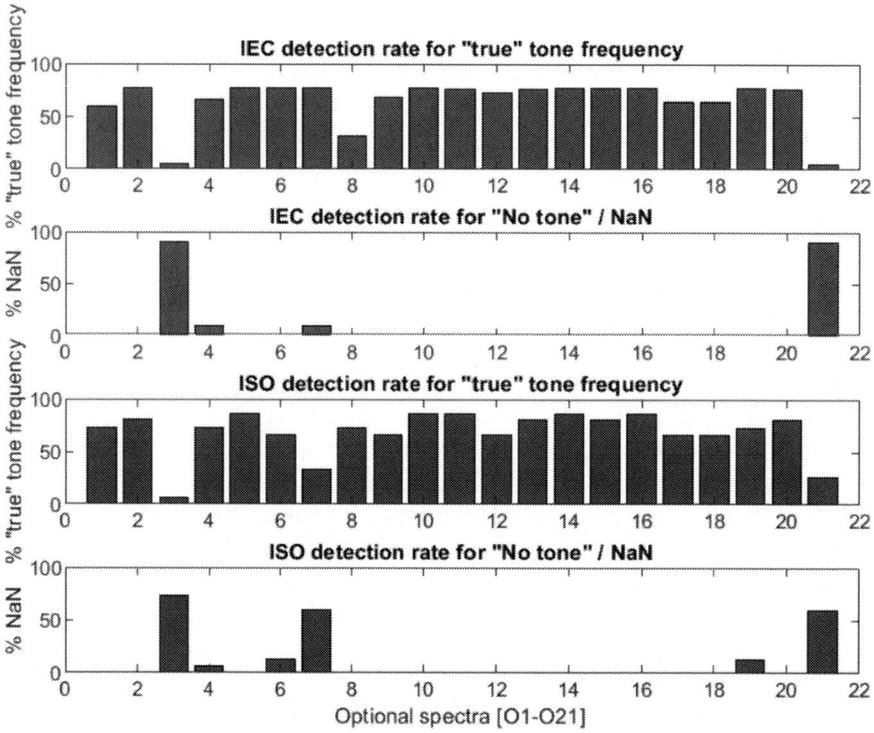

Figure 5. Detection rate for "true" tone frequency and "no tone" for optional spectra

5.2 No tone

Two spectra were included with no tones, O03 and O21. The detection rate for "no tone" or NaN are also shown in Figure 4 and Figure 5. The detection rate for "no tone" is high both for participants using the IEC and for participants using the ISO method. Apparently, the detection rate for "no tone" is higher for the ISO method than for the IEC method for spectra intended to contain a tone, for example M02 and O07. See also Table 2. It should be remarked that participants were not informed that spectra without a tone was included as well.

5.3 Tonal audibility

The average tonal audibility together with the standard deviation for each method is shown in Figure 6 calculated with all results. As it can be seen the deviation is quite large. When closer examined (not shown) some quite large deviations/outliers are found.

When outliers are removed, the average tonal audibility and the standard deviation for each method is shown in Figure 7. Compared with Figure 6 the deviation is significantly smaller now.

As described above the majority of the participants detected the same tone frequency. If the results are filtered so only those with the "true" tone frequency (regarding "true" tone frequency see section 5.1) are analysed (i.e. removed a second set of outliers), the average audibility and standard deviation are shown in Figure 8. Compared to Figure 6 and Figure 7 it is seen that the deviation is even smaller.

Figure 9 shows the standard deviation as a function of "true" tone frequency. It can be seen that variation is clearly largest in the 1-2 kHz frequency range, for spectra with multiple tones in the critical band. The reason for this has not been determined.

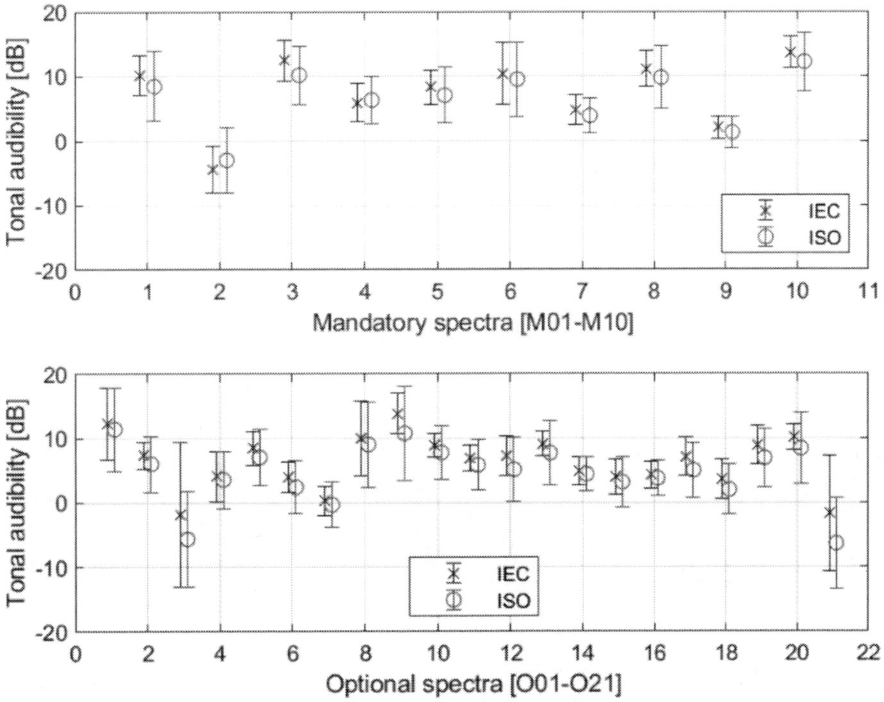

Figure 6 Average tonal audibility and standard deviation calculated with all results

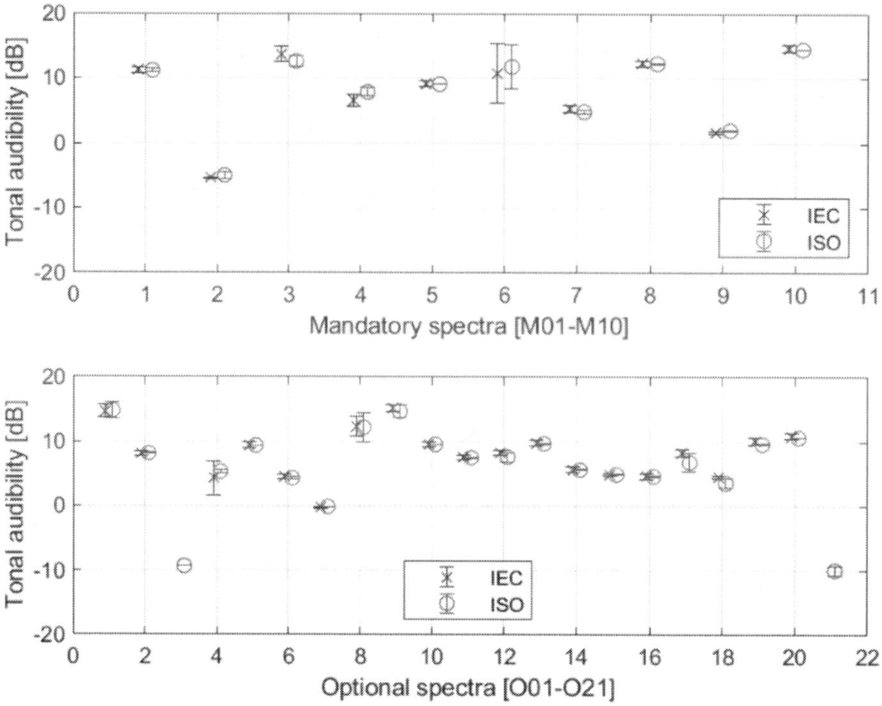

Figure 7 Average tonal audibility and standard deviation calculated without outliers

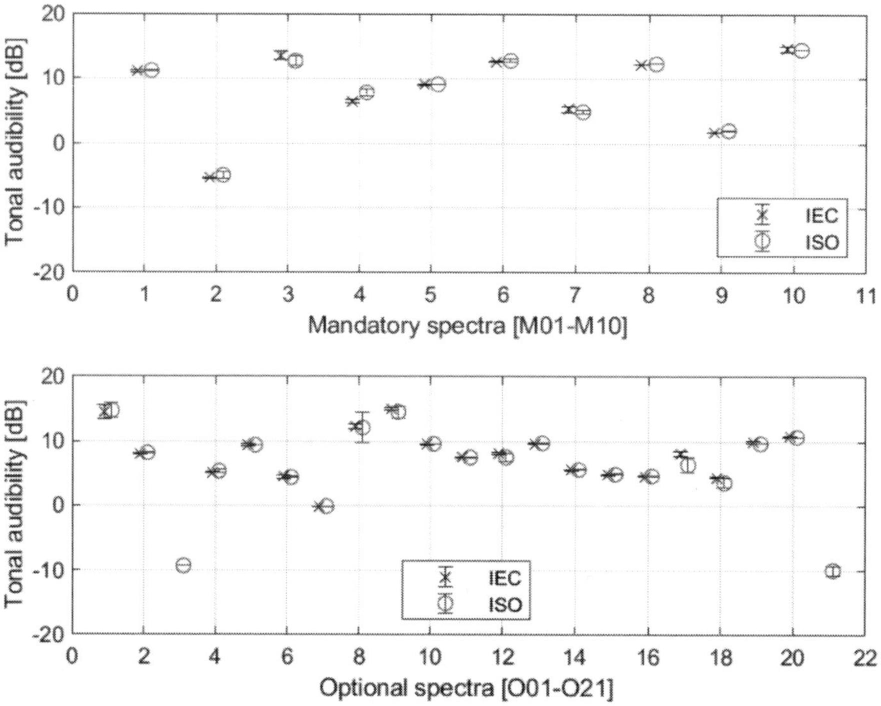

Figure 8 Average tonal audibility and standard deviation calculated only for "true" tone frequency

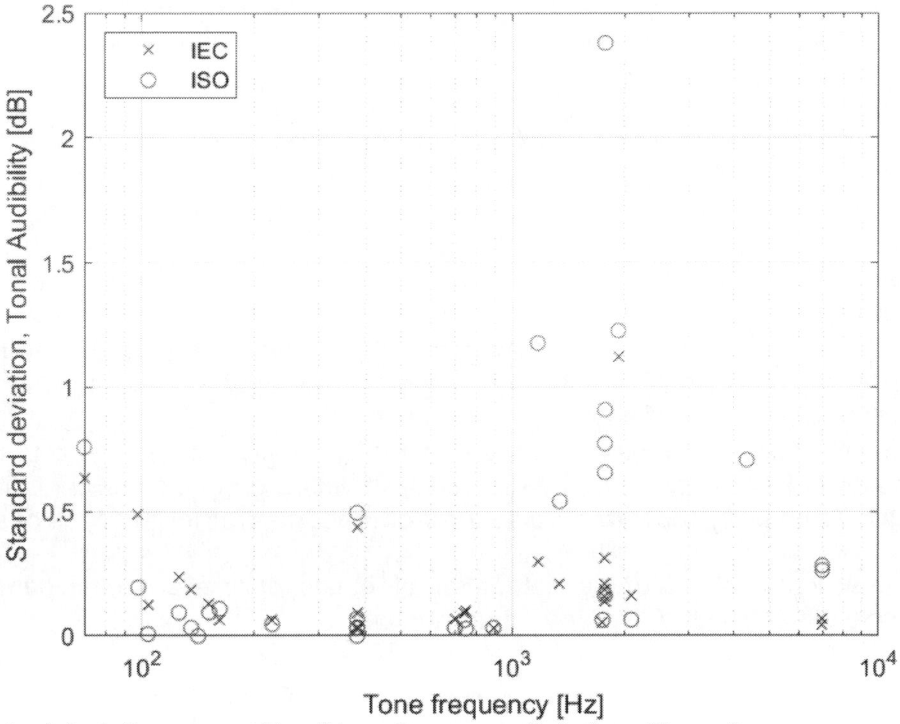

Figure 9 Standard deviation versus "true" tone frequency. Data from Figure 8.

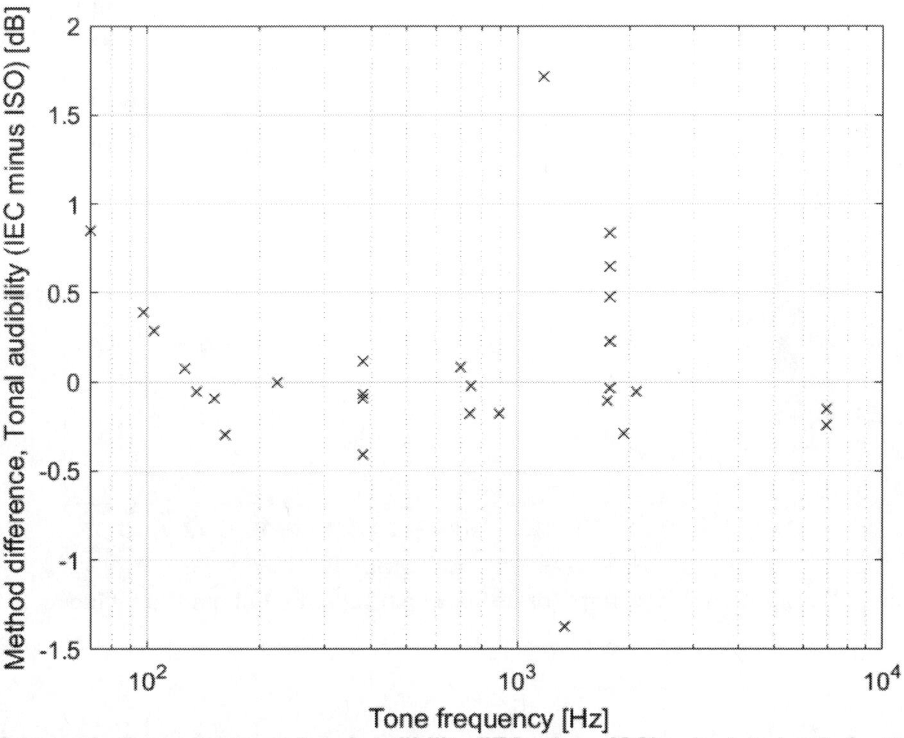

Figure 10 Difference between average tonal audibility (IEC minus ISO) calculated only for "true" tone frequency, shown versus "true" tone frequency

Figure 10 shows the difference between the average tonal audibility with the two methods. For most spectra the difference is small and below 0.5 dB, however in the 1-2 kHz area the difference is larger, which should be seen in comparison with the large deviation in standard deviation shown in Figure 9, and the spectra shown in Figure 1 - Figure 3.

If we choose an acceptable error of ± 0.5 dB, we note four outliers:

- M03 (70 Hz): The difference of 0.8 dB is in an area of the spectrum with a fairly steep downward slope below the 70 as well as very asymmetric skirts for the tone. Whether this is an inherent difference in the two methods has not been studied or determined.
- M04 (1344): This is a very complex multi-tone spectrum. The difference of 0.7 dB should be investigated as to whether it is related to this type of spectrum, and if it is a bias or random error.
- O17 (1172 Hz): This spectrum also contains multiple tones in the critical bands. This difference of 1.7 dB is significant. It should also be investigated if this a bias error or random error for this type of spectra.
- 018 (1776 Hz): This main tone has is surrounded by an asymmetrical spectrum, with a dip on the lower frequency side. The difference of 0.8 dB can be significant, and it should also be investigated if this a bias error or random error for this type of spectra.

In Figure 11 is shown an example of the distribution of results for different participants, shown as the difference to the calculated average Tonal Audibility.

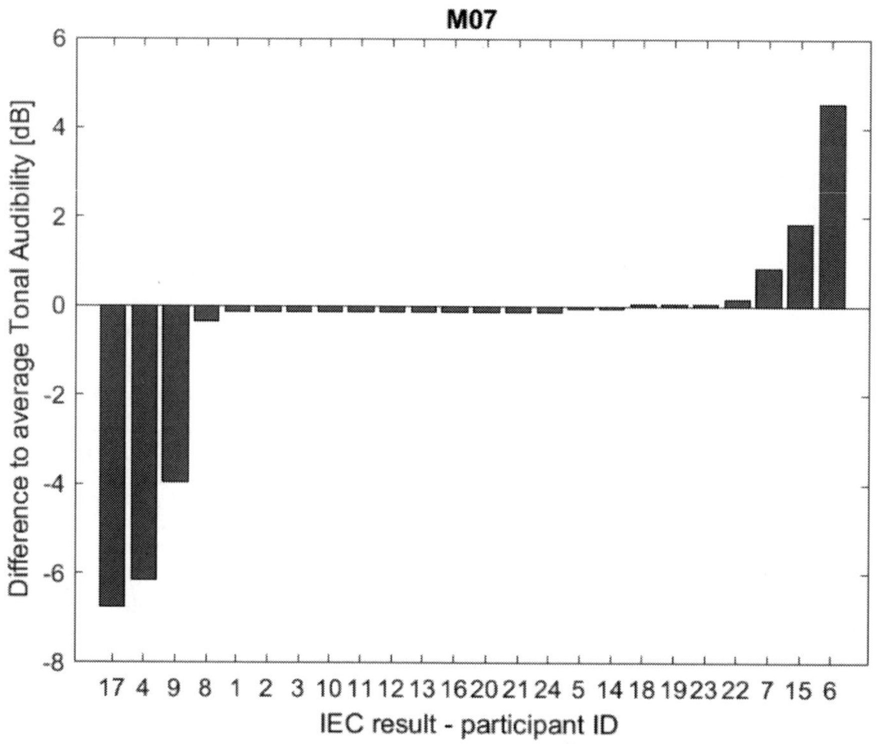

Figure 11. Example of Distributions of Results for different participants (all results included, also outliers)

The main results are shown in Table 2.

	(all participants) "True" tone frequency		Detection rate (all participants)				Tonal audibility (selected data)			
			Tone frequency		"No tone"		Average		Standard deviation	
	IEC	ISO	IEC	ISO	IEC	ISO	IEC	ISO	IEC	ISO
	[Hz]	[Hz]	[Number (%)]				[dB]	[dB]	[dB]	[dB]
M01	1776	1776	18 (75 %)	12 (71 %)	0 (0 %)	0 (0 %)	11.1	11.2	0.2	0.2
M02	380	380	17 (71 %)	4 (24 %)	5 (21 %)	11 (65 %)	-5.4	-5.0	0.1	0.5
M03	70	70	18 (75 %)	15 (88 %)	1 (4 %)	1 (6 %)	13.5	12.7	0.6	0.8
M04	1344	1344	18 (75 %)	12 (71 %)	0 (0 %)	0 (0 %)	6.5	7.8	0.2	0.5
M05	748	748	19 (79 %)	14 (82 %)	0 (0 %)	0 (0 %)	9.1	9.1	0.1	0.0
M06	6986	6986	14 (61 %)	13 (76 %)	2 (9 %)	0 (0 %)	12.6	12.7	0.0	0.3
M07	98	98	19 (79 %)	14 (82 %)	0 (0 %)	0 (0 %)	5.2	4.9	0.5	0.2
M08	888	888	13 (57 %)	11 (65 %)	0 (0 %)	0 (0 %)	12.1	12.3	0.0	0.0
M09	162	162	19 (79 %)	15 (88 %)	2 (8 %)	0 (0 %)	1.7	2.0	0.1	0.1
M10	380	380	19 (79 %)	16 (94 %)	0 (0 %)	0 (0 %)	14.6	14.5	0.4	0.0
O01	1930	1930	13 (59 %)	11 (73 %)	0 (0 %)	0 (0 %)	14.5	14.7	1.1	1.2
O02	740	740	17 (77 %)	12 (80 %)	0 (0 %)	0 (0 %)	8.0	8.2	0.1	0.1
O03	976	142	1 (5 %)	1 (7 %)	20 (91 %)	11 (73 %)	-	-9.3	-	0.0
O04	6984	6984	14 (67 %)	11 (73 %)	2 (10 %)	1 (7 %)	5.1	5.3	0.1	0.3
O05	136	136	17 (77 %)	13 (87 %)	0 (0 %)	0 (0 %)	9.4	9.4	0.2	0.0
O06	126	126	17 (77 %)	10 (67 %)	0 (0 %)	2 (13 %)	4.5	4.4	0.2	0.1
O07	380	380	17 (77 %)	5 (33 %)	2 (9 %)	9 (60 %)	-0.2	-0.1	0.0	0.0
O08	1776	1776	7 (32 %)	11 (73 %)	0 (0 %)	0 (0 %)	12.3	12.1	0.3	2.4
O09	1776	1776	15 (68 %)	10 (67 %)	0 (0 %)	0 (0 %)	15.0	14.6	0.2	0.9
O10	380	380	17 (77 %)	13 (87 %)	0 (0 %)	0 (0 %)	9.5	9.6	0.0	0.0
O11	224	224	16 (76 %)	13 (87 %)	0 (0 %)	0 (0 %)	7.5	7.5	0.1	0.0
O12	1776	1776	16 (73 %)	10 (67 %)	0 (0 %)	0 (0 %)	8.2	7.5	0.2	0.7
O13	2096	2096	16 (76 %)	12 (80 %)	0 (0 %)	0 (0 %)	9.7	9.7	0.2	0.1
O14	152	152	17 (77 %)	13 (87 %)	0 (0 %)	0 (0 %)	5.5	5.6	0.1	0.1
O15	1750	1750	17 (77 %)	12 (80 %)	0 (0 %)	0 (0 %)	4.8	4.9	0.1	0.1
O16	380	380	17 (77 %)	13 (87 %)	0 (0 %)	0 (0 %)	4.6	4.7	0.0	0.1
O17	1172	1172	14 (64 %)	10 (67 %)	0 (0 %)	0 (0 %)	8.1	6.4	0.3	1.2
O18	1776	1776	14 (64 %)	10 (67 %)	0 (0 %)	0 (0 %)	4.4	3.6	0.1	0.8
O19	104	104	17 (77 %)	11 (73 %)	0 (0 %)	2 (13 %)	10.0	9.7	0.1	0.0
O20	698	698	16 (76 %)	12 (80 %)	0 (0 %)	0 (0 %)	10.8	10.7	0.1	0.0
O21	150	4340	1 (5 %)	4 (27 %)	20 (91 %)	9 (60 %)	-	-9.9	-	0.7

Table 2. The "true" tone frequency is found on the basis on the results for all participants.
The detection rate columns show the number of participants submitting a result and is also calculated as a percentage of all participants for each particular spectrum. Notice that the number of results is not the same for all spectra.
The average and standard deviation for tonal audibility is only calculated with outliers removed and filtered for "true" tone frequency.

6. Listening tests

Formal listening tests were performed to assess the perceived tonality compared to the computed tonality.

6.1 Methodology and procedures

The SenseLabOnline Internet-based listening test software from DELTA – a part of FORCE Technology [5] was used for the tests. 31 sound samples mainly representing wind turbine noise with tones of different prominence and frequencies were prepared for presentation to the assessors. The samples were uploaded to SenseLabOnline which arranged the files in a random order in a double-blind paradigm for each assessor.

23 assessors from DELTA's expert listening panel (ages 22-52, mean 32 years) all with normal hearing and four acoustic experts (ages 33-64) with normal hearing for their age participated in the listening test. The assessors made the test as a "home test" via their own PC with soundcards and headphones (Sennheiser type HD 449) supplied from DELTA. The assessors were instructed to make the test in a silent room and time at home. The experts made the tests in DELTA's listening rooms with calibrated headphones (Sennheiser HD 650).

The participants had no prior information about the test signals. The total duration of the listening test was estimated to approximately 1 hour, including one repetition of all sound samples and including a training session where the assessors were trained in assessing tones in noise and scale usage with 38 other sound samples.

After the test, the assessor performance was checked with DELTA's eGauge tool [6] and five assessors with poor replication, discrimination or panel agreement were excluded before the final statistical analysis of the results. The methods used for these formal listening tests are based on best practice as described in "Sensory Evaluation of Sound" [7].

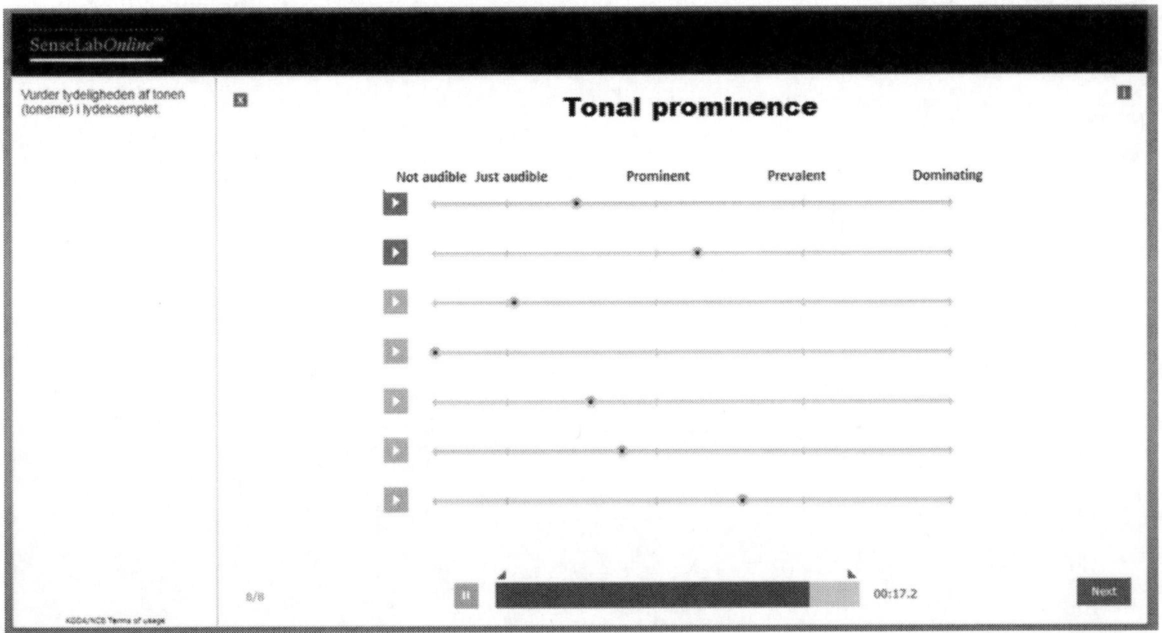

Figure 12. ***The user interface for the assessors in the SenseLabOnline test on prominence of tones in noise.***

The sliders shown in Figure 12 are used for the assessment of the stimuli. Beside the sliders are play buttons for the samples (sounds). The assessor is allowed to switch forth and back at will between any of the sounds during the test.

6.2 Level calibration

Each assessor started the listening test by adjusting the playback volume of an audio reference file with male speech, so the voice had a natural volume of a man talking at 1 m distance (nominal sound pressure level 64 dB(A)). From previous tests we have found out that the mean levels and deviations are within acceptable limits for this kind of test, see Table 3.

	L_{Aeq}, *dB*
Mean	63.9
Stand. Dev.	4.5
CI 95 %	1.8
Maximal difference	15.0

Table 3. ***Mean values, standard deviation and 95% confidence intervals (CI) for the level adjustment of male speech at 1 m distance made by 24 persons.***

6.3 Comparing the results from the round-robin with the listening test

The average tonal audibility from the round-robin are then compared with the results of the listening test, and the comparison is shown in Figure 13 and Figure 14.

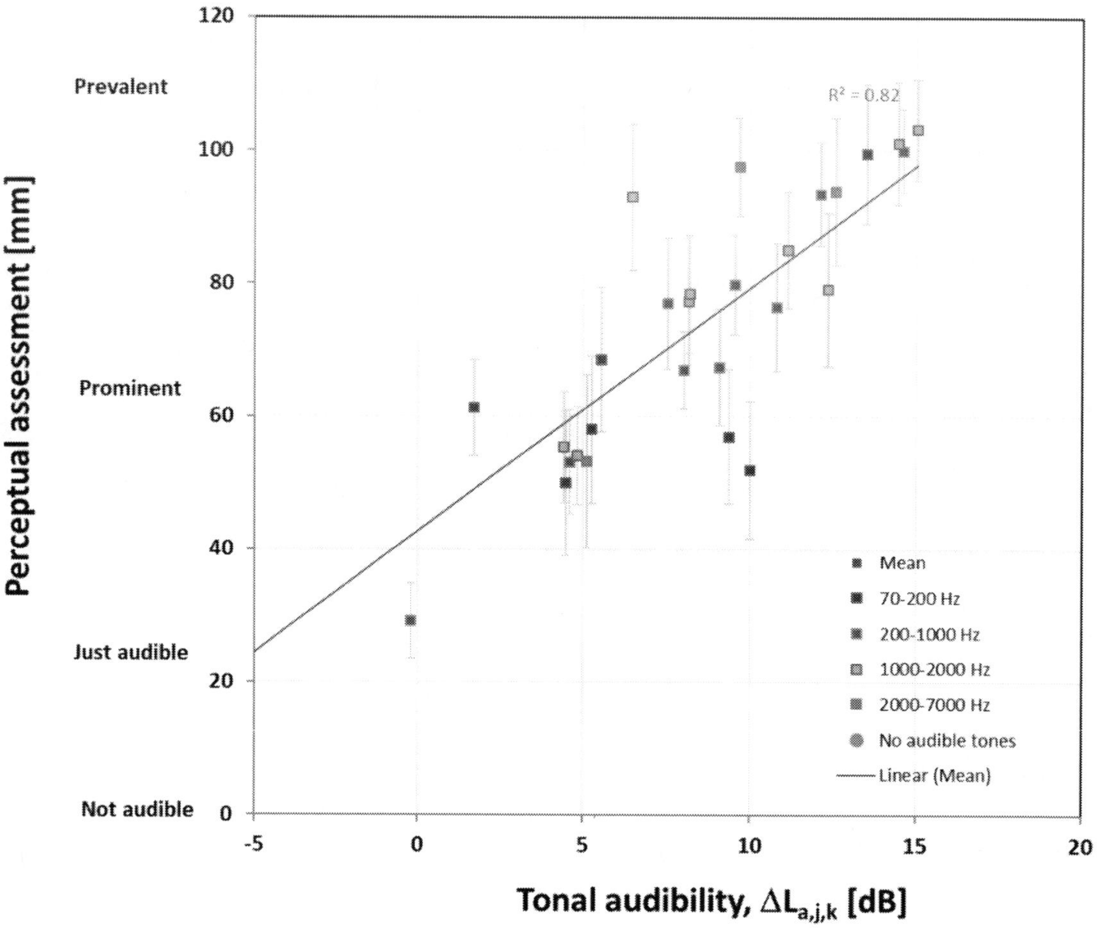

Figure 13. ***Tonal audibility according to IEC 61400-11 method versus the results of the listening test on an answering scale calibrated in mm. The vertical bars indicate the 95% confidence intervals. The percentage of explained variance of the linear regression is R^2 = 82 %.***

Figure 13 shows that the percentage of explained variance of the linear regression for the IEC 61400-11 is good (R^2 = 82%). The equivalent results for ISO PAS 20065 are essentially the same, where Figure 14 shows R^2 = 83 %.

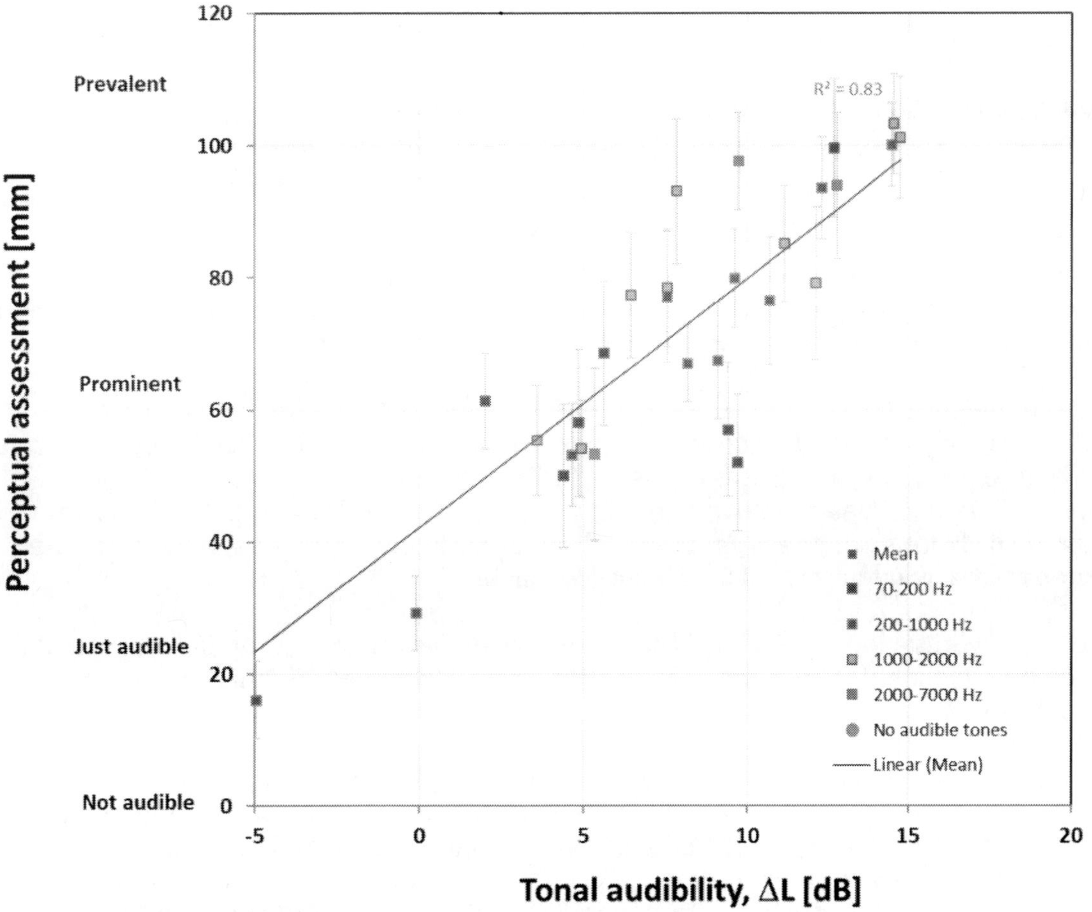

Figure 14. ***Tonal audibility according to ISO PAS 20065 method versus the results of the listening test on an answering scale calibrated in mm. The vertical bar indicates the 95% confidence intervals. The percentage of explained variance of the linear regression is R^2 = 83 %.***

7. Conclusions

A round-robin test has been performed comparing objective tone analysis with two methods, the method described in IEC 61400-11 and the new ISO PAS 20065 method.
Results were received from 23 different persons/laboratories, most submitting results with both methods.

This round-robin has focused on the first step of both tonality methods, investigating only the most significant tonal audibility for each spectrum. With this in mind, when comparing the results with the two methods, very similar results were obtained.

For both methods the clear majority of participants has determined the exact same tone frequency (except for the two spectra without a tone), and this tone frequency (here referred to as the "true" tone frequency) are the same for both methods.
When outliers have been removed and only results with the "true" tone frequency the average tonal audibility for each spectrum are very similar for the two methods, and where the difference is only larger than 0.5 dB for a few cases.

When the average tonal audibility is compared with a listening test the percentage of explained variance of the linear regression is R^2 = 82 % and 83 % for the IEC and ISO method respectively.

For both methods the majority of the participants clearly detected "no tone" for the spectra with no tone. However, where the majority of the participants with the IEC method detected a tone for the spectra with tones, the majority of the participants with the ISO method detects "no tone" for the two spectra with a 380 Hz tone.
However, additional research is recommended in the areas where significant differences have been found in this round-robin test (Section 5.1, 5.2 and 5.3). This research should attempt to determine if the differences are random or systematic errors, their root cause, and also investigate their potential impact on the final tonal audibility for the different spectrum types. In addition, it is important to recall that the scope of the round-robin is limited as described in Section 3.1, and hence is not a comprehensive comparison between the two standards.

Acknowledgement
The following persons/laboratories/companies (in alphabetic order) has kindly participated in this round-robin:

- Payam Ashtiani, Adam Suban Lowen, Kevin He, Allan Munro, Duncan Halstead **Aercoustics Engineering Limited**, Canada
- Mikel Amatriain **ACCIONA Energía S.A.**, Spain
- Guillaume Dutilleux **Acoustics Research Centre, NTNU/IES**, Norway
- Lars S. Søndergaard & Carsten Thomsen **DELTA - a part of FORCE Technology**, Denmark
- Philip Wocken **DNV GL (GL Garrad Hassan Deutschland GmbH)**, Germany
- Isabelle Schmich-Yamane **EDF DTG**, France
- Thomas Schroth **Enercon GmbH**, Germany
- Dr Kristy Hansen **Flinders University**, Australia
- Duc Phuc Nguyen **Flinders University**, Australia
- **Hayes McKenzie Partnership Ltd**, UK
- Tomohiro Kobayashi **Kobayasi Institute of Physical Research**, Japan
- Mrs. G. Arivukkodi **National Institute of Wind Energy (Ministry of New and Renewable Energy, Government of India)**, India
- Patrick Pannwitt **Nordex Energy GmbH**, Germany
- Arlinda Huskey **NREL**, USA
- **Peutz**, Netherlands
- Eugene McKeown **RPS**, Ireland
- Marko Arežina & Ben Coulson **RWDI**, Canada
- Lars Enggaard **SGRE**, Denmark
- Bo Søndergaard **Sweco**, Denmark
- Petri Välisuo **University of Vaasa**, Finland
- **Vestas**, Denmark
- **WG 45 Japan members**, Japan
- **Xi Engineering**, Scotland

References

[1] IEC 61400-11:2012+AMD1:2018 *Wind turbines - Part 11: Acoustic noise measurement techniques.*

[2] ISO 1996-2:2007 *Acoustics – Description, assessment and measurement of environmental noise. Annex C: Objective method for assessing the audibility of tones in noise.*

[3] ISO/PAS 20065:2016 *Acoustics -- Objective method for assessing the audibility of tones in noise -- Engineering method.*

[4] Søndergaard, L. S., Pedersen, T. H. (2013) *Tonality in wind turbine noise. IEC 61400-11 ver. 2.1. and 3.0 and the Danish/Joint Nordic method compared with listening tests.* 5th International Conference on Wind Turbine Noise (Denver, USA)

[5] www.senselabonline.com

[6] Lorho, G., Le Ray, G,. Zacharov, N. (2010). eGauge - *A Measure of Assessor Expertise in Audio Quality Evaluations.* AES 38th (Piteå, Sweden).

[7] Zacharov, N (Editor) (2019) *Sensory Perception of Sound.* CRC Press, Taylor & Lord Group

[8] Broneske, S., Søndergaard, B., (2019) *Measurement of wind turbine noise characteristics in receptor position – A new IEC Technical Specification.* 8th International Conference on Wind Turbine Noise (Lisbon, Portugal)

INCE*EUROPE*

8th International Conference
on
Wind Turbine Noise
Lisbon – 12th to 14th June 2019

Wind Farm Environmental Noise and Vibration Impact Analyses: Background Sound, Project Construction, and Assessment of Final Project Vibration Immissions

Ethan Brush – Acentech Incorporated, Cambridge, MA USA: ebrush@acentech.com

Summary

The siting of major wind energy developments requires numerous environmental factors to be studied, not only for the finished project but for the entire life cycle of a proposed project. The final noise immissions from a wind farm project are of course important, but the noise and vibration from construction and ongoing maintenance are important to consider as well. This paper will cover the many aspects of noise and vibration that must be considered in a wind energy developer's application. A brief review of wind energy developments in the Northeast United States will be covered.

We will also discuss how the background noise level at a given site is determined among a host of factors, including hub height wind speed, precipitation, and temperature. Ground vibration measurements at varying distances from an operating wind turbine will also be presented, providing additional information on the propagation of ground vibration surrounding wind farms.

1. Recent Wind Energy Developments in the Northeast United States

The political climate has slowed the growth of land based wind farm developments in the north-eastern United States in recent years. Many of the early stage developments after the 2016 elections have been put on hold. Further delays have occurred due to the emergence of white-nose syndrome affecting bats. Any additional threats to bats are now being more carefully examined.

Offshore wind energy in the north-eastern US is poised for growth. This region offers some of the best wind resources in the world. In 2018 the state of Massachusetts awarded a contract to develop 800 MW of offshore wind. Rhode Island also started the process for development of 400 MW of offshore wind. Other coastal states such as Maine, Connecticut, New York, and New Jersey are following suite with large proposals aimed at meeting future clean energy production goals. Maritime towns and ports are vying to be hubs for construction and maintenance to this newly forming coastal industry.

2. New York State – Article 10

In the state of New York, major electric generating facilities of any type larger than 25 MW are sited according to the state's Article 10 law. This law streamlines the application process for developers, while providing a rigorous process for local input and ensuring environmental and public health laws are followed. The process begins during the early planning of a facility and continues throughout the siting review, construction, and its operation.

When this law was enacted, local communities were worried that a state board could decide matters relating to their towns. To address these concerns, and to encourage community and stakeholder participation, developers in the Article 10 process are subject to an intervenor fee equal to $1,000 for each 1 MW of generating capacity of the subject facility, but no more than $400,000. These fees can be used by the local governments and other eligible people to pay for things like expert witnesses, consultants, and certain legal expenses.

2.1 Noise and Vibration Impact Studies

Exhibit 19 of New York's Article 10 law specifies the noise and vibration impacts that need to be considered in order for an application to be approved. The following list summarizes the required studies to be conducted.

1. Measurement and evaluation of ambient pre-construction baseline noise conditions at representative potentially impacted receptors that includes:
   - Summer Daytime/Night time Ambient Sound Level (Leq, Ln)
   - Winter Daytime/Night time Ambient Sound Level (Leq, Ln)
   - Combined Summer/Winter Ambient Sound Level (Leq, Ln)
   - The ambient pre-construction baseline sound level should be filtered to exclude seasonal and intermittent noise.

2. Evaluation of future noise levels during construction and operation of the facility at potentially impacted receptors using computer modelling to include:
   - A-weighted sound levels
   - Prominent discrete (pure) tones
   - Amplitude modulated sound
   - Analysis of whether the facility will produce significant levels of low frequency noise or infrasound

3. Identification and evaluation of reasonable noise abatement measures for construction activities and in the final design and operation of the facility

4. Evaluation of the following potential community noise/vibration impacts:
   - Hearing damage (as addressed by OSHA)
   - Indoor and outdoor speech interference
   - Interference in the use of outdoor public facilities and areas
   - Community complaint potential
   - Potential for structural damage
   - Potential for interference with technological, industrial or medical activities that are sensitive to vibration or infrasound

Because these impact studies apply to all electric generating facilities, some of the required evaluations are perhaps less intended for wind farms. Nonetheless, they still need to be addressed in a developer's application for a project in the state of New York. The rest of this paper will cover a few parts of these studies that may be useful to the wind turbine noise community.

3. Measurement of Ambient Pre-construction Baseline Noise Conditions

Measurement of site ambient sound levels is already a widely studied topic, yet the rules and metrics for arriving at an agreed upon ambient level can vary. For this reason, whenever possible it is expedient to measure and record sound data such that statistical metrics can be post processed in any number of ways. This requires the saving of 1/3 octave band Leq data in at least one-second increments.

3.1 Filtering Ambient Measurements

Figure 1 presents measurement data at a proposed wind farm during a two week summertime period. The overall A-weighted sound level is plotted for every second, which totals well over 1 million data points. The daytime (7:00 am – 10:00 pm) and night time (10:00 pm – 7:00 am) periods are delineated by the blue and pink data series. Ground and met tower wind speeds are also included on a secondary axis.

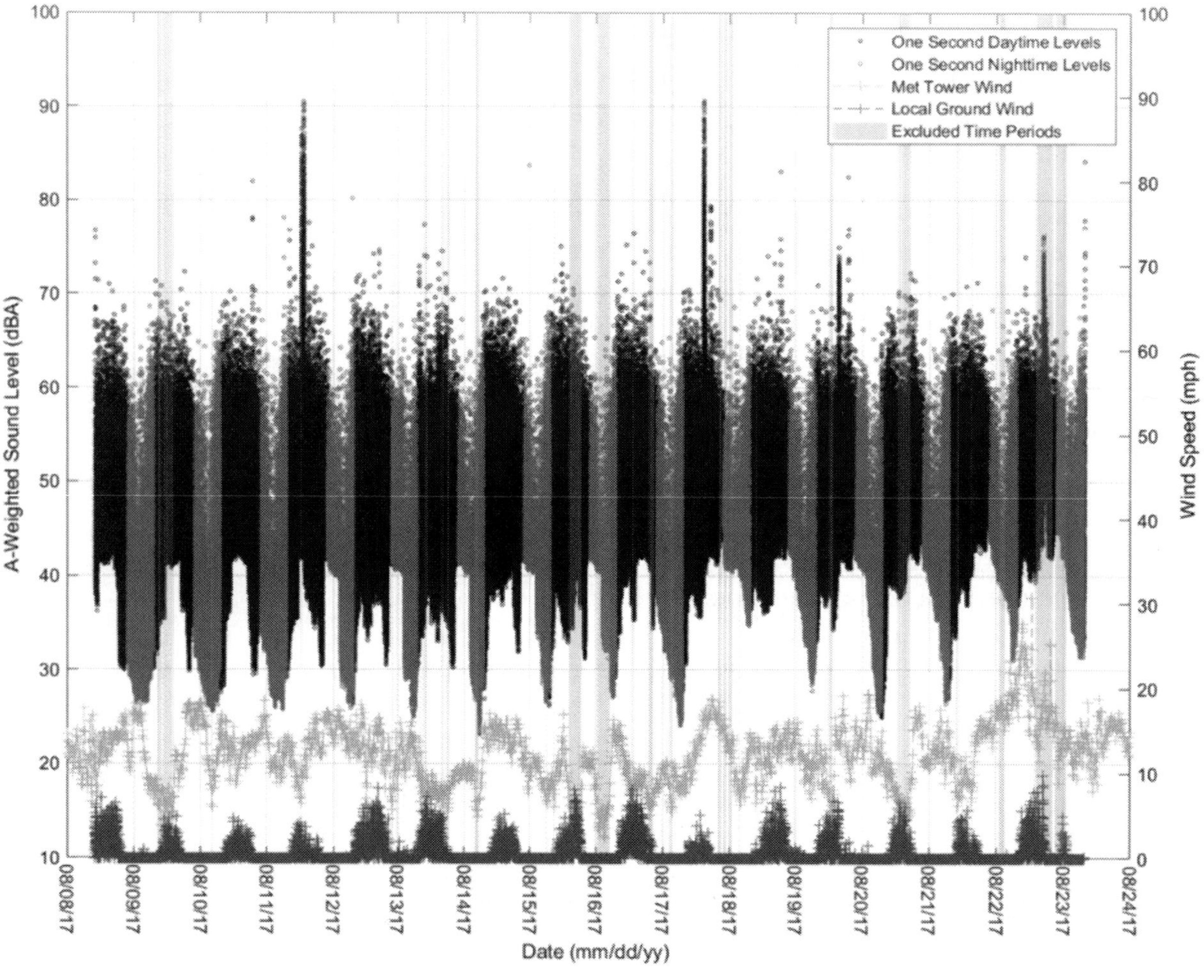

Figure 1: Representative Summer Ambient Noise Measurement Data

For this project it was required to filter days and nights for times without possible rain, snow, or temperatures outside the sound level meter specifications (-10 C to 50 C). Times when the met tower wind speed measured below the wind turbines' cut-in speed were also excluded. The shaded green areas on the chart in Figure 1 indicate the excluded time periods from the set of measured ambient data.

Figure 2 presents the ambient sound measurements at the same site during winter season. A much larger portion of the data set is excluded because of extreme low temperatures. The met tower also did not report data for some time, which required more exlusion of data.

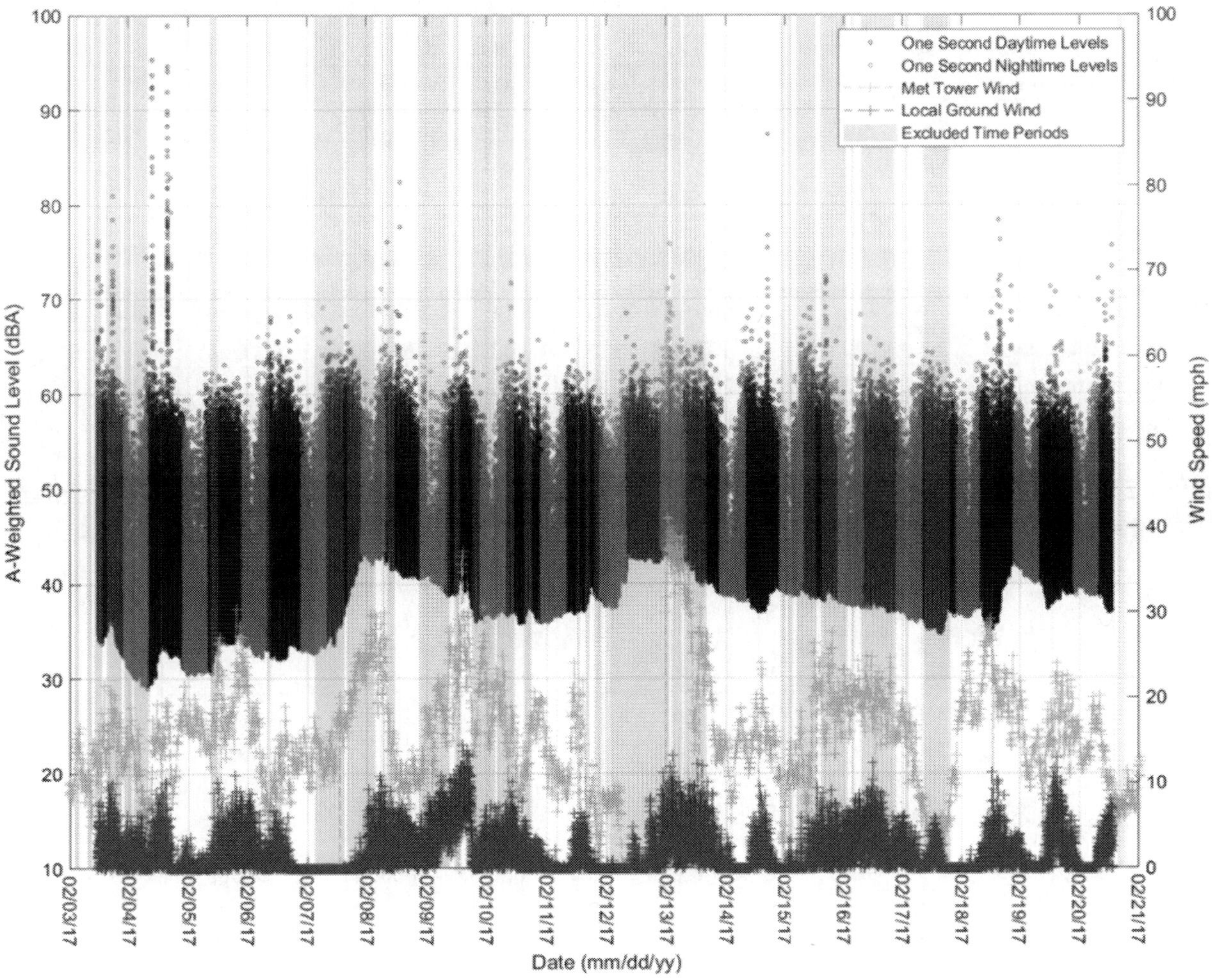

Figure 2: Representative Winter Ambient Noise Measurement Data

After filtering the summer and winter noise measurements the seasonal Leq and statistical metrics were calculated based on the remaining one-second data. Then the data from both seasons were combined in order to arrive at a year-round estimated ambient sound environment. There was roughly twice the amount of valid summer data as there was winter. Because of this the year-round statistical levels were calculated in two ways. First, the valid data points from each season were put together and the noise metrics calculated accordingly. Second, so that there was an equal number of data points for each season, the valid winter data was duplicated for the final noise metrics.

4. Construction Vibration

As discussed previously, the application for a wind development permit requires a study of the expected vibration impact that construction may have on nearby receptors. The nearest receptors could be residential or commercial buildings, each with their own limits for acceptable floor vibration. A list of equipment that might be used in the construction of a wind farm is provided in Table 1, along with the amount of ground vibration that could be expected at various distances using generic distance attenuation factors.

Table 1: Estimated Ground Vibrations Due to Various Large Construction Equipment [6]

Equipment	Distance from Activity (m)								
	10	15	30	60	90	120	150	180	215
	Estimated Ground Vibration Level (micro-m/sec)$_{rms}$								
Pile Driver (impact) (Upper range)	10,112	3,575	1,264	447	243	158	113	86	68
Pile Driver (impact) (Typical)	4,026	1,423	503	178	97	63	45	34	27
Pile Driver (sonic) (Upper range)	4,517	1,597	565	200	109	71	50	38	30
Pile Driver (sonic) (Typical)	1,135	401	142	50	27	18	13	10	8
Vibratory roller	1,273	450	159	56	31	20	14	11	9
Hoe Ram	569	201	71	25	14	9	6	5	4
Large bulldozer	569	201	71	25	14	9	6	5	4
Caisson drilling	569	201	71	25	14	9	6	5	4
Loaded trucks	507	179	63	22	12	8	6	4	3
Jackhammer	226	80	28	10	5	4	>3	>3	>3
Small bulldozer	20	7	>3	>3	>3	>3	>3	>3	>3

Figure 3 lists relative vibration sensitively limits for a number of sensitive and not-so-sensitive buidling uses. These can be used for comparing to the estimated vibration levels in Table 1.

Designation	Velocity Limits (μm/s)$_{rms}$	Applications of Limits
Potential damage to residential structures	5,000	
Offices	400	
Residences - Day	200	Average threshold of human perception / microscopes up to 40x
Residences - Night	140	
Operating Rooms	100	Perception threshold for the most sensitive persons
VC-A	50	Research labs, microscopes up to 400x
VC-B	25	Optical equipment on isolation tables
VC-C	12.6	Electron microscopes up to 30,000x / NMR, MRI, AFM
VC-D	6	Electron microscopes over 30,000x
VC-E	3	Unisolated optical research equipment
VC-F	1.56	Appropriate for extremely quiet research spaces; generally difficult to achieve in most instances, especially cleanrooms. Not recommended for use as a design criterion, only for evaluation.
VC-G	0.78	Appropriate for extremely quiet research spaces; generally difficult to achieve in most instances, especially cleanrooms. Not recommended for use as a design criterion, only for evaluation.

Figure 3: Typical Vibration Criteria [5]

Vibrations from blasting activities are not included in these charts. Blasting has the potential of producing considerable vibrations and noise. Both depend on the "charge" – the total amount of explosive set off at one time – and the "delay" (where the charge is divided into several smaller amounts and these are detonated with a brief, controlled, time delay from one to the next). At the usual distances from turbine installations to the nearest receptors the noise and vibration from blasting is unlikely to produce structural damage or cause adverse health effects. The vibration limits for residences or offices in Figure 3 may be briefly exceeded, but this is often mitigated by proper communication with potentially affected receptors well in advance of blasting activity. Blasting is usually only expected to happen a few times, whereas the other activities listed in Table 1 can happen throughout the construction process and some over the entire life of the wind turbine's operation.

5. Ground Vibration Near an Operating Wind Turbine

Exhibit 19 of New York State's Article 10 law requires an impact assessment to technological, industrial or medical activities that are sensitive to ground vibration. Very little reputable information could be found to support an assessment pertaining to wind turbines. Therefore, Acentech recorded vertical ground vibration measurements near operating utility scale wind turbines in a rural area. The results of one such measurement are presented in Figure 4.

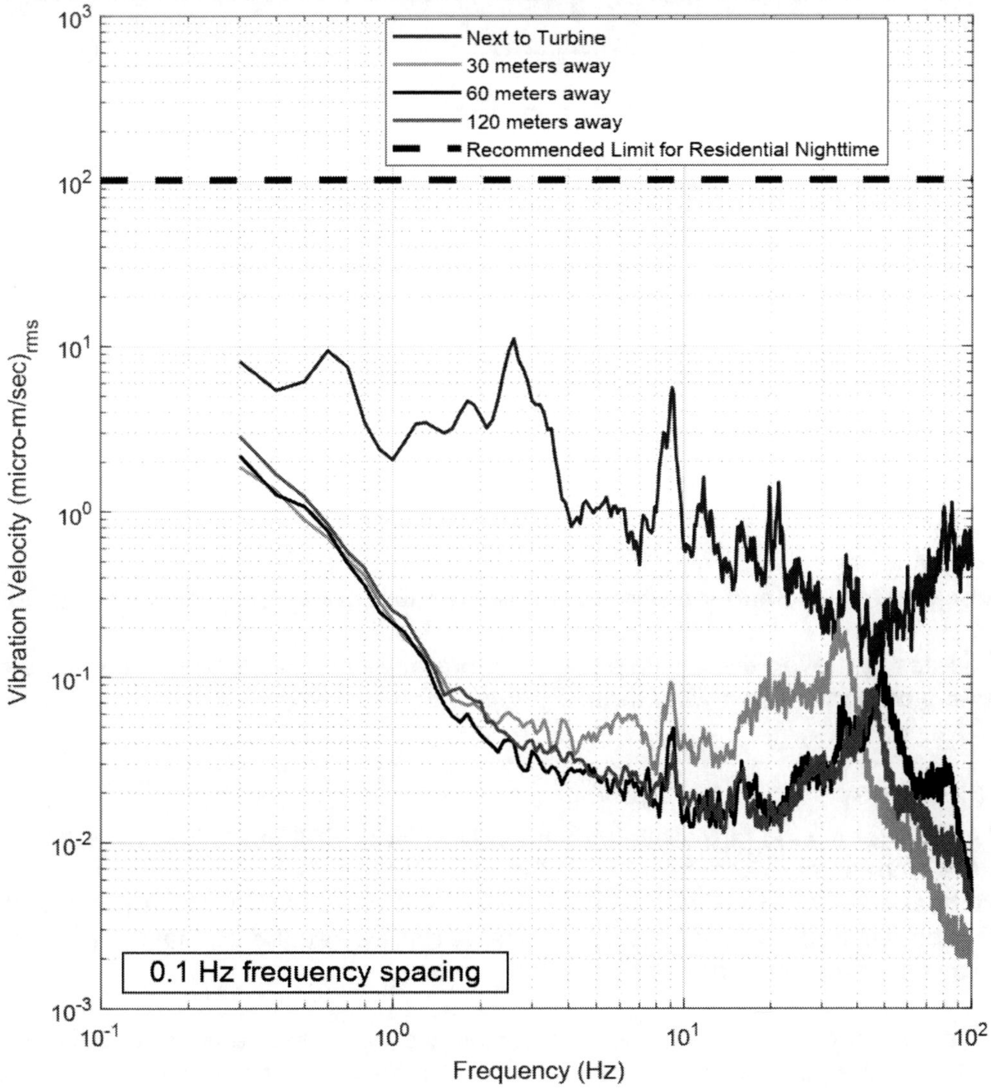

Figure 4: Narrow Band Vibration Data near an Operating a Utility Scale Wind Turbine

This chart shows the average vertical ground vibrations measured over a 10 minute period at each distance from an operating turbine. The results right next to the turbine show a peak at the blade pass frequency around 0.6 Hz as well as other harmonics and frequencies. At distances of 30, 60, and 120 meters away from the tower the vibration amplitude at the blade pass frequency is down at least an order of magnitude and perhaps even below the measurement noise floor. The dashed line on Figure 4 represents the recommended residential night time vibration limit listed in Figure 3.

For more perspective on how these measured vibrations compare to the most sensitive facilities, the data from Figure 4 has been converted to 1/3 octave band amplitudes for suitable comparison to the vibration criterion (VC) curves listed in Figure 3.

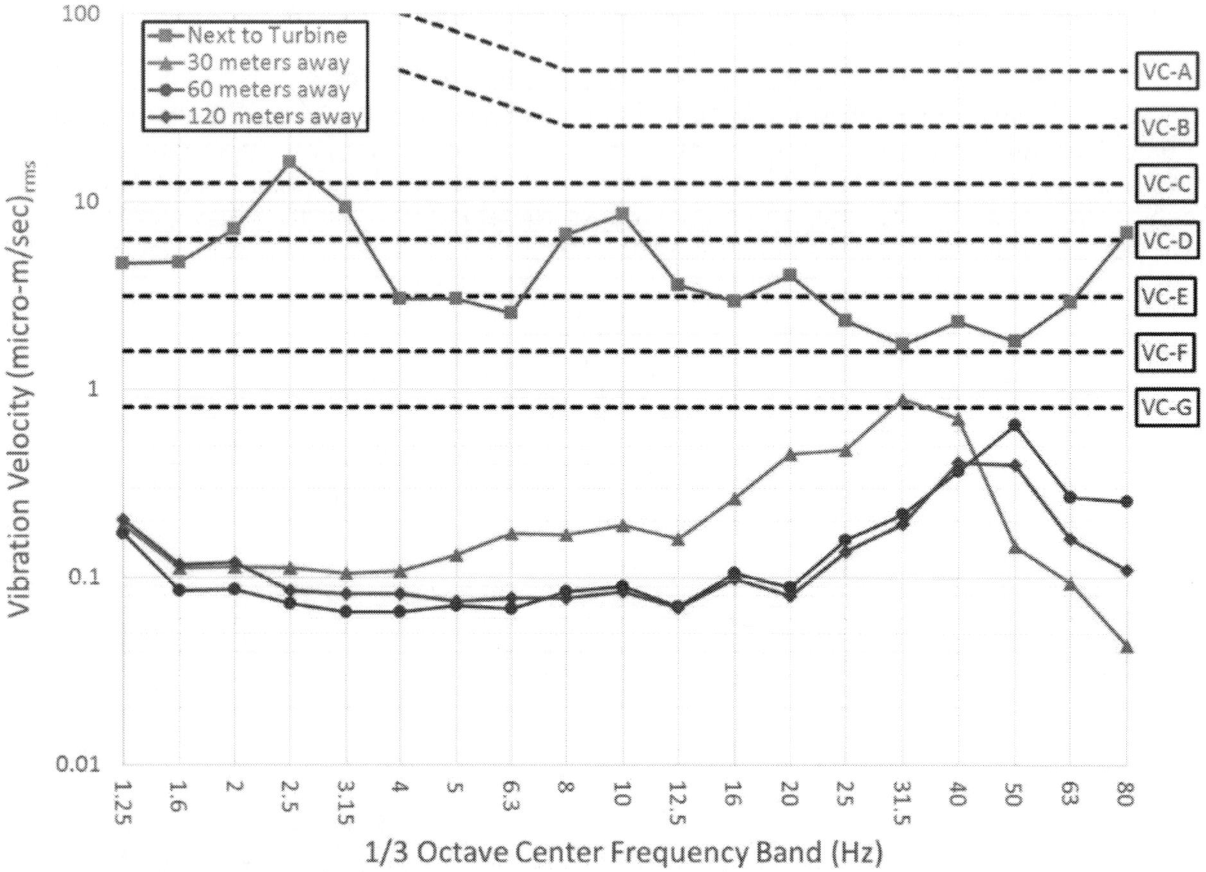

Figure 5: 1/3 Octave Band Vibration Data near an Operating Utility Scale Wind Turbine

Figure 5 indicates that even the most sensitive laboratories and hospital functions are not a risk due to ground vibrations from a wind turbine in operation.

6. Conclusions

The requirements for a wind farm development in New York State to follow the Article 10 law likely provides more work to acoustical and vibration consultants because of the wide ranging analyses that need to be considered. It is suggested that during ambient measurements the 1/3 octave and overall sound levels be recorded at least once a second so that sound metrics can be calculated (and re-calculated) based on many factors.

Guidance has been provided on the levels of ground vibration that could be expected during the construction and maintenance of wind turbines. We have also presented ground vibration measurements near operating utility scale wind turbines. This data shows that the levels of ground vibration around a single turbine are extremely small at even short distances away from the base of the tower. It is expected that Acentech will continue to gather ground vibration data near turbines in the future to supplement this observation.

References

[1] ISO 2631-2 Evaluation of human exposure to whole-body vibration – Part 2: Continuous and shock induced vibration in buildings

[2] New York State Article 10 Regulations (http://www3.dps.ny.gov/W/PSCWeb.nsf/All/143595FA3BE36AEA852579D00068B454?OpenDocument)

[3] Pellerin, Tricia (2017) *Background Noise Variability Relative to Wind Direction, Temperature, and Other Factors* 7th International Conference on Wind Turbine Noise

[4] Sondergaard, et al. (2017) *Variation of wind induced non-turbine related noise due to position, shelter, wind direction and season* 7th International Conference on Wind Turbine Noise

[5] Ungar, Sturz, Amick (1990) *Vibration Control Design of High Technology Facilities* Sound and Vibration, July 1990

[6] United States Federal Transit Administration (2018) *Transit Noise and Vibration Impact Assessment Manual* September 2018

INCE*EUROPE*

8th International Conference
on
Wind Turbine Noise
Lisbon – 12th to 14th June 2019

Session B

Low Frequency and Amplitude Modulation

Organised by INCE-Europe
mail@inceeurope.org
www.inceeurope.org

Supported by Portuguese Acoustical Society

spa Sociedade Portuguesa de Acústica

INCE*EUROPE*

8th International Conference
on
Wind Turbine Noise
Lisbon – 12th to 14th June 2019

How Critical is Low Frequency Noise for Micrositing?

Thomas Sørensen, EMD International A/S, Niels Jernes vej 10, 9229 Aalborg Ø, Denmark: ts@emd.dk
Adit Nand Kishore, EMD International A/S, Niels Jernes vej 10, 9229 Aalborg Ø, Denmark: ak@emd.dk.

Summary

Many national and regional Environmental Impact Assessment (EIA) guidelines state that low frequency noise impact from wind turbines must be examined, but only a few have described a specific procedure for this examination. The result is that low frequency noise is often either ignored, shrugged off as irrelevant or examined using a randomly selected approach that may or may not produce useful results.

Three countries, Finland, Sweden and Denmark, have described detailed procedures for calculating low frequency noise impact and, while not valid outside their respective countries, these procedures can be used for testing whether low frequency noise is potentially relevant for consideration.

This study presents and discuss the three guideline approaches and compares the low frequency impact across guidelines at dwellings located on the boundary for regular noise targets by testing if low frequency noise is critical according to the guidelines when regular noise targets are met.

This is done by using a number of anonymized noise datasets from actual commercial wind turbine types available to EMD. The datasets are tested by applying them to each of the three national noise codes in as many situations as they describe, such as onshore/offshore settings, day and night conditions and light dwellings versus regular dwellings. A narrow margin or an exceedance of guideline target criteria for low frequency noise means that low frequency noise would be a problem using this guideline and therefore could potentially reveal a problem in third countries without a specific guideline.

By using as many noise datasets from as many turbine types as possible, the extent of low frequency problems is uncovered to verify if the general assumption that low frequency noise is becoming less of a problem in planning as modern turbines are shifting noise emission away from the low frequency range, is true.

1. Introduction

A decade or so ago low frequency noise from wind turbines was a hot topic with concerns raised on health issues and on how to regulate the impact of low frequency noise from wind turbines. Since then the topic has to some extent cooled down, which begs the question if the low frequency issue is now a thing of the past and, secondly, how can we take low frequency noise into account in the planning process?

For planning it is useful to have a code with an operational methodology and tangible targets. Such targets may be based on a health consideration, but from a planning point of view we are mostly interested in the target itself. If we can stay within the target, we are good. Some countries have written low frequency noise into the national noise codes, either as a formalized model or in a loose formulation on that low frequency noise must be considered. The first of these are useful for planning, the latter less so. It would be desirable to get specific targets rather than fluffy ones, but only if low frequency noise is actually critical. If we could work on the assumption that low frequency noise from wind turbines is a thing of the past and that modern wind turbines will always comply with low frequency noise targets when they comply with regular noise targets, maybe there is no point to include low frequency noise in planning.

Colloquially there is a number of rules of thumbs going around in the wind energy community

1. Modern turbines do not have low frequency issues. Those were problems on older models. If you can comply with the standard noise target, you will not have problems with low frequency noise.
2. Reducing noise on wind turbines in the form of noise reduced operation reduces regular (normal range frequency) noise, but does little for low frequency noise
3. The larger the wind farm the more important low frequency noise becomes, because the target moves away from the wind farm and low frequency sound is attenuated less with distance than higher frequency sound.

The purpose of this study is to test these claims.

To achieve this goal we use the Danish, Swedish and Finnish codes for low frequency noise from wind turbines to test if low frequency noise becomes critical at a distance from the wind farm where the standard noise target is just met.

This test is conducted with a selection of anonymized modern turbines to get a representative, but not exhaustive sample group. The point here is not to point fingers at specific turbine types, but to answer the general question on whether low frequency noise can be critical for planning.

2. Noise codes

Denmark, Sweden and Finland have each defined a code for low frequency noise in planning of wind farms. The codes are different from each other and it is therefore useful to test each of these against the wind turbine types.

2.1 The Danish noise code

The currently valid Danish noise code is “Bekendtgørelse nr 135 af 07/02/19” (Miljø og Fødevareministeriet, Denmark, 2019). It defines noise targets as well as methodology in assessing the noise impact.

It operates with two types of planning zones: open land and areas of increased sensitivity with normal noise targets at 6 and 8 m/s at 10 m height as in table 1. The low frequency target is 20 dB(A), summed up in the frequency range 10 to 160 Hz. Recently a distinction has been made between standard dwellings and light dwellings (sommerhuse), where reduced noise insulation values are used based on a study from the Danish EPA (Søndergaard, 2016). Light dwellings are only considered as such if they are in an area of increased sensitivity.

Table 1. Wind turbine noise targets, Denmark

Zoning	6 m/s	8 m/s
Open land	42 dB(A)	44 dB(A)
Area of increased sensitivity	37 dB(A)	39 dB(A)
Low frequency: Any	20 dB(A)	20 dB(A)

The methodology for both regular and low frequency noise is a logarithmic model. For low frequency calculations the calculation point is indoor while it is outdoor for regular noise. For offshore wind turbines a reduced ground attenuation is used as well as multiple reflections.

2.2 The Swedish noise code

The Swedish noise code for wind turbines is administrated by Naturvårdsverket. For the present the guideline concerning regular noise is described in "Ljud från vindkraftverk" (Naturvårdsverket, 2010), while the low frequency guideline is only described on the home page of Naturvårdsverket (Naturvårdsverket, 2019). However, a new guideline is expected in 2019 at which point the methodology described here may be rendered invalid.

There are several noise targets in place for regular noise, but the most commonplace is 40 dB(A) at 8 m/s at 10 m height. The guideline methodology is a logarithmic model and the calculation point is outdoor. However, it is possible to use alternative calculation methodologies such as Nord2000.

For low frequency noise, a dwelling must be considered if the difference between C and A weighted noise at the receptor is higher than 20 dB (range 32-200 Hz). The noise target is a frequency curve supplied by Folkhälsmyndigheten (Socialstyrelsen, 2008) in the same range. The noise is calculated indoor. There is no specific model advised, though there seem to be a preference for Nord2000. Nor are there any official noise insulation values available.

The Swedish low frequency noise target is presented in figure 1.

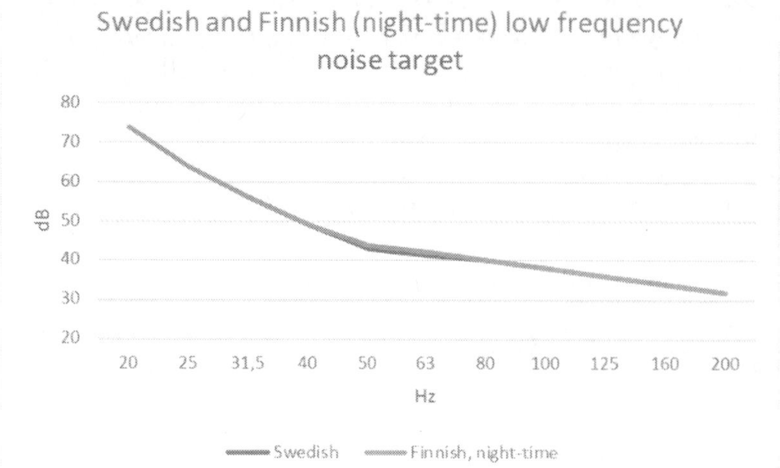

Figure 1. The Swedish and Finnish (night-time) low frequency noise targets are almost identical.

2.3 The Finnish noise code

The Finnish target for regular noise impact is described by statutory order (Miljöministeriet (Ministry of the Environment), 2015) to 40 dB(A) at night and 45 dB(A) at day for most receptors of interest. The low frequency noise target is not covered by the statutory order, instead the national residential health guideline (Sosiaali- ja terveysministeriön (The Finnish Ministry of Social Affairs and Health), 2003) is commonly used. This guideline sets the target as an indoor day and night frequency curve from 20 to 200 Hz, of which the night spectrum is the stricter.

The methodology is described in a guideline (Miljöministeriet (The Finnish Ministry of the Environment), 2014), advising a method for early stage planning and one for later stage. The early stage used here is a variety of the ISO 9613-2 (International Standards Organization, 1996) model, whereas for low frequency noise a logarithmic model similar to the Danish model is recommended. While the regular noise receptor is outdoors, the low frequency receptor is indoor. No advice is given concerning noise insulation values.

The Finnish low frequency noise target for night-time is presented in figure 1.

3. The test setup

The test setup is fairly simple in principle. For each permutation of the codes, a calculation point is found at the distance to a wind farm where the regular noise code is met. A noise target x with the methodology of noise code y results in a point z meters from the closest wind farm where the target is exactly met. For this point the low frequency noise is calculated with the guideline methodology and noted. If the low frequency target is exceeded, low frequency noise is the critical parameter. The margin to the target can be the margin in dB to the combined target (Danish code) or the margin to the target frequency curve at the most critical frequency (Sweden and Finland).

The tests made are listed in table 2. For the Danish tests, a distinction is made between onshore and offshore and between open land zoning or areas of increased sensitivity zoning. The later with a distinction between standard dwelling and light dwelling.

In the Swedish case only the 40 dB(A) target is used, being the most common target (though targets as low as 35 dB(A) is sometimes used). Since the low frequency target does not change, restricting the regular noise level will make low frequency noise less critical. The night low frequency target is used. As for low frequency propagation model, a Nord2000 model is set up using flat, very hard terrain (class G) and typical temperature and humidity settings. Noise insulation values are adopted from the Danish codes, using the standard insulation values for regular dwellings and light dwelling values to represent wooden houses.

For the Finnish case, the night 40 dB(A) target is used for regular noise and the night frequency curve as low frequency target. Standard values are used for the terrain. As for noise insulation the Swedish considerations are applied.

Table 2. Test scenarios

Code	On- or offshore	Zoning	Noise insulation
Danish code	Onshore	Open land	Standard dwelling
		Areas of increased sensitivity	Standard dwelling
			Light dwelling
	Offshore	Open land	Standard dwelling
		Areas of increased sensitivity	Standard dwelling
			Light dwelling
Swedish code	Onshore	40 dB zoning	Standard dwelling
			Light dwelling
Finnish code	Onshore	40 dB zoning	Standard dwelling
			Light dwelling

Calculations are done in windPRO for three different turbine configurations: 1, 5 and 21 wind turbines. The 5 and 21 turbine configurations are presented in figure 2

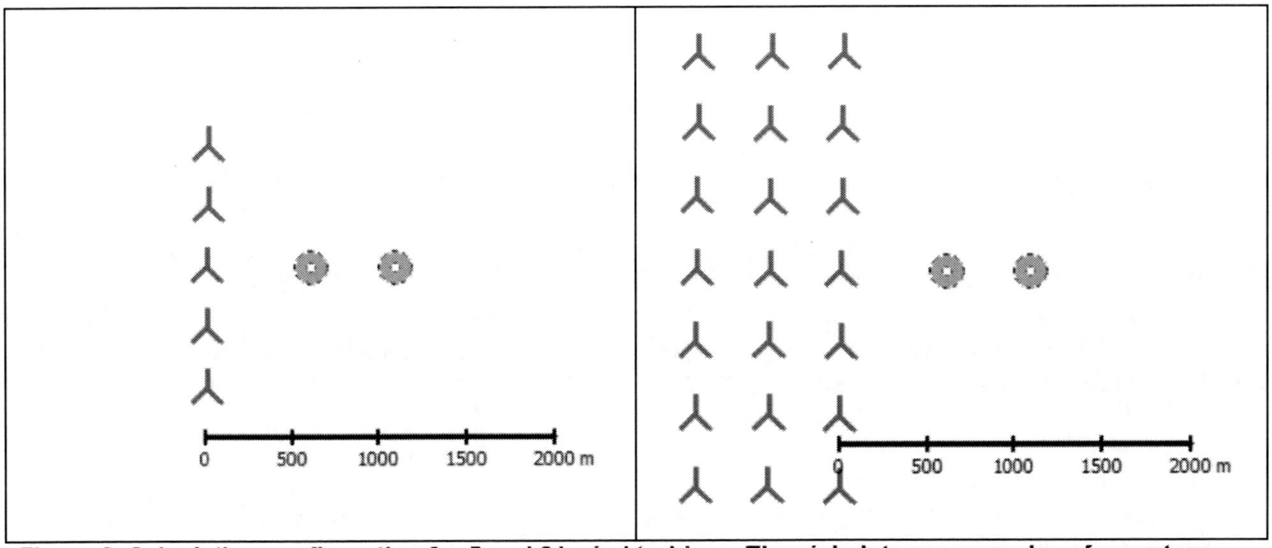

Figure 2. Calculation configuration for 5 and 21 wind turbines. The pink dots are examples of receptors.

4. Selection of wind turbine samples

The selection criteria for the wind turbine types to be used for the test were the following

1. The wind turbines had to be modern types in use for current developments
2. The wind turbines had to be comparable in size. The size range is 3.5 to 4.5 MW and 115 to 145 m rotor diameter.
3. The wind turbines had to be representative across suppliers, hence models were used from five different suppliers
4. Low frequency data had to be available to EMD
5. Each turbine type had to be able to be operated in a standard and in a noise reduced mode.

Most importantly the data had to be anonymized, as we have no intention of pointing fingers at specific wind turbine models. Furthermore, the data used are the data that suppliers make available for planning purpose (sales data) as these data are, if not public, then fairly easy to obtain.

An early consideration was to use measured data as supplement. However, these have a much higher confidentiality sensitivity and those in our possession are often from the prototype phase and are consequently not necessarily representative of the commercial version of the wind turbine. Therefore, only supplier provided data was used.

If available, data for serrated blades are used as these will typically be employed near settled areas. The standard mode will therefore, if available include serrations. No particular criteria are set for selecting a noise reduced mode, except that the noise reduction must be significant. This may be achieved through reduced rotational speed or reduced power or a combination of the two. It is possible that low frequency noise responds differently to these methods of reducing regular noise, but it was outside the scope to investigate this.

In total 10 turbine types from 5 suppliers were used, labelled 1 through 11 (model 4 was cancelled). Standard operation mode carries the suffix a and noise reduced mode the suffix b.

5. Results

5.1 Compliance

The low frequency noise impact on dwellings located exactly where the regular noise target is met was calculated for each of the scenarios in table 2 with each of the wind turbine types in two operation modes. The results are presented for a selection of the scenarios below.

For the Danish code case, the calculated low frequency noise impacts at the dwelling are presented with the orange columns, up towards the target of 20 dB(A). The black lines represent the distance from the dwelling to where the low frequency target would be met. In most cases there is a comfortable margin up to the target and the dwelling may be a safe distance away from where the target is met, but not for all turbines.

For onshore sites a single turbine type will be in trouble in the single turbine layout at the receptor in open land. At the area of increased sensitivity, the stricter regular noise target means that the low frequency impact is well beneath the target, but a change to light dwelling noise insulation gives a result very similar to the open land scenario.

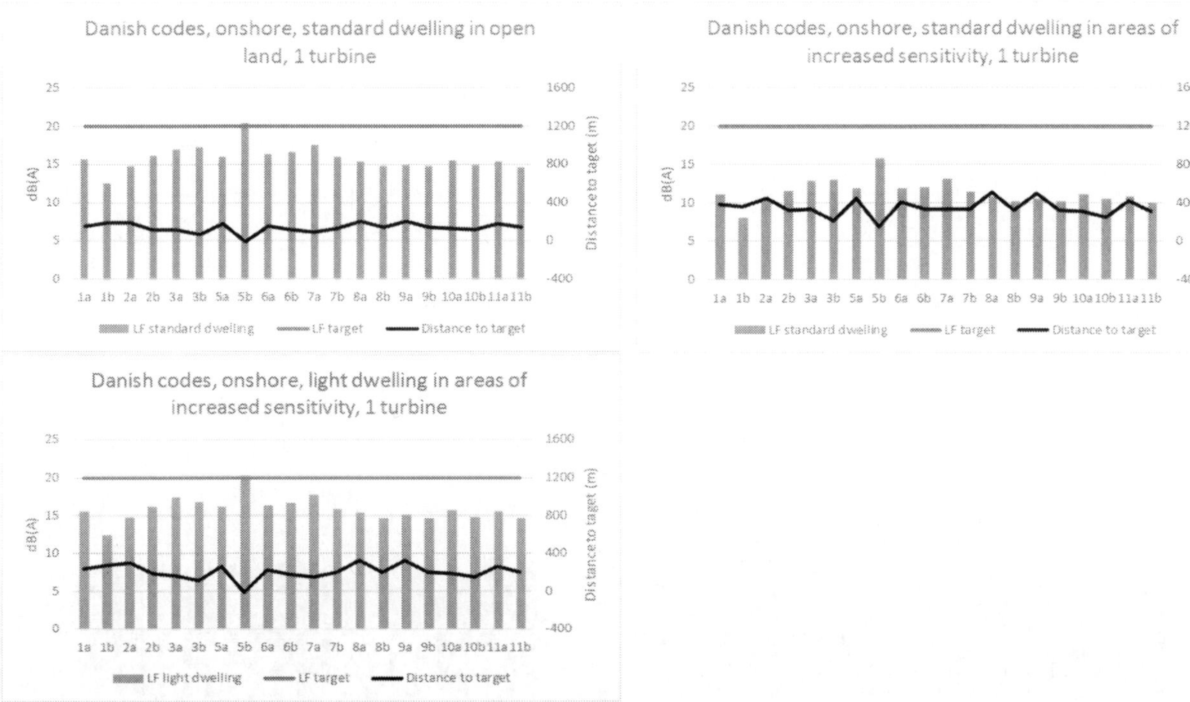

Figure 3. Results for three types of receptors using Danish codes, onshore with a 1-unit layout. Orange columns are the combined, A-weighted low frequency noise from 10 to 160 Hz. The black lines are the distance from the dwelling to the point where the low frequency target is met. A positive value means that this point is closer to the wind farm.

Increase the wind farm to 21 wind turbines and all the wind turbine types are pushed closer to the target, some of them critically close and, in one case, above the target. In the light dwelling scenario, there are now two types, 3a and 5b, that are above the target and a third, 7a right on the target.

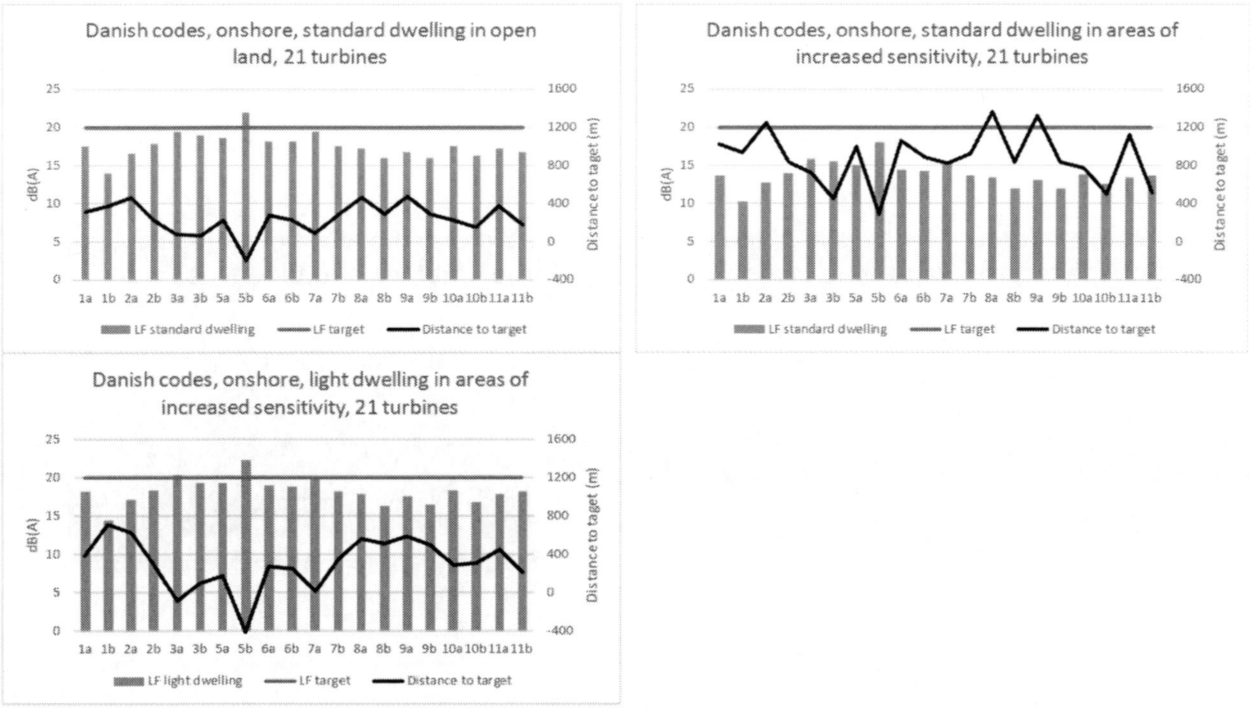

Figure 4. Results for three types of receptors using Danish codes, onshore with a 21-unit layout.

Moving offshore gives a marginal increase in the 1-unit scenario, but a larger increase in the 21-unit layout. Several wind turbine types are now very close to or exceeding the target.

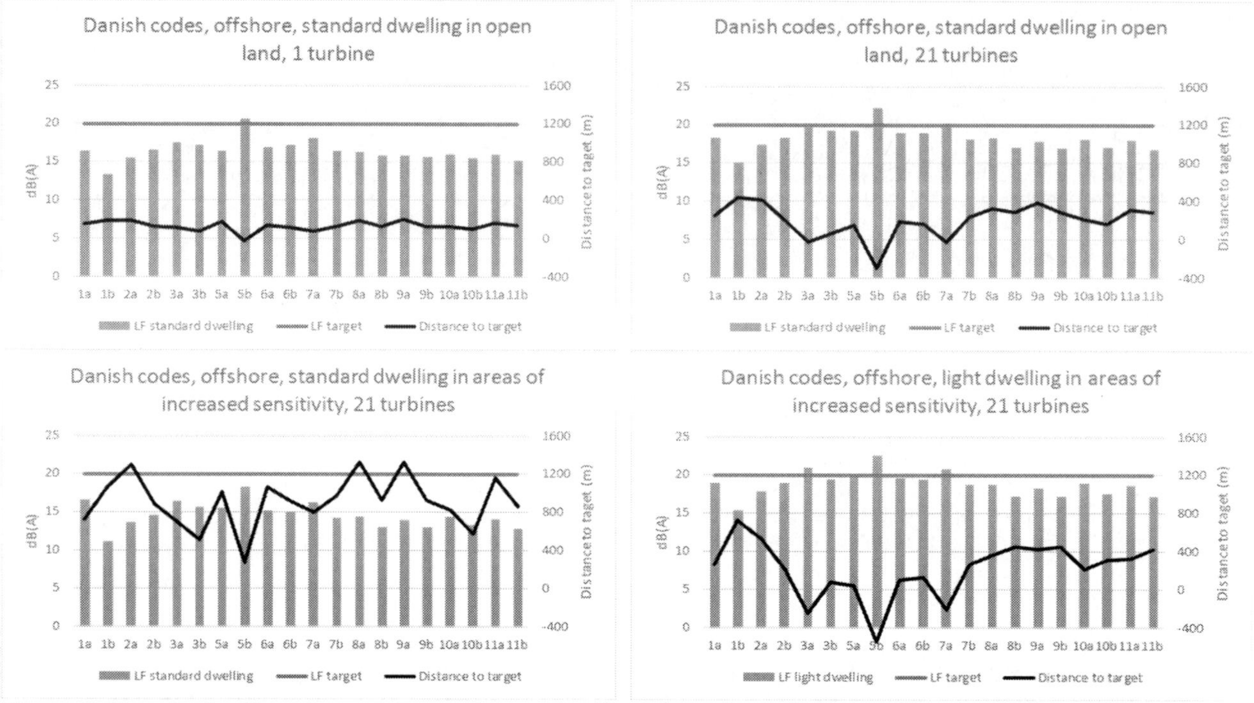

Figure 5. Results for three types of receptors using Danish codes, offshore with a 1- and 21-unit layout

The results using the Swedish codes are presented in figure 6, with orange and green columns representing the margin in dB between the impact and the target at the most critical frequency. Orange is for light dwellings and green for standard dwellings. The black graph is the C-A weighting criteria which in all cases are above 20 dB (meaning low frequency noise must be considered).

In the single wind turbine case, most wind turbine types will have a safe margin to the target, especially with the noise insulation of a standard dwelling, but change to a light dwelling and two types, 3 and 5 will exceed the margin. If the impact of 3b is plotted against the target spectrum we can see that the impact is actually safely within the target except for a single frequency (figure 7).

Considering the larger 21-unit layout, the problem is more pronounced with type 3 and 5 clearly exceeding the target, even with a standard dwelling. Three other wind turbine types are close to the margin.

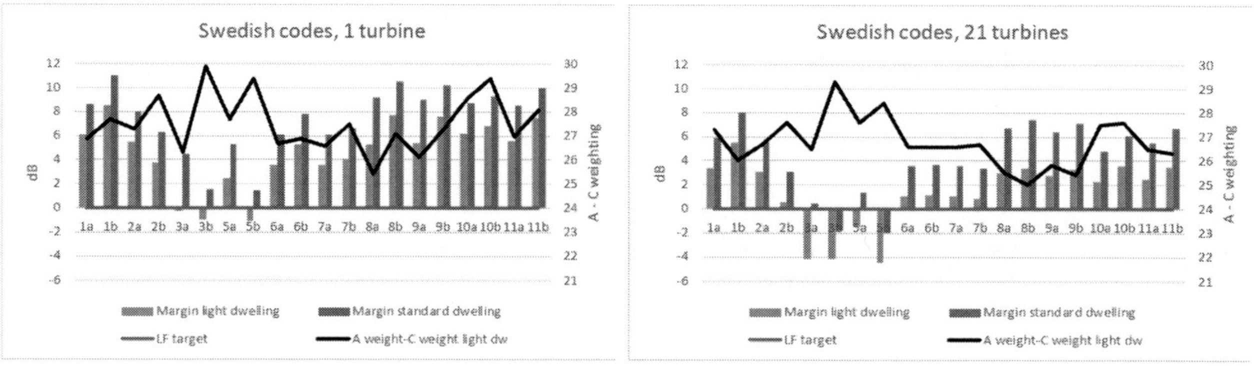

Figure 6. Results with Swedish codes, 1- and 21-unit layouts. Orange and green columns are the margin from the noise impact to the target value at the most critical frequency for light and standard dwelling respectively. A positive value means that the noise impact is within the target. The black lines are the difference between C- and A-weighting. A value above 20 dB means that low frequency noise at the dwelling must be considered.

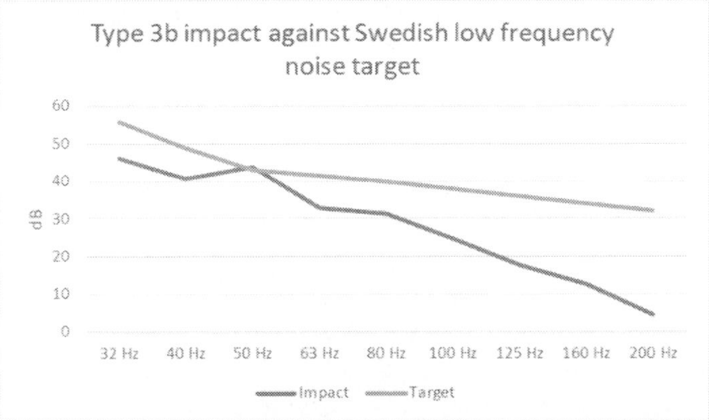

Figure 7. Impact of type 3a against Swedish low frequency target.

In the Finnish case, the pattern is similar, but not identical to the Swedish case despite the almost identical target spectrum. The critical margin is larger for the large wind farm case and the type that had trouble with the Swedish code are still in trouble but less so.

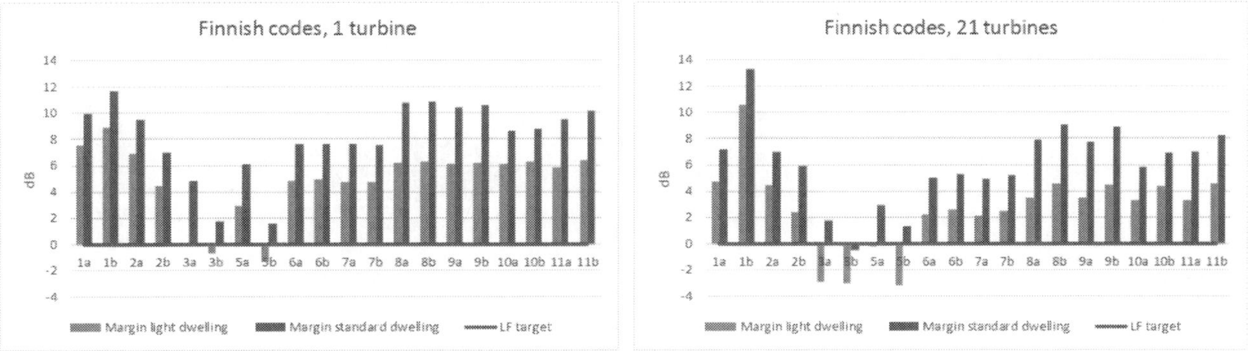

Figure 8. Results with Finnish codes, 1- and 21-unit layouts

5.2 Compliance as a function of wind farm size

Considering that attenuation is less with distance for noise at low frequencies than at high frequencies, it is expected that as the receptor points move further away from the wind farm, low frequency noise will become more critical. This has already been hinted at in section 5.1.

In the diagrams of figure 9, the margins from low frequency impact to targets are presented as a function of wind farm size (by units). A positive margin means that the impact is within the target.

For the Danish code only onshore, standard dwelling in open land and offshore, light dwelling in areas of increased sensitivity are presented, being the most interesting results. Only standard operation mode is included for the ten wind turbine types. The function is practically identical for all turbines, which is not really a surprise. In standard mode, onshore, open land, even the worst performing type will stay within the target even for large wind farms. In the other scenario with offshore codes and light dwellings there are many types that will exceed the target as the wind farm grows larger. Considering that offshore wind farms tend to be large this issue is expected to be relevant, though only for light dwellings near the shoreline as offshore codes in

Denmark are only to be used until the shoreline after which there is a 200 m transition zone before onshore methodology takes over.

For the Swedish codes the gradient is so flat from 5 to 21 units for half the types that it is doubtful if they will ever exceed the target even for very large wind farms. That is the case for both light dwellings and standard dwellings. Those types that are close to the target however have a steeper gradient and are prone to exceed the target at a point as the wind farm grows. That is the case even for standard dwellings.

In the diagrams of the Finnish results, there seems to be no such difference in gradient. The graphs are simply parallel, which means that eventually any turbine type will eventually exceed the target, it will just require a very large wind farm. For standard dwellings even the poorest performing type will take a very large wind farm to exceed the target.

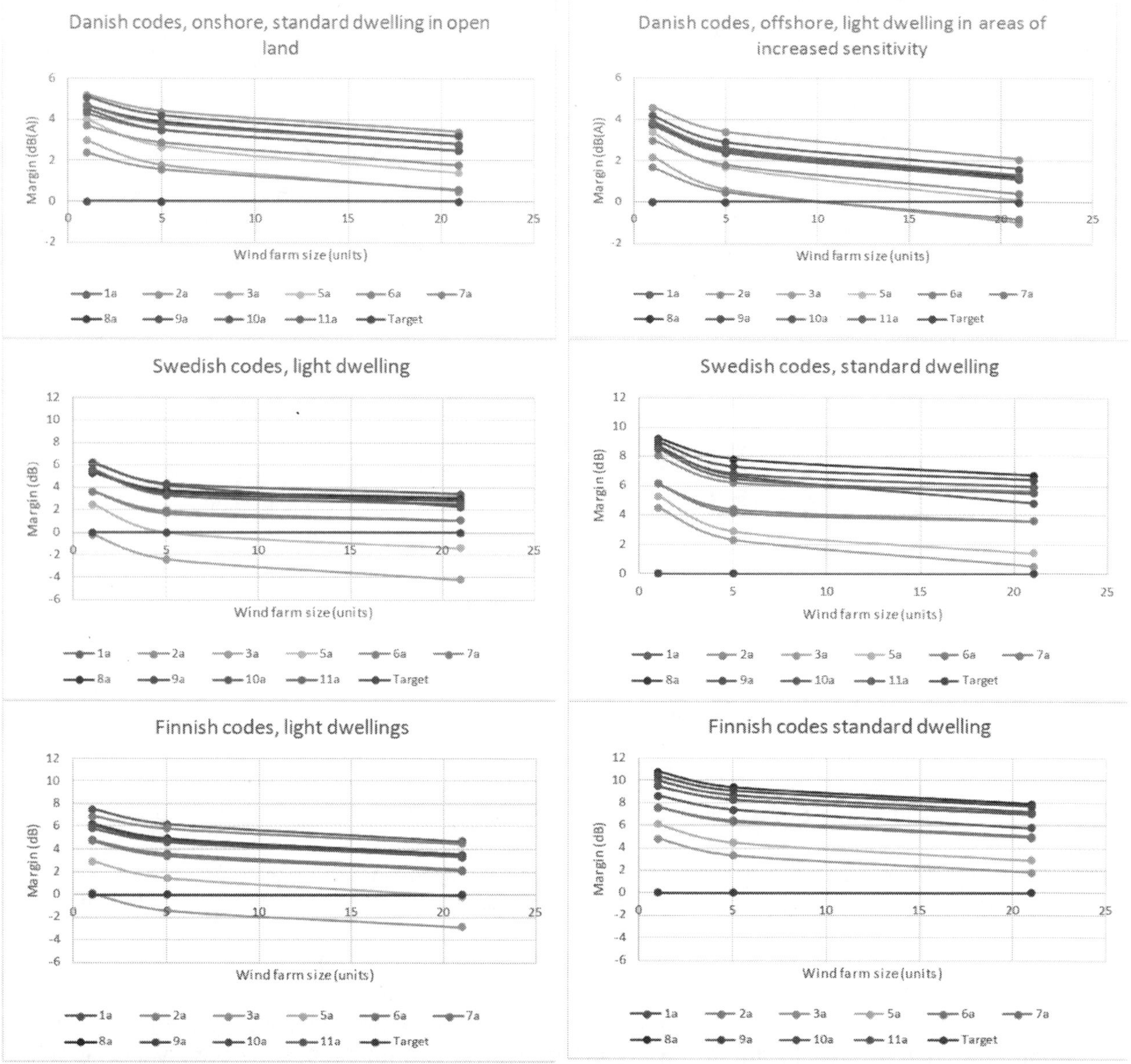

Figure 9. Low frequency noise as a function of wind farm size. The margin presented is the margin from low frequency noise impact to the target. In case of the Swedish and Finnish code, at the most critical frequency. A positive value means that the impact is within the target.

5.3 Changing operation mode

What is then the effect of changing noise mode?

Figure 10 presents the consequence of changing operation mode from standard operation mode to a noise reduced mode. The values presented are the change (increase) in margin from standard operation mode to noise reduced mode. An increase in margin means that low frequency noise is getting less critical while a decrease in margin means that low frequency noise is becoming more critical.

The direction of the change is uniform across the codes. An improvement will be an improvement using any code. But the magnitude is different and more importantly the change varies both in direction and magnitude from type to type. Some types get more critical while others get less critical, meaning that the low frequency response to a change in noise mode cannot be generalized.

A fairly large group of types shows almost no difference using the Finnish code, meaning that a reduction in regular noise is accompanied by a similar reduction in low frequency noise. That same group performs noticeably better using the Swedish code. Here those same types will reduce the noise relatively more for the low frequency test than for the regular noise test. Considering that the noise target is practically the same, this illustrates the difference in propagation model.

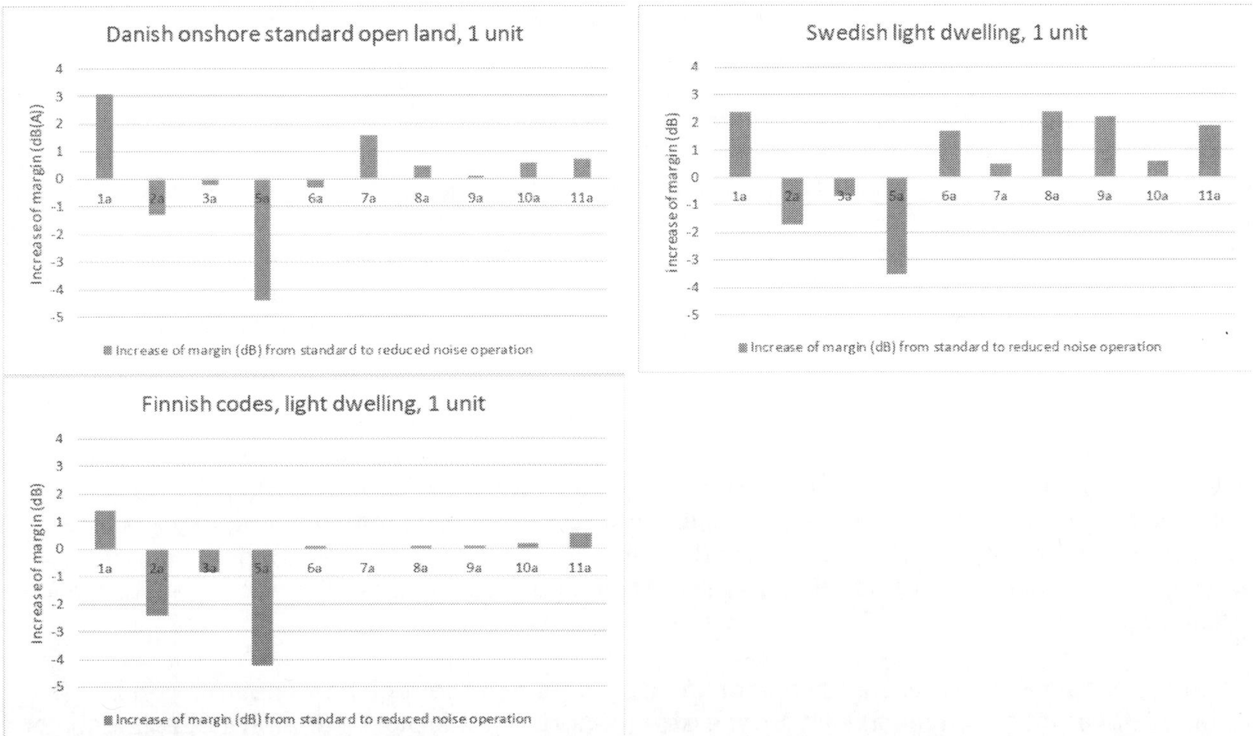

Figure 10. Effect on low frequency noise of changing operation mode from standard to reduced noise. The columns show the change in margin between low frequency noise impact and target as the operation mode changes. A positive value means that the margin is becoming larger while a negative value means that the low frequency noise criteria becomes more critical.

6. Conclusions

So, is it still relevant to test against low frequency noise codes or are we beyond that?

The answer to that is yes, it is still relevant.

While this study is not questioning if the targets in Denmark, Sweden and Finland are meaningful or serves their purpose of protecting the public, they do make up a practical and considered test to avoid excessive low frequency noise in the planning phase.

While some wind turbine types appear to keep comfortably clear of the low frequency targets as long as they comply with the regular (normal frequency) targets, others do not and because of that, it is meaningful to run a test against a low frequency code.

Wind farm size has an obvious influence on how critical low frequency noise gets and for large wind farms low frequency problems are not restricted to a few bad eggs but may be a problem for a larger group of wind turbine types.

Regulating the way out of low frequency trouble requires that reduced noise operation modes actually reduce low frequency noise along with the regular noise. We cannot rule out that the selection of reduced operation mode for some of the types was poorly considered and that there are operation modes available that actually does decrease low frequency noise along with regular, but it is clear from this investigation that selecting a reduced operation mode can go either way. There is no fundamental rule that low frequency noise is a bigger or lesser issue in reduced noise operation. We can only recommend being aware of this issue when considering the use of reduced noise operation mode.

What makes the low frequency noise critical then?

One reason may be found in the codes. If the codes allow a higher noise level for regular noise, but maintain the low frequency target, then the receptors may be allowed closer to the wind farm and the low frequency impact increases. The same goes for the reverse. If the noise target or methodology is tightened for low frequency noise without changing the regular noise, the low frequency issue becomes more critical, as exemplified with the new noise insulation values for light dwellings in Denmark.

The focus of the code may also be an issue. While the Danish code sums the impact over a range of frequencies, the Swedish and Finnish codes requires compliance at each frequency within the spectrum range considered. That means that while the Danish code considers broadband low frequency sound level, a single or two frequencies may be what triggers that low frequency becomes critical in the Swedish and Finnish code. We are not questioning whether this is a good or a bad thing, merely that they are targeting a different aspect of the low frequency sound.

From the wind turbine side, we have seen in later years that the regular noise from wind turbines have not grown much with turbine size (generator and rotor), certainly not as much as expected in earlier studies (Møller, 2011). It appears that particularly for onshore turbines even in standard operation mode attention has been given to reduce noise. Serrations of various sorts have become a standard feature on the blades across the range of suppliers and combined with general efforts to reduce noise that means that the regular noise impact from wind farms is relatively low. If the low frequency emission is not similarly reduced, low frequency noise becomes critical. Boiled down, you can create a low frequency issue by focusing on reducing regular noise. From this study it would appear that some suppliers have been reducing low frequency noise along with regular noise reduction (or have not reduced the regular noise in the first place), while others may not have managed to reduce the low frequency noise along with the regular noise.

All wind turbine noise data used in this study was obtained from material made available by suppliers for wind farm developments in general. We cannot verify if the values are correct and 1/3 octave data are rarely if ever warranted. Therefore, we cannot rule out that some of the data may be biased. However, we suspect that where that may be the case, it would rather be from simplifications in the calculation methods that provide the source noise data than from deliberately providing misleading values. There is very little to gain from massaging data from suppliers' side. The exercise could be repeated with measured source noise data, but this would likely require a large sample size to counter measurement uncertainty, and EMD does not have such data readily available.

In countries and regions where a low frequency noise methodology is not available in the local code, concerns regarding low frequency noise can be addressed with the Scandinavian models, at least to demonstrate whether or not low frequency noise could be a concern for the wind farm in question.

References

International Standards Organization. (1996). *Acousstics - Attenuation of soundduring propagation outdoors - part 2: General method of calculation.* International Standards Organization.

Miljø og Fødevareministeriet (The Danish Ministry of Food and Environment). (2019). *Bekendtgørelse om støj fra vindmøller, BEK nr 135.* The Danish Ministry of Food and Environment.

Miljöministeriet (The Finnish Ministry of the Environment). (2014). *Ympäristöhallinnon Ohjeita 2, 2014, Modellering av buller från vindkraftverk.* The Finnish Ministry of the Environment.

Miljöministeriet (The Finnish Ministry of the Environment). (2015). *Statsrådets förordning om riktvärden för utomhusbuller från vindkraftverk.* The Finnish Ministry of the Environment.

Møller, H. e. (2011). *Lavfrekvent støj fra store vindmøller - opdateret 2011.* Aalborg University.

Naturvårdsverket. (2010). *Ljud från vindkraftverk, Rapport 6241, koncept 20 april 2010.* Naturvårdsverket, Sweden.

Naturvårdsverket. (2019). *Buller från vindkraft.* Retrieved from https://www.naturvardsverket.se/Stod-i-miljoarbetet/Vagledningar/Buller/Buller-fran-vindkraft/

Socialstyrelsen. (2008). *Buller, Höga ljudnivåer och buller inomhus.* Socialstyrelsen, Sweden.

Søndergaard, B. (2016). *Udvidet datagrundlag for danske boligers lydisolation mod lavfrekvent støj (Extended databasis for noise insulation of Danish dwellings against low frequency noise).* The Danish EPA.

Sosiaali- ja terveysministeriön (The Finnish Ministry of Social Affairs and Health). (2003). *Asumisterveysohje.* The Finnish Ministry of Social Affairs and Health.

INCEEUROPE

8th International Conference
on
Wind Turbine Noise
Lisbon – 12th to 14th June 2019

Amplitude Modulation analysis - now and in the future

Matthew Cand, Hoare Lea, Bristol, UK: matthewcand@hoarelea.com

Summary

Over the past 10 years, significant progress has been made in the analysis of amplitude modulation in wind turbine noise. Several numerical analysis techniques have emerged and been refined by a number of authors. In particular, in 2016 a UK Institute of Acoustics working group published a proposed analysis methodology based on a hybrid frequency/time domain approach, which has been applied by a number of authors in the UK and abroad. The international community is now considering the assessment of this noise feature as part of an effort to standardise the measurement of wind farm noise in the far-field. It is therefore a good time to review the IOA method and its parameters, its application to a number of practical test cases, and perspectives for developing its scope further.

1. Introduction

Amplitude modulated (AM) sound can sometimes arise in proximity or in the far-field of wind turbines, the latter phenomenon being complex and sometimes pronounced to the extent that disturbance can arise. Following increased interest in this subject, different practitioners developed a number of comparable techniques to rate the level of modulation. However, the lack of a standard and robust assessment method made the research on this subject more difficult. Any such method needs to be useable on real-world recordings which are often corrupted by a number of other sources, particularly in the far-field.

The AM analysis method published in 2016 by the Institute of Acoustics [1] represents the outcome of the development of such a method by a UK-based working group, following a review of previous work on the subject. In summary, the procedure in Ref. [1] involves undertaking a frequency analysis of a band-filtered $L_{eq,100ms}$ signal separated in 10s blocks: if a clear peak is identified in the resulting modulation spectrum, at a rate expected for the turbines considered, then this indicates the presence of some AM. The identified modulation signal and its first two harmonics in the spectrum are retained and used to reconstruct a filtered modulated signal, whose variation is then rated using a simple statistical metric.

Results for individual 10s are then combined in groups of 10 minutes and a rating is given based on the top 10% of the individual 10s results (effectively representing a typical worst-case). Importantly, this rating is only given if sufficiently clear modulation (as defined using a "prominence" rating) is present for at least 50% of that 10-minute period. This last requirement in particular has been found to be very effective at automatically identifying wind turbine noise, which tends to be relatively persistent in time, compared to other sources which may trigger apparent "modulation" which corrupts the analysis. Such is the effectiveness of that method that it often eliminates a large proportion of erroneous data in the analysis, even without restricting analysis to the quieter night-time periods. Nevertheless, a residual number of erroneous results can still remain in the processed dataset and further data-validation procedures will be discussed below based on our experience.

The IOA method has already been applied by a number of other practitioners [2,3] with some success. As part of the use of this method at a number of UK sites, the present paper illustrates practices that have emerged in the application of these techniques. The paper also considers specific application cases and possible extensions of these techniques that could be considered as part of international development work for AM techniques.

2. Applications

2.1 AM Data analysis

In practice, due to a number of different factors, levels of AM in the far-field of wind turbines can experience a wide range of variability. Furthermore, the specific weather conditions associated with specific disturbance may not occur regularly. This can therefore require noise measurements over long periods of weeks or months in order to capture representative data. The resulting large data volumes can be daunting to analyse directly through review of the instantaneous noise levels time history; in any case this rarely provides directly meaningful information and provides little indication of the prevalence of any particular data.

The first step in the analysis using the IOA method involves producing a modulation spectrum for each 10s period. By plotting this spectrum as a function of time, the presence of modulation and its variation is clearly apparent, as in the example of Figure 1 below. This allows to quickly identify trends in the data over the short to medium term. These spectrograms also represent a rapid tool for data validation which, with some experience, largely precludes in practice the time-consuming review of audio samples or measured noise frequency spectra.

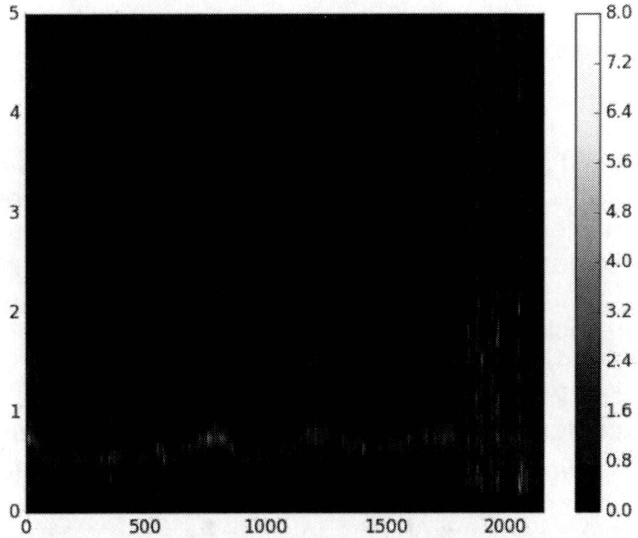

Figure 1 - Modulation spectrogram – scaled power spectral density of the modulation spectrum used in the IOA method analysis. Vertical axis: modulation frequency, horizontal axis: time, 10s blocks, representing a 6-hour period. Modulation from the turbines at variable rates is apparent as the varying but broadly horizontal spectral lines between 0 and 1 Hz (and harmonics); vertical lines represent corrupted data due to bird noise in this case.

2.2 Systematic analysis in relation to weather parameters

The short-term variations in noise or AM levels can be difficult to explain in themselves, but considering long-term results of the AM analysis against weather and operational data allows further analysis. Trends then often become more apparent which allows evaluating the situation at a particular site, either to consider the prevalence of this feature against any adverse comments from wind farm neighbours or to help consider mitigation of the feature if required.

Hoare Lea developed the use of representing the variation of AM through a type of "wind rose" showing the distribution of AM as a function of wind speed and direction, as shown in the example of Figure 2. This shows a particular case of AM present over a relatively wide range of conditions. This type of plot requires a single wind speed and wind direction, representative of the site, to be determined as in most wind farm assessments: either from a site anemometer mast or as measured from one or more turbines. In this case, the wind speed was "standardised" to 10 m height as per IEC 61400-11:2006.

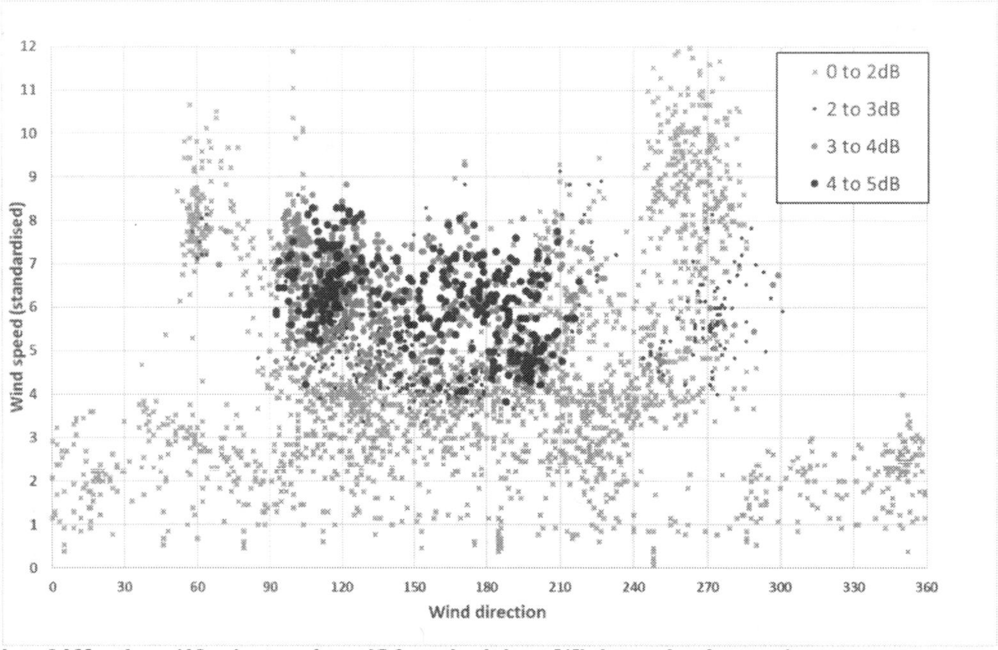

Figure 2 – Plot of AM ratings (10-minute values, IOA methodology [1]) determined over a long-term survey at a residential location in the far-field of a wind farm – values were separated into different bands and related to a representative wind speed and direction (average 10-minute values) for the site.

Provided a representative wind speed/direction is used, a clear pattern is often apparent. Furthermore, any outliers in this pattern will usually become clearly apparent in the chart and can be selected for further review and analysis: this can show if it is an atypical event or due to corruption from noise events not related to the wind farm. As this data is plotted as function of wind direction, a *radial* plot can also represent a useful graphical representation of the situation: this and other examples are included in Ref. [4].

The IOA method requires analysis in three frequency ranges: 50-200 Hz, 100-400 Hz and 200-800 Hz, to represent different turbine/site characteristics that may occur in different cases. The band that *"tends to give the highest modulation values over a representative range of valid data measured"* [1] should be use, and Ref [1] suggests a scatter plot is used. Additional experience suggests that this may be more complex on certain sites, for example if one band gives higher results for low modulation levels but the opposite is true for higher values.

In some cases, different frequency bands may give different results depending on the wind direction: for example, for the data in Figure 2, a different band-pass analysis gives much lower values in the 90-150 wind direction sector, but comparable values in the 150-220 sector. This suggests different AM generation effects at play, corresponding to other generation and/or propagation characteristics. This should be the focus of further research as this complex aspect of AM behaviour is currently poorly understood.

2.3 Binned analysis

Another way of represent the data such as that shown in Figure 2 is to undertake a "binned analysis". Table 1 shows the result of averaging the ratings obtained by the IOA method in each bin, with the data split in bins of 30 degrees for wind direction and 1 m/s for wind speed.

Table 1 - Average AM rating as a function of binned wind speed (1m/s) and wind direction (30°) – Figure 2 data.

Wind speeds	Wind directions (bin centred on...)											
	0	30	60	90	120	150	180	210	240	270	300	330
1	0.0	0.0	0.0	0.0	0.0	0.0	0.0	0.0	0.0	0.0	0.0	0.0
2	0.0	0.0	0.0	0.0	0.0	0.0	0.0	0.0	0.0	0.0	0.0	0.0
3	0.0	0.0	0.0	0.0	0.0	0.0	0.0	0.0	0.0	0.0	0.0	0.0
4	0.0	0.0	0.0	0.2	0.3	1.1	0.8	0.5	0.1	0.3	0.0	
5		0.0	0.0	1.6	2.2	2.7	2.0	2.1	1.0	1.1	1.5	
6			0.0	2.7	3.3	3.6	4.0	1.3	0.0	1.6	2.6	
7			0.7	2.6	3.6	3.9	3.6	1.9	1.0	0.6	2.5	
8			0.7	2.0	3.8	3.4	2.9	1.3	0.0	0.2	1.1	
9			0.2	0.5	3.2			1.1	0.1	0.0		
10			0.0	0.0					0.0	0.0		
11			0.0	0.0					0.0	0.0		
12				0.0					0.0	0.0		

Calculating an average AM value in each of the bins is only illustrative in this case. Depending on the local applicable regulations and/intended use, other approaches may be more relevant. In the UK, a potential penalty regime has been proposed for AM. Several UK practitioners have concluded [5] that, to determine the rating adjustment applied to wind turbine noise to reflect the presence of AM, the corresponding penalty value (up to 5dB for AM levels above 3dB) is determined for each 10 minute periods, and an average penalty is calculated over the range of conditions of interest. Some example applications of this approach are set out in Ref. [4].

Ref. [4] also notes that although a finer resolution with 10 degree-wide bins were also investigated, this resulted in scarcity of data in many bins and excessive variability which made the analysis less clear. The use of 30 degrees/1 m/s bins as in Table 1 therefore represents a good starting point for a systematic analysis of the statistical behaviour of the data, in the absence of more specific information, and can indicate trends which may not be obvious at first glance. It is important to acknowledge that lack of data in some conditions does not mean that AM would not potentially be present in these conditions, as sufficient data may not be present.

Figure 4 shows the results of the variation in average AM with wind speed for different wind direction sectors, from very wide to very focused. This illustrates that averaging over too wide or too narrow a range of wind directions can understate the average levels in other conditions. This may also suggest a more detailed statistical analysis could be of value. Caution should also be taken when averaging over limited quantity of data, which can throw spurious trends.

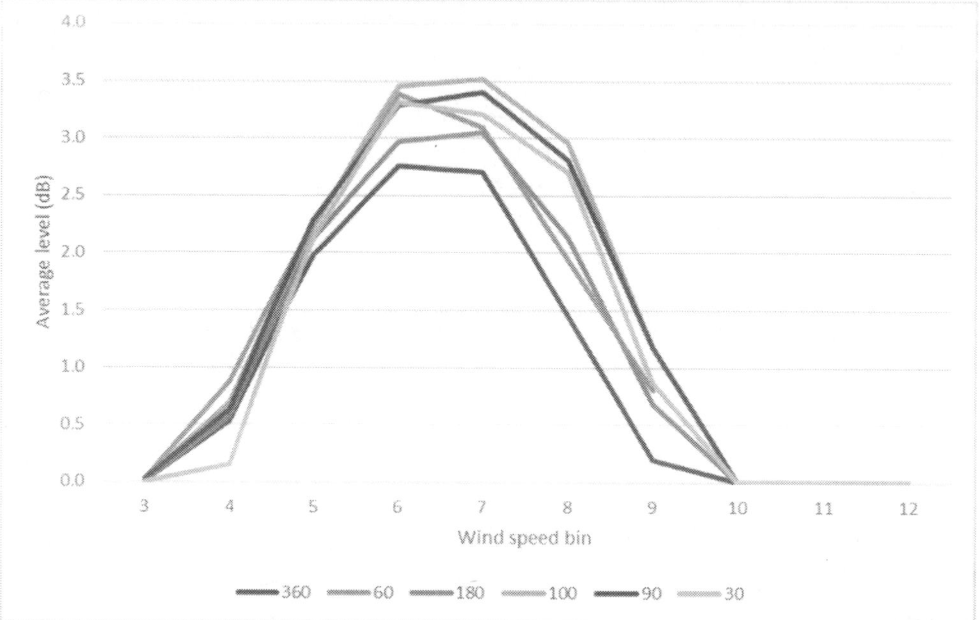

Figure 4 - variation of calculated average AM rating (IOA method) as a function of the width (degrees) of the wind direction sector for averaging: 360 degrees = all directions, 180 = 180 degree sector centred on DW conditions, etc. in In this case each direction sector was centred on the downwind direction.

Any distribution of AM obtained from this analysis, such as that of Table 1, is obviously only representative of the conditions obtained during the survey period, and whilst efforts should be made to make it as representative as possible, the wind conditions may differ from those observed in the long-term for any particular site. However, these results can allow identifying a particular set of wind conditions corresponding to relatively elevated AM (for example, in Table 1, wind directions of 105-195 degrees and wind speeds of 6 m/s and above). Reference can then be made to a long-term wind rose for the site which would allow determining the average prevalence (% of the time) in which these conditions arise, and therefore more elevated AM could occur. This can also help to target operational mitigation measures if relevant, either through targeted shutdowns or more advanced measures [6].

2.4 Use of turbine parameters

It is also useful to consider the AM analysis data against operational data from the wind farm control system (SCADA). In a first instance, of course, controlling for periods when turbines were not operating represents a good test.

In addition, the rotational rate of each turbine is generally available. The analysis of the AM modulation rate against this data also allows further filtering of the data. The IOA method requires as input a range of modulation frequencies which are potentially related to the turbines being considered (f=3*RPM/60 for a three-bladed turbine). However, this remains a broad criteria, particularly for modern variable-speed machines which can have strong variations in rotational rate at moderate wind speeds. So, even if a modulation "peak" is identified, this may not be valid if the AM frequency observed is inconsistent with that produced by the turbine(s).

Whilst determining a rotational rate is trivial for a site comprising a single turbine, this can be more complex for a larger wind farm. Our experience is that several turbines may be contributing to the AM in different individual 10s periods. This may therefore be at slightly different rates, particularly in sites with complex terrain where significant variation in the wind speed and therefore rotational speed can be experienced between different turbines. It may therefore be relevant to consider a range of RPMs for the relevant turbines potentially contributing to noise at the receiver rather than a single value. The case of different turbine types is considered below.

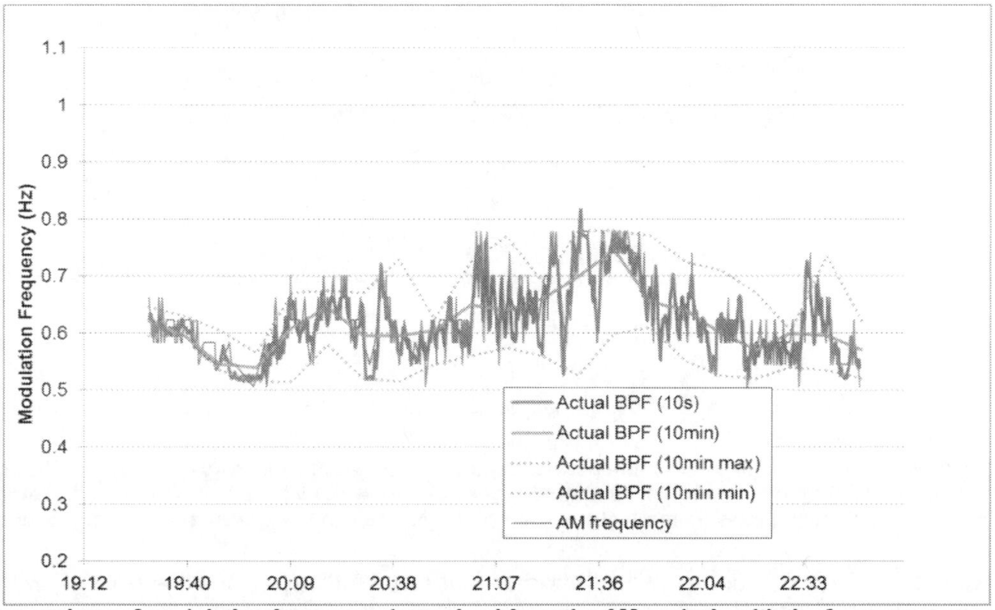

Figure 5 - Comparison of modulation frequency determined from the AM analysis with the frequency corresponding to the rotational speed of the closest turbine to the measurement location (both in 10s resolution and min/max/average over 10 minutes) – from [7] (based on a precursor of the IOA method)

Furthermore, the RPM data is generally available at 10-minute resolution rather than 10s, although in some cases a minimum and maximum value is recorded in addition to the 10-minute average. See Figure 5 for example: in this study [7], more detailed 10s data was recorded on one of the turbines and this showed an excellent match with the measured modulation rate (in cases of clear AM).

Another consideration is that the IOA method [1] defines a modulation frequency for each 10s period not over 10 minutes, as this frequency can vary slightly. There is no single frequency value as output from the algorithm directly usable for comparison with the turbine information. Possible candidates can be selected from a statistical analysis of the frequencies obtained over each of the 10s periods: in our experience the mode (most common value over each 10-minute) is usually representative, but alternatively the range of values may be considered instead. See next section for further discussion.

A data validation check is then to compare the frequency value or range determined from the AM analysis with the range of operational values determined from the rotational rates of the turbine(s). If the absolute difference is more than the tolerance of the method (0.1 Hz in the case of the IOA method) then these periods are flagged for further review.

2.5 Presence of turbines operating at different rates

The potential performance of the IOA method in complex modulation scenarios has been questioned. An example period was studied in which measurements were influenced by noise from two wind farms, with different turbine technologies: one with smaller turbines operating at 24-30 RPM (fast modulation around 1.3Hz) and one with larger machines operating at 13-18 RPM (slow modulation around 0.8 Hz). In the example period shown in Figure 6, modulation at both frequencies is apparent over a 30-minute period (with 3 separate 10-minute blocks of data with valid results).

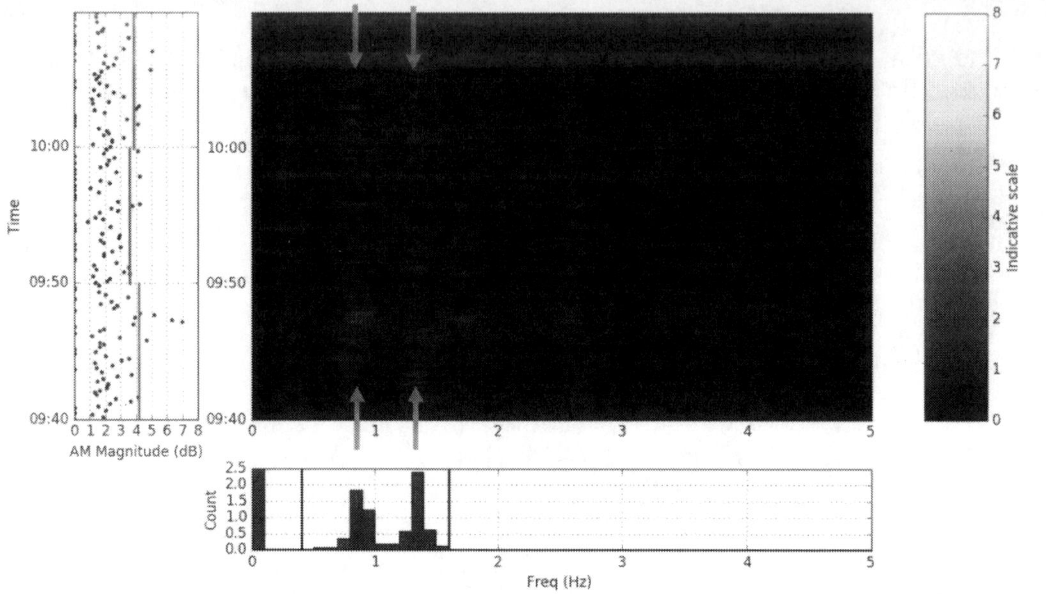

Figure 6 – Modulation spectrogram for a 30-minute period (central chart); left: time history of 10s AM rating values (blue) and 10-minute rating (red lines): this shows 3 x 10 minute blocks of data with valid AM ratings; below: histogram of the individual (10s) modulation frequencies determined. Arrows highlight the two frequency trends at 1.3Hz and 0.8Hz.

Figure 6 shows two clear trends for modulation at both rates throughout the period, with a double distribution in the histogram of frequencies obtained. To allow this analysis to work, the input range of valid frequencies for the method was set to 0.4 to 1.5 Hz. Whilst, in theory, a double modulation could create difficulties for the method, in practice, the variability of AM means that, for most of the periods analysed, only the modulation for one site is effectively dominant at any one time. In these cases, the strongest modulation peak in the input range specified was selected by the method.

A more detailed analysis showed that for some individual periods, the faster modulation was taken as the first harmonic of the lower modulation, or only one modulation component was analysed. But overall, the output of the method provided a good representation of the modulation values experienced throughout, despite the complex situation.

The question of determining a single output modulation frequency for each 10-minute period can then be considered in this case. Two clearly separate distributions are apparent in the graph below Figure 6. Calculating an average of the frequencies would yield a value in the middle of the two separate distributions which would not correspond to a real frequency for either site measured. If the most frequently occurring frequency (or mode) is used instead, excluding zero values, the result is 1.3Hz for all three periods in the example shown above. However, this does not necessarily represent the highest AM values which were experienced over this period. A more complex alternative is selecting the frequency of the 10s block with the AM rating *closest* to the 90th percentile value determined (red lines). This yields frequencies of 1.3Hz, 0.8Hz and 0.8 Hz respectively for each of the three 10-minute blocks shown. This is therefore more representative of the worst-case periods in each 10-minute block.

Analysis over 30s data blocks (see below) did not seem to yield good results as the modulation peaks were often too spread out, due to variable modulation, leading to poorly resolved peaks with poor apparent “prominence”, even if increasing the prominence criteria to account for the increased duration.

3. Future development

3.1 Varying the analysis parameters

The IOA method was developed with a focus on modern large commercial-scale turbines, with rotational rates (in full operation) corresponding to BPFs of 0.7 to 1.2 Hz (RPM of 14 to 24). Consideration can be given to expanding the remit of the method to faster or slower turbines. This means considering the parameters and approach of the method and if it could be extended.

In any Fourier-type analysis, there is a necessary compromise between frequency and time resolution that need to be balanced, but always involves trade-offs. In the IOA methodology [1], it was decided to separate the data into 10s blocks as this provided a good representation of the variability of far-field AM which was experienced from different site measurements, with individual “bursts” of modulation sometimes lasting only a few seconds. This corresponds to a frequency resolution[1] of 0.1Hz for the modulation spectrum. As discussed below, an increased analysis period could increase the frequency resolution some of the temporal detail can be lost.

Simplicity of approach is also a consideration: it was determined [8] that by using a 10s block, a simple measure of average modulation in each block provided a robust representation of individual modulation peaks and troughs over a 10-minute period. The working group concluded that it was reasonable to characterise the level of AM in each 10-second interval by a single value of the difference between L5 and L95 of the filtered dataset, which effectively represents the average of the different peak-to-trough values.

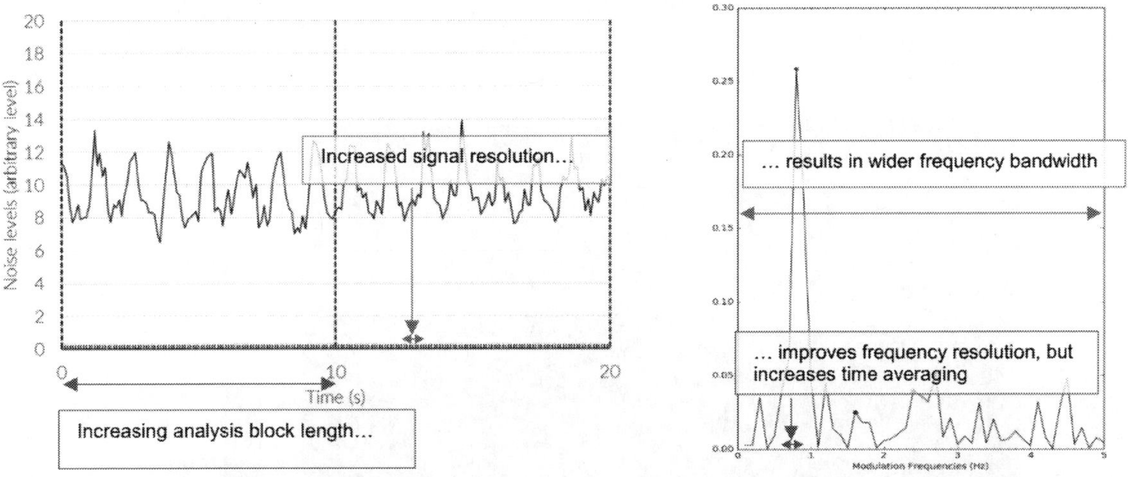

Time domain **Frequency domain**

Figure 7 – representation of the main parameters affecting the time and frequency resolution of the analysis.

Practical considerations are also a factor: noise levels, particularly with 1/3 octave band data, are typically only available on most sound level meters at 100 ms resolution (10 Hz). This means the modulation spectrum can only extend up to 5 Hz. As the method is based on accounting for the first three harmonics of the signal, this limits the frequency of the main peak to 1.6 Hz [1], which corresponds to a turbine operating at 32 RPM for a three-bladed turbine. A

[1] Although zero-padding the data can appear to increase the frequency resolution, this is only an artefact and does not represent a true increase in the actual resolution of the analysis.

higher signal sampling rate would be required in order to represent faster-rotating turbines, which will tend to be smaller models. Alternatively, less harmonics could be retained.

More generally, a balance has to be made when deciding how many harmonics to include – using too many will increase the noise floor while including too few will result in missing the true amplitudes of the peaks and troughs when harmonic energy is present. The use of the first three harmonics in the IOA method was found to provide a good representation of the original time-series in the large majority of measured data test cases, whilst maintaining a reasonably low noise floor.

3.2 Analysis over longer periods

During the development of the IOA method, consideration was given to a longer analysis time (30 s to one minute). This would provide not only more frequency resolution but also error reduction, through additional averaging of the turbine signal (which tends to persist in time), and averaging out spurious sources such as wind and animal noises which tend to be more short-lived. It was determined that 1-minute could be too long for variable-speed machines whose rotational speed can vary even over this period, but a 30s-period represented a potential alternative.

Figure 8 shows an example of the resulting spectrogram for the same period as Figure 1, demonstrating the cleaner signal and increased frequency resolution obtained. To obtain a comparable analysis, the parameters of the IOA method were modified as follows:

- Where one spectrum line was used before when inverting the Fourier analysis, three are used instead;
- In a similar way, the prominence of the spectral peak is calculated by ignoring three lines either side of peak and then calculating the average over six lines either side of this;
- The prominence criteria was increased from 4 to 12;
- The search for harmonics was done over wider area (2-3 lines either side of the exact multiple of the main spectral peak)

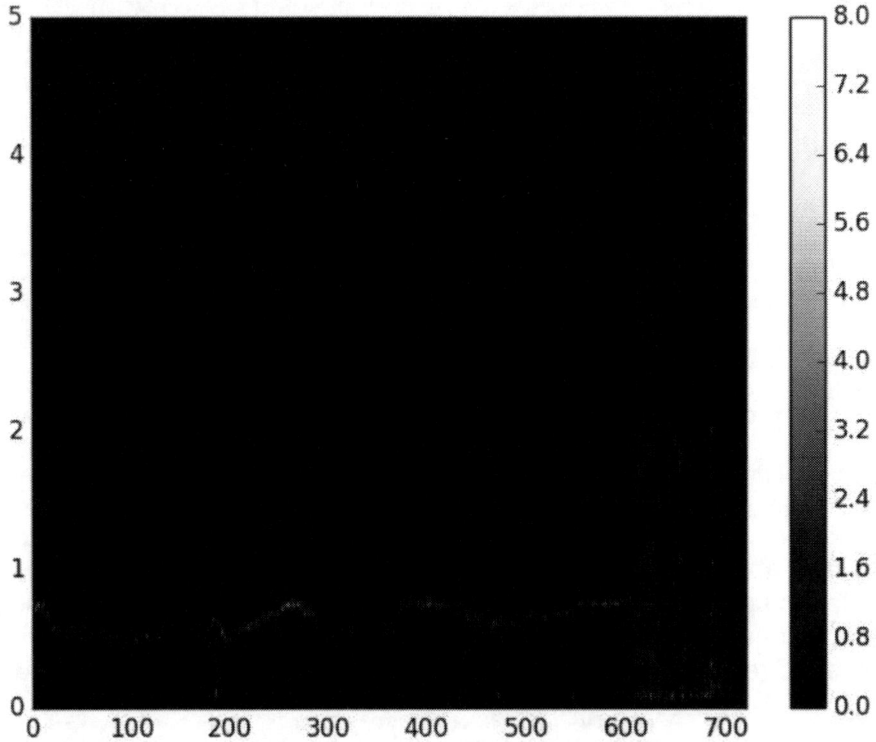

Figure 8 - Modulation spectrogram – same data and presentation as Figure 1 (6-hour period) but with the analysis undertaken in 30s blocks.

Figure 9 illustrates an example of analysis of one individual 30s period: this shows a relatively clear spectrum but illustrates two points: the variability of the reconstructed signal is much larger over the 30s, which was found to be often the case. This means that the simple L5-L95 variability metric used is too coarse to represent the signal variation over 30s. However, the filtered signal generated through inverse Fourier transform does include this time history element, and a more detailed analysis of the peaks/troughs is possible (as discussed in the next section). This however means the method then becomes more complex.

Figure 9 also illustrates a case where higher harmonics of the signal are more clearly visible, and it would be possible to consider these upper signal harmonics when reconstructing the signal. This tended however to represent a marginal effect on the analysis, with the exception of very isolated periods with strong harmonic content. For these reasons, the working group concluded that a 10s analysis period represented a better approach overall (at least for the modulation rates considered).

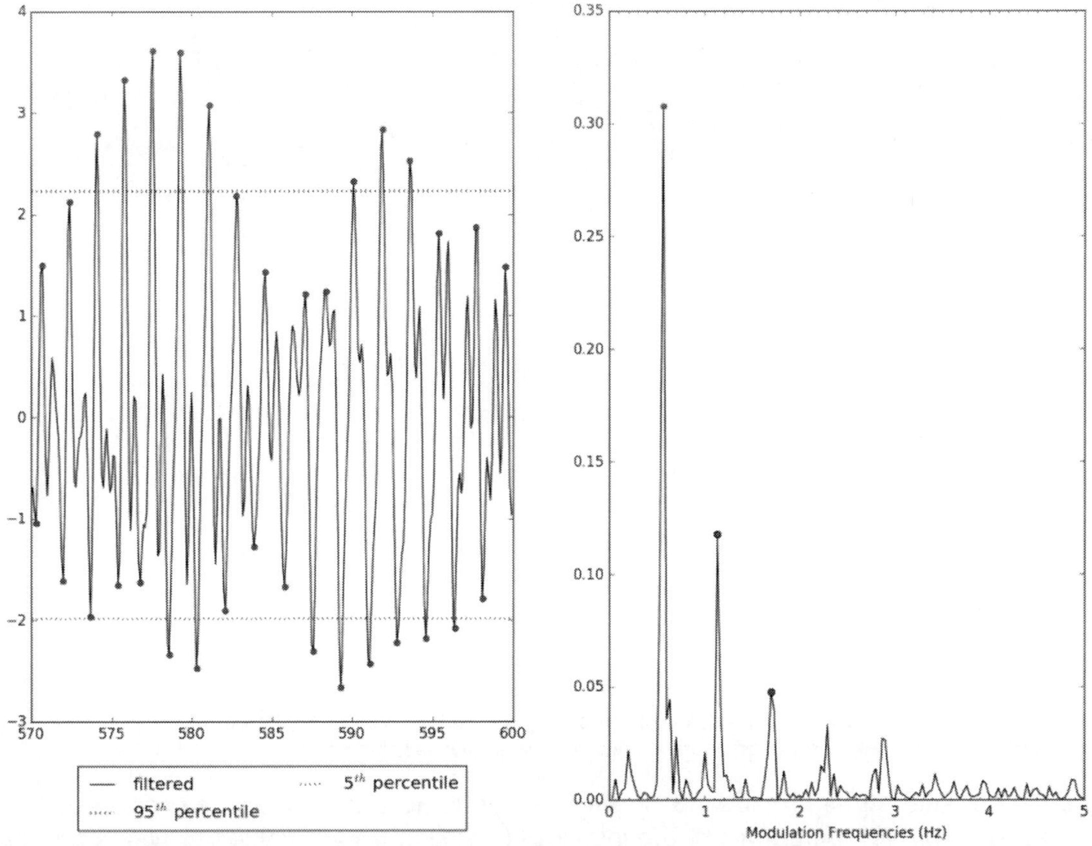

Figure 9 - Modulation signal and spectrogram analysed over one 30s period – left: filtered signal generated, 95th and 5th percentile values and individual peak/troughs marked by red and green circles respectively. Right: corresponding modulation power spectral density, with the red circles highlighting the three harmonics identified.

3.3 Individual peaks/troughs

In cases such as that of Figure 9 where the analysis period becomes relatively long compared to the modulation period T (which depends on the main modulation frequency), then the variability in modulation over the data "block" becomes significant. In these cases, considering the overall statistics (5 to 95th percentiles) of the signal may not necessarily provide a sufficiently robust estimate of the AM variability. But since the time series is reconstructed using knowledge of its frequency content, this allows the peaks and troughs to be identified and characterised individually, as illustrated in Figure 9. This means that a more detailed statistical analysis of variations of individual consecutive peak-trough differences is possible.

Ref. [9] also used a similar approach of identifying peaks/troughs directly, but the advantage of the hybrid approach used in the IOA method is that the frequency analysis undertaken guides the detection of peaks and troughs in the reconstructed signal. This was in fact the approach used in one of the earlier draft methods [10] proposed by the IOA working group, but not retained in the final document, to limit the complexity of the method. Additional steps such as these can also introduce errors and "edge cases" which makes the method less robust.

Specifically, a version of the reconstructed signal is obtained based only on the fundamental modulation peak in the spectrum (see Figure 4.5.2 in [1] or Figure 10 below). It is trivial to identify local maxima and minima in that signal (which has a fixed period) and these provide a guide as to the likely locations of peaks and troughs in the full reconstructed signal. The fundamental modulation frequency also provides the basic modulation period: a window of half this length is used to seek for a local maximum/minimum in the reconstituted (filtered) signal, starting from each peak/trough in the fundamental.

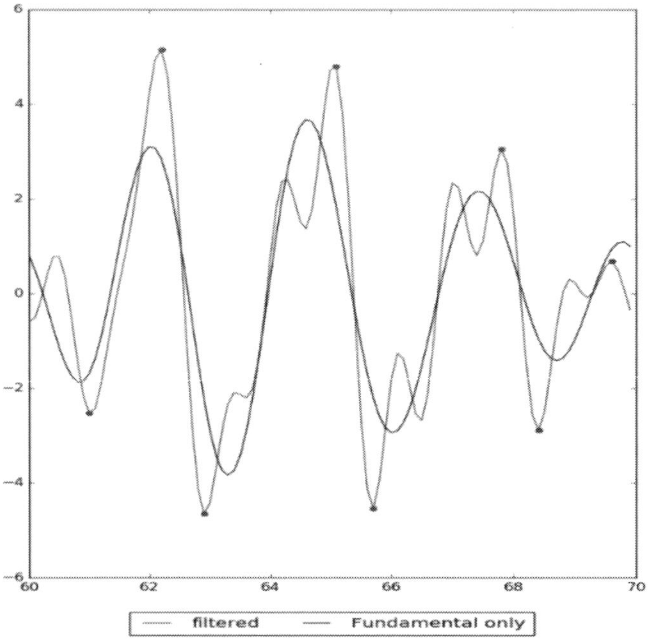

Figure 10 – filtered signal generated by the IOA methodology (red line), as well as first harmonic signal only (black). Individual peak/troughs identified are marked by red and green circles respectively.

Consecutive peaks/troughs provide individual AM results which can then be processed for further analysis. The 90th percentile of the individual peak-to-trough values provides a useful metric. It can be shown that, for data such as that in Figure 9, the average of the 90th percentile of peak-to-trough values in 30s blocks over a 10-minute period is equivalent to the IOA metric, *i.e.* the 90th percentile of (what is effectively) the average AM in each 10s block.

3.4 Slower turbines

As the move to larger and larger turbine rotors continues apace, consideration should perhaps be given to turbines operating at a slower rate (less than 10 RPM) and how any AM method could be extended to deal with this.

The first requirement would probably be to increase the time period for analysis, to capture more modulation periods and better capture this slower signal. This would also increase the frequency resolution, which would help to resolve the lower frequency area of the modulation spectrum (below 0.5 Hz) where the main modulation peak is likely to be found.

Despite this, initial investigations suggest that the prominence calculation in the standard IOA method can run into difficulties: the calculation of a "masking level" relies on averaging frequency lines either side of the main modulation peak and this could in theory decrease below the lowest spectral lines. In these cases, the calculation of the masking level can be modified to only consider spectral lines at higher frequencies (above the main peak) and still result in an effective procedure. Initial investigations using simulated data suggest this approach could be effective but the dynamics of actual 5-10 RPM turbines may be very different in practice and this would require further investigations.

References

[1] Institute of Acoustics (IOA) Amplitude Modulation Working Group, Final Report, A Method for Rating Amplitude Modulation in Wind Turbine Noise, June 2016. https://sourceforge.net/projects/ioa-am-code/

[2] Akinori Fukushima, Hideki Tachibana: Comparison of the IOA method and Japanese F-S method for quantitative assessment of amplitude modulation of wind turbine noise – A study based on the field measurement results in Japan. 7th International Conference on Wind Turbine Noise, Rotterdam – 2nd to 5th May 2017.

[3] Ian Bonsma, Nathan Gara, Brian Howe, Nick McCabe: An investigation into short-term fluctuations in amplitude modulation of wind turbine noise, preliminary results. 7th International Conference on Wind Turbine Noise, Rotterdam – 2nd to 5th May 2017.

[4] Matthew Cand, Malcolm Hayes, Gavin Irvine, Tom Levet: site experience with wind turbine noise AM analysis, Proceedings of the Institute of Acoustics (UK), Vol. 40. Pt. 1. 2018.

[5] Andy McKenzie, Matthew Cand, Dick Bowdler et. al. A planning condition for wind turbines, Acoustics Bulletin November/December 2017, Institute of Acoustics.

[6] Matthew Cand, Andrew Bullmore, Measurements demonstrating mitigation of far-field AM from wind turbines, 6th International Meeting on Wind Turbine Noise Glasgow 20-13 April 2015.

[7] Wind Turbine Amplitude Modulation: Research to Improve Understanding as to its Cause & Effect - Work Package D (WPD) - Measurement and Analysis of New Acoustic Recordings - Renewable UK, December 2013.
http://www.renewableuk.com/default.asp?page=IndustryStatementOAM

[8] T. Levet, Application of the AM Metrics – Case Studies, Proceedings of the Institute of Acoustics, Vol. 37. Pt. 2, 2015.

[9] J. Cooper and T. Evans, Automated detection and analysis of amplitude modulation at a residence and wind turbine, Proceedings of Acoustics 2013 – Victor Harbor, Australia, 17-20 November 2013.

[10] Institute of Acoustics (IOA) Amplitude Modulation Working Group, Discussion Document Methods for Rating Amplitude Modulation in Wind Turbine Noise, April 2015.

INCE EUROPE

8th International Conference
on
Wind Turbine Noise
Lisbon – 12th to 14th June 2019

Long-term noise monitoring of wind turbine amplitude modulation

Christopher Feist, University of Minnesota, St. Anthony Falls Lab: feist020@umn.edu

Matt Lueker, University of Minnesota, St. Anthony Falls Lab: luek0004@umn.edu

William Herb, University of Minnesota, St. Anthony Falls Lab: herb0003@umn.edu

Jeff Marr, University of Minnesota, St. Anthony Falls Lab: marrx003@umn.edu

Peggy Nelson, University of Minnesota, Speech-Language-Hearing Sciences: nelso477@umn.edu

Summary

A noise logging system capable of measuring 100 ms 1/3 octave bands was deployed 600 m north from a 2.5 MW wind turbine at the University of Minnesota's wind energy field research station for one year. The field station includes a 130 m meteorological tower located 170 m south of the turbine, which has wind, temperature, and humidity sensors at six elevations between 7 and 130 m. High-resolution wind speed and turbine SCADA data were recorded in conjunction with the noise measurements. Using these data we investigate correlations between meteorological data, turbine operating conditions, and the presence of amplitude modulation (AM). Analysis of the long-term turbine noise is performed using the IoA (Institute of Acoustics) method for quantifying AM rating.

AM rating was found to have moderate correlation with wind shear exponent, incoming velocity at elevation of 10 m, and turbine power output. The r correlation coefficient is equal to -0.37, 0.44, and 0.65 for the parameters, respectfully. The negative correlation with wind shear is opposite of what researchers have hypothesized to be the cause of enhanced AM and may indicate far-field AM is influenced by the sound propagation path rather than incoming flow conditions for a wind turbine.

1. Introduction

Wind turbine noise has been a significant obstacle to new wind projects in the United States. Wind projects are evaluated prior to installation to verify noise regulations are met, which is intended to reduce noise related annoyance. However, current regulations do not account for the sometimes variable noise characteristic of wind turbines that can cause annoyance even at low sound levels (Pedersen, 2004). This variability in wind turbine generated noise is termed amplitude modulation (AM) and is typically described as a "swishing" or "thumping" noise. The wind turbine sound is amplitude modulated in time by the blade passing frequency (BPF). It has been proposed that there are two types of AM, each having there own mechanisms. Near-field AM, or normal AM, is caused by the directivity of the trailing edge sound produced by wind turbine blade. Oerlemans demonstrated an increase in sound level as the blade moves toward the receptor, and a decrease as the blade moves away (Oerlemans, 2007). This near-field AM is typically observed in the cross-wind directions and at distances of up to 300 m from the wind turbine generator (WTG). At distances greater than a few rotor diameters and in the upwind and downwind directions, normal AM is diminished.

The second type of AM has been termed enhanced AM (EAM). EAM is defined as a variation in sound level that is greater than normal AM and is perceived in the upwind and downwind directions at distances in the far-field, 400 m and greater (Oerlemans, 2013). The mechanism for EAM is not fully understood; however, a number of theories have been presented. Synchronizing of sound between different WTGs (van den Berg, 2004) is one possibility. Oerlemans performed simulations to study the effects of wind shear on sound characteristics of WTG. It was reported that under certain operation conditions, strong wind shear can lead to local stall on a wind turbine blade during the upper part of the revolution. The simulation showed the sound characteristics due to local stall are similar to EAM measured in the field.

Alternatively, studies have shown little correlation between AM rating and meteorological conditions (Bonsma, 2017 and Cand, 2012) and suggest AM might be explained by the sound propagation path to the receiver. Larsson demonstrated wind direction and sound speed gradient are key parameters explaining the occurrence, or probability of occurrence, of AM immission but not emission (Larsson, 2014).

This paper discusses measurements made at a single utility-scale WTG for approximately one year. During the measurement period, acoustical, meteorological, and SCADA measurements were made nearly continuously. WTG sound was measured at a distance of 600 m north of the WTG. Meteorological data were measured at a met tower 170 m south of the WTG. Of primary interest was the observed AM rating at the acoustical measurement location and correlations with meteorological data. The Institute of Acoustics AMWG method is used for rating AM. The paper also investigates meteorological conditions which are most likely to produce AM imission.

2. Field Measurement Site

Field measurements were performed at the University of Minnesota's Eolos Wind Energy Research Station. A single 2.5 MW Clipper Liberty C96 WTG with a hub height of 80 m and 96 m rotor diameter is operated on the site. The landscape is relatively flat farm land with some groups of trees in the area. Trees were cleared in the predominant wind directions, south and northwest, up to 600 m from the turbine with a width of 300 m centred on the WTG. To the west of the turbine, starting at approximately 200 m from the WTG, are a significant number of trees. A highway is located approximately 1 km to the east of the site and carries moderate traffic.

Due south of the turbine is a research grade meteorological tower with a height of 130 m, spanning the swept area of the WTG blades. Ten elevations between 7 m and 128 m are instrumented with sensors for characterizing the atmospheric boundary layer (ABL).

3. Measurements

3.1 Acoustic measurements

Acoustic measurement were made at a height of 1.5 m using a Brüel & Kjær Noise Logger Type 3659. The noise logger uses a hand-held analyser Type 2250 and outdoor microphone Type 4952 using a free-field ½" microphone and a shield protecting the microphone against the effects of wind, rain, and perching birds. The noise logger was positioned 600 m north of the WTG in a farm field near a small grove. Sound spectra in 1/3-octave band L_{eq} for the frequency range of 6.3 Hz to 20 kHz were measured at 100 ms time intervals. The 1/3-octave band fast L_{eq} were recorded without weighting so that post-processing could be performed on unweighted values as well as A-weighted. The measurements were made continuously from April 24, 2018 to March 8, 2019.

In addition to the 1/3-octave band L_{eq} measurements, instantaneous sound pressure was recorded at 6.6 kHz with 24 bit resolution. These data were recorded to assist in identifying periods of high AM ratings.

The noise logger was deployed with and powered by photovoltaic solar panels and battery combination. Regular site visits were performed to calibrate the microphone and download data.

3.2 SCADA measurements

The WTG operating parameters were measured via the SCADA system simultaneously with the acoustic measurements. Parameters were logged at 1 Hz and time synchronized via GPS with meteorological measurements. Table 1 gives the parameters that are used in the analysis methodology for this paper.

Table 1. SCADA parameters used.

Parameter	Description
P	Wind turbine generator power output [kW]
γ	Nacelle direction w.r.t. north [deg]

3.3 Meteorological measurements

Atmospheric boundary layer (ABL) parameters were collected from the meteorological tower located 170 m south of the WTG. The 130 m high tower has sensors at elevations spanning from 7.3 m to 127.9 m. Table 2 outlines the sensors at each elevation. Campbell Scientific CSAT3 sonic anemometers are located at 4 elevations (9.9, 29.6, 79, and 127.9 m above ground level) which provide three dimensional wind velocity and temperature fluctuations at 20 Hz. Cup anemometers, wind vanes, temperature sensors, and relative-humidity sensors are located at 6 elevations (7.3, 27.1, 51.5, 76.7, 101.5, and 125.9 m) which provide 1 Hz data. The WTG rotor swept area has a bottom elevation of 32 m and top elevation of 128 m. Meteorological measurements span this entire height.

ABL measurements from the met tower are averaged over 10 minute periods. This averaging period corresponds with the 10 minute averaging period for the AM rating algorithm outlined in Section 4. The 10 minute meteorological window coincides with the beginning and end time of the 10 minute AM analysis.

Measurements from the sonic anemometers provide wind speed in all three directions; U_n – wind speed from the north, U_w – wind speed from the west, and U_z – vertical wind speed. All velocities presented in this paper were measured at 20 Hz. The horizontal wind components are rotated such that U, the instantaneous streamwise velocity, is aligned with the wind direction and V is the instantaneous spanwise velocity component. The instantaneous vertical velocity is defined as W. These rotated wind velocities are used to characterize the ABL. An average velocity for each 10 minute period was calculated and are defined as $\overline{U}$, $\overline{V}$, and $\overline{W}$.

Table 2. Meteorological instruments and measurement heights. WS, wind speed; WD, wind direction; T, temperature; P, atmospheric pressure; RH, relative humidity.

Height [m]	Parameter	Instrument	Rate [Hz]
1	P	Met One 092	1
7.3	T, WS, WD, RH	Met One 083E, Met One 011E, Met One 024A, Met One 083E	1
9.9	U, V, W, T	Campbell Scientific CSAT3	20
27.1	T, WS, WD, RH	Met One 083E, Met One 011E, Met One 024A, Met One 083E	1
29.6	U, V, W, T	Campbell Scientific CSAT3	20
51.5	T, WS, WD, RH	Met One 083E, Thies First Class, Met One 024A, Met One 083E	1
76.7	T, WS, WD, RH	Met One 083E, Thies First Class, Met One 024A, Met One 083E	1
79	U, V, W, T	Campbell Scientific CSAT3	20
101.5	T, WS, WD, RH	Met One 083E, Thies First Class, Met One 024A, Met One 083E	1
125.9	T, WS, WD, RH	Met One 083E, Thies First Class, Met One 024A, Met One 083E	1
127.9	U, V, W, T	Campbell Scientific CSAT3	20

A number of parameters are used to define the ABL for correlation with AM rating. All values are calculated for a 10 min period. Turbulence intensity, TI, is the amount of turbulence in the flow and is defined by

$$TI = \sigma_u / \overline{U}, \tag{1}$$

where σ_u is the standard deviation of U and is calculated for the same 10 min period as $\overline{U}$. Vertical wind shear defines the velocity difference between elevations in a boundary layer. This paper uses the wind shear exponent, α, from the power law equation to define wind shear. The wind shear exponent is determined by solving for α

$$\overline{U}_{128\text{m}} = \overline{U}_{10\text{m}} \left(\frac{128\text{ m}}{10\text{ m}}\right)^{\alpha}, \tag{2}$$

where velocity subscripts indicate elevation. Reynolds stress, or the turbulent transport of momentum, is an indicator of the amount of mixing happening in the boundary layer (BL). An unstable boundary layer has larger values of Reynolds stress as compared to a stable BL. The Reynolds stress term for the vertical kinematic eddy flux of U-momentum is defined as

$$\tau_{\text{Reynolds}} = \overline{u'w'}, \tag{3}$$

where the prime values are the turbulent velocity components, or the variation from the mean, and are defined by $u' = U - \overline{U}$ and $w' = W - \overline{W}$. Vertical kinematic eddy heat flux is a term used to quantify the turbulent mixing of heat in the BL and is defined as $\overline{w'\theta'}$ where θ' is the fluctuation of potential temperature from the mean. Negative values of heat flux occur during stable BLs, larger and positive values occur during unstable BLs (Stull, 1988).

4. AM Rating

The method used to quantify AM was proposed by the Institute of Acoustics (IOA) Amplitude Modulation Working Group (AMWG) and is used in this study. The method is designed to identify AM from wind turbine noise and not fluctuations in the background sound level. This is achieved by identifying the blade passing frequency (BPF) in the variation of 100 ms L_{Aeq} over a 10 s period. 10 s periods where no dominant frequency can be identified are rejected. Three frequency bands are analysed, 50 to 200 Hz, 100 to 400 Hz, and 200 to 800 Hz for AM. This paper focuses analysis on the 100 to 400 Hz bandwidth. The resulting series of 10 s AM ratings, which is calculated from the variation in the 100 ms L_{Aeq} over 10 s $(L_5 - L_{95})$, are further processed to identify 10 min windows where AM is identified at least 50% of the time and the 90th percentile of the sequential 10 s AM results is taken. The method produces a series of 10 min AM ratings.

The analysis in this paper uses both the 10 min and 10 s AM ratings for comparison with ABL parameters. The 10 min averaged ABL parameters coincide with 10 min AM ratings. The 10 s AM ratings occur within the start and end time of the ABL parameters, but are not centred in time as that would have significantly increased computational times.

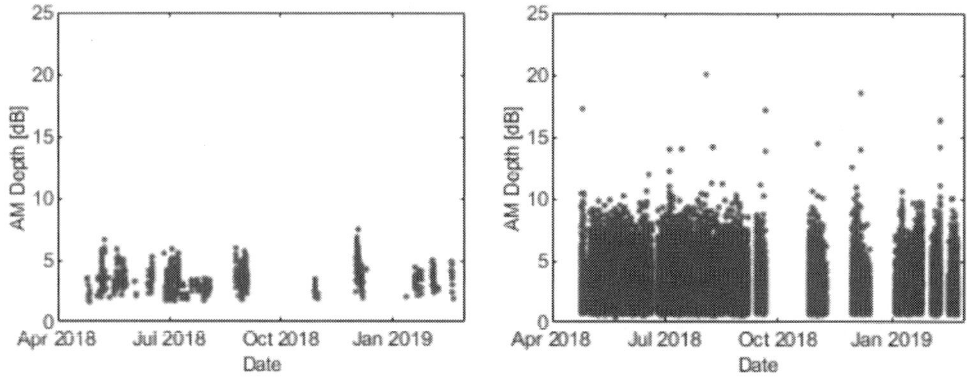

Figure 1. AM modulation rating for 10 min (left) and 10 s (right) analysis

Figure 1 gives the AM rating results for all acoustic measurements from April 24, 2018 to March 8, 2019. Both 10 min and 10 s results are shown, significant AM from the 10 s periods do not meet the criteria for the 10 min AM rating. From the measurement campaign there are 811 ten min periods where AM is identified. The average AM rating is 3.6 dB. There are 43,798 ten min periods between the start and end of measurements where instrumentation was working at the site. This corresponds to identifiable AM occurring 1.85% of the time. Filtering 10 min AM data for wind direction, so that the met tower measurements are not affected by the turbine wake, and instrumentation working leads to 305 ten min AM periods. This is further filtered by looking at data only when the turbine is generating greater than 200 kW, resulting in 183 ten min AM periods. The total number of 10 min ABL measurements is filtered down to 14,918 when the wind direction and turbine power output filters are applied.

5. Results

Measurements are filtered for wind direction such that the met tower is not in the wake of the WTG by filtering out times when wind is +/- 30 deg from due north. This filtering removes conditions where the noise logger is upwind from the turbine which is a limitation on the dataset in this paper. All ABL measurements use this filter unless otherwise stated. Data are further filtered for WTG power production greater than 200 kW for the periods of interest. This paper does not consider the different operating regions of the WTG controller.

Results from correlating ABL parameters with 10 s and 10 min AM ratings are given in Table 3. The reported correlation coefficient, r, uses data where the turbine is operating and generating at least 200 kW over the 10 min period of interest.

Table 3. Correlation coefficients for ABL parameters and AM rating. alpha, wind shear; $\overline{U}_{\mathrm{hub}}$, average hub height streamwise velocity; $\overline{U}_{\mathrm{10m}}$, average streamwise velocity at 10 m; TI, turbulence intensity at 10 m and 128 m; P, turbine output power.

Parameter	r (10 s AM)	r (10 min AM)
α	-0.19	-0.37
$\overline{U}_{\mathrm{hub}}$	0.26	0.40
$\overline{U}_{\mathrm{10m}}$	0.30	0.44
TI_{10m}	0.11	0.08
TI_{128m}	0.15	0.21
$\overline{u'w'}$	-0.20	-0.37
$\overline{w'\theta'}$	0.04	-0.27
P	0.31	0.65

Correlations between ABL parameters and AM rating for the 10 s data, show weak or no correlation. Weak, positive correlations are observed for the streamwise velocity at hub height, $r = 0.26$, and 10 m, $r = 0.30$. Figure 2 is scatter plot where each point is coloured by the spatial density of nearby points as a probability of occurrence in the dataset, termed scatter density plot. The figure shows 10 s AM rating vs hub height velocity (right) and 10 m velocity (left). As the correlation coefficient suggests, velocity at these two elevations are weakly correlated with AM rating as no clear trend can be observed in the scatter. However, the plots show higher probability of AM occurring at lower velocities.

Figure 3 investigates whether this is a phenomenon caused by more data being measured at lower velocities. The plots show the probability distribution of $\overline{U}_{\mathrm{10m}}$ occurring when AM is detected (red) versus during the entire measurement period (blue) for turbine power output greater than 200 kW. Comparing the distributions shows a higher probability of AM occurring at lower 10 m wind velocity as compared to the wind speed distribution for the whole measurement period. Likewise, figure 3 also compares distributions for hub height velocity with AM (blue) and for all measurements. AM is more likely to occur at higher hub height velocities.

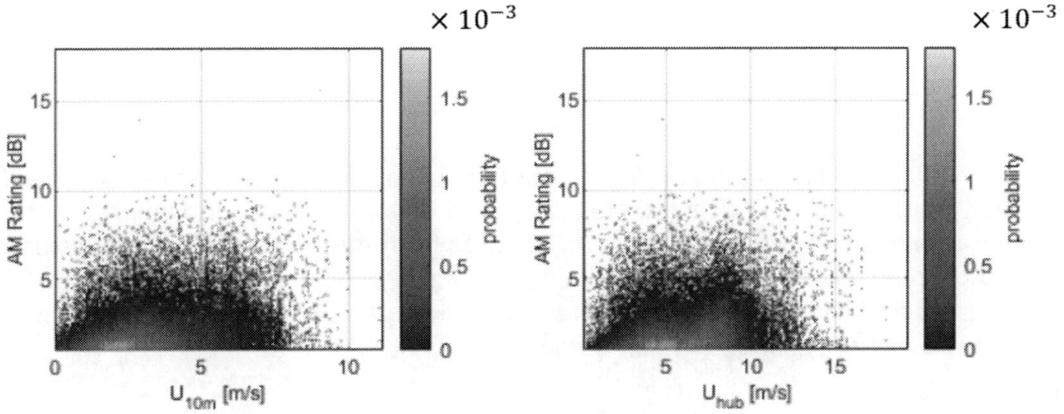

Figure 2. Scatter density plot of $\overline{U}_{10m}$ vs 10 s AM rating (left) and $\overline{U}_{hub}$ vs 10 s AM rating (right).

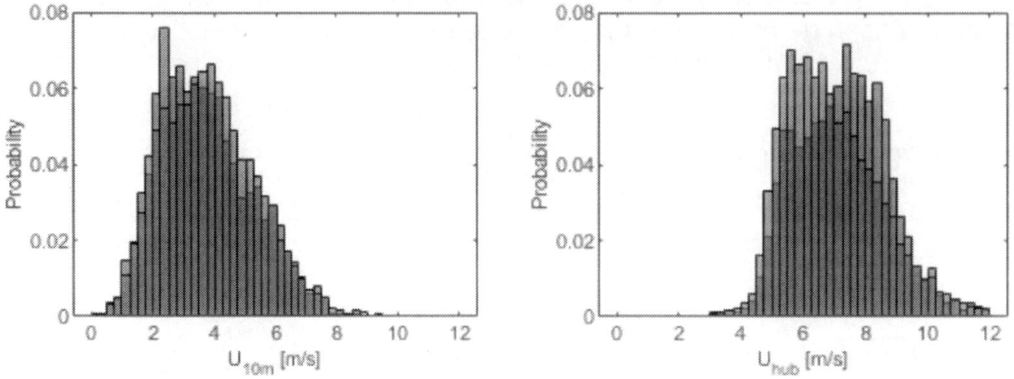

Figure 3. Probability distribution of $\overline{U}_{10m}$ for 10 s AM (blue) and entire measurement period (red). Similarly, probability distribution is shown for $\overline{U}_{hub}$.

The same correlations of velocity at hub height and 10 m are also made with 10 min AM ratings. Here, moderate and positive correlations are found with $r = 0.40$ for $\overline{U}_{\text{hub}}$ and $r = 0.44$ for $\overline{U}_{\text{10m}}$. The stronger correlations are likely a result of less noise in the AM ratings. Figure 4 shows scatter plots of $\overline{U}_{\text{10m}}$ and $\overline{U}_{\text{hub}}$ vs 10 min AM rating. The moderate positive correlation with velocity is seen as a trend in the scatter plot.

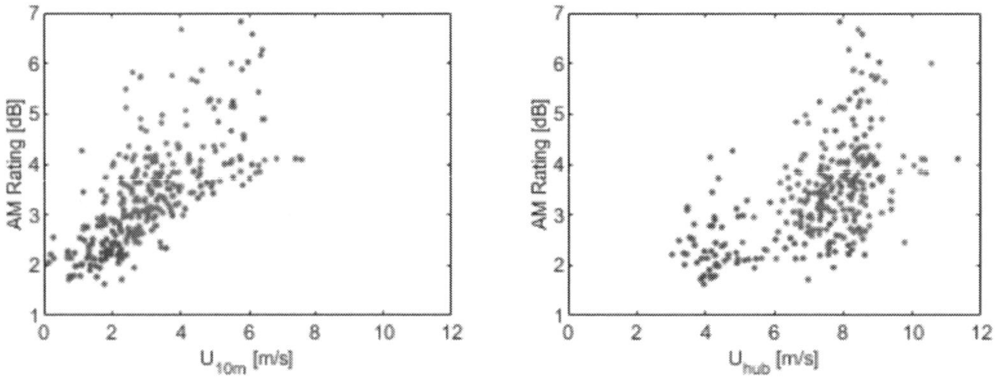

Figure 4. Scatter plot of $\overline{U}_{10m}$ vs 10 min AM rating (left) and $\overline{U}_{hub}$ vs 10 min AM rating (right). Blue points indicate AM ratings where turbine power is $\leq$ 200 kW, red points turbine power is $>$ 200 kW.

As with the 10 s data, the distribution of hub height and 10 m wind velocities are examined. Figure 5 gives the distribution. The blue histogram is the distribution for wind velocity occurring with 10 min AM, the red histogram is of all (filtered) data from meteorological measurements. Comparing the distributions for $\overline{U}_{\text{10m}}$ shows a difference in probability for 2 – 4 m/s with 10 min AM have a larger probability of occurring for this velocity range. Likewise, $\overline{U}_{\text{hub}}$ in the range of 7 – 9 m/s produces a higher likelihood of AM occurring.

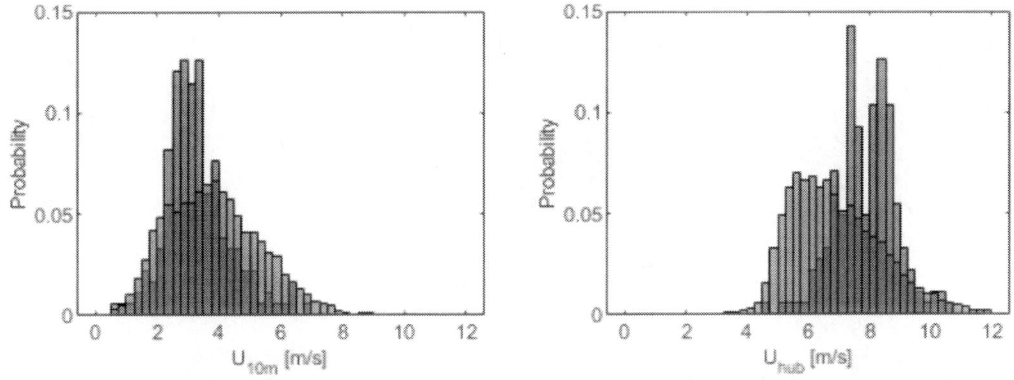

Figure 5. Probability distribution of $\overline{U}_{10m}$ for 10 min AM (blue) and entire measurement period (red). Similarly, probability distribution is shown for $\overline{U}_{hub}$.

Wind shear has been theorized to be a significant factor on the occurrence of AM, where high wind shear during a stable BL creates favourable conditions. i.e. low background noise at the ground due to low wind speeds and wind speed gradient across the WTG rotor, potentially causing partial flow separation at the top of the blade rotation. However, correlation between wind shear exponent and 10 s AM rating found a very weak relationship with r value of -0.19. Figure 6 is scatter density plot of wind shear exponent and turbine power output vs 10 s AM rating. There isn't a clear relationship between wind shear exponent and AM rating, however, higher values of AM occur at lower values of wind shear, or a mixed BL.

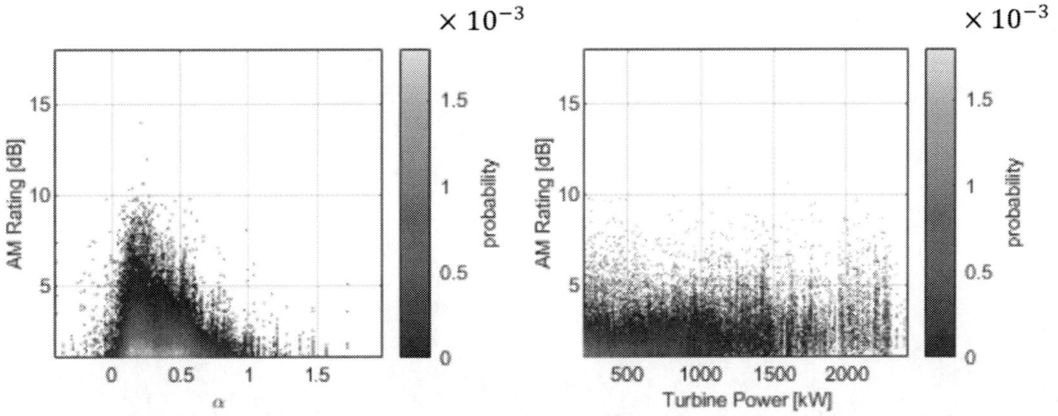

Figure 6. Scatter density plot of α vs 10 s AM rating (left) and turbine power vs 10 s AM rating (right).

Figure 7 shows the probability distribution of wind shear for periods with 10 s AM (blue) and all meteorological measurements (red), the later represents the baseline meteorological conditions for the site. There is a higher probability of AM occurring at larger values of wind shear as compared to the baseline, however, the difference is small at only a few percentage points.

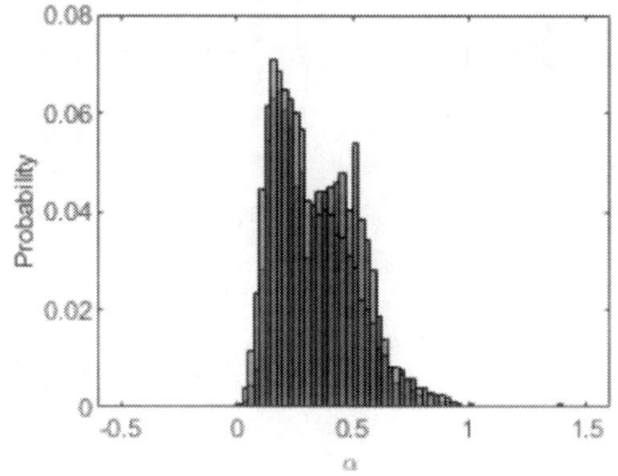

Figure 7. Probability distribution of wind shear exponent for 10 s AM (blue) and entire measurement period (red).

The 10 min AM ratings compared with wind shear exponent shows a stronger relationship as compared to 10 s AM values. A correlation coefficient of -0.37 is calculated. Figure 8 gives a scatter plot of wind shear exponent and turbine power for 10 min AM ratings. The inverse relationship between wind shear and AM rating can be seen. The moderate correlation between turbine power and AM rating is also seen, r value equal to 0.65.

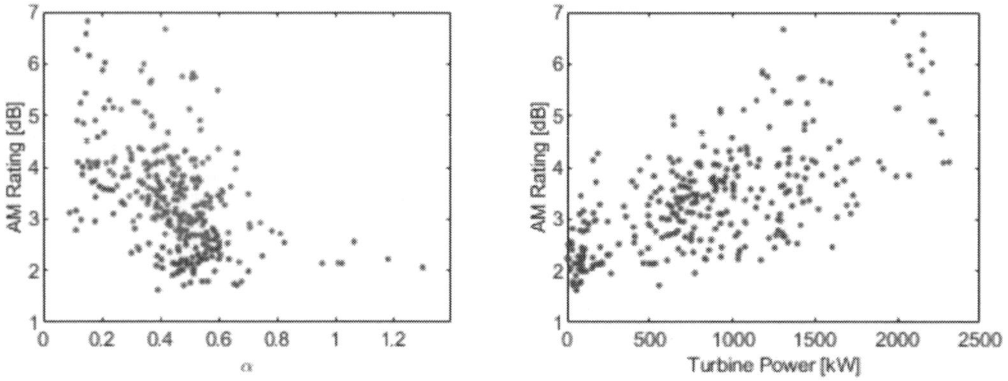

Figure 8. Scatter plot of alpha and turbine power vs 10 min AM rating. Red indicates turbine power > 200 kW.

From figure 8 it appears AM is more densely distributed around higher values of wind shear exponent. This is investigated in figure 9 with a probability density plot of α for periods with 10 min AM (red) and the site baseline condition (blue). Comparing the two distributions we can conclude AM is more likely to occur at higher values of wind shear, or under a stable BL.

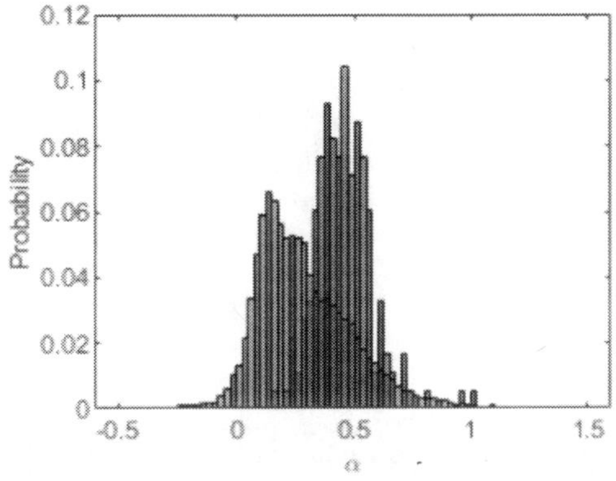

Figure 9. Probability density of wind shear exponent for 10 min AM ratings

Turbulence intensity, heat flux, and vertical momentum flux are also compared against 10 s and 10 min AM ratings. As with the streamwise wind velocity and wind shear exponent parameters, weak or no correlations are made with the AM ratings using the 10 s results. Stronger correlations are typically made with these parameters and the 10 min AM ratings likely due to the 10 min method filtering out contaminated AM ratings with background noise fluctuations.

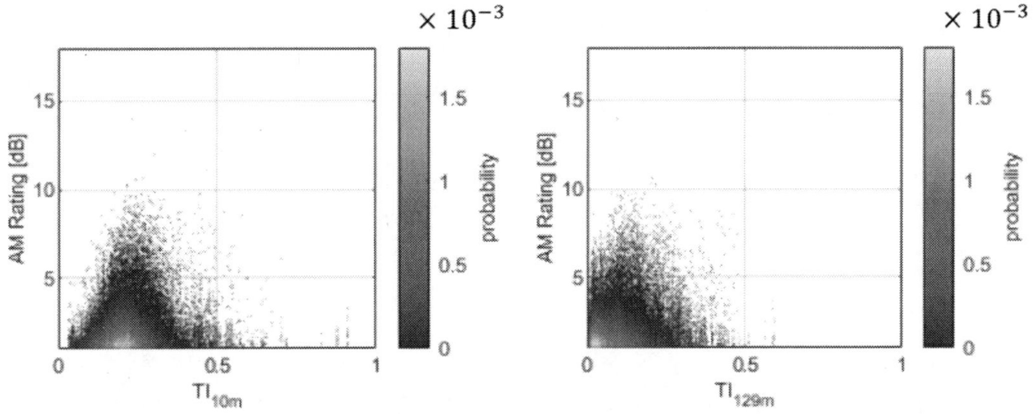

Figure 10. Scatter density plot of turbulence intensity at 10 m and hub height against 10 s AM ratings.

The distribution of the ABL parameter during periods with 10 min AM compared with the site baseline ABL characteristics show values for the ABL parameters at which AM is likely to occur. For example, figure 10 gives scatter density plots for turbulence intensity at 10 m and 129 m vs 10 s AM rating. The distribution of turbulence intensity with AM matches the site turbulence intensity well, figure 11. However, figure 13 shows the higher probability of AM occurring at lower *TI* values which is indicative of a stable boundary layer. Data used in the distribution of figure 13 are given in figure 12.

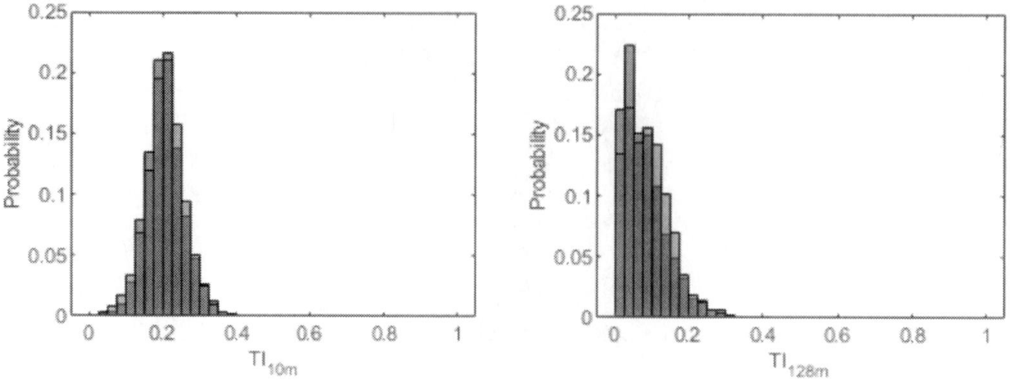

Figure 11. Probability distribution of turbulence intensity for 10 s AM ratings (blue) and baseline site meteorological conditions (red).

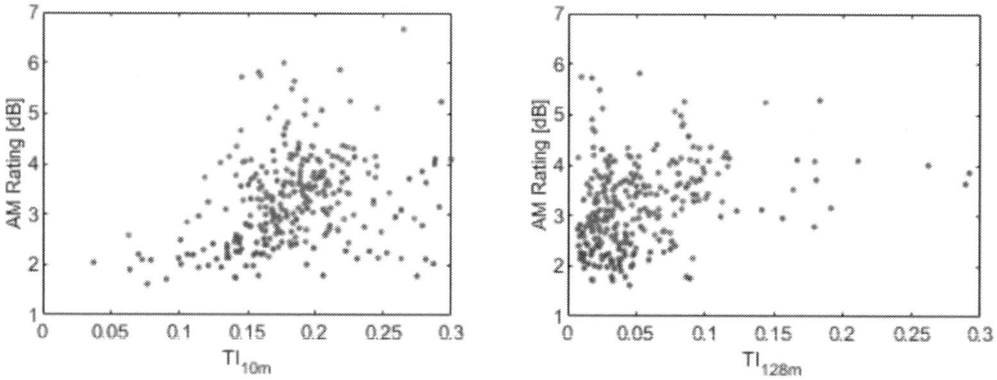

Figure 12. Scatter plot of turbulence intensity vs. 10 min AM rating for 10 m and 128 m elevations. Red indicates turbine power > 200 kW.

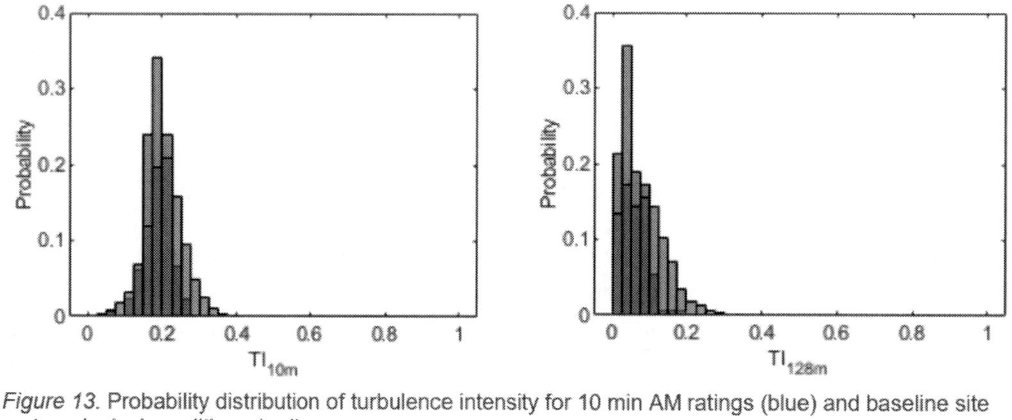

Figure 13. Probability distribution of turbulence intensity for 10 min AM ratings (blue) and baseline site meteorological conditions (red).

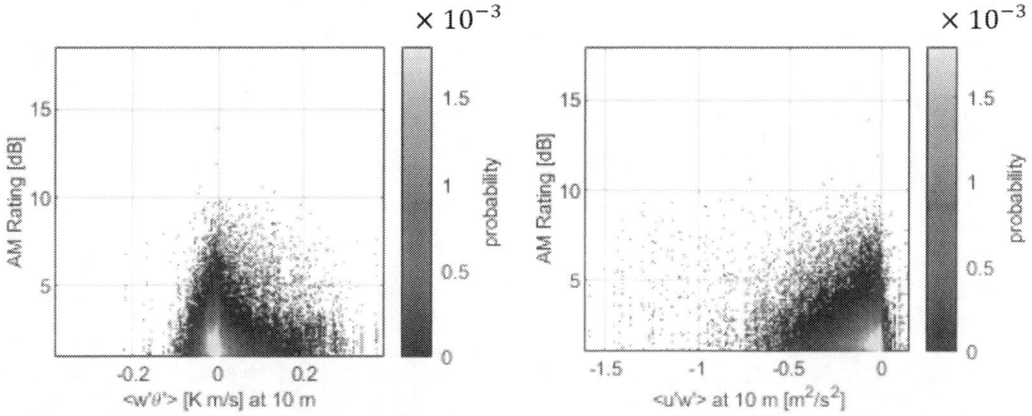

Figure 14. Scatter density plot turbulent heat flux vs 10 s AM ratings (left) and turbulent momentum flux vs 10s AM ratings (right).

The turbulent heat flux parameter, $\overline{w'\theta'}$, is an indicator of the stability of the BL. Negative values indicate a stable boundary layer, positive values an unstable BL. The scatter density plot in figure 14 shows higher values of AM occurring, as well as more often, near turbulent heat flux values around zero. This is confirmed by figure 15 where AM occurs more often at $\overline{w'\theta'} < 0$.

No obvious relationships are seen for $\overline{u'w'}$ in figures 14 and 15.

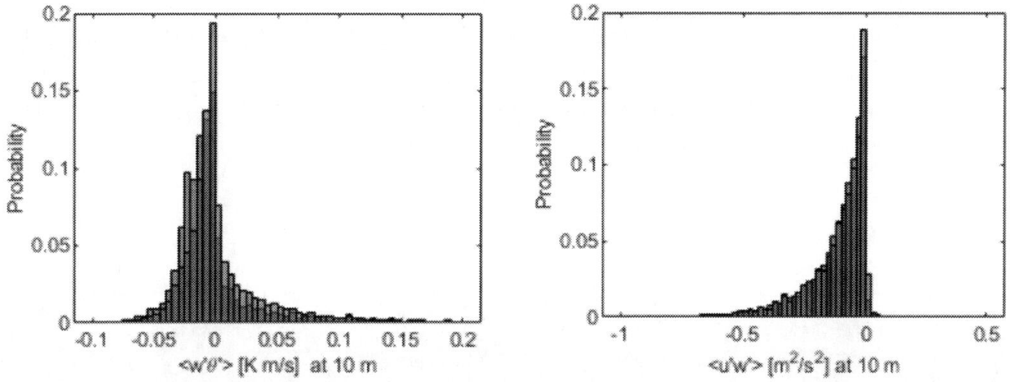

Figure 15. Probability distribution for turbulent heat flux (left) and turbulent momentum flux (right) during 10 s AM shown in blue. Red is the distribution for the site baseline turbulent fluxes.

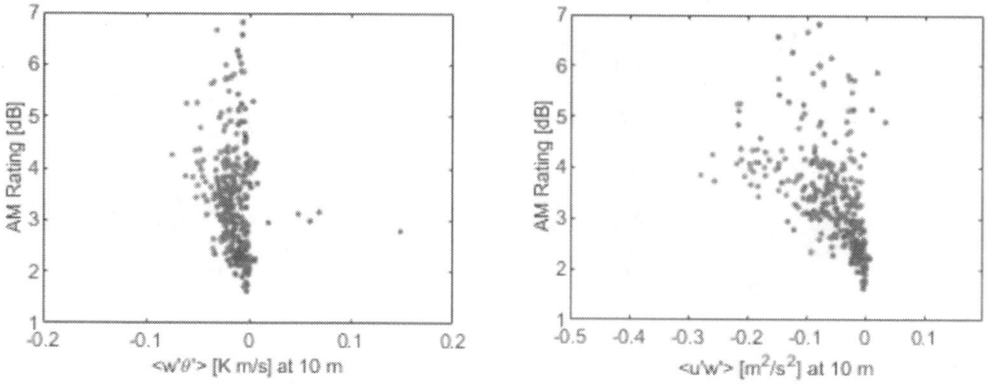

Figure 16. Scatter plot of turbulent heat flux vs. 10 min AM rating (left) and turbulent momentum flux vs 10 min AM rating (right). Red indicates turbine power > 200 kW.

Lastly, turbulent flux parameters are compared with 10 min AM ratings in figure 16. The turbulent heat flux shows a strong relationship between a stable BL, $\overline{w'\theta'} < 0$, and the presence of 10 min AM. This is further seen in figure 17 which shows the distribution of turbulent heat flux with AM vs the site baseline distribution. There is a weak relationship between decreasing heat flux and increasing 10 min AM rating, $r = -0.27$; however this is the only ABL parameter investigated in this study which shows a relationship between a stable BL and increasing levels of AM.

Larger absolute values of turbulent momentum flux, $\overline{u'w'}$, indicate a more turbulent BL. It is not clear at what values a stable BL or unstable BL occur, however, figure 17 indicates AM is more likely to occur at smaller absolute values of $\overline{u'w'}$ which would suggest a stable BL and is consistent with all other ABL parameters.

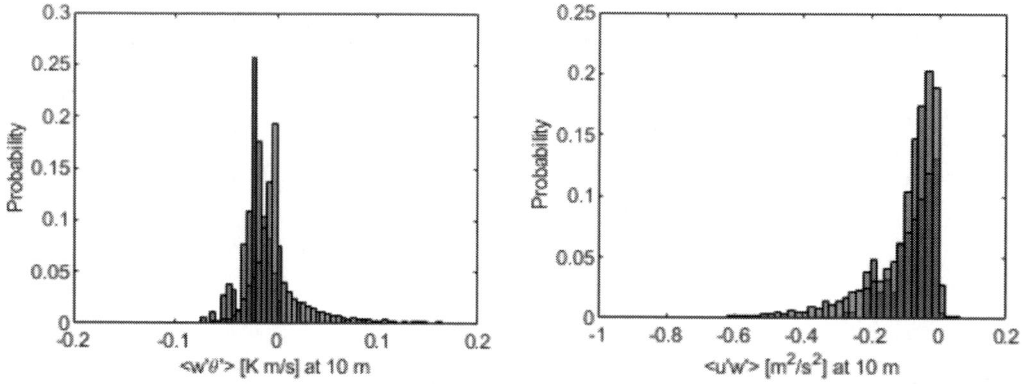

Figure 17. Probability distribution of turbulent fluxes with 10 min AM ratings (blue) and all meteorological measurements.

6. Conclusions

Correlations between ABL parameters and AM rating were investigated over a year-long measurement campaign at a single WTG site. Weak or moderate correlations were made. Wind shear exponent was negatively correlated with AM rating indicating higher values of AM typically occurring during an unstable BL. However, 10 m and hub height streamwise velocity and turbine output power had positive correlations with AM rating which may suggest it is a function of background noise fluctuations at the receiver. Further, the probability distributions for the ABL parameters with AM indicated higher probabilities of AM occurring when the meteorological conditions suggest a stable BL. The only ABL parameter to show correlations between a stable BL and higher levels of AM was the turbulent heat flux parameter.

As no strong correlations between meteorological conditions and AM rating were found, it is likely that AM in the far-field is an immission problem, perhaps amplified by inflow conditions at the source. A stable BL is more likely to produce AM, especially in the downwind direction. The cause may be the refraction of the sound waves down, towards the receiver. Further research would need to be performed to study this phenomenon.

References

AMWG, A. M. (2016). *A Method for Rating Amplitude Modulation in Wind Turbine Noise.* Institute of Acoustics.

Cand, M., Bullmore, A., Smith, M., Von-Hunerbein, S., & Davis, R. (2012, April). *Wind turbine amplitude modulation: research to improve understanding as to its cause & effect.* In Acoustics 2012.

Larsson, C., & Öhlund, O. (2014). *Amplitude modulation of sound from wind turbines under various meteorological conditions.* The Journal of the Acoustical Society of America, 135(1), 67-73.

Oerlemans, S., Sijtsma, P., & López, B. M. (2007). *Location and quantification of noise sources on a wind turbine.* Journal of sound and vibration, 299(4-5), 869-883.

Oerlemans, S., & Schepers, J. G. (2009). Prediction of wind turbine noise and validation against experiment. *International journal of aeroacoustics*, *8*(6), 555-584.

Oerlemans, S. (2013). *An explanation for enhanced amplitude modulation of wind turbine noise.* Wind Turbine Amplitude Modulation: Research to Improve Understanding as to its Cause and Effect. RenewableUK.

Pedersen, E., & Persson Waye, K. (2004). *Perception and annoyance due to wind turbine noise—a dose–response relationship.* The Journal of the Acoustical Society of America, 116(6), 3460-3470.

Stull, R. B. (1988). *An introduction to boundary layer meteorology.* Kluwer Academic Publishers.

Van den Berg, G. P. (2004). *Effects of the wind profile at night on wind turbine sound.* Journal of sound and vibration, 277(4-5), 955-970.

INCE EUROPE

8th International Conference
on
Wind Turbine Noise
Lisbon – 12th to 14th June 2019

Time dependent changes in sound pressure levels caused by wind turbines at long distances

Till Kühner, Dr. Kühner GmbH, Germany, kuehner@kuehner-it.de

Summary

Sound pressures levels caused by wind turbines typically show a lot of time-dependent changes. One obvious cause is the source itself, where the sound power of the wind turbine depends directly on the wind speed. Another obvious cause for changes in sound pressure levels at long distances is the wind direction, since the sound propagation depends on it.

Conditions for the source as well as the propagation and attenuation of sound can be classified to better understand the statistics of the sound pressure levels. But even choosing time sections with fairly constant wind speeds and wind directions at the wind turbine, there still is a lot of variability in the sound pressure levels.

On the very short time scale, this is partly due to the type of sound emitted by wind turbines. The aerodynamic sound from the rotor blades is band-limited noise, which results in fluctuations in sound pressure levels for purely statistical reasons.
Under some conditions, amplitude modulations (AM) occur, causing sound described with terms like whooshing or thumping. With 100ms-Leqs, the levels go up and down in synch with the rotor blades.
If the focus is on determining average sound pressure levels, and not on studying amplitude modulations, Leqs with integration times longer than the period of the AM can be used to suppress the effects of AM on the statistical distribution of the levels.

The variability and statistical properties of sound pressure levels caused by wind turbines at about 1000m distance for a two month long measurement project are presented.

1. Introduction

The sound pressure levels caused by wind turbine sound at long ranges vary for a number of reasons. The wind turbine itself emits sound that depends mostly on wind speed and the power produced, but may also change with turbine blade pitch angle, with the turbulence of the wind field, and the shape of the wind speed profile.

The attenuation due to the sound propagation also depends on the wind direction, the turbulence and shape of the wind field as well as the air temperature profile. The wind turbines have an effect on the wind field itself that may also be relevant for the propagation of sound for several hundred meters.

And there may be many more reasons for variations in the wind turbine as a source of sound as well as the propagation of the sound. The present paper discusses some of the problems that are encountered when evaluating the varying and fluctuating sound pressure levels caused by wind turbines.

2. Sound pressure levels

2.1 100ms-equivalence levels

The sound from wind turbines mostly is a band-limited noise – with noise being used in the statistical sense. For a noisy signal, a calculated sound pressure level with a limited integration time is also noisy, i.e. it shows statistical variations. The shorter the integration time, the more the sound pressure levels fluctuate, even if the sound pressure is a stationary signal – stationary indicating that it has a constant amplitude.

The longer the integration time, the smaller the statistical fluctuations become. But using long integration times reduces the time resolution. For wind turbines, the sound pressure signal often is not stationary, but has a modulated amplitude. If the amplitude goes up and down each 1.5s – for example – the integration time needs to be much shorter to still see the amplitude modulations in the sound pressure levels.

In studies on the sounds from wind turbines, it is very common to look at $L_{Aeq,100ms}$, which are A-weighted levels, with a rectangular integration window of 100ms, which seems to be a good compromise between time resolution and statistical noise.

2.2 Long term sound pressure levels

A typical use of the measured levels might be the calculation of an equivalent continuous A-weighted sound pressure level [ISO9613-2], also called time-averaged sound pressure level [ISO1996-1].
Since ISO9613-2:1996, there were several revisions of ISO1996-2. The text for the source was changed, where it used to say (6.1):

> "*The source operating conditions shall be statistically representative of the noise environment under consideration.*",

the current text is much more fuzzy, and the *statistically representative* was dropped. But the requirement of determining the uncertainties was added.

And for the meteorological conditions, the requirement in 3.6 for a meteorological window had the minor change from

> "*set of weather conditions during which measurements can be performed with limited and known variation in measurement results due to weather conditions.*"

to

> "*set of weather conditions during which measurements can be performed with limited and known variation in measurement results due to weather variation.*"

For wind turbines, the source depends on the meteorology, as do the propagation properties that affect the measured sound pressure levels. Depending on national regulations, old or new versions of ISO 1996-2 are to be used. But in both cases, the requirement that the variations in results due to meteorological conditions need to be limited and known stands.

And whether paragraph 6 is taken from the old or new version, single measurements can not fulfill the requirements.
For the old version of ISO1996-2, a possible way to fulfill the *statistically representative* requirement would be to perform measurements for a very long time. The time would have to be long enough to be statistically representative – which might be a year, two years, or even three years.

The new version of ISO1996-2 does not require a *statistically representative* measurement time, but it requires the determination of uncertainties. And this goes hand in hand with the *limited and known* requirement for the effects of the meteorology.

An approach to satisfy these requirements would be to form classes of conditions for the source and sound propagation, and then use known statistics for these classes to calculate a representative long-term averaged sound pressure level.
For each of these classes, the measurement results need to have "limited and known variation". The following is a discussion on such classes and meteorological windows.

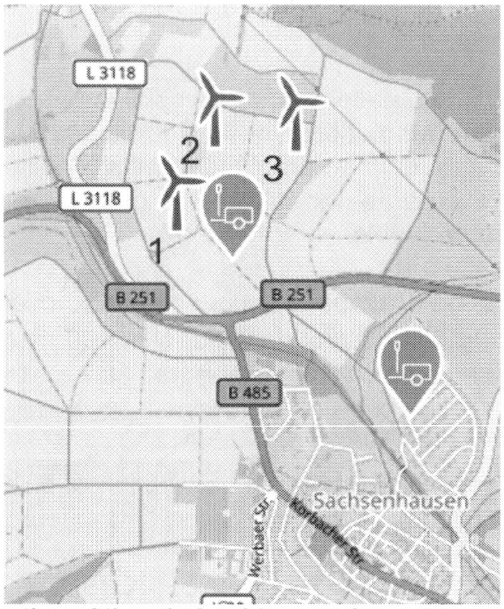

Figure 1 Location of the wind turbines and measurement positions

3. The measurements

The data used was measured in a project on the impact of sound from wind turbines funded by the German Environment Agency (Umweltbundesamt – UBA). It should be noted that this project does not focus on sound pressure levels, and that the following is not part of the project.

Outside of a small place called Sachsenhausen in Nordhessen, Germany, three wind turbines are about 1000m away from the residential area (see Figure 1). The wind turbines are:

- Wind turbine 1 and 2: Enercon E-82 with hub height 137 m,
- Wind turbine 3: Enercon E-101 with hub height 133 m.

Two measurement systems were installed by deBAKOM GmbH.
At 180m distance to wind turbine 1, meteorological data was gathered, including wind at several heights up to 10m. Operational data of wind turbine 1, including power and wind speed at hub height, was recorded at 1s intervals.

Figure 2 Measurements close to the wind turbines

Figure 3 Measurements at edge of residential area

A second measurement system was installed at the edge of the residential area, recording wind at two heights, and acoustical data at 7m height. All sound pressure levels discussed in the following were recorded in this location.
The measurement systems recorded from April 27th to June 26th 2018. Due to technical problems, not all systems ran for the whole time, but the acoustical measurements at the edge of the residential area are almost complete for the two month period.

4. Statistical classification

4.1 Background noise

As a very first step of looking at the measured levels, the distribution of levels is shown as a histogram in Figure 4. The vertical axis is labeled "relative density", the values shown are just the number of levels in each bin, divided by the total count of levels. In histograms, this is often called "frequency", the term "relative density" is used to avoid confusion.

Indicated in the histogram are also the 95, the 80, and the 50 Percentile. For the L50, the 50-Percentile, half of the measured levels are higher than this value, and half of the levels are lower. The L95, for which only 5% of the levels are lower than this value, is sometimes referred to as a background level.
The distribution shown in Figure 4 includes all the noise measured, which is not just from the wind turbines, but also from sources ranging from traffic to birds and mooing cows.

In the measurement location, the most quiet time of the night is the time from 23:00 to 04:00. Figure 5 shows the distribution of levels for this time. The shift of the distribution to smaller levels is very obvious.

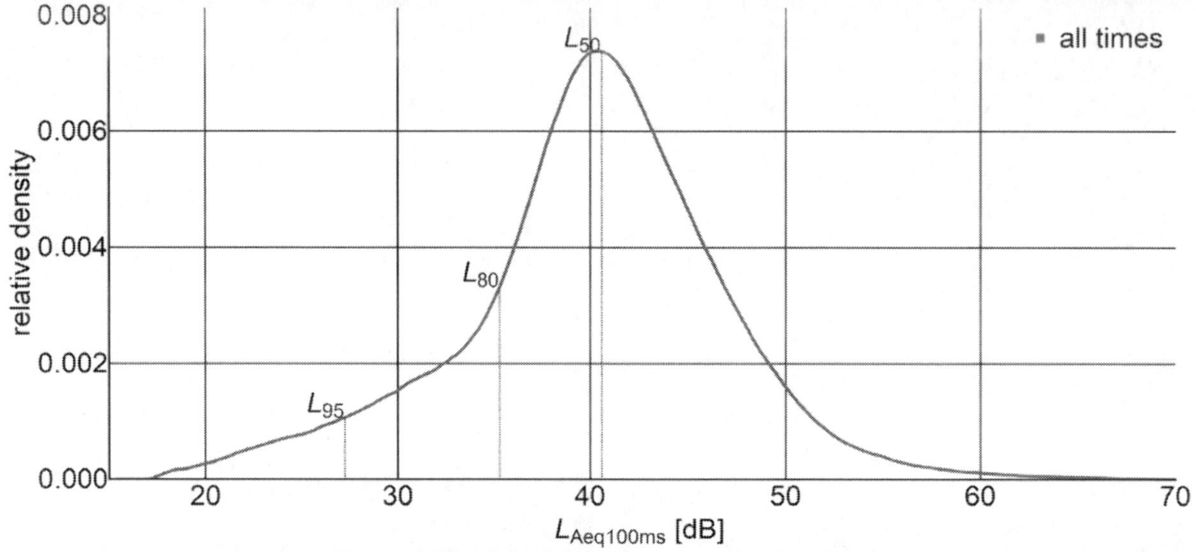

Figure 4 Distribution of $L_{Aeq,100ms}$ for all times

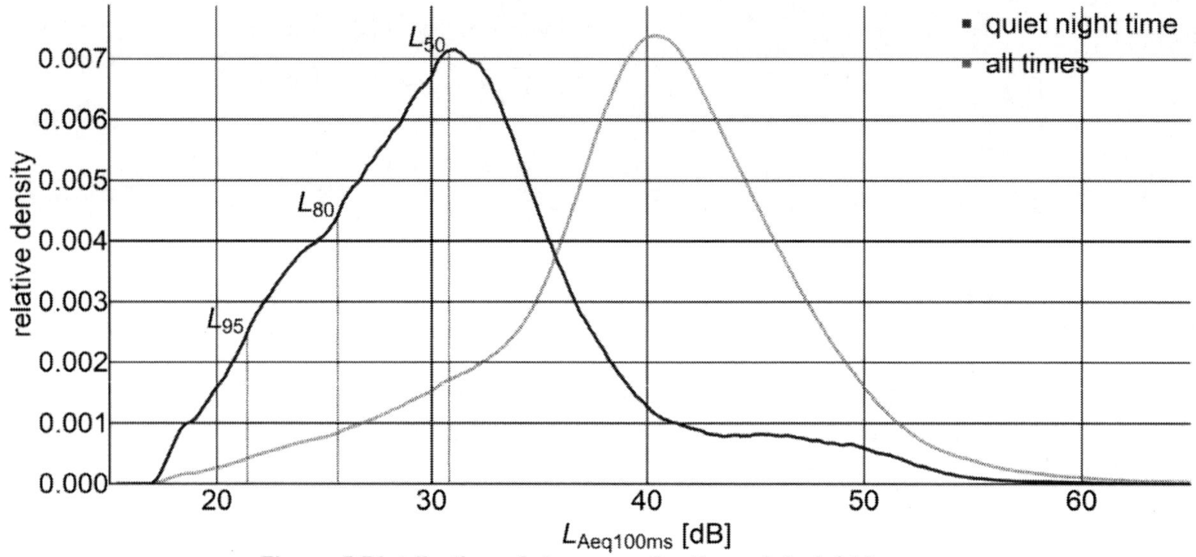

Figure 5 Distribution of $L_{Aeq,100ms}$ for the quiet night hours

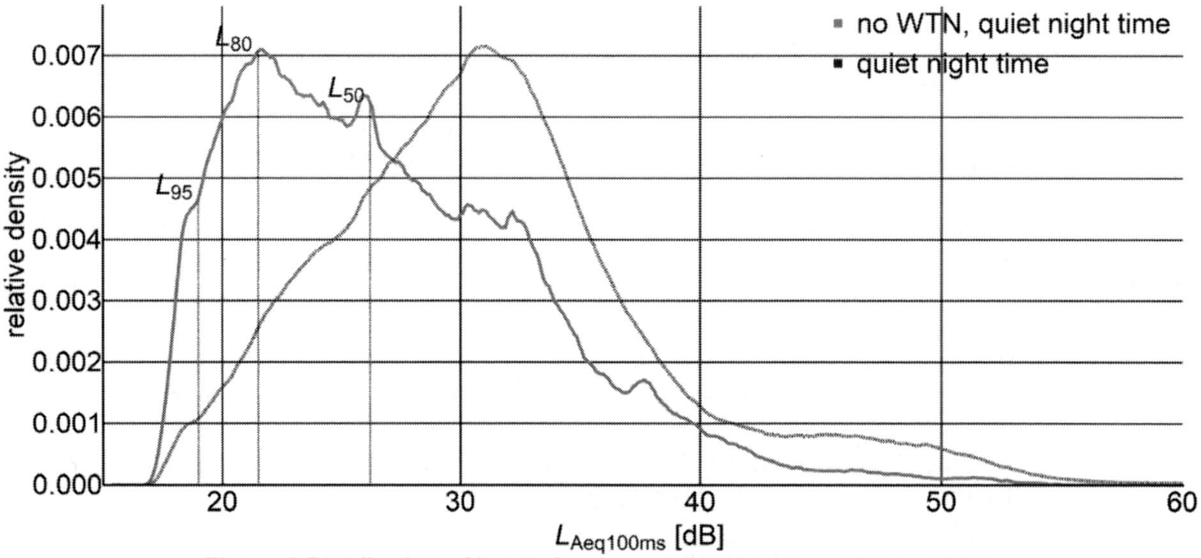

Figure 6 Distribution of levels for the quiet night hours without WTN

The next step is to reduce the times taken into account yet again, and only look at the quiet night hours where the wind turbines did not run, or - to be more specific - the times where wind turbine 1 produced less than 100kW power. Figure 6 shows this distribution, which should be a good approximation for what the night levels would be like it there were no wind turbines. With an L50 at about 26 dB(A), and an L95 at about 19 dB(A), the area is fairly quiet.

The levels in the absence of WTN mostly are so low, that for the following the measured levels under different conditions are looked at as if they were caused only by the wind turbines. Note that this is a very rough approximation, that is used to illustrate qualitative effects in this paper.

In the following, only data from the quiet time of the night from 23:00 to 04:00 is used.

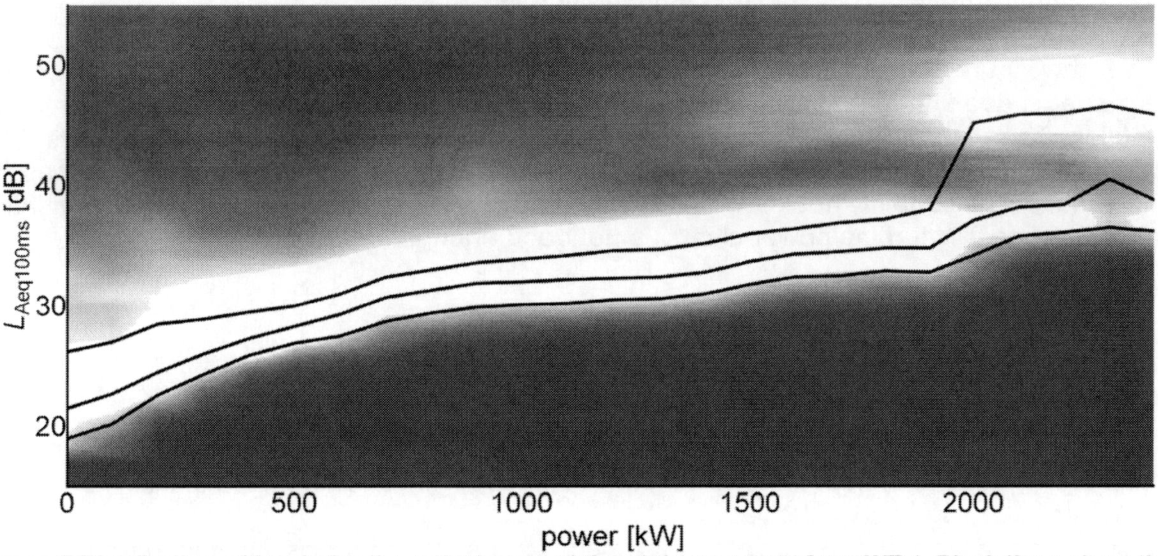

Figure 7 Distribution of levels for the quiet hours of the night vs power from WT 1. Black lines show the L95, L80 and L50.

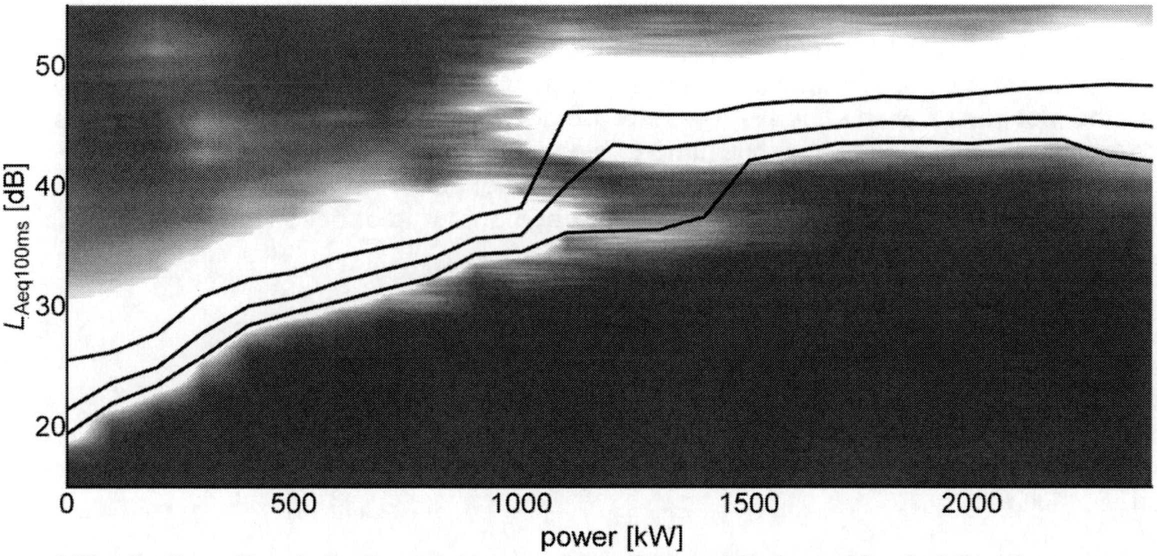

Figure 8 Distribution of levels for the quiet hours of the night and for favorable wind directions vs power from WT 1. Black lines show the L95, L80 and L50.

4.2 Levels vs wind turbine power

Some correlation of the sound pressure levels with the power from the wind turbines can be expected. Figure 7 show the distribution as 2D-plot for levels and power from wind turbine 1. The density of values is indicated by the color, with lighter colors for higher densities.
Up to about 1900kW the distribution of levels is smoothly shifted to higher levels, basically meeting the expectation that the levels depend on the power of the wind turbine.

Ignoring the discontinuity at 2000kW for now, another simple expectation that can be verified is the dependence of levels on the wind direction. Figure 7 shows levels for all wind directions, a distribution for favorable wind directions is expected to look similar, but with the distribution shifted to higher values. In the actual distribution for favorable wind directions shown in Figure 8, the levels are indeed distinctly shifted up – but the discontinuity that was at 2000kW for all wind directions is now between 1100 and 1200kW. For now, the reason for the discontinuities is unclear, but the distinct difference between Figure 7 and Figure 8 can be taken as warning on how deceptive the statistics for such measurements can be.

Figure 7 and Figure 8 also show L95, L80 and L50. Since the distribution of levels might be a bit asymmetric due to other noise, 2*(L50-L95) might be taken as a measure for the width of the distribution. This width of the distribution is about 8dB or higher for all wind turbine powers. So the width of the distribution is not limited in the sense of ISO1996 requirement, and the reason for the variations is certainly not known as per the same paragraph in ISO1996-2. Classification by wind direction and turbine power clearly is not good enough.

4.3 Levels vs wind at the wind turbines hub height

The distributions shown in Figure 7 and Figure 8 are plotted versus the power from wind turbine 1, and there is the inherent assumption that the other two wind turbines have similar power, and hence similar sound power levels.
Another approach would be to look at the wind speeds as the main classification parameter, which is also what is done in IEC 61400-1 for the sound pressure levels close to the wind turbines.

Figure 9 shows the distribution of $L_{Aeq,100ms}$ vs the wind speed at the hub of wind turbine 1, during the quiet hours of the night and with favorable wind directions. Similar to the plots versus power, the plot versus wind speed shows how the distribution is shifter to higher levels as wind speed increases, and a discontinuous jump to higher levels at 9 m/s wind speed.
The level distributions vs wind speed still show a width of 8dB or more, so a classification of source and propagation conditions just by wind speed and wind direction would be insufficient.

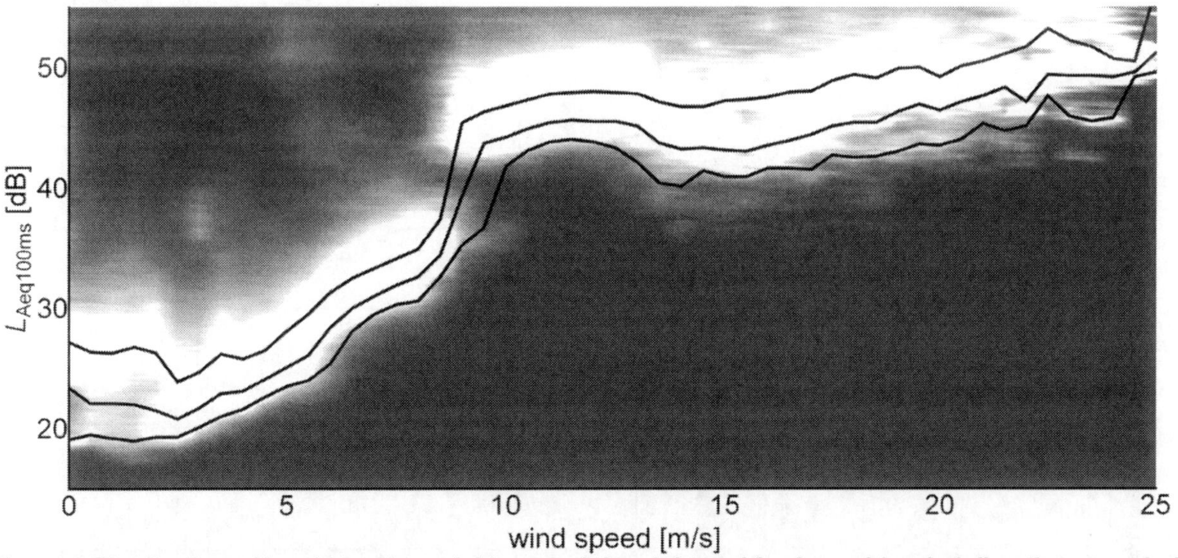

Figure 9 Distribution of levels for the quiet hours of the night and for favorable wind directions vs wind speed at hub height of WT 1. Black lines show the L95, L80 and L50.

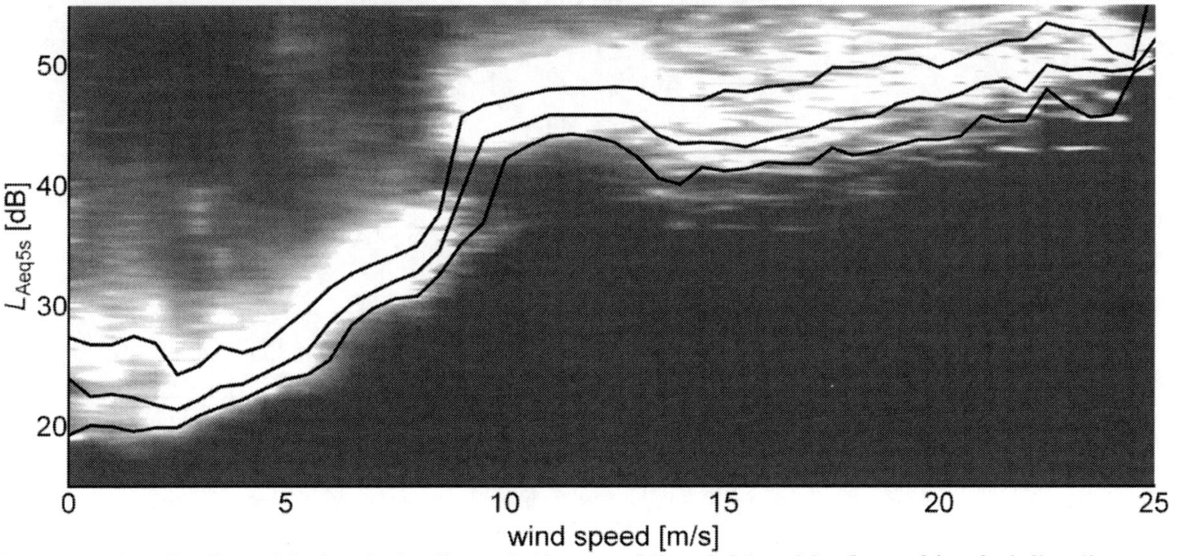

Figure 10 Distribution of 5s-levels for the quiet hours of the night and for favorable wind directions vs wind speed at hub height of WT 1. Black lines show the L95, L80 and L50.

5. Time dependent changes

5.1 Sound pressure levels

The statistical distributions shown in the previous section had width of 8dB or more for the sound pressure levels. A possible reason for the width of the distributions might be found in the levels themselves. The $L_{Aeq,100ms}$ are very sensitive to short time variations which might be of no interest when determining a long-term average. Figure 11 shows a sound pressure level plot for a time with amplitude modulations (AM). If one is interested in finding the AM, the $L_{Aeq,100ms}$ nicely shows how the level goes up and down. For a long-term average, these short-term variations may be considered a distraction. And they can be included with the levels assigned to the wind turbines, since they clearly are caused by the wind turbines.

Increasing the integration time of the sound pressure levels to 5s evens out the AM, as can be seen in the $L_{Aeq,5s}$ shown in Figure 11. The 5s are long enough that the plot does not follow the amplitude of the AM, and short enough to show the slower fluctuations on the scale of 5 to 15 seconds. Choosing an integration time is always a compromise, for the following 5s will be used.

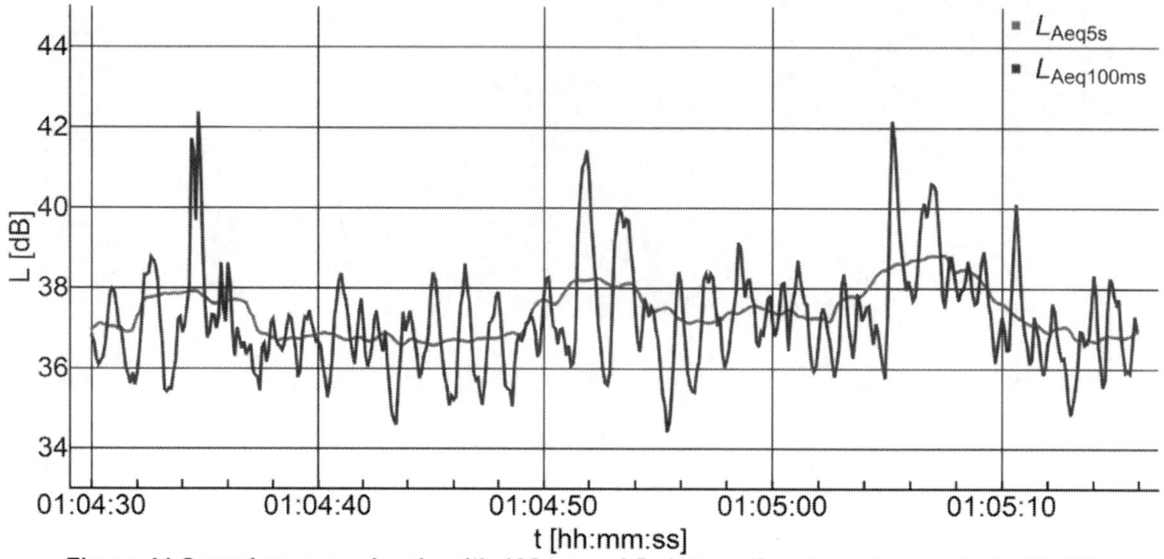

Figure 11 Sound pressure levels with 100ms and 5s integration times in a period with AM

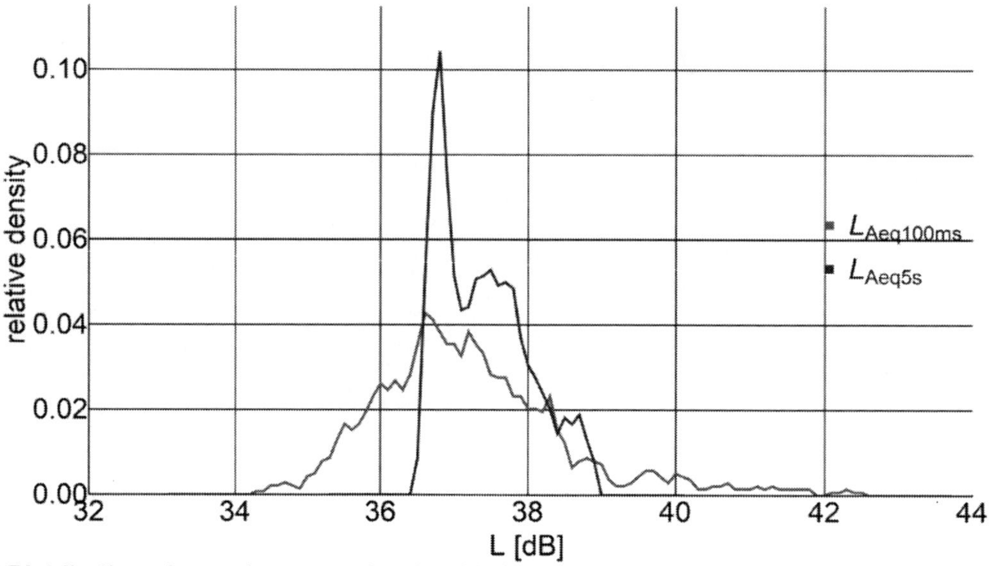

Figure 12 Distribution of sound pressure levels with 100ms and 5s integration times in a period with AM

Figure 12 shows the distribution of sound pressure levels for the same time as shown in Figure 11. As expected, for the $L_{Aeq,5s}$, the distribution is much more narrow than for the $L_{Aeq,100ms}$. But it still reflects that the sound pressure levels fluctuate even when AM are averaged out. Note in Figure 11 and Figure 12 that even within this 50s time-window, the $L_{Aeq,5s}$ varies by over 2dB.

Going back to the sound pressure level distribution vs wind speed, Figure 10 shows the distribution for the $L_{Aeq,5s}$. Comparing Figure 9 and Figure 10, some changes in the details can be seen due to the change from 100ms to 5s integration times in the levels, but overall the difference is small, and the width of the distribution is very similar. There is still a lot of variation in the sound pressure level, even with 5s integration time.

5.2 Wind speed

The discontinuity visible in the level distributions in Figure 7 to Figure 10 is an indication of a classification problem. Wind speeds and the potential influence on the measurement results need to be looked at closer.

Before going into more detailed classifications, the wind speed as hub height as the main parameter needs to be looked at more closely, as the wind speed data shows a lot of very quick fluctuations.
A possible way of quantifying this is shown in Figure 13. It is a standard deviation of the difference between two wind speeds with a time difference of τ in between:

$$\sigma(\tau) = \sqrt{E\left|\left(v_{t_i} - v_{t_i+\tau}\right)^2\right|}$$

(This is directly related to the autocorrelation function, but has the advantage of showing the difference to be expected between two values.)

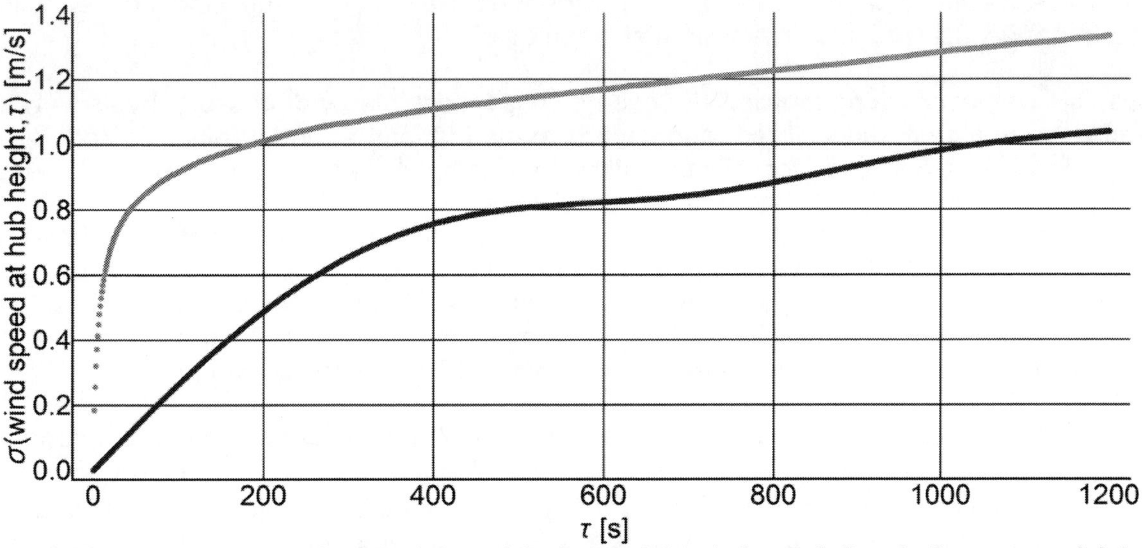

Figure 13 Time dependent standard deviation of the hub-height wind speed. Gray is the measured data, blue is the signal conditioned with a low-pass filter.

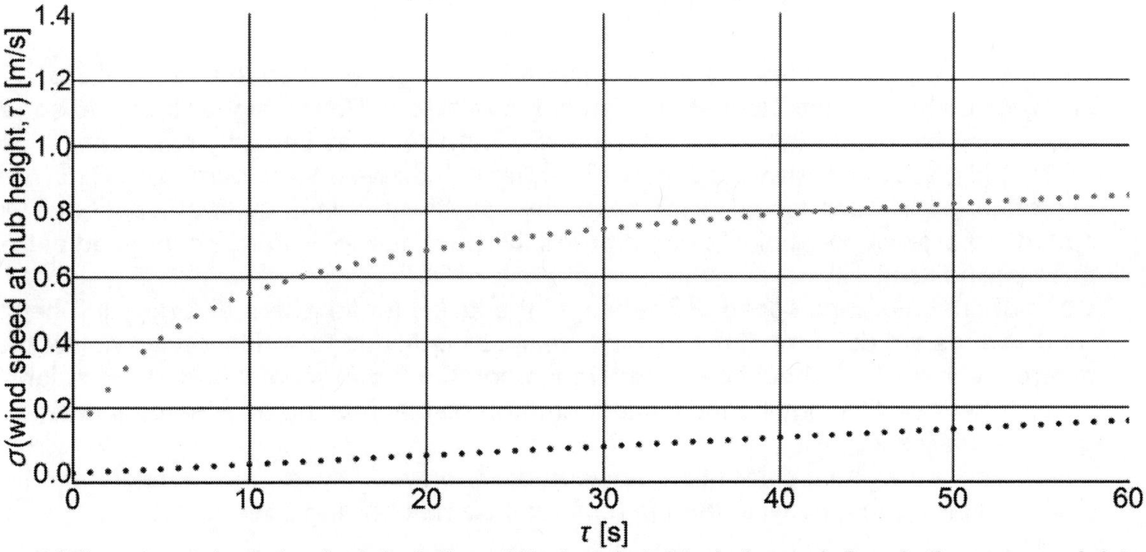

Figure 14 Time dependent standard deviation of the hub-height wind speed. Gray is the measured data, blue is the signal conditioned with a low-pass filter.

Figure 14 shows the standard deviations for small times. After less than 10s, 0.5m/s is exceeded. This indicates that the wind speed as recorded at hub height will usually change by 0.5m/s within less than 10s. The shape of the curve in Figure 13 indicates that these short term fluctuations mask the long-term correlations. They might be noise in the measurement, or they might be actual short-term fluctuations that contain relevant information on the turbulence of the atmosphere. In any case, they overlay the longer-term trend of the hub-height wind speed.

A possible approach to obtaining a more stable quantity is to apply a low pass filter to the wind speed signal, hopefully getting a better classification parameter.

6. Wind induced noise – or not?

A possible cause of increased measured sound pressure levels in the context of wind turbine noise is wind induced noise. As the wind speed at the hub of the wind turbine increases, so does the wind close to the ground. This wind causes noise itself, when blowing over or through vegetation for example, and that noise is not due to the wind turbine. Care has to be taken not to include this wind induced noise with the wind turbine noise.
Even worse, as wind speeds increase, the microphone itself may pick up wind induced noise, and that not be included in a environmental noise at all.

A possible way of checking for wind induced noise is to look a sound pressure levels vs wind speed close to the ground for fixed wind speeds at the hub of the wind turbine. To do this and have a reasonable resolution for effects at different hub height wind speeds, the wind speeds at hub height are conditioned as shown in the previous section.
With the conditioned hub-height wind speed, a classification by ground height wind speed is possible. As examples, three wind speed ranges are set:

- Conditioned hub-height wind speed 7m/s (6.75m/s to 7.25m/s)
- Conditioned hub-height wind speed 10m/s (9.75m/s to 10.25m/s)
- Conditioned hub-height wind speed 13m/s (12.75m/s to 13.25m/s)

As a second classification parameter, the wind speed at 7m height – next to the microphone - is used. Figure 15, Figure 16 and Figure 17 show the distributions of $L_{Aeq,5s}$ for the three hub-height wind speeds vs 7m height wind speed. The width of the distribution is roughly 5dB, and while that may not be narrow enough, it certainly is an improvement.

Figure 18 shows the L80 for the three distributions. Some qualitative information can be gathered:

- The measured sound pressure levels at small 7m height wind speeds is smallest for 7m/s at hub height, and biggest for 13m/s at hub height. So the higher the wind speed at hub height, the higher the levels from the wind turbines – as certainly expected.
- At (roughly) about 4m/s wind speed at 7m height, the levels start increasing. The microphone has been tested not to have any sensitivity to wind up to at least 6m/s wind speed, so at least some of this effect probably is not due to wind noise induced at the microphone itself.
- Up to about 8m/s wind speed at 7m height, the levels for the three different hub-height wind speeds are different. If the rise in levels was only due to wind induced noise, the difference in the L80 should be much smaller for 4 to 8 m/s wind speed at 7m height.
    - The statistics may be too limited, and the results may be an artifact of bad statistics, or
    - There may be a correlation between between wind speed at 7m height and the levels due to the wind turbine for fixed hub-height wind speeds.

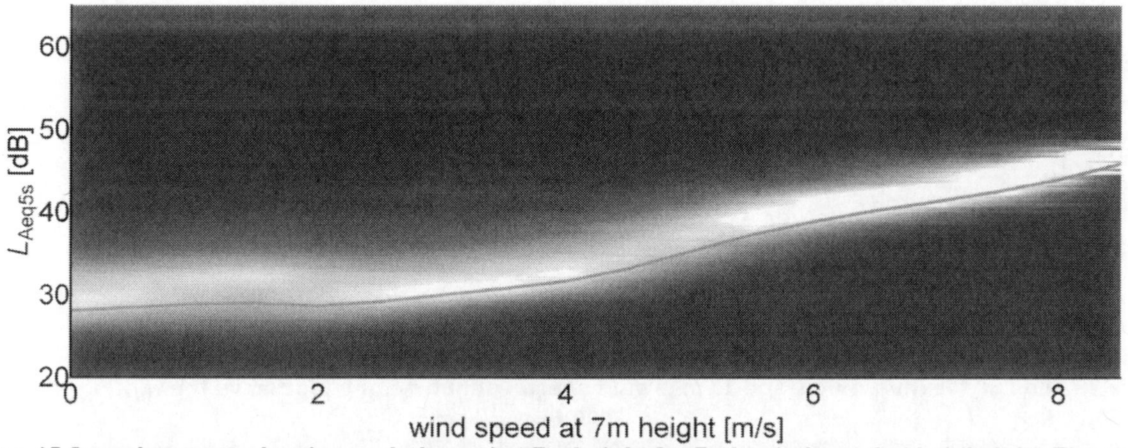

Figure 15 Sound pressure levels vs windspeed at 7m height for 7m/s windspeed at hub height. The green line shows the L80.

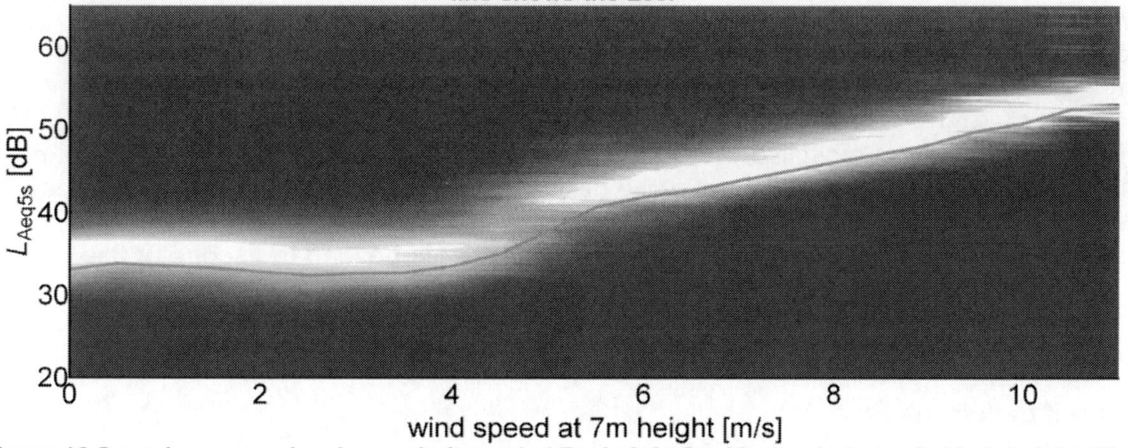

Figure 16 Sound pressure levels vs windspeed at 7m height for 10m/s windspeed at hub height. The orange line shows the L80.

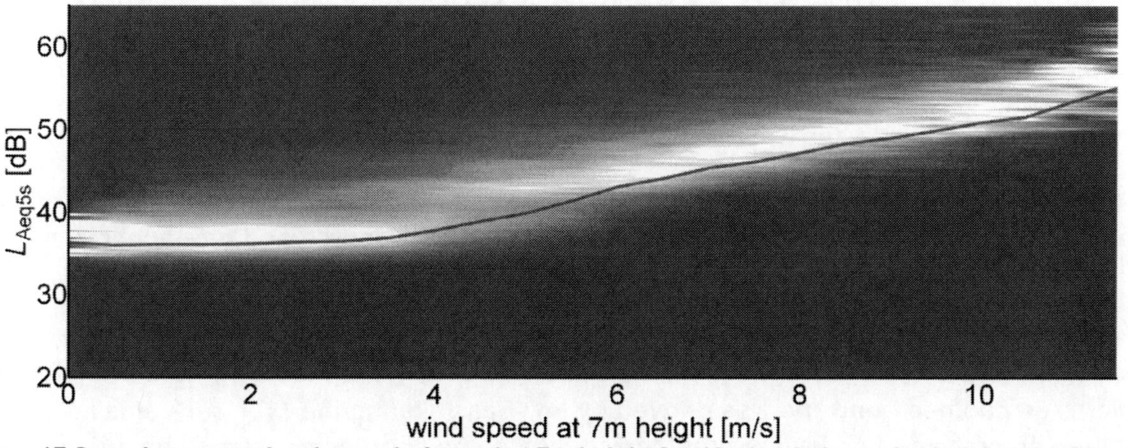

Figure 17 Sound pressure levels vs windspeed at 7m height for 13m/s windspeed at hub height. The red line shows the L80.

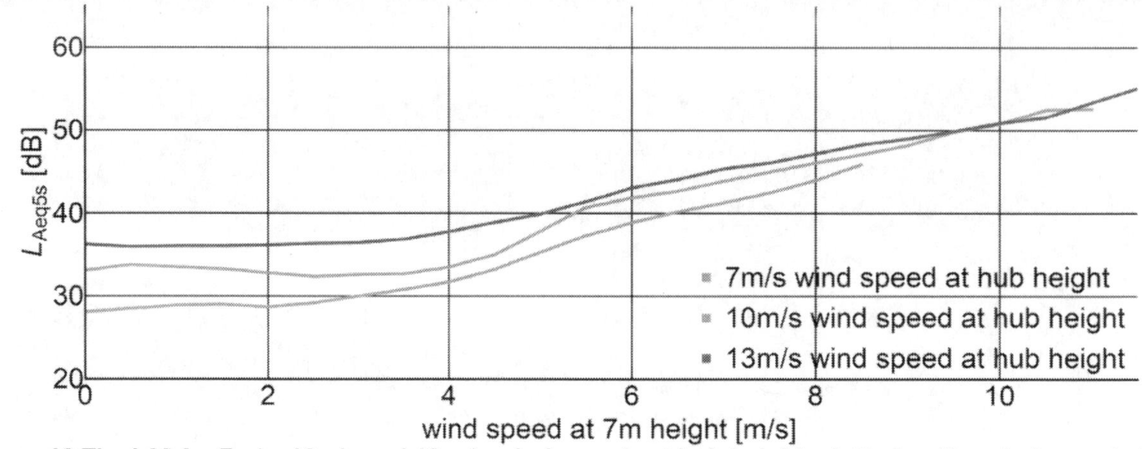

Figure 18 The L80 for 7m/s, 10m/s and 13 m/s wind speeds at hub height, plotted vs the wind speed at 7m height.

The data indicates that the sound pressure levels due to the wind turbines are dependent on the wind speeds at low levels. Data with high wind speeds close to the ground needs to be included in the evaluation of the long term equivalent sound pressure levels due to wind turbines, otherwise the levels would be underestimated.

On the other hand, there most likely is wind induced noise included in the measured sound pressure levels. And this wind induced noise needs to be subtracted to get to sound pressure levels due to the wind turbines. Doing this properly by just looking at A-weighted levels may not be possible, it may not be precise enough – in any case, it is beyond the scope of this paper.

7. More time dependent changes

7.1 Classification

As an example, assume that the conditions where wind turbine 1 is at 2100kW are to be studied. Figure 19 shows the distribution of hub-height wind speeds for 2100kW. The 15 and 85 percentile of this distribution are 11.0 m/s and 12.4m/s. Use these values as limits for the conditioned wind speed at hub height to find time periods that might be in the same class. Within the two month measurement time, 119 such time periods can be found, the longest being 41 minutes long. Figure 20 shows the distribution of length of the time periods, the average is just over 8 minutes.

A basic idea for time dependent classifications might be to take such time periods, take average sound pressure levels for each, and classify them by source and propagation conditions. The following is an illustration why this probably is too limited.

Figure 21 shows sound pressure levels vs time for a time period chosen because the wind speeds close to the microphone were low. On this time scale, even the $L_{Aeq,5s}$show a lot a variation. A smoothed trend line was derived by low-pass filtering the $L_{Aeq,5s}$. A small advantage of the smoothed levels is that they cut off some short noise events that do not belong to the wind turbines. The main purpose is to visualize how the levels change on the scale of several minutes.
Figure 22 shows the distribution of sound pressure levels in the example period. For the whole time, the hub-height wind speed was in a narrow window. The wind direction did not change. And still the level distribution is not narrow.

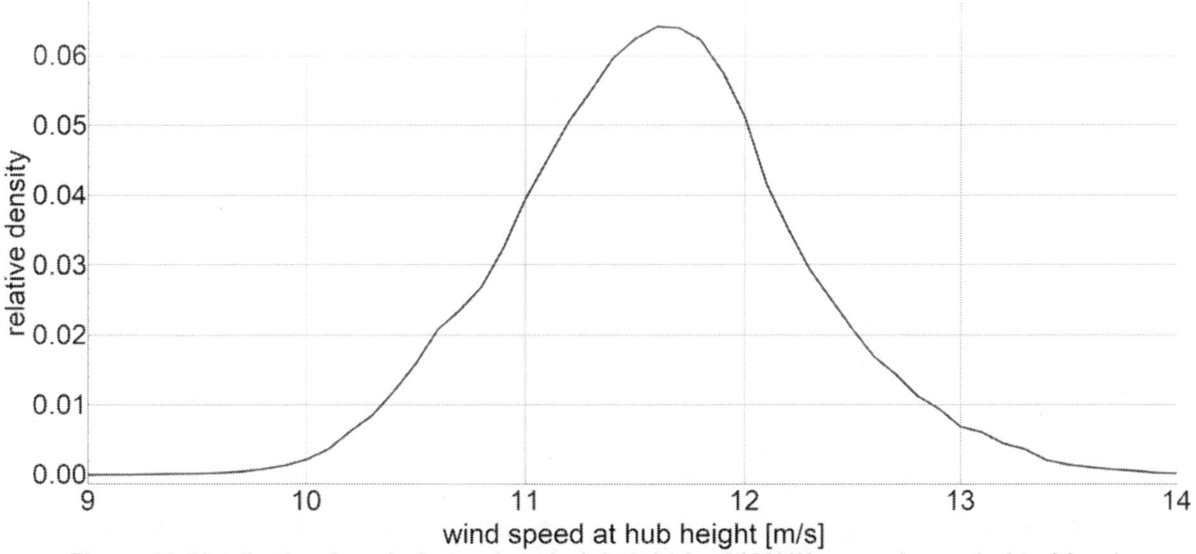

Figure 19 Distribution for wind speeds at hub height for 2100kW power from wind turbine 1

Shorter time intervals will result in narrow level distributions in each interval, if the intervals are chosen properly. At this time, it is not obvious how to choose the best intervals. Since wind speed profiles are very relevant for the sound source as well as the propagation, changing points may be visible in the recorded wind speeds.

Figure 23 shows the wind speeds recorded at different heights during the example period. These are unconditioned signals, and the presentation is not very useful.

Figure 24 shows the same wind data, but conditioned with a low pass filter. "MP1 7m" is at 7m height close to the microphone position, "MP2 4m" is at 4m height in the same position. Both wind speeds are well below 4m/s, so no or very little wind induced noise is expected.

In the level plot (Figure 21), vertical lines are used to mark the maxima in the smoothed levels. The same vertical lines are used in the wind data plots. A correlation between wind speeds and sound pressure levels is not obvious.

Figure 25 shows the variance of the wind speeds around the smoothed values. With wind measurements in non-ideal conditions, the variance of wind speeds might be a more robust indicator for the stability of the atmosphere and the wind speed profiles than the actual wind speed measurements. In the presented example, the correlation of wind speed variance with sound pressure levels is not apparent. More work on this is required.

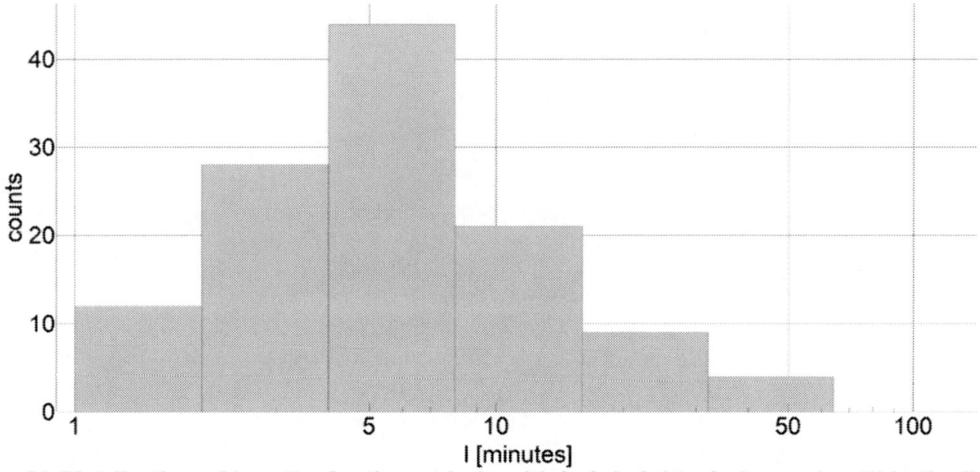

Figure 20 Distribution of lengths for the periods with hub height wind speeds within the limits.

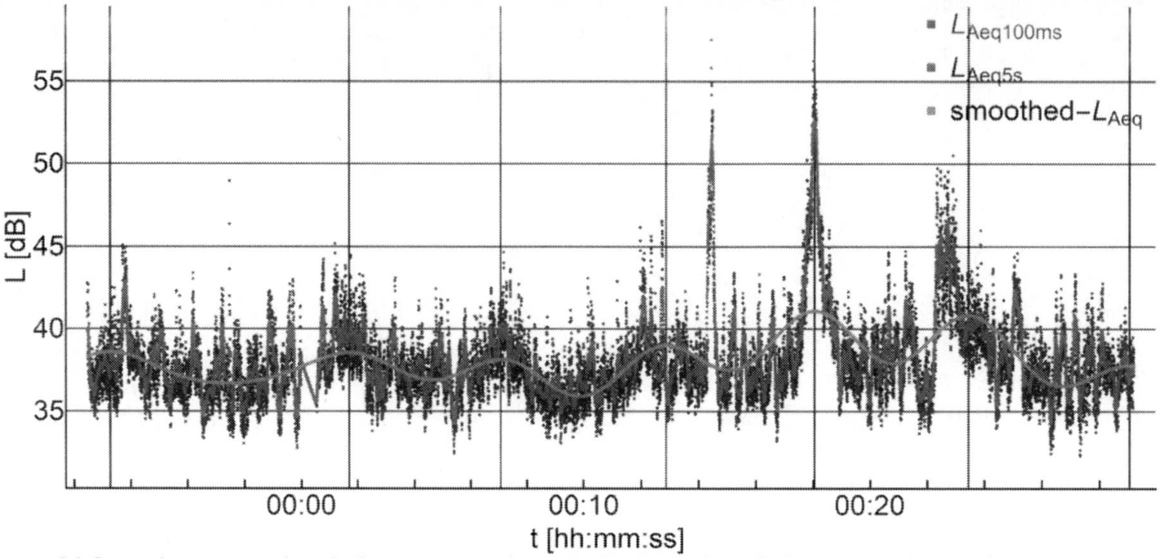

Figure 21 Sound pressure levels for an example period where the wind speed at hub height is within the limits

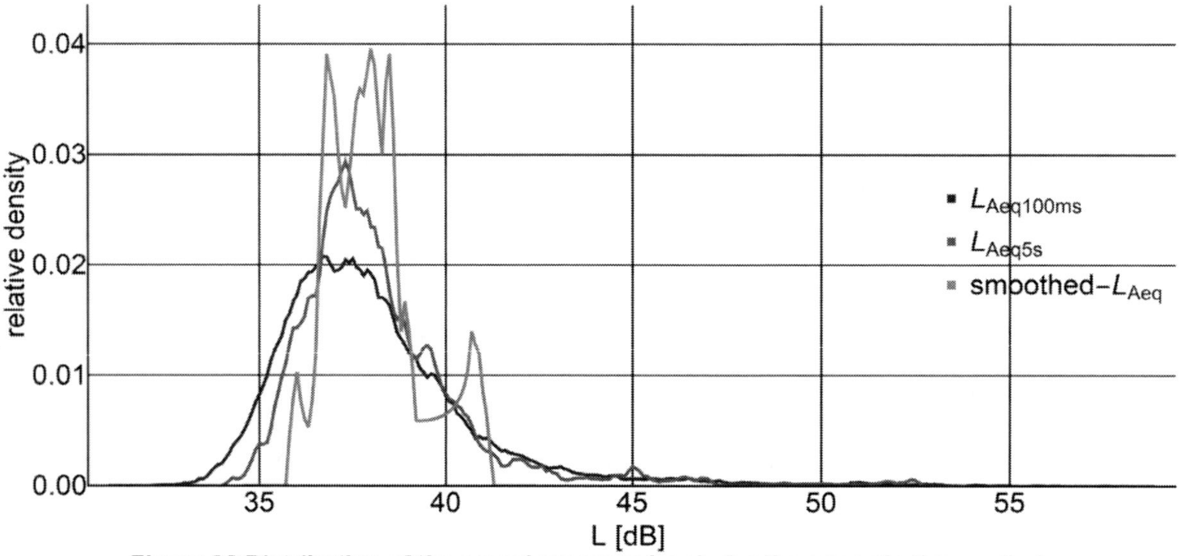

Figure 22 Distribution of the sound pressure levels for the example time period

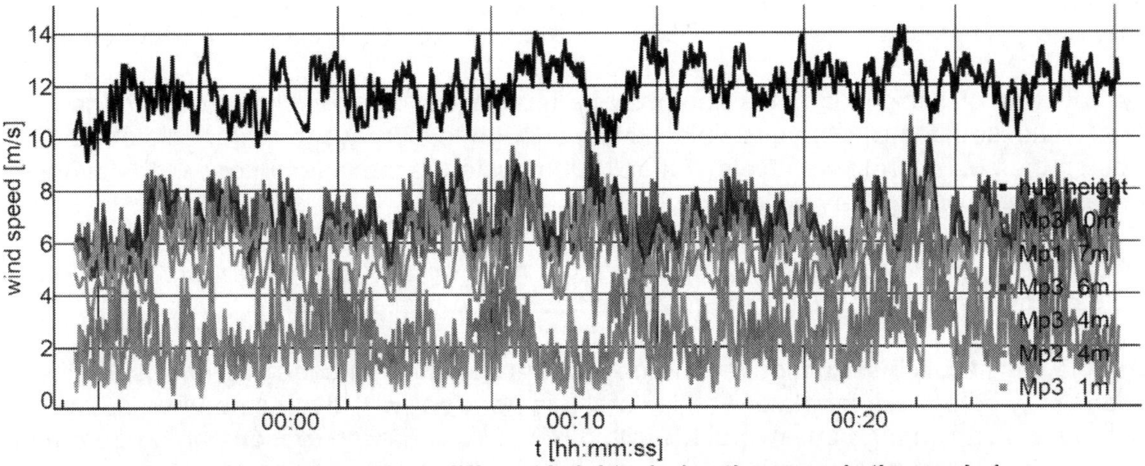

Figure 23 Wind speeds at different heights during the example time period

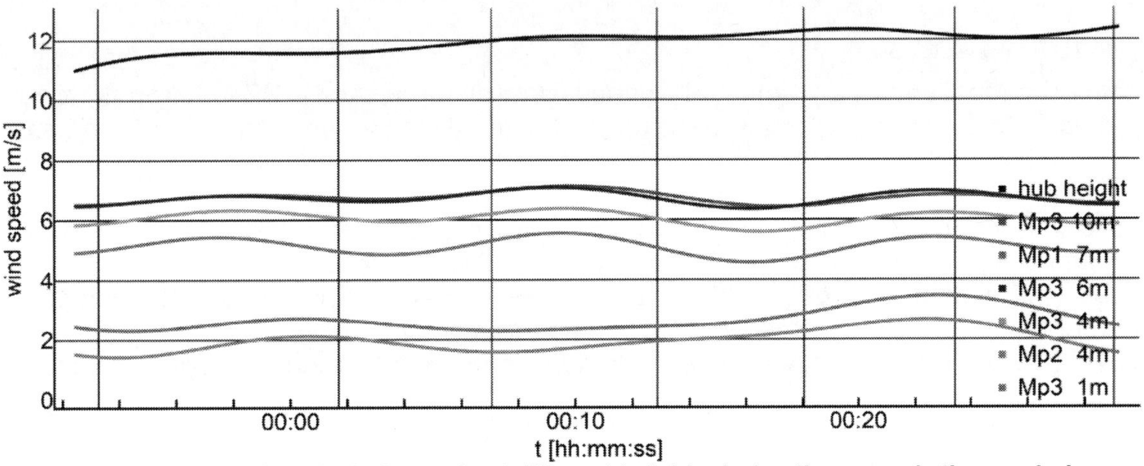

Figure 24 Conditioned wind speeds at different heights during the example time period

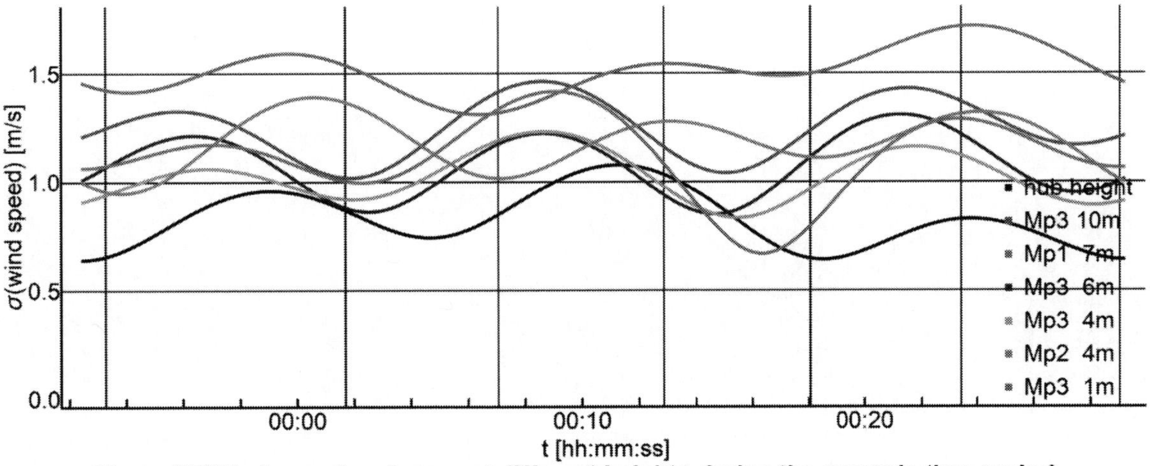

Figure 25 Wind speed variances at different heights during the example time period

8. Conclusions

It is desirable to find classes for source and propagation conditions that result in narrow distributions of the sound pressure levels within the classes. For the determination of equivalent continuous sound pressure levels, ISO 1996-2 actually makes such classes a requirement.

Purely statistical classification of the data may be too limited to find such classes. In the example discussed, the background noise is low enough to clearly see the effects of the wind turbines on the sound pressure levels, but at 1000m distance the distributions in the simple classifications shown are very wide.

Conditioning signals in the time-domain can help in improving the resolution within classes. Examples shown include the wind speed as well as sound pressure levels.

Beyond the statistical classification, a time-dependent detection of changing atmospheric conditions may help in sorting data. At this time, it is not clear why even smoothed sound pressure levels can change by several dB within minutes, and if the reason can be determined by studying the atmospheric conditions.

Acknowledgements

With thanks to the German Environment Agency (Umweltbundesamt – UBA) for authorizing the use of data from the project “Geräuschwirkungen bei der Nutzung von Windenergie an Land”.

References

[ISO1996-2] ISO 1996-2:2017 Acoustics -- Description, measurement and assessment of environmental noise -- Part 2: Determination of sound pressure levels

[ISO1996-1] ISO 1996-1:2016 Acoustics -- Description, measurement and assessment of environmental noise -- Part 1: Basic quantities and assessment procedures

[ISO9613-2] ISO 9613-2:1996 Acoustics -- Attenuation of sound during propagation outdoors -- Part 2: General method of calculation

INCE*EUROPE*

8th International Conference
on
Wind Turbine Noise
Lisbon – 12th to 14th June 2019

Session C

Propagation - 1

Organised by INCE-Europe
mail@inceeurope.org
www.inceeurope.org

Supported by Portuguese Acoustical Society

Sociedade
Portuguesa
de Acústica

INCEEUROPE

8th International Conference
on
Wind Turbine Noise
Lisbon – 12th to 14th June 2019

Measuring and Analysing the Sound Propagation of Wind Turbines

Susanne Martens, Tobias Bohne and Raimund Rolfes
Institute of Structural Analysis, Leibniz University Hannover, Germany

Summary
The influence of atmospheric conditions on the propagation of sound emitted from wind turbines is investigated on the basis of measured data. Effective sound speed profiles and gradients are calculated considering atmospheric stability and wind direction. At distances of 154 and 249 m to the wind turbine no meteorological effects on sound propagation but directivity effects of the sound source are observed. Due to the dipole characteristic of the dominant aerodynamic noise source, sound levels at gradients close to zero are lower than at higher absolute gradients. At a horizontal distance of 479 m, the effect of different sound speed gradients on measured sound levels is recognisable. With increasing sound speed gradients, sound waves bend downwards which results in a decrease of propagation loss. Upward bending occurs at strongly negative gradients, where the propagation loss is 6 dB higher than with highly positive gradients. The effect on refraction is less pronounced with weaker gradients.

1. Introduction
From 2007 to 2017, the capacity of onshore wind turbines has increased worldwide from 113 to 497 GW. Further growth is expected, reaching 830 GW by 2023 [5]. With the expansion of wind energy and wind turbines moving closer to local residents, the noise emission of wind turbines and its propagation has come into focus. This work focuses on the influence of meteorological conditions on sound propagation based on measured data.
Earlier studies already showed a large effect of the meteorological conditions on the sound emission and sound propagation. On the basis of long-term measurements van den Berg [11] shows that the

noise levels in the area of a wind farm are higher than expected at night. If the common description of the logarithmic wind profile is used to extrapolate the wind speed from the reference height to the hub height, the nightly wind speeds at hub height are underestimated. In a stable atmosphere at night stronger winds exist at hub height than those predicted by the logarithmic wind profile. As a result, a wind turbine produces more sound than expected from the wind speed at reference height. As described in van den Berg [11], due to the higher noise emissions, the levels at 400 m and 1000 m distance to the wind farm were up to 15 dB higher than expected. Consequently, when calculating wind profiles, it is important to consider not only surface roughness, but also atmospheric stability.
Öhlund and Larsson [9] carried out acoustic and meteorological measurements near to two wind turbine sites in Sweden. No meteorological effects were observed at a distance of 400 m from the nearest wind turbine. At shorter distances the directivity of the turbines predominates. At distances of 1-2 km to a wind farm, variations in sound pressure levels of 6-14 dBA were measured, depending on the sound speed gradient and correspondingly on the refraction. These effects were observed in total sound levels and in all 1/3 octave bands from 63 to 1000 Hz. According to Öhlund and Larsson [9], for high sound sources a certain distance is needed until sound waves are refracted at the strong velocity gradients near the ground. The meteorological effects in relation to refraction increase with distance and become significant at a distance between 400 m and 1000 m.
Based on the studies of van den Berg [11] as well as Öhlund and Larsson [9], the meteorological effects on sound emission and sound propagation are investigated using long-term measurements with three microphones. For this purpose, the theory of sound propagation with focus on the calculation of sound speed profiles will be discussed first. Subsequently, the performed measurement campaign and the evaluation method of the acoustic data are presented. Finally, the measurement results in relation to meteorology effects will be presented and discussed. The contribution ends with a conclusion and outlook.

2. Sound Propagation Theory

Sound propagation in the atmosphere is influenced by air absorption, atmospheric refraction and turbulence. Sound loses energy due to atmospheric air absorption, which is converted into heat. The amount of air absorption strongly depends on the sound frequency, the temperature, and the relative humidity of the air as well as increases with distance and frequency. The calculation of sound attenuation is described in DIN ISO 9613-2 [3]. Thus, depending on temperature and humidity, the attenuation at 4 kHz can be 22.9 to 88.88 dB/km.
As described in detail in the literature, the vertical gradients of the sound speed lead to atmospheric refraction downwards or upwards. The sound waves are bent in the direction of the low speeds. If the wind speed increases with height, and/or a positive temperature gradient is present the sound waves are bent downwards. This leads to higher sound pressure levels at the point of immission. When wind speed decreases with height, and/or there is an adverse temperature gradient, sound waves are bent upwards. As a result, shadow zones with low sound levels can occur. Turbulence scatters sound waves into the shadow zone. The described effects of refraction and scattering are shown in figure 1. Turbulence is described as temperature and wind fluctuations in the atmosphere. They not only lead in scattering of sound, but also cause a displacement of phases and amplitudes.
The focus of this contribution is on the effect of refraction on sound propagation. Since refraction is directly related to sound speed profiles, the calculation of profiles is described. The speed of sound is dependent on temperature and humidity and is also affected by the wind speed. In the direction

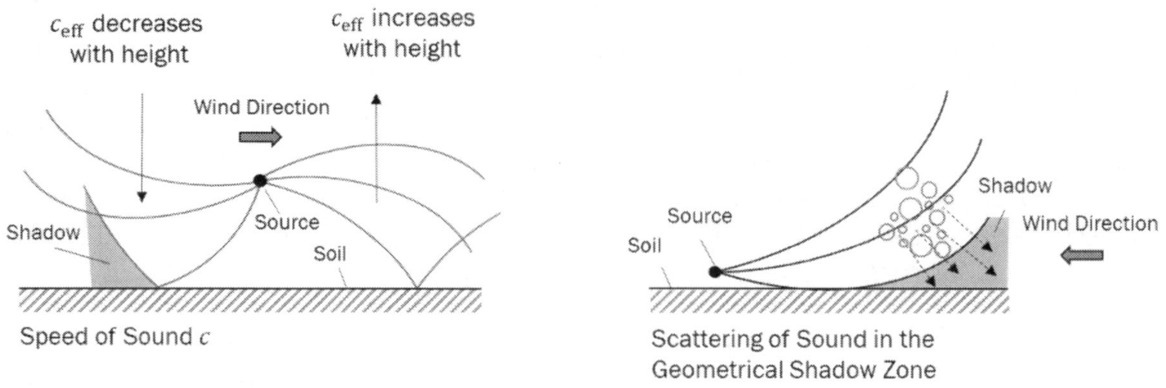

Fig. 1 Left: Atmospheric refraction due to vertical gradients in sound speed
Right: Scattering due to turbulences

of sound propagation, the relation between the sound speed $c(z)$, temperature profile $T(z)$ and wind speed profile $u(z)$ in dependence of height z is given by Rossing [10] with

$$c(z) = c_0\sqrt{\frac{T(z)}{T_0}} + u(z), \tag{1}$$

where T is in °C and u is in m/s. As explained before, wind and temperature gradients are significant for the refraction of sound. The influence of humidity is of minor importance.
If no measurement data in heights up to 100 m is available, the vertical profiles of wind speed and temperature can be determined using the Monin-Obukov similarity theory. According to this theory, the wind speed components in source-receiver direction and the temperature at height z are determined with

$$u(z) = \frac{u^*}{\kappa}\left[\ln\left(\frac{z}{z_0}\right) - \Psi_M\left(\frac{z}{L}\right)\right] \tag{2}$$

and

$$T(z) = T_0 + \frac{T^*}{\kappa}\left[\ln\left(\frac{z}{z_0}\right) - \Psi_H\left(\frac{z}{L}\right)\right] + \Gamma. \tag{3}$$

The definitions and the parameter values used in this work are summarized in table 1. The first term of the equation [$\ln(z/z_0)$] represents the turbulences due to mechanic friction. The second term [$\Psi_{M,H}(z/L)$] takes into account turbulences due to thermic friction, and thus the atmospheric stability is considered. This term includes the Monin-Obukhov lenght L, which depends on the heat flow at the ground surface and is particularly important for stability. Turbulent flow quantities such as the Monin-Obukov length are based on measured data and are determined using the Eddy-Corociance method. In order to smooth out random fluctuations in measurement data, atmospheric parameters are calculated with an averaging time of 30 minutes. The determination of L and other parameters listed in table 1 are described in Maas [7]. As shown in table 1, positive values of L represent a stable atmosphere and negative values an unstable one. Under neutral atmosphere $\Psi_M = \Psi_H = 0$

and $L \to \infty$ is assumed. In this contribution, the boundary layer stability conditions are classified in stability classes based on L. For stable and unstable atmospheric conditions, the correction value for the wind profile is calculated using

$$\Psi_M\left(\frac{z}{L}\right) = \begin{cases} 2\ln\left(\frac{1+x}{2}\right) + \ln\left(\frac{1+x^2}{2}\right) - \arctan(x) + \frac{\pi}{2} & \text{for } L < 0 \\ -5z/L & \text{for } L > 0 \end{cases} \tag{4}$$

with

$$x = \left(1 - \frac{16z}{L}\right)^{1/4}. \tag{5}$$

For the correction term of the temperature profile the following equation is applied:

$$\Psi_H\left(\frac{z}{L}\right) = \begin{cases} 2\ln\left(\frac{1+x}{2}\right) & \text{for } L < 0 \\ -5z/L & \text{for } L > 0 \end{cases} \tag{6}$$

The equations 4 to 6 are called Businger-Dyer profiles and correspond to measurements from a real atmosphere up to a height of 100 m. By using the calculated wind and temperature gradients (equations 2 to 3), the vertical profile of the sound speed can be determined from equation 1. Considering the horizontal component of the wind speed in a certain sound propagation direction, Larsson and Öhlund [6] introduced u_{comp} as wind speed parameter in equation 2. The horizontal component u_{comp} can be calculated by

$$u_{comp}(wd) = -|U| \cdot \cos(wd - dir), \tag{7}$$

where the wind direction in ° is marked with wd and the direction of sound propagation is dir. A sound propagation in northerly direction is indicated with 0°, $|U|$ is the magnitude of the wind speed. Conrady et al. [2] and Öhlund and Larsson [9] use the vertical gradient of the effective sound speed

$$\Delta c_{\text{eff}} = \frac{c(z_2) - c(z_1)}{z_2 - z_1} \tag{8}$$

as a measure of refraction. In equation 8 respective sound speeds at two chosen heights z_1 and z_2 are used.

Table 1 Definition and parameter values used in equations 2 and 3

Symbol	Denotation and Unit	Value
u^*	Friction velocity in m/s	Calculated by Maas [7]
z_0	Roughness length in m	0.08 (see Maas [7])
T^*	Scaling temperature in K	Calculated by Maas [7]
T_0	Air temperature at zero height	283
κ	Von Kármán constant	0.41
Γ	Adiabatic correction factor in °C/m	-0.01 for dry air
L	Monin-Obukhov length in m	Calculated by Maas [7]
		> 0 stable atmosphere
		< 0 unstable atmosphere
Ψ_M	Diabatic momentum profile correction (mixing) function	See equation 4
Ψ_H	Diabatic heat profile correction (mixing) function	See equation 6

3. Measurements and Data Processing

3.1. Site Description and Experimental Setup

Over a period of seven weeks, a measurement campaign was conducted in summer of 2018 next to a wind farm in northern Germany. The focus was on a 2 MW wind turbine with a 100 m hub height and 3-blade propellers of 50 m blade length. The terrain of the site is about 2 m above sea level and the landscape is homogeneous and flat. The area consists mainly of meadows divided by ditches, reeds, and rows of trees. Only one tree with a height of approximately 8 m is located around the acoustic measurement site. Acoustical measurement stations are considered to be 10 m away from natural disturbances such as trees or reeds. The area in which measurements were taken and the contribution of altitude levels in propagation of microphones are illustrated in figure 2.

In the campaign, three sound level meters were placed in 154 m, 249 m and 479 m distance to the wind turbine. The distance of the receiving point at 154 m is equal to the reference distance, as described in IEC 61400-11:2012 [4]. Sound pressure levels, 1/3-octave bands and audio were recorded at each measurement point. Each station consisted of a solar panel for external power supply and a sound level meter including windscreens, as shown in figure 2. Using a standard 90 mm windscreen as well as a 220 mm self-developed secondary windscreen, wind-induced noise was reduced effectively during the measurements. The sound level meters were fixed at a height of 1.70 m. Simultaneously to the acoustic recordings, extensive meteorological measurements were performed to describe the atmospheric conditions. In order to detect wind vectors, temperature and humidity, measurement devices were installed at a height of 2 and 10 m on a meteorological mast (see figure 2). Using Ultrasonic, 3D wind vectors and acoustical virtual temperatures are measured high-resolved. As a result, relevant turbulence parameters described in section 2 can be determined (see Maas [7] for detailed information). The meteorological mast was located 200 m from the wind turbine. Synchronously to acoustic and meteorological parameters, wind turbine operational data such as output power and rotational speed were monitored.

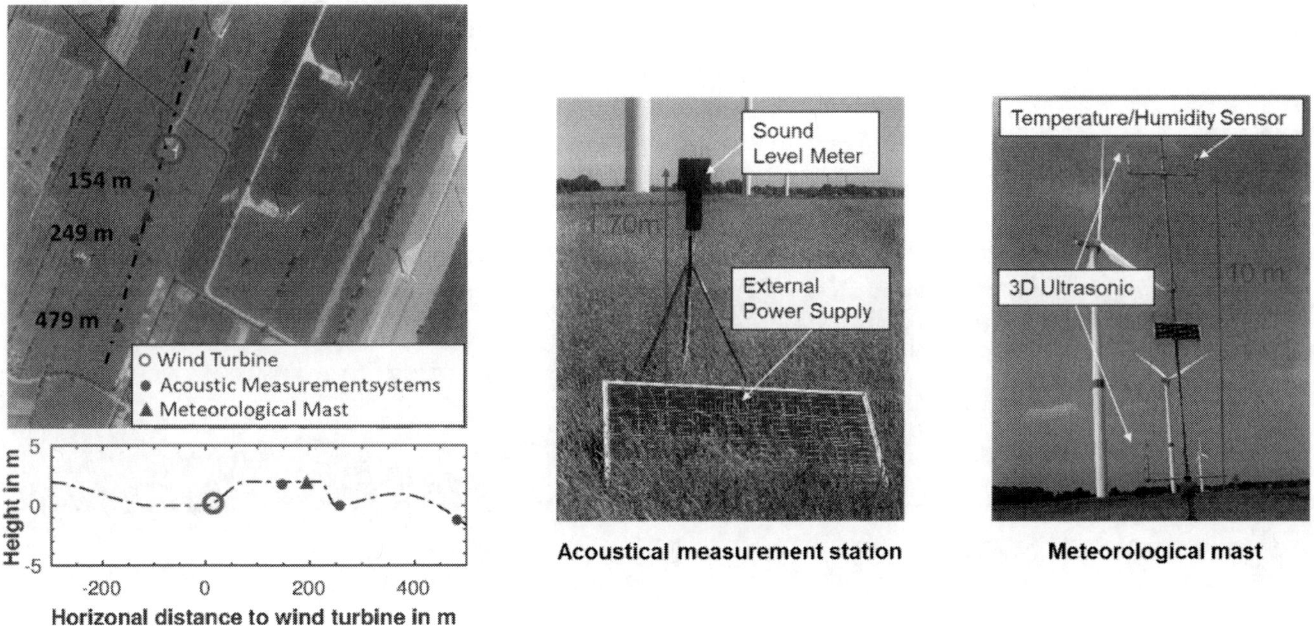

Fig. 2 Overview of the experimental setup

3.2. Selection of Data

While evaluating the acoustic measurements it is important to select data sets with dominant wind turbine noise, without disturbances from human activities, flyovers or from natural sources. Natural sources are mostly wind-induced sound from vegetation, and sound from animals such as bird-singing or frogs. Previous authors, especially named by van den Berg [11] and, Larsson and Öhlund [6], introduced criteria to avoid disturbances and to select data sets when the wind turbine sound is dominant. The criteria are described as follows:

$$L_5 - L_{95} \leq 4 \text{ dB(A)} \tag{9}$$
$$L_1 - L_{95} \leq 15 \text{ dB(A)} \tag{10}$$
$$L_{eq} - L_{1/3>3150Hz} \leq 1.5 \text{ dB(A)} \tag{11}$$

In criteria (1) and (2), statistical A-weighted percentile sound levels (L_n with n=1,5,95) are used. Those two criteria are based on the fact that disturbing sound events significantly affect the measured sound level during a short time period. To describe the criteria, 1 minute intervals of measured statistical sound levels are plotted for about two days in the left figure of 3. As illustrated, excess background noise shows high scattering in the sound level. In the case of dominant wind turbine noise, the sound levels fluctuate less and are more constant.

Criterion (3) represents the sound energy component in the frequency spectrum of a wind turbine. The high frequency components of wind turbine noise are strongly absorbed over distance. Based on investigations from Bolin [1], Öhlund and Larsson [9] show that wind-induced sound from vegetation at higher frequencies has more sound energy than the spectrum of a wind turbine. Typical background noises at the site, such as those from vegetation and bird-singing, dominate at high frequencies. This is also seen in right figure of 3, where an example of measured A-weighted 1/3 octave band during bird singing is shown. Using the A-weighted equivalent sound pressure level L_{eq}, criterion (3) insures that, for example, bird-related noise does not influence the measurement data.

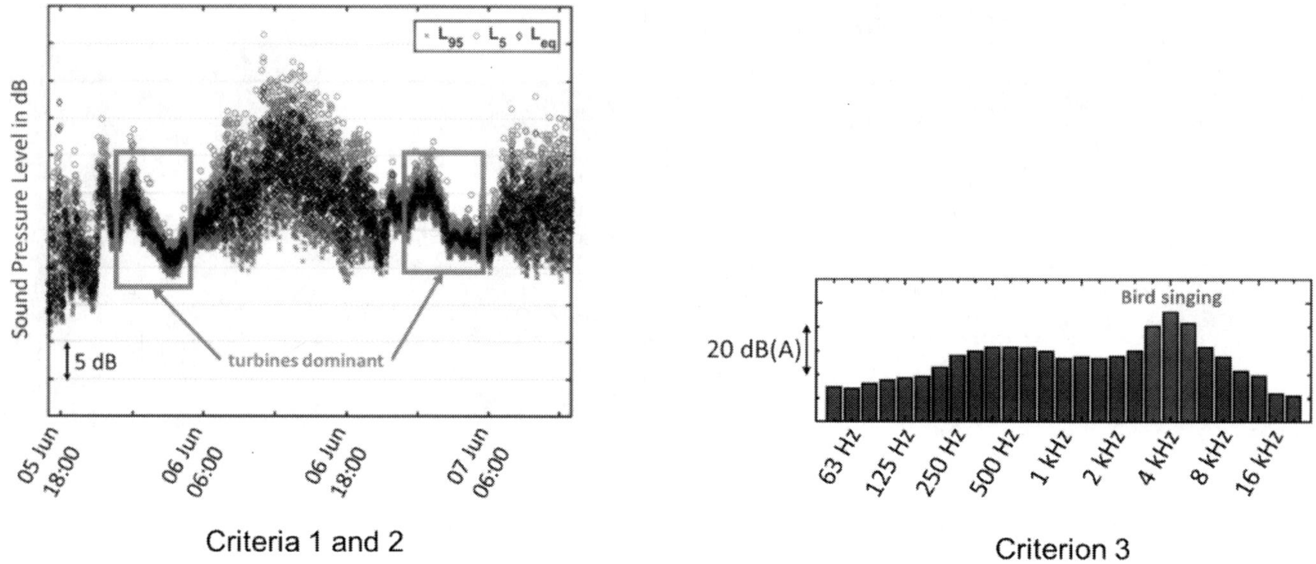

Fig. 3 Criteria 1 and 2: Measured statistical sound pressure levels per minute at 150 m from the wind turbine
Criterion 3: Measured A-weighted 1/3 octave band per minute at 150 m from the wind turbine while bird singing

For approximately 40% of the measuring time, the wind turbine sound was dominant. It is noticeable that the noises are mainly dominant at night and less during the day. This trend was also found by van den Berg [11], and corresponds with the complaints of residents in the area next to wind turbines. The complaints mainly relate to noise in the late evening and at night.

3.3. Method for Evaluating Sound Propagation

As shown in figure 3, the influence of ambient noise is large during the day at unstable atmospheric conditions. During the day the environment is more active then at night. Moreover, the wind speed at the ground is higher in an unstable atmosphere than in a stable one. Accordingly, the danger of measuring wind-induced noises is greater. Consequently, only data sets at stable atmopheric conditions are used for evaluating the sound propagation. Moreover, only data sets with rotor speeds of 12 rpm and above are selected. Since the sound propagation is investigated in all wind directions, the data is divided in four wind direction bins - Downwind, Crosswind 1, Crosswind 2 and Upwind. The division of wind direction is illustrated in figure 4.
Calculating the wind profiles the wind component in sound propagation direction u_{comp} is considered (see section 2). The high relevance of this component is visualized in figure 4. This figure shows the calculated effective sound speed for different wind directions in a stable atmosphere. If u_{comp} from equation 12 is not used to calculate the sound speed profile, it is assumed that the propagation direction of the sound is equal to the wind direction. As shown in the left picture, positive sound

velocity profiles are derived without directional correction - even in upwind situations. If the direction is corrected so that the propagation direction of the sound is not the same as the wind direction, a realistic negative sound velocity profile is obtained in upwind situations. In this contribution the sound propagation direction dir, for the calculation of u_{comp}, is 192.52°.

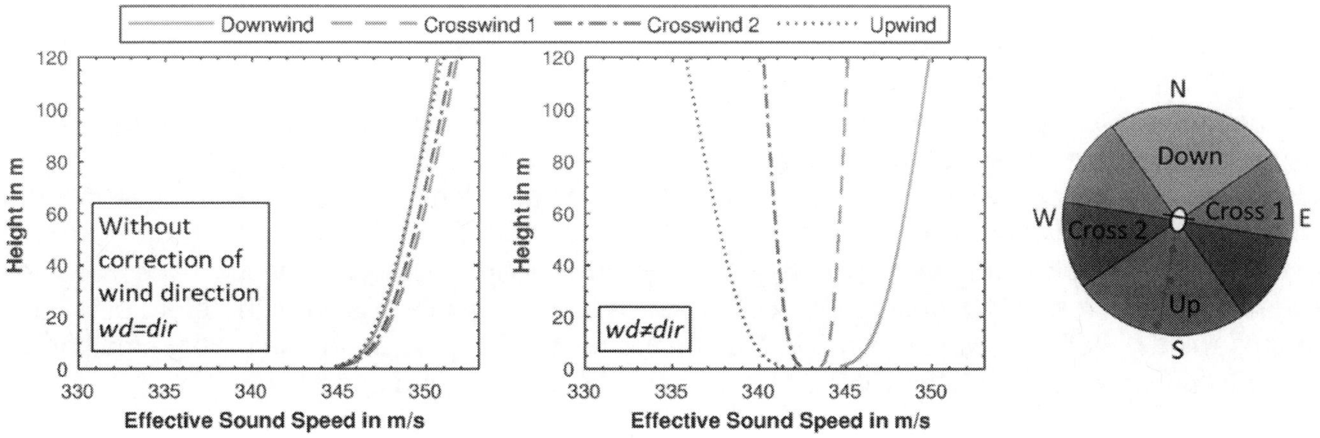

Fig. 4 Calculated effective sound speed profiles for different wind directions and during stable atmospheric conditions
left figure: without correction of wind direction
right figure: with correction of the wind direction using equation 12

The sound generation and radiation of wind turbines depends on many factors. For example, the turbine conditions and the environmental conditions have a direct effect on sound generation. In order to avoid additional error impacts due to inaccurate emission data of the wind turbine, the measured values at the first microphone position are used as a reference to examine sound propagation. The propagation loss is therefore estimated by

$$\Delta L_i = L_1 - L_i, \tag{12}$$

where L_i is the averaged A-weighted sound pressure level in dB at the microphone position i. Hence, L_1 is the averaged A-weighted sound pressure level at the first microphone position at 154 m distance to the wind turbine.

4. Results

In this section the environmental conditions during the measurement period are summarized. The sound propagation is investigated as a function of different gradients of the sound speed and hence, different sound speed profiles. Herein, only data sets during stable atmopheric conditions and rotor speeds of 12 rpm and above are used.

4.1. Meteorological and Wind Turbine Operational Conditions

Wind turbines emit sound as a function of operating parameters, which are dependent on meteorological conditions. During the measurement campaign, wind speed and wind direction were measured with an anemometer located on the nacelle behind upwind rotor. The rotating wings might have an influence on the measured meteorological data at hub height. A distribution of 10-minute averaged values of wind speed and nacelle position is shown in figure 5. In addition, the acoustic measurement positions and the location of the wind turbine are marked. During the period of acoustic measurements wind mainly flew from north-west to the south-east. A lot of acoustic data was therefore collected in the crosswind direction to wind turbine. In the downwind direction, the database is very small and the highest wind speed at hub height is 9.8 m/s, which is below the rated speed.

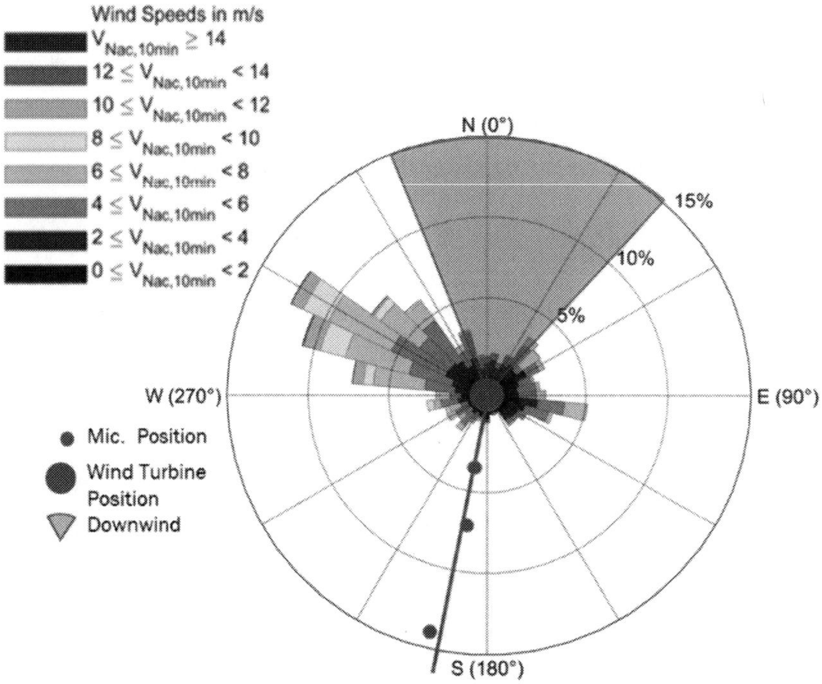

Fig. 5 Contribution of measured wind speed at hub height and nacelle position during the measurement campaign (10 minutes averaged data)

4.2. Sound Propagation

As in [2] and [9], the vertical gradient of the effective sound speed Δc_{eff} is used to evaluate the refraction of sound waves (see equation 8). In this contribution, the gradient Δc_{eff} between 120 and 1 m is calculated and divided into 0.01 Δc_{eff}-bins. A strong positive gradient is calculated in downwind conditions and a strong negative gradient is calculated in upwind conditions. Gradients close to $\Delta c_{\text{eff}} = 0$ indicate crosswind conditions.

Before the refraction of sound waves at different distances to the wind turbine is examined, an overview of the measurement uncertainties at the first microphone is given. For this purpose, the mean value and standard deviation of the sound-level data-sets per Δc_{eff}-bin are illustrated in figure 6. In addition, the number of 1-min data sets, which were selected according to the method described in section 3.2, is summarized in a table. As already known from figure 5, the data sets in crosswind direction predominate. The standard deviation of all data sets of the first microphone is less than 2.5 dB(A). As the distance to the wind turbine increases, the number of data sets decreases and the standard deviation becomes comparatively higher.

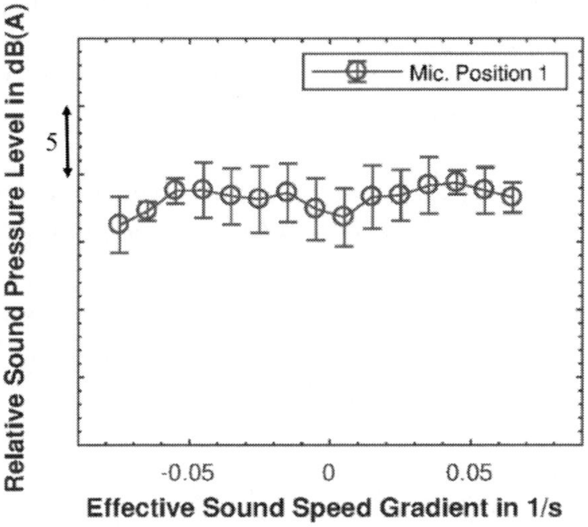

Δc_{eff}	No. of Data
-0,075	204
-0,065	59
-0,055	29
-0,045	**328**
-0,035	322
-0,025	297
-0,015	536
-0,005	**439**
0,005	624
0,015	**296**
0,025	270
0,035	275
0,045	61
0,055	60
0,065	**120**

Fig. 6 Mean value, standard deviation and amount of the sound-level data-sets per Δc_{eff}-bin at the first microphone

The average sound levels at all microphone positions for calculated effective sound speed gradients are shown in figure 7. In addition, the distance between the measuring position and the wind turbine is given. At the first two microphones with a distance of 154 and 249 m, lower levels are measured at Δc_{eff} close to zero than at high absolute gradients. This is not due to refractions, but represents the directivity of the wind turbine. Due to the dipole characteristic of the aerodynamic source, less sound is emitted in crosswind direction than in upwind and downwind direction. Based on the same measurement campaign, this behaviour is explained in Martens et al. [8].

At the 3rd measuring position at a horizontal distance of 479 m from the turbine, higher levels are measured with highly positive gradients than with more negative gradients. The difference between $\Delta c_{\text{eff}} = 0.065$ and $\Delta c_{\text{eff}} \approx 0$ is about 5 dB. This property is not only due to the directivity of the wind turbine, but also due to a downward bending of the sound waves. There are no differences between the levels at negative Δc_{eff} and $\Delta c_{\text{eff}} \approx 0$. The level at negative Δc_{eff} has accordingly decreased with distance in comparison with data at $\Delta c_{\text{eff}} \approx 0$. This indicates an upward refraction of the sound waves.

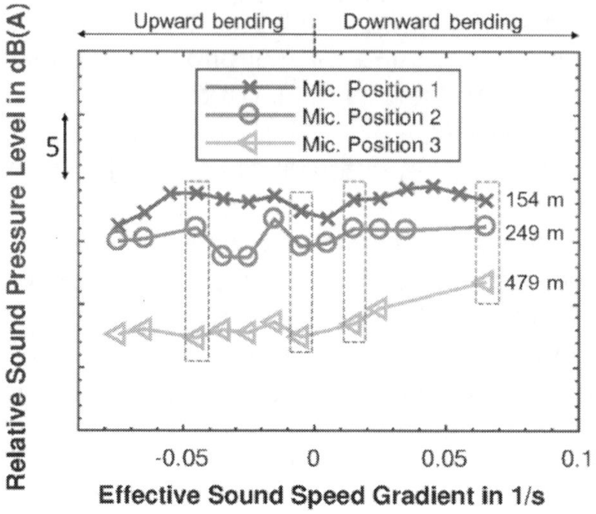

Fig. 7 Averaged A-weighted relative sound pressure levels for different effective sound speed gradients and distances to the wind turbine

Further discussion on the effect of sound speed profiles on sound propagation is based on four examples. The data at $\Delta c_{\text{eff}} = -0.045, -0.005, 0.015$ and 0.065 are used as examples, because those data sets are available for all measuring positions and have a comparatively large set of data (see marking in figure 6 and 7). In figure 8 the calculated sound speed profiles of the selected Δc_{eff}-bins are illustrated (right) and the respective propagation losses are plotted over the slant distance (left). In addition, the propagation losses due to geometric spreading is shown. This data is calculated by

$$\Delta A_{div} = \left(20log\left(\frac{d_1}{d_0}\right)\right) - \left(20log\left(\frac{d_i}{d_0}\right)\right), \tag{13}$$

where d_i is the distance from the source to the microphone i and d_0 the reference distance, which is set to 1. Effects of the ground or sound absorption are not taken into account.

The measurement results are consistent with the theory presented in section 2 and confirm the trend shown in figure 7. With increasing effective sound speed gradients and correspondingly growing positive sound speed profile, less sound is attenuated over distance. The more negative the sound speed gradient, the higher propagation loss over the distance. In figure 8, this tendency is especially seen at a distance of 490 m, where the difference between the first and last microphone is plotted. With strongly negative gradients, the propagation loss is 6 dB higher than with strongly positive gradients.

In comparison with the calculated propagation loss using geometrical spreading, an additional 3 dB attenuation of the sound is recognizable with strongly negative sound speed gradients. In comparison, 3 dB higher average levels are measured for strongly positive gradients. The same tendency, but less pronounced, is observed at the propagation losses of the gradients close to zero. Here, the slope of the sound speed profiles is smaller, so that the sound waves are bent less upwards or downwards.

As a result, the propagation loss deviates less from the theoretical value. The effect of refraction is particularly pronounced with strongly positive and strongly negative gradients and increases with distance.

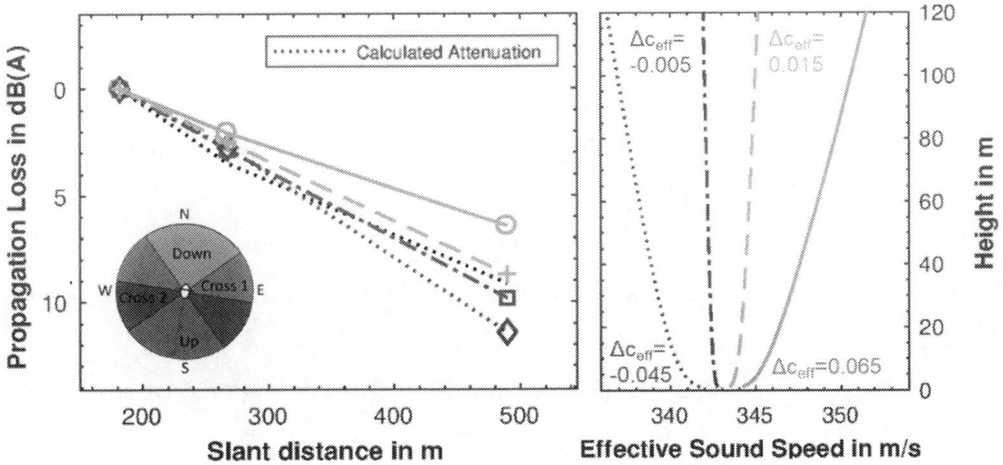

Fig. 8 Relative sound pressure level measured 154 m, 249 m and 479 m distance to the wind turbine during times of different effective sound speed gradients

5. Conclusion and Outlook

The sound propagation of wind turbines is affected by effective sound speed gradients, which are determined by temperature and wind speed profiles considering atmospheric stability. In theory, upward bending occurs with negative sound speed gradients while sound waves bend downwards with positive sound speed gradients. Hence, higher sound levels are expected with positive sound speed gradients. Measured acoustical data in 470 m distance to the wind turbine agree with the theory. A-weighted sound pressure levels with strongly positive sound speed gradients are 6 dB higher than with strongly negative gradients. This effect is less pronounced with weaker gradients and is not observed at shorter distances. At distances of 150 and 240 m to the wind turbine low sound levels were measured with gradients close to zero, which indicates crosswind situations. The measured values at those distances represent the source directivity of the wind turbine.

In future work, the results of this contribution will be verified with acoustical and meteorological data of a second measurement campaign. In this campaign, microphones are located in the statistical main wind direction, and larger distances up to 1150 m from the wind turbine to the receiver are realized. With a larger data base in downwind direction, the effect of the atmospheric stability on sound emission and sound propagation will be investigated in detail. A focus will also be on the influence of the turbulence intensity in the atmosphere.

Acknowledgements

Within in the project "WEA-Akzeptanz", the research at Leibniz University of Hannover is carried out together with the project partner Senvion GmbH and is funded by the Federal Ministry for Economic Affairs and Energy by an act of the German Parliament (project ref. no. 0324134A). The Institute of Structural Analysis is part of the Center for Wind Energy Research For-Wind. The authors gratefully acknowledge the financial support from the research funding organization, the determination and provision of meteorological data by the Institute of Meteorology and Climatology, and the great support from the operator of the wind farm and the local citizens. For further information about the project "WEA-Akzeptanz", please visit the project homepage at www.wea-akzeptanz.uni-hannover.de.

References

[1] Bolin, K. (2009). Prediction method for wind-induced vegetation noise. *Acta Acustica united with Acustica*, 95(4):607–619.

[2] Conrady, K., Sjöblom, A., and Larsson, C. (2018). Impact of snow on sound propagating from wind turbines. *Wind Energy*, 21(12):1282–1295.

[3] DIN ISO 9613-2 (1999). *Acoustics - Attenuation of sound during propagation outdoors - Part 2: General method of calculation.* Ed. 10 edition.

[4] IEC 61400-12 (2012). *Wind turbines: Acoustic noise measurement techniques*, volume IEC 61400-12 of *International standard.* International Electrotechnical Commission, ed. 3.0 edition.

[5] International energy agency (2019). Renewables - wind energy.

[6] Larsson, C. and Öhlund, O., editors (2014). *Wind turbine sound - metric and guidelines.*

[7] Maas, O. (2019). Detaillierte Vermessung der atmosphärischen Grenzschicht im Einflussbereich von Windenergieanlagen.

[8] Martens, S., Bohne, T., and Rolfes, R. (2019). Analyzing the sound propagation of wind turbines based on measured acoustical and meteorological parameters. *Proceedings of the 45th annual conference of the german society for acoustics, Rostock, Germany, 2019.*

[9] Öhlund, O. and Larsson, C. (2015). Meteorological effects on wind turbine sound propagation. *Applied Acoustics*, 89:34–41.

[10] Rossing, T. D. (2007). *Springer Handbook of Acoustics.* Springer - Verlag New York.

[11] van den Berg, G. P. (2004). Effects of the wind profile at night on wind turbine sound. *Journal of Sound and Vibration*, 277(4-5):955–970.

INCE*EUROPE*

8th International Conference
on
Wind Turbine Noise
Lisbon – 12th to 14th June 2019

Noise propagation at short range farfield position investigated by simultaneous measurements in the nearfield and the farfield.

Niels F. Christensen, Sweco Denmark A/S: nielsfrederik.christensen@sweco.dk
Rune Egedal, Sweco Denmark A/S: rune.egedal@sweco.dk
Bo Søndergaard, Sweco Denmark A/S bo.sondergaard@sweco.dk

Summary

To improve the accuracy of propagation models for long distance sound propagation it is important to compare measurements and predictions for different factors that affect the sound propagation, such as atmospheric conditions and terrain. A common problem in performing long distance measurements on a sound source is obtaining a good signal-to-noise ratio. This paper compares measured and calculated values at 500 m distance from a single wind turbine located at a test site. Comparisons are made for a range of wind speeds under monitored atmospheric conditions.

1. Introduction

Part of the environmental noise assessment of wind farms involves using propagation models to predict the sound pressure level (SPL) at neighbour dwellings at distances located at distances from around 500 m from the wind farm. This paper compares measured and calculated SPL at a position 500 m downwind from a wind turbine located at Test Centre Høvsøre, Denmark. The advantage of the test site is the possibility to stop all other turbines in the farm for a long period of time while performing measurements on a single turbine. Further, the test site at Høvsøre has meteorological masts which record details about atmospheric conditions at several heights above ground. Finally, test sites are generally chosen in places with low background noise which is critical for obtaining a good signal-to-noise ratio when measuring at a long distance from a sound source.

This paper presents comparisons between measured and calculated results for the SPL at 500 m downwind from a single turbine. Prediction were made using a simple transmission loss method described in the Danish statutory order for wind farms (BEK 1736), as well as the NORD2000 propagation model. For commercial reasons, no turbine specific details are provided in this paper. For this reason, only level differences are shown.

2. Wind Turbine Test Centre

2.1 General

Measurements were carried out from 17:15 on December 18. 2018 to 01:20 on December 19. 2018 on a wind turbine at Test Centre Høvsøre, Denmark. The site is in flat, rural farm land close to the seashore far from major roads and other noise sources. There are 4 additional turbines on the site, which were stopped during the measurements.

2.2 Weather conditions during measurements

Weather data for the site were provided by the Department for Wind Energy at The Technical University of Denmark (DTU). These include readings of temperature and relative humidity in both 2 m and 100 m height in addition to wind speed measurements at 10, 40, 60, 80, 100 and 116.5 m height. Data with a resolution of 10 minutes exists for the entire measurement period. The range of weather conditions during measurements is shown in Table 1.

Table 1. Recorded weather conditions during measurements.

December 18. 2018 – December 19. 2018					
Height above ground	Temperature [°C]	Relative humidity [%]	Atmospheric pressure [kPa]	Wind speed [m/s]	Wind direction [Degrees]
2 m	4.5 – 5.4	89.7-96.4	99.5-99.9	-	-
10 m	-	-	-	4.5 – 8.1	127 - 158
100 m	3.8 – 4.7	91.5-98.0	100.2-100.6	7.6 – 11.6	132 - 161

2.3 Measurement setup

Two microphones were placed downwind of the turbine, one in the standard IEC 61400-11 ed. 3 (IEC) measurement position, and the other 500 m from the turbine at 1,5 m height. Operational parameters were streamed from the turbine and used to determine the wind speed at hub height according to IEC. The turbine was stopped 3 times to measure background noise. An anemometer placed at 10 m height was used to determine the wind speed during background noise measurements. Weather data with 10 min resolution was assumed constant over each 10 min interval. All measurements were synchronized. Periods with high transient background noise at either the reference or the 500 m position were discarded from the analysis. Time periods were identified as total noise or background noise in the analysis. It was too dark to take pictures during measurements; however, examples of the setup from similar measurements on other sites are shown in Figure 1.

Figure 1. Examples of microphone setup at 500 m position (left) and reference position (right).

2.4 Measured Sound Power Levels

The sound power level of the turbine was determined according to IEC. A microphone was mounted on a reflective plate on the ground and placed in the reference distance downwind from the turbine. A secondary wind screen was used. Measured SPL were corrected for the influence of the wind screen in 1/3-octave bands and the +6dB reflection from the measurement board.

2.5 Measured Sound Pressure Levels at 500 m position

The microphone 500 m from the turbine was mounted on a tripod at 1.5 m above the ground. The microphone was placed in open, ploughed farmland with no nearby vegetation or reflecting structures. A secondary wind screen was used. At the time of writing this paper the insertion loss of the secondary wind screen has yet to be accurately determined. Consequently, the results are not corrected for the influence of the wind screen.

3. Results

Results from the measurements are averaged over 1-minute periods. All SPL's are A-weighted. In general, the simple propagation model (BEK 1736) was used to predict the SPL, $L_{pA,BEK\ 1736}$, at 500 m from the turbine based on the sound power level, L_{WA}, of the turbine.

3.1 Background noise

Figure 2 shows a time series plot of the wind speed and background noise during the measurement period. The measured background noise was lower at the 500 m position than at the reference position. This is probably due to noise from the turbine cooling fan which is less pronounced far from the turbine. However, there is a better signal-to-noise ratio (SNR) at the reference position, than at the 500 m position since the turbine noise is more prominent close to the turbine. For both positions the background noise increases with wind speed. Particularly at high wind speeds there is a poor SNR at the 500 m position. To rule out the influence of wind speed in the comparisons, results were split into 0.5 m/s wide wind speed bins.

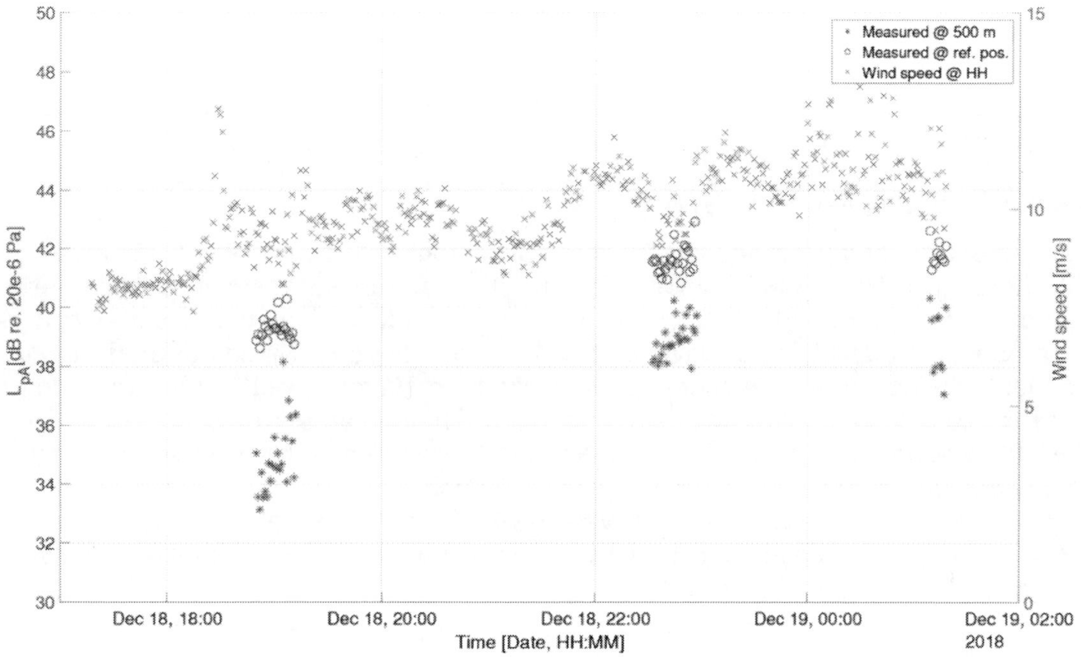

Figure 2. Time series plot of hub height wind speed and background noise levels.

3.2 Comparing 1/3-octave band spectra

For each wind speed bin all 1-minut periods of total noise were corrected for the bin average of the background noise in 1/3-octave bands. The average 1/3-octave SPL in each bin was determined. Figure 3 shows the difference in measured and predicted SPL at 500 m, $L_{pA,meas,f}$ - $L_{pA,BEK\ 1736,f}$, as a function of frequency. Above 3 kHz the predicted SPL at 500 m is small compared to the directly measured SPL. This is because air absorption increases drastically above 2 kHz. This results in predicted levels far below background noise in 500 m measurement at high frequencies. Therefore, 3.15 kHz is chosen as the upper limit for the comparisons. For hub height wind speeds below 10 m/s there is a fairly good agreement between $L_{pA,meas,f}$ and $L_{pA,BEK\ 1736,f}$. At 11 m/s $L_{pA,meas,f}$ is dominated by background noise, resulting in an overestimation of the turbine noise. Based on experiences with the insertion loss of secondary wind screens like the one used in the 500 m measurement, it is thought that the dip in ΔL between 1 to 4 kHz can be explained by the lack of a wind screen correction. However, this needs further investigation.

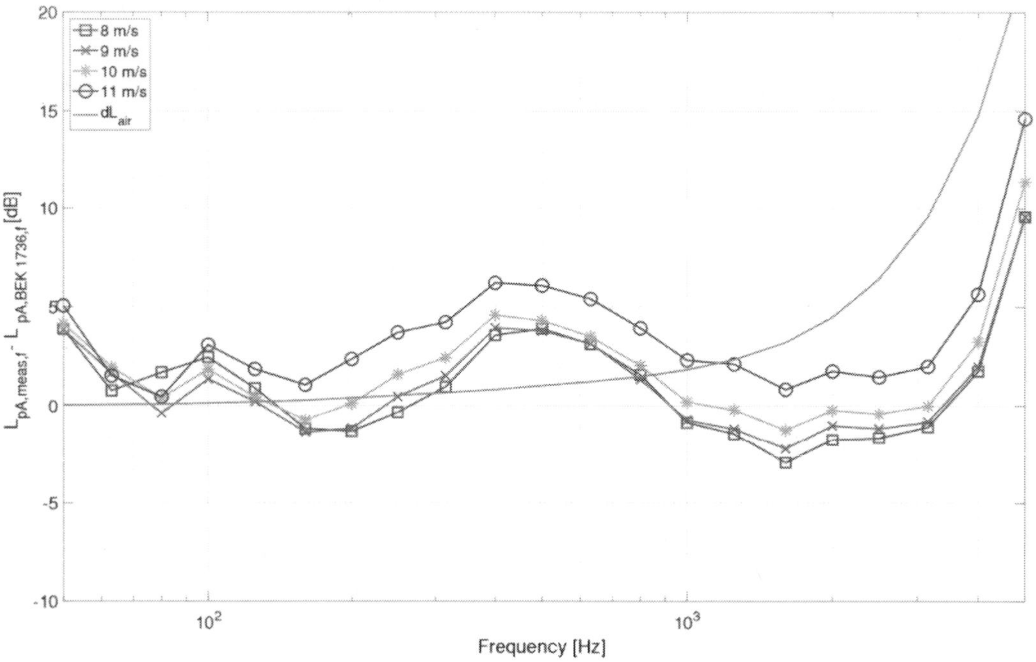

Figure 3. Difference between measured and calculated (BEK 1736) 1/3-octave SPL as a function of frequency. The predicted air attenuation, dL_{air}, is shown for reference.

3.3 Comparison between total levels

The total measured and predicted SPL's were calculated as the sum over 1/3-octave bands from 50 Hz to 3.15 kHz for each 1-minut period. A time series plot of the difference between the measured and calculated total SPL's, in addition to hub height wind speed, is shown in Figure 4. We can observe that ΔL is sensitive to changes in wind speed. Since the prediction model is independent of wind speed, the changes are likely the result of increasing background noise at the measurement position as wind speed increases. There is an abrupt change in ΔL around 21.00-22.00 o'clock when the wind speed exceeds 10 m/s. Since the wind speed was low during daytime and high during night-time it is not possible to do a day-night comparison of the level difference in wind speed bins. However, based on the increase in level difference during high wind periods during daytime, it is more likely that the change is caused by the increasing wind speeds than day-night differences.

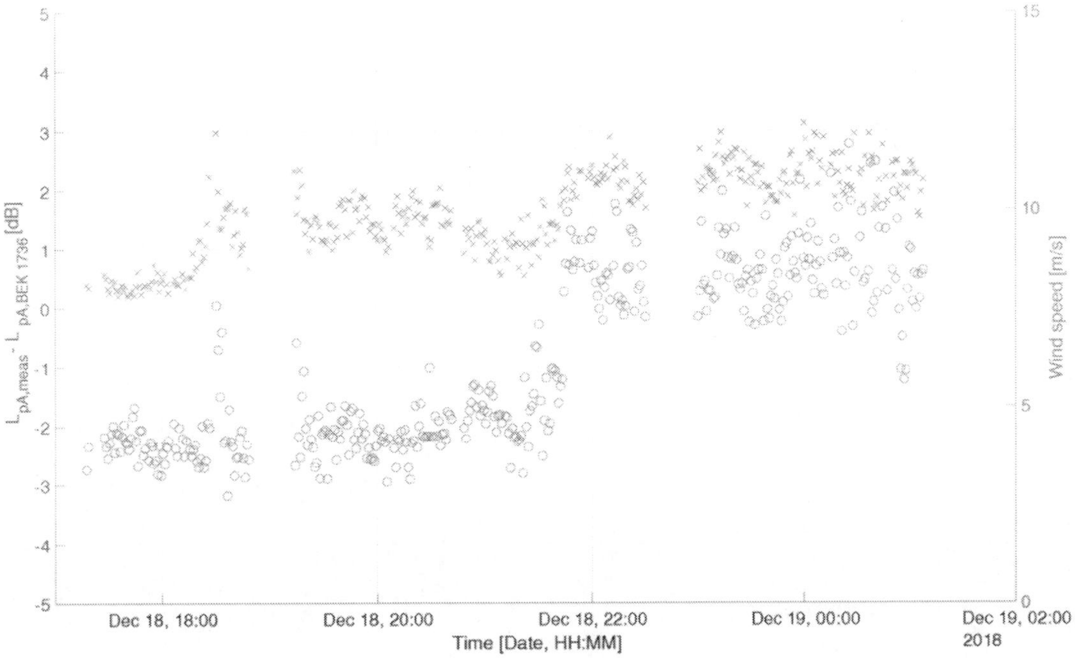

Figure 4. Time series of the difference between measured and predicted SPL and hub height wind speeds.

The air density was calculated from the from the measured temperature and pressure at 100 m height, using the ideal gas law. ΔL as a function of the calculated air density for wind speed bins from 8 m/s to 11.5 m/s is shown in Figure 5 and Figure 6.The results indicate that there could be a tendency towards lower values of ΔL with increasing air density.

The wind shear factor was calculated from the difference between the hub height wind speed and the wind speed measured at 10 m. ΔL as a function of the wind shear factor for wind speed bins 8 m/s to 11.5 m/s is shown in Figure 7 and Figure 8. Increasing wind shear factor means relative higher wind at hub height compared to wind at 10 m height. Thus, one would expect to see increasing values for ΔL with increasing wind shear due to low background noise near the ground and increased sound emission from the turbine. However, there is no observed influence of the wind shear factor on ΔL.

The relative humidity (RH) was measured in 2 m height above ground by the test centre meteorological mast. ΔL as a function of the RH for wind speed bins 8 m/s to 11.5 m/s is shown in Figure 9 and Figure 10. The variations in RH during measurements was small. There is no clearly observed correlation between ΔL and RH.

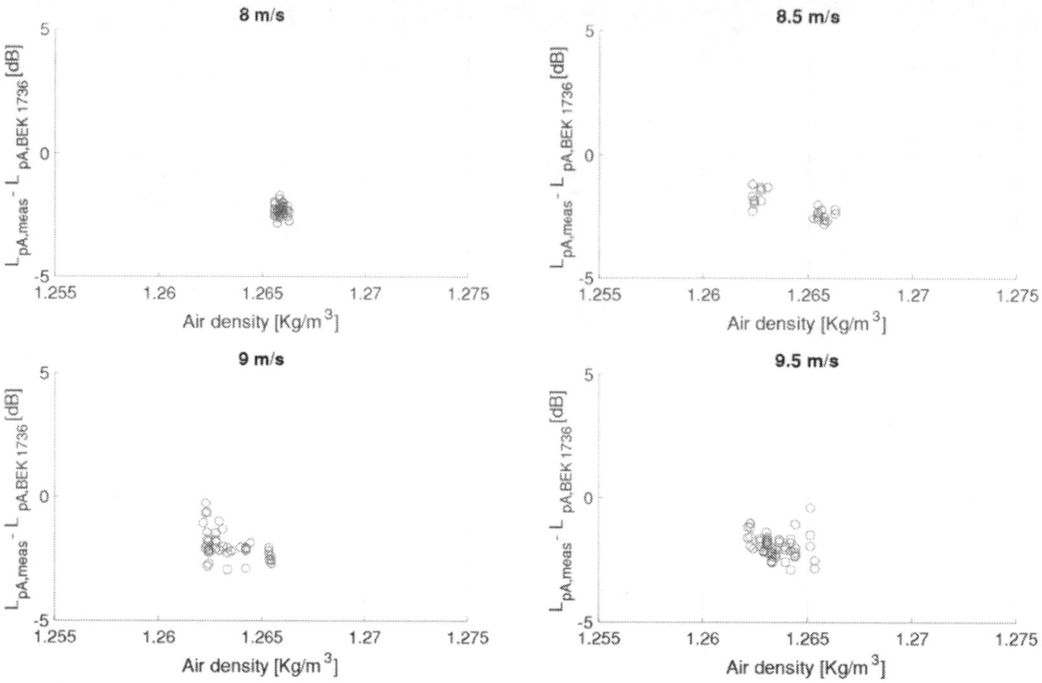

Figure 5. Difference between measured and predicted SPL as a function of air density for hub height wind speeds 8 m/s to 9.5 m/s.

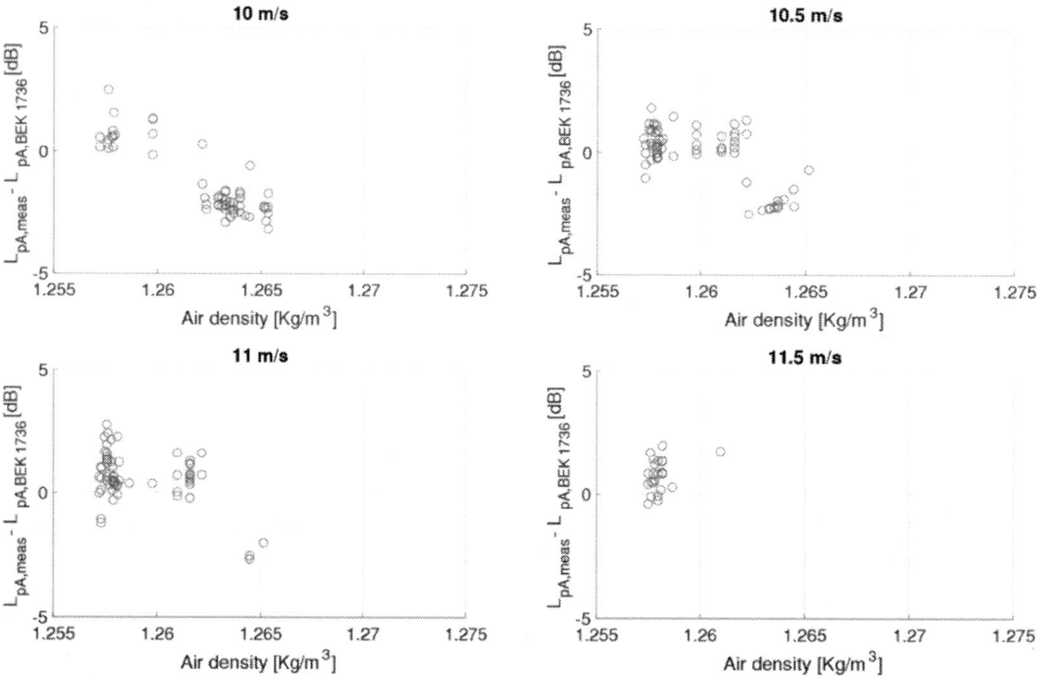

Figure 6. Difference between measured and predicted SPL as a function of air density for hub height wind speeds 10 m/s to 11.5 m/s.

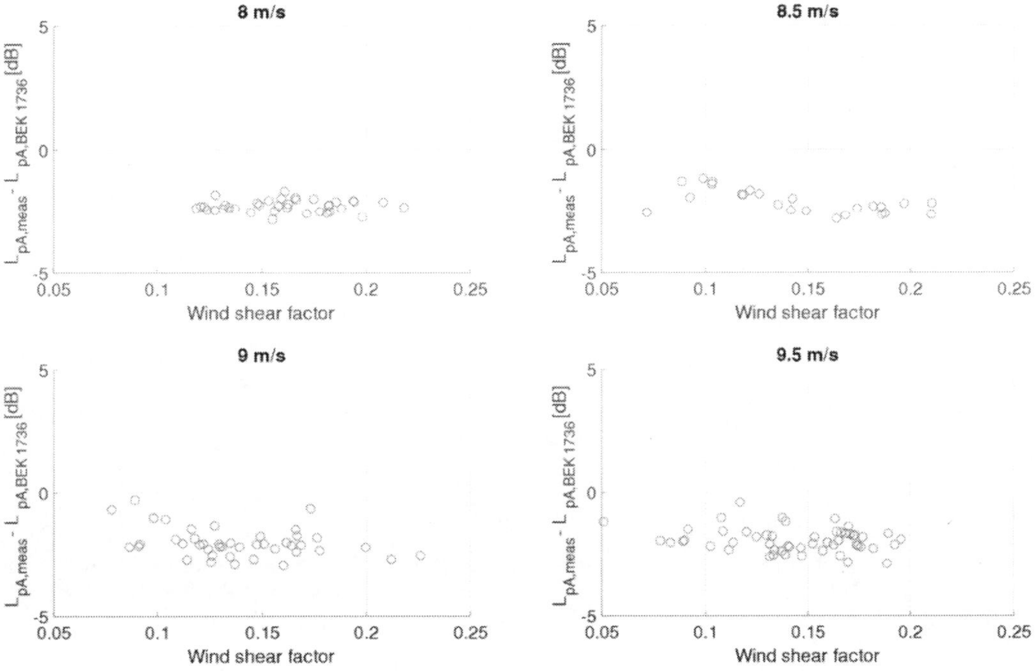

Figure 7. Difference between measured and predicted SPL as a function of wind shear factor for hub height wind speeds 8 m/s to 9.5 m/s.

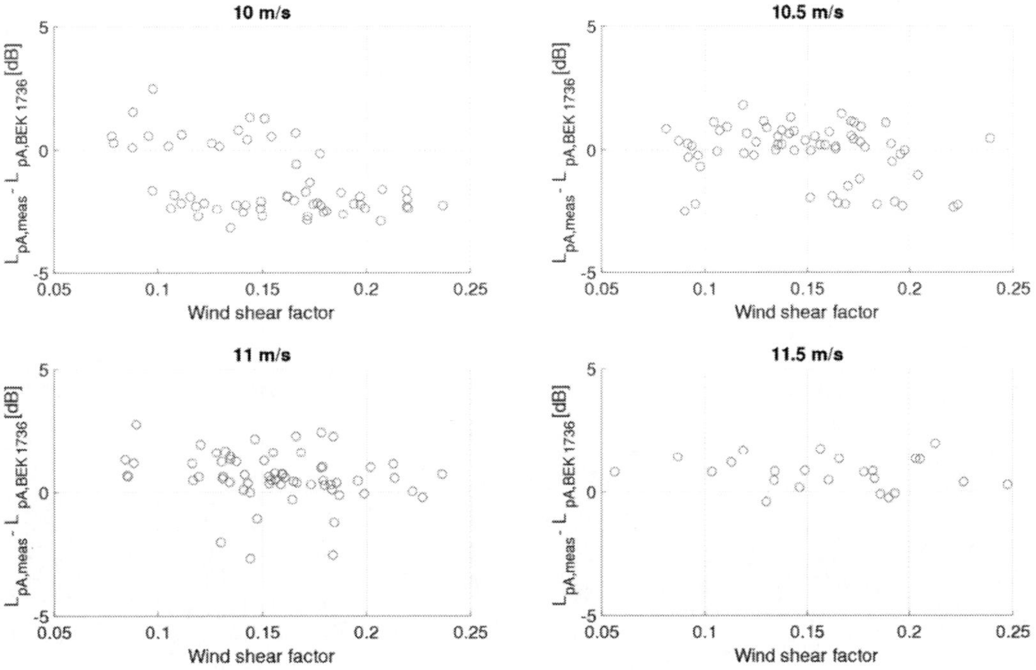

Figure 8. Difference between measured and predicted SPL as a function of wind shear factor for hub height wind speeds 10 m/s to 11.5 m/s.

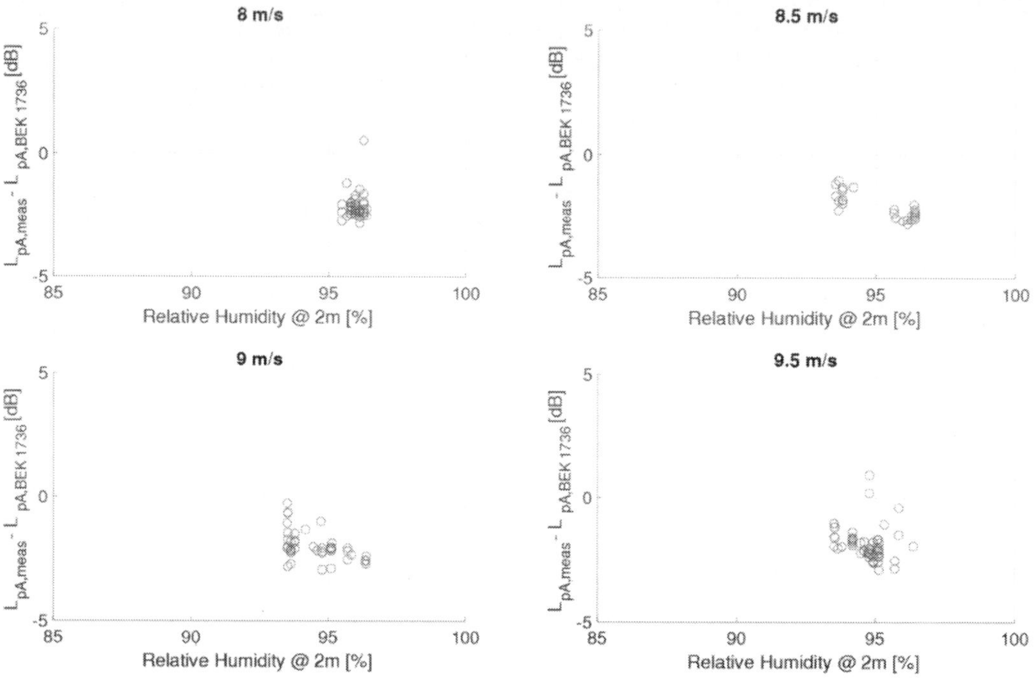

Figure 9. Difference between measured and predicted SPL as a function of relative humidity at 2 m for hub height wind speeds 8 m/s to 9.5 m/s.

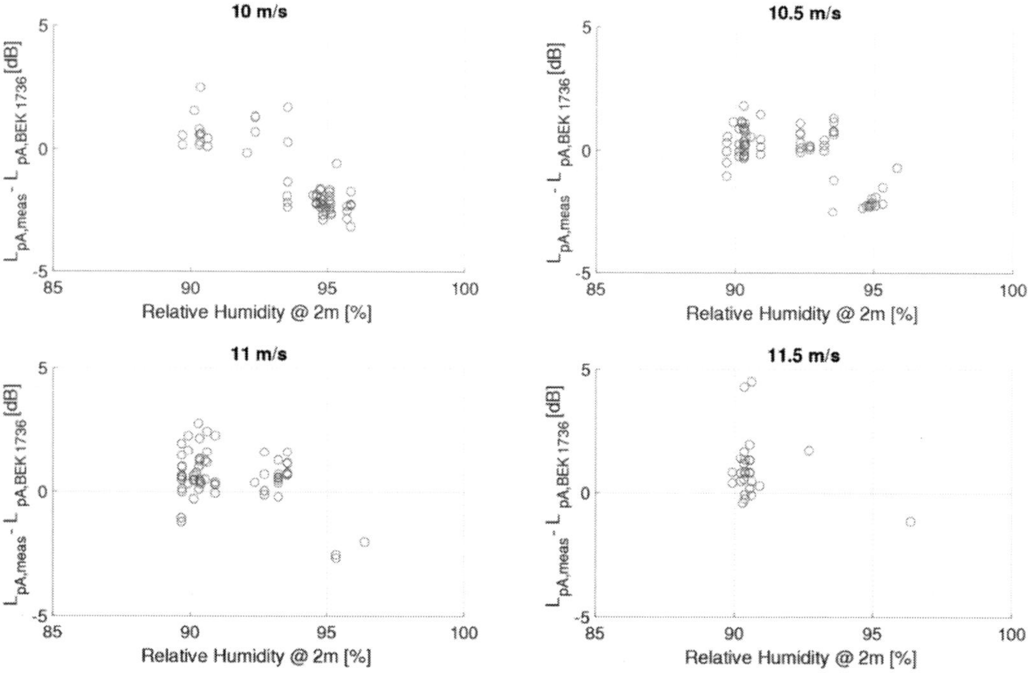

Figure 10. Difference between measured and predicted SPL as a function of relative humidity at 2 m for hub height wind speeds 10 m/s to 11.5 m/s.

3.4 Comparing BEK 1736 and NORD2000 results

A SoundPlan 8.0 model implementing the NORD2000 method was set up. The wind turbine was input as a point source placed at hub height. To compare the prediction method to the measurement with the best SNR, calculations were performed for 8 m/s hub height wind speed. The average values of temperature, atmospheric pressure, relative humidity and temperature gradient were calculated for the 8 m/s bin and input to the model. The input parameters are shown in Table 2.

Table 2. Input parameters for the NORD2000 predictions model.

Wind speed [m/s]	Temperature [°C]	Relative humidity [%]	Atmospheric pressure [kPa]	Relative humidity [%]
8	5	93	99.7	93
Roughness length [m]	Flow resistivity [kNsm^{-4}]	Temperature gradient [°C]	C_w^2 [m$^{4/3}$s^{-2}]	C_t^2 [Ks^{-2}]
0.05	80	-0.07	0.12	0.008

The differences between measured and predicted results using BEK 1736 and NORD200, respectively, are presented in 1/3-octave bands in Figure 11. The disagreement of the models around 200 Hz can be explained by the fact that BEK 1736 has a simple constant terrain correction of +1.5 dB, whereas the NORD2000 model predicts a decrease in SPL caused by interference with the ground reflection, depending on the relative height of the receiver above terrain. The NORD2000 method seemingly overestimates the destructive interference caused by the ground reflection.

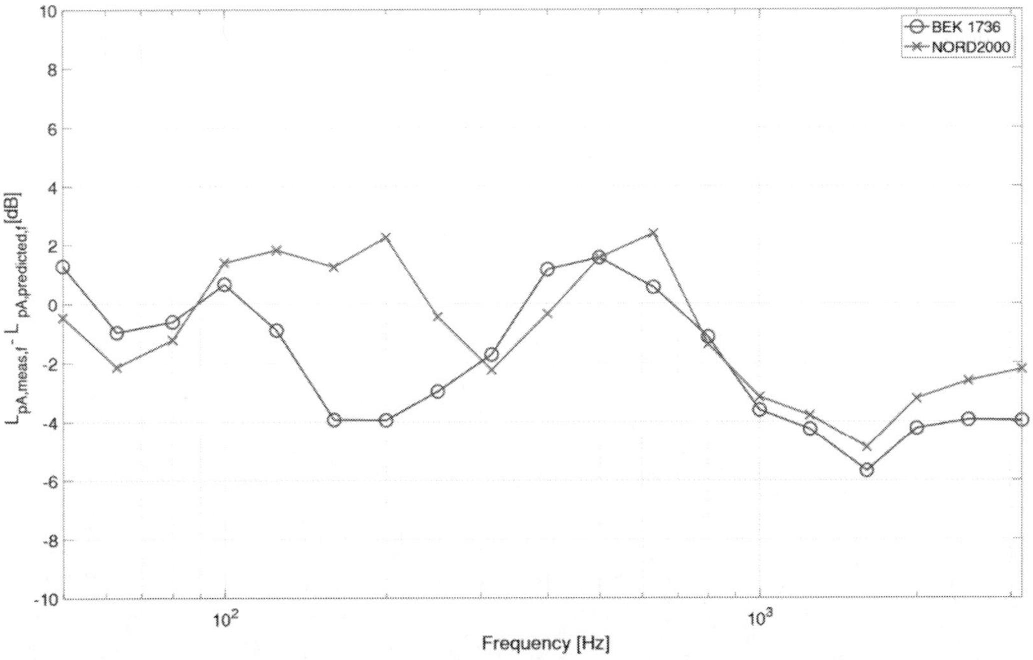

Figure 11. Difference between measured and predicted (BEK 1736) SPL (circled line) and Difference between measured and predicted (NORD2000) SPL (crossed line) as a function of frequency for 8 m/s hub height wind speed.

Both predictions models calculate slightly lower total levels than those measured directly. Part of the explanation is probably the difficulty of obtaining an adequate SNR at 500 m distance from the source, which leads to overestimation of the turbine noise. Another likely explanation is the lack of correction for the secondary wind screen for the 500 m measurement.

4. Conclusions

Comparisons were made between measured and calculated sound pressure levels at 500 m from a single turbine at Høvsøre Test Centre. It was shown that it is possible to get useful measurements by placing a microphone on a tripod at 500 m distance from a turbine. The level difference was sensitive to changes in wind speed due to decreased SNR at the 500 m position for high wind speeds. For future comparative studies it is therefore recommended to choose a far field position which is shielded from the wind. The far field position should be removed from vegetation and other sources of wind generated background noise. Additionally, it is important to use a secondary wind screen to limit wind generated noise in the microphone. The comparisons showed poor agreement between measured and predicted levels above 3.15 kHz due to air absorption. For calculating total levels, it is therefore recommended to limit the upper frequency to between 3-4 kHz. The measured 1/3-octave SPL from 1-4 kHz is lower than the predicted SPL. This is believed to be caused by the lack of correction for secondary wind screen for the 500 m measurement. Future work will necessitate determining the insertion loss of the secondary wind screen. The difference in measured and predicted total SPL's showed either no or very small correlations to the atmospheric conditions. Future work should aim to acquire data with a wider range of atmospheric conditions to draw conclusions on the accuracy of the propagation models for varying meteorology.

5. Acknowledgements

We would like to thank Siemens Gamesa Renewable Energy A/S for allowing us to use data from the measurement.

We would like to thank DTU Wind Energy for providing weather data from the test site.

References

[1]. DELTA (2007, revised 2014) *Proposal for Nordtest Method: Nord2000 – Prediction of Outdoor Sound Propagation* AV 1106/07.
[2]. The Danish EPA, 2015 BEK 1736 Bekendtgørelse om støj fra vindmøller. (Statutory Order on Noise from Wind Turbines; Statutory Order no. 1736).

INCE*EUROPE*

8th International Conference
on
Wind Turbine Noise
Lisbon – 12th to 14th June 2019

Using Measured Sound Power Inputs for Comparison of Measured Immission Noise Levels and Sound Levels Predicted with ISO 9613-2 for Various Ground Factors.

Kohl Clark, B. Eng., Aercoustics Engineering Limited: KohlC@Aercoustics.com,
Payam Ashtiani, B.A.Sc., P.Eng, Aercoustics Engineering Limited: PayamA@Aercoustics.com

Summary

In order to minimize the risk of adverse effect from a proposed wind facility, predictive modelling is often used as a pre-construction assessment method. In Ontario and elsewhere in the world, ISO 9613-2 is used to predict the noise impact from industrial noise sources, including wind facilities.

ISO 9613-2 makes use of several modelling parameters which affect the predicted attenuation over distance. The ground factor, G, dictates the degree of attenuation provided by the interaction of the wave front and the ground between a source and a receiver. Values between 0 and 1, representing perfectly reflective and perfectly absorptive ground, respectively, have been applied in the context of industrial noise modelling. There remains a lack of consensus within the industry on the appropriate ground factor to be used during the modelling for the permitting process for a wind farm. Studies have been performed assessing the applicability of various ground factors by comparing the sound levels predicted with ISO 9613-2 and those measured on site.

This study seeks to add further context to the comparison of these levels by using the sound levels measured according to IEC 61400-11 for inputs as noise source power levels in the model and making predictions at five locations across four wind farms. The cumulative impact from all relevant turbines has been assessed at each of the five locations using ground factors between 0.0 and 0.8. These predicted sound levels are then compared to average measured sound pressure levels from high-quality datasets collected at the same locations. These noise measurements are conducted such that the worst-case impact from the wind facility is assessed.

Comparison of the measured and predicted sound levels across all sites shows a trend indicating that lower values of G, between G=0.2 and G=0.4, may be most appropriate to accurately predict the sound levels on site, providing that the actual turbine emission levels and spectra closely match those of the specification. To further reduce the degree of uncertainty in these calculations, additional work to expand the assessment dataset is planned. Additionally, steps will be taken to more accurately quantify the uncertainty inherent to the immission measurements.

1. Introduction

In order to minimize the risk of adverse effect from a proposed wind facility, predictive modelling is often used as a pre-construction assessment method. In Ontario, and elsewhere in the world, ISO 9613-2 [1] is used to predict the noise impact from industrial noise sources, including wind facilities.

Although it has been shown that more complex modelling methods such as Parametric Equation Modelling can make predictions with a higher accuracy, the ISO 9613-2 standard is still applied widely for the permitting of wind facilities. The applicability of the ISO 9613-2 standard for wind turbine noise propagation prediction has been the subject of much research. Many studies have compared sound levels measured from wind farms to predicted noise levels in modelling, and specifically to those employing ISO 9613-2 [2] [3] [4].

When applying ISO 9613-2 in this context, there has been some variation in the literature regarding the most appropriate ground factor ("G") to ensure that the noise modelling does not under-predict the noise impact from the facility.

Previous research that was conducted in order to determine the most appropriate ground factor to use have largely been based on comparing the result of a measurement campaign with the predicted level from the model. [2] [3] [4]. However, a major shortcoming of these studies is the assumption that the turbines were emitting sound levels and spectra equal to that of the specifications used in the model. From experience, the sound emission output of the turbines have an inherent variation based on the turbine model, and may be subject to differences when compared to the manufacturer specified maximum sound emission. Consequently, any conclusions from those previous studies may have been biased by whether the turbines in the study were louder or quieter than their manufacturer specification used in the model, as well as the associated differences in spectral levels.

As part of this study, far field measurement datasets have been identified where the signal-to-noise ratio is sufficiently high, and the sound emission levels have been measured in-situ at one or more turbine from the site. This is often the closest turbine to the measurement location.

To assess the most appropriate ground factor, a comparison has been made between the noise levels predicted using measured sound power levels according to ISO-9613 and the noise levels measured during immission measurements at the same locations.

2. Background

The following section describes the nature of the acoustical immission measurements as well as the selection criterion for which sites were assessed in the study. Further, additional context is provided regarding the ISO 9613-2 modelling parameters. Finally, the methodology for the comparison of the measured and predicted sound levels is described.

2.1 Immission Measurements

In Ontario, it is often a condition of the approval for a wind farm that audits of the noise impact be conducted at a selection of the worst-affected points of reception around the wind farm. The primary requisites for a receptor being classified as worst-case are as follows:

- High predicted noise impact
- Situated downwind of the farm in the prevailing wind direction

The regulatory framework [5] necessitates assessment of the worst-case impact, only allowing the assessment of data satisfying the following criteria:

- Nearest turbine is producing above 85% of its rated power output.
- All turbines within 1.5 km are operating according to the scenario:
  - Total Noise: All turbines are spinning and producing power
  - Background: All turbines are parked and not producing power
- Measurement location is downwind of nearest turbine – within +/- 45° of the nearest turbine's measured yaw angle.
- Significant transient contamination is not present.

Night-time (22:00 – 05:00) noise measurements are carried out in two phases, with one taking place in the fall/winter (typically September-December), and the other taking place in the spring (typically March-May). Measurement locations are often between 550m and 850m from the nearest turbine and are sufficiently far removed from any reflective surfaces or foliage. Using a microphone with a primary and secondary wind screen at a typical height of 4.5 meters, audio data is recorded in one-minute intervals between frequencies of 20Hz to 10,000Hz. The insertion loss of the wind screen has been tested and accounted for in the processing. One-minute LA_{eq} and L_{90} values are recorded for determination of average sound levels and for transient event filtering, respectively. Weather data including wind speed, direction as well as atmospheric conditions are recorded simultaneously at a height of 10 meters above ground level (10m-AGL). Measurement datasets from two of the assessed sites are included below and generally demonstrate the quality of the data being used in this assessment.

Figure 1: Site A Measurement Data - Spring Measurement Period

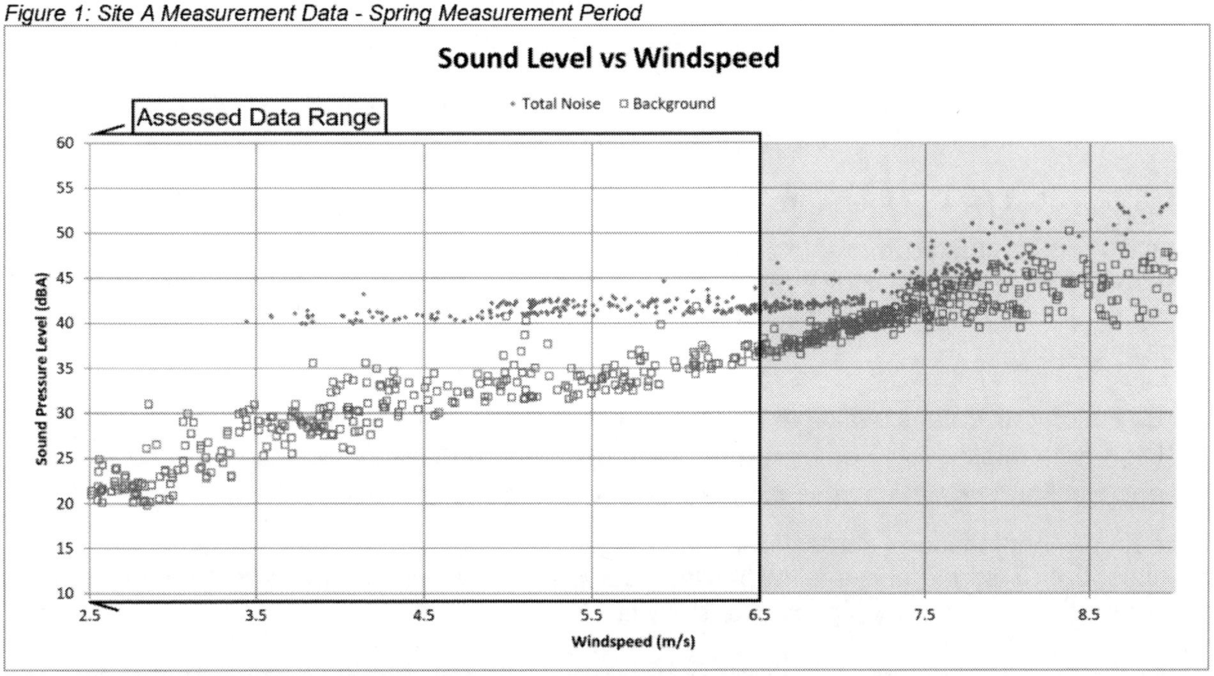

Figure 2: Site D2 Measurement Data - Fall Measurement Campaign

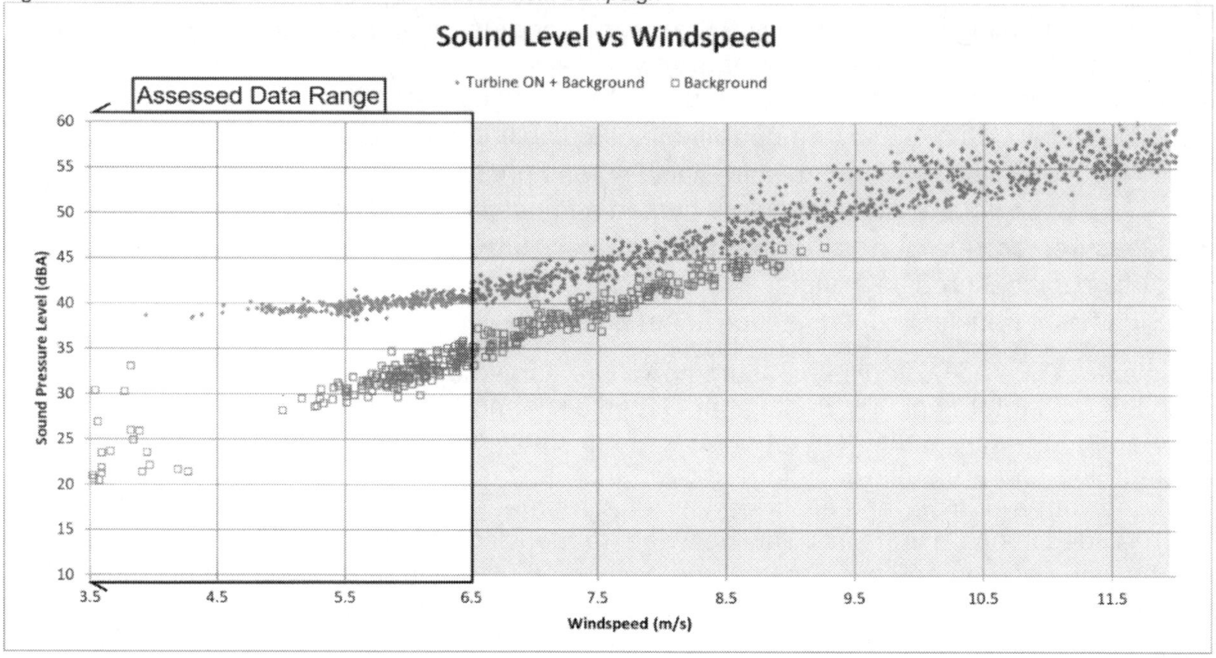

To account for the influence of wind speed on the acoustical measurements, data are sorted into integer wind speed bins ranging from 1-7 m/s (0.5-1.49 m/s, 1.5-2.49 m/s, etc).
The assessed project noise impact level is an average Turbine-Only sound level, which is taken to be the result of logarithmically subtracting the average measured Background sound level from the average measured Total Noise sound level, for each wind speed bin. The standard imposes data filtering to ensure representative conditions between these Total Noise and Background periods, such that the subtraction of windbin-averaged values for these periods results in the noise impact from the turbine operation only.

2.2 Sites Selected for Comparison

Aercoustics has conducted immission measurements at numerous sites across Ontario for which comparisons could be made. However, in order to remove as much uncertainty as possible from the noise level comparisons, the following considerations were made regarding the selection of sites for this study:

- Emission test data is available for turbines near the monitor; ideally the closest turbine.
- High data quality was demonstrated, demonstrating separation between the Total Noise and Background data across multiple wind speed bins.

Five measurement locations across four wind farms were selected for the assessment. For two of the sites, both a spring and a fall/winter dataset was assessed. For three of the sites, only a fall/winter measurement campaign was assessed.

All of the turbines in the study were pitch controlled and were uniformly in the 2 MW class with hub heights of approximately 100 meters. The assessed measurement locations are described in Table 1.

Table 1: Measurement Location Summary

Site			A	B	C	D1	D2
Measurement Period	Spring	Duration	Apr. 2017	Mar. – Apr. 2018	-	-	-
		# Days	8	58			
	Fall / Winter	Duration	Sept. – Nov. 2018	Oct. – Nov. 2018	Apr. – May. 2018	Nov. 2018 – Jan 2019	Nov. 2018 – Feb. 2019
		# Days	29	36	23	72	73
Wind Farm Size	Nameplate Capacity [MW]		10	8	8	>250	
Measurement Location Details	Distance from Nearest Turbine [m]		570	385	675	724	630
	Ground Cover between Turbine and Measurement Location		Crop / Grassland	Forest / Crop	Crop / Grassland	Crop	Crop
	Topography between Turbine and Measurement Location		Flat/Hilly, +/- 15m	Flat, +/- 10m	Flat, +/- 5m	Flat, +/- 5m	Flat, +/- 5m

2.3 Noise Level Prediction Using ISO 9613-2

Sound propagation according to ISO 9613-2 is calculated according to several attenuation parameters. In the context of a typical wind facility, the dominant factors are usually A_{div}, (geometric divergence), A_{gr} (ground effect), and A_{atm} (atmospheric absorption). The divergence term is based on spherical spreading of the wave front and varies solely with distance from the source. The ground attenuation term varies in frequency over distance and is dictated predominantly by the ground effect constant, G, which ranges between values of 0 and 1. These extremes are intended to represent perfectly reflective (G=0) and completely porous ground (G=1), with the value G=0.5 representing ground of 'mixed porosity'. The atmospheric term varies spectrally with the attenuation-per-distance depending on temperature and humidity values. The permitting guideline for Ontario [6] dictates that propagation be modelled accounting for a temperature of 10 degrees Celsius and a relative humidity value of 70%. Average meteorological values for each measurement campaign at each site are included in the table below. These values often shift throughout these measurement campaigns, however the average for the measurement periods generally agree with the modelled conditions.

In this data subset, five of the seven datasets were collected in the fall/winter months, and two of the seven were collected in the spring months.

Table 2: Measured Atmospheric Conditions at Each Measurement Location

Site	Period	Atmospheric Value	Relative Humidity [%]	Temperature [°C]
A	Spring	Average	44	8.9
		Maximum	66	17.8
		Minimum	25	4.6
	Fall	Average	85	16.0
		Maximum	96	20.9
		Minimum	62	2.2
B	Spring	Average	71	0.5
		Maximum	89	4.2
		Minimum	41	-4.8
	Fall	Average	70	2.6
		Maximum	89	9.9
		Minimum	62	-1.8

C	Fall	Average	60	1.5
		Maximum	96	9.4
		Minimum	44	-0.6
D1	Fall	Average	73	0.2
		Maximum	94	6.2
		Minimum	52	-12.4
D2	Fall	Average	73	-3.1
		Maximum	93	6.5
		Minimum	46	-15.8
Overall Average			68	4

The acoustical modelling software CadnaA, which utilizes ISO 9613-2, was used to make worst-case noise predictions at the five sites considered in the study. All wind turbines considered in the permitting of the wind farm were accounted for in the noise modelling, including those from adjacent facilities. Predictions of the cumulative noise impact from all relevant turbines were made at noise receptors corresponding to the real-world measurement locations at ground factors ranging from 0.0 - 0.8 in increments of 0.1.

Wind turbine noise sources were modelled as point sources at heights corresponding to the hub-height of the turbine. The worst-case sound power spectra measured according to IEC 61400-11 [7], [8] were used for each turbine where measurement data was available. Measurement data was available for some but not all turbines on a given site; for turbine noise sources that were not measured, turbine sound power spectra for were taken to be the energy-average of the tested values from other turbines on-site, for all turbines having the same model and rated sound output as those tested. Where no suitable test data was available for a given turbine model or rated sound power output, sound power spectra from the turbine specification were used. This was only the case for one model of turbine at Site D1/D2, and, at all sites, for the turbines from adjacent facilities. The table below compares the manufacturer specified sound power levels with those measured on site and provides context regarding the predicted acoustical contribution from measured sources compared to the overall predicted noise level.

Table 3: Measurement Location Noise Impact Comparison

Site	A	B	C	D1	D2
Nearest Turbine Rated Sound Power [dBA]	103.2	103.2	103.2	102.0	104.0
Measured Turbine Max IEC Sound Power [dBA]	102.6, 103.2	103.2, 103.2	102.9, 103.7	102.0	103.7
Predicted Sound Level[1] from All Noise Sources [dBA]	39.5	42.7	38.0	39.5	37.8
Predicted Sound Level[1] from Measured Noise Sources [dBA]	34.3	42.3	35.2	35.0	31.4

1 – Sound level Predicted at Measurement Location using Measured Values and G=0.7.

2.4 Comparison of Measured and Predicted Noise Levels

Data had been collected at a range of wind speeds at the sites mentioned above. A high-quality subset of this data was selected for assessment based on a number of criteria. Specifically, the windbin-average Turbine-Only sound pressure level for a given measurement location was only considered where:

- A minimum 1-minute data count of 20 was achieved for both Total Noise and Background Noise intervals
- A minimum signal-to-noise ratio between average Total Noise and Background sound levels of 5 dBA was achieved
- The wind speed bin was below 7 m/s, in order to reduce the influence of wind-related noise on the measurement data.

Applying these filters resulted in a dataset with a population of N=16. By virtue of the signal-to-noise threshold, minimum turbine power threshold, and downwind criterion for all data, these final Turbine-Only average sound levels represent the most accurate assessment of a worst-case impact from the nearby wind farm in the far field.

3. Results

The filtered selection of data has been grouped into wind speed bins for assessment of the average difference between measured and predicted noise levels for different values of G across multiple sites. Figure 3 below shows, for various wind speeds, the difference between measured levels and modelled levels with different ground factors. The grading is in 0.25 dBA increments, and the white area represents deviations of -0.25 to +0.25 dBA, i.e. most accurately reflecting the measured levels. Areas below the 'white belt' represent conditions where the measurements were less than the modelled levels (i.e. the model overpredicted) while the area above the white line represents times where the measured levels where higher than the modelled level (i.e. the model underpredicted).

Figure 3: Difference between Measured and Predicted Sound Level for Different values of G at Different 10m-AGL Wind Speeds

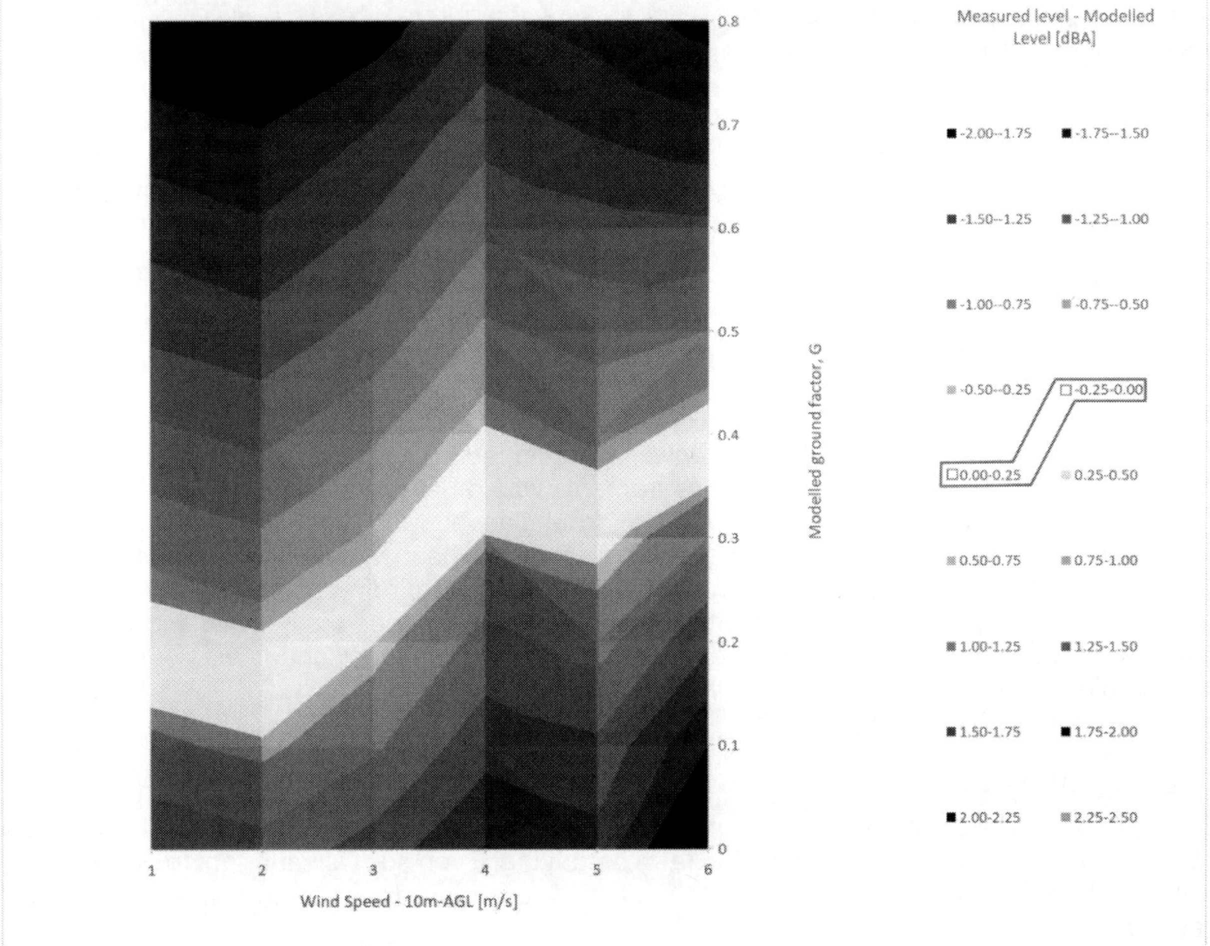

This same data is also presented in a more granular format in **Error! Not a valid bookmark self-reference.**

Values falling around the y-axis represent those most accurately reflecting the measured levels.

Figure 4: Difference between Measured and Predicted Sound Level for Different values of G at Different 10m-AGL Wind Speeds

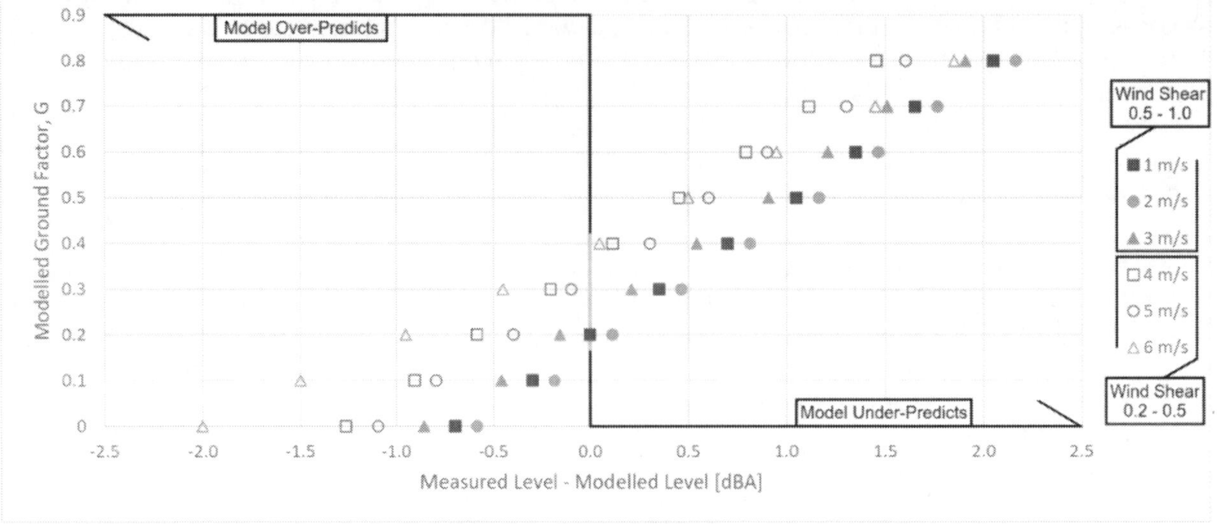

On average it was found that application of a ground factor of about G=0.3 was required in order to align the model with the measured value. This value is lower than findings from other studies [3]; potential reasons for these findings are presented in Section 5 of this study. These findings should be considered preliminary based on a limited sample size of 16. Further measurements and site locations are planned to be added to this data set in order to confirm the observed trends. Furthermore, the data was also analysed by the measured ground level wind speed. It should be noted that because the analysis only considers datapoints where the electrical power being generated is at least 85% of rated output, it would therefore necessitate that lower ground level wind speeds represent higher wind shear events, and higher wind speeds represent comparatively lower wind shear events.
That is, for the case of 10m-AGL wind speed of 1m/s, the hub height wind speed used in the analysis was still ~10m/s, making the wind shear coefficient approximately 1.00; for the case of 3m/s the wind shear coefficient would be approximately 0.5.

An interesting trend appears to be that at lower wind speeds (i.e. higher shear conditions) the ground factor to match the model to the measurements would have to be considerably lower than originally anticipated – roughly G=0.2. As the ground level wind speed increases and the associated wind shear decreases, the ground factor that best matches the measurements increases to G=0.4. This could be due to more extreme downward refraction conditions when the ground level wind speed is below 3m/s and the hub height wind speed is ~10m/s. However, the difference in measured level between extreme and moderate level shear was found to be relatively minor, in the order of 0.5 – 1 dBA.

4. Limitations of Study

Differences of fractional decibels are not practically relevant in terms of environmental acoustics but become important in the context of wind farm permitting. Even so, it is important to understand that the degree of uncertainty in this context often exceeds the magnitude of the value being assessed.

Compared to other studies regarding the applicability of various ground factors in wind turbine noise modelling, this study represents a lower degree of uncertainty in the results by virtue of using measured sound power levels combined with high quality immission data.

Regardless, there remain numerous sources of uncertainty involved in this assessment which arise from multiple sources. Specifically;

1. Wind Turbine Sound Emission Levels used for Modelling
    a. *IEC measurement uncertainty*
    b. *Variation in Turbine sound emission for turbines of the same model within a wind farm*
2. ISO-9613-2 Modelling Uncertainty
    a. *General uncertainty of model*
    b. *Downwind condition from all sources simultaneously*
3. Calculated Turbine-Only Sound Pressure Levels (Immission) - Unknown
    a. *Variation in environmental weather conditions across campaign*
    b. *Uncertainty related to the signal-to-noise ratio*
    c. *Uncertainty related to the measurement chain*

Items 1a, b, 2a, and 3a, b, c may bias values either above or below the real value, where item 2b is expected to have the effect of over-predicting the sound pressure level in situations where the receptor is surrounded by wind turbines and therefore cannot be downwind of all turbines simultaneously. This situation is most relevant at sites D1/D2, and less relevant at the other sites, where the measurement location was downwind of the majority of the closest turbines.

The uncertainty associated with the ISO 9613-2 methodology is stated to be +/- 3.0 dB in situations where the average height between source and receiver is below 30 meters, and where noise sources are within 1,000 meters of the receptor. In the context of most wind farms, this is not the case, and so the uncertainty may exceed the stated value.

Additionally, all of the measurement scenarios in this study generally correspond to flat ground. Variations with respect to hilly and complex terrain was not part of the scope of this study; the appropriateness of various values of G has been assessed to vary based on topography [4].

A detailed assessment of uncertainty was not possible at the time of this study. A simplified uncertainty for the results was calculated based on the various discrete uncertainties mentioned above, and was found to be in the order of at least +/- 1.6 dB

As mentioned previously, the results presented here represent a relatively small set of samples with high quality far-field data, and with field tested turbine emission data. Future work on this subject will include the addition of further sample sets in order to improve the quality of the conclusions.

5. Implications of Study - Appropriateness of G<0.5 Ground Factors

The results of the study indicate that a value of G=0.3 or lower is appropriate for conservatively and accurately predicting the sound level at a point of reception for a wind farm. Higher values of G, up to G=0.5, may be appropriate for lower-shear conditions, and may be valid for sites where high-shear conditions occur rarely. Conversely, for worst-case assessment of sites with a greater prevalence of high-shear conditions, values of G down to G=0.2 may be applicable. As stated earlier in this study, the applicability of these lower ground factor values is valid for situations where the sound power spectra used in modelling are known or expected to closely match the real-world turbine outputs.

For most wind facilities, the ISO 9613-2 methodology is inherently conservative in that it considers the point of reception to be down wind of all noise sources. This will not be the case for any receptor that is surrounding by turbines and therefore will only be down-wind of a selection of those. Conservatism can be added in different areas during the noise impact assessment, including in the manufacturer's turbine noise specifications, in the selection of ground factors, and in the application of explicit safety factors on the noise output level.

Higher values of G may be applicable in situations where the representative or average noise impact of the facility is being assessed, rather than the absolute worst-case.

In consideration of the findings of this study, it is important to be sufficiently conservative in the modelling process for wind facilities, particularly so considering the limitations of the ISO 9613-2 methodology. Modelling using ground factors as high as G=0.7 has somewhat consistently resulted in compliant immission sound levels in Ontario, providing that the turbine sound power output is within specification; this effect may not be the result of the applicability of that ground factor and may be due to other factors regarding sound output spectrum of the turbine(s) in-situ as compared to the assumed values in the pre-construction noise studies. However, this should not be an indication of the validity of using ground factor of G=0.7 in the future.

References

[1] ISO 9613-2, *Acoustics – Attenuation of Sound during Propagation Outdoors – Part 2: General Method of Calculation*, International organization for Standardization, 1996.

[2] Bullmore, A.J., Adcock, J.G., Jiggins, M, Cand, M, *Wind Farm Noise Predictions and Comparisons with Measurements*, Third International Meeting on Wind Turbine Noise, Aalborg, Denmark, 2009.

[3] Delaire, C, Griffin, D & Walsh, D 2011, *Comparison of predicted wind farm noise emission and measured post-construction noise levels at the Portland Wind Energy Project in Victoria, Australia*, Fourth International Meeting on Wind Turbine Noise, Rome, Italy.

[4] Evans, T. and Cooper, J, *Comparison of predicted and measured wind farm noise levels and implications for assessments of new wind farms*, Proceedings of Acoustics 2011, 2011.

[5] Ontario Ministry of the Environment, Conservation, and Parks, *Compliance Protocol for Wind Turbine Noise*, 2017.

[6] Ontario Ministry of the Environment, Conservation, and Parks, *Noise Guidelines for Wind Farms*, 2016.

[7] IEC 61400-11 Edition 2.1, *Wind Turbine Generator Systems, Part 11: Acoustic Noise Measurement Techniques,* 2016.

[8] IEC 61400-11 Edition 3.0, *Wind Turbine Generator Systems, Part 11: Acoustic Noise Measurement Techniques,* 2002.

8th International Conference
on
Wind Turbine Noise
Lisbon – 12th to 14th June 2019

Sensitivity analysis of influential parameters for wind turbine noise

Kayser B. and Gauvreau G.
Ifsttar, Cerema, UMRAE, CS4 4129, Bouguenais cedex, 44344, France

Écotière D.
Cerema, Ifsttar, UMRAE, 11 rue Jean Mentelin, Strasbourg cedex 2, 67035, France

Summary

Wind turbine noise is directly concerned by physical phenomena related to long-range sound propagation, i.e. ground effects (roughness, impedance), and micrometeorological effects (mean refraction, intermittency fluctuations, small-scale turbulence scattering).
In order to estimate the relative influence of these environmental and influence parameters on acoustic propagation, we present a sensitivity analysis based on a propagation model and on a specific numerical experiment design produced by a Morris screening method. The model used is a Parabolic Equation model (MW-WAPE) which takes into account ground effects (absorption through acoustic impedance diffusion through surface roughness) and micrometeorological effects (mean refraction through the vertical gradient of effective celerity atmospheric turbulence through the random Fourier modes technique).
The results allow the quantification of the global variability and uncertainties associated with main phenomena occurring in the emission-propagation-reception process. The results will allow quantifying the overall variability and uncertainties associated with the emission-propagation-reception sound chain. A better knowledge of these uncertainties will provide a better control of the quality of wind turbines noise prediction in an inhomogeneous outdoor environment.

1. Introduction

The noise emitted by wind turbines in operation can be the cause, in certain situations, of an annoyance reported by residents of wind farms. This sometimes leads to opposition against the development of these renewable energies, or to the implementation of degraded operating modes (acoustic curtailments) of the concerned wind farms, at the detriment of energy and environmental efficiency. Satisfactorily estimating the noise pollution potentially induced by wind turbines is a socio-economic and health challenge that requires the definition of reliable methods for predicting and estimating noise levels in the environment. Although the wind energy sound source has been the subject of many studies in recent years (sound emissions[9], directivity [24], sound power [1]), little work has focused on the influence of environmental parameters on the variability of the sound levels encountered. However, there is now a major challenge in estimating the accuracy and representativeness (spatial and temporal) of sound level measurements in non-homogeneous outdoor space. Estimating the global variability and uncertainties associated with the emission-propagation-reception chain thus appears essential for controlling the acoustic impacts of wind turbines in the environment.

Predicting the variability (deterministic and random) of sound levels at the output of prediction models will therefore require both (i) refining the models, (ii) estimating fluctuations in all input data, (iii) performing an optimized numerical experiment design, and (iv) using advanced techniques for uncertainty propagation and sensitivity analysis of complex systems. The final objective is to develop a method that allows to more accurately assess the uncertainties of the entire modelling chain due to deterministic variability and random uncertainties in influential parameters: wind farm operating regime, micrometeorological conditions (atmospheric stability and turbulence), ground acoustic properties (acoustic impedance), surface roughness (spatial spectrum), etc.

The purpose of this paper is to present the relative influence of parameters known to be the most influential on acoustic propagation: ground effect through absorption and roughness, and atmospheric stratification effect through the wind and temperature gradient. It is organized as follows: Section 2 introduces the model used and describes how it takes into account the above-mentioned effects; Section 3 describes the sensitivity analysis method followed, the parameters analyzed, and presents the results of the numerical experimental design.

2. Parabolic Equation model

2.1. Principles of PE model

The propagation model considered is a parabolic equation model with fractional steps called split-step Padé, based on higher order Padé (2,2) approximants [12] and solved with the method of Collins [13]. The model assumes an inhomogeneous propagation medium without any overall motion (mean-wind). In this approach, the real moving atmosphere is replaced by hypothetical motionless medium with the effective sound speed $c_{eff} = c + v_x$, where v_x is the wind velocity component along the direction of sound propagation between the source and receiver, and c is the adiabatic sound speed. It is based on the Mean-Wind Wide-Angle Parabolic Equation (MW-WAPE) derived from the Helmholtz elliptic equation for the sound pressure as described in [20]:

$$\left(\frac{\partial^2}{\partial r^2} + \frac{1}{r}\frac{\partial}{\partial r} + \frac{\partial^2}{\partial z^2} + k^2\right) p(r,z) = 0 \tag{1}$$

with $k = \omega/c_0$. Then the pressure $p(r,z)$ is split in two components:

$$p(r,z) = u(r,z)H_0^{(1)}(kr) \approx \frac{1}{\sqrt{r}} u(r,z)e^{jkr} \tag{2}$$

where $u(r,z)$ is an envelope fuction that varies very little with distance, and $H_0^{(1)}$ is a propagative cylindrical wave represented by the asymptotic expression of the Bessel function of the 1st kind and of the order 0. The evolution of $u(r,z)$ is governed by the equation:

$$\frac{\partial u(r,z)}{\partial r} = jk\left(Q_{pd} - 1\right) u(r,z) \tag{3}$$

with Q_{pd} a pseudo-differential operator defined as:

$$Q_{pd}^2 = \frac{1}{k^2}\frac{\partial^2}{\partial z^2} + n^2 \tag{4}$$

and $n = c_0/c_{eff}$ is the index of refraction.

As the MW-WAPE equation is solved at each spatial step (chosen here as $\lambda/20$ along the longitudinal axis), it is possible to modify the propagation effects (*i. e.* ground impedance or the effective celerity profile) at each forward step. Thus, the model can solve acoustic propagation problem above a mixed ground in a refractive and turbulent atmosphere. This elements makes parabolic equation based methods particularly suitable and accurate for acoustic simulations in long-range outdoor sound propagation [14] [25] [4][5] [3] [18] [6]. Furthermore, recent work has made it possible to take into account soil roughness by using effective admittance as a boundary condition for the ground in our PE model.

2.2. Ground effects

Acoustic properties of the ground (porous absorption, impedance and diffusion) are described by an effective admittance model [10]:

$$\beta_{eff} = \beta + \beta_{rough} = \frac{1}{Z} + \beta_{rough} \tag{5}$$

The effective admittance can be seen as an additive correction to the normalized admittance $\beta = 1/Z$ with Z the impedance of the ground described by the Miki's impedance model in our code [21]:

$$Z = Z_0\left[1 + 5.50\left(\frac{f}{a_{fr}}\right)^{-0.632} + i8.43\left(\frac{f}{a_{fr}}\right)^{-0.632}\right] \tag{6}$$

$$k = k_0\left[1 + 7.81\left(\frac{f}{a_{fr}}\right)^{-0.618} + i11.41\left(\frac{f}{a_{fr}}\right)^{-0.618}\right] \tag{7}$$

where $Z_0 = \rho_0 c_0$ is the specific impedance of air, $k_0 = \omega / c_0$ the acoustic wavenumber, ρ_0 the density of air, c_0 the celerity of sound in the air, $\omega = 2\pi f$ with f the frequency, and a_{fr} the ground airflow resistivity.

β_{rough} gives the average effect of surface roughness on acoustic propagation and it is modeled by the following equation [11] [8]:

$$\beta_{rough} = \int_{-\infty}^{+\infty} \frac{d\kappa'}{k_0 k_z(\kappa')} (k_0^2 - \kappa\kappa') W(\kappa\kappa') \tag{8}$$

with $\kappa = k_0 \sin\theta_i$ where θ_i is the incidence angle of the acoustic wave with the ground and $k_z(\kappa) = \sqrt{k_0^2 - \kappa^2}$. Considering that the probability density of the ground roughness heights follows a normal distribution, the roughness spectrum $W(\kappa\kappa')$ is defined in the wave numbers space by [8]:

$$W(k) = \frac{S_h^2 L_c}{2\sqrt{\pi}}^{-\frac{k^2 L_c^2}{4}} \tag{9}$$

with S_h the standard deviation of the ground roughness heights, and L_c the correlation length. S_h gives information on the vertical variations of the roughness, while the correlation length L_c reflects the horizontal variations of the roughness [11].

2.3. Atmospheric stratification effects

The atmospheric stratification effects on acoustic waves are taken into account *via* the assumption of effective celerity. In the case of a flat and clear ground, the atmosphere is considered homogeneous in the horizontal direction. It is then possible to neglect the evolution of the effective celerity according to distance. The average profile of effective celerity is then defined as follow :

$$\langle c_{eff}(z) \rangle = \sqrt{\gamma R \langle T(z) \rangle} + \langle V(z) \rangle \cos\theta \tag{10}$$

The vertical gradient of effective celerity is therefore expressed as:

$$\partial_z c_{eff} = \frac{1}{2}\sqrt{\frac{\gamma R}{T(z)}} \partial_z T(z) + \partial_z V(z) \cos\theta \tag{11}$$

$$\approx \frac{1}{2}\frac{\gamma R}{c_0} \partial_z T(z) + \partial_z V(z) \cos\theta \tag{12}$$

with $\gamma = C_p / C_v$ the heat capacity ratio of dry air at a constant pressure and volume, R the specific gas constant for dry air, and θ the angle between the wind direction and the direction of sound propagation. $V(z)$ and $T(z)$ are the vertical wind and temperature profiles, which are expressed as the sum of a mean part (symbol $\langle\ \rangle$) and a turbulent random part (symbol ')

$$V(z) = \langle V(z) \rangle + V'(z) \tag{13}$$

$$T(z) = \langle T(z) \rangle + T'(z) \tag{14}$$

(15)

these micrometeorological profiles can be approached by logarithmic profiles defined as follows:

$$\langle V(z) \rangle \approx a_u \ln\left(\frac{z-d}{z_0}\right) \tag{16}$$

$$\langle T(z) \rangle \approx T_0 + a_T \ln\left(\frac{z-d}{z_0}\right) \tag{17}$$

and their derivatives by the expressions:

$$\partial_z V(z) \approx a_u \frac{1}{z-d} \tag{18}$$

$$\partial_z T(z) \approx a_T \frac{1}{z-d} \tag{19}$$

$$\tag{20}$$

where a_u and a_T are coefficients (unitless) that determine the shape of the profiles, d the displacement height and z_0 the roughness height of the flux profiles.

3. Sensitivity analysis

3.1. Morris method

Sensitivity analysis is the study of the relative influence of different input factors on the model output. The simplest and most common way to study the sensitivity of a model is to modify its inputs one by one ("One-At-a-Time": OAT), the others remaining fixed at a nominal value [19]. The induced effect on the output is then quantified and attributed to the parameter in question. The main problem with this method is that it does not take into account the effects induced by interactions between input parameters. To do so, a complete factorial design method can be follow. It consists in evaluating the calculation code for all combinations of input parameters, which allows the estimation of all input effects and their interactions. In practice, the number of simulations required makes this plan often unfeasible because it increases according to a power law in the number of input parameters. The experimental design becomes too expensive in terms of calculation costs.

Morris method [22] [26] makes it possible to overcome the limiting assumptions of the OAT plan while keeping a cost that increases linearly with the number of input parameters. Morris' method consists in repeating r times ($r = 5$ to 10) [27] [17] an OAT plan randomly in the input space, discretizing each input into a suitable number of levels (depending on the number r of repetitions to be performed) which leads to a number of $r(k+1)$ runs, where k is the number of input variables. The method starts by sampling a set of start values within the defined ranges of possible values for all input variables and calculating the subsequent model outcome. The second step changes the values for one variable (all other inputs remaining at their start values) and calculates the resulting change in model outcome compared to the first result. Next, the values for another variable are changed (the previous variable is kept at its changed value and all other ones kept at their start values) and the resulting change in model outcome compared to the second run is calculated. This goes on until all input variables are changed. An example of the method is shown in figure 1 for $k = 2$ & $r = 2$.

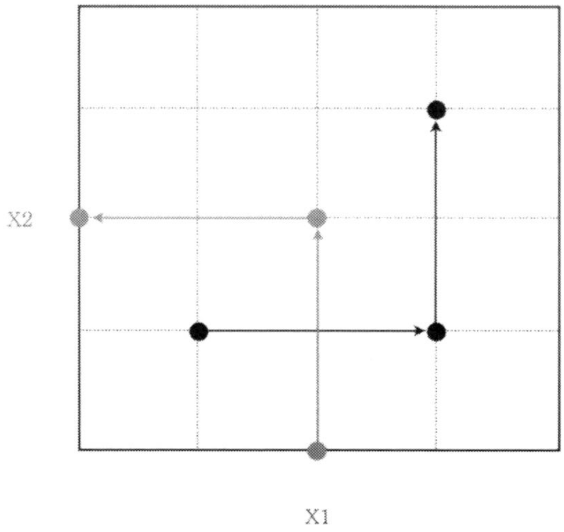

Fig. 1 Illustration of Morris' method with $k = 2$ and $r = 2$. The model has 2 inputs parameters X1 and X2 whose variation intervals are discretized into 5 levels. The calculations of the experimental design are represented by the points. The arrows represent the trajectories of the method (the value of a parameter is modified during a trajectory).

Each repetition i ($i = 1...r$) allows to evaluate an elementary effect (between 2 successive calculations) by input parameters. The entire experimental design provides a r−sample of effects for each input, from which the sensitivity indices are derived:

- μ^*, average of the absolute values of the effects,
- μ, average of the values of the effects,
- σ, standard deviation of the effects.

Thus, the more important μ^* is, the more the parameter contributes to the dispersion of the output. σ measures the linearity of the model: if the model output is linearly dependent on the X_n parameter studied, and X_n does not interact with other inputs, the effect of an elementary perturbation of X_n is the same regardless of its position in the input parameter space. In this case $\sigma = 0$. Therefore, the higher the σ is compared to μ^*, the less relevant is the assumption of linearity and non-interaction. Thence, Morris' method allows entries to be classified into 3 categories:

- inputs with negligible effects (low μ^* and σ)
- inputs with linear effects and no interaction (high μ^* and low σ)
- inputs with non-linear effects and/or interactions without distinction between the two (high σ compared to μ^*).

3.2. Study parameters

For this study we focus on the relative influence of the parameters known to be most influential in long-distance acoustic propagation: *i. e.* the ground effects (through roughness and absorption) and atmospheric stratification (through vertical wind gradient and vertical temperature gradients). Future work will lead to more exhaustive conclusions by taking into consideration other parameters, such as air humidity or vegetation height (which has an influence on the displacement height of vertical flux profiles) for example.

The first step in the sensitivity analysis is to determine the influence parameters that we want to study and their ranges of variation. For the ground effects we have chosen 3 parameters: a_{fr} (airflow resistivity of the ground), L_c (correlation length of the roughness) and S_h (standard deviation of roughness height). Their variation intervals for natural ground are visible in the table 1 [16] [23] [7] [15] [2]. For the refraction we have taken into account 2 parameters: the coefficients a_u and a_T. These coefficients are used to calculate the logarithmic wind and temperature profiles, as well as their derivatives over the height z. The range of variation of the a_u coefficient has been determined with respect to the operating ranges of the wind turbines; i.e. the lower bound corresponds to a wind of $3m.s{-}1$ at the nacelle, and the upper bound corresponds to a wind of $25m.s{-}1$ at the nacelle. For the thermal coefficient, the bounds were determined in accordance with the different profiles possibly encountered outdoor.

Then we chose to set up the Morris method with $r = 10$, and 4 levels of discretization for the intervals of variation of the input parameters. These values correspond to the values typically encountered during sensitivity analysis with this method (good compromise between the number of calculations and the coverage of the input space) [27] [17]. The table 1 shows the values that can be taken by the parameters studied. For each calculation of the experimental design, each parameter will be assigned a value (according to the Morris method) indicated in this table.

Parameters	min value	case 2	case 3	max value	Description
a_{fr} $(kNsm^{-4})$	50	500	5000	50000	airflow resistivity of ground
L_c (m)	0.05	0.37	0.68	1	correlation length of rough ground
S_h (m)	0.01	0.23	0.037	0.05	standard deviation of roughness heights
a_u	0.25	0.78	1.32	1.85	aerodynamic refraction coefficient
a_T	−0.5	−0.25	0	0.25	thermal refraction coefficient

Table 1 Description and values that can be taken by the 5 parameters into the sensitivity analysis.

3.3. Results

The study was conducted for an omnidirectional point source located at $100m$ height. The propagation domain is $500m$ high and $1500m$ long. The relative influence of the parameters has been quantified at two specific points in this domain: microphone1$(x = 500, z = 2)m$ & microphone2$(x = 1500, z = 2)m$. They correspond to a case of medium and long range sound propagation. The following figure shows an example of mapping obtained at $f = 50$Hz with the parabolic equation model, in the case of a fairly reflective ground (very marked interferences) and without refraction (unbent interference lobes).

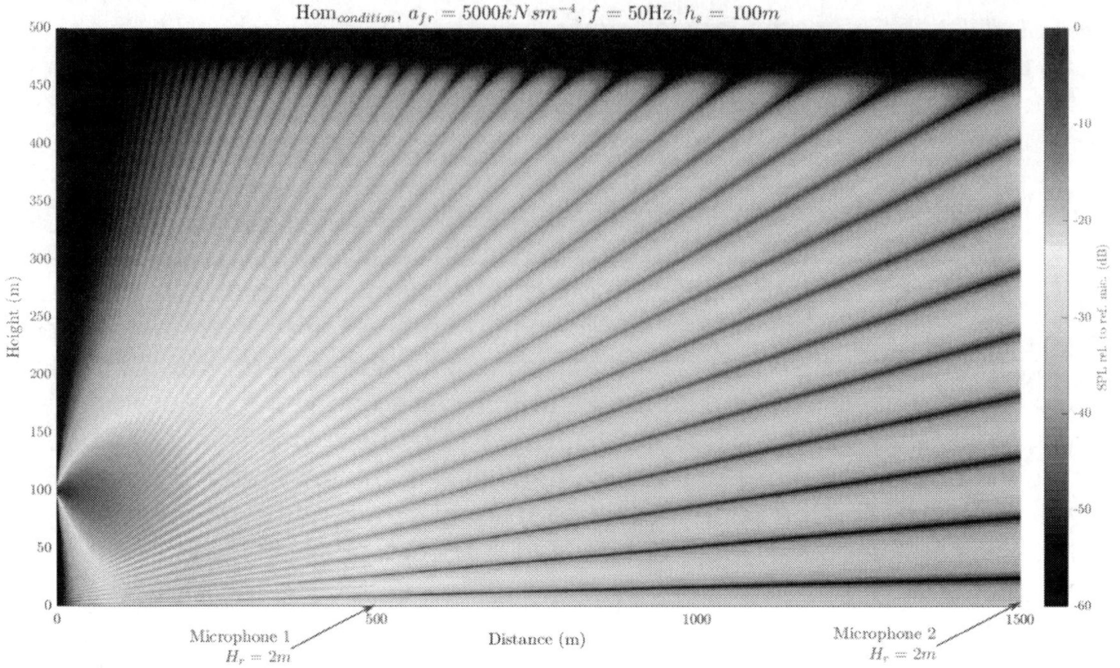

Fig. 2 Example of sound level mapping (attenuation) obtained with PE model.

The calculations of the experimental design will modify the mapping at the output of the model, and therefore the levels measured at microphones 1 and 2. As a result, the relative influence (μ, μ^* and σ) of each parameter is quantified. Some of the physical phenomena studied depend on frequency (i.e. absorption and roughness of the ground), so we carried out a sensitivity analysis per third octave band (50 to 1000Hz) in order to take this into account and not to bias the results. The figure 3 shows the results of the sensitivity analysis.

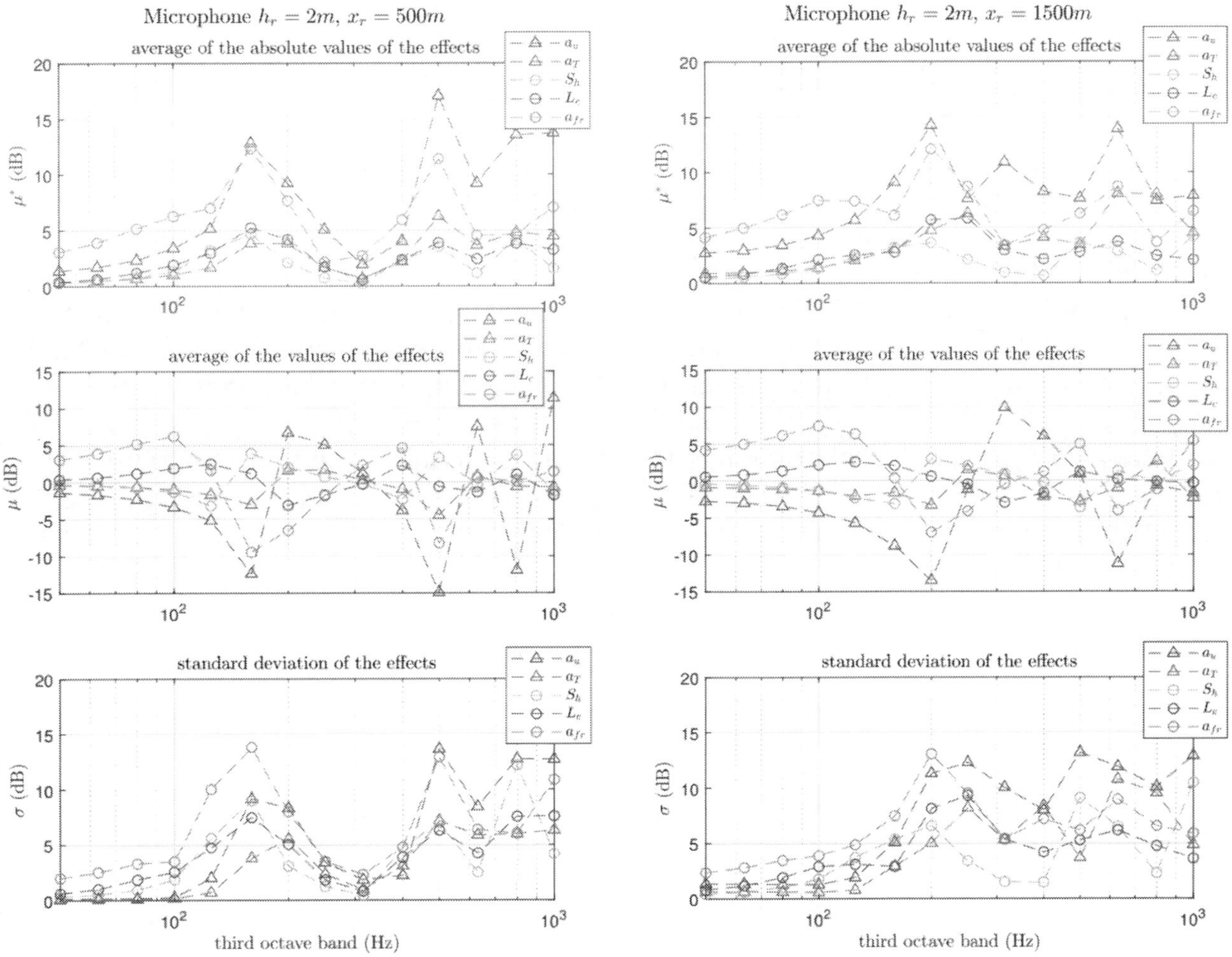

Fig. 3 Evolution of μ^*, μ and σ over the third octave band for: refraction effects (triangle) and ground effect (circle), in a medium range propagation case (left) and long range propagation case (right).

If we look at these figures as a whole, the wind profiles and airflow resistivity seem to be the most influential parameters (higher μ^* and σ) for almost all third octave bands. Nevertheless, the ground roughness and the temperature profile are not to be neglected regarding the values in dB of their respective μ and σ. These results are consistent because the 5 parameters studied are known to influence acoustic propagation.

If we look more closely at the phenomena at low frequencies (< 100Hz), it is ground absorption (a_{fr}) that emerges as the most influential parameters. In contrast roughness has little influence ($<$ 2dB for μ^* and σ). This is due to the fact that ground variations are negligible compared to the wavelengths considered in this frequency range. It can also be noted that the effects for all the 5 parameters are more important at microphone 2 ($1500m$) than at microphone 1 ($500m$). This is in line with the physics because the sound wave has propagated over a larger range and has therefore been more influenced by all the phenomena.

For frequencies above 100Hz, there is a fairly similar fluctuation of μ^* and σ for all parameters (whether at microphone 1 or 2): increase of the effects around the third octave band 200Hz and 630Hz. This fluctuation is related to the geometry of the system under study. Indeed, there are interference dips at the microphone position for the 200Hz and 630Hz third octaves band. As a result, when the environmental parameters are modified from one calculation to another, the interference patterns change. The microphones can alternatively be located in a minimum and maximum pressure, which leads to a very large dispersion of the results. This is why it is necessary to look at the relative influence of the parameters to conclude. In fact that doesn't mean that all the parameters are more influential at 200Hz and 630Hz third octave bands than at other frequencies, it is the system geometrical configuration that is more sensitive to effects at these measurement points.

Finally, when we consider the relative influence of the parameters, it therefore appears that above 100Hz, wind profile and ground resistivity are very influential parameters on acoustic propagation. The other parameters have also a non negligible influence on sound levels. Furthermore we note that μ^* and σ vary in a similar way regardless of the frequency and that σ is of the same order of magnitude than μ^*. This means that there are no interactions between variables and that the model's linearity assumption is valid for these 5 parameters.

4. Conclusion

In this study, the sensitivity analysis of a parabolic equation model was performed using the Morris method for a point source positioned 100m high above the ground. The relative influence of parameters known to have a strong influence on acoustic propagation (ground effects and atmospheric stratification) was determined at two points in space: a medium-range propagation case ($500m$) and a long-range propagation case ($1500m$) at $2m$ high. The conclusions of this paper are that wind profile and airflow resistivity are the most influential parameters for all third octave bands considered. Nevertheless, the influences of the other parameters (temperature vertical gradient and ground roughness) studied are even not negligible regarding the dispersion in SPL (dB).
Future work will allow other influence parameters to be taken into account, so that in the long term, the global variability and uncertainties associated with the emission-propagation-reception chain of sound will be quantified.

References

[1] (2004). NF EN 61400-11 Aérogénérateurs Partie 11 : techniques de mesure du bruit acoustique.

[2] Blaes, X. and Defourny, P. (2008). Characterizing Bidimensional Roughness of Agricultural Soil Surfaces for SAR Modeling. *IEEE Transactions on Geoscience and Remote Sensing*, 46(12):4050–4061.

[3] Blairon, N. (2002). *Effets de la topographie sur la propagation des ondes acoustiques dans l'atmosphère : modélisation avec l'équation parabolique et validation sur un site extérieur.* thesis, Ecully, Ecole centrale de Lyon.

[4] Blairon, N., Blanc-Benon, P., Bérengier, M., and Juvé, D. (2001). Calculation of sound propagation over non flat terrain using parabolic equation. volume 3, Rome, Italy.

[5] Blairon, N., Blanc-Benon, P., Bérengier, M., and Juvé, D. (2002). Outdoor sound propagation in complex environment: experimental validation of a PE approach. Grenoble, France.

[6] Blanc-Benon, P., Lihoreau, B., Pénelon, T., Gauvreau, B., Calmet, I., and Bérengier, M. (2004). Outdoor sound propagation in complex environments using the parabolic equation: A new PE code coupled with a micrometeorological code. Lake Morey, USA.

[7] Borgeaud, M. and Bellini, A. (1998). A database for electromagnetic scattering studies of bare soil surfaces. In *IGARSS '98. Sensing and Managing the Environment. 1998 IEEE International Geoscience and Remote Sensing. Symposium Proceedings. (Cat. No.98CH36174)*, volume 3, pages 1197–1199 vol.3.

[8] Bourlier, C., Pinel, N., and Kubické, G. (2013). *Method of Moments for 2D Scattering Problems: Basic Concepts and Applications.* John Wiley & Sons.

[9] Bowdler, D. and Leventhall, H. (2011). *Wind turbine noise.* Multi-Science Pub.

[10] Brelet, Y. (2008). *Diffusion des ondes électromagnétiques par une surface rugueuse monodimensionnelle sous incidences modérées et rasantes : application au domaine maritime.* thesis, Nantes.

[11] Brelet, Y. and Bourlier, C. (2008). Bistatic Scattering from a Sea-Like One-Dimensional Rough Surface with the Perturbation Theory in HF-VHF Band. In *IGARSS 2008 - 2008 IEEE International Geoscience and Remote Sensing Symposium*, volume 4, pages IV – 1137–IV – 1140.

[12] Chevret, P., Blanc-Benon, P., and Juvé, D. (1996). A numerical model for sound propagation through a turbulent atmosphere near the ground. *The Journal of the Acoustical Society of America*, 100(6):3587–3599.

[13] Collins, M. D. (1993). A split-step Padé solution for the parabolic equation method. *The Journal of the Acoustical Society of America*, 93(4):1736–1742.

[14] Craddock, J. N. and White, M. J. (1992). Sound propagation over a surface with varying impedance: A parabolic equation approach. *The Journal of the Acoustical Society of America*, 91(6):3184–3191.

[15] Davidson, M. W. J., Toan, T. L., Mattia, F., Satalino, G., Manninen, T., and Borgeaud, M. (2000). On the characterization of agricultural soil roughness for radar remote sensing studies. *IEEE Transactions on Geoscience and Remote Sensing*, 38(2):630–640.

[16] Embleton, T. F. W., Piercy, J. E., and Daigle, G. A. (1983). Effective flow resistivity of ground surfaces determined by acoustical measurements. *The Journal of the Acoustical Society of America*, 74(4):1239–1244.

[17] Faivre, R., Iooss, B., Mahévas, S., Makowski, D., and Monod, H. (2013). *Analyse de sensibilité et exploration de modèles: Application aux sciences de la nature et de l'environnement.* Editions Quae.

[18] Gauvreau, B., Bérengier, M., Blanc-Benon, P., and Depollier, C. (2002). Traffic noise prediction with the parabolic equation method: Validation of a split-step Padé approach in complex environments. *The Journal of the Acoustical Society of America*, 112(6):2680–2687.

[19] Iooss, B. (2011). Revue sur l'analyse de sensibilité globale de modèles numériques. *Journal de la Societe Française de Statistique*, 152(1):1–23.

[20] Lihoreau, B., Gauvreau, B., Bérengier, M., Blanc-Benon, P., and Calmet, I. (2006). Outdoor sound propagation modeling in realistic environments: Application of coupled parabolic and atmospheric models. *The Journal of the Acoustical Society of America*, 120(1):110–119.

[21] Miki, Y. (1990). Acoustical properties of porous materials-Modifications of Delany-Bazley models-. *Journal of the Acoustical Society of Japan (E)*, 11(1):19–24.

[22] Morris, M. D. (1991). Factorial Sampling Plans for Preliminary Computational Experiments. *Technometrics*, 33(2):161–174.

[23] Nicolas, J. and Berry, J. L. (1984). Propagation du son et effet de sol. *Revue d'Acoustique*, pages 191–200.

[24] Oerlemans, S., Sijtsma, P., and Méndez López, B. (2007). Location and quantification of noise sources on a wind turbine. *Journal of Sound and Vibration*, 299(4):869–883.

[25] Rasmussen, K. B. and Arranz, M. G. (1998). The insertion loss of screens under the influence of wind. *The Journal of the Acoustical Society of America*, 104(5):2692–2698.

[26] Saltelli, A., Chan, K., and Scott, E. M. (2000). *Sensitivity Analysis*. Wiley.

[27] Saltelli, A., Ratto, M., Andres, T., Campolongo, F., Cariboni, J., Gatelli, D., Saisana, M., and Tarantola, S. (2008). *Global Sensitivity Analysis: The Primer*. John Wiley & Sons.

INCEEUROPE

8th International Conference
on
Wind Turbine Noise
Lisbon – 12th to 14th June 2019

Session D

Propagation - 2

Organised by INCE-Europe
mail@inceeurope.org
www.inceeurope.org

Supported by Portuguese Acoustical Society

Sociedade
Portuguesa
de Acústica

INCE EUROPE

8th International Conference
on
Wind Turbine Noise
Lisbon – 12th to 14th June 2019

A proposal for the prediction of sound pressure levels due to wind turbine operation

Alice Elizabeth González: elizabet@fing.edu.uy
Pablo Gianoli Kovar
Luciana Olazábal Barrios

Summary

The prediction of environmental sound pressure levels related to the operation of wind turbines is still a concerning issue. During the last decade some different proposals have been made to enhance the results from the old prediction methods. We present the prediction model developed at the Environmental Engineering Department of the Faculty of Engineering (Universidad de la República, Uruguay), which has shown a good agreement with measured data at about ten different wind farms along the country. First of all, the atmospheric stability class according to Pasquill-Gifford and the wind velocity are taken into account for determining the acoustic power of the device. Then, a first module for describing noise generation along the blades due to turbulent phenomena (incoming wind turbulence and blades boundary layer release) is applied to compute sound pressure levels at 100 m far from the tower. Some coefficients have been experimentally obtained through wind tunnel tests. At last, a propagation module completes the prediction model to compute sound pressure levels at different distances from the tower. The propagation module considers not only geometric divergence but atmospheric absorption and turbulent dissipation (both phenomena depend on the frequency). The final sound pressure levels at a given reception point are obtained by superposing the sound pressure levels due to different wind turbines operation and the environmental noise due to the wind. Every computing is done in octave bands and the final results are expressed as L_{Aeq} values. The development of the prediction model has been done with measured data from 4 wind farms built on flat terrain. During calibration process some adjustment was needed for better prediction of noise propagation from wind farms built on complex terrain. This easy-to-use method has been programmed in Matlab®. The obtained results show almost 80 % cases within in ±3 dB range.

1. Introduction

One of the main problems that still make it difficult to predict the environmental sound pressure levels due to the operation of large wind turbines is the frequent disconnection between those who model aerodynamic phenomena and those who work in environmental acoustics.

In the development of this model, the Research Group on Noise Pollution of the Faculty of Engineering (UdelaR) has worked very closely to the Research Group on Wind Energy of the same Faculty. This second Group has been working on wind energy for more than 25 years. Actually, the strong growth of wind energy in Uruguay is closely related to their effort: in about only 10 years, the country has reached an installed wind energy power greater than 1500 MW. As Uruguay has about 3.400.000 inhabitants, what means less than 20 inh/km^2, it currently has one of the highest ratio of installed MW per capita in the world.

For developing our calculation procedure, the theoretical analysis of the different phenomena that are involved on noise generation, was complemented with tests at the University wind tunnel and, of course, with sound pressure levels measurements at some wind turbine farms in Uruguay.

This paper aims to present the main features of our prediction procedure and its results.

2. Noise sources

2.1 Mechanical noise

At the beginning of the development of large scale wind energy generation, there were many possibilities of improving the acoustic performance of the machines.

Currently, no significant problems are due to mechanical noises, which have been almost eliminated thanks to design improvement processes.

But aerodynamic noise generation during operation is inherent to wind turbines in nature. Their major acoustic emission is caused by the interaction between the wind and the blades. Most of these emissions occurs in low frequencies.

2.2 Aerodynamic noise

Aerodynamic noise generation at wind turbines can be classified into three types, according to the main process that causes the fluctuation in the pressure field (Van den Berg, 2003; Hernández Castellani et al., 2016):

I. The turbulence of the incoming wind, that causes a temporal variation of the pressure field around the blades.

II. The viscous forces in the boundary layer over the solid surfaces of the turbine, such as blades, tower and hub. Viscous forces on this layer are not negligible compared to the inertial forces (related to the medium air flow). The release of the boundary layer at the trailing edge causes a permanent releasing of eddies with negative gauge pressure at their cores.

III. Due to the power exchange between the wind and the rotor, two families of eddies linked to each blade are generated: one of them has helical motion and the other one is centred on the rotation axis and its length scale is about the length of the diameter of the rotor.

These phenomena are related to three different geometric scales (Van den Berg, 2003; Cataldo, 2016).

I. Macroscale: it is the scale related to the largest vortexes. If U, L and T are the scales of velocity, length and time associated, the Reynolds number of the biggest vortexes is the same as for the main flow.

II. Intermediate scale: it includes lower scales than the macroscale ones; there is still no power dissipation. The range of scales included here is called "inertial range".

III. Microscale: it is the smallest scale, in which the energy dissipation occurs. Unlike what happens in the macroscale, the smallest eddies are isotropic, as if the flow has "forgotten" where it comes from.

The turbulent cascade hypothesis is then to be considered. According to it, the larger vortexes are dissipating in smaller scale vortexes with increased kinetic energy. Thus, there is a spatial scale of vortexes that cannot continue to transfer power to another smaller scale of them. At this point, the cascade ends and the energy from the last vortex is finally dissipated. The smallest vortexes scale is known as the Kolmogorov scale. The so-called Kolmogorov frequency or dissipation frequency is the passing frequency or generation frequency of the smallest scale of eddies, which scale is the Kolmogorov scale (Cataldo, 2016; Hernández Castellani, 2016).

According to their frequency and energy, the released eddies can produce audible phenomena, i.e. they can become noise sources.

3. Noise propagation

Two phenomena are intended to be the most important ones during noise propagation: the atmospheric absorption and the turbulence energy dissipation.

3.1 Atmospheric absorption

To take into account the atmospheric absorption, the calculation method of ISO 9613 Part 1 has been used. For very low frequencies, it has been supposed to be negligible.

3.2 Turbulence dissipation processes

The released eddies are related to a certain energy level and a length scale. They can propagate along great distances while the turbulent cascade occurs (Cataldo, 2016).

We studied the scale of distances from the source at what the cascade is expected to stop, which is closely related to atmospheric stability. Figure 1 shows the two extreme cases of atmospheric stability.

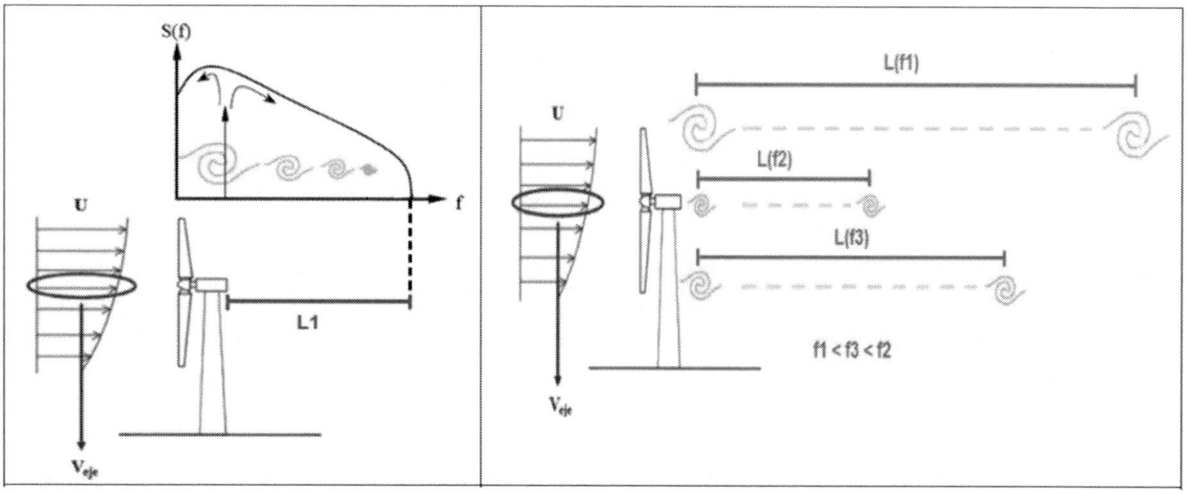

Figure 1: Atmospheric conditions for propagation (at left, strong instability; at right, extreme stability) From: Cataldo, 2016

As we have just integrated the atmospheric stability in the noise generation part of our prediction model (González et al., 2016a and 2016b), we could work here considering that the scale of distances mainly depends on the incoming wind velocity and the frequency of the released eddies.

As it is expected, we met different scales for different conditions. We grouped them into four classes according to wind velocity (lower or greater than 6.5 m/s) and distance to the wind turbine (lower or greater than 750 m). As a consequence, we obtained a set of depletion exponents for each one of these classes. We could also simplified the two sets of values for long distances, as they were equal or similar for most frequencies.

4. Our noise prediction proposal

As most of the research groups around the world did, we began working to understand ISO 9613:2 and CONCAWE Report 4/81 (Manning, 1981) limitations for the case of wind turbine noise (Van den Berg, 2006; González et al., 2011). Then, we begin working to overcome their weak faces.

The proposed model consists of two modules: the first one is related to aerodynamic noise generation; it allows to compute the sound pressure levels at 100 m from the wind turbine. The other module works with sound propagation; its input is the output of the first module and it propagates the levels of sound pressure up to the desired point.

The first module allows to calculate the acoustic power levels (in normalized third octave bands) associated with the aerodynamic phenomena of interaction between the wind and the blades. To do this, it takes the wind speed u in m/s from input data. If u is given at a height other than that of the wind turbine hub, the module itself computes the wind velocity at the hub height assuming that there is an atmospheric thermal inversion; the conversion is made following a potential curve with an exponent of 0.65 (Van den Berg, 2006).

Then, the module assumes that each of the blades is divided into n slices that behave as n point sources and for each of them calculates the sound emission associated with the noise of the incoming turbulence (noise at the attack edge), with the release of the boundary layer at the trailing edge and with the passage of the blade in front of the tower (these computing procedures are described in Deambrosi Papini et al., 2016; González et al, 2016a and 2016b; Hernández Castellani et al., 2016). The emission is propagated up to a distance of 100 m from the tower, supposing each segment of the blade is a point source; the contribution of every part of the blades are added to obtain the sound pressure levels in third octave bands that will serve as input data for the propagation module. For this computing, the values of drag and lift coefficients have been obtained from a set of tests at the wind tunnel of the Universidad de la República (Olazábal et al., 2018); these tests have been useful for improve the prediction results.

The propagation module takes the output of the generation module as an input and it propagates these sound pressure levels considering geometric divergence and atmospheric absorption. The absorption coefficients in third octave bands are obtained according to ISO 9613: 1.

The decay by geometric divergence is performed by using a depletion law in which the exponent n is not constant for all the bands, but n = n(f, d, u). The exponents are related to the distance scale at which eddies are expected to dissipate all their turbulence energy. The values of n(f, d, u) are given in octave bands according to whether the calculation distances are less than or greater than 750 m and that the wind speed at the hub height is less than or greater than 6,5 m/s (see table 1).

For distances greater than 750 m, the expected behaviour of noise is related to the frequency of the eddies but it is rather independent from the wind velocity. This means that the shorter distance scale phenomena are expected to have mostly ended.

For distances shorter than 750 m, the propagation phenomena are still conditioned by the wind velocity. Then the fact of not working within the usual general hypotheses (in particular because the phenomena under consideration are not adiabatic but dissipative (Núñez Pereira, 2013)) supports the position of assuming n as a variable. The eddies exhibit different behaviour for different wind velocities: they dissolve quicker for calmer situations but they can travel further for higher wind speeds. This is reflected by the values of n that our prediction model takes into account.

Table 1. Values of the exponent n = n(f, d, u) according to distance to the source and wind velocity

	16	31,5	63	125	250	500	1000	2000	4000	8000
d less than 750 m, u lower than 6,5 m/s	0,37	0,01	0,05	0,66	1,60	1,67	1,94	1,61	0,85	0,43
d less than 750 m, u higher than 6,5 m/s	0,52	0,02	0,01	0,22	0,93	0,79	1,04	1,07	1,06	0,65
d greater than 750 m	0,45	0,01	0,06	0,72	1,58	1,71	2,03	1,60	0,67	0,34

5. Results

The model was calibrated, validated and verified with measurements made in four wind farms. In the calibration and validation stages, a set of 59 data was used, while the subsequent verification included 49 cases from ten wind farms. The results obtained in the verification of the performance of the model are presented in Tables 2 to 4.

Table 2. Quality of results according to distance to the wind turbine

	Number of cases	Cases in ±3 dB	%
d shorter than 750 m	30	24	80.0 %
d greater than 750 m	18	14	77.8 %
TOTAL	48	38	79.2 %

Table 3. Quality of results according to wind velocity at the hub height

	Number of cases	Cases in ±3 dB	%
u lower than 6.5 m/s	12	10	83.3 %
u higher tan 6.5 m/s	36	28	77.8 %
TOTAL	48	38	79.2 %

Table 4. Quality of results according to distance to the wind turbine and wind velocity at the hub height

	Number of cases	Cases in ±3 dB	%
d shorter than 750 m, u less than 6.5 m	8	7	87.5 %
d shorter than 750 m, u greater than 6.5 m	22	17	77.3 %
d further than 750 m, u less than 6.5 m	4	3	75.0 %
d further than 750 m, u greater than 6.5 m	14	11	78.6 %
TOTAL	48	38	79.2 %

As important as the percentage of right predictions is to state that not only the levels in scale A are predicted in a reasonably adjusted way, but particularly that the spectra obtained with the proposed model are also rightly adjusted to the measured ones.

6. Conclusions

We have presented a proposal for the prediction of sound pressure levels related to the operation of large wind turbines. Our innovative point of view is to treat the exponent n used to compute the geometric depletion as a variable value depending on the distance to the wind turbine and on the wind velocity. The performance of this proposal is almost good, with about 80 % of the predicted results into an interval of ±3 dB from the measured values. It has an easy implementation and the needed data are easy to be obtained (geometrical and meteorological information).

Acknowledgments

This paper is the result of many years of research. The research team has been supported by funds from cooperation with the National Program of Wind Energy, ANII_FSE_10942 Project and CSIC I+D Research Groups. Many people have taken part of this team in its different steps; we thank all of them for taking part in this research. We are especially thankful to Dr. Eng. José Cataldo Ottieri for trusting on our team, for his permanent support and for generously sharing with us his wide and deep knowledge.

References

Cataldo, José (2016). *"Introducción a la Turbulencia"* (*Introduction to turbulence*). Postgraduate course notes. Universidad de la República, Uruguay.

Deambrosi Papini, Matteo; Hernández Castellani, Matías; Cataldo, José; González, Alice Elizabeth. *Prediction of environmental sound pressure levels due to large wind turbines.* ICA 2016, Buenos Aires, Argentina. September 2016.

González, Alice Elizabeth; Gianoli Kovar, Pablo; Deambrosi Papini, Matteo; Hernández Castellani, Matías; Paz Urban, Martín (2016a). Impacto acústico de aerogeneradores de gran porte. Modelo predictivo de niveles ambientales de presión sonora. Síntesis de Resultados del Proyecto ANII FSE_2013_1_10942 (*Acoustic impact of large wind turbines. Predictive model of environmental levels of sound pressure. Summary of Results of Project ANII FSE_2013_1_10942*). Montevideo, Uruguay, 2016.

González, Alice Elizabeth; Gianoli Kovar, Pablo; Deambrosi Papini, Matteo; Hernández Castellani, Matías; Paz Urban, Martín (2016b). *Environmental noise due to large wind turbines: what we have learnt.* ICA 2016, Buenos Aires, Argentina. September 2016.

González, Alice Elizabeth; Rezzano Tizze, Nicolás; Bianchi Falco, Fabiana. Algunas limitaciones de la Norma ISO 9613 – Parte 2 para el estudio de propagación de ruido de aerogeneradores de gran porte (*Some limitations of ISO 9613 - Part 2 for the study of noise propagation of large wind turbines*). Reunión Regional de Acústica, 11 y 12 de octubre de 2011, Montevideo, Uruguay.

Hernández Castellani, Matías; Paz Urban, Martín; Deambrosi Papini, Matteo; González, Alice Elizabeth (2016). *Propagation phenomena associated with noise due to the operation of large-sized wind turbines.* ICA 2016, Buenos Aires, Argentina. September 2016.

International Standard Organization. International Standard 9613. *Attenuation of sound during propagation outdoors- Part 1: Calculation of the absorption of sound by the atmosphere.*1993.

International Standard Organization. International Standard 9613. *Attenuation of sound during propagation outdoors- Part 2: General method of calculation.* 1996.

Núñez Pereira, Ismael. Elementos de Acústica (*Acoustic Elements*), Environmental Engineering Magister Program, UdelaR, Faculty of Engineering, Universidad de la República, Montevideo, Uruguay. 2013.

Manning, C. J., *The Propagation of Noise from Petroleum and Petrochemical Complexes to Neighboring Communities*, Report 4/81, CONCAWE, 1981.

Olazábal Barrios, Luciana; Gianoli Kovar, Pablo; Cataldo Ottieri, José; González Fernández, Alice Elizabeth. Estudio de generación de ruido asociado a la interacción entre un perfil aerodinámico y un flujo turbulento (*Noise generation study associated with the interaction between an aerodynamic profile and a turbulent flow*). XI Congreso Iberoamericano de Acústica; X Congreso Ibérico de Acústica; 49º Congreso Español de Acústica - TECNIACUSTICA'18. Cádiz, Spain, 2018.

Van den Berg, G. P. Effects of the wind profile at night on wind turbine sound. Journal of Sound and Vibration, doi:10.1016/j.jsv.2003.09.050, 2003.

Van der Berg, Godefridus Petrus. The sound of high winds: the effect of atmospheric stability on wind turbine sound and microphone noise (2006). Doctoral Thesis from the University of Groningen, Netherlands. Mayo, 2006.

INCE*EUROPE*

8th International Conference
on
Wind Turbine Noise
Lisbon – 12th to 14th June 2019

Low-level wind maxima and their effect on wind turbine sound

Kristina Conrady
Uppsala University, Uppsala, Sweden

Karl Bolin
Marcus Wallenberg Laboratory, Kungliga Tekniska Högskolan, Stockholm, Sweden

Anna Sjöblom
Uppsala University, Uppsala, Sweden

Anna Rutgersson
Uppsala University, Uppsala, Sweden

Summary

Refraction is an important factor for sound propagation from wind turbines. The simplified concept for the effects of refraction based on the logarithmic wind profile or other approaches where the wind speed is continuously increasing with height, is often not applicable in cold climates. The stably stratified boundary layer often features low-level wind maxima. This study was performed to analyse the effect of wind maxima below hub height on sound propagating from wind turbines. Long-term meteorological measurements in the vicinity of a wind farm in northern Sweden are used to identify low-level wind maxima. To analyse the effect of low-level wind maxima on sound propagation, sound measurements were conducted simultaneously to the meteorological measurements.

1. Introduction

Outdoor sound propagation, as wind turbine (WT) sound, is affected by meteorological conditions. One of the most important meteorological effects is refraction [6]. Refraction mainly occurs due to wind speed and temperature gradients and is characterised by the bending of the path of a sound ray towards regions of lower effective sound speed. A simplified concept illustrates the refraction pattern due to vertical wind speed gradients [e.g., 5, 6, 8]: refraction leads to downward bending and higher sound levels downwind of a sound source, and upwind of a sound source, upward bending occurs and results in lower sound levels (Fig. 1a). However, the stably stratified atmospheric boundary layer (SBL) is a common feature in cold climates and makes this simplified concept unapplicable. In the SBL, low-level wind maxima (LLWM) are often present. The reasons for the formation of LLWM are manifold [e.g., 7, 10].

LLWM above the WTs hub height were shown to increase the effect of downward refraction [1, 4, 9]. The maximum in wind speed above the sound source traps the sound waves between the LLWM and the surface, and therefore increases the level of WT sound near the surface at a distance much greater than the altitude of the LLWM [2].

LLWM can also occur below hub height, especially in very stable conditions. In the light of the ongoing development towards higher WTs, the probability for LLWM increases. Since in this case, the sound source is above the LLWM, a complete or partial trapping of the sound waves above the LLWM can be expected (Fig. 1b), which is contrary to the situation discussed above. Therefore, lower levels of WT sound close to the surface are assumed for downwind conditions. In order to test this assumption, long-term meteorological and acoustic measurements have been conducted in the vicinity of a wind farm in northern Sweden.

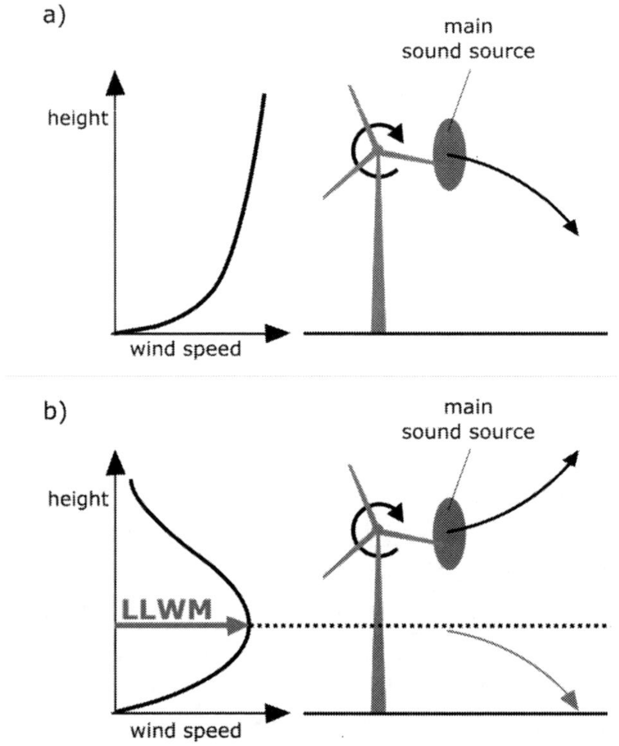

Fig. 1 Schematic illustration of the refraction patterns downwind of a wind turbine for a) a logarithmic wind profile and b) a low-level wind maxima (LLWM) below hub height [based on 3].

2. Measurements

Between August 2016 and June 2017, measurements were conducted in the vicinity of a wind farm in northern Sweden (Fig. 2). The surrounding landscape is undulating with hills and ridges up to 600 m above sea level. The sparsely populated area is predominantly covered by forests but several rivers, streams and lakes can also be found. An acoustic station was installed in a forest approximately 200 m away from the closest houses of a small settlement and ca. 500 m from a road, which bears little traffic. The wind farm is located in the northwest direction of the acoustic station on Nådagubbliden, which is a ridge with maximum elevations of 460 to 510 m above sea level. The wind farm consists of 22 Vestas V90 (2 MW) WTs with a hub height of 105 m. Meteorological measurements were conducted on a mast approximately 100 m southeast of the acoustic station and on a meteorological tower, close to the centre of the wind farm.

Wind speed and direction were measured at 1.3 m, 2.3 m and 4.6 m above the snow-free ground on a meteorological mast, which was placed in the middle of a field. The closest buildings and higher vegetation, i.e., trees and bushes, were around 50 m away to ensure as undisturbed measurement conditions as possible. Additionally, wind measurements at 60 m, 85.5 m, and 105 m at a meteorological tower were provided by the operator of the wind farm. 10-min averages of the equivalent sound pressure level, L_{Aeq}, were collected using a Norsonic NOR140 sound level meter and a Nor1214 outdoor microphone equipped with a rain hood and a dust mesh. The microphone was mounted 1.5 m above the snow-free ground. Simultaneous with the meteorological and the acoustic measurements, the wind farm operator provided 10-min averages of the rotational frequency of each wind turbine. All measurements were averaged over the same 10-min intervals.

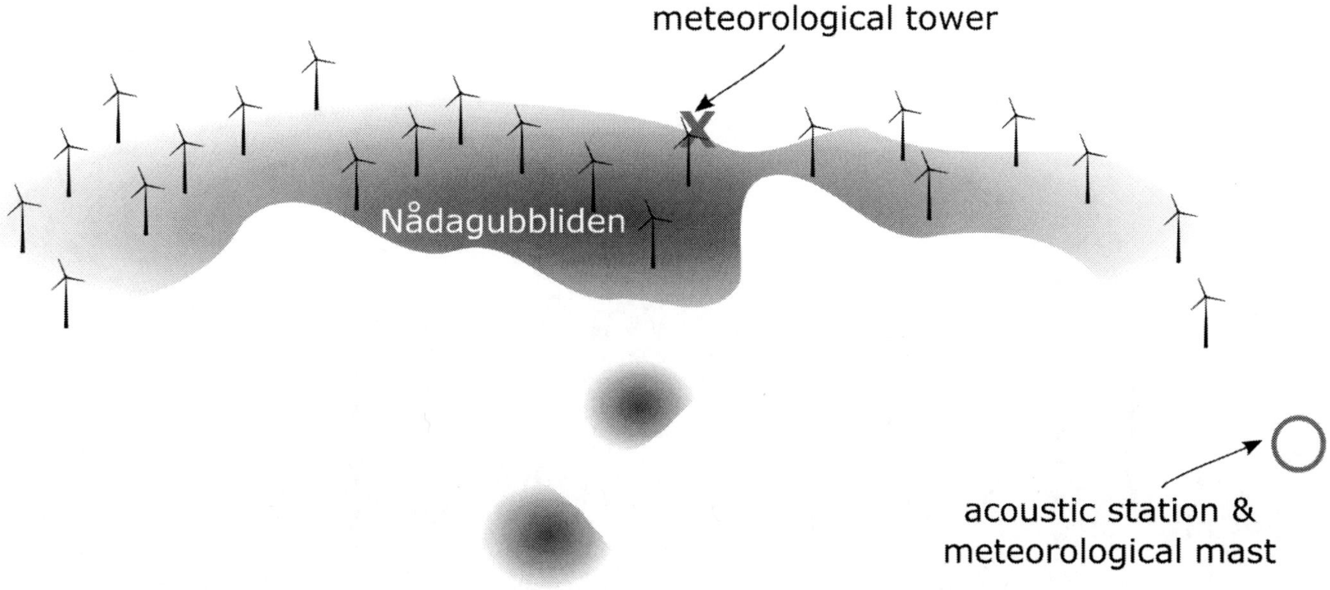

Fig. 2 Schematic illustration of the measuring site with the acoustic station and the meteorological mast (circle) and the meteorological tower (x).

3. Methods

Two types of LLWM have been chosen to analyse their effect on WT sound:

- **Type I**: wind speed is greatest at 85.5 m, second greatest at 105 m (hub height), and lowest at 60 m (tip of the lowermost blade) and
- **Type II**: wind speed is greatest at 85.5 m, second greatest at 60 m (tip of the lowermost blade), and lowest at 105 m (hub height).

Type I was expected to has a smaller effect on WT sound than Type II due to the stronger wind speed gradient between 85.5 m and 105 m for Type II. Two minimum differences, Δu, between the largest and second largest wind speed are chosen to analyse the impact of the strength of a LLWM: $\Delta u = 0.2\,\mathrm{m\,s^{-1}}$ and $\Delta u = 0.5\,\mathrm{m\,s^{-1}}$. Note that acoustic data was only selected for further analyses if the median of the rotational frequency of all WTs, RF, was $\geq 14.0\,\mathrm{rpm}$. This criterion was applied to ensure that WT sound was possible to be detected at the acoustic station if no LLWM below hub height was present.

4. Results and discussion

The occurrence of L_{Aeq} for occasions without LLWM and with LLWM (Type I or Type II) for downwind conditions and $RF \geq 14.0\,\mathrm{rpm}$ are shown in Figure 3, where a) shows LLWM meeting $\Delta u = 0.2\,\mathrm{m\,s^{-1}}$ and b) shows LLWM with an increased strength ($\Delta u = 0.5\,\mathrm{m\,s^{-1}}$). Compared to occurrences without LLWM, Type I mainly increases the occurrence of $L_{Aeq} < 32\,\mathrm{dBA}$, while Type II increases the occurrence of $L_{Aeq} \approx 27\,\mathrm{dBA}$ (Fig. 3a). $L_{Aeq} > 40\,\mathrm{dBA}$ seldom occur for either of the LLWM types. For LLWM with an increased strength ($\Delta u = 0.5\,\mathrm{m\,s^{-1}}$, Fig. 3b), the effect of LLWM on L_{Aeq} increases, especially for Type I.

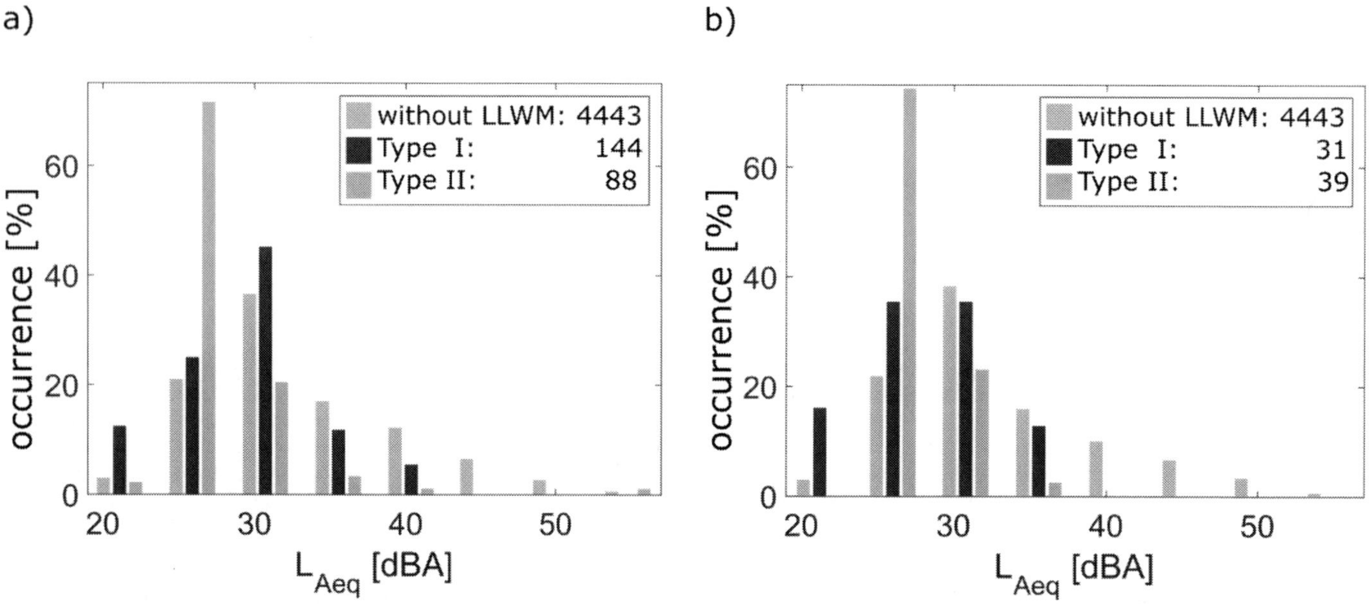

Fig. 3 Occurrence of L_{Aeq} for occasions without LLWM and with LLWM – Type I or Type II – for downwind conditions, $RF \geq 14.0\,\mathrm{rpm}$ and a) $\Delta u = 0.2\,\mathrm{m\,s^{-1}}$ or b) $\Delta u = 0.5\,\mathrm{m\,s^{-1}}$ [based on 3]. The numbers in the legends refer to the amount of data of each category.

The more pronounced effect of Type II compared to Type I can be explained by the definition of the two LLWM types. For Type II, the wind speed at the uppermost level is the lowest of the three levels, therefore, the gradient between the two highest levels is presumably larger than the one for Type I. Thus, more effective upward refraction is caused, leading to a more effective trapping of WT sound above the LLWM.

5. Conclusions

Analyses of long-term acoustic and meteorological measurements in northern Sweden indicate that LLWM below hub height reduce WT sound in the downwind direction. The effect is larger if stronger LLWM ($\Delta u = 0.5\,\mathrm{m\,s^{-1}}$) are present. Both LLWM types lead to lower L_{Aeq} when compared to occurrences without LLWM. The effect is larger if the lowest of the three wind speeds was measured at the uppermost level (Type II), which can be related to a more effective upward refraction above the LLWM. Since LLWM decrease the WT sound close to the surface in the downwind direction, the occurrence of LLWM can improve the conditions for residents. Furthermore, an increase in the height of the WTs at this site would increase the potential of LLWM below hub height.

Acknowledgments

This study was financially supported by the Swedish Energy Agency (project 40969-1). The authors are also grateful to Vasa Vind, especially to Per Olof Renling for sharing operational and meteorological data. We would like to thank Dr. Conny Larsson, who was the initial leader of the project.

References

[1] Bolin, K., Boué, M., and Karasalo, I. (2009). Long range sound propagation over a sea surface. *The Journal of the Acoustical Society of America*, 126 5:2191–7.

[2] Boué, M. (2007). Long-range sound propagation over the sea with application to wind turbine noise. Technical report, Swedish Energy Agency. Final report for the Swedish Energy Agency project 21597-3(TRANS), 68 pp.

[3] Conrady, K., Bolin, K., Sjöblom, A., and Rutgersson, A. (2019). Impact of low-level wind maxima on wind turbine sound propagation. in preparation.

[4] Johansson, L. (2003). Sound propagation around off-shore wind turbines – long-range parabolic equation calculations for Baltic Sea conditions. Licentiate Thesis, KTH, Stockholm.

[5] Lamancusa, J. S. (2008). *Noise Control – Outdoor Sound Propagation*. Penn State University, Pennsylvania. 19 pp.

[6] Larsson, C. and Israelsson, S. (1991). Effects of meteorological conditions and source height on sound propagation near the ground. *Applied Acoustics*, 33(2):109–121.

[7] Renfrew, I. A. and Anderson, P. S. (2006). Profiles of katabatic flows in summer and winter over Coats Land, Antarctica. *Quaterly Journal of the Royal Meteorological Society*, 132(616):779–802.

[8] Salomons, E. M. (2001). *Computational Atmospheric Acoustics*. Springer Netherlands, Dordrecht. 335 pp.

[9] Törnblom, K. (2006). Thermally driven wind modification in coastal areas and its influence on sound propagation with application to wind power. Licentiate thesis, Department of Earth Sciences, Uppsala University.

[10] Tuononen, M., Sinclair, V. A., and Vihma, T. (2015). A climatology of low-level jets in the mid-latitudes and polar regions of the Northern Hemisphere. *Atmospheric Science Letters*, 16(4):492–499.

INCE*EUROPE*

8th International Conference

on

Wind Turbine Noise

Lisbon – 12th to 14th June 2019

Wind turbine noise propagation in flat terrain for wind farm layout optimization frameworks

Matias Sessarego
Department of Wind Energy, Technical University of Denmark, Nils Koppels Allé, 2800 Kongens Lyngby, Denmark

Wen Zhong Shen
Department of Wind Energy, Technical University of Denmark, Nils Koppels Allé, 2800 Kongens Lyngby, Denmark

Emre Barlas
Goldwind Denmark, Delta 4, 8382 Hinnerup, Denmark

Summary

This paper describes the wind turbine noise propagation in flat terrain for use in wind farm layout optimization frameworks. Large-eddy simulations of a single wind turbine in flat terrain at varying wind speeds, shear and turbulence levels are performed. The wind turbine is modeled using an actuator line approach, while the wind turbine noise propagation is computed using the Technical University of Denmark's WindSTAR-Pro (Wind turbine Simulation Tool for AeRodynamic noise Propagation). The wind turbine noise propagation is computed in a quasi-three-dimensional manner by using a two-dimensional (2D) parabolic equation (PE) model at numerous 2D-planes around the wind turbine. In this study, the 2D, wide-angle, Crank-Nicholson PE model is used. The relative sound pressure level obtained from WindSTAR-Pro around the wind turbine is computed for varying wind speeds, shear and turbulence levels. Wind flow velocity and relative sound pressure levels for selected wind conditions are shown. The end goal is to create a noise propagation database that can be used as a lookup table in wind farm layout optimization frameworks to mitigate the noise impact in the development of new wind farms.

1. Introduction

According to the Nordic Energy Technology Perspectives 2016 [1], a five-fold increase of wind energy generation is proposed in the Nordic countries by 2050 and two-thirds, i.e., 2/3, of Nordic wind energy generation will be placed onshore. For the people who may be affected, such as farmers and people living nearby wind turbines, the large amount of onshore wind penetration emphasizes the importance of public acceptance. One of the main obstacles in the public acceptance of wind energy is wind turbine noise [2]. Many studies have been carried out on the topic, e.g. the human response to wind turbine noise as well as the reasons for the higher annoyance of wind turbine noise compared to other noise sources [3, 4]. Consequently, regulations have been put forth, such as the Danish wind turbine noise regulation [5] that limits the noise impact from wind turbines. Such regulations, however, may have an adverse effect on the energy yield of wind farms and limit the number of new wind farms being developed on land. Therefore, there is a need to develop wind farm design tools that include the noise emission to mitigate wind farm noise without compromising on energy yield and onshore wind farm development.

The following sections describe a wind turbine flow solver coupled to a noise propagation tool, which is then used for modeling wind turbine noise propagation in flat terrain for use in wind farm layout optimization frameworks. The relative sound pressure level (SPL) is computed for different inflow cases, i.e., varying wind speeds, shear and turbulence levels. The entire process of simulating the flow and noise is automated in a way that it becomes easy to generate a large number of relative SPL results for the different inflow cases. The end goal is to create a noise propagation database that can be used as a lookup table in wind farm layout optimization frameworks, e.g. [6], to mitigate the noise impact in the development of new wind farms. Instead of the table-lookup, a surrogate model could be constructed based on the set of simulated inflow cases and incorporated in the framework as well. Such an approach was used for quick assessment of site-specific lifetime fatigue loads of wind turbines [7, 8]. Nevertheless, the focus of the current article is in the simulation of the individual inflow cases rather than the structure of the database and its implementation in the framework.

In the current article, the noise propagation database within a wind farm layout optimization context is kept as general as possible, thus noise generation is not considered. For the moment, the idea is to have the noise generation calculations performed at each iteration in the wind farm layout optimization framework instead, since noise generation calculations can be calculated relatively quickly using engineering models such as classic BPM [9] or Amiet's model [10]. In the near future, noise generation may be incorporated into the database. The primary purpose of the noise propagation database is to include higher-order propagation effects to the noise emission in an efficient manner. The integration of the noise generation and propagation tools in a wind plant layout optimization framework is currently under development.

Last, noise is generally an onshore problem, which is commonly over non-flat or complex terrain. Since the noise database is based on flat terrain, there is an inconsistency if the database is used for wind turbine layout optimization on complex terrain. In this scenario, the database should be treated as an approximation. Still, a database approach strictly for complex terrain may not be possible because of the infinite amount of different terrain topographies which must be considered. It is assumed that the user of the wind turbine layout optimization framework is aware of the inconsistency when analyzing the final results for complex terrain scenarios.

The article is structured into five sections. Section 2 describes the wind turbine flow solver that provides the necessary flow input for the noise propagation model described in Section 3. Wind turbine noise propagation modeling under different inflow scenarios is described in Section 4. Finally, conclusions are provided in Section 5.

2. Wind turbine flow solver

Before the noise propagation computations are carried out, flow simulations are performed to provide the flow input to the noise propagation model. This section describes the simulation tool and models used to generate the flow inputs in two subsections. Subsection 2.1 describes the general purpose flow solver and subsection 2.2 describes the actuator line model used to represent the wind turbine within the flow domain. Subsection 2.3 describes the computational setup to execute the flow solver to obtain the flow fields.

2.1. Flow field

The flow is governed by the Navier-Stokes equations and simulations are performed using a computational fluid dynamics (CFD) solver typically based on either the Reynolds-Averaged Navier-Stokes (RANS) or Large-Eddy Simulation (LES). LES solvers are approximately three orders more computationally expensive than RANS solvers [11], since the large scale turbulence is resolved on a grid and only the smaller scale turbulence is modeled, e.g. using an eddy viscosity based sub-grid scale (SGS) model. RANS solvers are generally performed in a steady-state manner where all scales of turbulence are modeled [12]. The different turbulence scales in the flow are relevant because they are responsible for the noise generation in turbulent flow induced noise and have an impact on noise propagation. Flow fields from LES computations allows the noise propagation model to capture the effects of noise from some of the different turbulent scales.

The in-house EllipSys3D [13, 14] code developed at the Technical University of Denmark (DTU) is employed in the current work for generating the flow fields. EllipSys3D is a general purpose flow solver based on a structured grid topology, with a multi-block and cell-centered finite volume discretization. The Navier-Stokes equations are solved either steady or unsteady using the pressure-velocity coupling technique where the predictor-corrector method is used. In the predictor step, a second-order backward differentiation scheme is used as time discretization and a second-order central difference scheme is used as spatial discretization. The convective terms are discretized by the QUICK upwind scheme. Improved Rhie-Chow interpolation [15] is used in the corrector step to avoid numerical oscillations from the velocity-pressure decoupling. Besides the SIMPLE algorithm, the improved SIMPLEC scheme for collocated grids [16] is also implemented with the advantage that the solution is independent of the relaxation value. The EllipSys3D code is programmed with a multi-block topology and is parallelized using Message Passage Interface (MPI). Both the RANS and LES techniques are implemented in EllipSys3D, however only the filtered Navier-Stokes LES equations are being solved numerically in the current study. Last, only neutral atmospheric stability conditions are considered in the flow simulations.

2.2. Actuator line model

To include the effects of the wind turbine wake and unsteady moving sources on noise propagation, a wind turbine rotor model is required in the wind turbine flow solver. The actuator line (AL) technique as described in [17] is employed in the current work, where the wind turbine rotor blades are represented by rotating lines. By using the AL technique within EllipSys3D, complex flow

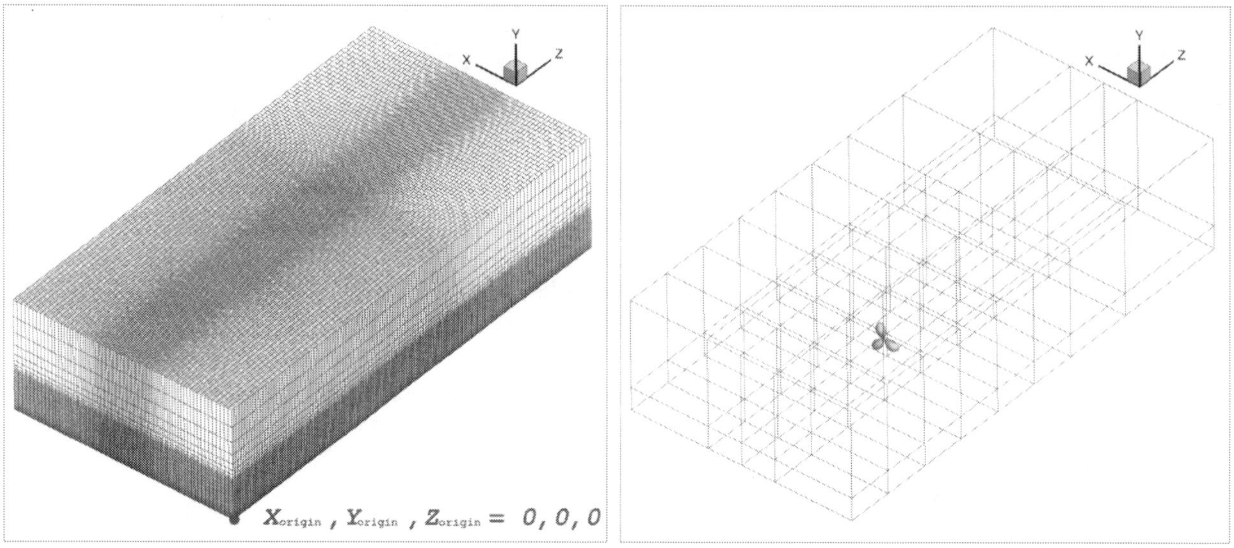

(a) (b)

Fig. 1 (a) Computational grid and (b) block structure used in the AL-LES computations. The three green spheroids in (b) represents the location of the actuator lines in the domain.

conditions on the wind turbine can be modeled, such as turbulent inflow, wind shear and yaw, etc. To compute the flow field over the wind turbine blades, a volume body force is added to the momentum equation. The body force is computed iteratively with the blade element approach with tabulated airfoil lift and drag data, e.g. lift (C_L) and drag (C_D) coefficients. The AL approach is coupled with the in-house developed aero-elastic code FLEX5 [18, 19]. The coupling with FLEX5 allows the addition of the wind turbine pitch and rotor RPM controllers as well as the flexibility of the turbine components, e.g. blades, tower, shaft, etc., within the simulation. However, for the purpose of this study, wind turbine control and flexibility are disabled. The focus is primarily on the noise propagation effects from different inflow conditions. The NREL 5 MW wind turbine [20] with a rotor diameter of 126 m and a hub height of 90 m is employed as the test turbine in the current study.

2.3. Flow solver setup

Figure 1 depicts the (a) computational grid and (b) block structure used in the flow-solver computations. In the flow solver setup, the mesh is a rectangular box with dimensions: width × height × length = $X_{\text{length}} \times Y_{\text{length}} \times Z_{\text{length}}$ = 1000 m × 500 m × 2000 m. The wind turbine rotor center is placed at X_{rotor} = 500 m, $Y_{\text{rotor}} = Y_H$ = 90 m, and Z_{rotor} = 800 m, where the origin of the mesh, X_{origin} = 0 m, Y_{origin} = 0 m, and Z_{origin} = 0 m, is defined in the bottom corner of the rectangular box as shown in Figure 1(a). The three green spheroids in Figure 1(b) represents the location of the rotor, or actuator lines, in the domain.

The mesh is comprised of 4 (horizontal) × 2 (vertical) × 8 (longitudinal) = 64 blocks with each block having 48^3 grid cells. The dimensions of all the 64 blocks are shown as black lines in Figure 1(b). Given an amount of computational resources to work with, since the amount of computational resources is limited, the computational mesh is designed such that there are more grid points upstream and downstream of the rotor location. Therefore, the sizes of the blocks are smaller in

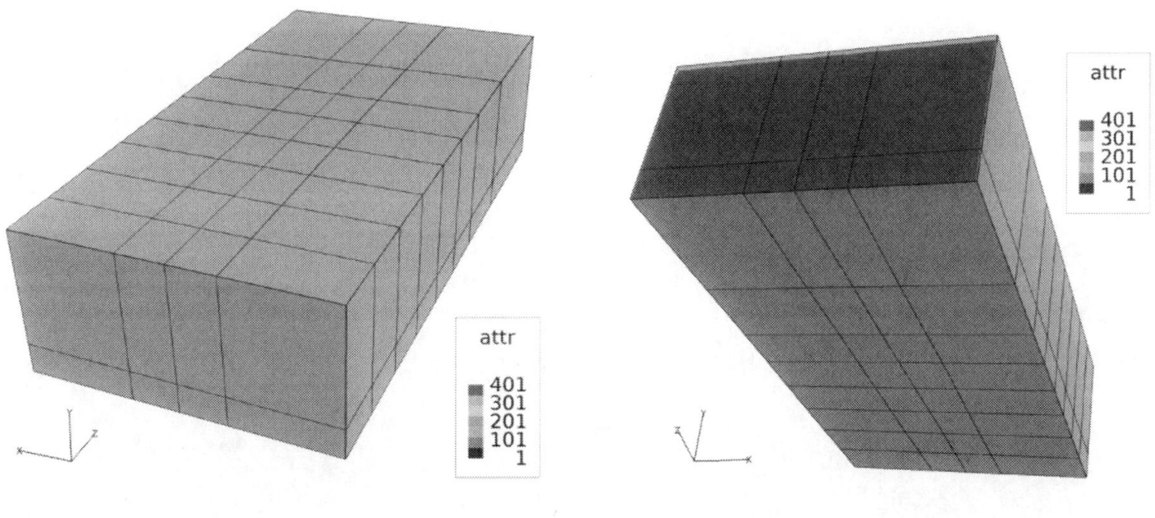

(a) (b)

Fig. 2 Two views of the boundary conditions applied on the mesh from the (a) front-top and (b) rear-bottom. No-slip, inlet, far-field, and outlet boundary conditions or attributes (attr) are color-coded as 101 (blue), 201 (green), 301 (orange), and 401 (red), respectively.

these regions as a means to increase the grid density. Upstream and downstream of the rotor, the velocity gradients will be the greatest and therefore the mesh should be refined accordingly. Further away from the rotor and towards the boundary of the mesh, a coarser grid is used to minimize the computational cost. There are approximately 50 (vertical) × 42 (horizontal) grid cells on the rotor swept area, which is similar to the number of grid cells used in Refs [18, 21]. The minimum/maximum grid cell lengths in the X, Y, and Z directions in the mesh are approximately 3.0 m/9.3 m, 1.4 m/40.0 m, and 4.2 m/11.4 m, respectively.

Figure 2 depicts two views of the boundary conditions applied on the mesh from the (a) front-top and (b) rear-bottom. No-slip, inlet, far-field, and outlet boundary conditions or attributes (attr) are color-coded as 101 (blue), 201 (green), 301 (orange), and 401 (red), respectively. The velocities for the inflow and far-field boundary conditions are prescribed. The fluid velocity for the no-slip boundary condition is zero.

Synthetic inflow turbulence is simulated by prescribing a plane, 1000 m (horizontal) × 500 m (vertical) in size, where the bottom-right corner of the plane lies at the origin of the mesh, i.e., $X_{\text{origin}} = 0$, $Y_{\text{origin}} = 0$, and $Z_{\text{origin}} = 0$. In other words, synthetic inflow turbulence is prescribed on the entire front-side of the mesh where the inlet boundary condition is also prescribed, see 201-green in Figure 2(a). Turbulence boxes are generated as a pre-processing step using the Mann model [22, 23] and slices from the box are gradually fed into the plane during the simulation. To capture the effect of each slice from the turbulence box in the AL-LES simulation, the resolution of the turbulence box should be slightly coarser than or equal to the resolution of the computational grid. A coarse resolution of approximately 8 meters/point (horizontal) × 8 meters/point (vertical) × 4.9 meters/point (longitudinal) was selected for the turbulence box. Due to the coarseness of the turbulence box, smaller turbulence scales that one could get from a finer turbulence box will not contribute to the flow solution. A finer turbulence box resolution will be used in the near

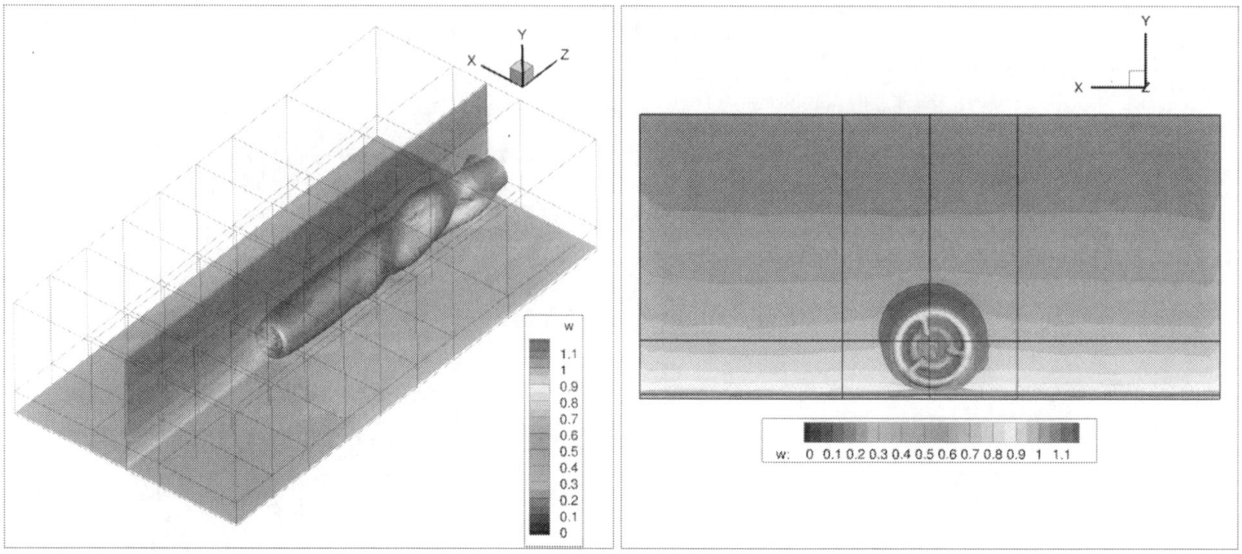

(a) (b)

Fig. 3 Two views of the iso-surface of vorticity from the actuator line large-eddy simulation in (a) three-dimensions and (b) from the front. Two-dimensional slices with normalized streamwise velocity, w, also shown.

future.

To simulate the wind shear, a prescribed wind profile power law is imposed everywhere in the domain at the first time-step in the AL-LES. Various shear exponents (ν) are used as defined by the parameter input range for the database. Using the coordinate system shown in Figures 1 to 2, the wind profile power law [24] is defined by Equation (1):

$$U(Y) = U_H \left(\frac{Y}{Y_H}\right)^{\nu} \tag{1}$$

where $U(Y)$ is the wind speed at height Y, U_H is the wind speed at hub height Y_H, and ν is a parameter giving the amount of shear.

Each AL-LES computation is performed using four nodes where each node contains twenty 2.8 GHz processors on a Linux cluster. The total number of processors is then 80, i.e., 4 nodes × 20 processors/node = 80 processors. However, for code execution efficiency, only 4 × 16 = 64 processors out of the 80 processors are used and each of the 64 processors is allocated one block out of the 64 blocks. The simulated real time for each simulation is 800 seconds including initial transients and requires approximately 19-24 hours for completion. Figure 3 depicts the result from an AL-LES computation in two views (a) and (b). In Figure 3, the iso-surface of vorticity in (a) three-dimensions and (b) from the front are shown. Two-dimensional slices with normalized streamwise velocity, w, are included in the figure as well. Figure 4 depicts the same as Figure 3 but for a simulation with an increased turbulence intensity. The iso-surface of vorticity in Figure 4 is adjusted to show the effect of the Mann turbulence box more clearly.

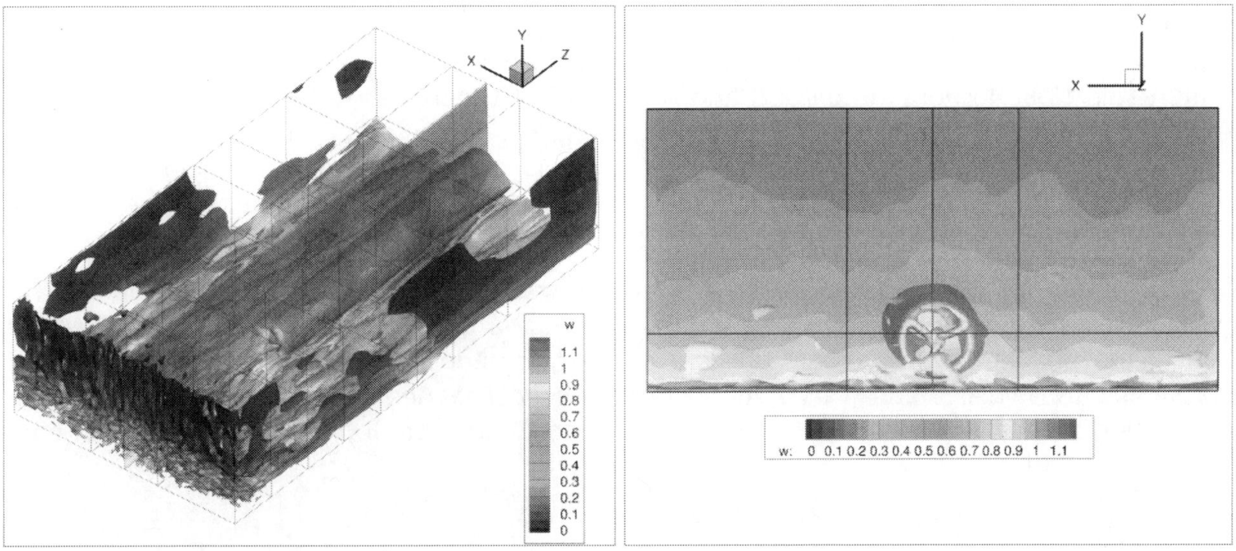

(a) (b)

Fig. 4 Two views of the iso-surface of vorticity from the actuator line large-eddy simulation with increased turbulence intensity in (a) three-dimensions and (b) from the front. Two-dimensional slices with normalized streamwise velocity, w, also shown.

3. Noise propagation solver

This section describes the noise propagation solver, WindSTAR-Pro [25, 26], in subsection 3.1 as well as the computational setup in subsection 3.2.

3.1. Noise propagation model

Predictions of wind turbine noise propagation can be made with various analytical and numerical modeling techniques, see [27] for a review. For example, there is a variety of ray tracing formulations available, which are analytical approaches based on geometrical acoustic theory. Due to the assumptions in geometrical acoustics, ray-tracing methods may not be the best method for low frequency noise propagation and generally cannot handle more complex atmospheric phenomena such as turbulence scattering, wind shear and diffraction effects. When considering more complicated terrain topography and/or atmospheric conditions, numerical approaches based on parabolic equations (PEs) are more suitable to the problem.

A two-dimensional (2D) PE model is employed in the current work as a compromise between simulation accuracy and computational cost. The PE method is a solution to the acoustic wave equation with approximations of harmonic wave propagations with a finite angle and a preferred direction of propagation. In this study, the 2D, wide-angle, Crank-Nicholson PE is used with the starter function and implementation details given in [28]. In this method the moving atmosphere is replaced by a hypothetical motionless medium with an effective speed of sound, $c_{eff} = c + v_x$, where v_x is the wind velocity component along the direction between source and receiver [29]. The wind velocity components are obtained from the flow solver described in Section 2. The solution of each PE simulation yields a steady solution at each frequency. Since a time dependent solution is desired in the present work, multiple PE simulations are performed successively to capture the relative sound pressure level as a function of time.

The PE model has limitations, but only some of them will be described here. First, the wide-angle implementation of the parabolic wave equation is only valid within ±35 degrees in the paraxial direction [25]. This limitation can be an issue for elevated sources like a wind turbine if the area of interest is close to both the ground and wind turbine. Second, the PE model is a one-way propagation method, which means that backscattering is neglected and results under multiple scattering are not reliable, e.g. a source between two noise barriers. Usually this is not the case for wind turbines though. Last, the 2D approach is another limitation. An axisymmetric approximation is employed here to reduce a 3D problem into 2D, which neglects azimuthal variations.

The PE model has been studied extensively by Barlas et al. [30–32] for single wind turbine noise propagation on flat terrain using a variety of inputs and simulation scenarios. In these studies, the effects of wind shear, different turbulence intensity levels and source modeling approaches are investigated. At DTU, the PE model is incorporated into a general purpose noise propagation tool called WindSTAR-Pro (Wind turbine Simulation Tool for AeRodynamic noise Propagation) [26].

3.2. Noise propagation setup

The PE model is coupled with a time varying sound source model. With this approach the flow solution and the location of the blades (or sources) at fifty time steps (or every 16 seconds of real-time) are extracted from the wind turbine flow solver. Preferably, the flow solution and blade locations should be extracted more frequently, e.g. every 0.2 seconds. However, such a small time step would generate a large amount of data and be more difficult to process. For the purpose of this study, a large time step is used which can be changed later on.

The flow field is extracted from the flow solution such as the ones shown in Section 2. Figure 5 depicts the extraction procedure for the streamwise wind velocity from the wind turbine flow solver for input to the noise propagation model. There are 801 slices from the domain normal to the streamwise direction that are extracted, but only twenty slices are shown in the figure for visibility. Since there are only 8 blocks × 48 cells/block = 384 cells in the streamwise direction, 801 slices might be about two times more than what is necessary and hence should be changed in the near future. The coordinate system shown in Figure 5 is different from Figures 1 to 4 because the EllipSys3D coordinate system labels during the extraction are modified to comply with the propagation model input coordinate system.

The sound emission from each wind turbine blade can be approximated by a lumped point source on the outer part of each blade, but not at the tip, e.g. at 80% of the blade span. This approximation is based on the source location studies in [2], which states that the sound is produced in the outer part of the blades. Therefore, the source locations are calculated based on 80% of the span of each actuator line. Figure 6 depicts the source locations shown on top of the streamwise wind velocity at time = 800 seconds in two views: (a) side view and (b) front view. The sources in Figure 6(a) are tilted because the NREL 5MW wind turbine has a shaft tilt of 5° [20].

Then, the relative sound pressure level is computed in a quasi-three-dimensional (quasi-3D) manner by performing 2D-PE simulations at numerous 2D planes within the domain and around the wind turbine. For the Crank-Nicholson PE calculations, the spatial resolution for both horizontal and vertical directions on the 2D-plane is set to one-eighth of the wavelength, i.e., $\delta x = \delta y = \lambda/8$, where λ is the wavelength of the solving frequency. Only flat terrain is considered and the ground impedance was defined using the theoretical and four-parameter model of Attenborough [33]. An

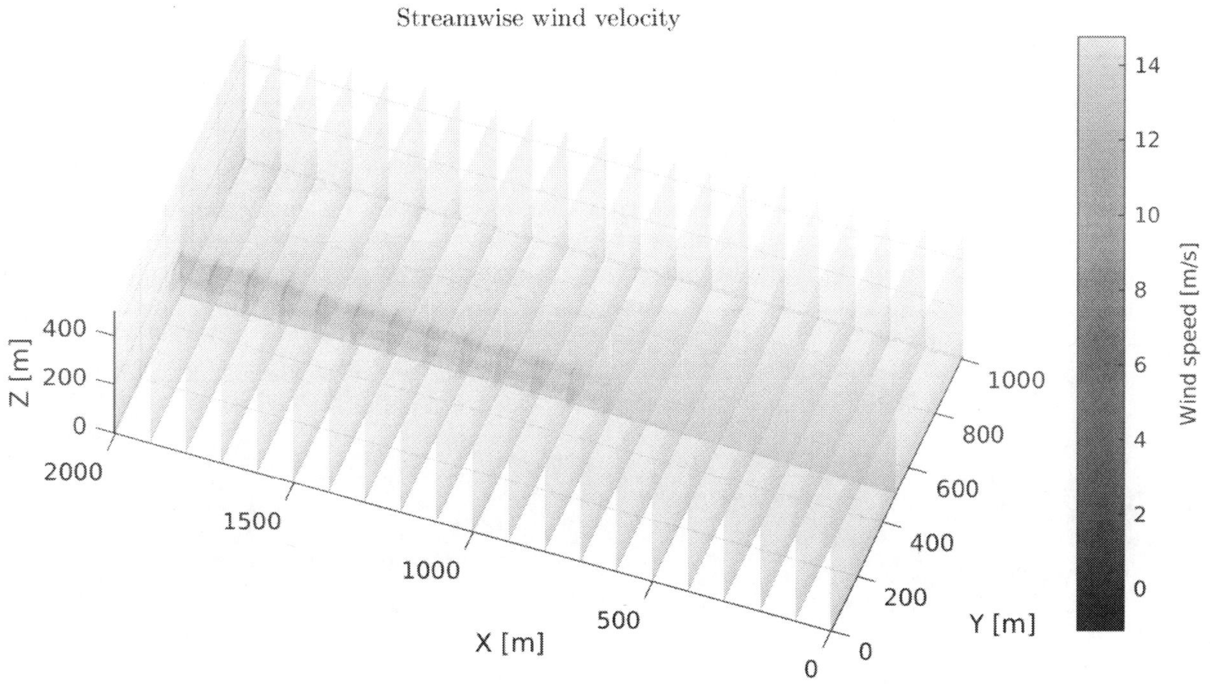

Fig. 5 Streamwise wind velocity extraction from flow solver for input to the noise propagation model. There are 801 slices from the domain that are extracted, but only twenty slices are shown for visibility. The coordinate system is modified for compatibility with the propagation model.

effective flow resistivity value of 200 kPa s/m^2 representative of grassland was chosen, which is the case where onshore wind turbines are commonly situated.

All simulations are carried out for 1/3 octave band center frequencies, f_i, from 50 Hz to 800 Hz. Frequencies higher than 800 Hz are assumed to have a negligible contribution to the overall SPL due to atmospheric absorption. Nevertheless, parameter settings can be easily changed and the simulations can be rerun if needed. The sound pressure level, L_p, is defined by Equation (2):

$$L_p(f_i) = L_W(f_i) - 10\log_{10} 4\pi r^2 - \alpha r + \Delta L \tag{2}$$

where L_W is the sound power level, α is the atmospheric absorption coefficient, r is the radial distance from the source, and ΔL is the relative sound pressure level. In the current study, only ΔL is considered in the analyses and the noise database. In other words, the sound power level, atmospheric absorption, and geometrical attenuation terms in Equation (2) are neglected. As mentioned in Section 1, noise generation or noise source, i.e., L_W, as well as the remaining terms in Equation (2) may be incorporated into the database in the near future. For now, only the relative sound pressure level in each band are summed logarithmically to obtain the overall relative sound pressure level, ΔL_{sum}, as defined in Equation (3):

$$\Delta L_{\text{sum}} = 10\log_{10}\left(\sum_{i=1}^{N} 10^{\Delta L/10}\right) \tag{3}$$

where N is the number of frequencies used.

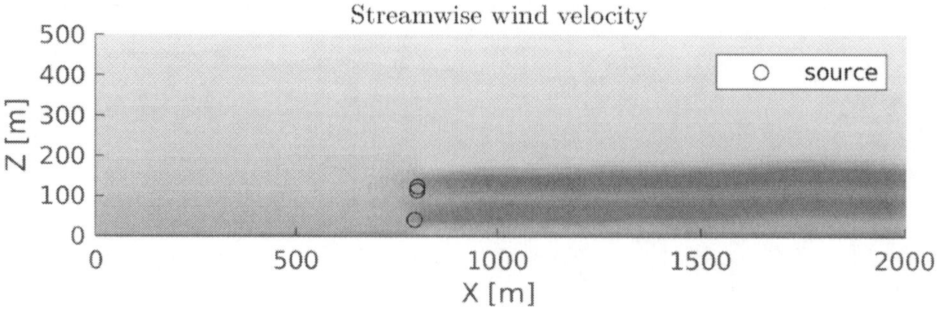

(a)

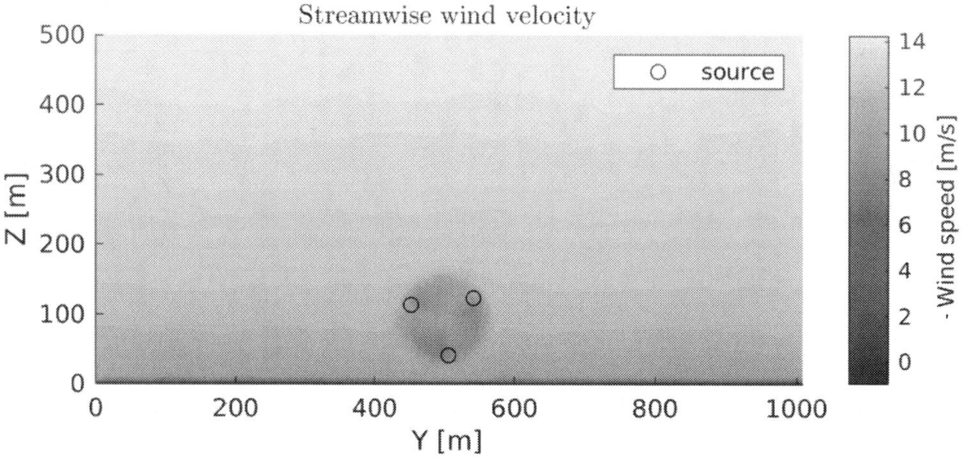

(b)

Fig. 6 Source locations shown on top of the streamwise wind velocity in two views: (a) side view and (b) front view.

4. Wind turbine noise propagation modeling

This section shows the computational results from the wind turbine noise propagation modeling based on the flow solution shown in Section 2. An excerpt from the noise propagation database is also shown in this section. Figure 7 depicts the two-dimensional planes from three sources (or three blades) to a receiver at X, Y, Z = (1950 m, 500 m, 2 m) with the (a) interpolated streamwise wind velocity flow field used in the PE model and (b) overall relative sound pressure levels computed from the PE model.

Figure 8 depicts the mean overall relative sound pressure levels, i.e. mean ΔL_{sum}, for the time period 640-800 seconds shown (a) on top of the receiver locations at 2 m height and (b) as a contour plot. The 640-800 second time period is a conservative choice to ensure that the wakes considered in the analyses have reached the end of the domain in the streamwise direction and is fully developed. Figure 9 depicts the wake at time = 80 seconds, which does not reach the end of the domain and is not fully developed like in Figure 6(a). There are a total of 1082 receiver locations that are used to construct a contour of rel. SPL for a given wind inflow condition. In the near future, the number of receiver locations might be halved since the contours are generally symmetric above and below the Y = 500 m axis in Figure 8(b).

As mentioned in Section 3.1, the PE model is only valid within ±35 degrees from the source in

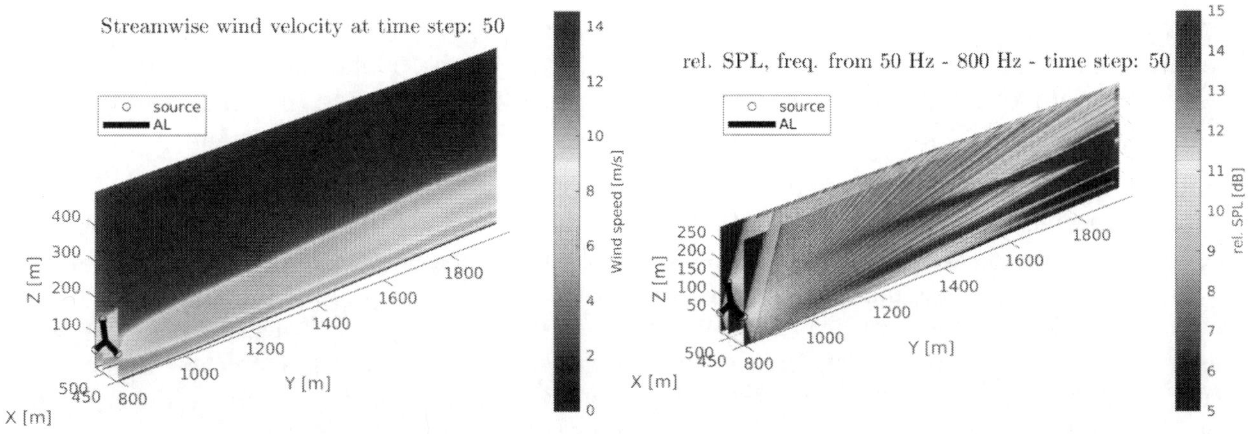

(a) (b)

Fig. 7 Two-dimensional planes from three sources (or three blades) to a receiver at $X, Y, Z =$ (1950 m, 500 m, 2 m) with the (a) interpolated streamwise wind velocity flow field used in the parabolic equation (PE) model and (b) overall relative sound pressure levels (rel. SPL) computed from the PE model.

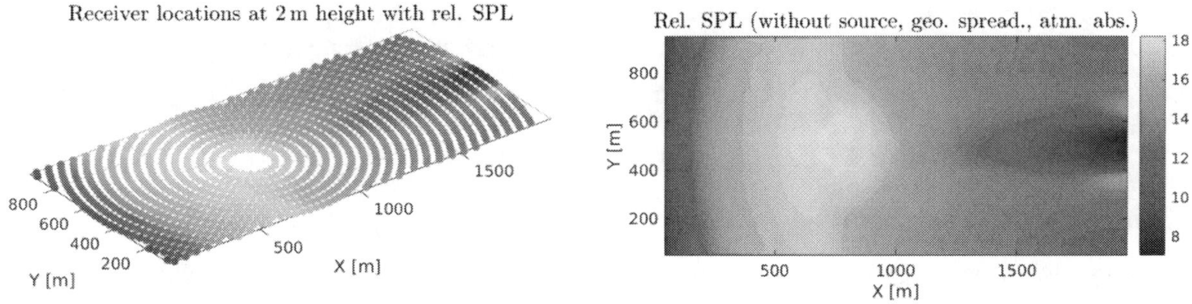

(a) (b)

Fig. 8 Mean overall relative sound pressure levels (rel. SPL) from time period 640-800 seconds shown (a) on top of the 1082 receiver locations at 2 m height and (b) as a contour plot.

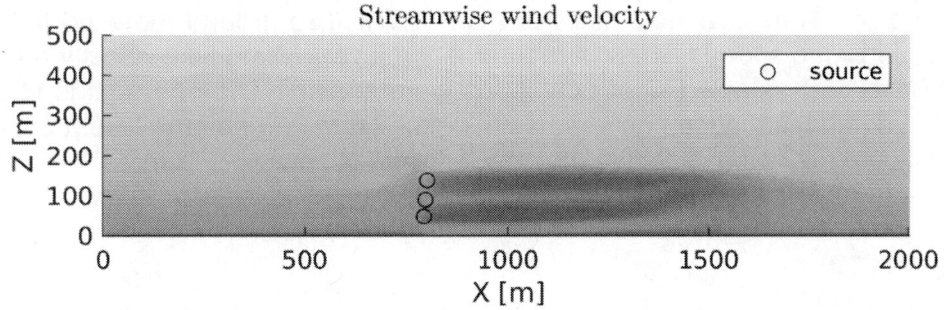

Fig. 9 Streamwise wind velocity at time = 80 seconds where the wind turbine wake does not reach the end of the domain and is not fully developed like in Figure 6(a).

the paraxial direction. The effect of the wide-angle implementation on the solution can be seen in Figure 7(b). For the NREL 5 MW wind turbine with a hub height of 90 m and a rotor radius of 63 m, the area at 2 m height with a radius of approximately [90 m + (0.8 × 63 m) - 2 m] × sin(35°) = 80 m around the turbine location in Figure 8(b) (and also Figure 10 later) should be disregarded. The contour data for all wind inflow conditions are stored in the database (or used to construct a surrogate model), which is then used in wind plant layout optimization frameworks.

Figure 10 depicts an excerpt from the noise propagation database. In Figure 10, contour plots of mean overall relative sound pressure level, i.e., mean $\Delta L_{\rm sum}$, from the time period 640-800 seconds are shown for a range of wind speeds (U_H) and wind shear exponents (ν): (a) U_H = 4 m/s, ν = 0.1, (b) U_H = 4 m/s, ν = 0.25, (c) U_H = 8 m/s, ν = 0.1, (d) U_H = 8 m/s, ν = 0.25, (e) U_H = 8 m/s, ν = 0.4, (f) U_H = 12 m/s, ν = 0.1, (g) U_H = 12 m/s, ν = 0.25, and (h) U_H = 12 m/s, ν = 0.4. The wind speed, U_H, is defined at hub height at the inlet, see subsection 2.3 and Equation (1). For all these cases, the turbulence intensity measured at the rotor center varies between 0.5% and 1.8%. The turbulence level in the flow can be scaled using the database parameter, K. For all cases shown in Figure 10, K = 1. Depending on the location of a particular wind turbine during a wind farm layout optimization run, the local wind conditions, e.g. U_H and ν, are extracted and used to determine the relative sound pressure level around this turbine based on the database.

Figure 11 depicts the relative sound pressure level spectra for one source/blade at time = 800 seconds upstream and downstream from the wind turbine position X, Y = (800 m, 500 m) at 2 m height for (a) U_H = 4 m/s, ν = 0.1 and (b) U_H = 12 m/s, ν = 0.1. The rel. SPL in general becomes less negative from upstream to downstream of the wind turbine. Downstream from the wind turbine, i.e., $X > 800$ m and Y = 500 m, the peaks and troughs in the spectra suggest constructive and destructive interference of ground reflected and direct waves. Upwind of the wind turbine, i.e., $X < 800$ m and Y = 500 m, the rel. SPL decreases smoothly for increasing frequency, f, suggesting that there is no wave interaction.

5. Conclusion

This paper described wind turbine noise propagation in flat terrain for use in wind farm layout optimization frameworks. Large-eddy simulations of a single wind turbine in flat terrain at varying wind speeds, shear and turbulence levels were performed. The wind turbine is modeled using an actuator line approach, while the wind turbine noise propagation is computed using the Technical University of Denmark's WindSTAR-Pro (Wind turbine Simulation Tool for AeRodynamic noise Propagation). The wind turbine noise propagation was computed in a quasi-three-dimensional manner by using a two-dimensional (2D) parabolic equation (PE) model at numerous 2D-planes around the wind turbine. In this study, the 2D, wide-angle, Crank-Nicholson PE model was used. The relative sound pressure level obtained from WindSTAR-Pro around the wind turbine was computed for varying wind speeds, shear and turbulence levels. Wind flow velocity and relative sound pressure levels for selected wind conditions were shown. The end goal is to create a noise propagation database that can be used as a lookup table in wind farm layout optimization frameworks to mitigate the noise impact in the development of new wind farms.

Acknowledgments

This research was funded by the Department of Wind Energy, Technical University of Denmark. Computations were performed on Jess, the high performance computing cluster at Risø Campus,

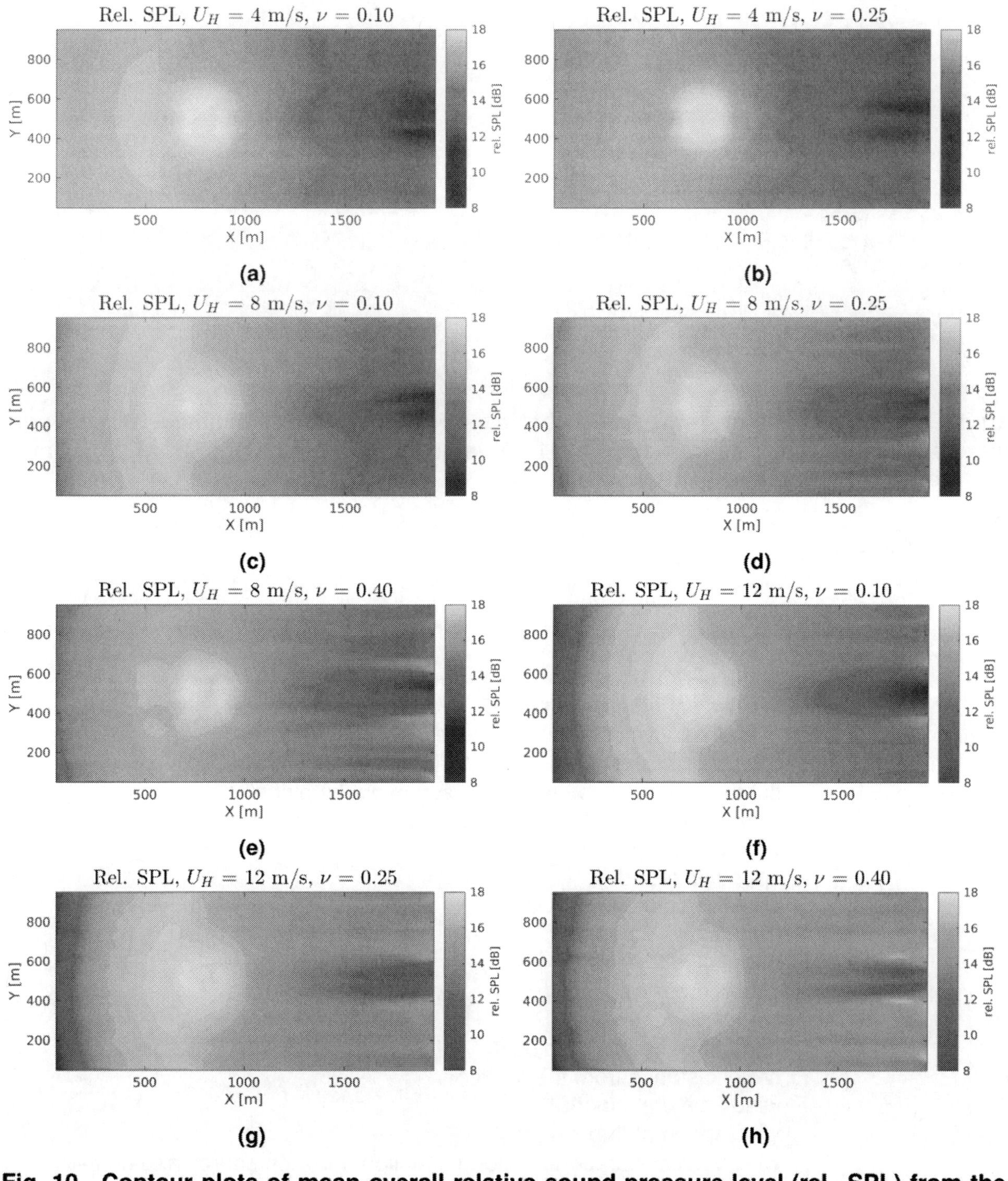

Fig. 10 Contour plots of mean overall relative sound pressure level (rel. SPL) from the time period 640-800 seconds for a range of wind speeds (U_H) and wind shear exponents (ν): (a) U_H = 4 m/s, ν = 0.1, (b) U_H = 4 m/s, ν = 0.25, (c) U_H = 8 m/s, ν = 0.1, (d) U_H = 8 m/s, ν = 0.25, (e) U_H = 8 m/s, ν = 0.4, (f) U_H = 12 m/s, ν = 0.1, (g) U_H = 12 m/s, ν = 0.25, and (h) U_H = 12 m/s, ν = 0.4.

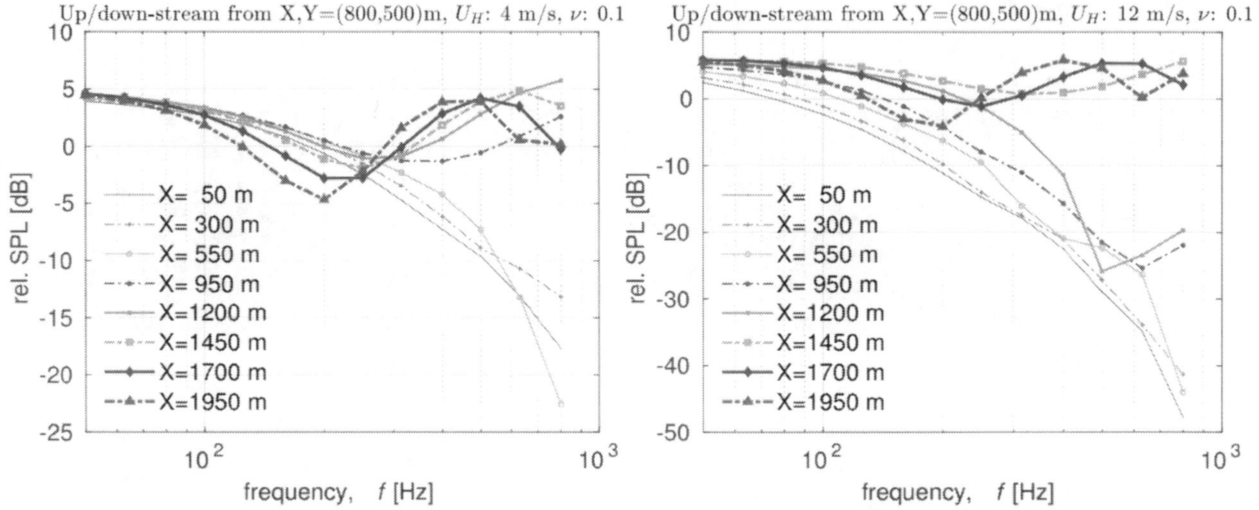

(a) **(b)**

Fig. 11 Relative sound pressure level spectra for one source/blade at time = 800 seconds upstream and downstream from the wind turbine position at 2 m height for (a) U_H = 4 m/s, ν = 0.1 and (b) U_H = 12 m/s, ν = 0.1.

Technical University of Denmark.

Nomenclature

α	atmospheric absorption coefficient
ΔL	relative sound pressure level
ΔL_{sum}	overall relative sound pressure level
δx	spatial resolution for horizontal direction on 2D-plane
δy	spatial resolution for vertical direction on 2D-plane
λ	wavelength of the solving frequency
ν	wind shear exponent
c	speed of sound
c_{eff}	effective speed of sound
f	frequency
K	turbulence level scale parameter
L_p	sound pressure level
L_W	sound power level
N	number of frequencies used
r	radial distance from the source
$U(Y)$	wind speed at height Y
U_H	wind speed at hub height Y_H
v_x	wind velocity component along direction between source and receiver
X, Y, Z	location in a 3D Cartesian coordinate system
$X_{\text{length}}, Y_{\text{length}}, Z_{\text{length}}$	domain lengths in a 3D Cartesian coordinate system
$X_{\text{origin}}, Y_{\text{origin}}, Z_{\text{origin}}$	origin location in a 3D Cartesian coordinate system
$X_{\text{rotor}}, Y_{\text{rotor}}, Z_{\text{rotor}}$	rotor location in a 3D Cartesian coordinate system
2D	two-dimensional

3D	three-dimensional
AL	actuator line
BPM	Brooks, Pope, Marcolini
CFD	computational fluid dynamics
DTU	Technical University of Denmark
LES	Large-Eddy Simulation
MPI	Message Passing Interface
PE	parabolic equation
RANS	Reynolds-Averaged Navier Stokes
rel.	relative
RPM	revolutions per minute
SGS	sub-grid scale
SPL	sound pressure level
w	normalized streamwise velocity
WindSTAR-Pro	Wind turbine Simulation Tool for AeRodynamic noise Propagation

References

[1] French-Brooks, J., Smith, M., and Wiesel, J., editors (2016). *Nordic Energy Technology Perspectives 2016: Cities, flexibility and pathways to carbon-neutrality*. Nordic Energy Research, International Energy Agency, Stensberggata 25, NO-0170 Oslo, Norway.

[2] Oerlemans, S., Sijtsma, P., and López, B. M. (2007). Location and quantification of noise sources on a wind turbine. *Journal of Sound and Vibration*, 299(4):869 – 883.

[3] Pedersen, E. and Kerstin, P. W. (2004). Perception and annoyance due to wind turbine noise—a dose–response relationship. *The Journal of the Acoustical Society of America*, 116(6):3460–3470.

[4] Persson Waye, K. and Öhrström, E. (2002). Psycho-acoustic characters of relevance for annoyance of wind turbine noise. *Journal of Sound and Vibration*, 250(1):65 – 73.

[5] The Danish Ministry of the Environment (2011). Statutory order no. 1284 on noise from wind turbines.

[6] Feng, J., Shen, W. Z., and Li, Y. (2018). An optimization framework for wind farm design in complex terrain. *Applied Sciences*, 8(11).

[7] Dimitrov, N., Kelly, M. C., Vignaroli, A., and Berg, J. (2018). From wind to loads: wind turbine site-specific load estimation with surrogate models trained on high-fidelity load databases. *Wind Energy Science*, 3(2):767–790.

[8] Schröder, L., Dimitrov, N. K., Verelst, D. R., and Sørensen, J. A. (2018). Wind turbine site-specific load estimation using artificial neural networks calibrated by means of high-fidelity load simulations. *Journal of Physics: Conference Series*, 1037:062027.

[9] Brooks, T. F., Pope, D. S., and Marcolini, M. A. (1989). Airfoil self-noise and prediction. *NASA Reference publication 1218*.

[10] Amiet, R. (1975). Acoustic radiation from an airfoil in a turbulent stream. *Journal of Sound and Vibration*, 41(4):407 – 420.

[11] van der Laan, M. P., Sørensen, N. N., Réthoré, P.-E., Mann, J., Kelly, M. C., Troldborg, N., Schepers, J. G., and Machefaux, E. (2015a). An improved k- ϵ model applied to a wind turbine wake in atmospheric turbulence. *Wind Energy*, 18(5):889–907.

[12] van der Laan, M. P., Sørensen, N. N., Réthoré, P.-E., Mann, J., Kelly, M. C., and Troldborg, N. (2015b). The k-ε-f_p model applied to double wind turbine wakes using different actuator disk force methods. *Wind Energy*, 18(12):2223–2240.

[13] Sørensen, N. N. (1994). *General purpose flow solver applied to flow over hills*. PhD thesis, Technical University of Denmark.

[14] Michelsen, J. A. (1992). Basis3D - a platform for development of multiblock PDE solvers. Technical

report, Technical University of Denmark.

[15] Shen, W. Z., Michelsen, J. A., and Sørensen, J. N. (2001). Improved Rhie-Chow interpolation for unsteady flow computations. *AIAA Journal*, 39(12):2406–2409.

[16] Shen, W. Z., Michelsen, J. A., Sørensen, N. N., and Sørensen, J. N. (2003). An improved SIMPLEC method on collocated grids for steady and unsteady flow computations. *Numerical Heat Transfer, Part B: Fundamentals*, 43(3):221–239.

[17] Sørensen, J. N. and Shen, W. Z. (2002). Numerical modeling of wind turbine wakes. *Journal of Fluids Engineering*, 124(2):393–399.

[18] Sørensen, J. N., Mikkelsen, R. F., Henningson, D. S., Ivanell, S., Sarmast, S., and Andersen, S. J. (2015). Simulation of wind turbine wakes using the actuator line technique. *Philosophical transactions. Series A, Mathematical, physical, and engineering sciences*, 373(2035):20140071.

[19] Øye, S. (1996). FLEX4 simulation of wind turbine dynamics. In *Proceedings of 28th IEA Meeting of Experts Concerning State of the Art of Aeroelastic Codes for Wind Turbine Calculations*, pages 71–76, Lyngby. International Energy Agency.

[20] Jonkman, J., Butterfield, S., Musial, W., and Scott, G. (2009). Definition of a 5-MW reference wind turbine for offshore system development. Technical report.

[21] Sessarego, M., Shen, W. Z., Van der Laan, M. P., Hansen, K. S., and Zhu, W. J. (2018). CFD simulations of flows in a wind farm in complex terrain and comparisons to measurements. *Applied Sciences*, 8(5).

[22] Mann, J. (1994). The spatial structure of neutral atmospheric surface-layer turbulence. *Journal of Fluid Mechanics*, 273:141–168.

[23] Mann, J. (1998). Wind field simulation. *Probabilistic Engineering Mechanics*, 13(4):269 – 282.

[24] Hansen, M. O. L. (2008). *Aerodynamics of wind turbines*. Earthscan, London, 2nd edition.

[25] Barlas, E. (2017). *Development of an advanced noise propagation model for noise optimization in wind farm*. PhD thesis, Technical University of Denmark.

[26] Zhu, W. J., Shen, W. Z., Barlas, E., Bertagnolio, F., and Sørensen, J. N. (2018). Wind turbine noise generation and propagation modeling at DTU wind energy: A review. *Renewable and Sustainable Energy Reviews*, 88:133 – 150.

[27] Bérengier, M. C., Gauvreau, B., Blanc-Benon, P., and Juvé, D. (2003). Outdoor sound propagation: A short review on analytical and numerical approaches. *Acta Acustica united with Acustica*, 89(6):980–991.

[28] West, M., Gilbert, K., and Sack, R. (1992). A tutorial on the parabolic equation (PE) model used for long range sound propagation in the atmosphere. *Applied Acoustics*, 37(1):31 – 49.

[29] Ostashev, V. E. and Wilson, D. K. (2015). *Acoustics in Moving Inhomogeneous Media*. 2nd Edition. CRC Press.

[30] Barlas, E., Zhu, W. J., Shen, W. Z., Kelly, M., and Andersen, S. J. (2017a). Effects of wind turbine wake on atmospheric sound propagation. *Applied Acoustics*, 122:51 – 61.

[31] Barlas, E., Zhu, W., Shen, W., Dag, K., and Moriarty, P. (2017b). Consistent modelling of wind turbine noise propagation from source to receiver. *Acoustical Society of America. Journal*, 142(5).

[32] Barlas, E., Wu, K. L., Zhu, W. J., Porté-Agel, F., and Shen, W. Z. (2018). Variability of wind turbine noise over a diurnal cycle. *Renewable Energy*, 126:791 – 800.

[33] Attenborough, K. (1985). Acoustical impedance models for outdoor ground surfaces. *Journal of Sound and Vibration*, 99(4):521 – 544.

INCE*EUROPE*

8th International Conference
on
Wind Turbine Noise
Lisbon – 12th to 14th June 2019

Session F

Assessment and Compliance

Organised by INCE-Europe
mail@inceeurope.org
www.inceeurope.org

Supported by Portuguese Acoustical Society

Sociedade
Portuguesa
de Acústica

INCE*EUROPE*

8th International Conference
on
Wind Turbine Noise
Lisbon – 12th to 14th June 2019

Measurement of wind turbine noise characteristics in receptor position – A new IEC Technical Specification

Sylvia Broneske, Innogy Renewables UK Ltd.: sylvia.broneske@innogy.com
Bo Søndergaard, Sweco Danmark A/S: bo.sondergaard@sweco.dk

Summary

Following an initially rejected proposal for a new international standard on the measurement of wind turbine noise characteristics at receptor positions in 2017, the subsequently submitted revised 2nd proposal for a technical specification (TS) received approval by the IEC/TC88 member committees in April 2018. The International Electrotechnical Commission (IEC) Technical Committee TC88 tasked Mr. Bo Søndergaard from Denmark with organizing and convening the new project team, PT61400-11-2 (PT11-2), and the first meeting of the group took place in Hamburg in June 2018. Further meetings were since held in the US and in Denmark. At the second meeting in October 2018, the preliminary structure of the TS was defined, and the text of the individual chapters and appendices is currently being written in smaller project groups until the next update meeting in June 2019.

The TS comprises measurement and evaluation methods relating to topics such as sound pressure level, amplitude modulation, tonality, impulsivity, low frequency, and the determination of wind speed and other non-acoustic measures describing the details of the implementation of such measurements under conditions relevant to wind turbine noise. The project team currently includes 29 experts from 11 national committees, traditionally strongly represented from Europe and North America, but also a growing number of experts from Asia are interested in collaborating within the project team. The TS will be of particular interest in countries, where no guidance for the measurement of wind farm noise is available, but also in countries where guidelines and legislation for environmental noise measurements may not be sufficiently adapted to the specifics of immission measurement and characterization of sound from wind turbines under windy conditions.

This paper will cover the background of and intentions for the initiative, as well as a description of the items included in the TS and the current status of writing, including the estimated timeline of publication.

1. Introduction

In most of the countries known to the authors, wind turbines are structures that require a planning permission (permission for construction, operation and deconstruction at the end of the life-time), usually involving an environmental impact assessment which also includes the assessment of wind farm noise at the nearest residential properties. Subsequently, the building permit or permit to operate a wind farm is often issued with a number of conditions relating to the noise level and noise characteristics from wind turbines, the wind farm immissions impacting on the nearest residential neighbours. In order to show compliance with such conditions or in the event of complaints about wind farm noise and/or its characteristics, measurements at the receptor position may be required. Some countries have specific guidelines for the measurement of wind farm noise, many have only general guidance for the measurement of industrial noise and others have no guidance at all. Unless wind farm specific guidelines are available, national legislation or guidance often require adaptation to the peculiarities of wind turbine noise. Available guidance is usually created for wind speed independent sound sources and is therefore often not suitable for wind turbines which exhibit a wind speed dependent sound emission. In the absence of a well described guidance, the quality of measurements and reports is often poor, and/or not reproducible by another expert.
In order to simplify and standardize their work, wind turbine manufacturers, internationally operating energy providers, wind farm developers and operators prefer uniform and internationally recognized measurement and evaluation methods, especially in countries where there are no suitable regulations adequately adapted to wind turbines. Internationally it can be observed that many permitting authorities use ISO 1996-2, which should only be used for wind speeds up to 5 m/s at the location of the measurement.
In order to have a basis for wind farm measurements and assessments all over the world, and to fill a 'vacuum' in various countries, experts of wind turbine acoustics and stakeholders from the wind industry came together to discuss in which way an international standard would be suitable and under the guidance of which international standards organization (ISO or IEC) this standard should be placed. This process will be described in the following chapters.

2. History of the IEC/TS 61400-11-2 with the working title "Wind energy generation systems - Part 11-2: Measurement of wind turbine noise characteristics in receptor position"

2.1 IEC and ISO Standards

A collection of currently 32 published standards relating to wind turbines is in the care of the approximately 800 experts from 29 countries under the umbrella of the IEC[1] in the Technical Committee TC88 "Wind energy generation systems". Another 27 standards are under development (revision or new documents in progress). Amongst others, the IEC 61400 series describes topics from individual components of the wind turbine (mechanical and electronical), wind measurements, test procedures, stability, wind park regulation, site suitability and occupational safety for onshore and offshore wind turbines and wind power plants as a whole (including supply to the power grid). Further information can be found on the IEC website under 'TC 88 Dashboard'[2].

Standards dealing with sound measurements of sources and propagation calculations of industrial noise have traditionally been written and maintained by ISO[3]. An exception is the standard for the determination of sound power levels of wind turbines, which was included in

[1] International Electrotechnical Commission: International Standards and Conformity Assessment for all electrical, electronic and related technologies
[2] https://www.iec.ch/dyn/www/f?p=103:7:5599856330528::::FSP_ORG_ID,FSP_LANG_ID:1282 25 (last accessed 20/03/2019)
[3] International Organization for Standardization

the IEC series because it is strongly linked to the operation of this particular plant and therefore fits into the IEC series which describes this product. The IEC 61400-11 is an expansion of the ISO standards for the determination of sound power levels (especially ISO 3744), but which includes as a special feature the correlation of the sound power level with (hub height) wind speed and in particular describes measurements at higher wind speeds. The first version (Edition 1) was released in September 1998, with the latest version (Edition 3) published in December 2012 and Amendment 1 in June 2018.

2.2 A new IEC standard/technical specification

The existing ISO 1996 standard series has been written for the description, measurement and evaluation of environmental noise, which includes the sound sources road, rail and air traffic, as well as industrial plants. However, it is unsuitable for wind power plant immissions because it defines a maximum allowable wind speed range at which the wind turbines may not have reached their highest sound power level yet. It is required to avoid the influence of meteorological parameters on the measurements. Recently, there was a proposal to include wind turbines as a sound source in the ISO 1996 series, but this has not been followed up so far. Now that the IEC has taken this task on, it is expected that there will be reference to the new IEC document rather than their own creation. From the beginning, experts from the relevant acoustics committees at ISO collaborated with the with the IEC experts. This is to avoid doubling the work and possible contradictions between an ISO and an IEC standard.

In March 2017, the Chinese National Committee submitted a proposal for writing a new standard titled "Wind turbine aero acoustic noise measurement techniques (proposed IEC 61400-11-2)", which was accepted by the majority of P-members[4] but rejected owing to lack of experts from P-members (these two criteria must be met for a successful application). This was then taken as an opportunity to revise the Chinese proposal and submit it in a second proposal to the IEC, written by members of the MT11 maintenance team and submitted by the Danish National Committee. This included the removal of overlaps of the new proposal with the existing IEC 61400-11, a more integrated approach that included more parameters than just amplitude modulation and low frequency noise, changing from an international standard (IS) to a technical specification (TS) to speed up the process and the intention to collaborate with the relevant ISO working group. This revised proposal was adopted by the P-members and sufficient experts were put forward by the National Committees, so that at their annual meeting, TC88 mandated the formation of a new project team in April 2018.

Documents published by the IEC are classified based on their maturity and experience with the methods described. Some of the publication types produced by the IEC are listed here:

- International Standard (IS)
- Technical Specification (TS)
- Publicly available specifications (PAS - publicly available specifications)
- Technical reports (TR)

MT11 opted for a TS in this case, as it was felt that e.g. the topic amplitude modulation still requires further research and therefore standardization as an IS was considered as premature. A TS refers to international standards with respect of detail and completeness. Furthermore, a TS is similar to an IS in that it is normative and developed by consensus. A TS is approved by a two-thirds majority of the participating members of a technical committee of the IEC (TC) or a subcommittee (SC). The approval process for the TS is similar to that of the IS, but saves one round of voting and therefore goes through the IEC process faster for publication.

A TS goes through the following stages as shown in Figure 1 and further explained below.

[4] P-member: participating National Committees in TC88

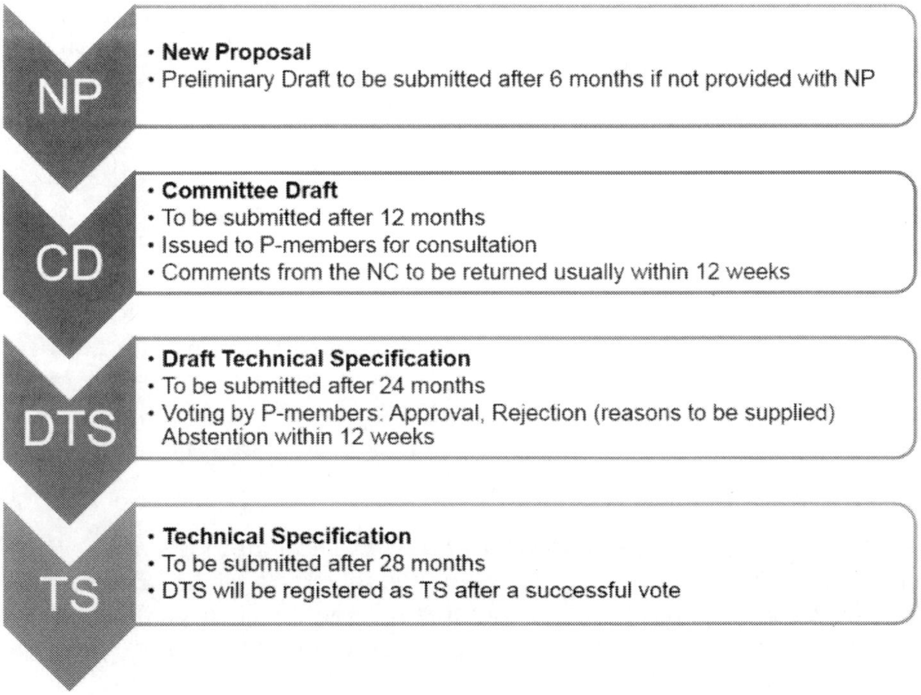

Figure 1: Milestones of an IEC/TS

A New Proposal (NP) is submitted by a National Committee and is usually written by experts with the required background in the topic. The expectations of the IEC on deliverables are as follows:

- If not present at the time of submission of the NP, a 'working draft' must be submitted within half a year. This can be done in the form of a table of contents or the schematic representation of the work processes.
- One year after accepting the NP and assigning the project team, the Committee Draft (CD) is expected. This is distributed to the P-members of the TC88 with the requests, to submit comments within 12 weeks, which will eventually lead to TS acceptance by that Member State.
- The project team must now consider all the comments received and decide whether they are justified and therefore passages need to be changed, whether it is a misunderstanding that can be resolved by a better description of the text, or whether it is a topic that may be known, but cannot be considered at the time and is therefore postponed until the next revision.
- Subsequently, the Draft Technical Specification (DTS) is prepared. This is sent out to the P-members again with a 12-week voting period. Each member state has three options: acceptance of the TS (positive), rejection of the TS (negative, reasons must be presented) and abstention. A case for the latter option could be if a National Committee does not have its own experts on the subject and therefore cannot form an opinion on the subject.

Ideally, the DTS is accepted by the National Committees and is then registered and published by the IEC as TS. If the criteria are not fulfilled at the election, the project team can rework the document. The text may then be reissued as a revised DTS, or, which would be one step before, being circulated as CD again. And the last option is to present it to TC88, and if well argued, it may be approved by TC88, however may also be rejected if there are good reasons for it.

2.3 Motivation

Writing a document that includes the characterization of the noise from wind turbines at the receptor location into an IEC standard is a consequence of the fact that sound power level measurements at the wind turbine do not describe the nature of far-field sound as perceived by the residents, and the need for a description of the measurement and evaluation of amplitude modulation (AM). It was decided that such a document should include the analysis of all wind farm sound characteristics, not as originally intended, just AM. It is not the aim to invent new methods e.g. for tonality or impulsivity, but to use existing standards and methods wherever possible.

It has already been found that the TS cannot consist of just one well-defined method for carrying out measurements and assessment per topic, as there are already many different methods in existence in various countries. The aim for the TS is therefore to describe available and reliable methods and give instructions on how to work with the relevant method for the required task. A toolbox will be created in which the expert can pick out the best fitting approach for a meaningful assessment.

It will be stated very clearly that the different measurement and evaluation methods are independent of each other and that they do not have to be applied all in one report. Before each measurement it should be defined which purpose the measurement will serve and the appropriate method should be selected appropriately. It is expected that only those assessments will be carried out which are purposeful for the task at hand.

2.4 The Project Team

The PT61400-11-2 (PT11-2) project team currently consists of 29 experts, representing eleven National Committees (shadow committees to TC88), and numbers are still growing. Only countries, where their national standards committee is a member of TC88, can send experts on their behalf. Experts of other countries can only take part in meetings of the project team in an advisory manner at the discretion of the Convenor. Traditionally strongly represented are Germany and Denmark, which together already made up half of all participants at the inaugural meeting and, with the exception of participating Goldwind from China, also have all the major international wind turbine manufacturers in their ranks. Project management for PT11-2 has been assigned to Mr. Bo Søndergaard of Sweco Danmark A / S, who is already in charge of the MT11 maintenance team. A project manager can be assisted by a secretary which in this case has been taken over by Ms Sylvia Broneske of Innogy Renewables UK Ltd.

3. State of Progress

The members of the Danish NC came prepared to the first meeting of PT11-2 with a proposal for the chapter structure and topics to be included in the TS. This was generally accepted by participants of the meeting and is shown in Figure 1 in a revised version representing the work in progress after the 3rd meeting.

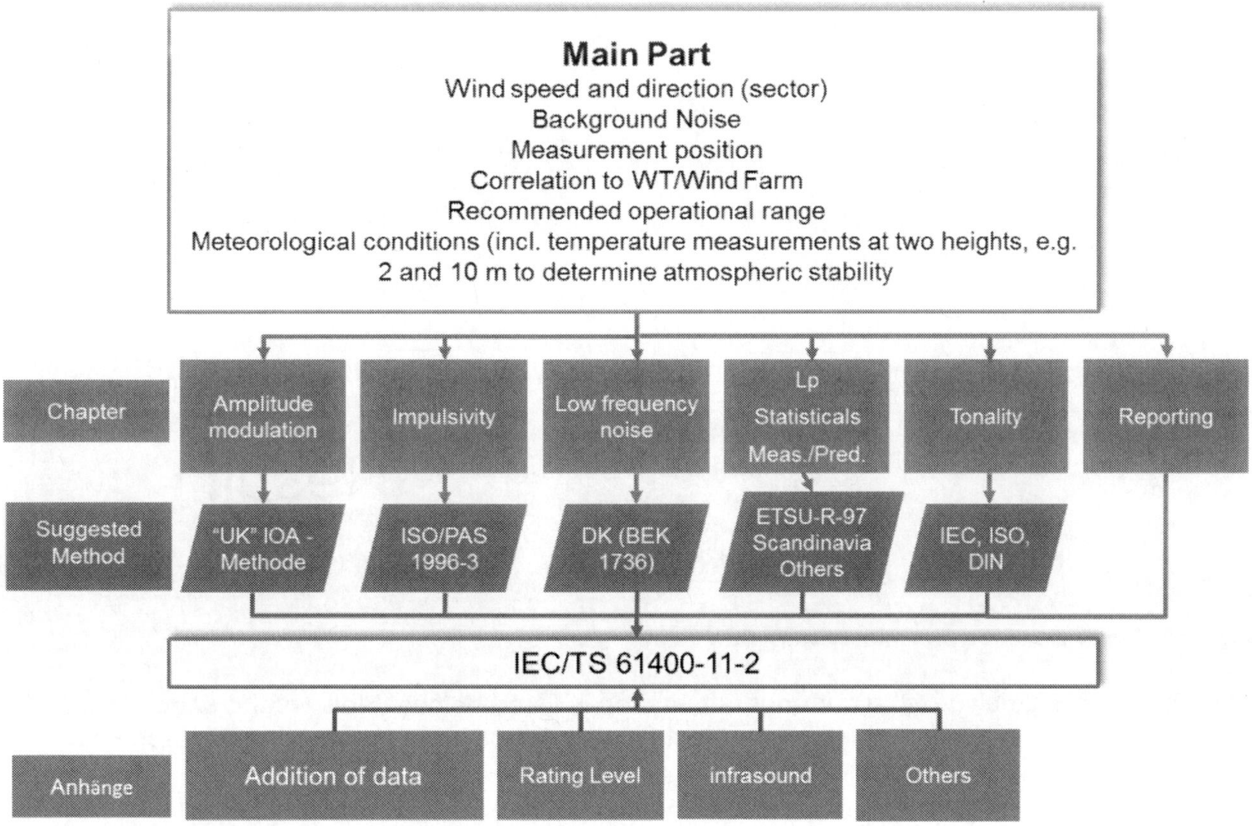

Figure 2: First draft of structure of IEC/TS 61400-11-2

It has also become very clear that it will not be possible to decide on only one method of measurement and evaluation per topic, since the tasks assigned to such measurements may vary considerably. For example, the effort and the requirements for a complaint measurement will differ from those for compliance testing or troubleshooting. Therefore, the goal is to offer a choice of possibilities that can be put together for the measurement depending on the task. The buzzword commonly used here is toolbox. Depending on the job you need a certain approach. If the task is not defined, a proposal for the minimum requirement will be defined so that the measurements provide a meaningful result.

The following schematic structure of the TS was provided by a member of PT11-2 and is included here as Figure 3.

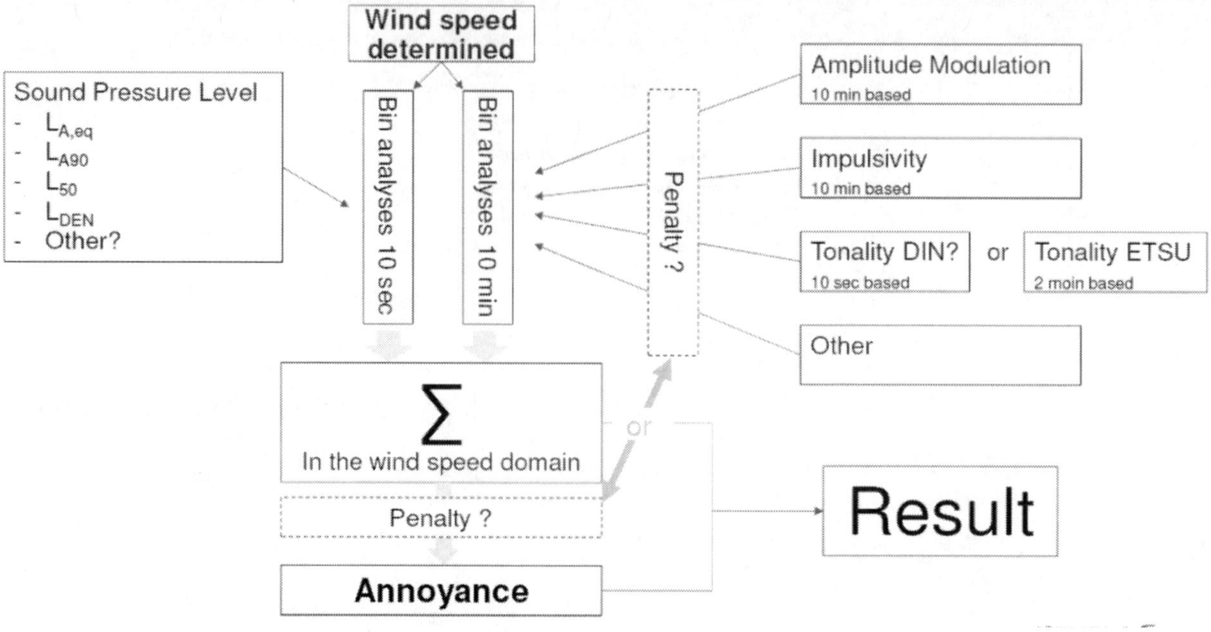

Figure 3: Overview of the preliminary work flow

The chapter breakdown of the current draft version is listed below. It follows the specifications of the IEC, as far as the introductory, non-technical chapters are concerned. Each technical chapter is subdivided as follows:

- Outline
- Instrumentation
- Acoustic measurements
- Non-acoustic measurements
- Data reduction
- Reporting

If similarities of all described measurements and evaluations are found, these are described in the main introductory chapters. For topic-specific explanations that are not useful in any other case, there will be a description in each individual chapter. Care must be taken to ensure that no repetitions of the same facts of the main text occur in the individual chapters.

The preliminary list of contents is provided in Figure 4 below.

CONTENTS

Figure 4: Preliminary list of contents

For the evaluation of the tonality, no final decision has yet been made yet, which of the available methods represents the preferred method of the PT11-2. Søndergaard et al. carried out a round robin in which individual spectra were evaluated according to different standards, namely IEC 61400-11 and ISO/PAS 20065. At the last session of the PT11-2 it was also suggested to include the evaluation according to ETSU R-97. After a complete evaluation of the round robin, the results will be presented at the Wind Turbine Noise Conference in June 2019 and subsequently PT11-2 may be able to decide on just one method to assess tones in wind turbine noise immissions.

The description of amplitude modulation (AM) is currently based on the publication of the British Institute of Acoustics, though the expert group on AM also discusses improvements and extensions of the evaluation procedure and, as far as possible in the short time, incorporates this into the text of the TS. There is another large study on AM currently conducted in Germany under the sponsorship from the German Umweltbundesamt[5] (UBA) and results are eagerly awaited[6]. However, their project plan extents to 2021 which will be too late for the anticipated publication of this TS.

For the assessment of impulsivity, the TS will refer to the future ISO/PAS 1996-3, which is expected to be published within the next three years. This ISO/PAS is based on the publicly available Nordtest NT ACOU 112: 2002-05.

4. Challenges

The challenge of writing an international norm or even report of a working group, is to balance the interests of the various stakeholders at the table. In PT11-2 all major wind turbine manufacturers from Europe are represented and for the first time there is also an Asian manufacturer present. Another large group in the project team are the experts from the measuring institutes from various countries, while operators / project developers and universities / research institutes are present only in a small number. All these various groups, and indeed individual members, aim to bring forward and defend their own interests. As representatives of the National Committees, experts are also expected to represent the opinion and interests of their home countries. An IS / TS is the result of long discussions based on consensus. This is not always easy, as can be imagined.

Consensus is defined as follows:

> ***consensus**: General agreement, characterized by the absence of sustained opposition to substantial issues by any important part of the concerned interests and by a process that involves seeking to take into account the views of all parties concerned and to reconcile any conflicting arguments.*
> *NOTE: Consensus need not imply unanimity.*

Even the selection of topics to be covered in the TS 11-2 led to lively discussions, which are not yet completed. Should e.g. the evaluation of a noise characteristic be complete with the determination of a value (example: tonal perceptibility or depth of the amplitude modulation)? Should a typical "normal" value be defined? Does the PT11-2 want to attempt a definition for penalties and a rating level? The latter would be of particular interest to countries in which there are no regulations with regard to immission standards and pollution control. The text would have to be phrased in such way that it does not cancel out existing rules and regulations about the allocation of penalties such as tonality, impulsivity and AM.

Details must be clarified, such as, which of several existing nearly equivalent evaluation methods would be listed as the preferred method, how are uncertainties to be calculated, how and where should wind speed be measured or determined, or where should the microphone be placed for the measurement.

There are also administrative challenges. Working with a group of international experts across the world brings challenges in finding suitable time slots for webconferencing. Attending face to

[5] Federal Environment Agency
[66] Research Project "Geraeuschwirkungen bei der Nutzung von Windenergie an Land" carried (
Gesellschaft fuer sensorische Messtechnik mbH

face meetings is not an option for every expert as this depends on their availability to take out 2 days for each meeting and additional travel time, but also on the willingness of their company to pay for expenses and allow them to take the required time out of a usually busy work schedule. Writing a TS or IS is time intensive, so that availability of 'free' time to carry out the required work and attend meetings could easily have a major impact on meeting the deadlines set out by IEC as described in Section 2.2 above.

5. Outlook

It had already emerged at the first meeting of the new project team that the original idea of referring exclusively to existing standards and only occasionally writing explanatory words to adapt existing methods for wind farm noise was too simplistic. Therefore, the original timetable has turned out to be too ambitious. The plan was to hand over the CD to IEC in January 2019, so that after passing through the necessary processes, the TS could be published in June 2020. PT11-2 has requested an extension of the deadline by a year, so that the new deadline is to finalize the Committee Draft by the end of January 2020 and if everything goes smoothly[7], to have the TS available for mid-2021.
Again, this is a tight schedule, which requires a lot of engagement from the participants, not only at further meetings, but also in the current phase of working out individual chapters in the different subgroups away from the larger PT11-2 meetings.

It will be interesting to see if there are any more presentations at the Wind Turbine Noise 2019 that need to be considered in the TS. Furthermore, we invite the discussion with members of PT11-2 to either confirm that we have taken the right approach so far or, if we missed out on important topics, to make us aware at this stage where topics can still be discussed and included if required.

It looks like the results of the German research project will be too late to be considered for this version of the TS, but their findings may then be incorporated into a revision of the TS or, if results confirm previous research, into an IS.

References

Bauerdorff, A, (2019) *Entstehung und Minderung des Lärms von Windenergieanlagen – Aktivitäten des Umweltbundesamtes*. Conference proceedings 10. Rheiner Windenergie-Forum (Rheine, Germany)

Søndergaard, L S., Thomsen, C and Pedersen, T H (2019) *Prominent tones in wind turbine noise - Round-robin test of the IEC 61400-11 and ISO/PAS 20065 methods for analysing tonality content*. 8th International Conference on Wind Turbine Noise (Lisbon, Portugal)

IEC 61400-11:2012+AMD1:2018 CSV *Wind turbines - Part 11: Acoustic noise measurement techniques*. IEC, Geneva/Switzerland.

Institute of Acoustics (2016) *AMWG Final Report: A Method for Rating Amplitude Modulation in Wind Turbine Noise.* IOA, Milton Keynes/United Kingdom of GB and NI.

ISO 1996:2017 *Acoustics - Description, measurement and assessment of environmental noise -- Part 2: Determination of environmental noise levels.* ISO, Geneva/Switzerland.

[7] ... and when does that ever happen???

Not yet published: ISO/PAS 1996-3 *Acoustics — Description, measurement and assessment of environmental noise — Part 3: Objective method for the measurement of prominence of impulsive sounds and for adjustment of* L_{Aeq}. To be published within the next three years.

ISO/PAS 20065:2016 *Acoustics – Objective method for assessing the audibility of tones in noise – Engineering method.* ISO, Geneva/Switzerland.

ETSU-R-97 *Assessment and Rating of Wind Turbine Noise*

INCE EUROPE

8th International Conference
on
Wind Turbine Noise
Lisbon – 12th to 14th June 2019

Danish experiences with measuring wind turbine noise at neighbor dwellings.

Rune Egedal, Sweco: rune.egedal@sweco.dk
Niels F. Christensen, Sweco: nielsfrederik.christensen@sweco.dk
Bo Søndergaard, Sweco: bo.sondergaard@sweco.dk

Summary

This paper investigates measurements conducted in Denmark at neighbour dwellings. The results are based on three studies, two short term and one long term measurement. The paper will summarize the results and experiences gained through the studies. The focus is drawn to the differences in noise level seen at different neighbour dwellings, and hence the influence of the surroundings e.g. the vegetation. Furthermore, the windspeed measured at different positions compared to the measured noise levels are investigated. It is indicated in the studies that it in some cases is possible to measure the noise from wind turbines at the neighbours, but there are several factors that must be accounted for, in order to ensure that the measured noise inherit from the wind turbines.

1. Introduction

In connection with criticism of the Ministry of the Environment's guidelines for documentation of noise from wind turbines through sound power measurements and noise predictions, the Danish Environmental Protection Agency (EPA) has launched a series of studies based on direct measurement of noise from wind turbines at neighbors to wind turbines.
Sweco has conducted two separate studies, where measurements have been conducted during the day period for one site and day and night period for the second site.
The purpose of these studies is to gain knowledge about whether it is realistic to implement documentation of noise from wind turbines through direct measurements at neighbors. Furthermore, whether special noise issues occur outside the wind speed range of 6 m/s to 8 m/s, that the current regulations include, and which may give rise to reassessment of the regulations. Where the first study focuses on more general conditions during the day the second study includes measurements during night time.
Beside the studies conducted for the Danish EPA a third study at a, for Danish circumstances, large windfarm has been conducted.

This paper focuses on the knowledge gained through these three studies. The main points will concern the following:

- Is it possible to measure the noise from the wind turbine at the neighbor dwellings?
- The difficulties experienced during the measurements and data handling.
- Correlation between measured noise and measured windspeed at different positions.

2. General setup and site description

In the following the main characteristics of the three studies will be presented. Starting with the two studies conducted for the Danish EPA, which has similar setups and motivation. Ending with the third study which aim was long term surveillance of noise levels at different dwellings. The project report for study number 2 has been released from the Danish EPA [1] the first study remains to be officially released.

2.1 Study no. 1

For this study a site in western Jutland was investigated. The wind farm consisted of 6 wind turbines with a hub height of 90 meters and a rotor diameter of 120 meters, the combined effect of the wind farm is 36 MW. The turbines are placed on a straight-line oriented north to south. For this study 3 neighbours in different directions were selected, respectively south, north and west of the wind farm. The turbines are situated in a flat terrain surrounded by agricultural land, and scattered vegetation like small and large groups of trees, a site overview can be seen in Figure 1. The distance to the neighbour dwellings varied between 650- to 760 meters from the nearest turbine in the wind farm.

Figure 1: Site overview for study 1.

The measurements were conducted over three different campaigns. Each campaign was conducted at similar times during the day from around 09:00 in the morning to 16:00 in the afternoon. The setup consisted of 3 different positions.

- One position approximately between the nearest wind turbine and the dwelling in question.
- One position outside the dwelling.
- A setup indoors consisting of 3 microphones and measured according to the guidelines described in [2].

Beside the noise measurements the following non-acoustic parameters were obtained:

- Wind speed at one of the neighbours at 10-meter height.
- Operational parameters from the wind farm e.g. produced power, and wind speed at hub height.

2.2 Study no. 2

For this study a site in mid Jutland was investigated. The wind farm consisted of 3 wind turbines with a hub height of 94 meters and a rotor diameter of 112 meters, the combined effect of the wind farm is 9 MW. The turbines are situated on a straight-line oriented north to south. Three neighbours in different directions were selected, respectively east and west of the turbines. The turbines are situated in flat terrain surrounded by agricultural land and scattered vegetation like small and large groups of trees, all three neighbours have larger trees around their dwelling. A site overview can be seen in Figure 2. The distance to the dwellings varied between 650- to 920 meters from the nearest wind turbine.

Figure 2: Site overview for study 2

The measurements were conducted like the measurements in study number 1, with the similar positions and data sampling. However, the measurements were conducted over night and until the afternoon the day after, from around 21:00 in the evening until 16:00 the afternoon the day after.

2.3 Study no. 3

This study was conducted near a large wind farm in western Jutland. The wind farm consists of 22 wind turbines with a combined effect of 72,6MW. The turbines are situated in two parallel lines with 11 turbines in each line. For this study eight neighbours were selected located around the wind farm in different directions. The terrain around the wind farm is flat and a large part of the area is agricultural land.

The study was conducted as long-term measurements over a period of 17 days. Microphones were placed at eight different neighbours outside their dwellings. If the residents had special requests on the position of the microphone this was followed. The total A-weighted sound pressure level was recorded at all the neighbours. Beside the sound pressure level, the windspeed at 10-meter height was recorded at two of the neighbours. Operational parameters from the wind farm was not recorded during the measurement. However, operational data from one of the wind turbines has been obtained afterwards as 10 minutes averages and used in this analysis. The wind turbine is situated in the outer limit of the park.

3. Experiences & Results

In the following the results and experiences from the two Danish EPA studies are presented. For comparison two neighbours has been selected showing two different scenarios of neighbour measurements.

3.1 Short term noise measurements (study 1 and 2)

Common to the two studies is that the measurements are conducted over a short period (less than 24 hours). The purpose of dividing the study into several short term measurements is to cover as varying meteorological conditions as possible. The measurements are performed from December to June, so the measurements cover winter, spring and summer.

In Figure 3, one of the measurements campaigns from study no 1 is shown. The light blue (P5) represents the position between the wind turbine and the neighbour. The light green (P4) is the position near the neighbour, and finally the black, red and orange are the three microphones at the indoor position. The pink and blue respectively represents produced power and the normalized windspeed derived from the power curve at hub height

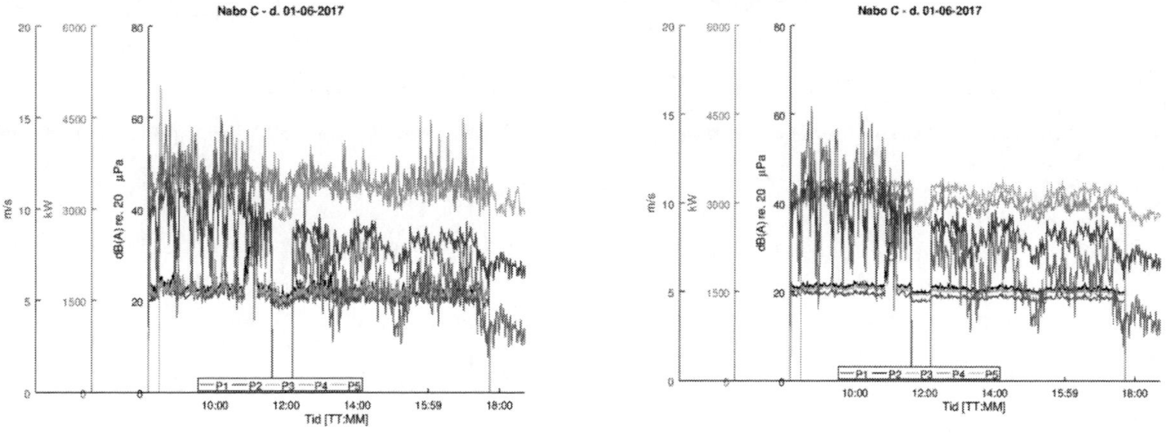

Figure 3: Results of the measurements at neighbor C for study 1. Left L_{pAeq} right L_{A90} 1-minute averages.

It is seen from the left graph in Figure 3 that the noise measured at the neigbour position outdoor is comparable and in some cases higher than the noise measured close to the windturbine. This indicates that the noise measured near the neighbour is contaminated by other spurious noise sources in the area. In order to exclude as much as possible of the spurious noise sources e.g. car passings or activities in the dwelling the noise indicator L_{A90} is chosen as a suitible indicator of the steady noise measured at the nieghbour. On the right side in Figure 3 a time slice is shown with L_{A90} as the indicator. As opposed to the left side the outdoor levels are now more clearly separated. However it is seen that the level at the neighbour and closer to the tubine in some cases are the same. During the measurement the turbines were shut down for a short period, and a drop in the measured noise level at all positions can be seen indicating that the noise from the wind turbines are contributing to the measured noise in all positions.

Noise from wind turbines have a strong correlation with wind speed, therefor the noise measured at the neighbour are compared to the windspeed observed at the wind turbine. In order to show the complexity of the measurements, measurements with high background levels have not been excluded from the plots. The plots are seen in Figure 4, Figure 5 and Figure 6. Kanal 1-3 is indoor, Kanal 4 is outside the house and Kanal 5 is close to the wind turbines.

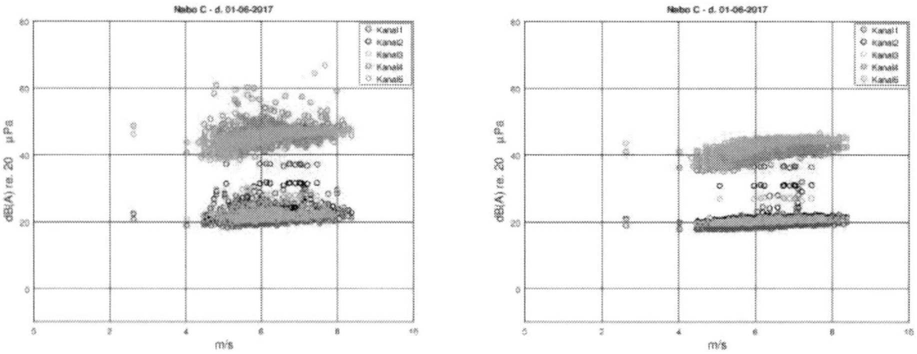

Figure 4: Results of the measurements at neighbor C for study 1. Left L_{pAeq} right L_{A90} 1-minute averages.

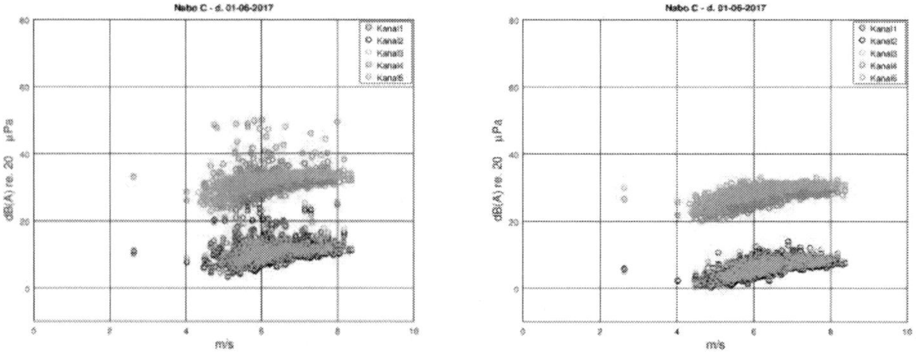

Figure 5: Results of the measurements at neighbor C for study 1. Low frequent levels left L_{pALF} right L_{A90LF} 1-minute averages.

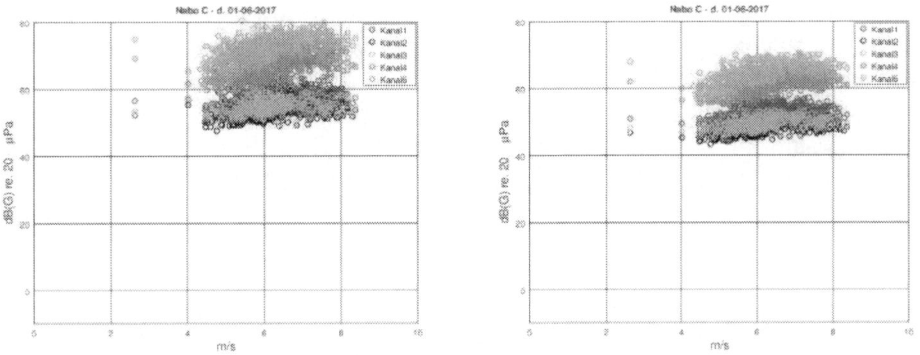

Figure 6: Results of the measurements at neighbor C for study 1. Infrasound (G-weighted) left L_{Geq} right L_{G90} 1-minute averages.

It is seen that there is a correlation between the measured noise levels and the wind speed for all the measurements. For the measurements outdoor light green and light blue, the correlation is visible for all three indicators. The pattern of the measured noise does follow the regular pattern for a pitch regulated wind turbine indicating that the noise is affected by the wind turbine. For the indoor measurements the correlation is weak when using the L_{A90} indicator (probably due to influence from background noise). However, when the L_{pALF}[1] indicator and to some extend L_{pG} is used a correlation is seen even at low levels. This indicates that the background noise in these frequency ranges is low.

To see if the noise is merely a result of increased background noise due to the rise in windspeed and hence more noise from vegetation, the measurement is scrubbed from outliers and afterwards bin analyzed. In Table 1 the results from neighbor C is shown. It is seen that there is a correlation between windspeed and the measured noise level, and that there is a difference between the turbines running or shut down. For the indoor low frequent level and the outdoor L_{A90} level the distance to the background noise is good enough to count the results as liable for the measure of the wind turbine noise at the neighbor, when corrected for background noise.

Windspeed	4 m/s	5 m/s	6 m/s	7 m/s	8 m/s	Background noise
L_{A90} Indoor P1-3	19.3	20.4	21.0	24.2	20.7	19.6
$L_{pA,LF}$ Indoor P1-3	7.5	12.2	12.7	13.4	13.9	8.5
L_{A90} Outdoor P4	36.5	38.5	40.3	41.7	42.3	37.8
L_{pG} Indoor P1-3	53.0	56.8	56.7	58.3	56.7	55.0

Table 1: Bin-analyzed results at neighbor C for study 1, in dB [re 20µPa]. The background noise is representative for 7-8 m/s

For the second study a similar analysis as in study 1 has been conducted. For comparison with the results obtained study 1, and to show the difficulties that can arise for a measurement with a higher degree of influence from background noise.

[1] L_{pALF} is the low frequent part of the A-weighted sound pressure level, given in dB [re 20µPa] for the 1/3 octave bands from 10 Hz to 160 Hz.

As for study 1 the correlation between wind speed and measured noise is investigated. In Figure 7, Figure 8 and Figure 9 the measured levels are shown with light blue for the position closest to the turbine light green for the outdoor position at the neighbour and red, orange and black for the indoor positions at the neighbour. One of the first differences compared to study 1 is that the pattern recognized from a pitch regulated turbine is not clear for the different positions except for L_{A90} for the position closest to the turbine. Instead the measured levels seem to follow a more linear tendency for all the indicators. It is also seen that the level measured outdoor at the neighbour is higher than the measured level close to the turbines for all wind speeds indicating that the measured noise does not inherit from the turbines but from other noise sources.

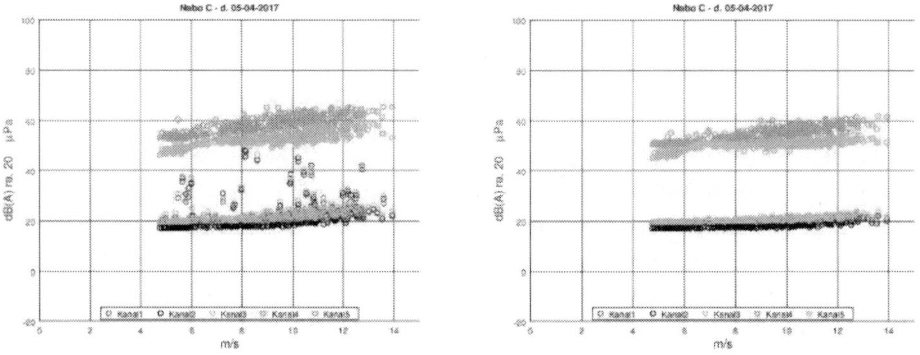

Figure 7: Results of the measurements at neighbour C for study 2. Left L_{Aeq} right L_{A90} 1-minute averages.

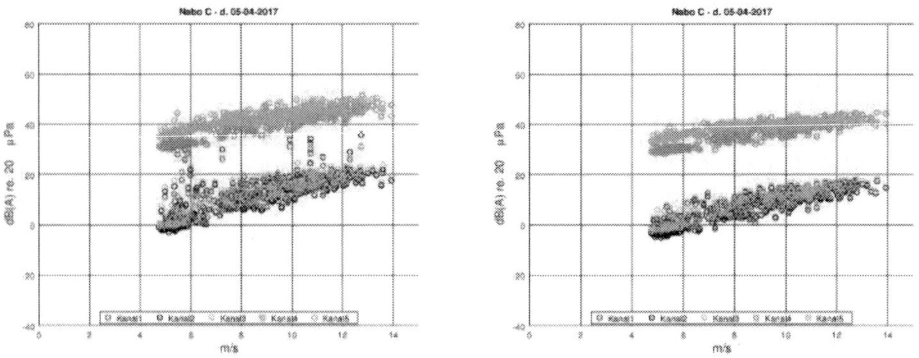

Figure 8: Results of the measurements at neighbour C for study 2. Low frequent levels left L_{pALF} right L_{A90LF} 1-minute averages.

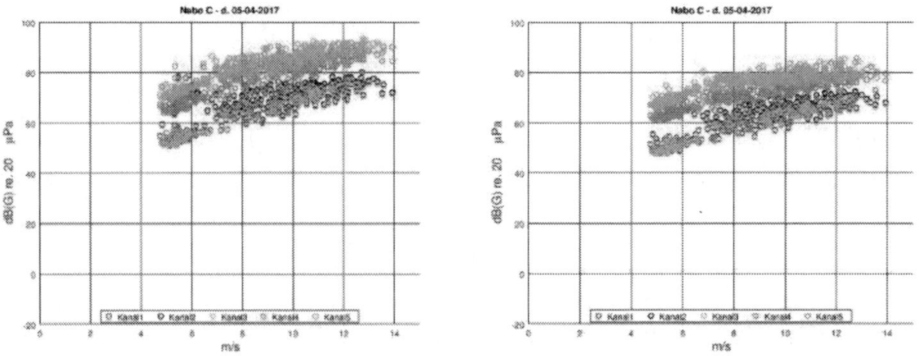

Figure 9: Results of the measurements at neighbour C for study 1. Infrasound (G-weighted) left L_{Geq} right L_{G90} 1-minute averages.

As for study 1 the measured data has been scrubbed for outliers and bin sorted. In Table 2 the result of the analysis is shown. For the indoor indicator L_{A90} the noise with the turbines turned on or off is approximately the same, and hence it is not possible to conclude anything about the noise from the turbines. For the indoor L_{pALF} indicator a difference is seen for low windspeeds. However, as the windspeed rises the difference between the background level and the total level diminishes, and it is difficult to deduct if the noise arises from the turbines, the same occurs for the G-weighted level. For the outdoor L_{A90} level a difference is seen for low wind speeds, but as the windspeed increases the background noise levels reaches the same level as the total level indicating that other noise sources at high windspeeds dominates.

Windspeed	L_{A90} indoor		L_{pALF} indoor		L_{pG} indoor		L_{A90} outdoor	
	Total noise	Bg-noise	Total noise	Bg-noise	Total noise	Bg-noise	Total noise	Bg-noise
3 m/s	18.1	-	0.2	-	46.1	-	33.4	-
4 m/s	18.0	17.9	10.9	-5.6	61.3	43.7	39.1	33.2
5 m/s	18.0	18.0	9.8	-2.7	56.3	46.7	36.4	35.1
6 m/s	18.0	18.0	1.3	4.4	54.7	55.8	42.6	35.6
7 m/s	18.1	18.0	7.8	-3.7	60.0	44.4	47.0	34.1
8 m/s	18.2	18.5	7.0	7.6	60.0	61.0	48.6	52.1
9 m/s	18.5	18.6	10.3	9.3	62.8	62.4	52.1	53.0
10 m/s	18.7	18.7	11.0	9.2	64.4	62.5	53.6	53.9
11 m/s	18.8	18.9	11.3	10.9	65.2	63.9	54.3	54.4
12 m/s	19.0	19.0	11.6	12.6	65.3	66.5	54.9	55.5
13 m/s	19.3	-	13.0	-	66.2	-	56.5	-

Table 2: Bin-analysed results at neighbour C for study 2, in dB [re 20µPa].

As it is seen from the table above it can be difficult to extract valid information on the noise caused by the wind turbine from the background noise. In Figure 10 the background noise is marked with a black cross and total noise with light green in order to compare the levels experienced at the neighbour outdoor position. It is seen that the noise levels for total noise and background noise merges, indicating the complexity of recognizing the turbine at this neighbour for the total noise level.

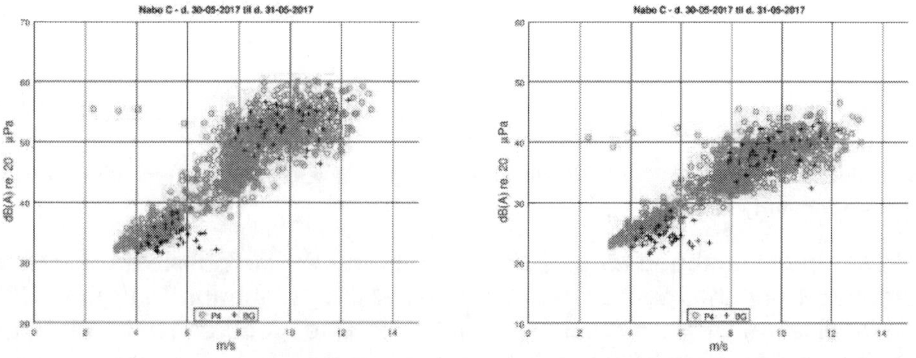

Figure 10: Results from neighbour C for study 2, left L_{A90} right L_{pALF}

From the measurements conducted in the two different studies for the Danish EPA, it is shown that it is, to some extent, possible to measure the noise from the wind turbine at a neighbour position. However, it is also seen that the environment around the neighbour can have an impact on the ability to measure the noise from the turbine at the neighbours. Another parameter that is of high importance is windspeed. As it is seen in the above the noise is correlated with the windspeed, and hence it is important to use the right windspeed in order to assure that the noise measured is due to the wind turbine and does not inherit from e.g. vegetation or wind induced noise in the house.

3.2 Concerns regarding determination of windspeed (Study 1)

When measuring noise at the neighbour dwellings the windspeed becomes a parameter that can be discussed. Following the Danish EPA statutory order [3] for measuring noise from wind turbines the wind speed is defined through the measured power output of the wind turbine. The determination of the wind speed is under Danish circumstances very important as the noise limits are connected to the wind speeds, 6 and 8 m/s. When measuring at neighbour distance from the wind turbines there are several possibilities for determining wind speed. It can be deduced from one of the turbines, it can be as the average of the wind experienced by the entire wind farm, or it can be measured at the neighbour dwelling at a height of 10 meter or more.

In Figure 11 two different measurements are shown. If there is a good correlation between the measured wind speed at the neighbour and det deduced wind speed from the power produced by the wind turbine, ideally they should form a line with a slope equal to 1. To the left it is seen that the wind experienced at the neighbour does not correspond very well with the wind measured at the turbine, and to the right a tendency is seen towards correlation between the two measured wind speeds.

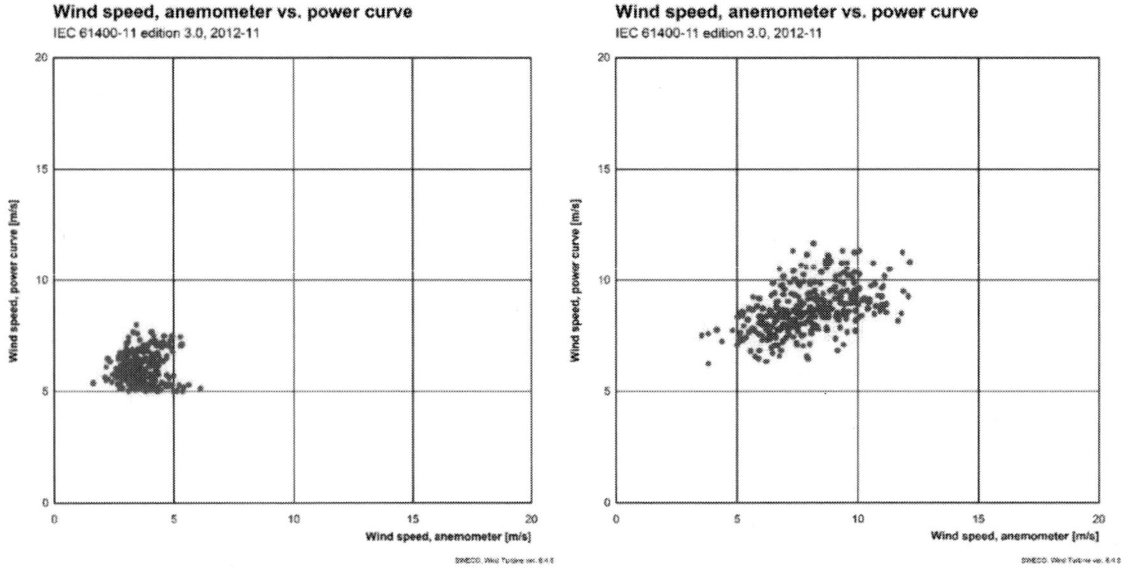

Figure 11: Left, Windspeed at neighbour B, right windspeed at neighbour A. (the measurements were obtained during two different campaigns for study 1)

Moving forward with the windspeed measured at neighbour A, it is possible to link the determined windspeed with the measured noise level at the neighbour. In Figure 12 the measured noise level is shown as a function of the local windspeed at 10 m height at the neighbour (left) and as a function of the windspeed at the wind turbine (right).

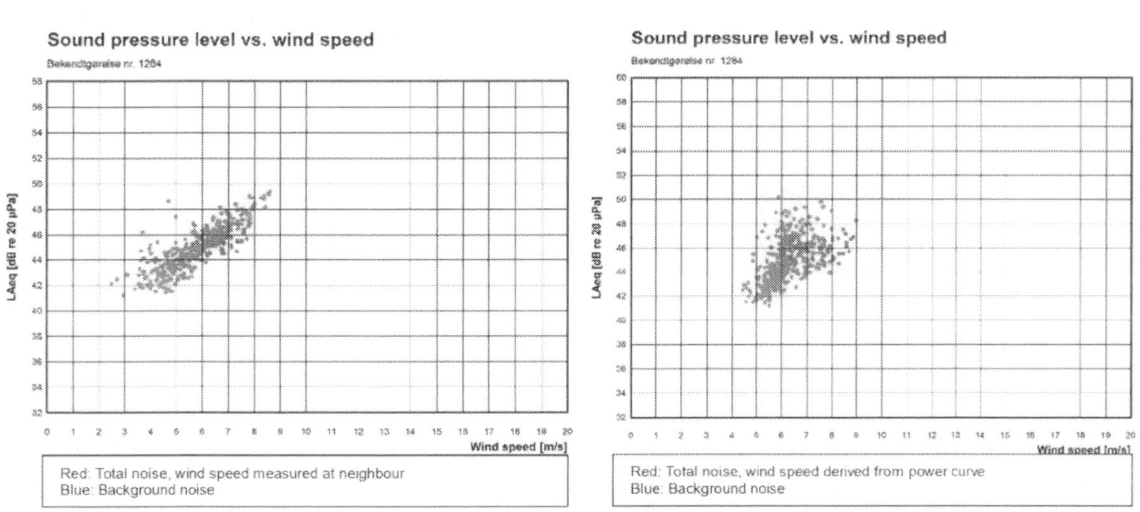

Figure 12: Left, noise as a function of local windspeed. Right, noise as a function of wind turbine windspeed.

It is seen that the noise measured at the neighbour dwelling correlates better with the windspeed measured at the dwelling. This indicates that the noise inherits from other sources than the wind turbines, and most likely from noise caused by the vegetation and/or the surroundings. If the windspeed only was measured at the dwelling it would be possible to draw the wrong conclusion concerning the noise correlation with the wind turbines production. This indicates the necessity of measuring the windspeed at the neighbour and in parallel measure the windspeed at the wind turbines. This will ensure that that the conclusion drawn is fair to the noise produced by the turbines.

4. Long term measurements (study 3)

The third study differs from the two above in duration and data available. For this paper the focus for the third study is on measurement of windspeed and concerns regarding placement of the wind mast. In the full study the noise measured at the eight neighbours have been correlated with the windspeed measured at the neighbours. All averages presented in the following is 10 minutes averages, and the noise indicator is in all cases L_{A90}. The measurements were conducted in April (spring). In the following, measurements at 3 neighbours have been selected. All measures are synchronous over 17 days.

4.1 Wind speed determination.

In Figure 13 the correlation between the wind measured at two different neighbours and the wind turbine is shown. The data from the wind turbine is deduced from hub height to 10 meters. The data is sorted in day/night-bins with night hours defined from 22:00 until 04:00 in the morning. As opposed to the large difference in correlation seen for study 1 this comparison shows a tendency in both cases. However, it is also seen for both cases that the higher windspeeds at the neighbours yields a better correlation with the windspeeds measured at the wind turbine. For neighbour F16 the night data is seen to be higher for the wind turbine than for the measured wind speed at the neighbour which is caused by some periods of high wind shear. For Neighbour No4 it is not as pronounced, however, the tendency is also visible there.

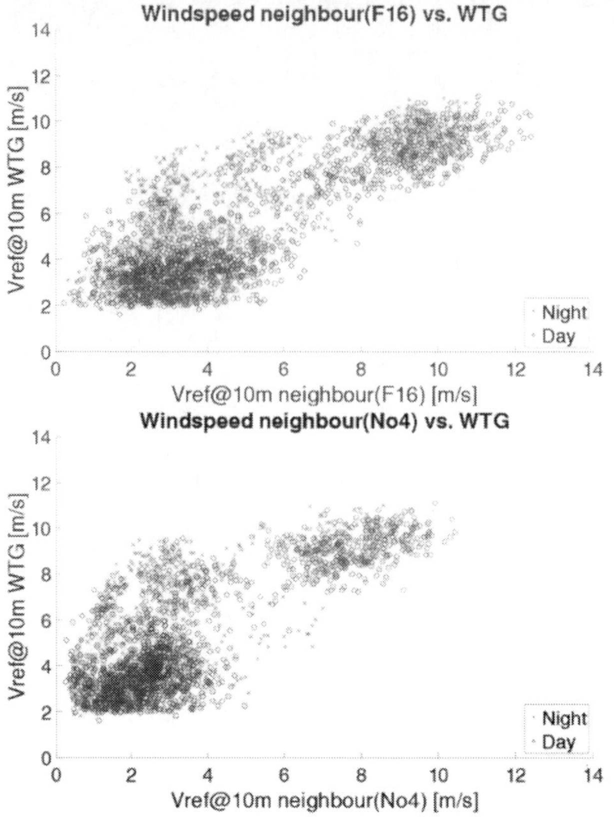

Figure 13: Upper windspeed at neighbour F16 correlated to the turbine wind speed. Lower windspeed at neighbour No4 correlated to the windspeed measured at the same turbine.

To further investigate the difference experienced in Figure 13, the wind direction is considered. In Figure 14 the relative difference in windspeed for the two neighbours is shown together with wind direction. Here 0° designates that upwind is north and 270° that upwind is west.

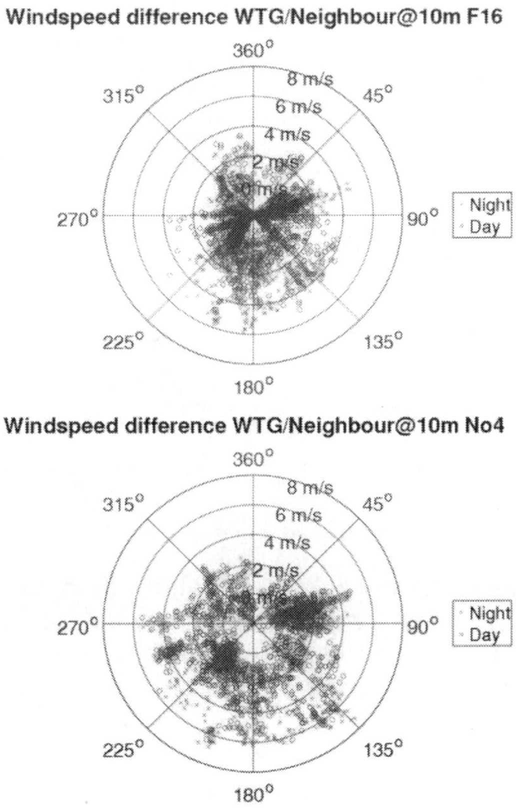

Figure 14: Upper: Relative difference in windspeed neighbour F16. Lower: Relative difference in windspeed neighbour No4.

It is seen for neighbour F16 that the highest differences are reached at night due to high wind shear but for a large part of the day the differences are within 2 m/s. For neighbour No4 the picture is different as the differences are higher, and it is not mainly due to high wind shear as it is seen during both night and day and from approximately 90° and until 270°. The explanation for the higher differences is in this case due to the placement of the wind mast. The wind mast is placed at the neighbour dwelling which is surrounded by small trees. Even though the wind mast is placed free of the trees, their influence is seen in the lower part of Figure 14.

As for study 1 the measured noise at the dwelling is shown as a function of wind speed measured at the wind turbine and at the neighbour. In this study two neighbours with different siting is chosen for comparison (Neighbour F16 and No20). At neighbour F16 there are some low trees and a beech hedge, at neighbour No20 there is low vegetation and an old machinery house. In Figure 15 the measured noise level is shown for the two neighbours as a function of respectively the windspeed at the turbine and at the neighbour. For neighbour F16 it is seen that the noise correlates better with the wind measured at the neighbour dwelling as in study 1. It is also seen that the noise at low windspeeds has an upper level, however when the windspeed increases above 7 - 8 m/s the noise continues to increase. For comparison the noise curve for the wind turbine is introduced in the plot. It is seen that the noise from the wind turbine reaches its maximum at around 7 - 8 m/s, which suggests that the noise measured above 7 m/s most likely does not inherit from the wind turbine, but from other noise sources e.g. the beech hedge.
For the other neighbour, No20, the same comparison is made. Here it is seen that the noise measured tends to follow the pattern of the wind turbine noise to some extent at the higher windspeeds compared to neighbour F16. This might indicate that the noise level measured at neighbour No20 is influenced by noise generated by the wind turbines.

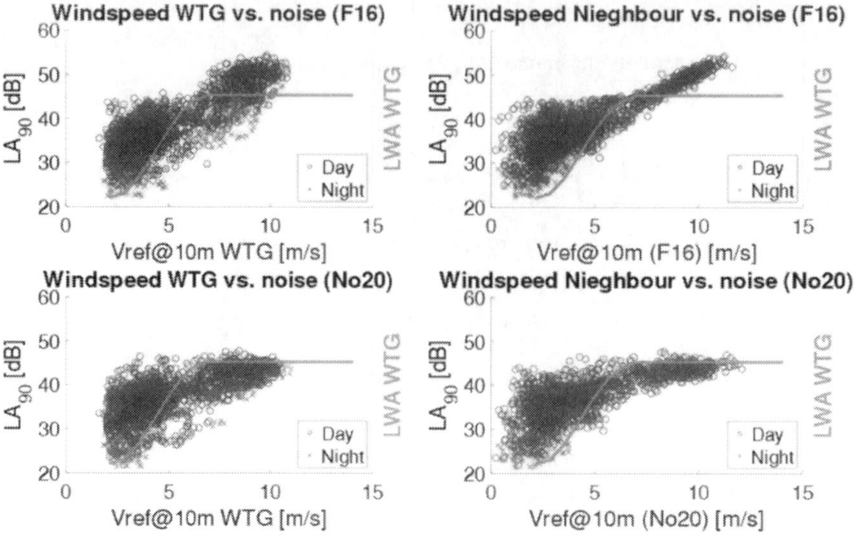

Figure 15: Left, noise as a function of wind turbine windspeed. Right, noise as a function of local windspeed.

Due to the occurrences of high wind shear it is known that the wind turbine in some periods during night hours has experienced higher windspeeds than what has been measured in 10 meters height at the neighbours. In these periods the noise generated at the wind turbine has been at a maximum and the background noise induced by vegetation must be assumed to be on a minimum. In Figure 16 the same data as in Figure 15 is shown, however, only data where the wind turbine has been producing more than 85% of rated power is shown. This is inspired by Payam Ashtiani, Aercoustics Ltd. For the measured noise as a function of the wind speed measured at the neighbour it is seen that the night hours show higher noise levels at low windspeeds, during high wind shear. These measurements could indicate the maximum wind turbine noise level at the neighbour, however due to the lack of background noise measurements from the same period it is not certain that the measured noise inherit from the turbine.

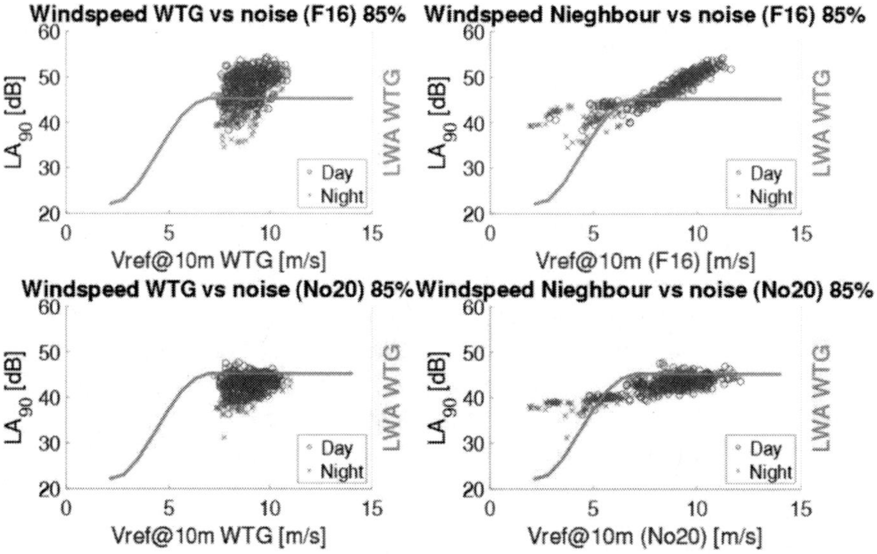

Figure 16: Left, noise as a function of wind turbine windspeed. Right, noise as a function of local windspeed. During a minimum production of 85% of rated power.

From the above it is seen that there are several critical issues when measuring at neighbour distance. The surroundings of the neighbour are decisive of what one can expect to measure at high windspeeds. A small amount of vegetation such as a beech hedge can introduce high levels of background noise at higher windspeeds and might already introduce it before the wind turbine reaches maximum noise generation. It is also seen that placement of the wind mast is critical in order to assure that there is a good correlation between wind speed measured at the turbine and at the neighbour and at all wind directions. The simultaneous measurement of wind speed ensures that periods with high wind shear and other phenomena's can be detected and accounted for, when drawing conclusions on the noise measured at the neighbour.

5. Conclusions

The two studies conducted for the Danish EPA, shows the difficulties that comes with measuring at neighbour dwellings. The results of the first study show that it is possible to evaluate the noise from a wind farm at a neighbour. The second study on the other hand show that it is or can be difficult to evaluate the noise from a wind farm through measurements at the neighbour dwellings, especially when there is vegetation around the dwellings. In order to secure a good measurement, there are some conditions which should be met before and during measurement:

- The background noise must be evaluated at each dwelling of interest, to secure that a satisfying signal to noise ratio can be achieved (6 dB). This should be investigated both for the outdoor measurement but also for the indoor measurements. Different appliances indoor can give significant contributions to the background noise, e.g. fridges, ventilation, oil furnaces and even clocks.
- During measurements it is important to have several and longer periods where the wind farm is shout down, so that the actual background noise during measurement can be evaluated. It is important to cover the same meteorological conditions for the background noise.
- Operational parameters from the wind farm shall be available during measurement together with windspeed measured at the neighbour dwelling. This is to secure that the measured noise levels can be compared to the expected noise from the wind turbine and vegetation at the dwelling.
- In order to exclude spurious noise sources statistical noise indicators should be used. Knowledge must be established of how to apply the statistical levels especially for low frequency measures. It is however, important to notice that the statistical levels do not exclude continuous noise such as noise caused by vegetation.

For the third study the necessity of windspeed measurements at several positions is shown. The measurements can when correlated with the measured noise levels give valuable information on what may cause the noise measured at the neighbour dwelling. Especially for neighbour F16 the correlation with the local measured windspeed indicates that the noise is induced by vegetation rather than the wind farm. The measurements performed under high levels of wind shear can indicate the level of wind turbine noise at the neighbour. However, it is necessary to have thorough knowledge of the background noise and the representative windspeed and direction for the entire wind farm to ensure that the shear conditions are common for the wind farm.

From what can be derived from the 3 studies there is a strong indication that the Danish method with sound power measurements and a simple prediction method complies with the measurement at relevant Danish distances.

References

[1]. Søndergaard, B. Heiden, J. and Christensen, N.F (2018) Måling af støj hos vindmøllenaboer, mindre vindmøllepark målinger i dag og nat perioden. (*Measurement of noise at neighbors to wind farms, smaller wind farm measurement at day and night*) Miljøportalen nr. 2048.

[2]. Orientering fra miljøstyrelsen, nr. 9 1997, Lavfrekvent støj, infralyd og vibrationer I eksternt miljø. *(Orientation from the Danish Environmental Agency, no. 9 1997. Low frequent noise, infrasound and vibrations in external environment)*

[3]. The Danish EPA, 2015 BEK 1736 Bekendtgørelse om støj fra vindmøller. (*Statutory Order on Noise from Wind Turbines; Statuary Order no. 1736.)*

INCE EUROPE

8th International Conference
on
Wind Turbine Noise
Lisbon – 12th to 14th June 2019

Atmospheric absorption in sound power measurements of wind turbines

Tomas Rosenberg Hansen, SGRE (DK), tomas.hansen@siemensgamesa.com
Lars Enggaard, SGRE (DK), lars.enggaard@siemensgamesa.com

Summary

Attenuation of sound due to atmospheric absorption varies for different meteorological conditions and will cause variability in the measured apparent sound power level for a wind turbine according the IEC 61400-11 standard. But the attenuation is not considered in the measurement method described in latest version of the standard.

This document describes:

- theoretical effect of variations in atmospheric absorption on a measured sound power level 1/3 octave spectrum from a modern wind turbine
- a method to take the atmospheric absorption into account in the sound power measurement method described in IEC 61400-11
- the described method applied on real turbine measurements

The study shows convincing results giving lower variability of measured apparent Sound Power Level after applying the proposed correction for atmospheric absorption.

1. Introduction

Attenuation of sound due to atmospheric absorption varies for different meteorological conditions and is well described and standardized. But attenuation caused by atmospheric absorption is not considered in the wind turbine noise measurement method described in the IEC 61400-11 standard. Within typical northern European weather conditions, mainly cold and dry weather can result in significantly higher attenuation at frequencies in kHz ranges. The influence on the IEC sound power results can be significant on large modern turbines due to increased measurement distances. At Siemens Gamesa Renewable Energy we also observe that the main energy in the noise frequency spectrum has changed in the last decades from approximately 200-500 Hz to 2-3 kHz. Therefore, atmospheric absorption becomes more and more important when calculating the apparent Sound Power Level.

The present study proposes a correction method for atmospheric absorption and shows convincingly that this method gives a lower variability of the measured apparent Sound Power Level.

2. Air absorption effects on measured IEC noise levels

Attenuation of sound due to atmospheric absorption varies for different meteorological conditions and is sensitive to the composition of the air. Air absorption in general will not be discussed in this document, but the following investigations are based on the well described and standardized formulas given in ISO 9613-1:1993 [3].

2.1 Variations of atmospheric absorption

The molar concentration of water vapor in the air varies a lot over time close to ground level. The variation is related to seasonal changes at a specific geographical point and can even vary rapidly within a few hours. Day and night shifts or just rapid changes in weather will affect the molar concentration of water vapor in the air. Based on ISO 9613-1 these changes have a major influence on the atmospheric absorption.

The air absorption can be calculated from equation 3-5 in ISO 9613-1:1993 [3]. Below is equation 5 in the ISO standard given as equation 1 in this document:

α is the pure tone attenuation coefficient, given in dB/m for atmospheric absorption, and is calculated using equation 1 below:

Eq. 1

$$\alpha = 8.686 f^2 \left(\left(1.84 \times 10^{-11} \left(\frac{P_a}{P_r} \right)^{-1} \left(\frac{T}{T_0} \right)^{1/2} \right) + \left(\frac{T}{T_0} \right)^{-5/2} \times \left(0.01275 \left(\exp\left(\frac{-2239.1}{T} \right) \right) \left(f_{rO} + \left(\frac{f^2}{f_{rO}} \right) \right)^{-1} + 0.1068 \left(\exp\left(\frac{-3352.0}{T} \right) \right) \left(f_{rN} + \left(\frac{f^2}{f_{rN}} \right) \right)^{-1} \right) \right)$$

Where:

p_a ambient atmospheric pressure [kPa]
p_r reference ambient atmospheric pressure [kPa]
T ambient atmospheric temperature [K]
T_0 reference air temperature [K]
f frequency on sound [Hz]
f_{rO} oxygen relaxation frequency [Hz], can be calculated using equation 2 below
f_{rN} nitrogen relaxation frequency [Hz], can be calculated using equation 3 below

Eq. 2

$$f_{r0} = \frac{p_a}{p_r} \left(24 + 4.404 \times 10^4 h \frac{0.02 + h}{0.391 + h} \right)$$

Eq. 3

$$f_{rN} = \frac{p_a}{p_r} \left(\frac{T}{T_0} \right)^{-1/2} \times \left(9 + 280 h \times \exp\left(-4.170 \left(\left(\frac{T}{T_0} \right)^{-1/3} - 1 \right) \right) \right)$$

Where:

h molar concentration of water vapor, as a percentage (calculated from relative humidity)

Equations 1-3 correspond to the equations 3, 4 and 5 in ISO 9613-1:1993 standard [3]. In the standard it is mentioned that *"these equations are all that is needed to calculate the pure-tone attenuation coefficient for atmospheric absorption for selected values of the variables … Humidity data … are rarely supplied in terms of molar concentration of water vapour. Annex B* ..[in the standard].. *provides information on conversation of humidity, dewpoint and other measures, to corresponding values of molar concentration."*

Temperature, relative humidity (RH) and atmospheric pressure (and frequency and distance) are the variables that change the air absorption levels. Since variation of atmospheric pressure for a given geographical position is relatively small and the atmospheric pressure has minor influence on the air absorption, variations in atmospheric pressure are generally negligible. Therefore, all calculations in this study are based on a fixed atmospheric pressure of 1 ATM.

2.2 Method for normalization of atmospheric absorption

To reduce the variability caused by variation in temperature and relative humidity, it has been considered to correct for the absorption and thereby calculate the "real" apparent sound power level based on pure results from the IEC61400-11 ed. 3.0:2012 [1]. Since the atmospheric absorption is very high at high frequencies, this will result in a dramatic change of the overall sound power level, mainly caused by high frequencies in the spectrum. Since the measured sound power spectrum in the high frequency range often has a large uncertainty due to poor S/N-ratio, the air absorption *correction* will cause an even larger variation in the results due to background noise.

For this reason, a method was developed which *normalizes* the measured sound power level for the effect of atmospheric absorption, i.e. calculating the spectrum that would have been measured under reference conditions.

The *corrected* sound power level, $L_{W,C}$, can be calculated by correcting the measured sound power level spectrum, $L_{W,m}$, for each 1/3 ovate band, i, using equation 4 below:

Eq 4

$$L_{W,m,i} = L_{W,C,i} - R_1 \alpha_{act,i}$$

$$\Updownarrow$$

$$L_{W,C,i} = L_{W,m,i} + R_1 \alpha_{act,i}$$

Where:

$L_{W,m,i}$ is the measured sound power in the 1/3 octave band with center frequency i

R_1 is the slant distance between rotor center and microphone according to IEC 61400-11

$a_{act,i}$ is the atmospheric absorption coefficient at frequency i, at actual atmospheric conditions

The *normalized* sound power in the 1/3 octave band with center frequency i, called $L_{W,N,i}$, at reference conditions is given by equation 5:

Eq 5

$$L_{W,N,i} = L_{W,C,i} + R_1 \alpha_{ref,i}$$

Where:

$L_{W,C,i}$ is the corrected sound power in the 1/3 octave band with center frequency i

Inserting the corrected sound power, $L_{W,C,i}$,from equation 4 in equation 5, the normalized sound power level, $L_{W,N,i}$, is then given by equation 6:

Eq 6 $$L_{W,N,i} = L_{W,m,i} + R_1(\alpha_{act,i} - \alpha_{ref,i})$$

Where:

$L_{W,N,i}$ is the measured sound power in the 1/3 octave band with center frequency i

R_1 is the slant distance between rotor center and microphone according to IEC 61400-11

$a_{act,i}$ is the atmospheric absorption coefficient at frequency i, at actual atmospheric conditions

$a_{ref,i}$ is the atmospheric absorption coefficient at frequency i, at reference conditions

It has been decided to use the following reference conditions for this study:

- Atmospheric pressure: 1013.25 hPa corresponding to 1 ATM
- Temperature: 15°C
- Humidity: 80% RH

These reference conditions are chosen because atmospheric pressure and temperature are equal to the reference conditions for the power curve used in the IEC 61400-11 ed. 3.0. A relative humidity of 80%, which correspond to the relative humidity used in the Danish legislation, BEK 135 [4], as standardized value for the propagation calculation. It seems to be a reasonable mean humidity for windy areas in northern Europe.

2.3 Calculated effect of atmospheric absorption for realistic turbine spectra

The possible effect of air absorption on the apparent sound level was calculated for a modern Siemens Gamesa turbine with a 130-m rotor diameter and equipped with the latest serrations and vortex generators. Table 1 shows the normalized overall sound levels for different atmospheric conditions, representative for a north European site.

The reference condition is indicated in red and errors within ±0.5 dB are shown as green text. It is interesting that the sound power level is never overestimated by more than 0.5 dB. However, depending on the conditions the apparent sound level may be underestimated by as much as 2 dB or more.

RH	Temp [Deg C] -15	-10	-5	0	5	10	15	20	25
30	0.90	-0.64	-2.35	-3.04	-2.74	-2.18	-1.64	-1.20	-0.89
35	0.50	-1.19	-2.63	-2.86	-2.39	-1.82	-1.31	-0.92	-0.68
40	0.10	-1.64	-2.72	-2.62	-2.08	-1.52	-1.04	-0.71	-0.52
45	-0.28	-1.96	-2.68	-2.38	-1.81	-1.26	-0.83	-0.53	-0.39
50	-0.63	-2.18	-2.57	-2.15	-1.57	-1.04	-0.65	-0.39	-0.30
55	-0.95	-2.30	-2.43	-1.93	-1.35	-0.86	-0.49	-0.28	-0.22
60	-1.22	-2.34	-2.28	-1.73	-1.16	-0.70	-0.36	-0.18	-0.17
65	-1.44	-2.34	-2.12	-1.55	-1.00	-0.55	-0.25	-0.11	-0.12
70	-1.62	-2.30	-1.97	-1.39	-0.85	-0.43	-0.16	-0.04	-0.09
75	-1.76	-2.23	-1.82	-1.23	-0.71	-0.32	-0.07	0.01	-0.06
80	-1.86	-2.15	-1.68	-1.10	-0.59	-0.22	0.00	0.06	-0.04
85	-1.92	-2.05	-1.55	-0.97	-0.48	-0.13	0.06	0.10	-0.03
90	-1.96	-1.96	-1.42	-0.85	-0.38	-0.05	0.12	0.13	-0.02
95	-1.97	-1.86	-1.30	-0.74	-0.29	0.02	0.17	0.16	-0.01
100	-1.97	-1.76	-1.19	-0.64	-0.21	0.08	0.21	0.18	-0.01

Table 1: Variation in the apparent sound power level under different atmospheric conditions for a modern Siemens Gamesa turbine, assuming the given source spectrum.

In order to show an example on a measured spectrum, the result of a measurement from a cold but wet winter day is shown in figure 1. The measured spectrum is shown as a bold black line.

The thin dashed magenta line shows a spectrum where only the correction has been applied and not the normalization (using only the correction calculated in equation 4). It is noticeable that there is a high correction at maximum level due to the relatively high peak frequency in the spectrum. The background noise level was low at high frequencies during the measurement, but it still results in unrealistic high levels from approximately 5 kHz and up. Due to too low S/N-ratio at the highest frequencies the overall level will probably be predicted wrong and is therefore not shown for the reference spectrum. The background noise levels dominate the levels at the highest frequencies.

Dotted bold curves are spectra normalized to different atmospheric conditions. These curves illustrate that differences up to more than 1 dB can occur in the apparent sound level, depending on the atmospheric conditions. These differences are currently not accounted for in the IEC standard.

Keeping in mind that the purpose of measuring sound power level is to predict sound pressure levels further away from the turbine, the real sound power level is irrelevant for longer distances and thereby for neighbor noise levels, because the air absorption affected frequencies will anyway be damped a lot in the propagation calculations towards the neighbor.

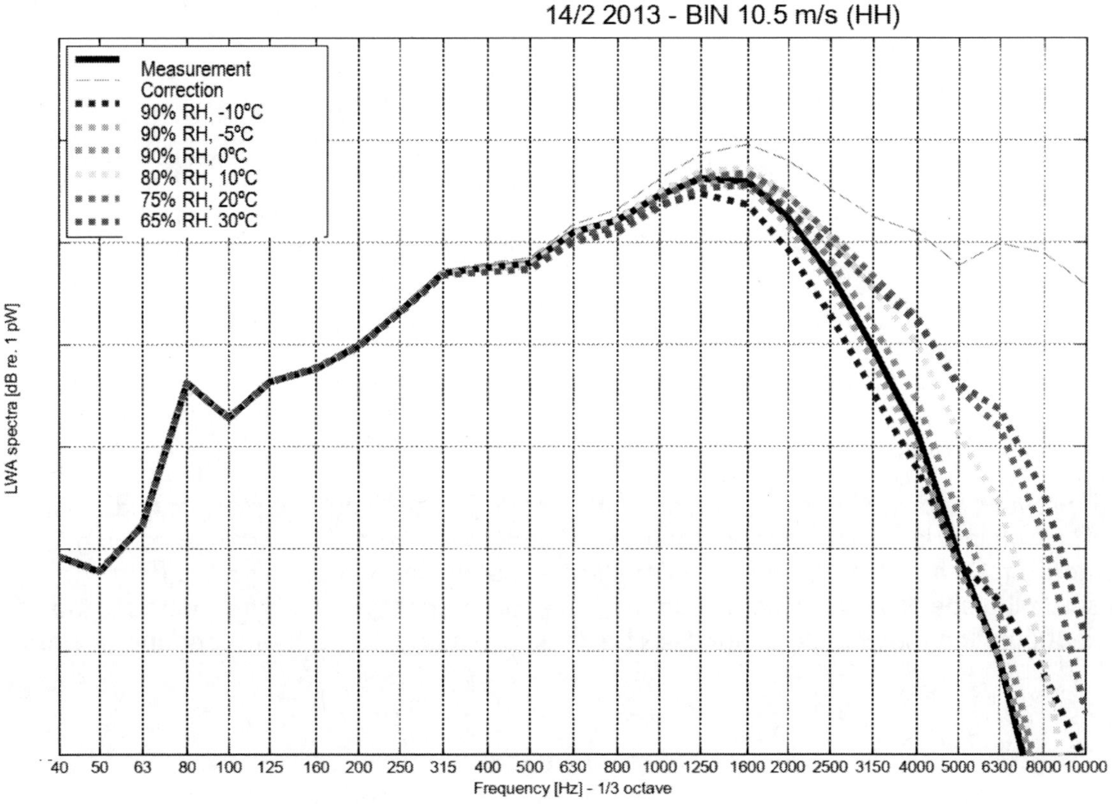

Figure 1: Atmospheric absorption correction and normalization applied to a measured spectrum of a SiemensGamesa 108 m rotor. There is 5 dB between horizontal lines in the plot.

To illustrate this effect on smaller and older turbine types, appendix 1 shows similar results for this and two other measured turbine spectra. The data shows that the influence of atmospheric absorption is more prominent for large modern turbines due to IEC measurement distance and shape of the sound power spectrum.

2.4 Experimental evidence for atmospheric absorption

Until now we have calculated the potential effects of air absorption by *assuming* the absorption values from the ISO standard. In this section we will show that these effects are real and can be observed in actual field measurements. Figure 2 below shows that normalization of air absorption results in a substantially better reproducibility of the measurement results. The data are results of measured sound power level according to IEC 61400-11 3d 3.0:2012 [1]. The turbine rotor has a diameter of 130-m, and is equipped with the latest serrations and vortex generators.

The red curve was measured on October 8th, 2015 on a day with atmospheric conditions very close to reference conditions (10°C, 90% RH). The red curve spectrum was measured a few months later a winter day, January 22nd, 2016 (-2.7°C,71% RH). The left plot shows measured data according to IEC 61400-11 ed. 3.0 and the right plot shows the same spectra but normalized to reference conditions (15°C, 80% RH).

The normalization shows that the apparent sound power level of the January measurement was underestimated by approximately. 1.2 dB on BIN averages due to atmospheric absorption, compared to reference conditions. 10 sec data points were underestimated by up to 1.8 dB.

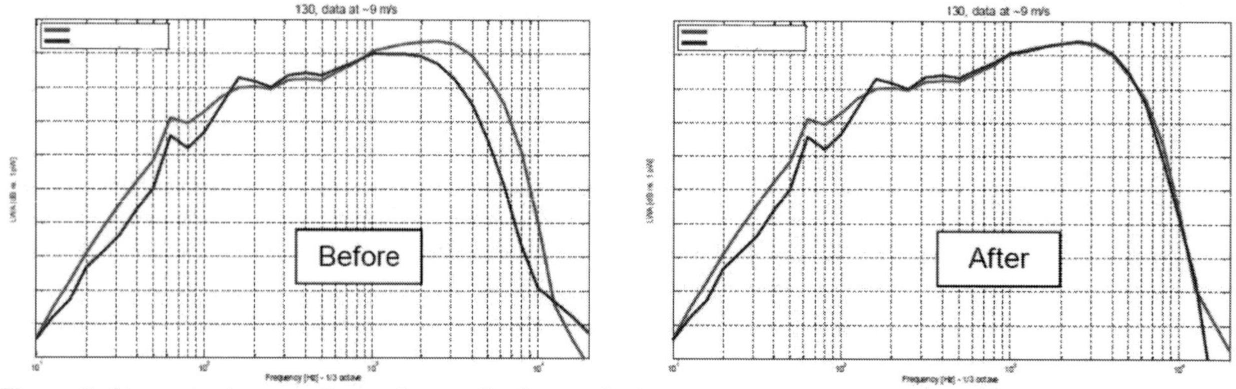

Figure 2: Atmospheric correction and normalization applied to a measured spectrum from two measurements of a SiemensGamesa 130 m rotor. There is 5 dB between horizontal lines in the plot. Left plot shows measured data according to IEC 61400-11, right plot shows results normalized to reference conditions.

Based on the results shown above, variations in atmospheric conditions, namely humidity and temperature, result in a large variation in apparent sound power level measured according to the IEC standard for a modern turbine. The variation is mainly seen in the high frequency range, but since the spectra for a modern turbine have more energy at high frequencies, the atmospheric absorption may cause a significant error in the measured apparent sound power level.

Figure 3 below show the sound power level obtained at three measurements on the same turbine at three different periods. The dashed curves are measured in October with higher air temperature. The blue is measured in January and the red curve is the same but normalized for atmospheric absorption. The normalization allows a higher repeatability. The difference in sound power level up to 9 m/s is due to other factors.

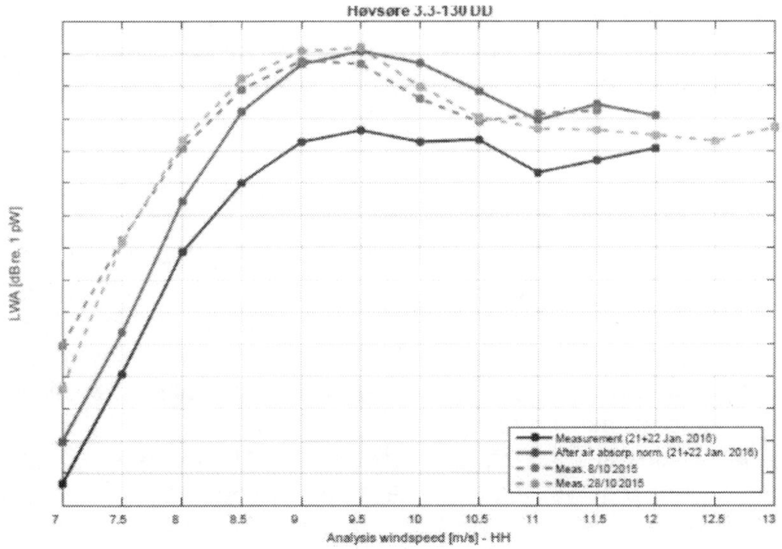

Figure 3: Atmospheric correction and normalization applied to a measured spectrum from three measurements of a SiemensGamesa 130 m rotor. The red curve shows the normalization of the measurement result shown as the blue curve.

There is 0.5 dB between horizontal lines in the plot.

2.5 Variation within one measurement campaign

But what happens if the measurement is carried out over a long period having significant variations in the weather conditions? E.g. a measurement carried out over 24 hours with change in weather conditions due to a front passage or a day/night shift causing temperature drop and change in the relative humidity.

If the 10 sec averaged measurement points are measured for different atmospheric conditions, this will result in large scatter of the points and air absorption normalization cannot be applied to the final bin-averaged noise curve.

According to the IEC 61400-11 ed. 3.0 standard, the spectra are calculated based on a bin analysis of all results. In the amendment from 2018 [2] some directions are given for cases where it is necessary to combine measurement series measured under different conditions. But it can be difficult to divide the measurement into series with equal conditions if the temperature or relative humidity changes slowly during the measurement.

The next chapter describes how this difficulty can be solved by correcting individual 10-s averages for absorption effects.

3. Sound power domain method

As mentioned at the end of the previous section, this chapter describes how 10-s averaged measurement points within one test period can be corrected for atmospheric absorption effects. The methodology in IEC 61400-11 ed. 3.0 is that the measured 10 sec values of background noise are binned in the wind speed domain. The 10 sec averages of total noise collected during turbine operation are binned separately from the background noise results. Background noise correction is carried out by correction bin center values of the total noise using bin center values of background noise.

Since background noise obviously is not affected by air absorption, normalization of uncorrected total noise values will result in significant errors due to the typical poor signal to noise ratio high frequencies.

3.1 Engineering method description

To reduce variability in internal measurements, it has been necessary to develop an engineering method based on the method given in IEC 61400-11 ed. 3.0. The idea is therefore to be able to normalize background corrected 10 sec data points separately.

The methodology deviates from the IEC methodology in the following points:

- Background noise is binned as described in the IEC standard.
- Total noise spectrum is background corrected in 10 sec values in the wind speed domain without binning, using calculated background noise spectrum from the bin analysis corresponding to the exact same wind speed.
- It is crucial that the binned background noise result is representative for the measurement period. This can be simply achieved by using different sets of background noise data, e.g. if microphone position was changed or other conditions changed during the measurement period.
- After background correction of the 10 sec total noise data, it is now possible to calculate sound power levels in terms of 10 sec data points. This opens also the opportunity to normalize for air absorption based on measured atmospheric conditions. Furthermore, any other correction can be added here.
- When data are highly affected by background noise, the air absorption normalization will result in obvious wrong results. To avoid this, we have introduced a requirement of minimum 5 dB S/N-ratio at each frequency to normalize for air absorption. This means that some 1/3 octave bands may not be normalized, but this occurs typically at very low or very high frequencies. For practical reasons we have furthermore decided not to normalize 1/3 octave bands above 8 kHz.
- Finally, the bin analysis is straightforward using the same principle as in the IEC standard, but now for the 10 sec sound power data points.

Because the sound power level data are now given in 10 sec data points, the method opens opportunities to study variability – even between different measurement series, time of the year, weather conditions, distance to turbine etc. An example is given in Section 3.2.

3.2 Example of results based on sound power domain method

A measurement was carried out in a period of approximately 8 hours on January 21st and 22nd. The measurement was done at daytime and night time with a significant shift in meteorological conditions. The temperature dropped from 0°C to -5°C during the measurement and the relative humidity changed between almost 100% and 65% RH.

The turbine with 130 m rotor was situated at DTU's prototype site in Høvsøre. Since the measurement was done at this site, there is access to a lot of meteorological data. The data from the meteorological mast was correlated in time with the acoustic data and data from the turbine. Figure 4 below shows temperature, relative humidity, atmospheric pressure during the measurement period.

The last plot in figure 4 shows the corresponding variation of normalization for air absorption calculated for the measured spectrum at each individual 10 sec periods. The air absorption normalizations on the overall noise levels range from approximately +0.5 to +1.8 dB.

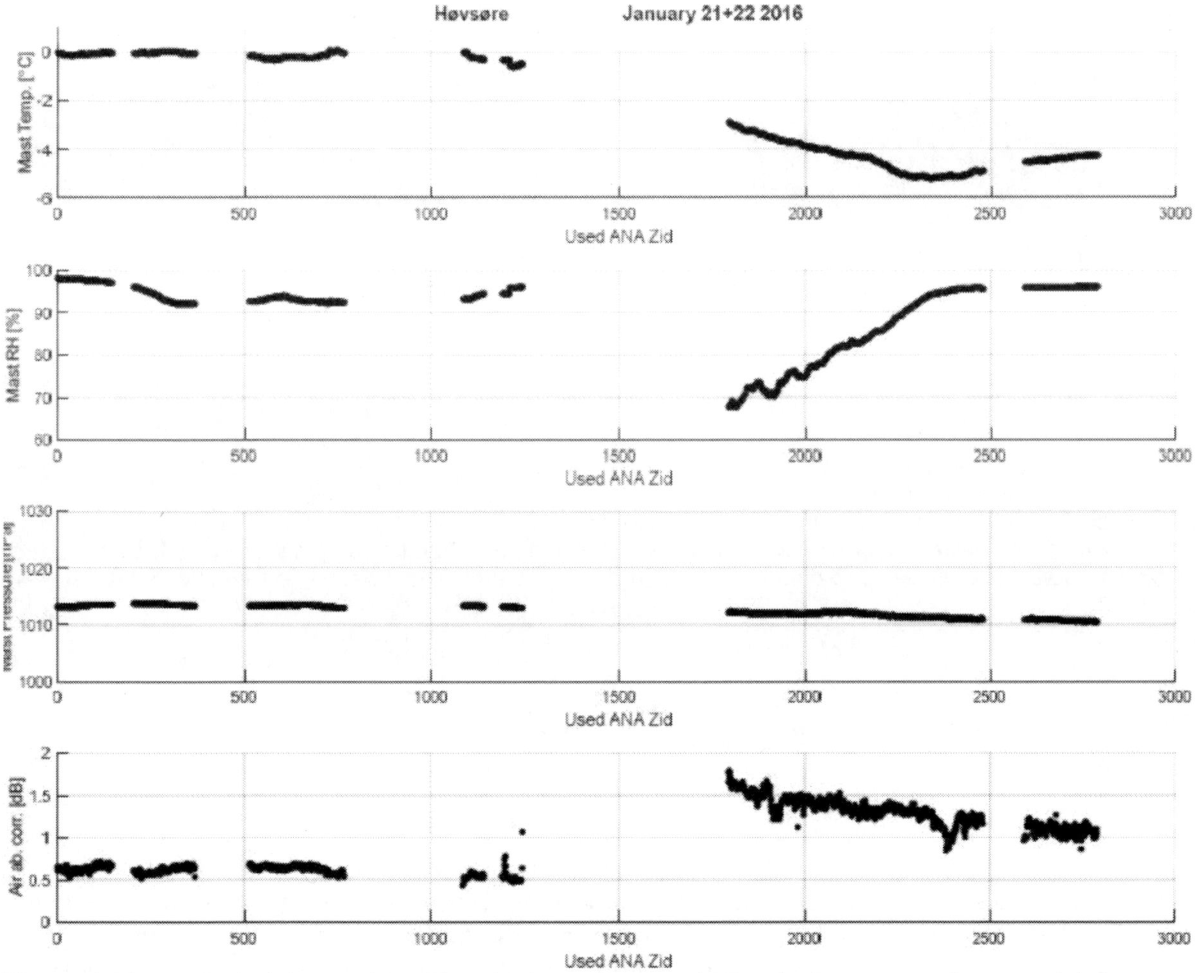

Figure 4: Meteorological data and resulting air absorption nomaization for the measured spectra duing a measurement with significant changses in the metrolodical conditions.

Figure 5 below show 10 sec data points in terms of sound power levels calculated using the methodology described above.

Blue circles are not normalized for air absorption, red dots are normalized. On the left plot measured and normalized data are shown as measured, with and without atmospheric absorption normalization. On the right plot a 1 dB offset is added to the measured data to show the difference in the scatter of the data. It is clear the scatter of the 10 sec sound power level data points has reduced after the data has been normalized for air absorption.

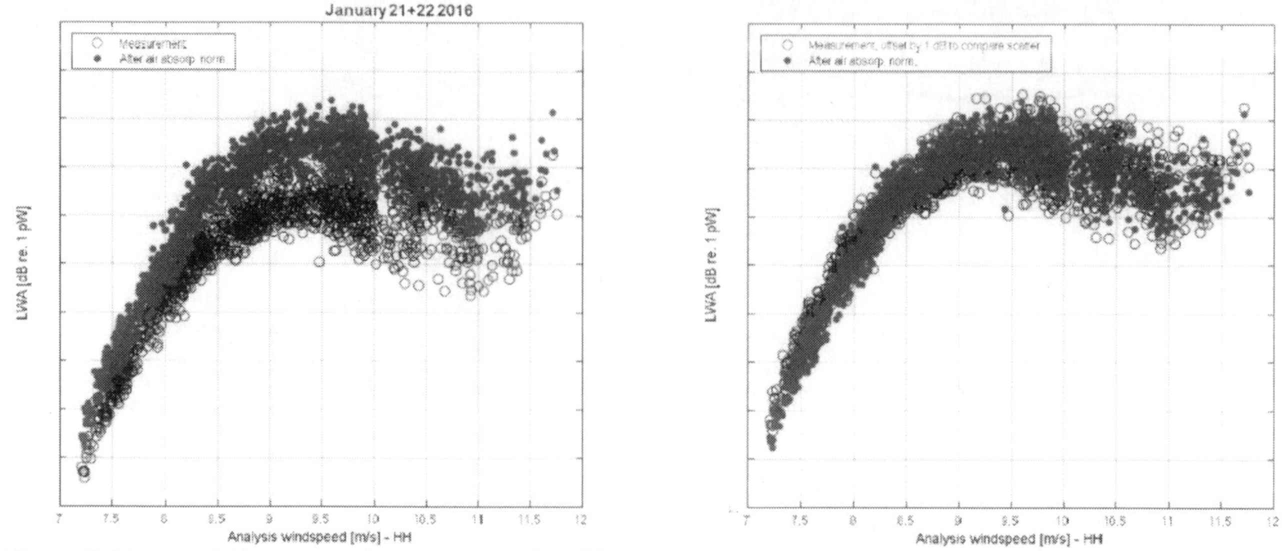

Figure 5: Measured 10 sec sound power data points. Blue circles are the measured data, red dots are data points where air absorption normalization is applied. Left is measured and normalized data points. In the right plot a 1 dB offset has been added to the measured data to show difference in scatter. The distance between horizontal lines is 1 dB.

4. Remarks about the normalization method

This section provides some remarks about the proposed correction method for atmospheric absorption effects.

Atmospheric absorption is given in dB/m. Large underestimation of the apparent sound level happens for spectra with most energy at high frequencies, where the atmospheric absorption is greatest. The suggested underestimation of sound power level from modern turbines at specific meteorological conditions is therefore negligible on predicted neighbor overall sound pressure level, because the atmospheric absorption will be even higher at large distances.

The results underline that it is important to compare noise performance of turbines based on noise propagation calculations using the correct spectra and the frequency dependent atmospheric absorption, and not only based on sound power levels.

There are several benefits of the proposed method.

- The methodology makes it possible to merge several measurement series recorded under different conditions in a consistent way.
- The method makes it possible to correct for other conditions that may vary during the measurement and can influence the results

Obviously, the methodology introduces some uncertainties:

- It is known that the major contributor to the noise from a modern wind turbine is the outer part of the blade. The slant distance used is the distance to the rotor center and not the distance to the primary noise source near the tip. This effect is not expected to be significant.
- Temperature and humidity are measured according to the IEC standard at 2 m height. The meteorological conditions may vary with height, which introduces some uncertainty in the normalization. This has been investigated based on one year of meteorological measurement on DTU's test site in Høvsøre. Atmospheric absorption is calculated at 2 kHz

based on pressure, temperature and humidity measured in 2 and 100 m height, see figure 6 below. Significant sessional variations are seen. But difference in resulting atmospheric absorption based on 2 m and 100 m data is less than 0.5, which is evaluated as an acceptable variation compared to the effort it takes to measure and correct for the error.

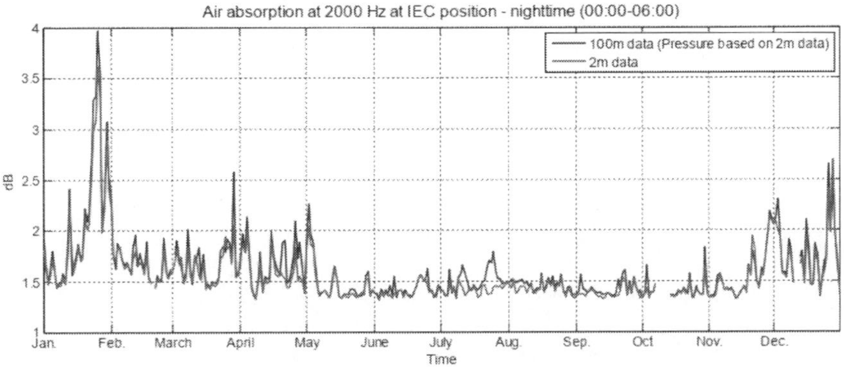

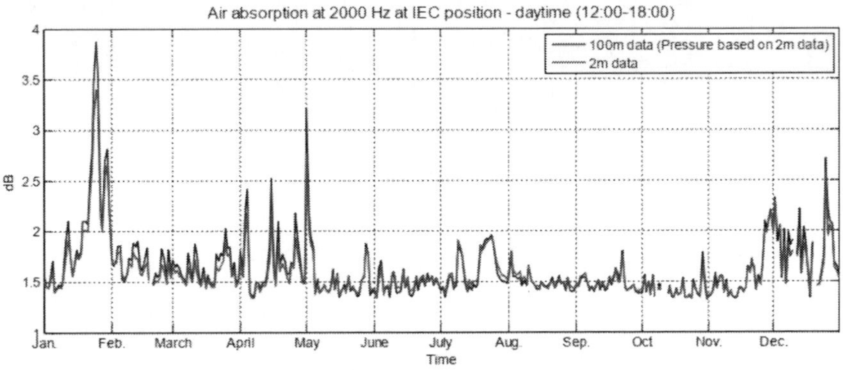

Figure 6: Atmospheric absorption is calculated at 2 kHz based on pressure, temperature and humidity measured in 2 and 100 m height at DTU's test site in Høvsøre. Upper plot is night time data, lower plot is daytime data.

Even though some uncertainties are introduced, this study suggests that the scatter of the data points is significantly reduced by introducing normalization for atmospheric absorption. Therefore, the methodology discussed in this document is proposed to be introduced in the next version of the IEC 61400-11 standard.

5. Conclusions

Studies of variability and repeatability of internal noise measurements according to IEC61400-11 performed by Siemens Gamesa suggest that one of the most important factors influencing variability is the attenuation due to atmospheric absorption. IEC results obtained in cold and dry conditions can be underestimated by up to 1.5 dB or more in overall A-weighted sound levels.

A procedure has been developed to correct measured sound levels for atmospheric absorption effects. The method can provide the sound power spectrum for each 10-sec data point. To minimize large errors introduced by the background noise level at higher frequencies, the developed analysis method is based on normalization to a chosen reference meteorological condition.

The study shows convincing results giving lower variability of measured apparent Sound Power Level after applying the proposed correction for atmospheric absorption.

6. References

[1] IEC 61400-11 ed. 3.0:2012
[2] IEC 61400-11 ed. 3.0 amendment A1:2018
[3] ISO 1996-2:1993
[4] BEK nr 135 af 07/02/2019 (https://www.retsinformation.dk/Forms/R0710.aspx?id=206666)

7. Appendix A

This Appendix presents calculations on three IEC measurements to show the possible influence of air absorption on the measured results. The three spectra have different shapes (energy distributions) to show that it matters for the influence of the air absorption on the overall levels. Results are only shown for hub height wind speeds of 10.5 m/s, which corresponds to the top of the noise curves.

The air absorption normalization is based on the method given in section 3.

Temperature, relative humidity (RH) and atmospheric pressure (and frequency and distance) are the variables that change the air absorption levels. All results shown are calculated with 1 ATM (1013.25 hPa).

The table below shows the three measurements:

Turbine	Rotor size [m]	Date	Hub height [m]	Slant dist. [m]	RH [%]	Temp. [°C]
1	93	7/11 2012	80	148	85	10
2	113	9/10 2014	90	173	88	16
3	130	14/2 2013	100	189	80	-1

Table 2: Data for measurements shown in this appendix

Turbine 1: this spectrum has most energy around 300 Hz
Turbine 2: this spectrum has most energy around 1 kHz, but is relatively flat
Turbine 3: this spectrum has most energy around 1.6 kHz

The three measurements are *corrected* to show a spectrum with "no" air absorption using equation 4, based on the corresponding temperature and relative humidity. Finally, this reference spectrum is recalculated to show the spectrum in different atmospheric conditions (temperature and relative humidity), based on equation 6, by *normalizing* to different atmospheric conditions. The results are shown in figure 6, 7 and 8.

The table below shows the calculated difference in sound power levels, that potentially would be measured in different combinations of temperature and relative humidity. The values are calculated based on equation 6, *normalizing* to different atmospheric conditions.

RH [%]	Temp. [°C]	Turbine 1 [dB]	Turbine 2 [dB]	Turbine 3 [dB]
90	-10	-0.29	-0.79	-0.82
90	-5	-0.16	-0.49	-0.37
90	0	-0.06	-0.22	0
80	10	0	0	0.30
75	20	-0.03	0.00	0.27
65	30	-0.14	-0.24	-0.08
Largest difference (-10°C to 30°C)		0.29	0.79	1.12
Largest difference (0°C to 20°C)		0.06	0.22	0.30

Table 3: Difference in calculated expected measured sound power levels at different atmospheric conditions

The data shows that the influence of atmospheric absorption is more prominent for large modern turbines due to IEC measurement distance and shape of the sound power spectrum.

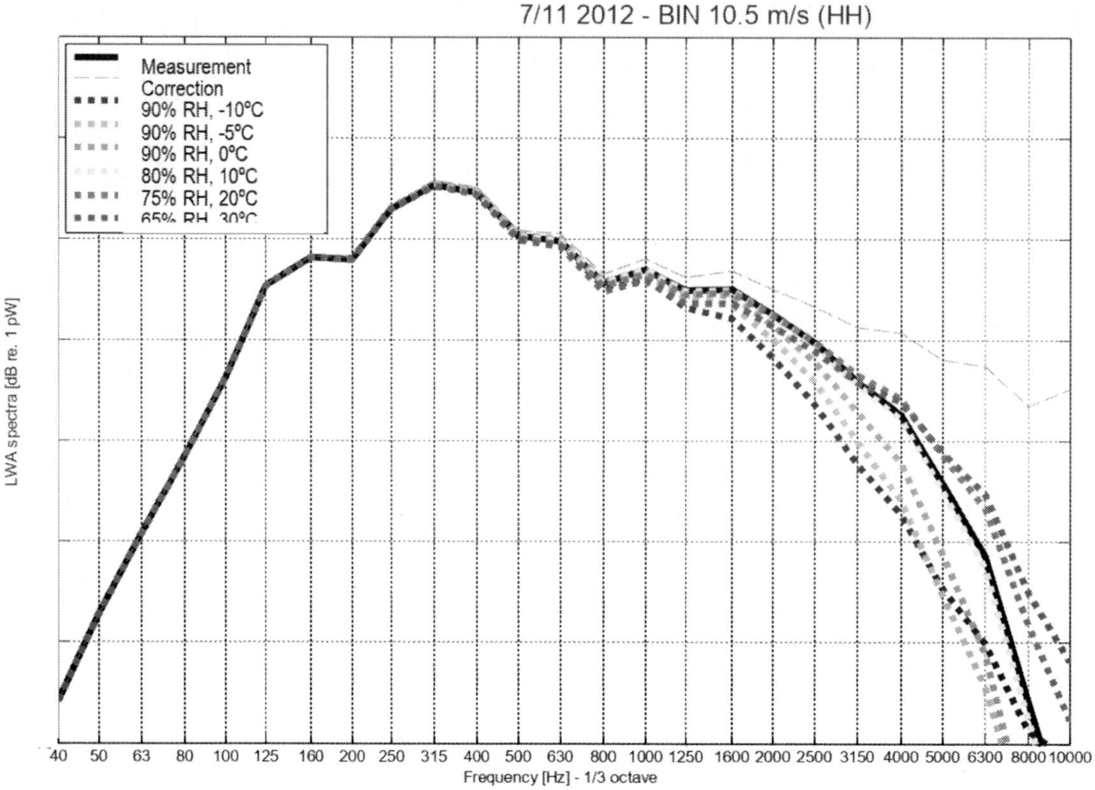

Figure 7: Atmospheric absorption correction and normalization applied to a measured spectrum of a SiemensGamesa 93 m rotor (turbine 1). There is 5 dB between horizontal lines in the plot.

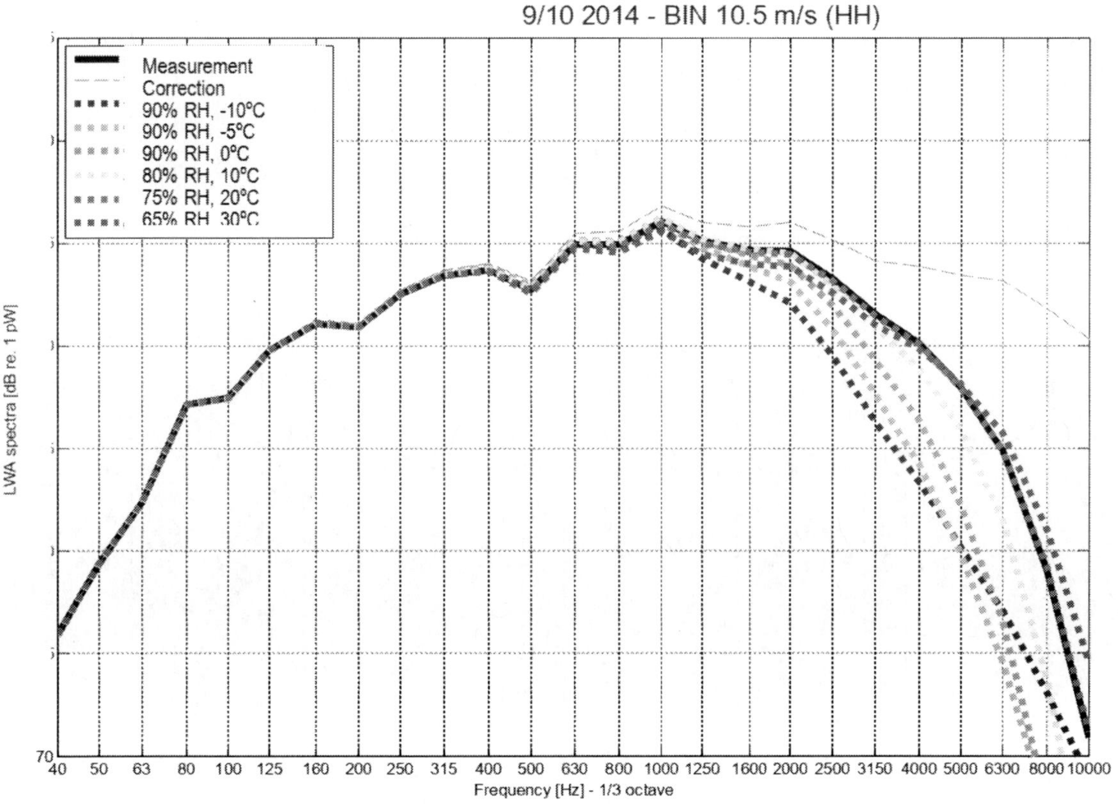

Figure 8: Atmospheric absorption correction and normalization applied to a measured spectrum of a SiemensGamesa 113 m rotor (turbine 2). There is 5 dB between horizontal lines in the plot.

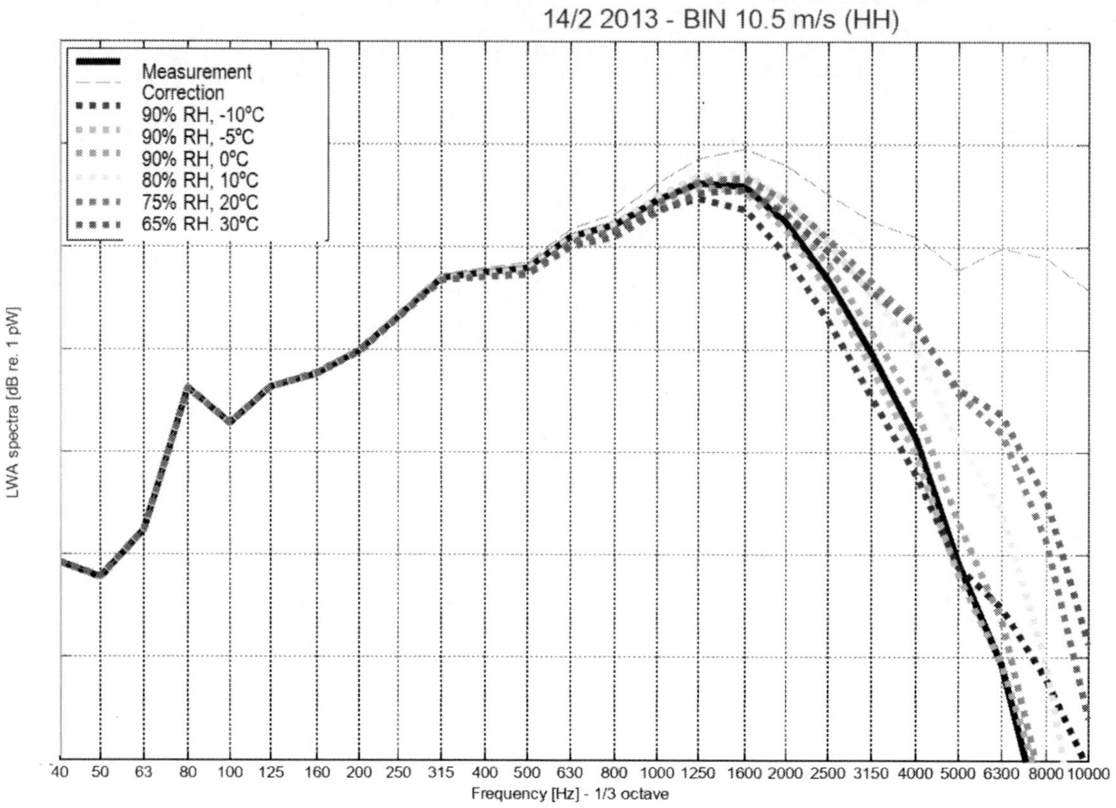

Figure 9: Atmospheric absorption correction and normalization applied to a measured spectrum of a SiemensGamesa 108 m rotor (turbine 3). There is 5 dB between horizontal lines in the plot.

INCE*EUROPE*

8th International Conference
on
Wind Turbine Noise
Lisbon – 12th to 14th June 2019

France - Germany: A Comparison of the Acoustic Assessment Procedures

Pierre Dutilleux, Evergy Engineering GmbH, Malsenstrasse 84, 80638 Munich, Germany
pierre.dutilleux@evergy.de

Summary

The French and the German regulation frameworks for the acoustic assessment of wind farm projects are briefly described. How far the interpretation of the same acoustic situation can differ when the French or the German regulation framework is considered, is shown on the basis of the observation of the same measurement data set. Several aspects of the assessment procedure are discussed: choice of the weather conditions, use of wind screens, assessment of the tonality, measurement and reproduction uncertainties. The consequences of the differing procedures are discussed from the point of view of the dweller, of the wind turbine manufacturer, of the wind farm operator, of the financing institutions and of the regulators. Recommendations for international projects are derived on the basis of the experience gained in France and Germany.

1. Introduction

As a French engineer providing advisory services to German investors, I am often required to explain how and why the assessment of wind farm noise are so much differing from one country to the other [4][12][13].

As an introductory example, please consider the scatterplot of the sound pressure levels which have been measured at a receptor location in the vicinity of a German wind farm (Figure 1).

The scatterplots of the total sound and of the residual sound level hardly differ from each other. The maximum total noise level during the daytime measurement periods remains within 55 dBA, a noise level which compares to the German daytime rating level for small residential areas. On the basis of such a data set, an acoustic consultant in Germany would however resent at giving any opinion about the compliance of the wind farm project. The following sections of this paper will explain why he would rather state that no qualified assessment of the wind farm noise can be made at this location because the signal/noise ratio of the wind farm noise is too low. He would further recommend performing other measurements at a location where the signal/noise ratio of the wind farm noise can be significantly improved.

Similar scatterplots for a wind farm in France would be interpreted differently. An acoustic consultant in France would stress the point that no sound emergence can be noticed. As a consequence, he would say that the wind farm is compliant to the French regulations.

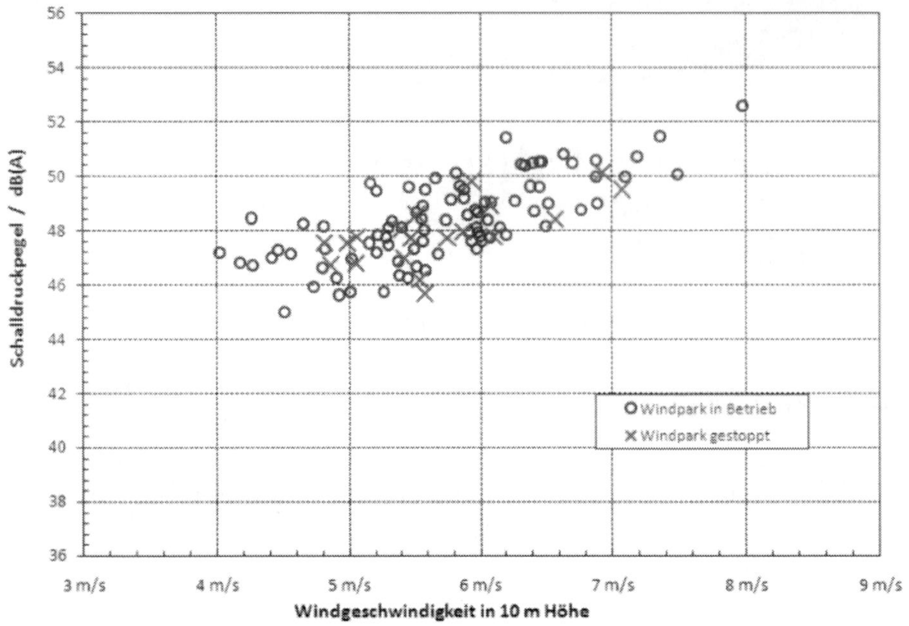

Figure 1: Measured total sound (blue circles) and residual sound (red crosses) levels in $L_{Aeq,1min}$ at the receptor location during daytime (Source: [11])

A wind farm operator is responsible for the compliance of his facility to the local regulations and strives at improving the electricity production. In case of complaints from the dwellers or the need for a noise curtailment plan in order to comply to the regulations, the operator will need a detailed analysis of the acoustic situation at the site in order to assess which corrective measures can be implemented. In order to provide qualified answers to the operator, an acoustic consultant often needs to investigate the acoustic situation far beyond the minimum requirements of the local regulations.

This paper is based on the experience gained while reviewing and assessing numerous wind farm projects mainly in France and Germany. The assessment methods are heavily dependent on the regulatory framework of the country of application. We will hence begin with a brief description of the French and the German regulatory frameworks. For a detailed description of the technical procedures please refer to the relevant national regulations and standards as well as to the international standards. We will then go into more details on selected topics, discuss a few issues and finally compare the advantages and drawbacks of the respective procedures for the different parties involved in a project.

In this paper we will mainly use the international terms defined in ISO 1996-1 but it might be occasionally appropriate to revert to the respective French or German terms.

International: ISO 1996-1	France: NF S 31-110	Germany: TA Lärm §2.4
Total sound	Bruit ambiant	Gesamtbelastung
Specific sound	Bruit particulier	Zusatzbelastung
Residual sound	Bruit résiduel	Vorbelastung
Sound emergence	Émergence	

2. Regulatory Frameworks

2.1 The French Regulatory Framework

The regulations which apply to wind farm projects since 2011 are derived from the general framework of the registered environmental facilities (Installations Classées pour la Protection de l'Environnement (ICPE)) [28][15].

An absolute acoustic threshold at the border of the facility is set to 70 dBA during daytime and 60 dBA during nighttime. The border is defined as a distance of 120 % the total height (hub height + rotor radius) of the wind turbine.

A relative threshold of 5 dBA during daytime and 3 dBA during nighttime is set with reference to the residual sound. The term "sound emergence" is often used, a definition of which can be found not only in the French standards NF S 31-110 and NF S 31-010 [31][29] but also in the ISO 1996-1 international standard as: "increase in the total sound in a given situation that results from the introduction of some specific sound" [22]. In our case the specific sound is the wind farm noise.

Assessing the sound emergence is typically based on the comparison of the sound levels when the wind farm is operating and when it is not operating. The sound emergence is only relevant in France if the total sound level is larger than 35 dBA. The term "sound emergence" is in itself worth a lengthy discussion but this will be addressed later in this paper.

The tonal character of the specific sound (tonalité marquée) must remain within given limits. More details in this issue will be also given below.

The minimum distance to the dwellings is set to 500 m [28].

The main steps of an acoustic acceptance measurement campaign are, in reference to the preliminary French standard PR NF S 31-114 [32]:

a) Prepare the measurement and wait for suitable weather conditions.
b) Measure during a time period typically ranging from a few days to a few weeks
c) During the measurement period, collect weather data at the site and noise level data at a selected set of dwellings
d) Operate the wind farm under test in a sequence of operation / stop phases. The practical duration of the phases typically ranges from one hour to a few hours depending on the operating and the weather conditions
e) Exclude the erroneous data sets from the evaluation
f) Build data classes corresponding to homogeneous conditions e.g.: daytime vs nighttime, standardised wind speed bin, wind direction, hours of the day etc...). A number of data classes is suggested in the preliminary standard but project specific classification methods may be proposed. The classification according to the atmospheric stability [10] is one example.

The acoustic descriptor is $L_{50,10min}$ of the $L_{Aeq,1s}$. The acoustic indicators in each data class are the medians of the residual sound level and of the total sound level. The sound emergence in each class is calculated as the difference between the total sound level and the residual sound level.

The noise limit depends on the residual sound level during the measurement. A detailed assessment of the various acoustic situations will hence typically require several measurement campaigns covering several seasons and/or meteorological conditions.

The residual sound might also evolve as the traffic on the surrounding roads increases from year to year or when additional industrial facilities or wind farms are commissioned. In agreement with the general regulations for the registered environmental facilities, each wind farm is considered individually, the other surrounding facilities contributing to the residual sound.

2.2 The German Regulatory Framework

The German local development plans define areas and their intended use. For each type of area, maximum immission noise levels are set according to the technical directive on noise (TA Lärm, [34][14]).

Type of area	Daytime: 6 am to 10 pm	Nighttime: 10 pm to 6 am
Industrial areas	70 dBA	70 dBA
Business areas	65 dBA	50 dBA
Cities, villages and mixed-use areas (farming)	60 dBA	45 dBA
Common residential areas	55 dBA	40 dBA
Exclusively residential areas	50 dBA	35 dBA
Health resorts, hospitals and nursing homes	45 dBA	35 dBA

The total sound level to be compared to the maximum allowed immission level is the addition of the maximum contribution of each of the existing and future facilities. The total sound level (Gesamtbelastung) which is relevant here is fairly independent from the weather conditions.

3. Verification of the Emission Characteristics of the Noise Sources

When the characteristics of the wind farm as a source of noise has to be characterised, a sound level measurement using a typical sound level meter setup at the border of the facility is sufficient in France. In Germany, an accurate measurement of the source is necessary. It must be performed according to a German adaptation [35] of the international standard IEC 61400-11.

4. Measuring Equipment

4.1 Acoustic Measuring Equipment

In France, the sound level measuring equipment must fulfil the class I requirements but a standard foam wind screen is usually installed on the microphone. Such wind screens are typically recommended by the manufacturers when the short-term wind speed at the microphone height does not exceed 5 m/s (NF S 31-110 §6.2.2). Under the assumption of a standard logarithmic wind-speed profile, a wind-speed at the microphone height (1,2 m) is sometimes derived from the hub-height wind speed and it is explained to remain under 5 m/s. This method forgets that the short-term wind-speed profile is variable and that the wind speed in the eddies around the microphone location can be much higher than the average wind-speed of the air flow. Experience shows indeed that the turbulent flow around the microphone can induce significant sound level errors [16] and a procedure for the estimation of the bias and the uncertainties due to the wind-induced noise has been proposed [8].

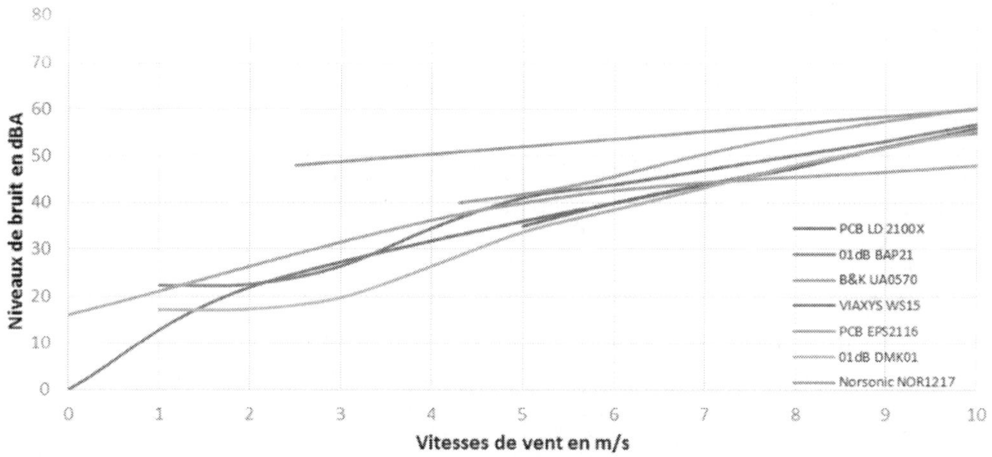

Figure 2: Wind-induced noise on a microphone depending on the type of wind screen (Source: [16])

Under the assumption that the total sound as well as the residual sound will be affected in the same manner, the sound emergence level is considered to be unaffected by the wind-induced noise at the microphone (NF S 31-110 §6.2.3).

In Germany, considering the importance of avoiding any significant bias, of low measurement uncertainties and of the reproducibility of the measurements, the use of a large secondary wind screen is required [35].

A few publications are available where the characteristics of the wind screens and the effects of the wind-induced noise have been investigated [1][2][21][8].

4.2 Meteorological Measurements

The French standards require a detailed measurement of the meteorological conditions with a focus on the acoustic propagation. NF S 31-110 requires i.a. measurements using 10-m masts and the description of the acoustic propagation conditions including cloud cover and solar radiation. Similar requirements are found in the Annex F of NF S 31-010. The wind farm specific standard NF S 31-114 refers additionally to the standardised wind speed which can be derived from the hub-height wind speed. The use of the standardised wind-speed instead of the measured wind-speed at 10 m height has been well accepted by the acoustic consultants and it significantly improves the reproducibility of the assessment procedure. We unfortunately notice that the documentation of the acoustic propagation conditions, although required by the French standards, is seldom fully available in the acoustic assessment reports. Beyond the aspect of acoustic propagation, we consider that the same information can greatly improve the assessment of the residual sound levels [10].

In Germany, the aim of relating the measured sound levels to the operating parameters of the wind turbines is well served by using a similar setup and similar evaluation procedures as required by the IEC 61400-11 standard.

5. Documentation of the Wind Turbine Operation

The particular sound level being the main relevant parameter for the compliance of a wind farm in Germany, the involved parties strive to control and document the operating conditions and the status of the wind turbines as accurately as possible. In critical situations, the operating mode and the rotational speed of the wind turbines have to be logged and made available upon request to the regulators. The involved parties have adapted their working procedures so that this information remains traceable over the past operation periods.

The sound emergence being the main indicator in France, the most important parameter for the wind farm operation is its status: in operation (ON) or stopped (OFF). While preparing an acoustic acceptance measurement, the acoustic consultants usually specify a sequence of ON/OFF operation periods and rely on the wind-farm operator to implement this sequence, and to the wind turbine manufacturer as to the wind turbine specifications.

According to the regulations on registered environmental facilities (ICPE), the legal representative of the project company is ultimately responsible for the compliance of the wind farm to the regulations. In the course of a number of project reviews, we had to figure out in what extent the acoustic assessment could be relied upon and in case not, to sort out which assumption(s) of the acoustic assessment might happen to be misleading. The investigations required addressing the acoustic consultant, the wind farm technical manager and the wind turbine manufacturer. After studying the topography, cross-checking the site-specific wind and weather data, a site visit and critical listening often provided the decisive cues.

6. Tonal Adjustments

Tonal sounds are known to be particularly annoying but modern wind turbines are expected not to emit objectional tonal components. In Germany, the tonal sounds have to be assessed on the basis of the narrow-band analysis according to IEC 61400-11 ed.2 standard in order to later derive a tonal adjustment according to the German standards [35][36][34]. The assessment method is well defined and can be effectively enforced. As a consequence, most wind turbines in Germany are free of objectionable tonal components. Some exceptions are however known but the problems can usually be investigated and solved when necessary.

The general method in France for the assessment of tonal sounds from registered environmental facilities is based on a 1/3rd octave analysis [25][28]. This method is much less sensitive than the narrow-band method according to IEC 61400-11 ed.2. The sensitivity of the method is further reduced by the ministerial order dated 26.08.2011. If tonal sounds are identified on the basis of the 1/3rd octave analysis, they are considered acceptable as long as they do not occur more than 30 % of each of the daytime and nighttime operation periods. We consider this situation very unfortunate because although these tonal sounds are not a necessary evil, they have a high nuisance potential. The dwellers on the countryside can especially feel annoyed, because such tonal sounds do not belong to the “traditional” soundscape. On the other hand, when the tonal sounds remain undetected, a chance is missed to identify and correct occasional malfunctions of the wind turbines.

In some cases, the tonal sounds even prevent newly implemented curtailment plans or other noise reduction measures to reach their objective. The noise-reduced operation modes usually achieve a reduction of the broad-band noise of the wind turbine. In the presence of a tonal sound, the broad-band noise might be reduced but the tonal sound usually remains unaffected. The tonal sound then becomes prominent in the noise spectrum; the objective of reducing the overall particular sound level is not achieved and the wind turbine sound is even more annoying because the masking of the tonal sound by the broad-band noise is less effective than in the former operation mode.

7. Uncertainties

The discussion of the measurement uncertainties is beyond the scope of this paper. We will mention here only a few simple guidelines for the interpretation of the results of a wind farm acceptance measurement campaign. Readers interested at the details of measurement uncertainties will revert to the applicable standards for more information.

The rating of wind farm noise based on sound emergence levels below 3 dBA is prone to a bias and to uncertainties which grow very high as the sound level difference between ambient and residual sound vanishes. Whereas the uncertainties due to the instrumentation and to the propagation conditions have being investigated in details [20][19][33], the investigations are very scarce when it comes to assessing the reproduction uncertainties of the residual sound levels.

The French standards [32] allow expressing the result of the assessment after deduction of the extended uncertainties from the calculated sound emergence levels. Using this procedure, a wind farm only faces the risk of being non-compliant if the allowed emergence level is exceeded with a high level of confidence, say 95 %. This procedure is in favour of the operator. The dwellers might rather wish an assessment giving them a 95 % confidence that the allowed emergence levels are not exceeded but under the current practice in France, such a requirement would not be realistic.

If a wind farm is found to be compliant and the remaining margin is used for the optimisation of the wind farm, the chances become very few, although they do exist, that the optimised wind farm is still compliant. During a financial transaction between a project developer and an investor, the wind farm developer will aim at increasing the energy output of the wind farm by avoiding any noise curtailment. The acoustic acceptance measurement campaign might turn out to be successful after application of the corrections for uncertainties. The wind farm would be declared compliant because the acoustic assessment has been performed in line with the French standard. The investor would however take over a wind farm with a 95 % confidence level that the acoustic thresholds are not exceeded by more than the extended uncertainty. This means also that he would have only 5 % confidence that the wind farm noise is within the acoustic threshold. This might not be considered to be a problem since the wind farm has already been declared compliant but if a verification measurement becomes necessary, the chances are by far lower than 50 % that the assessment results will be acceptable. The first verification measurement campaign might fail and identifying and waiting for the suitable measurement conditions might last for a long time. If several measurement campaigns have to be performed until acceptable result are available, the accumulated production losses during the stop phases might also become prohibitive. For this reason, we would rather rely on the emergence level results based on the raw emergence calculations, in the hope of having 50 % chances to have correctly assessed the acoustic situation. This means that in case of non-compliance, there would be reasonable 50 % chances of success in case of a new measurement campaign. We consider that this approach balances the risks between the parties.

Considering that the bias and the uncertainties of a method based on sound emergence levels at the receptor location are prone to be high and that the extended uncertainties can be deducted from the sound emergence level, the method does not provide incentive for reducing these uncertainties. In some cases, when the uncertainties grow too high, the French procedure recommends acquiring additional measurement data and / or to refine the classes of homogeneous conditions.

The German regulator is aware of the numerous sources of uncertainty but seeks nevertheless a 90 % confidence level that the particular sound will be lower than the threshold which has been set on the operating permit. To achieve this goal, the acceptance procedure has been tailored to reduce the sensitivity to the measurement and reproduction uncertainties [34][36]. To achieve this goal, the wind turbine noise is preferably measured from a short distance using a large secondary wind screen and the loudest wind turbine operating conditions are considered relevant. The assessments of the noise level at the dwellings rely however on a propagation calculation which itself is prone to uncertainties. The compliance of a wind farm is finally assessed on the basis of a calculated noise level at the receptor location after addition of the extended combined uncertainties which account for the measurement uncertainties, the deviations of wind turbine characteristics from the specifications as well as the meteorological and acoustic propagation uncertainties. Once a wind farm has been declared compliant, the operator has typically only 10 % risk that a new acceptance measurement would fail at providing confirmation. The 10 % risk means however that exceptions are possible.

Under consideration of the allowed immission noise level which has been set by the regulator, the particular noise of the wind turbine can be slightly increased if all the involved parties succeed at reducing the uncertainties at the various steps of the assessment process. This fosters the development of more accurate working procedures, a better optimisation and a tighter control of the wind turbine.

8. Discussion of Selected Topics

8.1 France: Discussion of the Definition of the Sound Emergence

According to NF S 31-110 the sound emergence refers to the temporal level variation which is caused when the wind farm noise appears or disappears i.e. when the windfarm is put in operation or stopped. The ministerial order of 31.08.2006 about community noise [26] and the French standard NF S 31-010 [29] defined the terms of global and spectral emergence but the assessment is not based on the temporal variation of the noise level but on the difference between the "usual" initial sound level and the noise level during normal operation of the specific sound source. The ministerial order of 05.12.2006 [27] further specified that the assessment within the framework of community noise had to be based on at least 30 minutes of measurements. This assessment procedure has been in use for some time in France and was unfortunately extremely sensitive because, if at any given time the wind farm noise was higher than the background noise for more than 30 minutes, the wind farm would have been declared non-compliant. Due to the fact that the application of this method to the newly commissioned wind farms lead to very unstable decisions, another method had to be proposed. At that time, the method of statistical evaluation of the noise levels after binning according to the wind speed was already under investigation by the French acoustic consultants. Initial attempts were based on a wind speed measured at 10 m height. The statistical evaluation after binning significantly improved the stability of the decision but the reference to a wind speed at 10 m height occasionally led to severe discrepancies depending on the measurement period. Further developments of the method including the use of a standardised wind speed lead to a draft standard [32] which has been referred to by the French regulator [28], at the time when the French administration decided to consider the wind farm projects as registered environmental facilities (ICPE). From this time on, the regulations about community noise were no longer applicable to the wind farm projects. The main principles set out by the preliminary standard PR NF 31-114 [32] are nowadays in application in France but in spite of lengthy discussions and several updates this preliminary standard could not yet be established as a standard.

Through the binning process, several noise level data sets are aggregated as long as they originate from the same binned wind speed but they might as well originate from acoustic situations which, for other reasons, significantly differ from each other.

The sound management plans which are derived from this method are typically static and ensure a deemed-compliance with reference to the acoustic acceptance measurement and evaluation report. It can however not be ruled out that under particular conditions, the curtailment of the wind farm is too severe or other situations where the emergence is indeed temporarily exceeding the permitted levels. Some acoustic consultants and wind farm operators do hope to design an automatic control system which would be able to adjust in real-time the operation modes of the wind turbines on the basis of the real-time assessment of the sound emergence at the dwellers locations [18]. This approach is technically challenging because it relies on signal analysis and processing techniques.

8.2 France: Discussion of the Significance of the Sound Emergence

There is a general agreement that the higher the wind farm noise, the more annoying it is for the dwellers. This formulation is fairly simplistic but it explains why many countries have adopted absolute noise thresholds. On the other hand, a few countries such as France have chosen to consider that the change in noise level is a more relevant criterion. We are personally sceptical because we have not read so far any survey demonstrating the relation between emergence levels and annoyance level. In the course of discussions about the topic of sound emergence we have noticed however what we would consider here misconceptions:

a) "The sound emergence reflects the perception of the wind farm noise by a human." If the wind farm noise is at the same level as the background noise, the total sound level is 3 dB higher than each of the specific sound and the residual sound. The sound emergence level is hence acceptable according to the French regulations although the particular sound is clearly audible as an independent source of noise and a trained listener could even describe it with a wealth of details.

b) "When the background noise and the particular noise are at the same level, they cannot be distinguished from each other." A wind farm noise which is at the same level as the background noise is usually not masked by it, by far. As a rule of thumb, a wind farm noise can be masked by the background noise if the wind farm noise is about 8 to 12 dB lower than the background noise [7]. A rule of thumb inherited from the field of speech and audio coding says that the noise of a coding and transmission channel does not disturb the listener if it lies about 10 dB below the speech or audio signal in each frequency band and can be masked by the speech or audio signal.

c) "The nighttime sound emergence level is allowed to grow up to 3 dB in France because such a sound emergence level is hardly perceptible." It is indeed difficult to identify absolute sound levels on a scale with 3 dB steps, but a level difference of 3 dB can easily be perceived. As an example, the level steps of a portable audio player are typically 2 dB each. This is large enough to be clearly perceptible but not too small so that the user can quickly adjust the level. According to Fastl and Zwicker, the just-noticeable level differences are below 1 dB [3]. Further, a trained ear is capable of perceiving relative level differences of a fraction of 1 dB in individual frequency bands. The level variations might not be perceived for themselves but as a tonal coloration.

8.3 Germany: High Stability of the Assessment Results

If a wind farm is found compliant in Germany, there is only 10% risk that it is not. If a new measurement is performed, the risk of a significantly differing result is low. As a consequence, the administrative decision is stable (a few exceptions do exist however).

The procedure is designed for the worst-case scenario. The sensitivity to the weather conditions is rather low (exceptions do exist however).

The structured set of standards and of the accreditation procedures (FGW TR-1, DIN, VDI, IEC, §26 BImSchG, MEASNET) contributes to the reliability of the procedure and of the personal as well as to a high stability of the assessment results.

The cumulative effects when additional facilities are commissioned are limited because the noise budget for all the wind farm projects and industrial facilities is defined by the noise level guidelines (TA Lärm). Exceptions do exist however, especially when the facilities have been permitted in the past using simpler procedures. Before permitting a new facility, a preliminary update of the planning documentation of the existing facilities might be necessary.

8.4 From the Point of View of the Various Involved Parties

We finally summarize in a tabular form the above discussions.

Involved Party	French Procedure	German Procedure
Wind turbine manufacturer	The French procedure is more flexible as to the operating modes. The lower constraints on tonality leave more flexibility for the choice of the individual components and for the operation procedures of the wind turbine. In case of acceptance problems, it can become difficult to identify which aspects of the project or of the wind turbines have to be improved.	The verification of the wind turbine noise characteristics is more straightforward. The wind turbine specifications and warranties have to be tighter. The introduction of alternative wind turbine versions or other operating modes might be difficult due to the burden of the certification and of the documentation of each modification.
Wind farm technical manager	More operation modes to choose from. The sound management plans are more flexible but also more complicated to implement, to check and to trace.	Fewer operation modes to choose from. Tighter specifications. Easier control and documentation of the operating conditions.
Wind farm operator	More latitude is available as to the sound optimisation strategy. High uncertainty of the acoustic acceptance measurements. The reproducibility risk due to adverse meteorological conditions can be high. The costs of an acoustic acceptance measurement and the associated production losses can be very high. The residual sound levels might increase along the project life, allowing to gradually reduce the noise curtailment measures and, as a consequence, to improve the electricity production.	The sound management plan is usually defined during the development phase of the project. Improvements are only possible in case of a significant upgrade or improvement of the wind turbine. The costs of a verification measurements and the associated production losses are limited.
Financing institution	The assumptions of the pre-construction acoustic survey might be too optimistic, especially as far as the residual sound level is concerned. A significant risk might remain that the assumptions cannot be confirmed by the acoustic acceptance measurements and that a more restrictive sound management plan has to be implemented.	The acoustic risk can be assessed and managed.
Regulator	The operator is responsible for the compliance of the wind farm project to the regulations. The possibilities for the regulator to have direct insight into the project are limited. The acoustic measurements at the receptor location provide the relevant information for compliance testing.	The maximum noise contribution from the wind farm can be set and controlled with a 90 % confidence interval. The regulator has possibilities to directly control several acoustic aspects of the project. The measurement of the source, followed by a propagation calculation, only provides an indirect information about the sound levels at the receptor locations. The administrative decision is very stable.

Dwellers	Thanks to the sound emergence criterion, the increase of the ambient sound level at low wind speeds is moderate. The same sound emergence criterion allows however the wind farm to contribute more noise at high wind speeds because the residual sound level also increases with the wind speed. The overall ambient sound level is subject to increase over time as the number of facilities increases. Tonal sounds are annoying but often remain undetected.	The overall maximum noise contribution from all the surrounding facilities can be controlled by the regulator. The wind turbines are expected to be "state of the art" i.e. free of objectionable tonal sound. Noise curtailment sets in at a given wind speed and the wind farm contribution does not increase beyond the level which has been allowed by the regulator.
Acoustic consultant	The sound emergence criterion is subject to large uncertainties. The standardisation procedure is lengthy. The reproducibility risk due to weather conditions is fairly high. A large variety of requirements and of expectations from the involved parties has to be fulfilled. The expertise of the acoustic consultant plays a key role during the evaluation of the measurement data and the interpretation of the measurement results.	An acoustic measurement of the source can be straightforward. Project specific measurement concepts and alternative measurement locations are sometimes necessary. The measurement and assessment procedures are very detailed and subject to round robins. Low uncertainty measurements are required.

9. Conclusion and Outlook

The comparison of the French and the German procedures has shown that the same acoustic situation can lead to largely differing assessments depending on the regulations of the country of application.

The French standard NF S 31-114 has been under discussion for a number of years now but could not be finalised yet. The notion of sound emergence is intrinsic to many assessment procedures in France but the discussion whether it should be determined on the basis of statistics of binned values, on time sequences or even better in real-time thanks to source-separation signal processing algorithms remains open.

One of the main problems of the French procedure is the reproduction uncertainty depending on varying meteorological conditions. This issue will hopefully be addressed in a future revision of the French standards.

The German procedure has proven to be effective and robust but as the number of wind turbines and the density of the wind farms in some areas increases, an update of the procedures is necessary. Modifications of the acoustic propagation calculation procedure has already been introduced (topic beyond the scope of in this paper) but the technical guideline for the measurement of the source has still to be updated in relation to the IEC 61400-11 ed.3.

Experience shows that reducing the uncertainties goes along with improved and more qualified evaluation methods.

This paper might be of interest for the project team on the measurement of wind turbine noise characteristics in receptor position [23].

We have focussed here on two countries but similar concepts might be applicable to other countries. The IFC EHS guidelines for wind farm [12] are potentially applicable in countries where no specific regulations are available. Their use requires mastering most of the concepts addressed in this paper but special care should be given to the implementation details and to the human, material and financial resources available in the country of application.

References

[1] R. A. Davis and Dr M. C. Lower (1996) *Noise measurement in windy conditions* ETSU W/13/00386/REP

[2] DELTA (1997) *Investigations of Wind Screens Insertion Loss and Attenuation of Wind Noise Note 1* JOR3-CT95-0065

[3] Hugo Fastl and Eberhard Zwicker (2007) *Psychoacoustics, Facts and Models, 3rd edition* Springer

[4] Pierre Dutilleux and Joachim Gabriel J (2008) *Recommendations for Improved Acceptance of Wind Farm Projects in France with Regard to Acoustic Noise* DEWI Magazin No.32, February

[5] Pierre Dutilleux, Joachim Gabriel (2008) *Assessment of the acoustic noise issues of wind farm projects in the light of the experience gained in Germany* DEWEK, November

[6] Michel Rumeau (2010) *Incertitude de mesurage du bruit en environnement par une méthode détaillée* SFA, Lyon

[7] Karl Bolin et al. (2010) *The Potential of Natural Sounds to Mask Wind Turbine Noise* Acta Acoustica united with Acustica, Vol. 96, p. 131-137

[8] David Ecotière (2012) *Estimation of uncertainties due to the wind-induced noise in a screened microphone* Nantes

[9] Joachim Gabriel, Steffen Vogl, Thomas Neumann, Gundula Hübner, Johannes Pohl (2013) *Amplitude Modulation and Complaints about Wind Turbine Noise* Denver

[10] Alexis Bigot et al. (2015) *Influence of vertical temperature gradient on background noise and on long-range noise propagation from wind turbines* Glasgow

[11] Joachim Gabriel and Steffen Vogl (2015) *Untersuchung der Beeinträchtigung von Anwohnern durch Geräuschimmissionen von Windenergieanlagen* DEWI RS14-00017-01

[12] Pierre Dutilleux (2017) *Noise Guidelines for Countries where no National Regulations are Available. A discussion of the IFC EHS Guidelines for Wind Energy* Rotterdam

[13] Simon Bénard (2017) *Impacts des parcs éoliens : balisage, impacts sonores et infrasons* OFATE, Paris, June

[14] Hermann Lewke (2017) *Noise protection Regulations for Wind Turbines in Germany Paris* 08.03.2017

[15] Claire Sauron (2017) *La prise en compte des impacts des parcs éoliens : Le cadre réglementaire français* MEEM, DGPR, Paris 08.03.2017

[16] Alexis Bigot (2017) *Wind farms noise impact: theory and regulations* Paris 08.03.2017

[17] Johannes Pohl, Joachim Gabriel and Gundula Hübner (2018) *Understanding stress effects of wind turbine noise – The integrated approach* Energy Policy (112) pp.119-128

[18] Baldwin Dumortier (2018) *Contrôle acoustique d'un parc éolien* Université de Lorraine, PhD 25.09.2018

[19] David Ecotière et al. (2018) *NF S 31-120 une nouvelle norme pour évaluer l'influence des effets météorologiques et des effets de sol sur la mesure du bruit dans l'environnement* CFA, Le Havre

[20] David Ecotière et al. (2019) *XP S 21-115-1 : une nouvelle norme pour évaluer les incertitudes liées à l'instrumentation lors de mesurages du bruit dans l'environnement* CFA, Le Havre

[21] IOA (2014) *A good practice guide for the assessment and rating of wind turbine noise – Supplementary guidance note 1: Data Collection*

[22] ISO 1996-1 (2016) *Acoustics — Description, measurement and assessment of environmental noise — Part 1: Basic quantities and assessment procedures*

[23] PT 61400-11-2 *Wind energy generation systems - Part 11-2: Measurement of wind turbine noise characteristics in receptor position* IEC

[24] World Bank Group (2015) *Environmental, Health, and Safety Guidelines for Wind Energy* 07.08.2015

[25] *Arrêté du 23 janvier 1997 relatif à la limitation des bruits émis dans l'environnement par les installations classées pour la protection de l'environnement*

[26] *Décret no 2006-1099 du 31 août 2006 relatif à la lutte contre les bruits de voisinage et modifiant le code de la santé publique (dispositions réglementaires)*

[27] *Arrêté du 5 décembre 2006 relatif aux modalités de mesurage des bruits de voisinage*

[28] *Arrêté du 26 août 2011 relatif aux installations de production d'électricité utilisant l'énergie mécanique du vent au sein d'une installation soumise à autorisation au titre de la rubrique 2980 de la législation des installations classées pour la protection de l'environnement*

[29] *NF S 31-010 (1996) Acoustique – Caractérisation et mesurage des bruits de l'environnement – Méthodes particulières de mesurage* AFNOR

[30] *NF S 31-010 (1996) Acoustique – Caractérisation et mesurage des bruits de l'environnement – Méthodes particulières de mesurage. Annexe F : Paramètres météorologiques* AFNOR

[31] *NF S 31-110 (2005) Acoustique – Caractérisation et mesurage des bruits de l'environnement – Grandeurs fondamentales et méthodes générales d'évaluation. Acoustics — Description and measurement of environmental noise — Basic quantities and general evaluation methods* AFNOR

[32] *PR NF S 31-114 (2011) Mesurage du bruit dans l'environnement avec et sans activité éolienne, version préliminaire* AFNOR July

[33] *NF S 31-120 (2018) Acoustics — Characterization and measurement of environmental noise — Soil and weather conditions influence* AFNOR December

[34] *TA Lärm (1998) Technische Anleitung zum Schutz gegen Lärm* 26:08:1998

[35] *Technical Guideline for Wind Turbines (2008) TG 1 – Determination of noise emission Revision 18*

[36] *LAI Hinweise zum Schallimmissionsschutz bei Windkraftanlagen (WKA)* 30.06.2016

INCEEUROPE

8th International Conference on Wind Turbine Noise Lisbon – 12th to 14th June 2019

Noise Source Location in Wind Turbines

Pablo Alloza, gfai tech GmbH: alloza@gfaitech.de
Benjamin Vonrhein, gfai tech GmbH: vonrhein@gfaitech.de
Olaf Bölke, gfai tech GmbH: boelke@gfaitech.de

Summary

Sound source localization has a direct practical application on wind turbines. Thus offering a new point of view that provides novel solutions to several challenges from the wind turbines sector. With this acoustic measurement method it is easy to record, identify and localize sound sources in different parts on wind turbines, analyzing in time and frequency domain. In addition, the results can be showed in the wind turbine 3D model, acquired using a laser-scanner.
This publication presents the basic properties and functionalities of sound source localization using beamforming and the Acoustic Camera, it introduces the principle of operation and the different array typologies and concludes with several results in different scenarios.

1. Introduction

Wind energy is one of the renewable energies widely used around the world. The wind is a source available almost anytime not depending of other factors like temperature or sunlight. Sometimes the best locations for wind turbines are close to cities. Hence noise emissions from the wind turbines, can be a problem; especially in wind farms with a lot of units. The relation between the noise exposure and health is well know and been analysed in several publications. In outdoor noise emissions the wind strength and his orientation, temperature or humidity affect directly the noise propagation in the air. In addition, the sound footprint from a wind turbine has low frequency content, which increases the noise problems in the reception points.
The noise in and from wind turbines during operation is a complex equation because there are several potential noise sources: Blades, gearbox, generator, brakes, tower, etc. In addition aerodynamic noise can be present in the external part like blades or nacelle.
Be able to identify this sources separately is the key for analysing the contribution of each one in the global level and to take the right decisions about how to improve the noise reduction in a wind turbine. In addition, the noise localization provides a new point of view for maintenance purposes, visualizing where the specific noise like chirps, squeaks or screaks are coming from.

The Acoustic Camera system can provide this kind of results, using a special multichannel microphone array, a data recorder and a computer running the software, all this analysis in time and frequency domain can be done on site.
This method can be used in research & development for new prototypes, control and monitoring for installed ones or comparison between different models or different workflows. One of the main features of this system is that it is not necessary to install any sensor on the wind turbine, neither to stop in operation.

2. Principle of operation

There are different techniques for localising sound using microphone arrays; For near field measurements, Nearfield Acoustic Holography is the best method. When the target needs to be independent from background noise, to measure in very low frequencies and in close distances, Intensity is the best solution. Focusing in our aim, wind turbines are big objects and are usually far away from the point of evaluation, for this reasons Beamforming is the best technique for this application.
The basic algorithm running behind the processing software is named Delay-and-Sum-Beamforming, the most widely known sound localisation technique. It is based on the following principle: The sound of different noise sources travels using different ways so it takes different amounts of time to reach the array microphones. This means different run-times from the sound sources to the different microphones. By calculating this time differences between a sound event and each microphone of an array, direction and strength of sound sources are determined. The calculated sound pressure is then mapped on the optical picture or video of the measurement object. Figure 1 shows an easy example with two sensors; the human spatial hearing is a good and easy comparison.

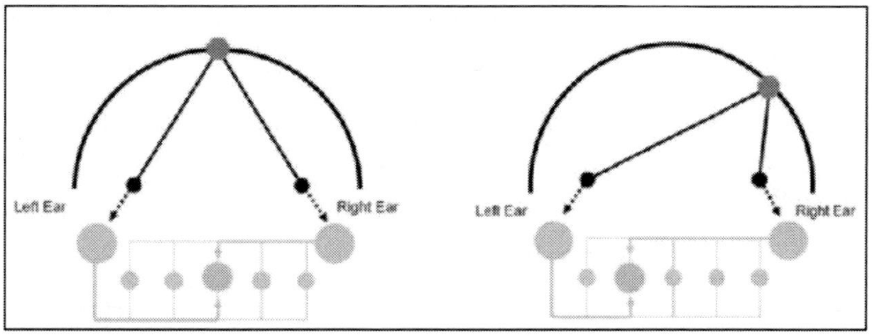

Figure 1. Run-time example using human earing.

The data from all the microphones are used to calculate the sound pressure level for a certain position using the following formula:

$$\hat{f}(\mathbf{x},t) = \frac{1}{M}\sum_{i=1}^{M} w_i f_i(t - \Delta_i)$$

Figure 2. Delay and sum formula.

In the formula (Figure 2) the f corresponding sound pressure level for a given t time and a x position on a reference plane using M number of microphones; w_i could be used as optional spatial shading weights or set to unity if does not occur. Figure 3 shows the process in detail:

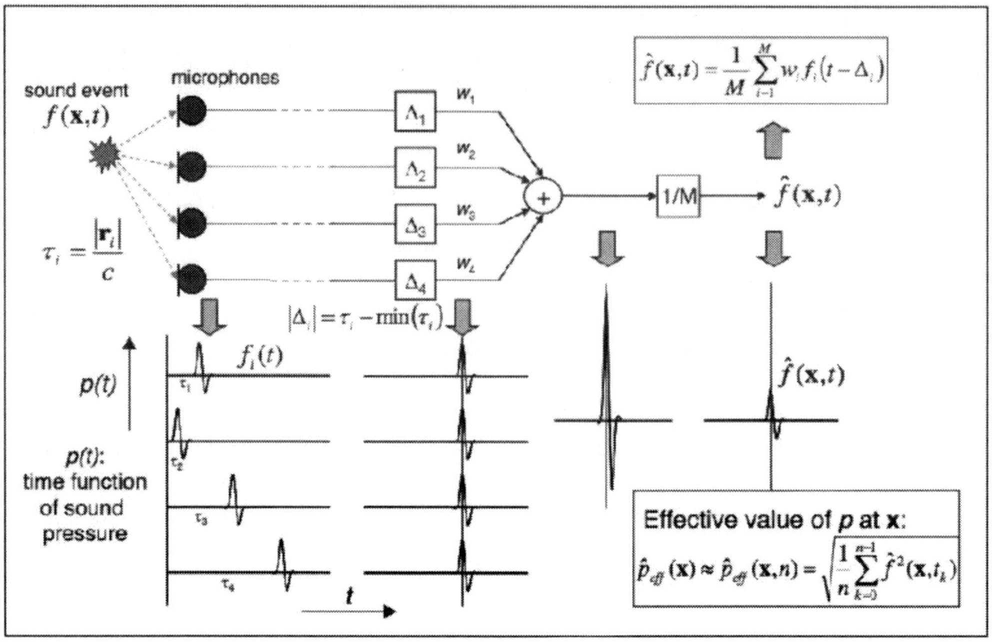

Figure 3. Time domain Beamforming operation principle.

Delay-And-Sum-Beamforming in the Frequency Domain allows acoustic evaluation via the frequency domain. The spectrum and the spectrogram show the different frequencies a signal is composed of. Using Fourier Transformation, a transition from time to frequency domain is possible and the spectrum and spectrogram can be calculated.
In the final step this process is applied to a number of picture points set by the user in a photo plane, in which the target is located. Figure 4 shows that to each picture point a corresponding set of distances r_i and run-time delays t_i can be calculated.

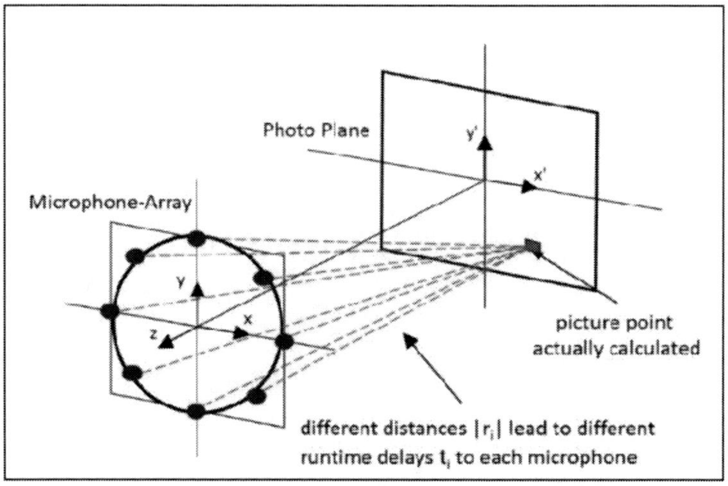

Figure 4. Picture points calculation.

Thus, the sound pressure level can be determined for all the picture points using delay-and-sum-beamformer. To represent the sound pressure levels, a colour scale is used in the picture points with the optical reference photo, generating an acoustic picture.

2.1 Microphone array typologies

Depending on the beamforming applications, different microphone arrays, varying in geometry, size and number of microphones are used. Ring type arrays are suitable for several applications, inside and outside, depending on the array diameter. Sphere type arrays are designed for 3D applications, inside cabins or rooms. For the measurement campaigns in this publication, a star microphone array (STAR 48 AC PRO) with 48 channels and 3,4 meters was used. This microphone array (Figure 5, centre) is designed for outdoor applications and specialized for low frequency localization.

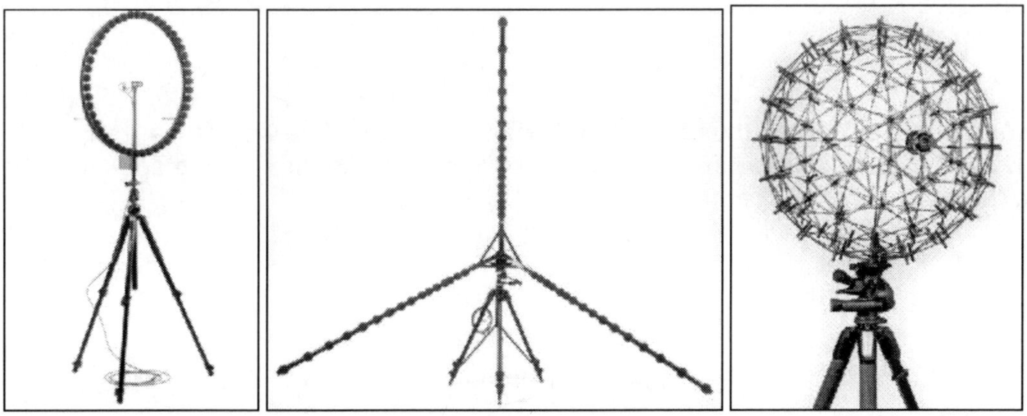

Figure 5. From left to right: Ring array with 48 microphones, Star array with 48 microphones and Sphere array with 120 microphones.

3. Measurement scenarios

The measurement tests were divided in 3 scenarios, in order to provide results from different locations and different wind turbine models (Figure 6).

- Scenario 1: Wind farm, several wind turbines in lines. Focus on the first one, the other ones are on the right.
- Scenario 2: Same point of view from scenario 2, focus on the right side.
- Scenario 3: Wind farm, several wind turbines inside a small forest. Point of evaluation in the middle of several wind turbines, focus on the right side.

Figure 6. From left to right: Evaluation points for scenarios 1, 2 and 3.

3.1 Scenario 1 results

In this measurement scenario, the Acoustic Camera is set 330 meters from the wind turbine selected as target. Other wind turbines are in the area, mainly on the right side. One first look to the spectrogram (Figure 7) provides us the possibility to identify continues frequencies in all the measurement or short time events.

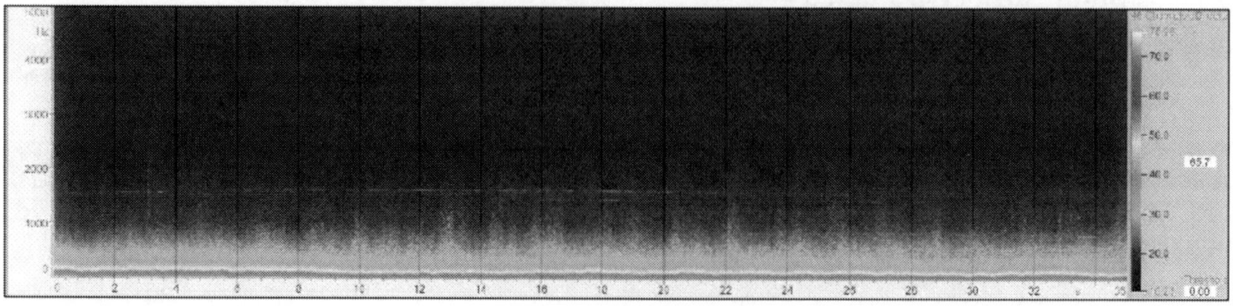

Figure 7. Spectrogram

Selecting when one blade is on the top, the 500 Hz third octave is always localized in the same position. The acoustic pictures are from different time selections. The blades on the right position are the noise source (Figure 8).

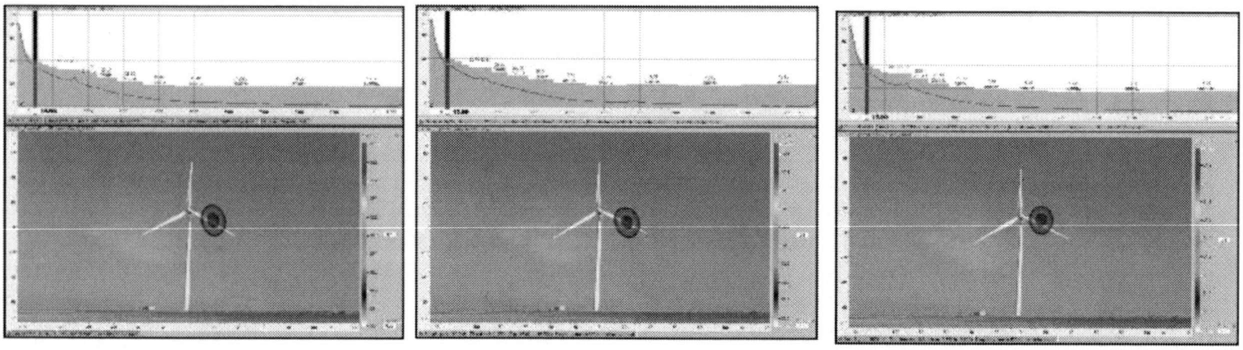

Figure 8. Acoustic pictures from different times selections, 500 Hz third octave

Going to the low frequency, 200 Hz third octave, the noise source is always localized on the wind turbine tower, no matter which blade is on the top position (Figure 9).

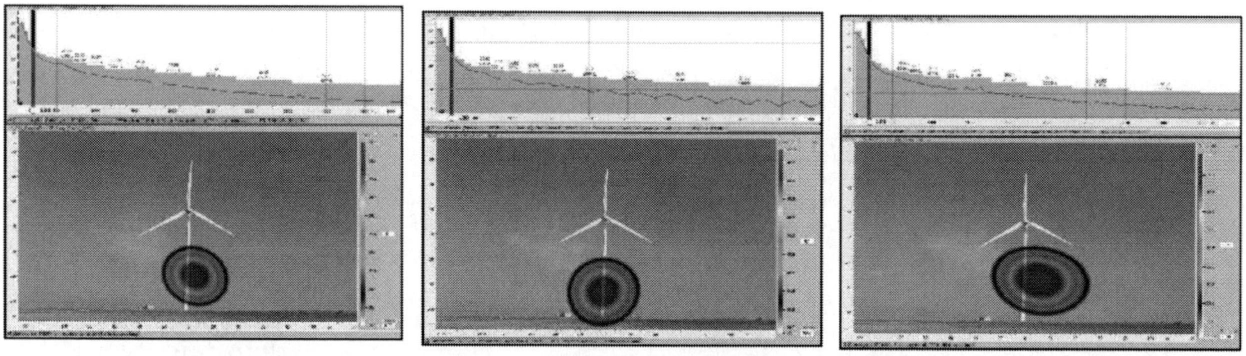

Figure 9. Acoustic pictures from different times selections, 200 Hz third octave

3.2 Scenario 2 results

From the same evaluation position as scenario 1, just rotating to the left of the microphone array. In this position the microphone array is not parallel aligned in front of the noise sources. Now the first wind turbine is located in 400 meters distance. In this measurement two wind turbines are in the view, 120 meters distance between them (Figure 10). The striking frequency peak on 1.594 Hz is localized on the blade.

Figure 10. Acoustic picture focused on the frequency peak on 1.594 Hz

The 2.000 Hz third octave is localized on the first wind turbine tower, but increasing the dynamic range (Figure 11, right), the same third octave is localized on the second wind turbine tower too.

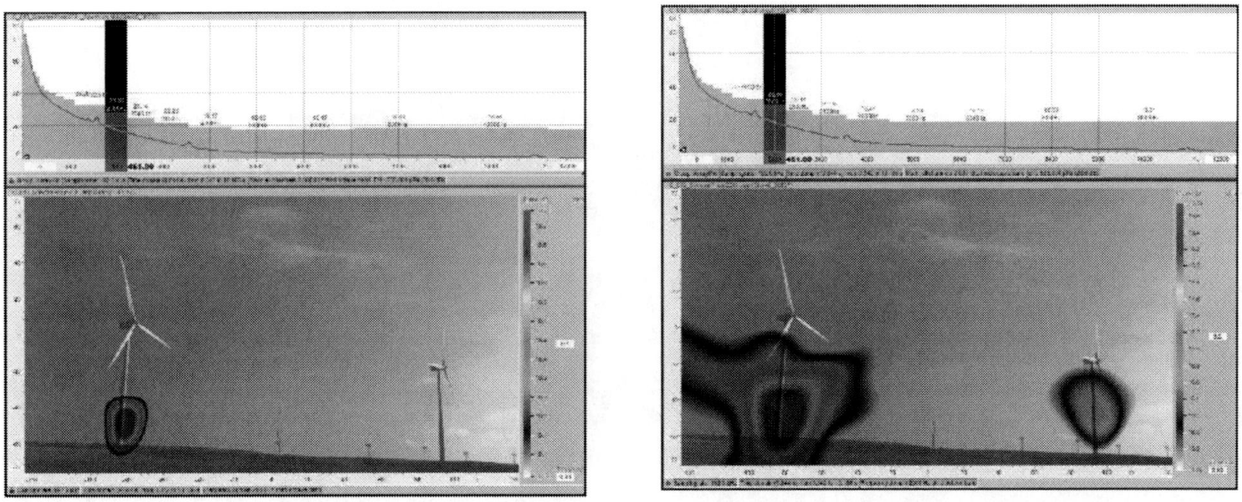

Figure 11. Acoustic pictures with different dynamic range, 2.000 Hz third octave

Looking in the low frequencies, 400 Hz third octave is localized in the bottom of the first wind turbine and again, increasing the dynamic range (Figure 12, right), the second noise source in the same frequency but less level appears in the second wind turbine tower.

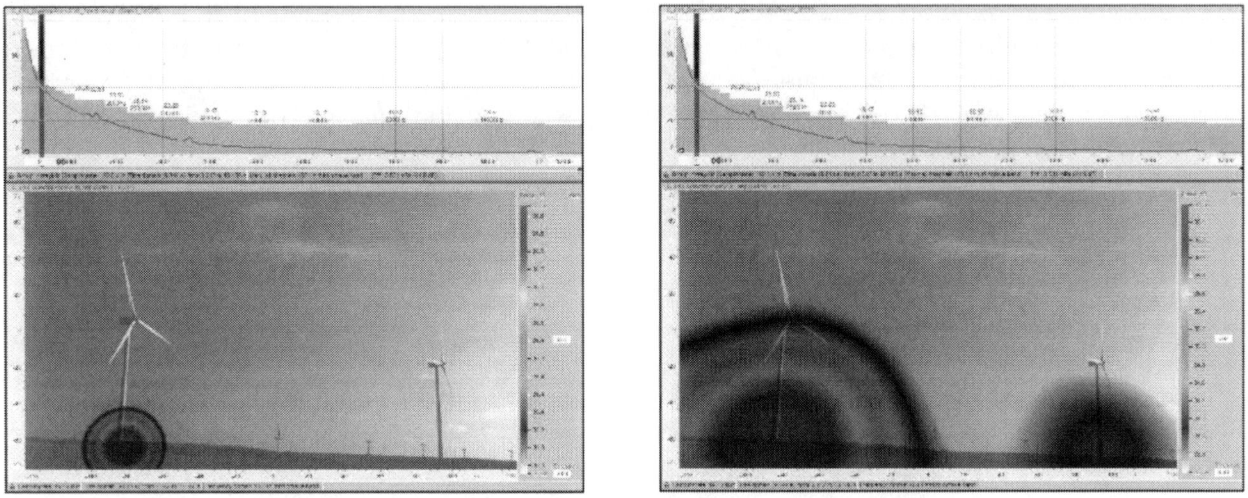

Figure 12. Acoustic pictures with different dynamic range, 400 Hz third octave

3.3 Scenario 3 results

The area for this test is a small forest on a mountain with several wind turbines around in the high part. The point of evaluation is in front of three wind turbines more or less aligned. The microphone array is 350 meters from the first one. In the spectrogram some fluctuations are showed between 2.000 Hz and 3.000 Hz. Selecting this fluctuations directly in the spectrogram (Figure 13), the noise source is pointed on the top blade.

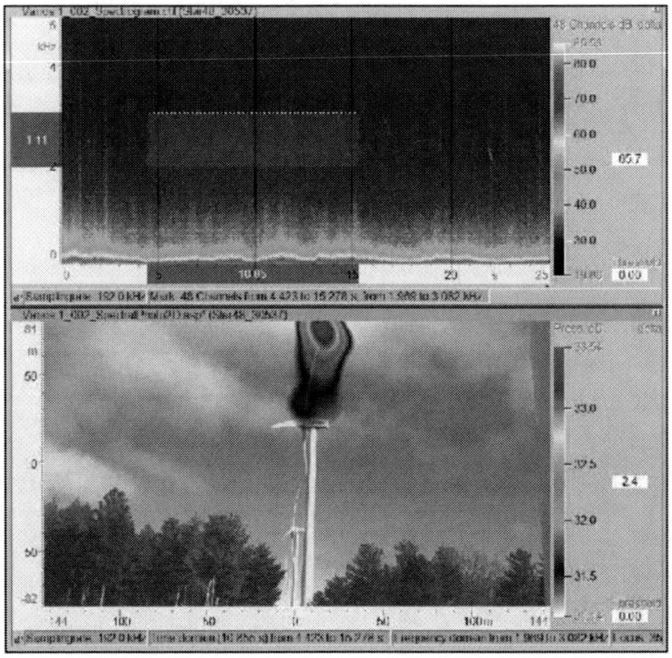

Figure 13. Spectrogram and acoustic picture from the selected area.

The 400 Hz octave band can be focused on the rotor from the first wind turbine. Independently of the blades position, during the whole workflow the noise source remains.

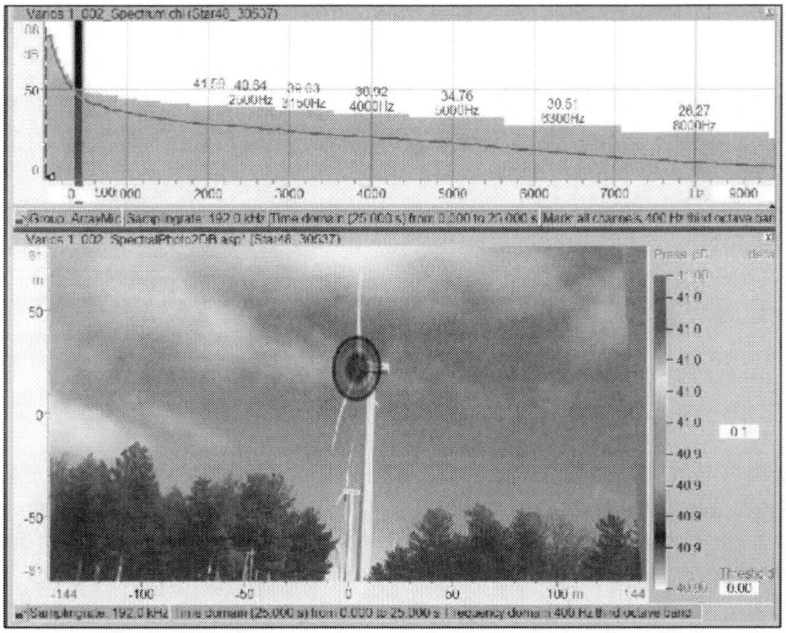

Figure 14. Acoustic picture, 400 Hz third octave

The low frequencies, like the 160 Hz third octave band showed, can be focused near to the ground (Figure 15).

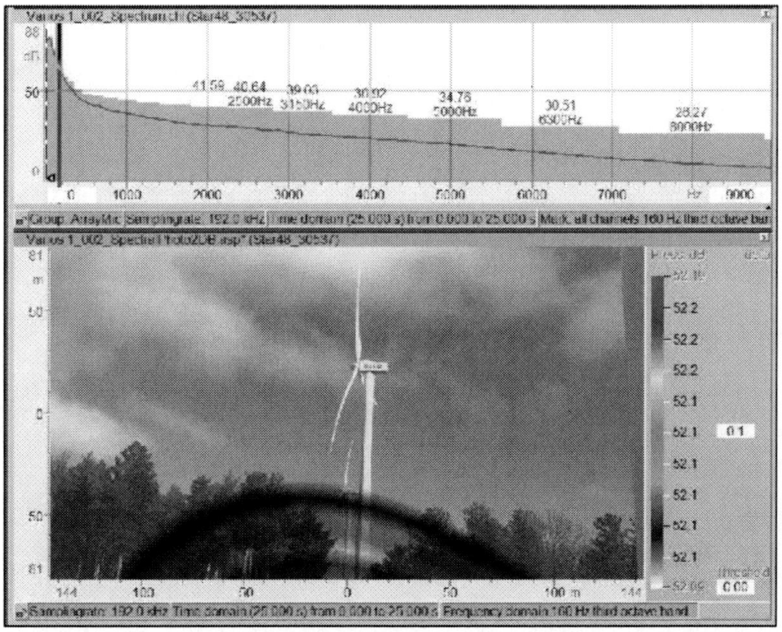

Figure 15. Acoustic picture, 160 Hz third octave

4. Conclusions

A lot of data has been generated in these measurement tests, but only the more interesting are showed on this publication. The results from the different scenarios show the different kind of data that can be used as basis for analysing, regardless if the wind turbine target is close to others. This data can be used as reference for comparing it with other workflow conditions, like different wind speeds or wind directions. Repeating the same points of evaluation in the same units and same conditions after a known number of working hours can provide useful information comparing the noise emissions regarding level and point or area of emission. Without stopping the wind turbine and without installing any sensors on it, noise results related with maintenance and malfunction can be acquired and processed on site.
The acoustic pictures can be used for validating predictive noise models, focusing on the noise emission from each part of the wind turbine or using global values for predictive noise propagation softwares, feeding them with real data from installed and working wind turbines.

References

Stuart Bradley, Michael Kerscher, Torben Mikkelsen (2017) *Use of the Acoustic Camera to accurately localise wind turbine noise sources and determine their Doppler shift.* Wind Turbine Noise

Michael Kerscher, Benjamin Vonrhein, Prof. Dr. Friedrich Ueberle, Dr. Dagmar Rokita (2016) *How Acoustic Camera Measurements can help to increase the Acceptance of Wind Turbines.* Wind Europe Submit.

Ofelia Jianu, Marc A. Rosen, Greg Naterer (2011) *Noise Pollution Prevention in Wind Turbines: Status and recent advances*

Oerlemans, S, Sijtsma, P and Méndez López B (2007) *Location and quantification of noise sources on a wind turbine.* J Sound Vibration 299, 869-883.

Zhang X-Z, Bi C-X, Zhang Y-B and Xu L (2017). *A time-domain inverse technique for the localization and quantification of rotating sound sources.* Mech Systems Sig Proc 90,15–29

www.acoustic-camera.com (2019)

INCE*EUROPE*

8th International Conference
on
Wind Turbine Noise
Lisbon – 12th to 14th June 2019

Session G

Background Noise and Measurement

Organised by INCE-Europe
mail@inceeurope.org
www.inceeurope.org

Supported by Portuguese Acoustical Society

Sociedade Portuguesa de Acústica

INCE EUROPE

8th International Conference
on
Wind Turbine Noise
Lisbon – 12th to 14th June 2019

A Low Wind-Noise Microphone for Wind Turbine Noise

Sabine von Hünerbein, University of Salford, UK. S.VonHunerbein@salford.ac.uk
Stuart Bradley, University of Auckland, New Zealand. s.bradley@auckland.ac.nz

Summary

A persistent challenge in characterizing wind turbine noise is that turbulent noise on the microphones is dominant in the same spectral area as the turbine noise. While baffles certainly help, they generally also modify the reception of signals of interest.
In response to this problem we describe a new approach where a closely-spaced array of microphones is mounted flush into a small-diameter surface. When wind is incident on this surface, a turbulent boundary layer forms, whose thickness and spectral properties depend on wind speed and distance downwind from the leading edge. The distance from individual microphones to the upstream leading edge of the surface varies depending on wind direction and microphone location. The result is that a number of independent measurements are made of the local wind noise, allowing for correction.
We present the theory for this new method and first results with our small microphone arrays under controlled conditions.

1. Introduction

Microphones respond to pressure fluctuations due to turbulence created in a boundary layer above the surface of the microphone (Bradley et al., 2003; Beranek and Vér, 2006). These wind-dependent fluctuations can degrade the estimation of external sound. The aeroacoustic noise peaks at low audio frequencies and, in many applications, can be removed via spectral filtering. However, in applications such as recording the noise from wind turbines, the external signal of interest and the locally generated wind noise on the microphone have overlapping spectral peaks of comparable strength (Heckl and Petersson, 2005). Directionality of microphones does not necessarily help. The usual approach is to use wind shields, but it may be necessary to have multiple, and physically large, shields. Alternative approaches, such as recording the external audio only during low-wind periods at the microphone, risk biasing data (von Hünerbein et al., 2017).
For the case of small microphones, dipole noise is the dominant turbulent noise, and Ffowcs-Williams and Hall (1970) showed that the mean-square sound pressure varies as

$$\overline{p^2} \propto U^5 l \frac{\overline{u^2}}{U^2} \qquad (1)$$

due to turbulence of intensity $\overline{u^2}/U^2$ of correlation length l, where U is the speed in the undisturbed flow. The turbulent scales are limited by the boundary layer thickness, which increases approximately as the square root of the distance d downstream of the leading edge. Putting these factors together, the mean-square voltage output from a microphone (which is proportional to the mean-square sound pressure) can be approximated by

$$\overline{v^2} = \overline{v_n^2} + \overline{v_{ext}^2} \tag{2}$$

where

$$\overline{v_n^2} = \frac{aRe^2}{1+bSt^5}. \tag{3}$$

and $\overline{v_{ext}^2}$ is due to the external sound source of interest. The term $\overline{v_n v_{ext}}$ doesn't appear because v_n and v_{ext} are uncorrelated, and circuit noise and microphone self-noise are generally negligible by design. The dimensionless parameters involved are the Strouhal number $St = fd/U$ and the Reynolds number $e = Ud/\nu$, where ν is the kinematic viscosity, and f the frequency. In general a and b are functions of d.
The wind-noise spectrum described by (3) depends on the distance of a microphone from the leading edge of the microphone mounting, unlike the spectrum from an external sound source such as a wind turbine. This potentially provides a method for separating the local turbulence-generated spectrum from the external sound source spectrum.

2. Design of a composite microphone

A number of circular composite microphones were designed, of the general form shown in Figure 1.

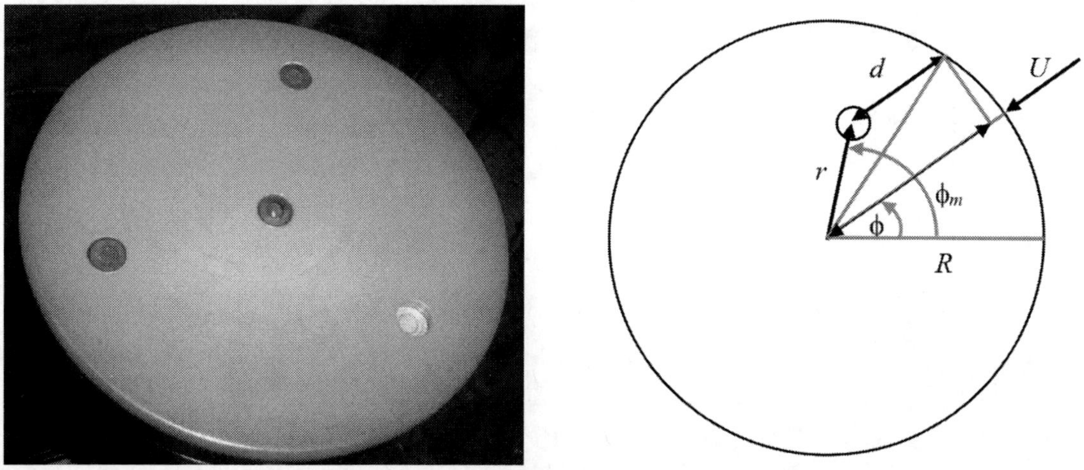

Figure 1. A circular composite microphone (left - one microphone is not mounted) and the geometry for computing d (right).

This gives

$$d = \sqrt{R^2 - r^2\sin^2(\phi - \phi_m)} - r\cos(\phi - \phi_m). \tag{4}$$

The design shown has four microphones and, for any given wind direction, there will be four distances d and four equations (2) in the two unknowns U and ϕ.

3. Proof of concept

The first step is to check the validity of (3) and to determine the dependence of a and b on d. To do this, a small, quiet, wind tunnel was used (Figure 2). Measurements of normalised spectra at fixed wind speed and at fixed wind direction are shown in Figure 3. The f^{-5} dependence at higher frequencies is evident. Also shown is the background noise spectrum from a microphone not exposed to the air flow, showing that the external sound pressure is negligible.

Figure 2. The composite microphone under test.

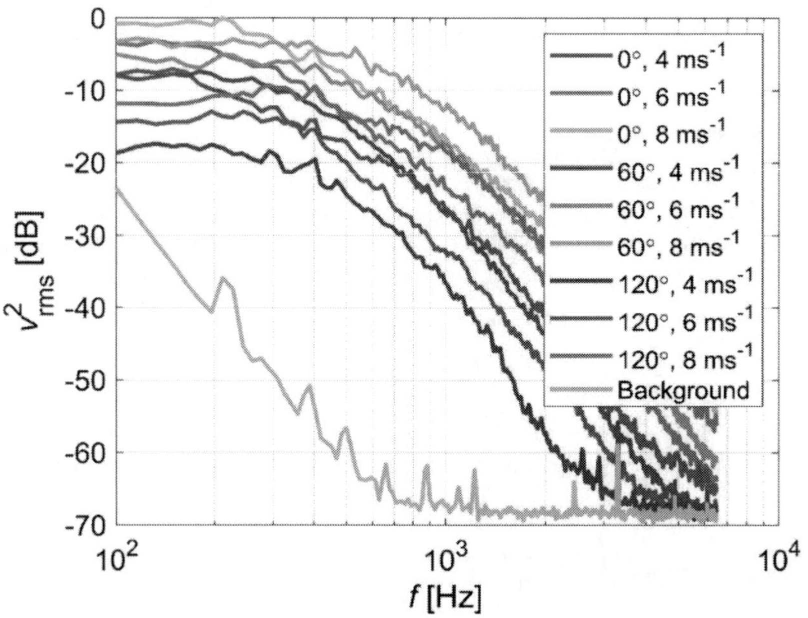

Figure 3. Spectra at U = 4, 6, and 8 m s^{-1} and three microphone angles 0°, 60°, and 120°.

Next, the simple scaling with a and b constant is tested. Figure 4 shows that the -3 dB frequency and the low-frequency intensity are both dependent on d. Figure 5 shows scaled spectra with $a \propto d^{-4}$ and $b \propto d^{-5/4}$ which captures the main dependence on wind speed and direction.

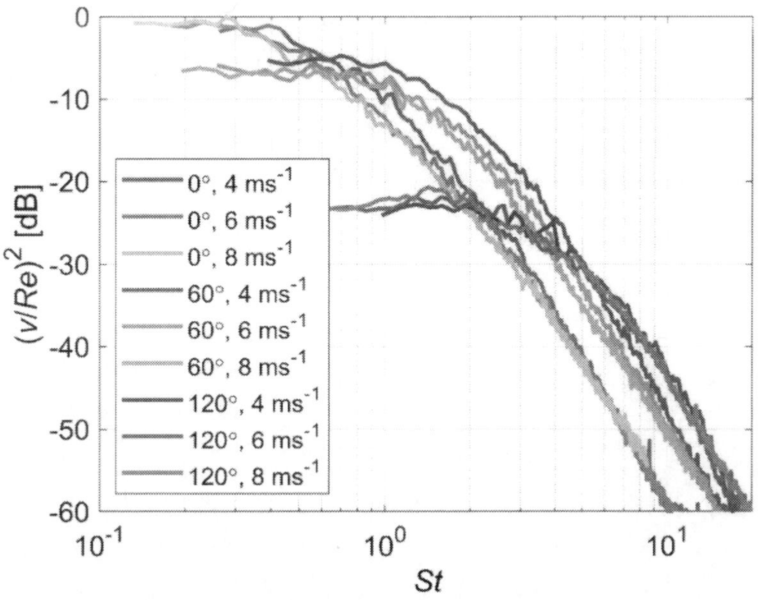

Figure 4. Scaled spectra versus Strouhal number.

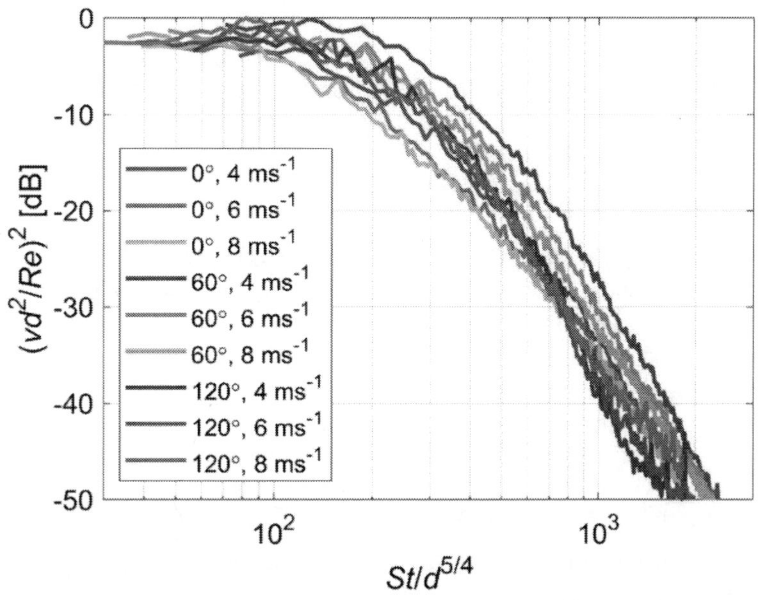

Figure 5. Spectra with a and b dependent on d.

4. Estimating wind noise

The spectra shown above did not include the central microphone for which $d = R$. However, an indication of the estimation of *U* and ϕ can be given from 3 microphones. Highest accuracy may be expected with the full fit over the frequency range, but a first estimate can be obtained with the simpler low-frequency plateau part of the spectrum, since the fit is then of the form

$$\overline{v^2} = A\left(\frac{U}{d}\right)^2 + \overline{v_{ext}^2} \tag{5}$$

where A is known from the above wind tunnel calibrations. For example, for U = 6 m s^{-1}, and using the measured spectra for the cases of 0°, 120°, and also using the 120° spectrum for -120°, a simple fit by searching parameter space gives the result in Figure 6. The estimated wind speed is 5.76 m s^{-1}, and the estimated value of ϕ is 0°. The rms error in this simple fit is 2% of the mean low-frequency $\overline{v_n^2}$

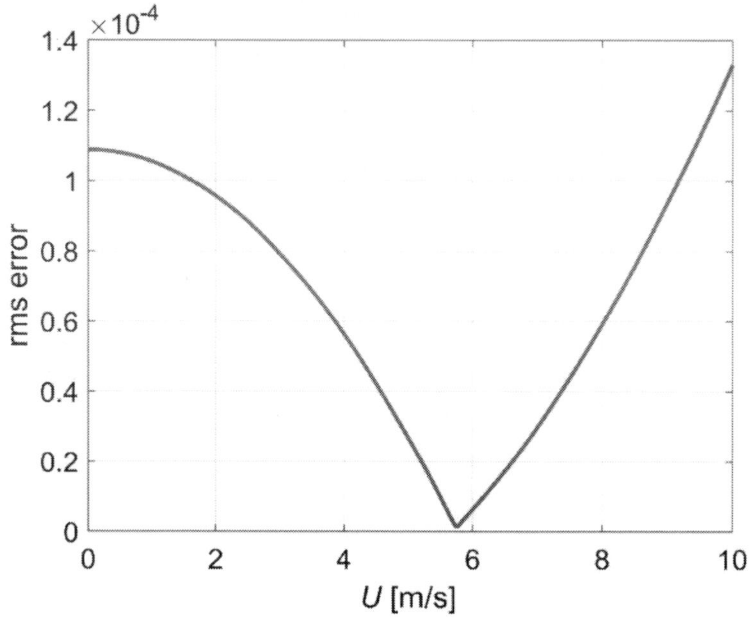

Figure 6. The rms error for fitting versus trial *U* values.

5. Conclusions

We have described a method by which locally-generated wind noise on microphones can be distinguished from sound from an external source. An simple example of estimation of wind speed and direction based on wind tunnel data has an estimation error of $0.02\,\overline{v_n^2}$, where $\overline{v_n^2}$ is the noise spectrum. If the method gives 2% error in estimated wind noise, and the intensity of the wind noise is comparable to the intensity of the external sound, then the error in estimation of the required external sound intensity is also 2%. Thus if the masking wind noise spectrum is of comparable magnitude to the external sound source spectrum, high quality spectral 'cleaning' should be possible.

References

Beranek L and Vér I (2006) *Noise and Vibration Control Engineering – Principles and Applications*, John Wiley & Sons, New York

Bradley S, Wu T, von Hünerbein S and Backman J (2003) *The mechanisms creating wind noise in microphones*, Audio Engineering Society Convention, Amsterdam, paper 5718, 9 pp.

Heckl C and Petersson B (2005) *Structure-Borne Sound - Structural Vibrations and Sound Radiation at Audio Frequencies,* Springer

von Hünerbein S, Kendrick P and Cox T (2017) *Extended simulations of wind noise contamination of amplitude modulation ratings*, Wind Turbine Noise conference, 2 -5 May 2017, Rotterdam, NL.

Ffowcs-Williams J and Hall L (1970) *Aerodynamic sound generation by turbulent flow in the vicinity of a scattering half-plane.* J. Fluid Mech., 40, 657-670.

INCEEUROPE

8th International Conference
on
Wind Turbine Noise
Lisbon – 12th to 14th June 2019

Measuring infrasound from wind turbines: the benefits of a wind-shielding dome

Sarah D'Amico, Information Technology Department, Ghent University: Sarah.DAmico@UGent.be
Timothy Van Renterghem, Information Technology Department, Ghent University: Timothy.VanRenterghem@UGent.be
Dick Botteldooren, Information Technology Department, Ghent University: Dick.Botteldooren@UGent.be

Summary

Measuring infrasound from wind turbines is a challenging task. A device for the signal-to-noise ratio improvement of wind turbine emitted acoustic pressure at low and infrasound frequencies has been designed and tested. The semi-spherical shape chosen aimed at maximizing the pressure averaging of large atmospheric turbulent eddies. In addition, the wind shielding dome should still be reasonably compact for easy handling. A computational fluid dynamic study was conducted with Fluent to find the optimal porosity of the windbreak textile used to constitute the outer surface of the dome. For a range of different inflow conditions, the most suited porosity was found to be 40% when minimizing the inside dome wind speed and related turbulence parameters. The dome fence's acoustic absorption in absence of flow was tested in a semi-anechoic chamber. The maximum absorption observed in the lower part of the test signal's spectrum was less than 0.1 dB, making the dome almost fully transparent for the frequency range of main concern. The dome was also tested outdoors with a free and a dome-covered low-frequency microphone. The test was performed under different wind speeds, measured turbulence intensities and wind directions. Reductions of wind-induced noise of 12 dB were observed for frequencies down to 0.1 Hz, while maximum reductions of more than 14 dB were measured for frequencies around 1 Hz. It can thus be concluded that the wind shielding dome is able to strongly improve the infrasound signal-to-noise ratio, opening possibilities to more accurately measure the infrasonic emissions from wind turbines.

1. Introduction

The need for renewable sources of energy is resulting in an increment in the construction of wind farms. While wind turbines have a good reputation as efficient devices for the generation of green energy, their noise emissions are often reported as the main negative impact on the environment. Health problems related to long-term noise exposure and the potentially long distances infrasound can propagate in the atmosphere are reasons for concern.

The scientific community agrees that the effect of wind turbine infrasound on human health is currently unclear. Infrasound exposure from wind turbines is typically lower than the threshold of hearing (van Kamp & van den Berg, 2018), while intrinsic human body pressure variations could create infrasound signals in the inner ear that are larger. Psychological factors are thought to have prominence on the genesis of annoyance and health disorders (Leventhal, 2013). On the other hand, physiological studies on the sensitivity of the human ear have revealed that although infrasound is not able to generate a hearing sensation, it could still produce neuronal signals whose impacts are unknown (Salt & Huller, 2010). Furthermore, the studies on prolonged exposure to low intensity infrasound are still sparse (Baliatsas, et al., 2016).

Characterizing wind turbines' infrasound also presents difficulties. Infrasound emissions from wind power generators occur simultaneously with other environmental noise sources. In particular, wind-induced noise will mask wind turbine infrasound. The large turbulent eddies convected by the wind produce pressure fluctuations at these very low frequencies. Fully discriminating between the two sources is often not possible. The standard methodology applied in relation to wind turbine measurements is insufficient to produce a good signal-to-noise ratio in the infrasonic frequency range. Large wind fences have been proposed for military applications still showing limited SPL reductions at the frequency range of interest. In addition, the dimensions of such devices make them impractical for day-to-day operation.

A large porous structure can reduce the mean wind speed and will affect the turbulence structures present in the flow. Studies from spherical wind screens suggest that they act as pressure-averaging devices providing pressure fluctuation reductions for turbulence eddies larger than their characteristic length (Hedlin & Raspet, 2003). A wind fence designed using this principle could therefore be a good candidate for efficient wind infrasound mitigation.

2. Methodology

2.1 Dome design

The optimal shape of the dome fence was based on designs applying the principle of spherical wind screens to reduce wind-induced noise in the infrasound range starting from frequencies of approximately 0.1 Hz. A retaining structure of 220 cm height and with a 360 cm base diameter was used.

The optimal porosity of the screening textile was found based on a CFD study with Ansys Fluent 17.1. The dome was discretized in a 2D domain as a surface area-equivalent hemisphere of 2 m radius and simulated under inflow wind conditions of 5 m/s and 10 m/s at 10 m height. The k-ε model was used. Pressure jump coefficients for different porosities were derived from actual measured values of this type of windscreens (Dierickx, 1998).

Reductions of the flow parameters with respect to the free field at the centre of the wind-shielding dome were compared for different porosities and flow conditions. More precisely, mean wind speed and turbulence intensities were investigated resulting in an overall optimal porosity of 40%. The textile was manufactured on demand. The wind-shielding dome was first tested in a semi-anechoic room to assess its acoustic absorption. The dome was shown to be fully acoustically transparent.

2.2 Field test

The dome's wind-shielding efficiency was tested in an open field. Simultaneous measurements were performed with a microphone inside the dome and compared to an uncovered microphone at close distance. Both microphones used a standard spherical windscreen.

The test was performed in a rural environment approximately 400 m away from a rail road, national secondary roads and an inhabited village. A wind farm composed of seven wind turbines of 2.3 MW was near the test location. Two GRAS 46 AZ low frequency microphones and preamplifiers were used and equipped with rain protection and wind screens of 10 cm. Wind flow parameters were monitored at 10 m height with a 3D sonic anemometer.

Long-term measurements were performed in two configurations. In the first one, the two microphones were positioned at 200 m from each other, while the free-standing microphone was 100 m close to the wind farm. In the second configuration, the two microphones were positioned at 30 m from each other and at 300 m from the wind farm. Rainy periods were excluded as well as periods with noise from agricultural machine or trains. In the second configuration, data relative to a wind direction parallel to the two microphones were also excluded to avoid the influence of the dome's wake on the response of the uncovered microphone.

Ten-minutes 1/3rd octave bands were sorted for nacelle-position, mean wind speeds and turbulence intensities. For each group, results were averaged and compared for the two receivers. Background noise measurements were also recorded; these were performed with the wind farm fully off at 2 m/s mean wind speed.

3. Results

Preliminary results of the aforementioned measurement campaign are discussed in this section. Figure 1 illustrates the values of the 1/3rd octave bands for the measurements acquired by the two receivers in the second configuration at different wind speeds. The measurements are presented over non-dimensional frequencies representing the ratio between the dome's characteristic length (L_{dome}) and the atmospheric turbulence's characteristic length scale.

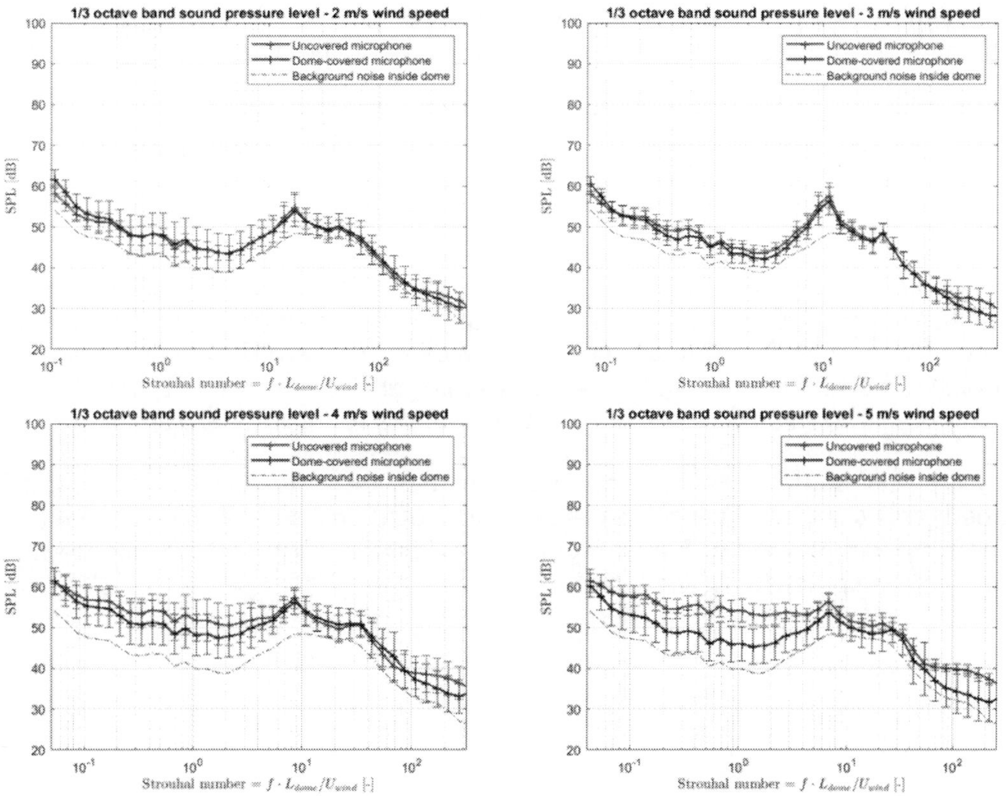

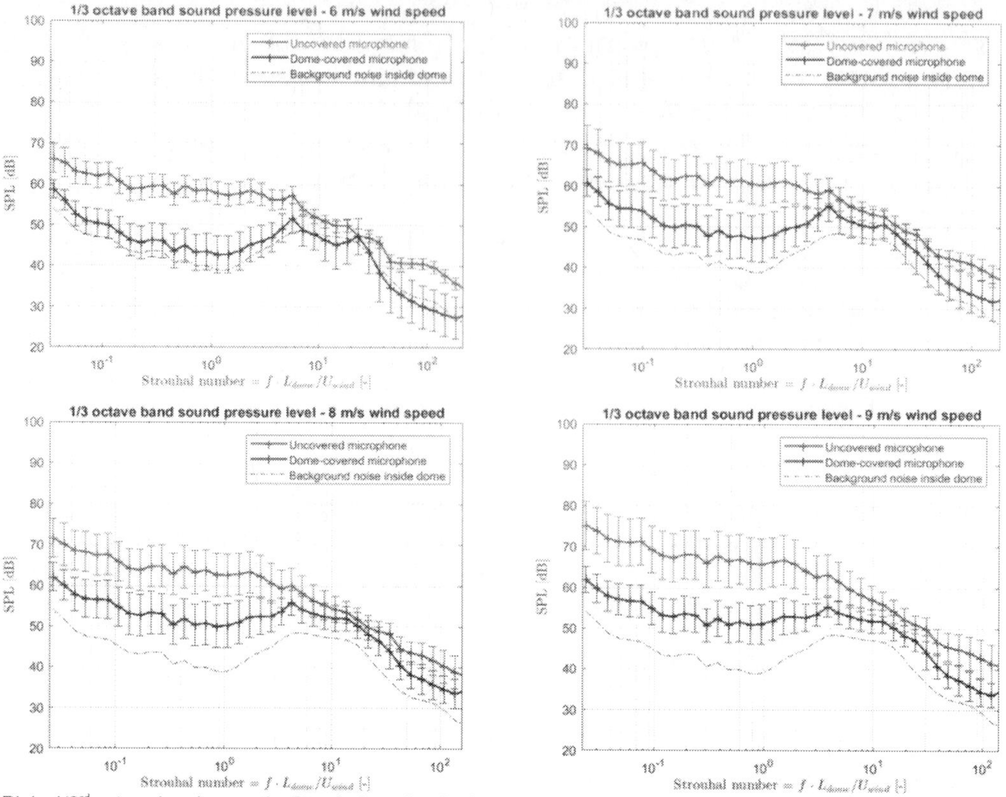

Figure 1: SPL in 1/3rd octave bands over the Strouhal number for free and dome-covered receiver under different mean wind speeds. f is the frequency of the acoustic wave and U_{wind} is the mean wind speed. Total error bars are double the standard deviations

Reductions in sound pressure levels by the presence of the dome in the infrasonic frequency range were found up to 14 dB. For low mean wind speeds, 2 and 3 m/s, SPLs inside and outside the dome are similar. As the mean wind speed increases, at 4 and 5 m/s, the levels at the uncovered receiver start increasing for Strouhal numbers (Sr) between 0.1 and 6. A maximum reduction of the wind induced noise inside the dome of 3 dB at 4 m/s and 7 dB at 5 m/s can be observed at Sr equal to 1. For higher wind speeds, from 6 to 9 m/s, absolute levels inside and outside the dome increase. However, this increase is much slower inside the dome, highlighting the wind-induced noise reduction produced by the dome. A maximum reduction of approximately 14 dB is observed at at Sr of 1. The dome reduces the mean wind speed of the flow and acts on the turbulence eddies as a pressure averaging device, mitigating the resulting pressure fluctuations at the microphone. The effect is maximum for turbulence length scales of the same order of magnitude as the dome's characteristic length. For smaller Sr numbers the reduction is almost constant down to Sr 0.1 with reductions of 12 dB or higher. The dome shows to reduce turbulence-induced noise from turbulence structures larger than its characteristic length scale. For Sr numbers higher than 1, thus smaller turbulences, the reductions are smaller.

Results from the first setup, with 200 m separation between the two receivers, confirm the aforementioned results. Figure 2 presents the behaviour of the dome at 5, 10 and 25% turbulence intensities for a low wind speed of 3 m/s. Figure 3 depicts the case of 6 m/s mean wind speed and turbulence intensities of 10, 15 and 25%.

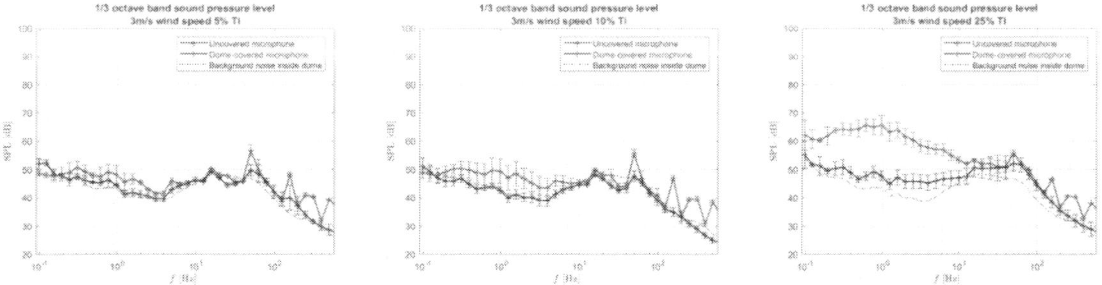

Figure 2: SPL in 1/3rd octave bands for free and dome-covered receiver for 3 m/s mean wind speeds and 5, 10 and 25% turbulence intensity; total error bars are double the standard deviations.

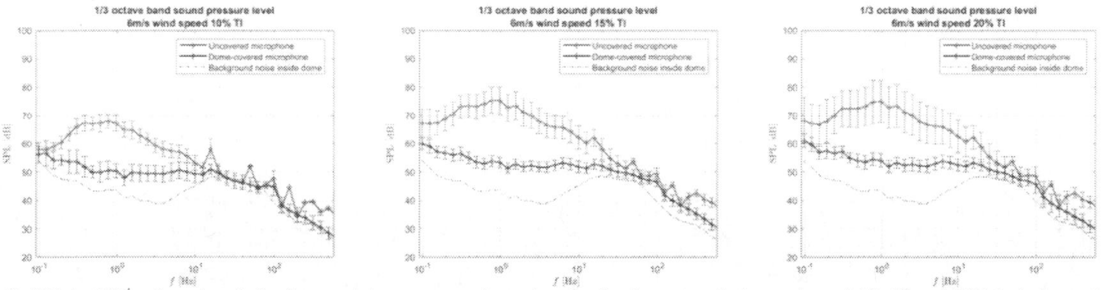

Figure 3: SPL in 1/3rd octave bands for free and dome-covered receiver for 6 m/s mean wind speeds and 10, 15 and 20% turbulence intensity; total error bars are double the standard deviations.

Figure 3 shows that the wind-induced noise does not simply correlate with the mean wind speed but it is also influenced by the turbulence intensity. For 3 m/s mean wind speed, the wind induced noise at the free receiver is quite low at 5 and 10% turbulence intensity in the infrasonic range. For the case of 25% turbulence intensity, the wind induced noise at the uncovered microphone undergoes an increment of approximately 15 dB at 1 Hz. SPLs inside the dome undergo a slight increase at high turbulence intensities, but this effect is limited to 5 dB at 1 Hz. For the highest turbulence intensity analysed here the dome gives a wind-induced noise reduction of 15 dB at 1 Hz at 3 m/s.

For the 6 m/s mean wind speed, a reduction of approximately 14 dB can be observed already for 10% turbulence intensity at 1 Hz inside the dome. This gives 20 dB at the same 1/3rd octave for the case of 25% turbulence intensity.

4. Conclusions

A wind-shielding dome for infrasound wind-induced noise was designed and tested. Its optimal geometry was based on wind screen theory and its porosity was optimized based on CFD simulations. The dome's performance was tested outdoors where pressure fluctuations recorded at a free-standing microphone where compared with a dome-covered microphone.

The dome shows the capability of reducing wind-induced noise in the infrasound spectral range up to 14 dB, having its maximum effect on atmospheric turbulences of comparable characteristic length. Wind noise reductions can be observed down to 0.1 Hz. The preliminary results show reductions correlated to mean wind speed and to turbulence intensity. For low mean wind speeds, the infrasonic sound pressure levels at a free-standing microphone show to be strongly influenced by the turbulence content in the flow resulting in an increase of circa 15 dB between low and high turbulence at the same low mean wind speed. For higher mean wind speeds and high turbulence intensities, reductions in sound pressure level by the dome reach 20 dB at 1 Hz. The dome, although still reasonably compact, allows to significantly increase the signal-to-noise ratio in the infrasonic range. This opens possibilities for accurately characterizing wind turbines infrasonic emissions.

References

Baliatsas, C., van Kamp, I., van Poll, R. & Yzermans, J., (2016). *Health effects from low-frequency noise and infrasound in the general population: Is it time to listen? A systematic review of observational studies.* The Science of the Total Environment, 557-558, 163-169.

Dierickx, W., (1998). *Flow Reduction of Synthetic Screens Obtained with Both a Water and Airflow Apparatus.* Journal of Agricaltural Engineering Research, 71, 67-73.

Hedlin, M. A. H. & Raspet, R., (2003). *Infrasonic wind-noise reduction by barrier and spatial filters.* The Journal of the Acoustical Soiety of America, 114(3), 1379-86.

Leventhal, G., (2013). *Infrasound at the ear.* Denver, 5th International Conference on Wind Tubine Noise.

Salt, A. N. & Huller, T. E., (2010). *Response of the ear to low frequency sound, infrasoud and wind turbines.* Hearing Research, 268(1-2), 12-21.

van Kamp, I. & van den Berg, F., (2018). *Health Effect Related to Wind Turbine Sound, Including Low-Frequency Sound and Infrasound.* Acuostics Australia, 46(1), 31-57.

INCE*EUROPE*

8th International Conference
on
Wind Turbine Noise
Lisbon – 12th to 14th June 2019

Is it possible to predict background noise levels from measured meteorological data with machine learning techniques?

Alexis Bigot, Sixense Environment: alexis.bigot@sixense-group.com
Guillaume Hochard, Quantmetry: ghochard@quantmetry.com

Abstract

Background noise levels play a major role when it comes to ensuring compliance with French noise regulation. Background noise levels are usually assessed based on scatter plots: noise levels vs wind speeds and/or wind directions.

In this article, we propose to model the background noise levels using machine learning techniques.

The datasets are built with 10 minutes meteorological data as well as background noise level measured in dBA.

We used gradient boosting which is a supervised machine learning technique used in classification and regression problems. The objective is to train the model for each dataset, and evaluate the accuracy of the regression algorithm.

The results are very promising: the mean absolute error (MAE) of the prediction are 1.07 dBA on dataset A and 1.71 dBA on dataset B. We are convinced that this technique will change the way we manage noise and meteorological data in acoustics and wind energy.

1. Introduction

Background noise levels play a major role when it comes to ensuring compliance with French noise regulation. Background noise levels are assessed using scatter plots: noise levels vs wind speeds. This statistical analysis is carried out for several "homogeneous conditions" as defined in the French normative [1]:

- Day/night
- Wind direction sectors
- Time of the day
- Human activities
- Meteorological conditions
- Seasons

The problem is that this methodology is not precise enough if we need an accurate model of the background noise vs meteorological conditions: the standard deviation is always greater than 3 dBA. First, we should take into account all the meteorological conditions available (not only wind speed and the wind direction as shown in figure 1). Secondly we need new techniques, with a better accuracy than the classical scatter plots.

That's why we decided to test machine learning techniques for a better assessment of noise levels around wind farms.

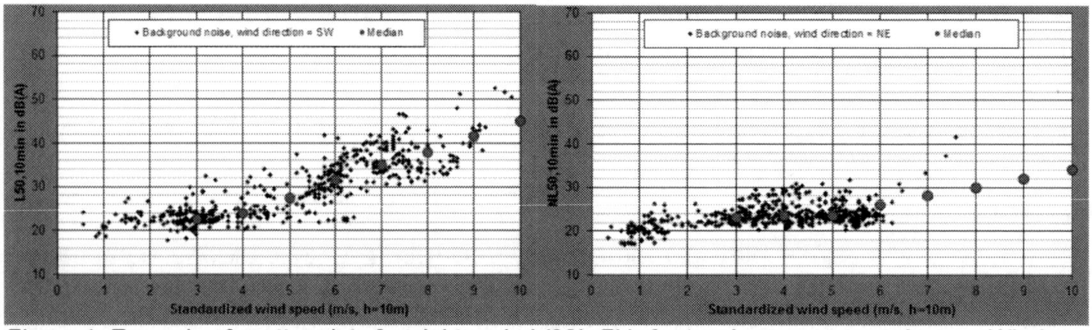

Figure 1: Example of scatter plots for night period (22h-7h), for two homogeneous classes: Wind direction = South-West [135° - 315°[and wind direction = North-East [315° - 135°[. In red = calculation of the median or extrapolation of the trend of background noise.

2. Dataset & methodology

2.1 Dataset

During previous WTN conferences we showed that the wind speed gradient and the temperature difference had an influence on the background noise levels [2]. Those parameters also drive the refraction during noise propagation of wind turbine noise [3].

Based on this observation we decided to choose two datasets that include wind speed and temperature measured for at least two different heights.

In the dataset A, the input features are:

- Datetime, from 11/23/2017 to 12/14/2017
- Temperature measured at a height of 1.5 m
- Temperature measured at a height of 10m
- Wind speed measured at a height of 10m
- Wind speed measured at a height of 100m (hub height of the future turbines)

- Wind direction
- Relative humidity
- Point of acquisition

and the measured noise level, the target to model.

In the dataset B, the input features are:

- Datetime, from 05/03/2017 to 06/02/2017
- Temperature measured at a height of 24.9m
- Temperature measured at a height of 99.5m
- Wind speed measured at a height of 24.9m
- Wind speed measured at a height of 99.5m (hub height of the future turbines)
- Wind direction
- Relative humidity
- Point of acquisition

and the measured noise level, the target to model.

In both cases the output data are background noise level L50,10min in dBA, measured at several locations around the wind farm project: there are 9 location of noise measurements in dataset A and 5 in dataset B.
We decided to use only the night time period (22h-7h) of the dataset, because French regulation is more restrictive during night time.

Datetime	Temperature at 1.5m (°C)	Temperature at 10m (°C)	Wind speed measured at 10m (m/s)	Wind speed measured at 100m (m/s)	Wind direction (°)	Relative humidity (%)	Measure point	Noise level
11/23/2017 00:00	12.2	12.4	9.0	14.2	164	67.9	1	37.7
11/23/2017 00:10	12.4	12.5	9.7	13.6	163	67.4	1	36.6
11/23/2017 00:20	12.5	12.5	9.4	13.6	164	67.1	1	36.4
11/23/2017 00:30	12.6	12.7	9.7	13.6	167	66.1	1	32.5
11/23/2017 00:40	12.7	12.8	10.2	14.2	167	66.1	1	31.4
11/23/2017 00:50	12.7	12.8	10.1	15.0	167	65.9	1	32.8

Figure 2: First rows of dataset A

2.2 **Methodology**

To model this dataset, we propose to use a supervised machine learning approach.

Supervised machine learning is part of artificial intelligence techniques which aims at learning from a set of features a decision pattern to predict target values. A supervised machine learning model thus implies two phases, the learning and the inference phase.

In the learning phase, a model is trained on a set of samples that includes prepared features and corresponding targets in order to make the most accurate predictions. Once the model is trained, we can use it to infer new target values based on a dataset of new samples with the same set of features used for the training phase.

Gradient boosting is a supervised machine learning technique used in classification and regression problems based on the concept of ensembling, i.e. combining weak learners to produce a prediction model. The model is built sequentially, first producing a first model which performance will be evaluated.
The prediction errors are then weighted in order for the next model to correctly predict the difficult samples that were incorrectly predicted by the previous model. This process is repeated iteratively for a given number of rounds.

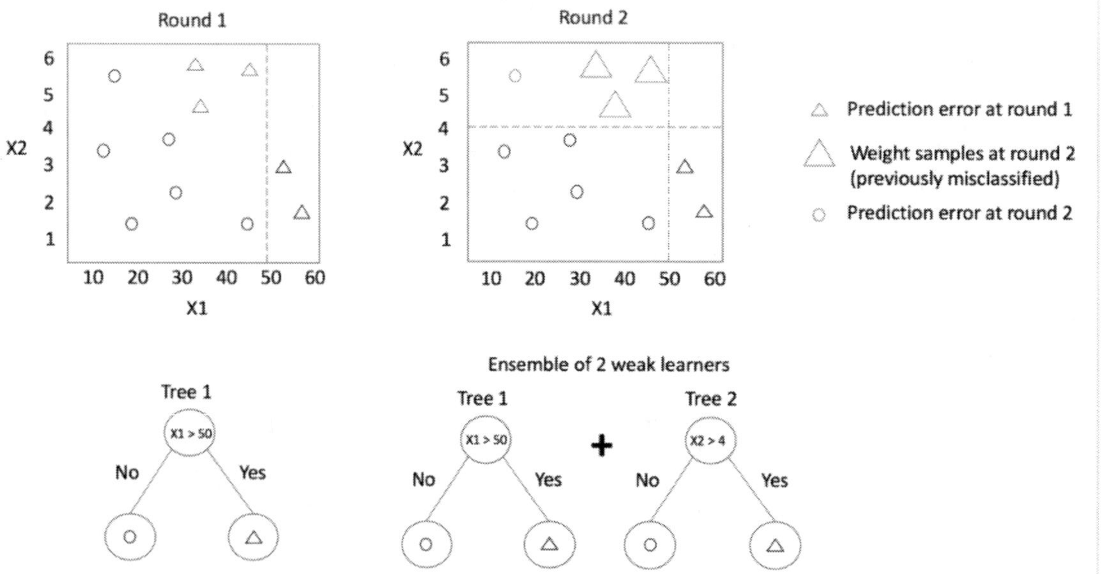

Fig. 3 Gradient boosting explained simply for a supervised classification example. Decision trees are iteratively built in order to make better predictions for the incorrectly predicted values by the previous tree.

For this paper, we have used XGBoost, a popular and efficient open-source implementation of gradient boosting [4]. XGboost is famous for winning machine learning competitions and because it adapts to a large variety of data types and offers a broad palette of hyperparameters that can be tuned to enhance model performance. This kind of model is thus a reliable choice for supervised machine learning regression problems, like the one presented in this paper.

One problem that could occur during the training phase is overfitting. This term means that the model complexity is too high and has learned "by heart" the training set. The effect of overfitting is straightforward on new data, as the performance measures on new predictions get lower than those on the training set. To overcome this problem, we split the data in two sets - a training and a validation set - with a 80/20% ratio. To assess model generalization and performance on new data, we will compare the predictions errors on these two sets.

One other way to control overfitting is by performing cross-validation on the training set. This process involves to randomly partition the training data into several folds. One of the folds will be left out the training set and will be used for performance evaluation. This process is repeated for every fold and validation results are averaged over the different validation rounds to get a better evaluation of the model's predictive capability.

To evaluate the performance, we use the mean absolute error and the standard deviation error. Given a predicted value $\hat{y}_i$ for and the corresponding ground truth value y_i, the mean absolute error over the n samples in the dataset is defined as:

$$MAE = \frac{1}{n}\sum_{i=1}^{n}|\hat{y}_i - y_i|$$

3. Results

3.1 Global results

The machine learning models trained on these two datasets offer good performances, as illustrated in Table 1, with low bias (1.10 dBA and 1.82 dBA on the test sets) and limited variance between the training and the validation sets (resp. 0.79 dBA and 1.16 dBA on dataset A and B), which indicates good generalization of the models on data unseen during the training phase, and thus limited overfitting.

	Train set		Test set		Test/train comparison
Dataset	Mean Abs Error (1)	Std Dev.Error	Mean Abs Error (2)	Std Dev. Error	Error difference (2) – (1)
A	**0.29 dBA**	0.38 dBA	**1.07 dBA**	1.45 dBA	**0.78 dBA**
B	**0.60 dBA**	0.83 dBA	**1.71 dBA**	2.42 dBA	**1.11 dBA**

Table 1. Performance of the two models on the training and test sets

Figures 4 and resp. 5 present the predicted values over the ground truth values for all the samples in the training and test sets. The difficult values to predict in the test set are clearly identified as outliers from the linear fit. Further exploration on the prediction errors could allow to better understand in which conditions the model is less accurate. We can also compare the generalization power of the two models by observing the distribution of errors for the train and test sets: more variance is observed on the training dataset A than on the training dataset B, and we observe the same pattern in the corresponding test sets.

On dataset B, we observe greater prediction errors for larger values. This could be explained by the fact that the model has less data to be trained on for large values, as we clearly observe lower density for large values on the test set evaluation scatter plot in Figure 5.

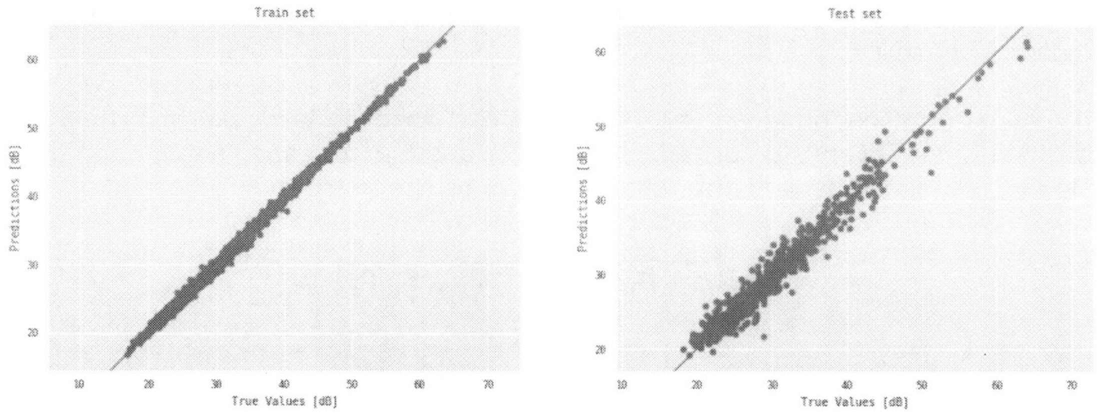

Figure 4: Predictions and ground truth values for training and validation set on dataset A

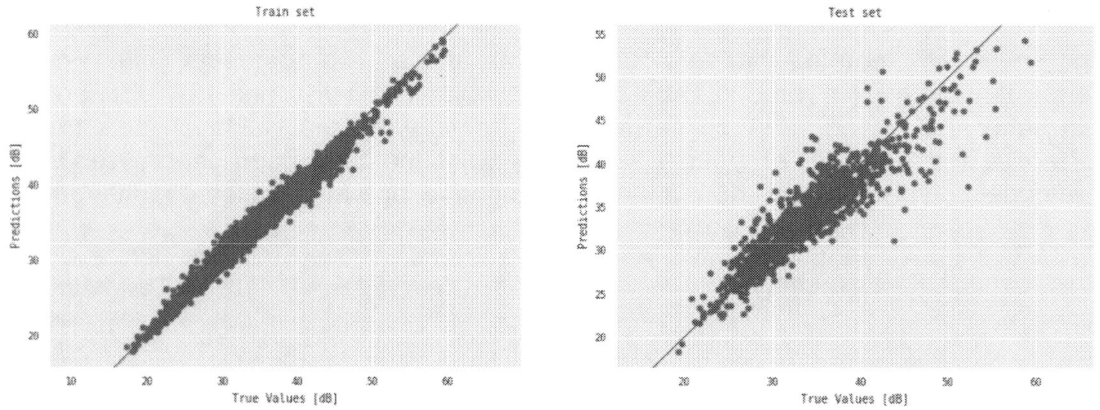

Figure 5: Predictions and ground truth values for training and validation set on dataset B

3.2 **Feature importance**

One advantage of using gradient boosting trees methods in the explainability of the decision making process performed by the model, both at a global scale (understanding which feature is important in the model) and at a local scale (which feature contributed quantitatively to drive the target value prediction in a specific direction). Local scale interpretability can be obtained using methods as shapley values [5].

At the global scale, the feature importance is an insightful tool to assess the predictive power of features. In the context of gradient boosting trees, it measures and computes the average reduction in impurity across all trees in the ensemble of weak learners due to each feature. Therefore, features that are used early in the tree construction (closer to the root node) get larger importance value.

By plotting the feature importance of all features in Figure 6 and 7 for the two models, we observe that the top most important features in the dataset A are the temperature at 1.5m, the mean wind speed at 10m/100m and the relative humidity. In the dataset B the important meteorological features are similar. The point of acquisition (area to predict the noise level) is also of prime importance, which confirms the fact that the noise level depends on the meteorological conditions but also the location of the measure itself.

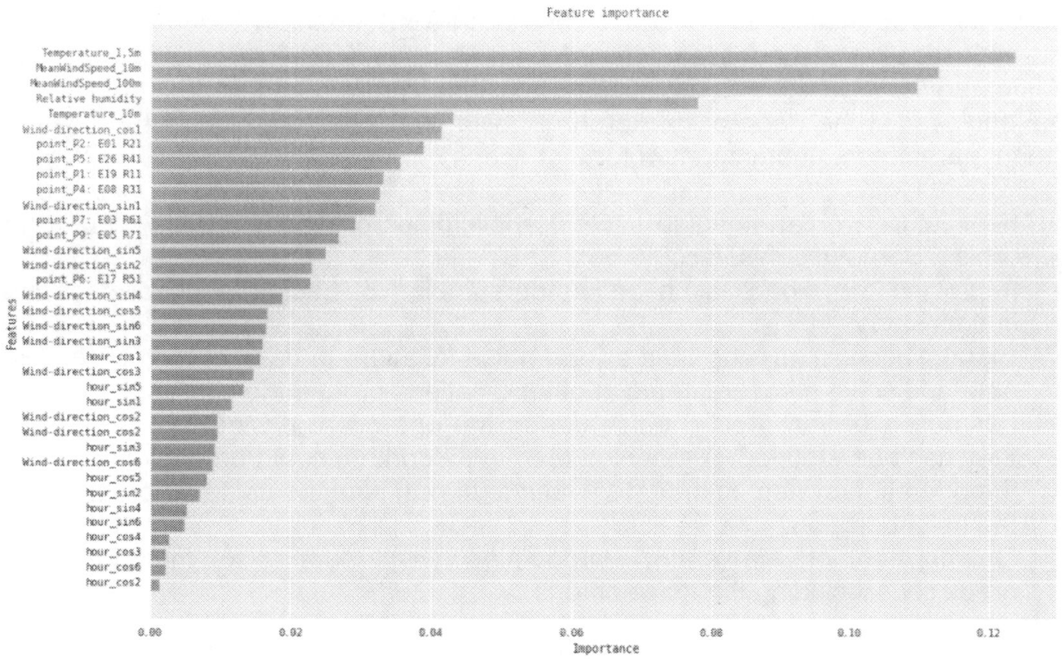

Figure 6: Features importance on dataset A

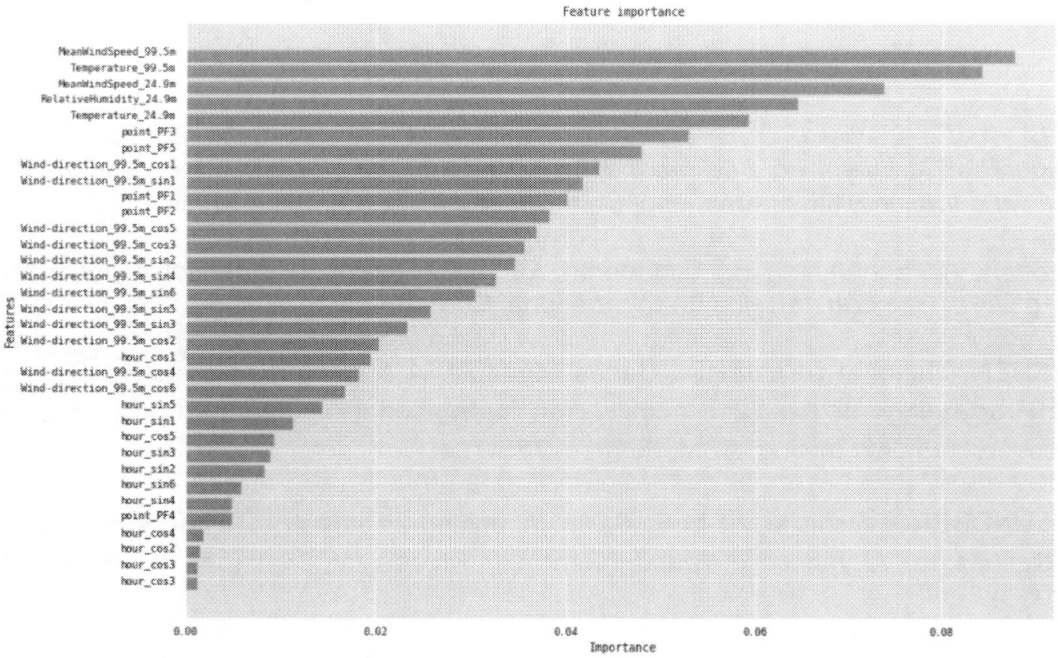

Figure 7: Features importance on dataset B

In parallel of this machine learning approach, we tried to benchmark these results to neural networks based models, but these models gave slightly less performance than the one obtained with the gradient boosting trees models. Nevertheless, they may give good results and should be tested again if the dataset is very large.

4. Discussion

Machine learning techniques give good results on wind turbine noise prediction, but further investigation are necessary in order to draw more reliable conclusions. The questions are: is it possible to extrapolate from one site to another? Is it possible to model and predict the trend of background noise over one year? and what are the minimum parameters and amount of for that scope?

In the near future we can imagine several applications of machine learning in wind turbine noise. Some use cases could be:

- Improving the modelization and understanding of background noise during impact study.
- Extrapolation of missing data at one noise location, in base of meteorological data and noise measurements at other locations around the site.
- Estimation of background noise during the operation of wind turbines, when they can't be stopped. A machine learning algorithm could be implemented in the operating system in order to evaluate in real-time the noise emergences in the neighborhood.
- Optimization of the energy production with respect to meteorological predictions including the noise criteria.

References

[1] Projet de norme AFNOR NF S 31-114 - Version du 7/7/2011 – v3

[2] Bigot, A.; Slaviero, D.; Mirabel, C.; Dutilleux, P. (2015). *Influence of vertical temperature gradient on background noise and on long-range noise propagation from wind turbines*. 6th International Conference on Wind Turbine Noise.

[3] Bigot, A.; Economou, P.; Economou, C. (2017). *Wind turbine noise prediction using Olive Tree Lab Terrain*. 7th International Conference on Wind Turbine Noise.

[4] CHEN, Tianqi et GUESTRIN, Carlos. *Xgboost: A scalable tree boosting system*. In : Proceedings of the 22nd acm sigkdd international conference on knowledge discovery and data mining. ACM, 2016. p. 785-794.

[5] LUNDBERG, Scott M. et LEE, Su-In. A unified approach to interpreting model predictions. In : Advances in Neural Information Processing Systems. 2017. p. 4765-4774.

INCE*EUROPE*

8th International Conference on Wind Turbine Noise Lisbon – 12th to 14th June 2019

Does background noise vary with seasons ?

Arthur PETIT (ENGIE Green) : arthur.petit@engie.com
Jaigh-Patrick DURIEUX : jaigh-patrick.durieux@external.engie.com
Arthur FINEZ (ENGIE Green) : arthur.finez@engie.com
Colin LEBOURDAT (ENGIE Green) : colin.lebourdat@engie.com

Summary

1. Introduction

Regarding wind turbine noise, compliance with French regulations depends in part on the background noise. One of the main criterion is indeed an emergence descriptor, which is the difference between the levels of total noise and background noise. Basically, if the total noise level exceeds 35 dB(A), the emergence must remain below a threshold of 5 dB(A) during the day and 3 dB(A) at night. Acoustic measurement campaigns are generally of one to two weeks duration, which makes it difficult to assess the seasonal variation of background noise. However, although their influence on noise levels is not well known, the fact remains that anthropogenic, floristic and faunistic activities do vary throughout the year. Besides, as for most wind farms, the acoustic compliance requires noise curtailments that are often based on a most constraining scenario, it is also of interest to assess the proportion of production losses that could be avoided by a seasonal refinement of the noise curtailment plans.
This study focuses on the particular case of a village in northern France where a long-term measurement campaign has been carried out. Acoustic measurements in two locations were conducted from late January to August 2014 alongside meteorological measurements. The

data were processed over a 2-week moving window in accordance with the best practices of the French profession, as described by the draft standard NFS 31-114. Then, considering the evolution of the background noise level obtained, an assessment of the potential earnings were carried out for typical wind farm layouts.

2. A quick reminder of French regulations regarding wind turbine noise

Depending on the country, both thresholds and indicators vary, making comparisons between regulations difficult. For example, Germany uses the LAeq indicator and England the LA90, with accordingly adjusted thresholds.
This study is based on French practices and French regulation.

2.1 French legislation on classified installations

In France, since the publication of the decree of August 26, 2011, wind farms enter the scope of classified installations for environmental protection (installation classée pour la protection de l'environnement) (ICPE) regime. They are subject to authorization. As such, there are three acoustic criteria to be respected to prevent neighboring areas from noise :

- Compliance with global emergence values in dB(A);
- Compliance with maximum ambient noise values in dB(A);
- Tonal noise absence.

Concerning the respect of global emergence values, the regulatory thresholds are of +5 dB(A) on day time (7am-10pm) and +3 dB(A) at night (10pm-7am). If the total noise level is less than 35 dB(A), these constraints no longer apply. The areas concerned include the indoor and outdoor living areas and the constructible areas defined in the current urban planning document.

Table 1 : Global emergence thresholds

Total noise level	Permissible maximum emergence	
	Day (7am-10pm)	Night (10pm-7am)
Lamb > 35 dBA	5 dB(A)	3 dB(A)

The maximum noise criterion is the adaptation to onshore wind power of a pre-existing standard that was intended for industrial installations. The smallest polygon embracing a buffer zone of 1.2 times the total height of the wind turbines symbolizes the property limit of the park, where the maximum noise criterion applies. The thresholds are 70 dB(A) on day time (7am-10pm) and 60 dB(A) night time. These rules do not apply if the background noise level is higher than the regulatory limit (70 or 60 dB(A)).

Finally, the tonal noise criterion concerns any peak in the wind turbine's emission spectrum. In regulatory terms, this criterion results in a maximum emergence not to be exceeded between each third octave band of the spectrum and the four adjacent bands (the two immediately lower bands and the two immediately upper bands). The limits to be observed are 10 dB for third octave bands centered from 50 to 315 Hz and 5 dB for third octave bands centered from 400 to 8000 Hz. This analysis should be conducted in residential areas, considering the spectrum of noise attributable to wind turbines. Frequency peak apparition as defined before shall not exceed 30% duration time.

2.2 Noise measurement standards

The draft standard NFS 31-114 is used by acoustics professionals to process wind energy measurements in France, in accordance with the decree of August 26, 2011.
Measurement campaigns generally last between 10 and 30 days. The measurement points must be carefully chosen in order to check the compliance of the wind farm with the most

exposed residents., The receivers must be installed in an ordinary acoustic environment of a living space.
Measurements are made with an integrating sound level meter or a class 1 measuring system (according to NF EN 60804 or NF EN 61672-1). The devices shall measure Leq every second. Sound level meters are placed at least 2m from the facades of the houses.

The standard recommends an analysis of noise levels as a function of wind speed. In addition to the distinction between day and night periods, it is advisable to work on homogeneous classes in wind direction, road traffic intensity, season, etc. Thus, within a homogeneous class, the only parameter influencing the noise level shall be the wind speed. The descriptor used by the standard is the median level L5010min based on measured Leq, 1s and each bin must contain at least 10 samples. Once the distribution of samples has been made, medians that are centered on integer wind speed values are extracted from the scatter plots. The centered medians characterize the measured noise levels.

3. A review of the method used for the measurement analysis NFS31-114

This study focuses on the particular case of a village in northern France where a long-term measurement campaign has been carried out. Acoustic measurements in two locations were conducted from late January to August 2014 alongside meteorological measurements. The data were processed over a 2-week moving window in accordance with the best practices of the French profession, as described by the draft standard NFS 31-114.

3.1 Experimental set-up

Acoustic measurement took place in two locations R1 and R2 close to each other in Pas-de-Calais, France. The devices measured from 01/23/2014 to 08/21/2014, i.e. 210 days. Recovery rates were 70% for R1 and 87% for R2. Figure 1 shows photographs of the measurement points taken in winter and summer. Several indicators were measured on a second scale, including global LA_{eq} and LZ_{eq} per third octave band.

Figure 1 : Pictures of the acoustic measurement points

The meteorological data were taken from a wind farm located approximately 4 km from the acoustic measurement points. The nearest wind turbine is 2.8 km away from R1, so it can be

assumed that the wind farm acoustic contribution is negligible. However, since the site roughness is homogeneous between the windfarm and the receivers, SCADA data from this power plant provide a good correlation with noise levels. Figure 2 allows to locate the different measurement sites.
SCADA data give for each wind turbine the average of the wind direction, wind speed and active power achieved on a 10-minute basis. Since the wind direction and wind speed data are very close from one wind turbine to another, the following analysis gives very similar results depending on the wind data source chosen.

Figure 2 : Location of the instruments

3.2 Data processing

First, acoustic data were aggregated to $LA50_{10min}$ according to French standard Pr-NFS 31-114 and synced with SCADA data.
The evolution of acoustic data according to the hours of the day led to the consideration of three distinct periods: day (7am-8pm), evening (8pm-11pm), night (11pm-7am). As for wind direction, it had no significant influence on sound levels. In addition, the main wind regime, from the south-western sector, was overwhelmingly represented (see Appendix 0 Wind rose from 2014/01/24 to 2014/08/21).
As some noise and wind speed pairs did not follow an increasing trend in noise level with wind speed, they were removed from the analyses. These samples are considered as disturbed because they are very high compared to representative values. For instance, dawn chorus needed special care and treatment because it is not representative of the daytime period. An overview of the data once filtered is available in Appendix 0 Data filtering.

After having been properly prepared, data have been processed over a 2-week moving window, following the same standard Pr-NFS 31-114. In the case of a too great lack of availability, that is, if the available data represent less than 10 cumulative days out of the expected 15 days, the statistics have not been calculated.

Beside the calculation of the recentered medians, the following values have been determined for each wind speed bin : the most frequent direction based on Kernel density estimation, median active power and uncertainty of the background noise level indicator.
In parallel with these analysis over reduced periods, an estimation of long-term background sound levels was carried out, allowing the seasonal variation around the median value to be calculated. This last processing step allows to free the operator from other parameters such as wind speed and focus only on the influence of the time of year.

4. Results of a 7-month acoustic measurement campaign

This section describes the results of the experiment of measuring background noise for several months at the same locations. Details concerning data processing can be found in the previous section.

4.1 Long-term background noise levels

Figure 3 shows long-term background noise levels calculated over the entire period of measurement, from 2014/01/24 to 2014/08/21. On the same subgraph, the difference between the curves comes from the wind speed data source : wind turbine E1, E2, E3 or E4. It appears that daytime and nighttime values are more reliable than evening period values, especially for low wind speed. This is due to the fact that the evening period is a transitional period where both night and daytime noise levels are present. In addition, there are fewer samples for this period because the time frame covered is smaller. It can therefore be seen that the study of the seasonal variability of evening noise levels will be constrained by the dispersion of the data.

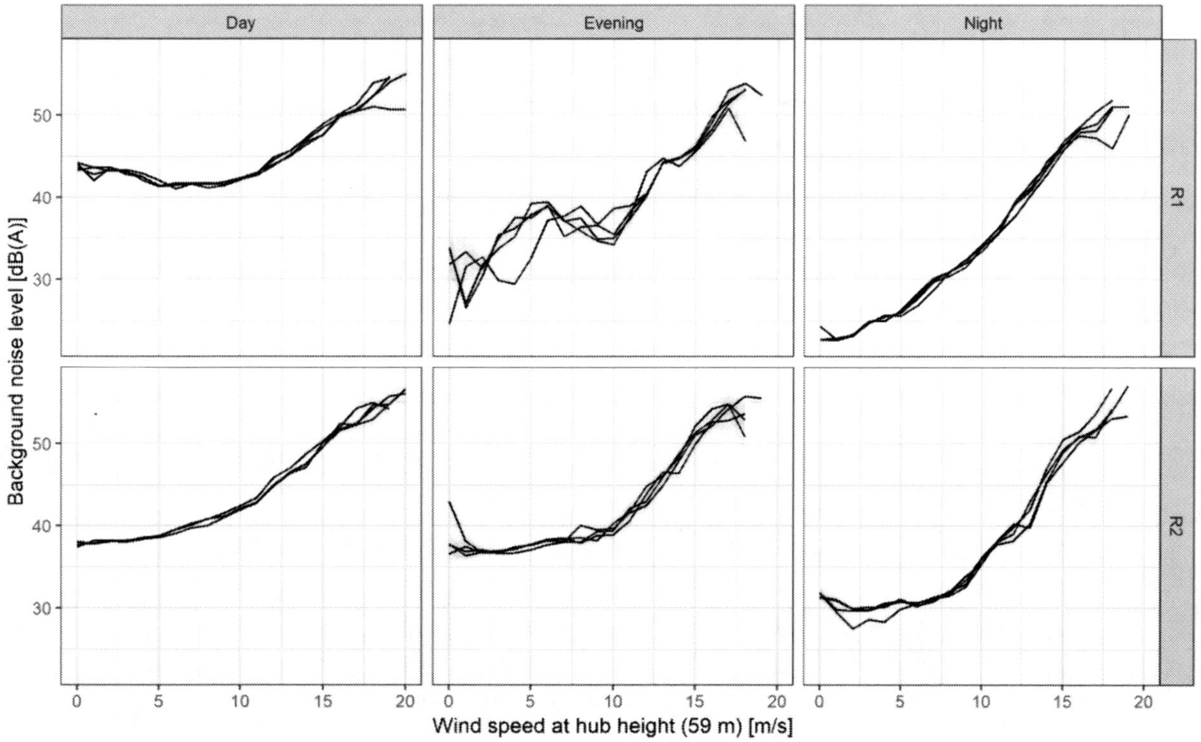

Figure 3 : Long-term background noise levels

4.2 Background noise variability

Figure 4 shows for wind speed bins from 1 to 15 m/s the background noise level variability around its long-term value. Red areas correspond to relatively louder situations, blue areas correspond to quieter situations.

The results show that background noise can vary with a 2 dB(A) amplitude around its median value because of seasonality. Both locations recorded a general increase of night-time noise levels during spring. It has to be noted that noise levels at low wind speed are more influenced by seasons than noise levels at high wind speed.

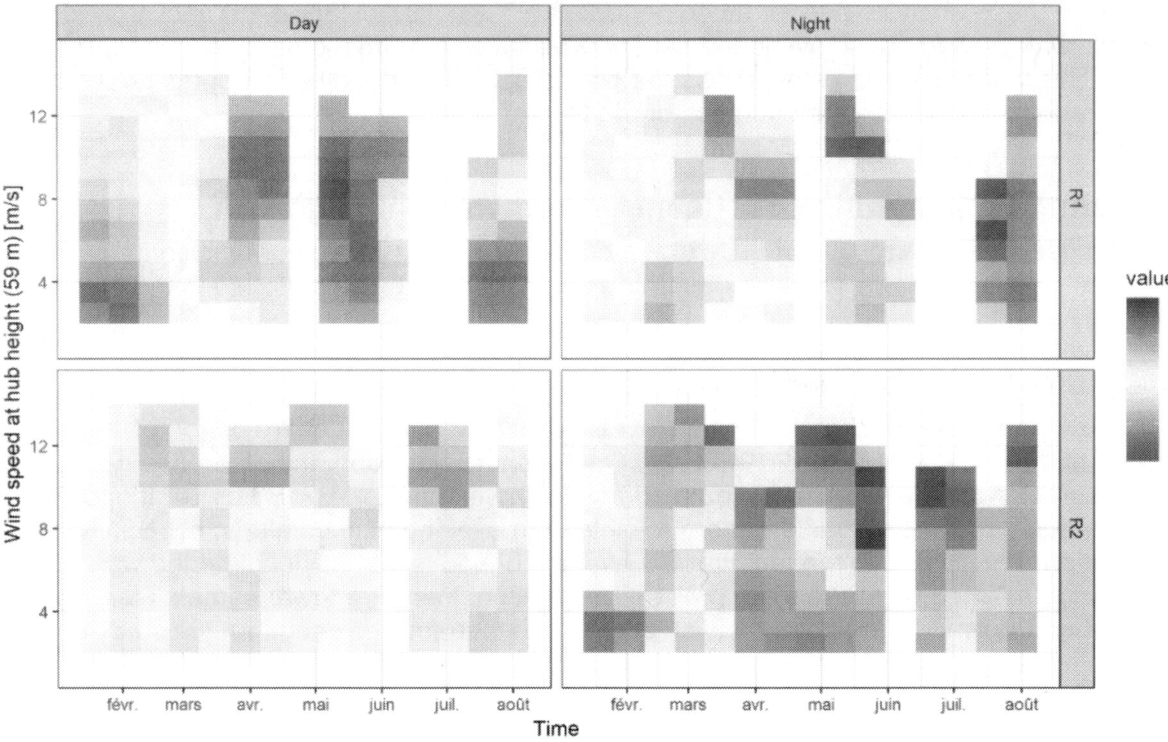

Figure 4 : Heat map of background noise variability regarding time period, with respect to wind speed

It has been noted that noise levels at low wind speed are more influenced by seasons than noise levels at high wind speed. However, to give a general order of magnitude of the variability of background noise level over the year, regardless of wind speed, averages of the variability weighted by the inverse of the uncertainty of the bins were plotted on Figure 5. Variability is significative and confirm a general increase during spring.

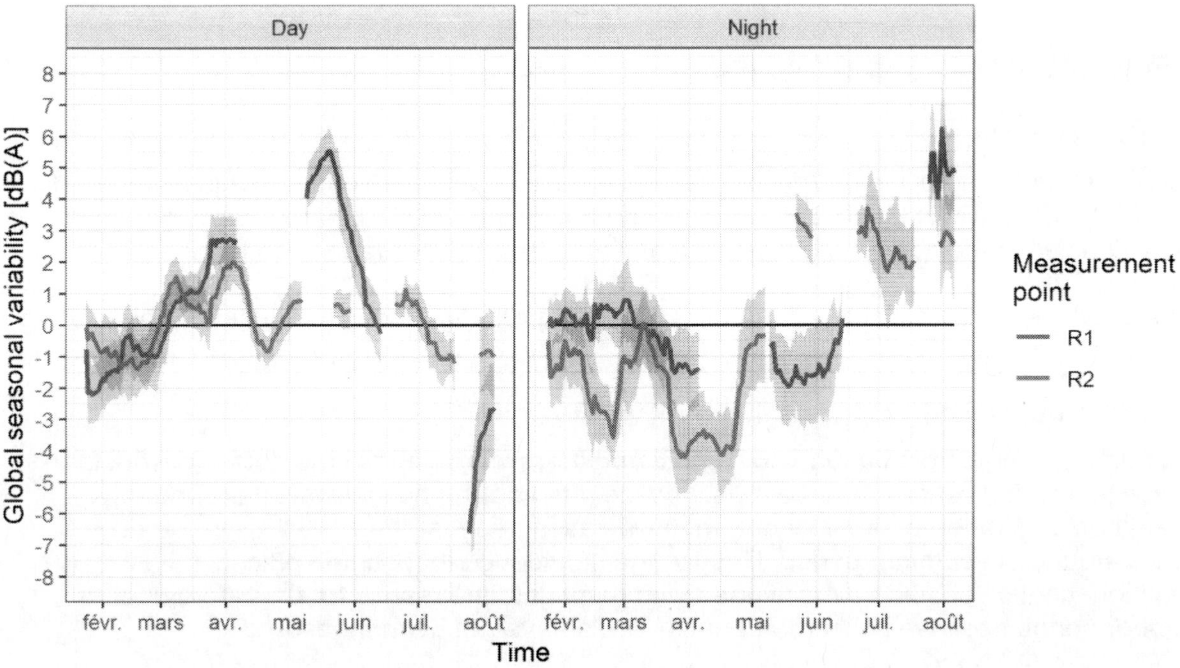

Figure 5 : Global seasonal variability

5. Presentation of the use case

5.1 Use case layout

A typical French onshore wind farm layout was considered as a use case. In this example, 4 wind turbines ENERCON E-103 EP2 with serrated trailing edge (STE) are surrounded by 8 receivers placed at the nearest houses. The minimal distance between wind turbines and receivers is 620 m.
The wind turbine model considered features 6 noise operating modes in addition with standard mode and stop mode.

5.2 Background noise hypothesis

Measurements lasted from January to August and there were gaps in the data. To work on a use case that would consider the year as a whole, the missing noise levels were roughly estimated using a random forest algorithm. It was assumed that there was a symmetrical behavior around the summer period. Despite the confidence that can be placed in this assumption, caution should be maintained with respect to the results for the September to December period because background noise levels have not been measured.

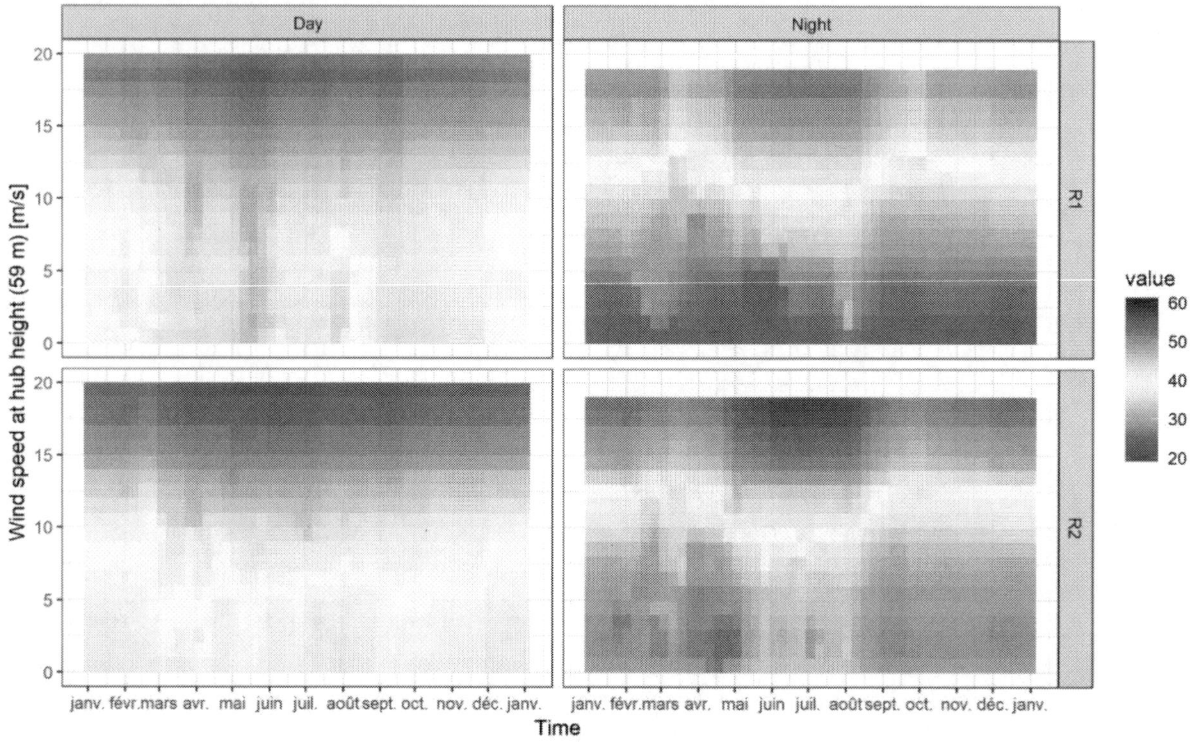

Figure 6 : Rough estimate of background noise levels throughout the year

Figure 6 illustrates the levels selected for the subsequent analysis.

The assignment of the noise levels R1 and R2 was made arbitrarily. Basically, R1 background noise curves were assigned to even receivers R'2, R'4, R'6 and R'8 and R2 curves were assigned to odd receivers R'1, R'3, R'7 and R'9.

6. Influence of seasonality on acoustic curtailment losses

6.1 Acoustic impact in standard operation

From the annual variations in background noise, it is possible to determine what conclusions would have been reached by acoustic studies whose measurement campaigns would have been carried out at different times. Figure 7 illustrates the evolution throughout the year of the acoustic impact of the wind farm under standard operating conditions. The impact is quantified by the number of non-compliances with regulatory thresholds and by the maximum acoustic exceedance.

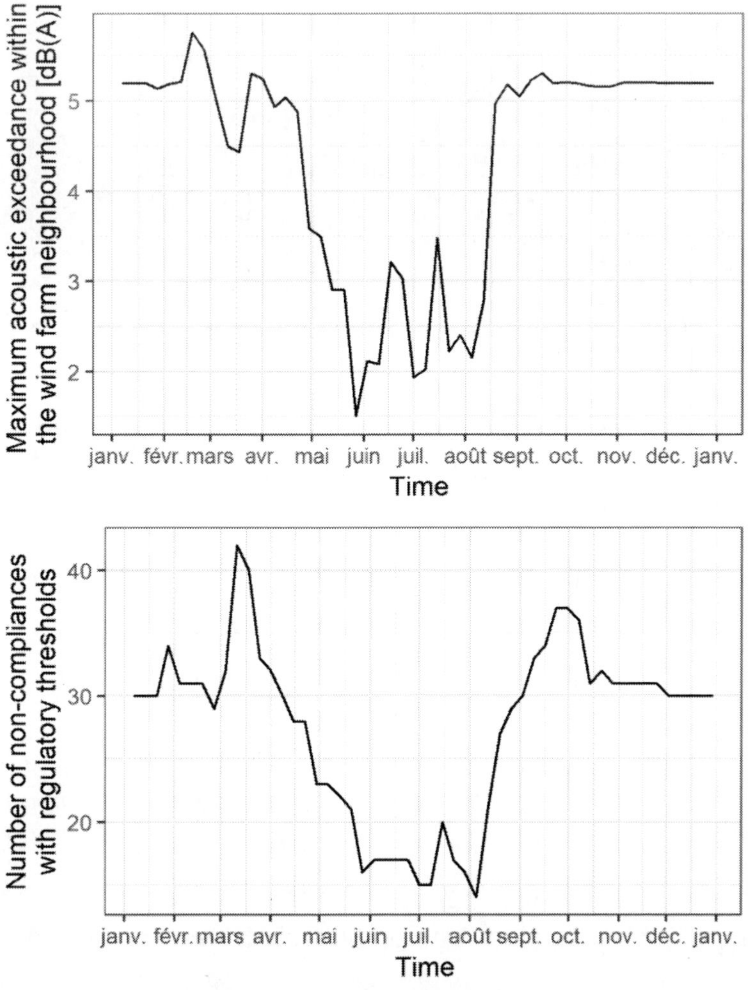

Figure 7 : Acoustic impact of the wind farm

Obviously, the impact is higher in winter than in summer because background noise levels are lower. On this particular use case, maximum exceedance vary between 2 dB(A) in summer and 5 dB(A) in winter. An acoustic curtailment plan is therefore necessary whatever the time of year considered for acoustic measurements. However, since the constraint is less severe in the summer, it is expected that the associated losses will be lower.

Nighttime background noise levels fluctuations have a major influence on the acoustic impact because night is the period of highest risk. Indeed, regulatory threshold is 3 dB(A) and background noise are lower than during the day.

6.2 Long-term production losses due to acoustic curtailments

When a wind farm in standard operation does not comply with acoustic requirements, a curtailment plan is disigned in order to sufficiently reduce its acoustic impact, without unnecessarily reducing production. Figure 8 represents the evolution of acoustic losses that could be expected if background noise levels measured during a given 15-days window were used to establish a yearly acoustic curtailment plan on this particular wind farm.

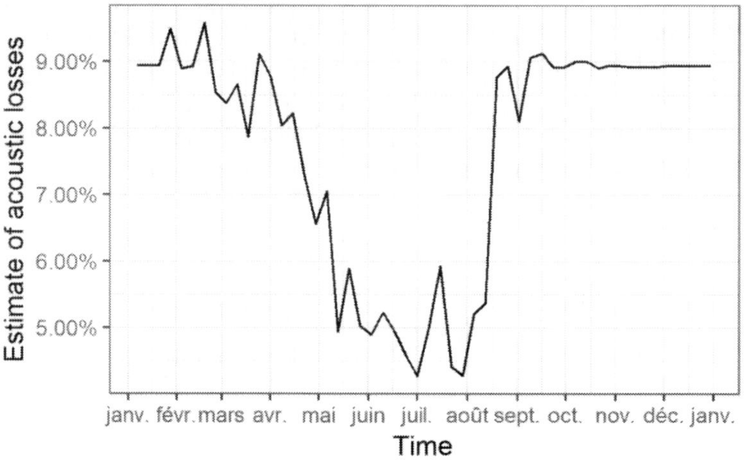

Figure 8 : Production losses due to acoustic curtailments

Long-term production losses, expressed in percent of the wind farm gross production, vary from 5% in summer and 9% during winter.

The results are heavily dependent on the assumptions regarding the wind farm layout and the turbine specifications. Therefore, production losses could be way more or less sensitive to background noise variations.
Further investigations should look at the actual acoustic impact of a wind farm under a given curtailment plan. Also, an estimate of the production gains and acoustic confort provided by a dynamic curtailment plan would be a great contribution.

7. Conclusions

The results show that background noise can vary with a 2 dB(A) amplitude around its median value because of seasonality. Both locations recorded a general increase of night-time noise levels during spring. This could have an outcome of several percent on wind farms Annual Energy Production (AEP).

References

AFNOR (2011) *Projet NF S 31-114 Mesurage du bruit dans l'environnement avec et sans activité éolienne*
Sébastien Garrigues (2016). *Maîtriser l'environnement sonore d'un parc éolien.* Gamba
Commission Electrotechnique Internationale (2005). *Cei 61400-12-1 :2005. AFNOR.*

Appendices

Wind rose from 2014/01/24 to 2014/08/21

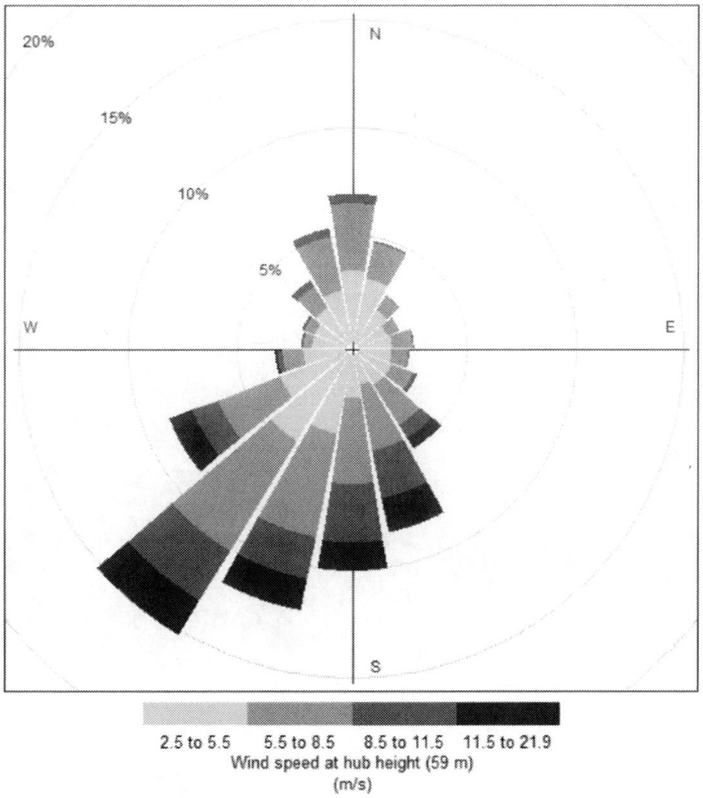

Figure 9 : Wind rose from 2014/01/24 to 2014/08/21

Data filtering - Location R1

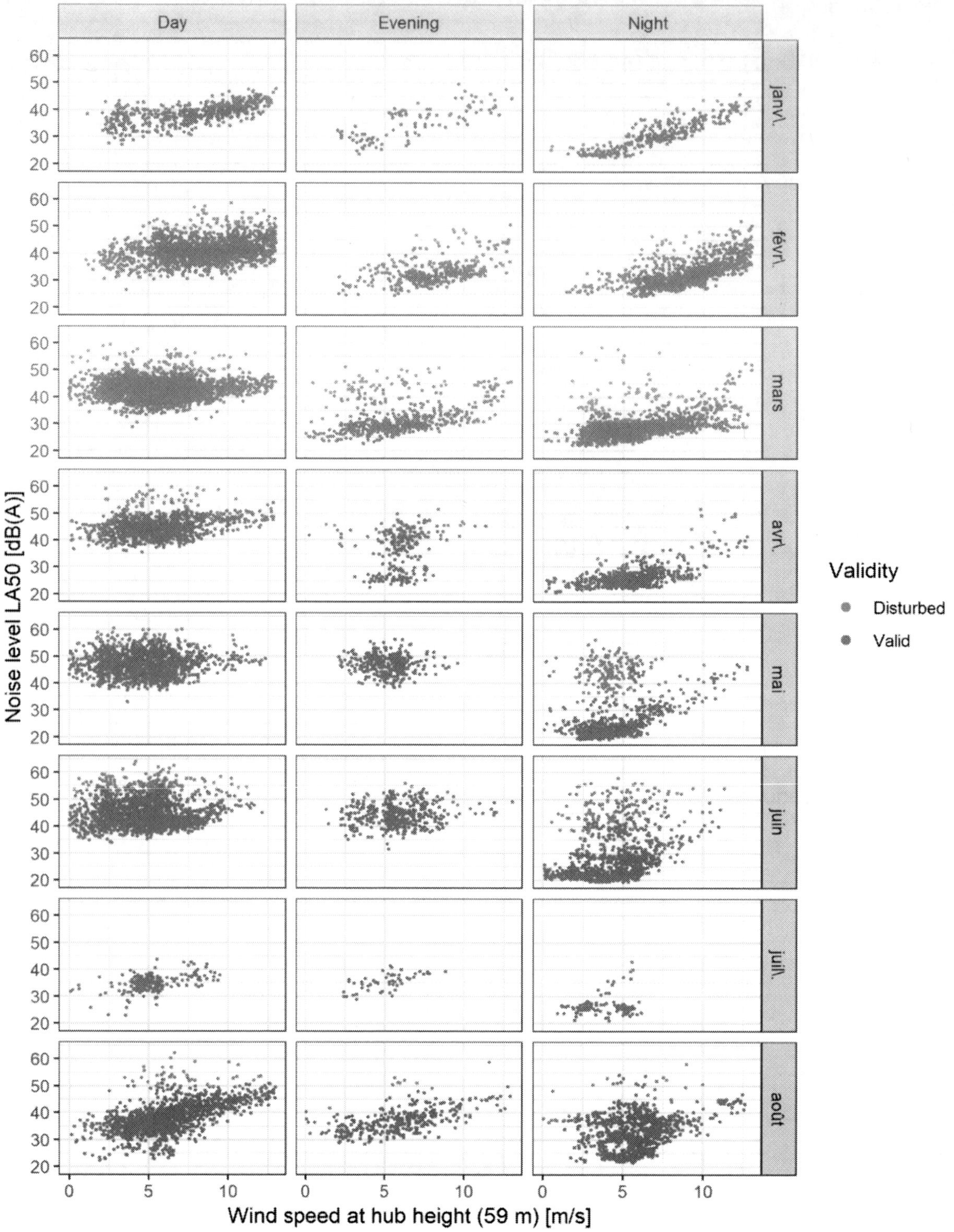

Figure 10 : Data filtering on measuring point R1

Data filtering - Location R2

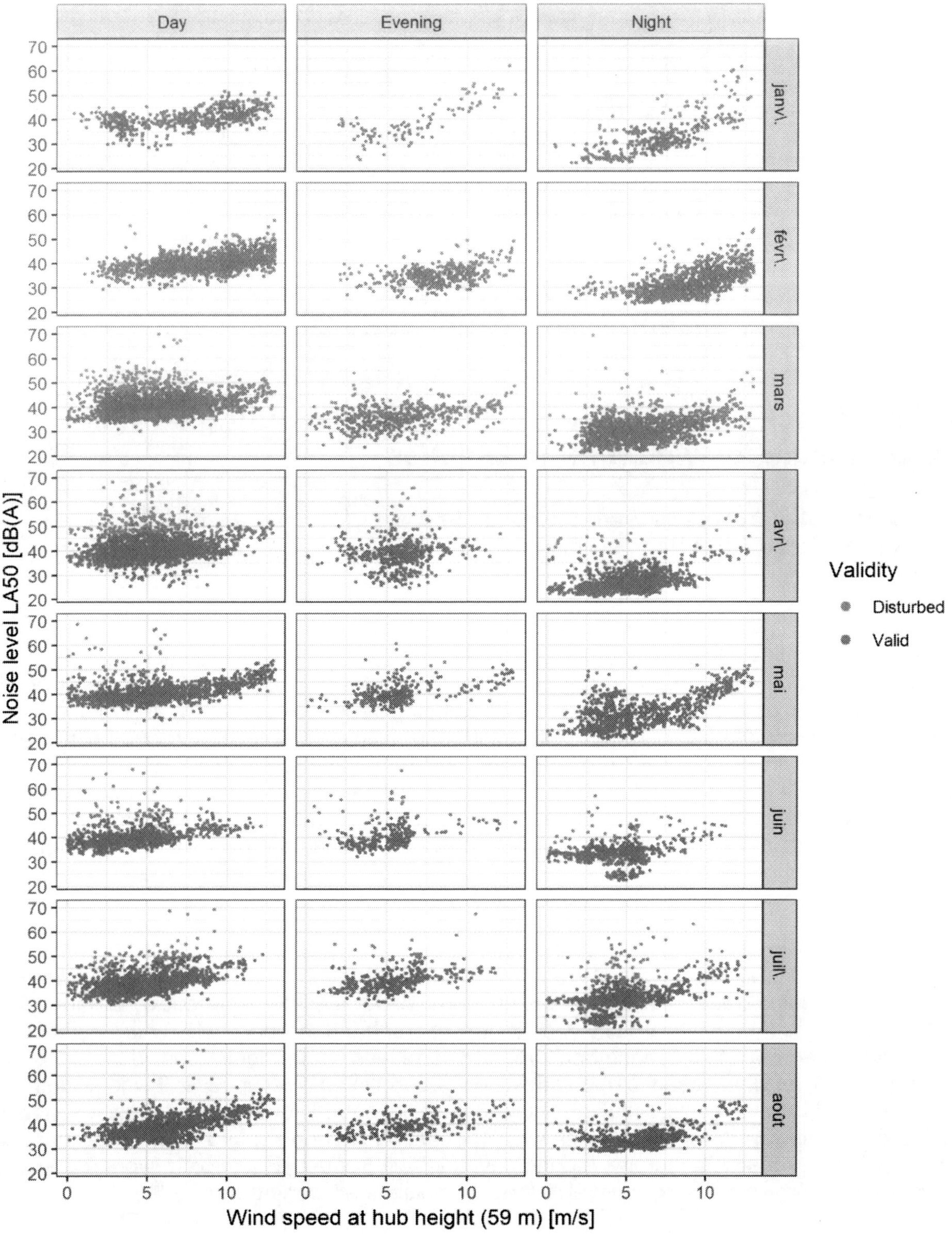

Figure 11 : Data filtering on measuring point R2

INCEEUROPE

8th International Conference
on
Wind Turbine Noise
Lisbon – 12th to 14th June 2019

A study of background noise levels measured during far-field receptor testing of wind turbine facilities

Duncan Halstead, Aercoustics Engineering Limited
Nicholas Tam, Aercoustics Engineering Limited

Summary

This study presents an assessment of the background noise levels measured across a range of wind facilities in southern Ontario. Background sound levels are measured during wind turbine shutdowns as part of the acoustic compliance auditing required for nearly all Ontario wind facilities. The background measurement data plotted against ground level and hub height wind speed to determine the impact wind speed has on ambient sound levels. An examination of the variation in sound level conducted at sites having nearby features such as trees and roadways to determine their effects on background sound levels. Finally, the average spectral content of background sound in different conditions is compared with a typical far-field wind turbine spectrum to determine the important frequencies of interest in far-field measurements, as well as the conditions that might make an accurate determination of facility-only sound level difficult or even impossible.

It was found that fluctuations in sound level of 10-20 dB are common, particularly in the presence of roadways or nearby trees. Forested areas also provide an elevated sound level at high wind speeds, even if the immediate vicinity around the microphone is open field. Hub-height wind speeds were determined to influence background sound levels and require consideration along with the ground level wind speed, particularly at low ground level wind speeds. It was also found that background sound levels regularly exceed turbine sound levels at frequencies above 1000 Hz, even at low wind speeds. At high wind speeds, background sound levels were found in some cases to exceed the turbine sound level at all frequencies, which would make a determination of turbine-only sound level in these conditions exceedingly difficult.

1. Introduction

An accurate measurement of any noise source requires a careful consideration of the background sound level present during the measurement. For wind facilities, this consideration of background noise is particularly important, because both the sound output of the wind facility and the background sound vary with wind speed. Increasing wind speed for both conditions corresponds to an increase in background sound level.

This consideration for background noise becomes even more important in the case of far-field noise measurements, which are often conducted at distances of 500 metres or more from the facility. These are typical conditions present during compliance testing or complaint evaluations performed on wind facilities in Ontario and other jurisdictions. Aercoustics has conducted extensive testing of wind facilities in Ontario, where the measurement protocol for compliance testing prescribes a requirement to periodically turn off the wind turbines and measure background sound. This has enabled Aercoustics to collect a variety of background sound measurements across several different wind facilities.

In this study, different background noise datasets are examined to determine important factors associated with background sound levels in the context of wind turbine noise measurements. Wind speeds at different heights are compared and the surrounding area of each monitor is categorized to determine which parameters are important in categorizing and filtering background sound and selecting an appropriate measurement location. Lastly, an assessment of the spectral content of background sound levels in different wind conditions is compared to a typical far-field wind turbine spectrum, to determine the impact of background noise compared to a 40 dBA wind turbine level.

Given the increasing prevalence receptor testing as the method of choice for many jurisdictions to evaluate wind facility compliance, the understanding of background sound levels and their impacts on the sound measurements is of great importance.

2. Measurement Methodology

The measurement data used for this study was obtained from long-term measurement equipment installed at a variety of locations across several wind farms in southern Ontario. Measurement campaigns typically last anywhere between 4 to 16 weeks and must follow a measurement protocol published by the Ontario Ministry of the Environment [1].

2.1 Measurement Equipment

Data in this study was collected using measurement equipment installed near a wind facility and configured to remotely record sound levels and meteorological (MET) information every night. The equipment synchronously captures acoustic and MET data at heights of 4.5 meters and 10 metres, respectively. The microphone is fitted with a primary and secondary windscreen to reduce the effect of wind generated self-noise. The 10-metre [1] weather station measures a range of weather parameters for each interval. A picture of a typical measurement apparatus is shown in Figure 1.

Figure 1: Measurement apparatus

[1] Wind speed at 10-metres is referred to as "ground level wind speed" in this paper.

These measurement stations are typically installed at distances ranging from 500 to 800 metres from the nearest wind turbines. All measurements are unattended, conducted nightly from 10 pm to 5 am with data downloaded wirelessly at regular intervals. A total of 1,700 hours of background measurement data was assessed in this study.

2.2 Measurement Parameters

The data presented in this study was captured in 1-minute intervals. The overall LA_{eq}[2], 1/3rd octave band level, and L90 was recorded for each interval. Weather data recorded includes wind speed, wind direction, temperature, pressure, humidity, and precipitation.

Acoustic and weather data are acquired synchronously by the same data acquisition system. Turbine operational information was obtained separately from the wind facility SCADA system; this data was available in 10-minute average intervals.

2.3 Measurement Environments

The measurement locations in this study represent a range of conditions typical of southern Ontario. Terrain surrounding most of the wind facilities is flat, open fields dotted with patches of trees. Roadways pass through and around most facilities, with traffic levels anywhere from small regional roads, major highways, and everything in between. Forested areas range in size from a few dozen metres to multiple kilometres across. Many of the houses in these areas are surrounded by narrow tree lines for shelter against the wind. Tree noise and roadway noise is prevalent and pervasive in most background measurements, particularly at high wind speeds.

Measurement locations were categorized based on the nearby features surrounding the microphone as well as the presence and proximity of forested areas and major roads. In total, data from 40 separate measurement campaigns measured at 21 different locations across 8 wind facilities were examined.

For the current discussion, a sample of these datasets is presented. The background measurement data from a selected 4 monitoring locations are compared more closely. A summary of the locations and their surrounding environment is presented in Table 1. The locations of the features presented in this table are with reference to the "front" direction pointing towards the nearest turbine.

Table 1: Description of Monitoring Locations and Surrounding Environment

Location ID	Immediate Surroundings	Distance to Nearest Forest	Distance to Nearest Roadway (and traffic level)
Site A	Front: open field Side: trees, 25m Back: open field	100m	340m (Infrequent traffic)
Site B	Front: open field Side: trees, 15m Back: house, 75m	130m	15m (infrequent traffic)
Site C	Front: open field Side: open field Back: forest, 50m	50m	40m (infrequent traffic)
Site D	Front: open field Side: house, 50m Back: open field	500m	15m (moderate traffic)

[2] Energy-equivalent sound level, A-weighted

The wind facilities measured in this study are all modelled to a 40 dBA limits (Class III environment), representative "areas dominated by natural sounds having little or no road traffic"[3]. It is not an uncommon notion in the acoustic community that rural ambient environments are very quiet, which would imply that receptor measurements of wind turbine noise shouldn't be adversely impacted by background noise. This notion is usually based on experiences of noise levels in rural environments during calm conditions (i.e. low wind speeds). Since wind turbines to not operate in calm conditions, it is important that background sound levels be measured during conditions when the turbines would otherwise be operating.

Background noise variation in this study is presented in the form of sound level histograms sorted by wind speeds or 1/3rd octave frequency. High concentrations of measurements are represented by darker colours in each plot. Figure 2 shows an example of this plotting technique for a site located 500 m from a forested area. The x-axis is ground level wind speed in this example.

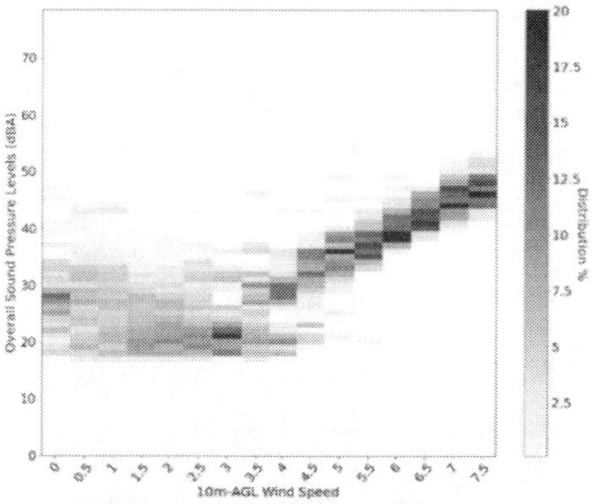

Figure 2: Sample Histogram

3. Results and Discussion

Background sound levels are filtered and categorized by different parameters and presented in this section. These plots show the typical variety of background sound levels observed at each measurement location with respect to wind speed, measurement season, and 1/3rd octave frequency. These comparisons illustrate the important parameters to consider when measuring background levels for wind turbine noise testing. Of particular interest are the background spectra comparisons to a typical far-field wind turbine spectrum. These plots show that in certain conditions the background sound levels have the potential to dominate noise from the wind turbine at all frequencies.

3.1 Overall Sound Level and 10-metre Wind Speed

The overall LA_{eq} (OASPL) is plotted against the measured ground level wind speed in Figure 3. Several different site conditions are plotted and compared. The horizontal line across each plot is 40 dBA, which represents the permitted sound level limit for all the wind facilities in this study. Each increment on the y-axis of each plot represents a change of 10 dB, and each increment on the x-axis represents a change of 0.5 m/s. The x and y axis ranges are the same for all plots.

[3] Per Ontario Ministry of the Environment document NPC-232 [2].

Figure 3: OASPL vs. Ground Level Wind Speed for Sites A-D

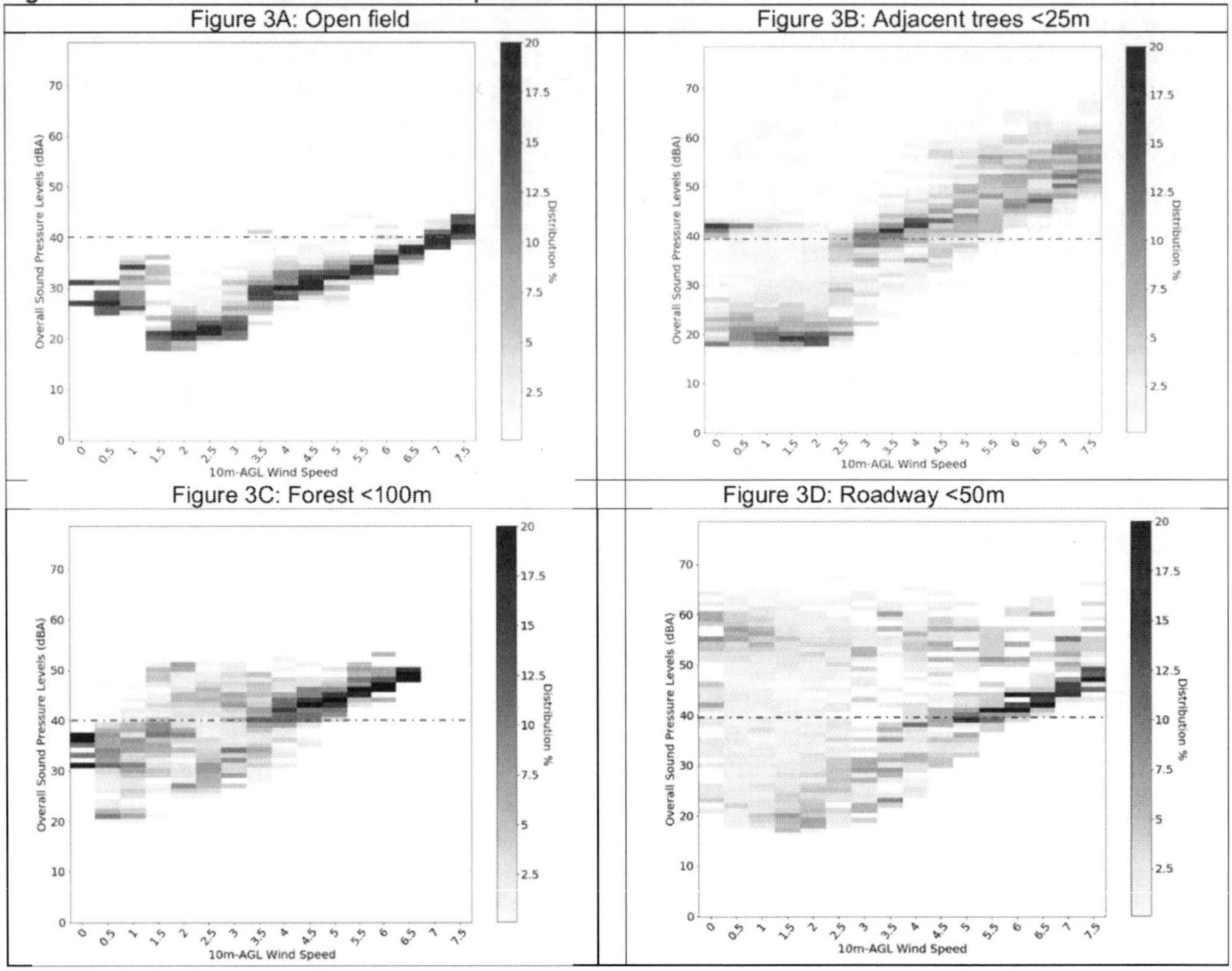

The figures above show a positive correlation between background sound level and ground level wind speed. This is no great revelation, as it has long been accepted that the sound levels increase with ground level wind speed at the monitoring location. Each monitor appears to have an increasing floor, or minimum noise level, as wind speed increases. However, the variation in sound level above this floor is quite pronounced at different sites, depending on the nearby features.

Data in Figure 3A is from an open site and shows a high concentration of sound levels near the sound floor. Figure 3D, on the other hand, is near a roadway that has intermittent but regular vehicle pass-bys, which cause significant fluctuations in sound level. Figures 3B and 3C represent sites that have trees adjacent the monitor (3B) or a forested area in the vicinity (3C). The impact of the tree noise presents both a more rapid increase in sound level with wind speed as well as an increase in the variation in sound level at lower wind speeds. This variation at low wind speed implies that there are other factors that are affecting the measured background sound level when the ground level conditions at the monitor are calm.

3.2 Overall Sound Level and Hub Height Wind Speed

To further investigate the effect of wind speed on sound level, the overall LA_{eq} is plotted against the measured hub-height wind speed in Figure 4. Each increment on the y-axis represents a change of 10 dB, and each increment on the x-axis represents a change of 0.5 m/s. The x and y axis range is the same for all plots. The horizontal line across each plot is drawn at 40 dBA.

Figure 4: OASPL vs. Hub Height Wind Speed for Sites A to D

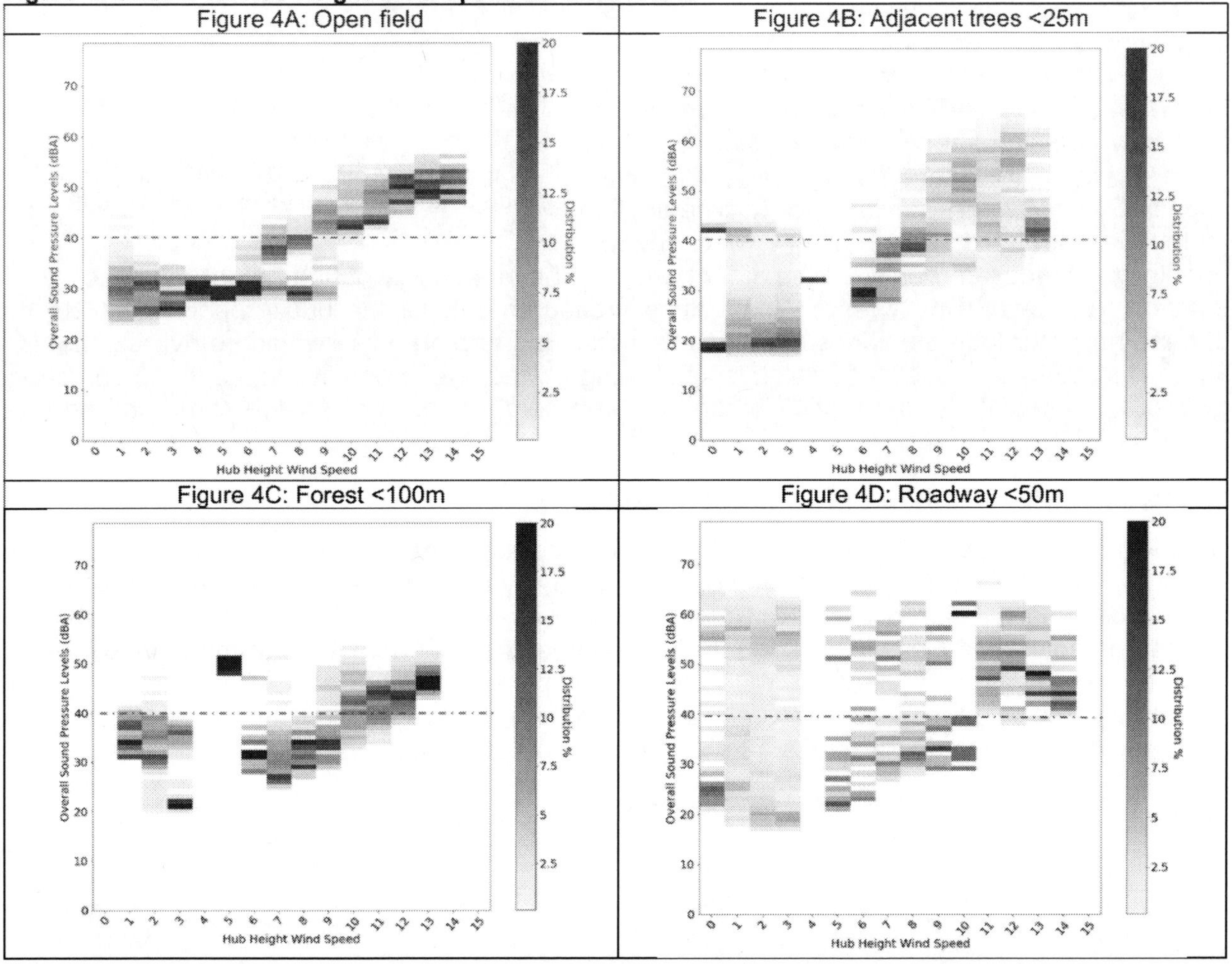

The above figures show similarities to the trends observed in 3.1, with increasing sound levels measured with increasing hub height wind speed. The wind speed at hub-height can change independent of ground level wind speed, expressed as variations in wind shear[4]. This means that high hub height wind speeds can be present at low ground level wind speeds during high wind shear conditions. Given the findings in Figure 4, the variation in background sound level observed at low ground level wind speeds in Figure 3A-D may be explained by different wind shear conditions at low ground level wind speeds.

For a more illustrative example, sound level is plotted against ground level wind speed in Figure 5, with the colour of each point representing the hub-height wind speed. This plot shows that

[4] Wind shear characterizes the difference in wind speed with elevation.

variation in hub-height wind speeds can explain much of the variation in background sound level observed at low ground level wind speeds. That is, increasing wind speeds at high elevations (hub height) also have the effect of raising the background sound level at the measurement location.

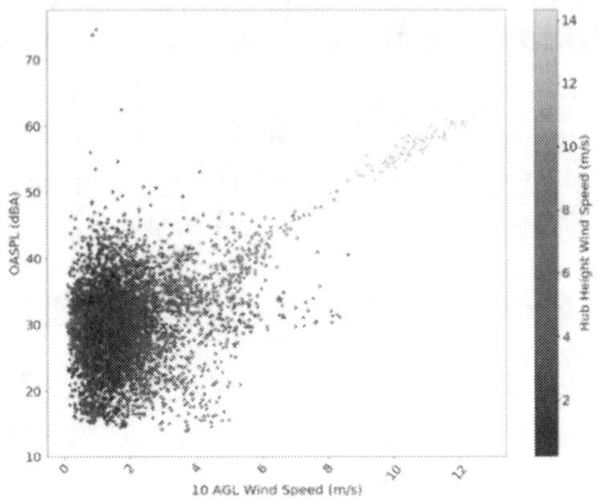

Figure 5: OASPL vs Ground Wind Speed (hub height wind speed indicated by colour)

This finding is particularly important in the case of high wind shear conditions, where turbine sound levels are compared to background sound levels at low ground level wind speeds near the monitor. Typically, measurements conducted when the wind turbines are operating will naturally have high hub height wind speeds, since turbine power output is correlated to hub-height wind speed. Background measurements, on the other hand, can have a variety of hub-height wind speeds for each ground level wind speed. When using measurements of the background to isolate the turbine noise through ambient subtraction, the resulting turbine noise would be overestimated if the background sound level contains low hub-height wind speed conditions, as those conditions are found in this study to be quieter than high hub-height conditions for the same ground level wind speed.

3.3 Seasonal Variation in Background Sound Level

The two figures below (Figure 6A and Figure 6B) represent measurement data from the same location, during different times of the year. Figure 6A was measured in the spring months and Figure 6B was measured in the fall months. The most notable difference between the two measurement periods was that during the fall months, the field adjacent the monitoring location (100 metres away) was planted with corn. It is suspected that the noise from the corn is the reason for the increased sound levels at high wind speeds between the two periods. The increase is not trivial, upwards of a 10 dB difference at some wind speeds.

Figure 6: Seasonal Comparison of Background Noise at Site A

Figure 6A: Open field

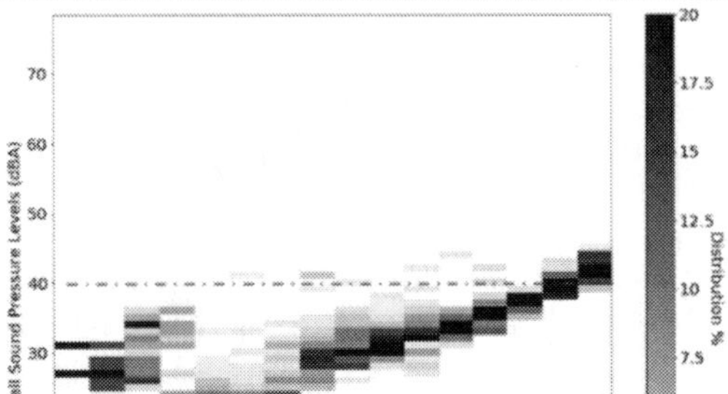

Figure 6B: Corn field <100m

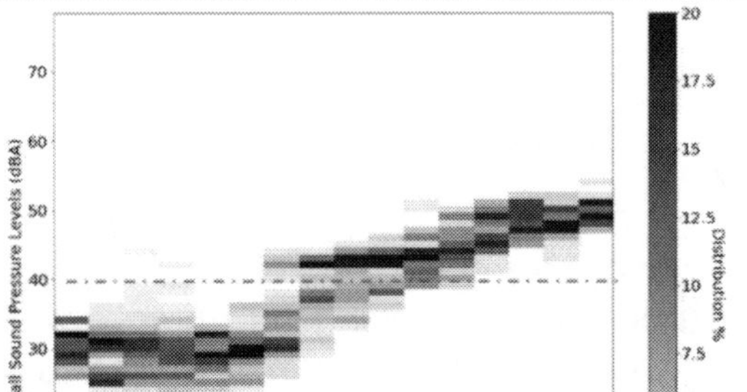

The sound levels in Figure 6B show background noise elevated above 40 dBA starting at ground level wind speeds of 3 m/s, in contrast to sound levels in Figure 6A which remain largely below 40 dBA until 7 m/s. These figures show that even at the same measurement location, the background sound levels can vary significantly depending on changes in the surrounding landscape between seasons.

3.4 Average 1/3rd Octave Spectra – Turbine ON vs. Background

The findings presented so far show that background sound levels are highly dependent on wind speed (ground level and hub-height), transient sources (such as road traffic) and nearby features such as crops and trees. Furthermore, it was found that in many locations the overall background sound level regularly exceeded the 40 dBA sound level of the wind turbines being measured. For further study, an evaluation of the spectral content of the 1/3rd octave band noise levels is performed. For comparison, a typical wind turbine spectrum is plotted for each graph, representing the noise from the facility that would be the subject of the measurement[5].

Figure 7 shows the 1/3 octave band spectra for Sites A to D with ground level wind speeds at 5 m/s. These figures illustrate a typical shape of background noise spectra at different sites in relation to the turbine noise. High frequencies are dominated by background in all cases, and in some cases even middle and low frequencies are dominated by background noise.

Figure 7: Spectral Comparison of Background Noise at Sites A to D at 5 m/s Ground Wind Speed. 40 dBA turbine spectrum plotted in blue.

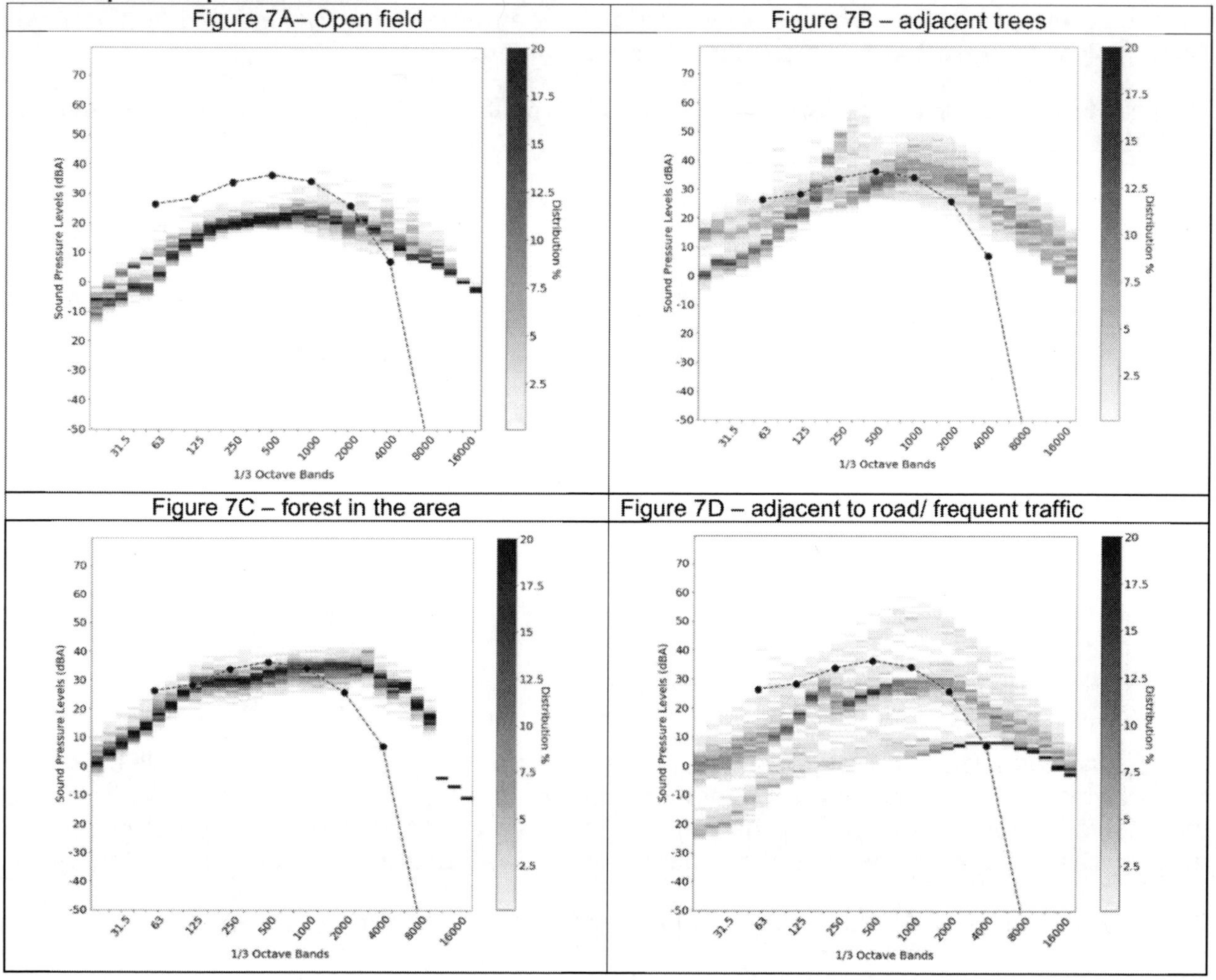

Of the locations above, only Site A (open field) has a good separation between the wind turbine noise and the background noise, but only at 1000 Hz and below. For Site B (adjacent trees), around 40% of the background data is higher than the turbine spectrum levels and around 70%

[5] Wind turbine spectrum at a location predicted to be 40 dBA.

of the data collected does not satisfy the 6 dB signal to noise ratio criteria for good quality data[6] at octave bands of 1000 Hz and below. Similar conditions were found in Site C, indicating that the presence of trees – whether a nearby tree line (Site B) or a distant forested area (Site C) – will elevate background sound levels to the point where the levels regularly exceed a 40 dBA turbine spectrum.

The location at Site D (adjacent roadway) shows greater variation in the distribution of the sound levels. This variation is likely attributed to varying traffic levels at different times of the night. There appears to be a concentration of sound levels in the middle this data set which may represent the average amount of road traffic noise experienced at this site. The impact of traffic is also shown in Figure 3D and 4D to cause large variations in measured noise level that are independent of wind speed.

For open locations like Site A (shown in Figure 7A), the background levels are generally quite low; allowing for a good signal to noise below 2000 Hz octave bands. This location represents a site having a very quiet ambient environment. However, even for low ambient sites, the quality of the signal drops with increasing wind speed. Figure 8 shows the background spectra at Site A for 6 m/s and 7 m/s. The signal to noise ratio between overall background and turbine levels drops to less than 6 dB at 6 m/s and less than 3 dB at 7 m/s.

Figure 8: Spectral Comparison at Site A at 7 m/s and 6 m/s ground wind speeds. 40 dBA turbine spectrum plotted in blue.

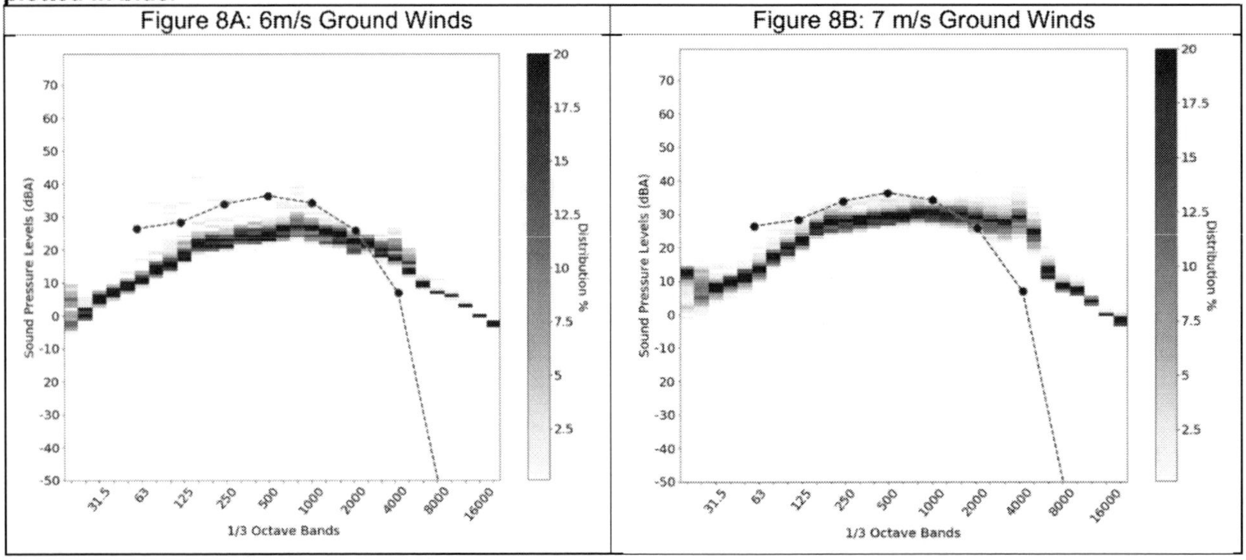

These findings show that low ground level wind speeds are most appropriate for collecting high-quality measurement data, and even then only for frequencies of 1000 Hz and below. This limits a suitable assessment to periods having high wind shear, which may or may not occur on a regular basis depending on the environmental conditions that are typical for the monitoring site.

Given the findings of 3.2, hub height wind speed must be considered during low ground level wind speed conditions. To that end, Figure 9 shows the background 1/3rd octave spectra at a ground level wind speed of 3 m/s for different hub height wind speed conditions. Site C was chosen for illustration, having open fields immediately adjacent the monitor and a forested area starting 50 metres behind it.

[6] Threshold for "good quality" data is taken from the IEC 61400-11 standard for wind turbine noise testing [3].

Figure 9: Spectral Comparison at Site C for 3 m/s Ground Winds with different Hub Height Winds. 40 dBA turbine spectrum plotted in blue.

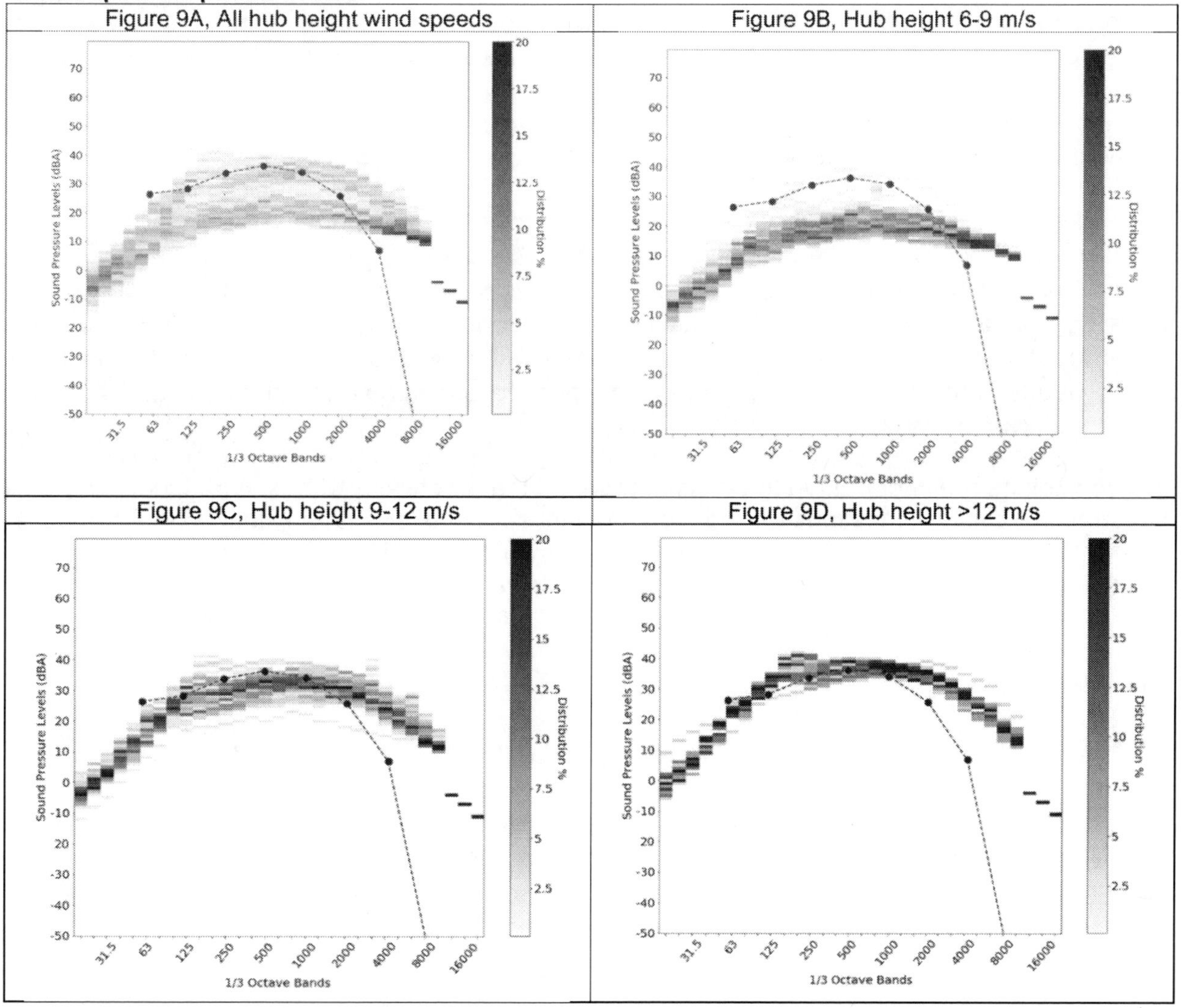

The above figures show a big influence of hub height windspeeds on the measured background spectra. When all hub-height data is plotted (Figure 9A), a large variation in spectral sound level is apparent. However, when this data is broken down by range of hub-height wind speed, the impact on the background is clearly visible. At medium (Figure 9C) and high (Figure 9D) hub-height winds, the 40 dBA turbine spectrum is almost completely overshadowed by background noise.

In Figure 9B, the turbine only sound levels could be resolved above the background spectrum. Comparing the average octave levels at this hub height wind speed bin, there is a >6 dB signal to noise ratio for the 250 Hz to 1000 Hz band. It is important to note however that in these conditions (6-9 m/s hub height wind speed) the wind facility will likely not be generating its maximum output (both power and noise). Most of the southern Ontario wind facilities have turbines that reach their rated power and noise output in the 9-12 m/s range. The turbine spectrum shown above represents the noise from the wind facility at this rated condition. Therefore the most representative background noise for this condition would be shown in Figure 9C.

4. Conclusions

This study has examined the variation in background sound levels across several wind facilities in southern Ontario. It was found that consideration should be given to both the ground level and hub height wind speed, the latter being particularly important to avoid underestimating the background sound level in low ground level wind speeds. An assessment of the spectral shape of background sound in different wind conditions was also examined. It was found that high wind speeds (both ground level and hub-height) increase the background sound level so much that the turbine sound level can be overshadowed at almost all frequencies, particularly when the sound level of the wind turbines is low (around 40 dBA).

High wind shear conditions and locations having large amounts of open area on all sides of the measurement location are ideal for an accurate determination of wind turbine sound level. These conditions may not always be achievable at some facilities, and in these situations it would be more appropriate to conduct far-field testing at locations closer to the wind facility in order to maximize the signal-to-noise level between wind turbines and background sound. Although, it is acknowledged that some regulators require compliance testing be done at the residence, and so this is not always possible.

These findings are important in the consideration of the practicality and cost of conducting post-construction audits in locations having high ambient sound level. At many locations, an accurate determination of turbine-only sound level at a nearby receptor will be hampered by the presence of background noise. The most appropriate solution in these cases would be to measure at a position closer to the wind facility in an area likely to generate low ambient sound levels.

References

[1] Ministry of the Environment and Climate Change, "Compliance Protocol for Wind Turbine Noise," Government of Ontario, Toronto, 2017.

[2] Ministry of the Environment, "NPC-232-Sound Level Limits for Stationary Sources in Class 3 Areas (Rural)," Government of Ontario, Toronto, 1995.

[3] IEC TC88, "61400-11 Wind turbines - Part 11: Acoustic noise measurement techniques," International Electrotechnical Commission, Geneva, 2012.

INCE*EUROPE*

8th International Conference on Wind Turbine Noise Lisbon – 12th to 14th June 2019

Session H

Impact on People - 1

Organised by INCE-Europe
mail@inceeurope.org
www.inceeurope.org

Supported by Portuguese Acoustical Society

spa Sociedade Portuguesa de Acústica

INCEEUROPE

8th International Conference
on
Wind Turbine Noise
Lisbon – 12th to 14th June 2019

Wind Turbines and Groundwater Contamination: An Analysis

W.D. Colby, Western University, London, Canada drdavidcolby@gmail.com

Summary

Allegations of water well interference, sediment infiltration and aquifer contamination due to ground borne vibrations from wind turbine construction and operation have been levied against a wind farm in Chatham-Kent, Ontario, Canada. Ground vibration measurements revealed nothing extraordinary. Baseline water analyses before construction revealed poor water quality overall. There were few complainants after construction began. Retesting of water post construction from the complainants revealed a smaller proportion of wells with turbidity than in the baseline samples. The data submitted to the Ministry of the Environment are summarized and analysed. These data do not support the notion that seismic vibrations from wind turbines are damaging wells nor causing contamination of the aquifer.

Introduction

In Chatham-Kent, Ontario, Canada, the North Kent 1 wind farm received Ontario *Ministry of Environment* approval (REA) on 29 June 2016 as a Class 4 wind facility up to 100 megawatts.[1] The project consists of 45 Siemens SWT series utility scale wind turbines rated as 104-106 dBA sound power. As objections to the project had been expressed based on the theoretical possibility of ground borne vibrations affecting the aquifer, the Ministry ordered extensive pre- and post-construction vibration monitoring and water testing as a condition of approval. On July 13, 2016, Kevin Jakubec appealed the REA to the Environmental Review Tribunal under s. 142.1(2) of the Environmental Protection Act on the grounds that the Project will cause serious harm to human health and serious and irreversible harm to the natural environment via groundwater interference from vibrations of operating wind turbines, but the appeal was withdrawn prior to the hearing[2]. Despite that, members of an activist group have complained about sediments in well water since the construction began and have alleged that the entire aquifer has been contaminated due to pile fracturing of the shale bedrock and mobilization of shale sediments. Shale is known to contain trace amounts of heavy metals. Ongoing concerns are expressed to Provincial and Municipal agencies and the media about heavy metal poisoning from drinking and even bathing in water from wells in the area and a health hazard investigation has been demanded. This paper summarizes the data submitted to the Ministry of Environment by the wind farm operator and discusses the implications to groundwater and health.

Results

1. Preconstruction Water Survey

An extensive background water survey was done by AECOM Canada.[3] 959 properties were identified within the project boundary, of which 827 were privately owned by 581 property owners (143 owned more than one property). Survey packages were mailed to the 581 property owners and 393 responses were received. Of the 393 respondents, 210 were identified by survey responses as having active water wells, 167 had no active wells and six respondents were unclear about whether they had active wells.

201 respondents agreed to a *Detailed Well Assessment* (DWA), but further analysis revealed that 10 of these did not actually have active wells and one respondent did not provide well information nor contact information for clarification. A total of 189 DWAs were completed and analysed for multiple chemicals and parameters based on the criteria of the *Ontario Drinking Water Standards, Objectives and Guidelines* (ODWSOG).

36 (20%) of wells failed on baseline bacteriology, 53 (29%) failed on turbidity and 155 (85%) failed on colour.

2. Postconstruction Complaints of Well Water Turbidity Alleged from Pile Driving[4]

There was a total of 17 complaints of well water turbidity during the study period, and 15 of these had been in the group of 189 with completed baseline DWAs. One of the complaints without a DWA was questioned as invalid since the turbidity allegedly began 35 days after pile driving was completed and the well was 9 Km from the turbine construction. Post construction water sampling was completed on 13 of the 15 complainants with DWAs which permitted comparison with baseline results.

11 wells passed turbidity testing postconstruction, two wells had measurable turbidity arising postconstruction and two were not resampled. Of 189 wells with DWAs completed, 185 (98%) either did not report turbidity problems postconstruction or were free of water turbidity upon retesting.

3. Vibration Monitoring

Phase 1, Test Pile Program[5] The pile driving hammer energy was selected to be as large as possible for the equipment. Accelerometer data from the monitoring of three wells indicated that the peak particle velocity was less than 0.035 mm/s during pile driving. At times when pile driving was not occurring, the maximum particle velocity was nearly double this value as related to other activities on and near the monitored wells. The character of the vibrations at the three wells as measured by the sensitive instrumentation was similar to or less than those induced by common day-to-day activities (e.g., traffic, walking nearby, farming equipment and light vehicles) and by walking or jumping near the well casing.

Phase 2[6]: Measured pile-driving vibration magnitudes at the well casings were significantly smaller than those identified at the turbine site where measurements in the bedrock were made, consistent with known vibration distance attenuation. The measured turbine site vibrations were less than those anticipated, based on the test pile program (Phase 1) and use of the CALTRANS (2004) model for predictive vibration magnitude estimation. Peak vibration velocities at the turbine sites were less than 14 mm/s. Well pumps were the largest sources of well casing vibrations measured during the monitoring program. The peak vibration velocities at domestic water well casings that were directly attributable to pile driving were less than 0.04 mm/s at the closest well to pile distances. These measured velocities are one or more orders of magnitude less than vibrations induced by typical pump operations where the pumps are connected in relatively close proximity to the wells and less than vibrations induced by local road traffic.

Discussion

Complaints about wind turbine sounds and sometimes shadow flicker are often expressed as the basis for objecting to wind farm projects but allegations of aquifer contamination and water well interference from ground vibrations arising from construction (specifically pile driving) or operation of wind turbines had not been encountered in Ontario.

Ground vibrations can be measured by peak particle velocity (PPV) in mm/s, root mean square velocity (RMS) or particle displacement. There is an abundance of published literature and standardized methods for estimating ground vibrations from pile driving and other sources at various distances; for example, the California State Department of Transportation formula (CALTRANS 2004). In soft ground Deckner[7] found PPV of magnitude 2-30 mm/s, 1-3 m from the pile hammer. In loose soils there is rapid attenuation of ground vibration waves; typically, 85-95% at 10 m from the pile and 99% attenuation by 30-50 M. No changes to well performance were noted at PPV 21-220 mm/s.[8-10] At normal wind turbine setback distances to homes (usually 550 M in Ontario) it is not plausible for structural damage to occur to wells from pile driving. The notion that extensive fracturing of bedrock could result from piles is ludicrous.

Ground borne vibrations from the operation of wind turbines are exceedingly small, requiring extremely sensitive measuring equipment to even detect. Even at the turbine base, Styles[11] found RMS velocities of approximately 0.07 mm/s, well below the threshold of human perception. Snow[12] found RMS velocities not exceeding 0.015 mm/s at 100 m distance, about 10 times less than the threshold of perception. Styles[11] and Schofield[13] looked at particle displacement 100 m from a wind farm (>6 turbines) and found that the maximum motion induced to be 120 nanometres, or 1.2 X 10^{7} m, about 10 times the diameter of a human hair. The measurements obtained at the North Kent 1 wind farm were consistent with these published data. This range of seismic motion could not result in structural damage to well casings etc. particularly when the vibrations from passing traffic and especially well pumps are much greater. The notion that particular kinds of sediment particles (black shale) could be translocated over long distances by ground borne vibration waves is scientifically implausible and is probably based on a fundamental misunderstanding of wave physics.

It was evident from the baseline assessment that well water quality in the study area of Chatham-Kent was poor from the outset. Of 571 private property owners, 17 (3%) registered complaints about deterioration of well water during the study. Only 15% (two wells) of the 13 complainants which could be resampled had significant postconstruction turbidity and 85% (11 wells) had no significant turbidity. These numbers are small, but the proportion of complainants with documented postconstruction well water turbidity was less than the proportion with turbidity in the pre-construction baseline study. This does not support the hypothesis that well water is being adversely impacted. Overall, 98% of wells studied had no documented deterioration in water turbidity postconstruction. These findings are inconsistent with widespread contamination of the aquifer. There is no evidence that water wells are being systematically affected by construction or operation of wind turbines. This evidence was examined by officials from Ministry of the Environment who agreed that there is no evidence implicating wind turbines to well problems. Furthermore, this has been recently tested at the Ontario Environmental Tribunal. In the case of Concerned Citizens of North Stormont v. Ontario (Environment, Conservation and Parks, 2019 Can LII 287[14], approval of a wind farm was appealed partially on allegations that turbine construction in Chatham-Kent caused turbidity in drinking water wells. This was refuted by experts retained by the Approval Holder and by the Director, Ministry of the Environment. The Tribunal accepted the evidence that there was no causal relationship between pile driving, wind turbines and well water issues and that sediment in wells is usually due to construction and maintenance of those wells. The appeal was dismissed.

Sediment in well water is due to structural issues with the well casing, screen or seals or sometimes because the well is being overpumped. These causes need to be investigated before implicating unproven theories of causation such as external ground vibrations. Bacteria and dissolved heavy metals etc. in water are health hazards at sufficient concentrations, but turbidity in the absence of other problems is considered an esthetic issue. Most people do not drink noticeably turbid water and filter or settle sediments out before consumption. Ingested inorganic particulates in suspension pass through the body unabsorbed.

Conclusion

There is no evidence that wind turbine construction or operation results in the contamination of groundwater and no scientifically plausible mechanism has been offered by which groundwater contamination with translocated sediments or associated health hazards could theoretically occur.

References

1. https://northkentwind.com/files/2414/6765/0819/NUMBER_5272-A9FHRL_accessible.pdf
2. http://www.ert.gov.on.ca/CaseDetail.aspx?n=16-076
3. https://northkentwind.com/files/1115/1127/7109/Rpt_2017-03-30_Baseline_Well_Survey_NKW1_60343599.pdf
4. https://northkentwind.com/project-documents/water-well-investigation-reports/
5. https://northkentwind.com/files/9215/1188/7980/NKW-Surface_and_Subsurface_Vibration_Monitoring_Report_-_Jun_16_17_REVISED_FINAL_NK1.pdf
6. https://northkentwind.com/files/7315/2122/3694/1668031-2000-R02_FINAL_Dec_7_17_Construction_Vibration_Monitoring_NK1.pdf
7. Deckner, F (2013) Ground vibrations due to pile and sheet pile driving – influencing factors, predictions and measurements. Thesis, KTH, Royal Institute of Technology, Stockholm, Sweden
8. Robertson, DA, Gould JA, Straw, JA and Dayton MA (1980) Survey of blasting effects on ground water supplies in Appalachia. (Contract Report J-0285029.) Washington, DC: U.S. Bureau of Mines
9. Rose, R, Bowles, B and Bender W (1991) Results of blasting in close proximity to water wells at the sleeper mine. In Proceedings of the Seventeenth Conference on Explosives and Blasting Technique. Cleveland, OH: International Society of Explosives Engineers.
10. Straw, JA and Shinko, JP (1994) Blast Vibration Effects Upon a Deep Injection Well and the Reduction of Ground Vibration Over Depth. Proceedings, Tenth Annual Symposium on Explosives and Blasting Research, Austin, TX, 31 – 40.
11. Styles, P, Stimpson, I, Toon, S, England, R and Write, M (2005) Microseismic and Infrasound Monitoring of Low Frequency Noise and Vibrations from Windfarms,

Recommendations on the Siting of Windfarms in the Vicinity of Eskdalemuir, Scotland. Applied and Environmental Geophysics Research Group, Earth Sciences and Geography, School of Physical and Geographical Sciences, Keele University
12. Snow, D (1997) Low frequency noise and vibrations measurement at a modern wind farm. ETSU W/13/00392/REP, Powergen - Ratcliffe Technology Centre, Nottingham UK, Department of Trade and Industry, London UK
13. Schofield, R (2010) Seismic Measurements at the Stateline Wind Project and a Prediction of the Seismic Signal that the Proposed Maiden Wind Project Would Produce at LIGO. University of Oregon. LIGO-T020104-00-Z.
14. https://www.canlii.org/en/on/onert/doc/2019/2019canlii287/2019canlii287.html

INCEEUROPE

8th International Conference
on
Wind Turbine Noise
Lisbon – 12th to 14th June 2019

Human Health Hazard – The Shirley Wind Story

Isaac Old: isaac.old@rsginc.com

Summary

In October 2014 the Brown County Board of Health declared the Shirley Wind project – located in Brown County, Wisconsin, U.S.A. – a Human Health Hazard, a finding that surprised many within the acoustics community. This paper tells the story of the Shirley Wind project, from the permitting phase, all the way through the current time. The paper will primarily rely on public documents, such as the Board of Health meeting minutes. Comparisons will be made between project characteristics and context, and factors that are known to contribute to human noise response.

1. Introduction

The Shirley Wind project ("Project") is a 20 MW, eight turbine wind power project, located in the Town of Glenmore, in Brown County, Wisconsin. The Town of Glenmore is a small town about 16 kilometers (10 miles) from the City of Green Bay, with a population of approximately 1,100 people [1]. A map of the project area, including turbine locations and surrounding residences, is shown in Figure 1. Constructed turbines are Nordex N100 2.5 MW units, with 80-meter hub heights. In many respects, including its size and locale, there is little unusual about the project.

Response to the project has been far from typical with at least 75 unique complainants and three families abandoning their homes [2]. This response resulted in the Brown County Board of Health passing a motion on October 14, 2014, declaring the Project a Human Health Hazard. This paper goes through the history of the project, with the hope that it will provide better understanding of some of the basis of the response.

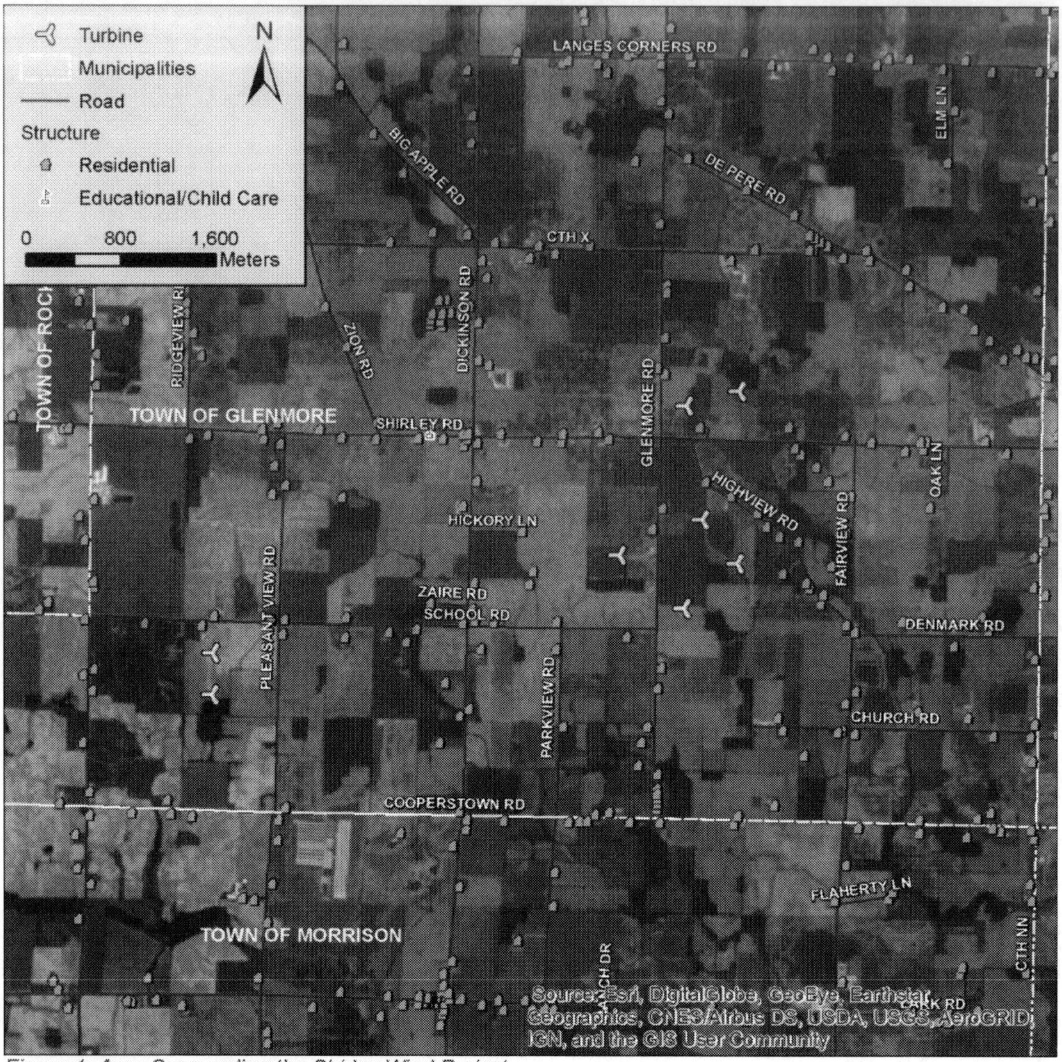

Figure 1: Area Surrounding the Shirley Wind Project

2. Project Permitting

The Shirley Wind Project permitting began around 2007 and was granted a Special Use Permit by the Town of Glenmore, Wisconsin the same year. The sound level limit from the special use permit was 50 dBA [2]. There was a provision in the permit, where a payment of $4,000 per megawatt ($80,000 total) is split between the Town of Glenmore, Brown County, and all landowners within a ½ mile radius of turbines on a yearly basis [3]. The project was initially permitted by a small developer, permitting their first project [4]. After completion the project was sold twice, the second time to Duke Energy [5] [4].

The author of this paper requested the permitting documents from the Town of Glenmore, but, as of the writing of this paper (March 22, 2019), he has not yet received them. Further information will be added to this section if/when these documents are received.

3. Between Permitting and Commissioning

3.1 Wisconsin PSC Rulemaking

The State of Wisconsin began rulemaking to develop wind turbine siting standards in 2009 following 2009 Wisconsin Act 40 [6]. Act 40 also specifically appropriated power to the state for regulating wind power projects, saying "A political subdivision that chooses to regulate shall enact an ordinance subject to sub. (6)(b), that is no more restrictive than the applicable standards established by the commission in rules promulgated under s.196.378 (4g)." The result of this rulemaking procedure was Wisconsin PSC 128 [7]. This set sound level limits of 45 dBA during nighttime hours and 50 dBA during daytime hours, along with an audible tone penalty. No specific metric is given for these limits and the duration it applies to is ambiguous.

3.1.1 Brown County Response to PSC 128

Acceptance to the rulemaking process in Brown County was decidedly negative. Residents commenting to the Brown County Board of Health (BCBOH) seemed to see it as an indication of the political power of wind developers. This sentiment was based on the makeup of the council that developed the legislation, which was viewed as slanted towards the wind power industry. The council also did not include "one member who is a University of Wisconsin System faculty member with expertise regarding the health impacts of wind energy systems," as was required by Act 40 [6].

Three Towns in Brown County, Glenmore, Morrison, and Wrightstown (locations shown in Figure 2) submitted comments on June 23, 2010 regarding the draft of PSC 128 [8]. The comments were reportedly the product of much research and input from the public in seven town meetings, including two that were joint meetings of all three towns. The comments advocate setbacks of at least 0.5 miles (0.8 km) and sound level limit recommendations from various sources of below 40 dBA, though the metrics specified varied wildly. The comments also requested low frequency and infrasound specific limits. The most interesting aspect of this is the reduction in requested sound level limits from what is in the permit for Shirley wind (50 dBA). Also included is a request that towns should be able to require compliance for their existing ordinances.

3.1.2 Most Recent Glenmore Ordinance

The Town of Glenmore most recently modified its Wind Energy Facility ordinance in October 2010 [9], about three years after Shirley Wind received its permit. This ordinance sets sound level limits of 35 dBA for "any period of time" at inhabited structures, along with the permitting requirement of sound propagation modelling that includes low frequency sound and infrasound. Low Frequency Noise is defined in the ordinance as "an ongoing noise that impairs the health of a resident emitted during periods of turbulence as the blades are buffeted by changing winds that can cause structure vibration. Low frequency noise refers to sounds with energy in the lower frequency range of 20 to 200 Hz."

3.2 Board of Health Involvement

The Board of Health's involvement in wind turbine issues seems to begin with an email from the County Supervisor to the Health Department, dated March 30, 2010. The email states, "from what I understand there will be real health issues affecting the people in southern Brown County if the wind turbines are installed.... I will be looking for the BC health department to research this subject and disclose their findings and recommendations in a written report. I would request that either the Health Department, the Corporation Counsel, the Board Attorney or all three parties file an injunction to halt the process of construction until the health assessment by our local governing body is determined." [10] This email resulted in a joint meeting of the Board of Health and Brown County Human Services Committee (BCHSC) on March 25, 2010. The purpose of the meeting was to provide information from Invenergy Wind (developer of the proposed Ledges Project, proposed for Brown County), the State of Wisconsin, and Brown County Citizens for Responsible Wind Energy (BCCRWE) [11]. The State of Wisconsin Department of Health Services representative indicated that there was little information available on the health and safety of wind turbines and that the state had not formally taken a position on the subject. He went on to say that other organizations had not found an association between wind turbines and health impacts at typical setback distances. Individuals claiming health impacts should see their doctor. The Invenergy representative spoke, pointing to their health and safety record, largely regarding occupational health. Then he points to the economic and environmental benefits of projects including their proposed Ledge Wind Farm. Invenergy also submitted a newspaper article called "Winds of Change Inevitably Get the Hot Air Stirring." BCCRWE has a presentation from several residents presenting a request for a moratorium on wind projects in Brown County pending completion of an epidemiological study. Other presentations included literature forwarding the view point that adverse health effects are inherent to wind turbine installations. Other residents presented interviews of those claiming health effects from other projects, claims of harm to livestock, reduced crop productivity, and concerns about groundwater. BCCRWE finally presented an 890-signature petition to the state to perform a health study in advance of more wind power project approvals.

After this meeting, discussion on wind turbine issues became a common theme of Board of Health meetings. At the October 12, 2010 meeting, Dr. Jay Tibbetts (a member of the Brown County Board of Health) distributes the minority report for the Wind Siting Commission, and explained that he found their position well written and objective. Particular among their concerns, were a lack of low frequency noise consideration [12]. The minority report was a document that specified considerably more restrictive siting rules than what was eventually included in PSC 128 and was written by dissenting members of the commission.

At the January 11, 2011 Board of Health meeting, BCCRWE introduced a draft legislature bill was introduced that increased setbacks to 1,800 feet (549 meters).

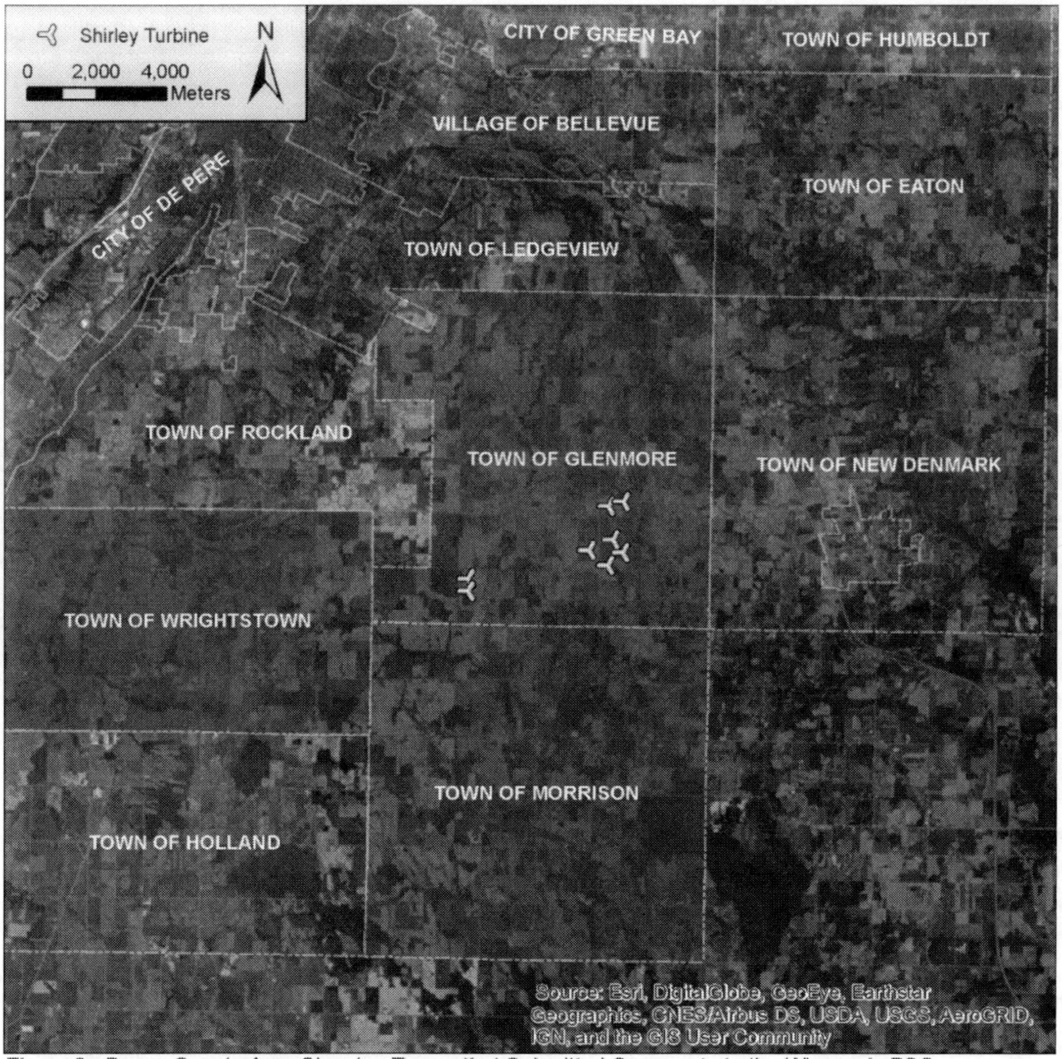

Figure 2: Brown County Area Showing Towns that Submitted Comments to the Wisconsin PSC

4. Post-commissioning

The Shirley Wind Project was commissioned in December 2010 [5].

4.1 Other Brown County Wind Power Projects

A second wind power project was proposed for the Town of Glenmore, by a developer called Prelude that would have included seven turbines. Most of the permits for the project had been obtained by the developer before 2010 [13]. A power purchase contract had also been signed, with the provision that all permits would need to be obtained before March 1, 2011. The developer learned around September 2010 that the project would require building permits for each of the seven turbines. After this, the developer submitted applications to the Town Board for these building permits, after which the Town Board requested further information. The developer provided this information by December 31, 2010 and informed the Board Chair of the impending deadline. At the January and February 2011 town meeting angry citizens opposed the plan and the Board Chair was receiving "threats to his physical safety should he approve the wind turbine project" [13]. The original developer (Prelude Energy) was purchased by CEnergy in a sale that closed in February 2011. The Town Board did not take up the permit issue at either the January or February meeting, citing need for the Town Attorney to review the newly-submitted information, though the developer later claimed it was due to threats the Board Chair was receiving [13]. The permits were finally approved on March 7, 2011, and the meeting was closed, but the meeting was re-opened due to pressure from some of those in attendance, and the permits were rescinded [13]. One week later the board had a special meeting, reinstating the revoked permits. The permits were never actually issued due to completeness concerns from Glenmore's lawyer.

The permit approval ended up being too late, since the power purchase agreement was cancelled on March 4, 2011. The developer ended up suing the Town of Glenmore citing that it was deprived of property without due process, requesting a judgment that would have cost each resident of Glenmore $6,000.

The appeal for the suit was decided against the developer on August 7, 2014.

The Ledge Wind farm was a 150 MW wind farm proposed by Invenergy for southern Brown County. This project was cancelled in March 2011, citing "regulatory uncertainty" at the state level [14].

4.2 Board of Health Proceedings

The first mention of Shirley Wind to the Board of Health after commissioning comes in the May 10, 2011 minutes, where residents from two households report having health issues that they blamed on the Project [15]. One mentions low-frequency noise and describes it as a "sound you can feel" and report that others within the town have had the same issues. Another reports that they did some internet research and attributes the issues to infrasound. One Board of Health member recommends they see a doctor, which is met with the response that they found it unlikely the doctor would find anything since symptoms are only present near the turbines. Another Board of Health member remarks, it's similar to what Nina Pierpont reported in her book and some of the stories told at the May 25, 2010 joint meeting. Dr. Tibbetts considers state health officials to be unsympathetic and wants a study of the issue. The following meeting on July 26, 2011 included a response from the State of Wisconsin indicating that it was unlikely that the State would be performing a study on wind turbine noise.

The next time the Shirley project was mentioned in the health board minutes is for November 15, 2011, where the board considers whether or not to support a bill that would require submittal of a study on wind turbine health effects by the State health board, before the new siting guidelines take effect [15]. Members of the Board of Health also went on a site visit to a farm where 13 of the 87 cows had died without apparent cause since operation of the wind farm. On the site visit, one of the residents who had abandoned their residence due to health effects they attributed to the turbines, invited the board members over to his house. One board member reported having pressure in their ears. Another board member reports that the Shirley project had been bought and sold three times since the turbines were installed (other sources indicate twice [4]).

The January 10, 2012 meeting brought a presentation by BCCRWE on stray voltage [16], which is what was blamed for the cow deaths mentioned in the November 15, 2011 minutes. After this presentation, the Board of Health passes a resolution to ask the state for financial aid for those that relocated due to the wind project. The board votes to support the "Wisconsin Citizens Safe Siting Guidelines." These guidelines were endorsed by the Towns of Glenmore, Holland, and Morrison, and drafted by a handful of Wisconsin Citizen groups. Dr. Tibbetts indicates that Richard James (an acoustician and Principal of E-coustic Solutions) had been hired to measure infrasound in the Shirley Project area, since he considered that to be the primary concern.

At the March 13, 2012 meeting, there is a report of several more complaints due to the Project. The Board of Health tasks the Health Department with measuring sound levels at a complainant's residence [17]. There is more discussion about having Mr. James revise the county ordinance and do a noise study of the area. At the next meeting on May 8, 2012, results from the noise complaint test are reported, which indicated an exceedance of the county noise ordinance [18]. The report was presented to Duke, which responded that the Town of Glenmore special use permit possessed by the Project superseded the County's jurisdiction, a conclusion which the counsel for the county agreed with. Effectively the County does not have the jurisdiction to regulate wind farms, since the Special Use Permit was granted by the Town of Glenmore and current legislation has moved regulation to the State of Wisconsin. Dr. Tibbetts puts forward the idea that the noise issues can be dealt with when Shirley's permit is brought up for a review, which would occur at a five-year interval. The next meeting in July 2012 indicates that there had been several other complaints about noise the Project [19]. Apparently, a study was performed by Mr. James indicating an "issue" with low frequency noise, though no further details are given. This study was distributed to Board of Health Members.

There is talk among the BCBOH of having Mr. James draft a low frequency noise ordinance during this period [19] [20] [21]. The final conclusions seem to be that the ordinance would not be enforceable for wind farms due to PSC rules. BCCRWE encourages the ordinance anyways, to send a message [20]. The board approves hiring Mr. James to write the ordinance in January 2013 and funding was obtained [22].

At the September 2012 meeting, a study regarding infrasound from wind turbines was distributed by Dr. Tibbetts, though no further information was provided in the minutes. Another Board member reports that three families have now abandoned their homes due to infrasound. Interestingly enough, one of these families specifically referenced audible sound in their complaint [23]. Further discussion about the low frequency noise ordinance leads to the conclusion that it might not be enforceable due to the restrictions on local regulations. The Board of Health also approves of Dr. Tibbetts sending a letter to municipalities to consider health effects of wind turbine noise that included the infrasound study mentioned above, as well as information about the three families that abandoned their homes. At the November 2012 meeting, there is extensive discussion about how the PSC took away local control of permitting. BCCRWE says that they are looking for the Board to facilitate low frequency sound studies but with the PSC's funding. The Board doubted that they would have sufficient expertise or the ability to pick anyone other than the lowest bidder, since there were concerns about picking an expert that might be "biased." They agree to send a letter by registered email to the PSC along with letters of those that claimed they were experiencing health effects caused by wind turbines, which were provided by BCCRWE.

The January 2013 minutes include a review of the Cooperative Measurement Program (described in Section 4.3) [21]. The next available set of minutes, is March 11, 2014, though wind power is not discussed. At the May 13, 2014 [24] meeting a letter was read that was sent to the medical community to "increase their index of suspicion." It was written by Dr. Tibbetts with assistance from two of the residents that abandoned their homes (one is a BCCRWE member), and Mr. James. This seems to be the meeting where the idea of declaring the Project a human health hazard is first put forward. There is a discussion about whether it more fit the definition of a "Public Nuisance," "Human Health Hazard," or "Immediate Human Health Hazard." There is also discussion of surveys from those near the Shirley and Fond Du Lac wind farms being sent to Carmen Krogh (a pharmacist) and an expected noise study from Mr. James. It's implied that Mr. James is assembling much of the literature collected by the Board.

A series of binders are passed out to the BCBOH members, assembled by Mr. James. These binders appear to include his report conclusions, along with a letter containing his request for the Human Health Hazard declaration [25]. A special meeting was scheduled for September 9, 2014 to discuss the information distributed. Minutes are not available for this meeting. Dr.

Tibbett's letter was also sent to area newspapers and BCCRWE distributed the information as flyers to about 400 people in the Glenmore area.

4.3 Cooperative Measurement Program

As part of the permitting process for the Highland Wind project, also located in Wisconsin, a measurement study at the Shirley Wind Project was performed by five acoustical consultants (George Hessler, David Hessler, Robert Rand, Bruce Walker, and Paul Schomer) [26]. The study entailed detailed sound level measurements, particularly in the low frequency and infrasonic range at three homes that had been abandoned by the residents. Measurements showed clearly visible blade passage frequency harmonics at two of the three monitored locations. Audible sound levels at the closest residence (1,100 feet or 335 meters) were as high as 50 dBA. The report included four individual reports by each consultant, as well as combined conclusions, including the statement:

> "The four investigating firms are of the opinion that enough evidence and hypotheses have been given herein to classify LFN and infrasound as a serious issue, possibly affecting the future of the industry. It should be addressed beyond the present practice of showing that wind turbine levels are magnitudes below the threshold of hearing at low frequencies."

This report has since been extensively cited, both for its measurement results and for the above-referenced conclusion. One of the difficulties of the study was that the project owner did not cooperate with the study, by providing data or by performing organized turbine shutdowns to allow for background sound level measurements. The Health Board reviewed this report, focussing on the above statement. Discussion of the report lead to discussion of developing a low frequency noise and infrasound ordinance for the county.

Paul Schomer formalized his hypothesis as was first elucidated in the cooperative report in a journal paper [27]. Bruce Walker later published a listening test pilot study, using measurements from the cooperative study at the 2015 Wind Turbine Noise conference [28]. Four of the five acousticians involved in the cooperative study (excluding Robert Rand) published a follow-up paper [1] in 2017. This paper had the following conclusions:

> "Our analysis illustrates that a wind turbine is not a classic LFN source; that is, one with excessive low-frequency spectral content. But a wind turbine is a unique power-generating source with spectral content down to the 1-Hz octave band, emitting measurable IS in addition to LFN. Infrasound (IS, 0-20 Hz) from wind turbines can almost be ruled out as a potential mechanism for stimulating motion sickness symptoms. But to be thorough and complete, we recommend that one or two relatively simple and relatively inexpensive studies be conducted to be sure no infrasound pathways to the brain exist other than through the cochlea. Pending the results of these studies, we feel that no other IS or LFN criteria are required beyond an acceptable A-weighted level."

While this does not fully rebut the statement from the cooperative report, it is much more sceptical of the idea that sub-audible infrasound is the cause of health effects near wind turbine installations.

4.4 Other Sound Studies

E-coustic Solutions (written by Rick James) was hired by one of Shirley Wind's neighbors to do a noise study. The study included microbarometer measurements in and around the project, up to a distance of about 10 kilometers (6 miles) [29] to measure infrasound. Most of the measurements were performed by two project complainants who had abandoned their homes [30]. The E-coustic summary report measured wind-turbine caused infrasound at each of these locations. Conclusions were:

"Based on the above, it is reasonable to conclude that the adverse health effects reported by member of the Shirley community are linked to the operation of the Shirley Wind project wind turbines. While there may still be debate about the precise mechanism that causes these sound to induce the symptoms; it is clear from this study, and others conducted in different parts of the world by other acousticians, that acoustic energy emitted by the operation of the modern utility scale wind turbines is at the root of the adverse health effects."

This conclusion was followed by a recommendation that either landowners near (within about 4 miles) the project should be bought out, or that turbine operations should be halted.

5. Board of Health Decision

The Board of Health Declaration was made on October 14, 2014. At this meeting, only four of six Board members were present [2]. The second item of the meeting was regarding additional information about wind turbine sound and included the request letter from Mr. James which reads.

"The purpose of this letter is to inform you that residents living in and near the Town of Glenmore, citizens of Brown County, Wisconsin, are being negatively impacted by sources that began after the construction of the Shirley Wind Project. This letter is an appeal and urgent request to the Brown County Board of Health and the Director of the Health Department to declare the Shirley Wind Project a Human Health Hazard in order to protect the health and well-being of the residents who are in extreme peril."

The statute read to support the request included the following definition of Health Hazard.

"Human Health Hazard means a substance, activity or condition that is known to have the potential to cause acute or chronic illness or death if exposure to the substance, activity or condition is not abated."

After this was read, a BCBOH member read the timeline of their involvement with wind turbine noise issues, beginning with the March 30, 2010 letter from the BCHSC through to 2014, including the cooperative study and concluding with the E-coustic study. This is followed by several testimonies by residents about their experience near the project, typically urging the BCBOH to declare Shirley wind a Human Health Hazard and recounting of stories worldwide about adverse effects on humans and animals attributed to wind turbines. Mr. James describes more of his studies in a recorded presentation. Patrick Evans (the author of the March 2010 letter to the BCBOH from the BCHSC) several times reminds the Board that it's the Health Board's mission to protect the public. Declaring Shirley Wind a "Human Health Hazard" is then brought to a vote and the motion is passed, though the vote count is not recorded in the minutes. At a later meeting, Dr. Tibbetts says the vote was unanimous [30].

6. Post-decision

At the first few meetings after the declaration there is considerable discussion by BCBOH about next steps to compel Shirley Wind to somehow address the issue. Most of these discussions occurred in closed sessions with legal counsel, so the details are not part of the minutes [31] [32] [33] [34] [35] [36] [36].

Shirley Wind's first interaction with the BCBOH seems to occur only after the decision, with a letter sent on October 31, 2014 requesting due process [32]. At the November 11, 2014 meeting, a lawyer representing Duke objects to the declaration on legal grounds. The Board then approves Duke presenting to the Board at the January 20, 2015. At this presentation, a lawyer presents Duke's version of events, a legal analysis indicating that the BCBOH does not have jurisdiction over Shirley Wind, and requests that the Board rescind the declaration [32].

BCBOH seems to conclude that they cannot compel Shirley Wind to do something, but the Health Director can, since one of their responsibilities is enforcement [37]. At the May 12, 2015 meeting, they hand all information on the subject that had been collected during previous BCBOH meetings to the Health Director, named Chua Xiong. Ms. Xiong indicates that she will review the collected information, including that sent by Duke, conduct meetings with interested parties, and do additional research before making a decision [38]. The deadline for document submission was set at September 30, 2015 [36]. This additional research ended up being a full literature review of available sources, with the assistance of a local graduate student [36]. At the October 13, 2015 meeting, Ms. Xiong presented the procedures she was using for the literature review with no objections from Board members or the public mentioned in the minutes [39].

BCCRWE takes increased interest in the BCBOH during this period, with multiple members present at each meeting, presenting new information and challenging the validity of information presented by others and/or the presenter's credibility [32] [34] [38] [40]. They asked to invite speakers to present to the Board, particularly Carmen Krogh, though this is challenged by a couple of Board members and citizens due to Krogh's lack of credentials and perceived bias [38].

There also seems to be more scepticism of the idea of infrasound causing health effects from some of those attending the meetings including criticisms directed at the Board that they were getting a single sided story [38] [40] [36].

6.1 Health Director Determination

The Health Director's determination was given on December 15, 2015 and is one of two sets of minutes that is presented verbatim instead of in a summarized format [37]. The meeting starts with public comments, mostly those praising the BCBOH's declaration or specifying health issues attributed to Shirley Wind. One interesting comment is from a resident near the proposed Highland Wind farm that indicates residents near that project received a postcard from Brown County warning them of the effects of wind turbines.

The Health Director then presents her findings, indicating that:

> "Presently, there is insufficient scientific evidence based research to support the relationship between wind turbine and health concerns. However, I will continue to keep an open mind and review any new scientific evidence based literature on an annual basis. As the Brown County Health Officer, I will accept and review any new, current, scientific evidence based literature and provide updates on an annual basis regarding significant findings."

In the short period between the Health Director's report and the end of the meeting, reaction from the public seems stunned, but upset. Most of the Board members seem to have accepted Ms. Xiong's decisions, though Dr. Tibbetts says,

> "Ah, I know that when we, when we, passed this, ah, October 14th, ah, it was unanimous, ah, and I for one stand by that. I'm afraid Chua and at least I, as a member, are at odds with, ah, declaring this, ah, human health hazard. Ah, I understand why she wouldn't want to because that demands action and I for one am very disappointed with the, ah, with the response of Chua."

7. Board Activity After Health Director Decision

The next meeting on January 12, 2016 is also a verbatim transcript [41]. The meeting opens with more public comments. Most of the comments are either expressing disappointment about the decision, or reporting more health effects. A member of BCCRWE went through the list of "experts" that they provided to Ms. Xiong that she did not contact. They take this as evidence that she did not do her due diligence in making the decision. Another BCCRWE member said, "You are now part of the problem and will guarantee more people, more families, more children, more parents will be harmed." One resident says, "Are you a leader or not? If you are not a leader step aside." Other comments accuse Ms. Xiong of being derelict of her duties. Dr. Tibbetts says he considers the literature review she performed to be a waste of time.

Ms. Xiong is not present at the next meeting on March 8, 2016 [3] and resigns on March 18, 2016 [42]. BCBOH receives criticism from other parts of the county government, calling their conduct towards Ms. Xiong unprofessional [3].

In following meetings there are considerable criticisms of Ms. Xiong's decisions by the public, BCCRWE, and even an acoustician, Robert Rand, who writes a letter condemning the decision. BCCRWE issues a formal rebuttal of some of the references cited by Ms. Xiong and asks for the decision to be reversed [42].

In March 2016, another Board member was added, who ended up being more critical of BCBOH's declaration and submits information to the Board to support criticism of the declaration, though these submissions are criticized by BCCRWE [42]. A motion was later brought to rescind the declaration which failed in two votes to four [43].

Duke offers a presentation on health effects due to wind turbine noise, presented by Dr. Robert McCunney in September 12, 2017 [44]. Dr. McCunney is a medical doctor who has contributed to and authored several papers on wind turbine noise. Best known is his 2014 literature review.

As of the writing of this paper, the declaration has not been rescinded, but there has not been a health director decision affirming the declaration. There has not been any action against the project by the State of Wisconsin.

8. Comparison with Human Response Factors

While sound exposure is correlated with human response, it often only explains 20 percent or less of the variation [45]. Several authors have looked at other factors that influence response both for wind turbine noise and other noise sources. Table 1 shows some of these factors along with an assessment of whether they were present for Shirley Wind.

For some of these factors, such as Noise Sensitivity, it would be difficult to impossible to assess the relevance for Shirley Wind, due to the types of evidence reviewed for this paper. In other cases, these factors may have been influential for some residents, but not for others (such as Project Audibility, Visual Annoyance, and Place Attachment). Consequently, a "yes" in the second column indicates that there is some kind of evidence that this was a factor for at least some residents, but it may not have been factor for all.

Table 1: Comparison of Human Response Factors with Context of the Shirley Wind Project.

Factor	Presence at Shirley	Source(s)	Notes
Noise Sensitivity	Maybe	[46] [47]	Not possible to determine from analysed information

Factor	Presence at Shirley	Source(s)	Notes
Perceived Stress	Y	[46]	
Concern for Physical Safety	Y	[46] [48] [49]	Concerns about infrasound and stray voltage.
Property Ownership	Y	[46]	All three of the cases where homes are abandoned were owned by the residents.
Self-reported Sleep Disturbance	Y	[46]	Several complaints mention sleep disturbance.
Visual Annoyance to Wind Turbines or annoyance to other non-acoustic characteristics	Y	[46] [49]	Shadow flicker is mentioned as an issue in some cases. Concern about stray voltage. Located in a flat area, so turbines are visible within a radius of several kilometres.
Audible Wind Turbine Noise	Y	[46] [47]	Some complaints in the minutes specified that the noise was audible. In other cases it was specified as inaudible.
Complaint About Wind Turbines	Y	[46]	Over 75 complaints
Attitude Towards the Noise Source	Y	[47] [49]	Developer/owner lack of response and mis-steps. It's not clear if those in the project area viewed the turbines as necessary.
Belief noise can be prevented	Y	[49]	
Place attachment	Y	[47]	Not possible to tell from the minutes for many cases. However, rural areas adjacent to larger populated areas often have a large number of people who moved there for the rural environment.
Trust/Distrust of Authorities	Y	[48]	Change of jurisdiction and lack of response

Factor	Presence at Shirley	Source(s)	Notes
			from the State or developers
Expectations	Y	[48] [50]	Consistent presentation of literature showing relationship between infrasound and health effects
Noise Exposure History (Character of area)	Y	[48] [49]	No previous multi-turbine projects

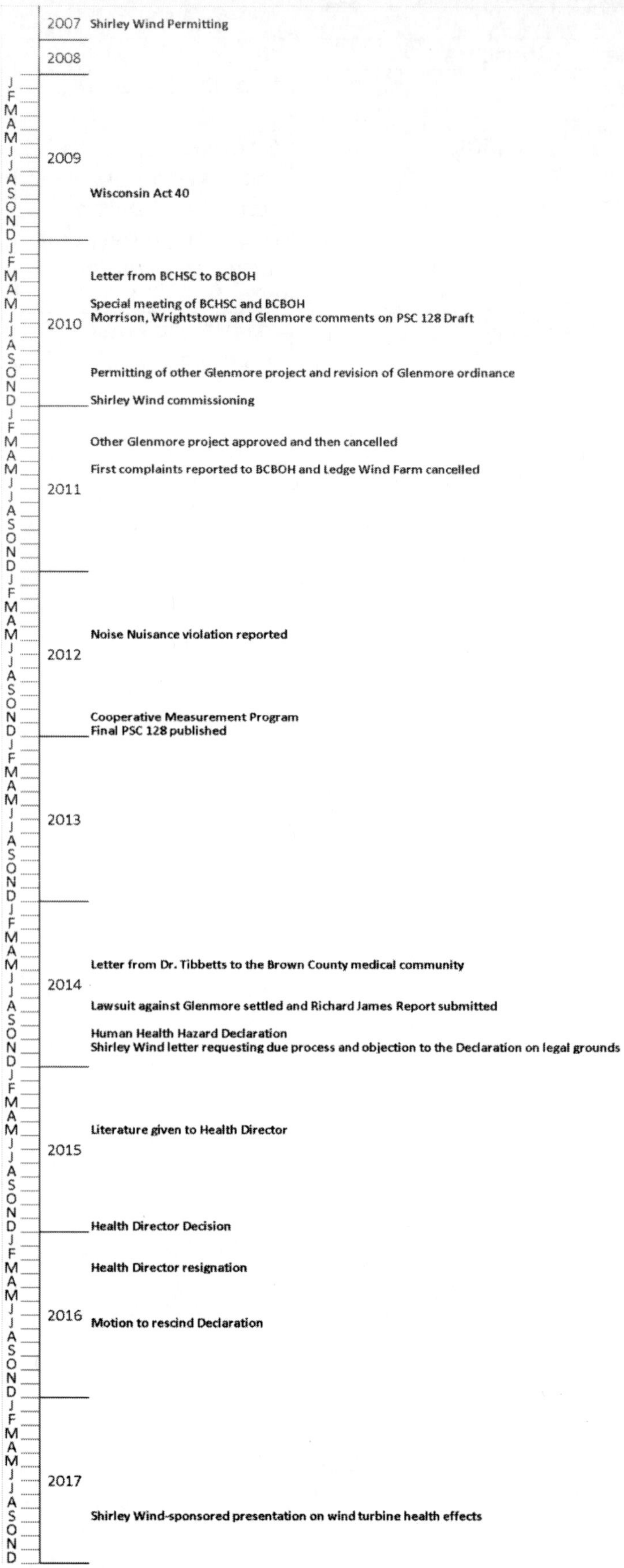

Figure 3: Shirley Wind Event Timeline

9. Discussion

Response to the Shirley Wind project has been extraordinary, particularly considering the relative size of the project (other projects in Wisconsin are larger). Although relatively high sound emissions (compared to WHO guidelines, for example) are a potential cause, sound level limits in excess of 50 dBA are not uncommon for the region. Similarly, the hypothesis that infrasound could be the cause of such a response is also unlikely, since one would expect a similar response from similar projects, with similar turbines installed, and similar setbacks. While there are other projects that have had strong adverse response, this is not a typical response. Furthermore, recent laboratory studies have shown a lack of response to wind turbine-produced infrasound at magnitudes typical near wind power projects in blinded tests [28] [50].

Other factors that can influence response to environmental noise sources have long been studied in acoustics. Particularly striking is the extent to which some cultural and situational factors were present, particularly negative expectations and events that are likely to have degraded trust in state authorities and the developer/owner of the Project. The presence of two of these factors are summarized below:

- Several events seem to result in a strong distrust of authorities, whether it was the state government or specific developers:
  - Wisconsin Act 40 removing permitting and regulator authority for local municipalities and inability of local authorities to address noise complaints.
  - A lack of response from the State of Wisconsin to requests that the State study the issue.
  - The lawsuit between a developer and the Town of Glenmore.
  - Lack of response by Shirley Wind to resident concerns, other than legal analyses. Similar is the 2010 presentation which contained information about economic benefits and an article which seemed to minimize any concerns.
  - Multiple sales of the project.
- Although the origin of the concerns about health effects from infrasound and other characteristics of wind turbines has not been determined in this paper, expectation was established and propagated:
  - Proliferation of information by BCCRWE in the community and elsewhere.
  - Dissemination of information by the BCBOH.
  - Use of those who claimed health effects due to the turbines as infrasound testers and the report provided by Mr. James.

Although some of the factors that may have eroded trust were not directly related to Shirley Wind, it may be that people of Brown County mentally grouped all wind power developers/owners as an external "they." Distrust of authorities may have also sped acceptance by some residents to accept the idea that characteristics of wind turbines (including infrasound) were harmful. It was several years between the first concern about wind turbine health effects and a strong response by either the state of Wisconsin or any developers presenting alternative health-based information. By the time there was any response, most BCBOH members seemed to have firmly taken a position and any alternative explanation was met with firm resistance, as can be seen through the response to the Health Director's decision. The two Board members that ended up questioning the declaration were only added after the declaration was made.

The root of the specific health effects indicated by some of the residents is an interesting topic, considering the factors listed in Section 8 are related to noise annoyance and not the specific symptoms (nausea, tinnitus, headaches, etc.) mentioned by the residents. It's beyond the scope of this paper to fully address this, except to mention some authors who have looked at the influence of high level annoyance to some of these effects [51] [52].

10. Conclusions

The Shirley Wind project (Project) has been noteworthy for the strong response of residents in the project area, resulting in the Brown County Board of Health (BCBOH) declaring the Project a "Human Health Hazard." This paper has looked through the history of the Project, using public information to assess which factors, that are known to result in adverse response to noise sources, were present in the project context. The paper shows that:

- Several events from project permitting through the Human Health Hazard declaration, likely resulted in a degradation of trust in local authorities and wind power project developers/owners.
- Extensive proliferation of information by several parties associating several characteristics of wind turbines with various health effects. Distrust of authorities likely helped speed acceptance of this information and reduced acceptance of other information that pointed to a different conclusion, once it became known.

Given the sheer density of information proliferated, the number of events that could have eroded trust of authorities, and relatively high noise exposure to some residents, the adverse response to the Shirley Wind project is understandable.

Acknowledgements

Thanks to Geoff Leventhall for discussions on this subject at Wind Turbine Noise 2017 and the encouragement to write up the topic.

References

[1] United States Census Bureau, "American Fact Finder," [Online]. Available: https://factfinder.census.gov/faces/tableservices/jsf/pages/productview.xhtml?src=bkmk. [Accessed 26 March 2019].

[2] Brown County Board of Health, *Proceedings of the Board of Health Meeting - Tuesday, October 14, 2014,* Green Bay, Wisconsin, 2014.

[3] Brown County Board of Health, *Proceedings of the Board of Health Meeting - Tuesday, March 10, 2016,* Green Bay, Wisconsin, 2016.

[4] M. Vickerman, "Shirley Wind: An Auspicious Debut for Emerging Energies," Renew Wisconsin, 10 November 2010. [Online]. Available: http://www.renewwisconsin-blog.org/2010/11/shirley-wind-auspicious-debut-for.html. [Accessed 15 March 2019].

[5] "Duke Energy," [Online]. Available: https://www.duke-energy.com/our-company/about-us/businesses/renewable-energy/wind-energy/shirley-windpower. [Accessed 14 March 2019].

[6] State of Wisconsin, *2009 Wisconsin Act 40,* Madison, Wisconsin: State of WIsconsin, 2009.

[7] Public Service Commission, *Chapter PSC 128 - Wind Energy Systems,* Madison, Wisconsin: Legislative Reference Bureau, 2012.

[8] G. Schwalbach, *Comments by the Towns of Morrison, Wrightstown, and Glenmore, Brown County Wisconsin on Draft Chapter 128 - Wind Energy Systems,* 2010.

[9] Town of Glenmore, *Wind Energy Facility,* Glenmore, Wisconsin, 2010.

[10] P. Evans, "Wind Turbine Communication - Health of BC Citizens," 2010.

[11] Brown County Board of Health and the Brown County Human Services Committee, *Proceedings of the Joint Meeting of the Board of Health and The Brown County Human Services Committee - Tuesday, May 25, 2010,* Green Bay, Wisconsin, 2010.

[12] Brown County Board of Health, *Proceedings of the Board of Health Meeting, Tuesday, October 12, 2010,* Green Bay , Wisconsin, 2010.

[13] *CEnergy-Glenmore Wind Farm #1, LLC v. Town of Glenmore Appeal Decision,* 2014.

[14] M. Gallucci, "Developer Pulls Plug on Wisconsin Wind Farm Over Policy Uncertainty," Reuters, 24 March 2011. [Online]. Available: https://www.reuters.com/article/idUS408832364820110324. [Accessed 15 March 2019].

[15] Brown County Board of Health, *Proceedings of the the Board of Health Meeting - Tuesday, May 10, 2012,* Green Bay, Wisconsin, 2011.

[16] Brown County Board of Health, *Proceedings of the Board of Health Meeting - Tuesday, January 10, 2012,* Green Bay, Wisconsin, 2012.

[17] Brown County Board of Health, *Proceedings of the Board of Health Meeting - Tuesday March, 13 2012,* Green Bay, Wisconsin, 2012.

[18] Brown County Board of Health, *Proceedings of the Board of Health Meeting - Tuesday, May 8, 2012,* Green Bay, Wisconsin, 2012.

[19] Brown County Board of Health, *Proceedings of the Board of Health Meeting - Tuesday, July 17, 2012,* Green Bay, Wisconsin, 2012.

[20] Brown County Board of Health, *Proceedings of the Board of Health Meeting - Tuesday, November 13, 2012,* Green Bay, Wisconsin, 2012.

[21] Brown County Board of Health, *Proceedings of the Board of Health Meeting - Tuesday, January 8, 2013,* Green Bay, Wisconsin, 2013.

[22] Brown County Board of Health, *Proceedings of the Health Board Meeting - Tuesday, May 28, 2013,* Green Bay, Wisconsin, 2013.

[23] Brown County Board of Health, *Proceedings of the Board of Health Meeting - Tuesday, September 11, 2012,* Green Bay, Wisconsin, 2012.

[24] Brown County Board of Health, *Proceedings of the Board of Health Meeting - Tuesday, May 13, 2014,* Green Bay, Wisconsin, 2014.

[25] Brown County Board of Health, *Proceedings of the Board of Health Meeting - Tuesday, August 12, 2014,* Green Bay, Wisconsin, 2014.

[26] B. Walker, G. Hessler, D. Hessler, R. Rand and P. Schomer, "A Cooperative Measurement Survey and Analysis of Low Frequency and Infrasound at the Shirley Wind Farm in Brown County, Wisconsin," Clean Wisconsin, Madison, Wisconsin , 2012.

[27] P. Schomer, J. Erdreich, P. Pamidighantam and J. Boyle, "A Theory to Explain Some Physiological Effects of the Infrasonic Emissions at Some WInd Farm Sites," *Journal of the Acoustical Society of America,* vol. 137, no. 3, pp. 1356-1365, 2015.

[28] B. Walker and J. Celano, "Progress Report on Synthesis of Wind Turbine Noise and Infrasound," in *6th International Meeting on Wind Turbine Noise*, Glasgow, 2015.

[29] R. James, "Sound Pressure Level (SPL) Measurements of Infrasound Inside Homes In and Proximate to the Footprint of The Shirley Wind Project (Duke Energy)," E-Coustic Solutions, Okemos, Michigan, 2014.

[30] Brown County Board of Health, *Proceedings of the Board of Health Special Meeting - Tuesday, December 15, 2015,* Green Bay, Wisconsin, 2015.

[31] Brown County Board of Health, *Proceedings of the Board of Health Meeting - Tuesday, November 11, 2014,* Green Bay, Wisconsin, 2014.

[32] Brown County Board of Health, *Proceedings of the Board of Health Meeting - Tuesday, January 20, 2015,* Green Bay, Wisconsin, 2015.

[33] Brown County Board of Health, *Proceedings of the Board of Health Meeting - Tuesday, March 3, 2015,* Green Bay, Wisconsin, 2015.

[34] Brown County Board of Health, *Proceedings of the Health Board Meeting - Tuesday, March 17, 2015,* Green Bay, Wisconsin, 2015.

[35] Brown County Board of Health, *Proceedings of the Board of Health Meeting - Tuesday, April 28, 2015,* Green Bay, Wisconsin, 2015.

[36] Brown County Board of Health, *Proceedings of the Board of Health Meeting - Tuesday, August 25, 2015,* Green Bay, Wisconsin, 2015.

[37] Brown County Board of Health, *Proceedings of the Board of Health Special Meeting - Tuesday, December 18, 2015,* Green Bay, Wisconsin, 2015.

[38] Brown County Board of Health, *Proceedings of the Board of Health Meeting - Tuesday, May 12, 2015,* Green Bay, Wisconsin, 2015.

[39] Brown County Board of Health, *Proceedings of the Health Board Meeting - Tuesday, October 13, 2015,* Green Bay, Wisconsin, 2015.

[40] Brown County Board of Health, *Proceedings of the Board of Health Meeting - Tuesday, July 28, 2015,* Green Bay, Wisconsin, 2015.

[41] Brown County Board of Health, *Proceedings of the Board of Health Special Meeting - Tuesday, January 12, 2016,* Green Bay, WIsconsin, 2016.

[42] Brown County Board of Health, *Proceeding of the Board of Health Meeting - Tuesday, May 18, 2016,* Green Bay, Wisconsin, 2016.

[43] Brown County Board of Health, *Proceedings of the Board of Health Meeting - Tuesday, July 19, 2016,* Green Bay, Wisconsin, 2016.

[44] Brown County Board of Health, *Proceedings of the Board of Health Meeting - Tuesday, September 26, 2017,* Green Bay, Wisconsin, 2017.

[45] R. Job, "Community Response to Noise: A Review of Factors Influencing the Relationship Between Noise Exposure and Reaction," *Journal of the Acoustical Society of America,* vol. 83, no. 3, pp. 991-1001, 1988.

[46] D. Michaud, S. Keith, K. Feder, S. Voicescu, M. Leonora, J. Than, G. Mireille, T. Bower, A. Denning, E. Lavigne, C. Whelan, S. Janssen, T. Leroux and F. van den Berg, "Personal and Situational Variables Associated with Wind Turbine Noise Annoyance," *Journal of the Acoustical Society of America,* vol. 139, no. 3, pp. 1455-1466, 2016.

[47] E. Pedersen, F. van den Berg, R. Bakker and J. Bouma, "Response to Noise from Modern Wind Farms in The Netherlands," *Journal of the Acoustical Society of America,* vol. 126, no. 2, pp. 634-643, 2009.

[48] R. Guski, "Personal and Social Variables as Co-determinants of Noise Annoyance," *Noise and Health,* vol. 3, pp. 45-56, 1999.

[49] J. Fields, "Effect of Personal and Situational Variables on Noise Annoyance in Residential Areas," *Journal of the Acoustical Society of America,* vol. 93, no. 5, pp. 2753-2763, 1993.

[50] R. Tonin, J. Brett and B. Colagiuri, "The Effect of Infrasound and Negative Expectations to Adverse Pathological Symptoms from Wind Farms," *Journal of Low Frequency Noise, Vibration, and Active Control,* vol. 0, no. 0, pp. 1-14, 2016.

[51] G. Leventhall, "Why do Some Believe that they are "Made Ill" by Wind Turbine Noise," in *7th International Conference on Wind Turbine Noise*, Rotterdam, Netherlands, 2017.

[52] S. Chapman, A. St. George, K. Waller and V. Cakic, "The Patter of Complaints about Australian Wind Farms Does Not Match the Establishment and Distribution of Turbines: Support for the Psychogenic, 'Communicated Disease' Hypothesis," *PLOS One,* vol. 8, no. 10, p. e76584, 2013.

[53] G. Hessler, G. Leventhall, P. Schomer and B. Walker, "Health Effects from Wind Turbine Low Frequency Noise and Infrasound: Do Wind Turbines Make People Sick? That is the Issue," *Sound and Vibration,* pp. 34-44, January 2017.

[54] Brown County Board of Health, *Proceedings of the Board of Health Meeting - Tuesday, November 15, 2011,* Green Bay, Wisconsin, 2011.

[55] D. Michaud, "Wind Turbine Noise and Health Study: Summary of Results," in *6th International Meeting on Wind Turbine Noise*, Glasgow, Scotland, 2015.

[56] D. Michaud, S. Keith, K. Feder, S. Voicescu, L. Marro, J. Than, M. Guay, T. Bower, A. Denning, E. Lavigne, C. Whelan, S. Janssen, T. Leroux and F. van den Berg, "Personal and Situational Variables Associated with Wind Turbine Noise Annoyance," *Journal of the Acoustical Society of America,* vol. 139, no. 3, pp. 1455-1466, 2016.

Eighth International Meeting

on

Wind Turbine Noise

Lisbon, Portugal, 12th-14th June 2019

ABSTRACT ONLY

Self-reported health in the vicinity of five wind power areas in Finland

Authors: Anu Turunen (National Institute for Health and Welfare), Pekka Tiittanen, Tarja Yli-Tuomi, Pekka Taimisto, Timo Lanki.

Introduction

Public discussion regarding the potential health effects of wind turbine infrasound has emerged in many countries but scientific evidence supporting an association is lacking. The aim of this study was to assess the association between exposure to wind turbine sound and the prevalence of self-reported symptoms, diseases and medications.

Materials and methods

A questionnaire study was conducted in the vicinity of five wind power areas (3–16 turbines with nominal power of 2.4–3.3 MW) in Finland in 2015–2016. Exposure to wind turbine sound was determined by the distance between the closest turbine and the respondent's home (<2.5, 2.5–5, and >5–10 km) and modelled sound pressure level outside the home (<30 dB, 30–32 dB, >32–34 dB, >34 dB). The questionnaire contained a large array of symptoms (headache, nausea, dizziness, tinnitus, ear fullness, arrhythmia, fatigue, difficulties in falling asleep, waking up too early, anxiety, stress), diseases (hypertension, heart insufficiency, and diabetes), and medications (analgesics for headache, joint/muscle pain or other pain, and medication for sleep disturbance, anxiety, depression, or hypertension). Logistic regression analyses were adjusted for age, sex, marital status, education, work situation, smoking, alcohol consumption, physical activity, BMI, and hearing problems.

Results

The response rate was 50 % (n=1,411). The shortest distance between the residence and turbine was 680 m, and the highest modelled sound pressure level outside the residence was 43 dB. The prevalence of symptoms, diseases and medications was essentially the same in all distance and sound pressure level categories, and no statistically significant associations were observed in the adjusted models.

Conclusion

There was no consistent association between the residential proximity of wind turbines and self-reported health problems. The results do not support the hypothesis that wind turbine sound or infrasound could cause the proposed health problems.

INCE*EUROPE*

8th International Conference on Wind Turbine Noise Lisbon – 12th to 14th June 2019

Testing the human response to wind turbine emissions

Peggy Nelson, University of Minnesota, Speech-Language-Hearing Sciences: nelso477@umn.edu

Andrew Bryne, University of Minnesota, Multi-Sensory Perception Lab: byrn0050@umn.edu

Matthew Waggenspack, University of Minnesota, Speech-Language-Hearing Sciences: wagge008@umn.edu

Matthew Lueker, University of Minnesota, St. Anthony Falls Lab: luek0004@umn.edu

Christopher Feist, University of Minnesota, St. Anthony Falls Lab, feist020@umn.edu

Bill Herb, University of Minnesota, St. Anthony Falls Lab: herb0003@umn.edu

Jeff Marr, University of Minnesota, St. Anthony Falls Lab: marrx003@umn.edu

Summary

Our human response laboratory tests participants for their detection, rating, and postural sway during laboratory exposure to wind-turbine generated emissions. Testing of ~70 adults indicated that these participants experienced few symptoms from re-created wind turbine audible (modulated) sound and infrasound emissions. The sample included subjects ages 21-73 years who attended to audible and infrasound signals generated from a wind turbine, recorded at 300 meters and re-created in a laboratory.

During Phase I, stimuli consisted of thirty second \ modulated and unmodulated audible turbine sounds presented at 50 dB SPL, as well as natural and peak-enhanced turbine infrasound at an overall level of approximately 85 dB SPL with peaks up to 100 dB SPL. Infrasound presented was ~40 dB higher than that recorded in the field. This was required as ambient infrasound levels in the lab were higher than wind turbine generated infrasound. Participants were tested for their postural stability, detection, and ratings of audible and infrasound emissions randomly presented in thirty second exposure intervals in the laboratory. Very few and minor adverse effects were noted, mostly ear fullness or pressure. Healthy participants showed no evidence of any change in postural sway in the presence of infrasound for the group tested.

Phase II of testing extended the duration of infrasound to seconds and excluded amplitude modulation to maintain total testing duration. The infrasound used in Phase II was more impulsive than in Phase I yielding clear blade passing and harmonic frequencies in the spectra while maintaining the same presentation level peaks of ~ 100 dB SPL.

Testing of participants who have symptoms of dizziness/imbalance as reported to their ENT specialist is underway. Results from postural sway remain inconclusive; however, these participants report greater symptoms following exposure and a few are reliable detectors of infrasound. Some negative symptoms have been reported even in the absence of infrasound. Testing continues with participants remaining in a seated condition, focusing on detection, sound quality judgments, and pre- and post-exposure symptoms.

Project funding provided by customers of Xcel Energy through a grant from the Renewable Development Fund.

1. Introduction

The noise produced by wind turbines is a major obstacle for wind energy projects (Oerlemans, Sijtsma, & Lopez, 2007). However, developers and regulators frequently cite that farms are within existing noise limits. A potential discrepancy between these two groups is that wind turbines emit noise that is not covered by typical regulation such as infrasound or amplitude modulation. Typical regulations only cover frequencies in the 20 to 20,000 Hz range. Infrasound is noise below 20 Hz that is generally not sensed through the auditory system. (Moller & Pedersen, 2004). Amplitude modulation (AM) is the sinusoidal modulation of the sound level (amplitude) of an audible carrier frequency. Neither of these types of noise are covered in typical regulations. In this study, participants are subjected to infrasound and amplitude modulation which is recorded and modified from a single turbine site. Balance (sway) is used as a quantitative cochlear or vestibular response to these types of noise along with detection, noise ratings, and symptom surveys.

2. Experimental Setup

Each individual was tested at the Center for Applied and Translational Sensory Science (CATSS) lab at the University of Minnesota with a testing protocol approved by the University of Minnesota Institutional Review Board. The testing room was a 6 foot wide by 15 foot long by 8 foot tall room (Figure 1) with reproduced/modified audible and infrasound recordings obtained from field measurements. The field measurements used in the current study were recorded at the University of Minnesota Eolos Wind Research Station. This site has which has a single 2.5 MW Clipper Liberty Turbine which is primarily used for wind energy research.

Postural stability and sway were measured by having individuals stand on an AMTI AccuSway Optimized™ balance forceplate which measures left-right and front-back sway continuously.

Infrasound stimuli were generated using an Eminent Technologies© rotary subwoofer with a frequency range of 0.01 to 30 Hz. A rotary woofer is a unique subwoofer which utilizes a fan with pitch controlled blades to create the infrasonic pressure waves. Audible stimuli were played through a custom subwoofer with a frequency range of 50-800 Hz simulating the audible turbine noise.

Audible and infrasound signals were recorded from the field site. They are reproduced/modified in the CATSS lab and noise levels are calibrated using an infrasound microphone (Brüel and

Kjær Type 4193 with UC0211 adapter), Infiltec microbarometer, and Brüel and Kjær 2250 sound level meter in the testing room.

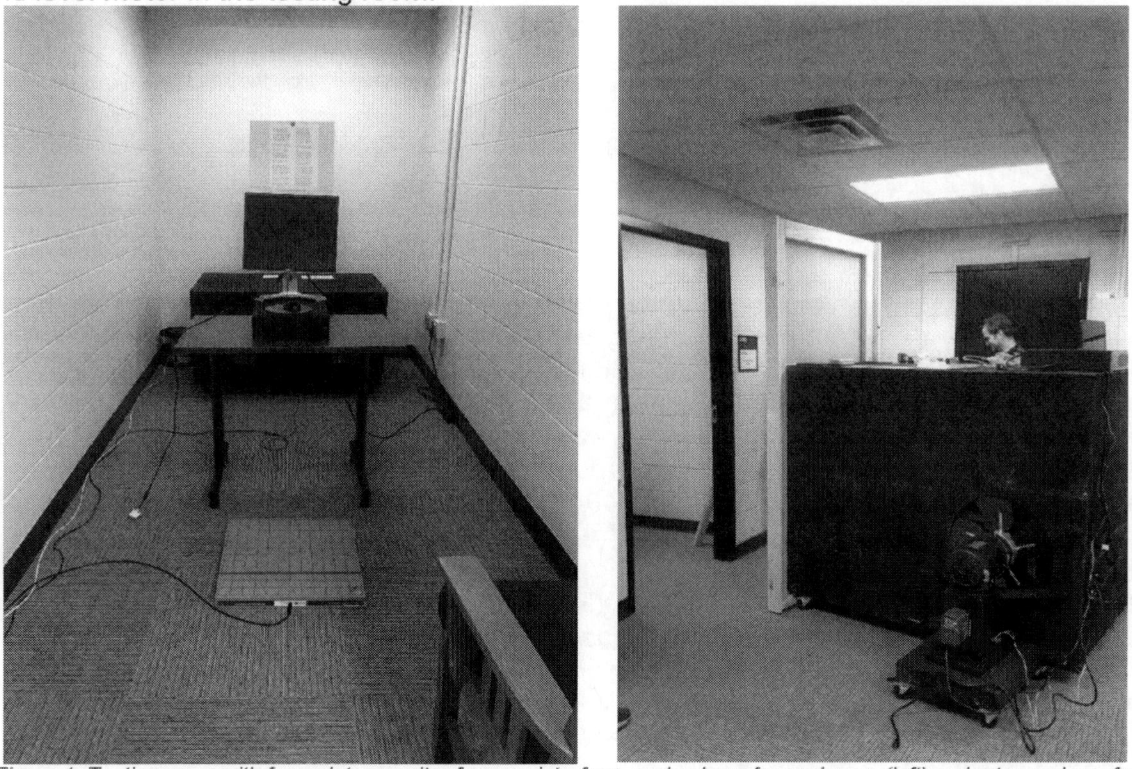

Figure 1: Testing room with forceplate, monitor for user interface, and subwoofer enclosure (left) and rotary subwoofer enclosure used to generate stimulus (right). Enclosure was pushed up to door frame after participant enters testing room.

2.1 Testing Protocol

A written general health survey was collected from each subject prior to testing and the generalized testing procedure is outlined below.

1. The subject/monitor/table are in the center of room, with the subject standing on the forceplate and facing the subwoofer with the rotary woofer to their back.

2. Testing
    a. Pre-Test: One set of eyes open and eyes closed conditions, random order.
    b. Main Experiment:
        i. Each stimulus is randomly played with the subject's eyes either closed or shut, the later order also randomly selected
        ii. Repeated for designated repetitions
        iii. Mid-session break halfway through main experiment
    c. Post-Test: one set of eyes open and eyes closed conditions, random order.

    The subject initiates each individual stimulus by clicking a button on the display.

    Subject is given the one of following auditory instructions through subwoofer.
    - "Please close your eyes."
    - "Keep your eyes open."

    The subject is asked to respond to the following questions after each stimulus:
    1. "Did you sense anything besides the sound?"
       "Yes" or "No" buttons for possible responses.
    2. "How would you rate the condition?"

Slider: "Very Negatively" (0) to "Neutral" (0.5) to "Very Positively" (1)

After testing, a post experiment survey was given on symptoms typical of wind turbine noise and the following scale for each.

- 0 = "not at all"
- 1 = "mild"
- 2 = "moderate"
- 3 = "severe"

3. Testing Overview

The original testing plan (Phase I) for human response testing involved a target of ~100 healthy individuals subjected to a combination of amplitude modulated audible and infrasound noise files for a short duration in a controlled laboratory environment. Three repetitions of nine, 30 second combinations of amplitude modulation and infrasound were presented to the individual during testing. Total recorded response time was 13.5 minutes excluding pre/post-test baseline measurements. This excludes time for user input after each condition and stimulus volume ramping. Total testing duration was of 30 to 40 minutes to prevent contamination of results due to fatigue for a standing test.

However, the testing plan was modified twice (Phase II and Phase III) over the course of testing due to lack of human response to stimuli. Phase II investigated longer duration infrasound with a more impulsive infrasound described below and Phase III investigated symptomatic individuals with the same stimulus as Phase II. Symptomatic individuals were told of connection to wind turbine noise for ethical reasons. An overview of the testing phases are summarized in Table 1.

The reasoning for the deviation for the testing plan to Phase II was the lack of human response to the stimulus, updated analysis of field data that revealed impulsive nature of infrasound, and feedback from colleagues about short stimulus duration in testing. To maintain the same approximate total testing time, the amplitude modulation was dropped and the updated "impulsive" infrasound duration was doubled to 60 seconds. Furthermore, amplitude modulation is more likely related to annoyance. Since there was not a significant negative rating to AM conditions from the Phase I population, it seemed unlikely that extending the short term durations to approximately one minute would cause increased response.

The reasoning for Phase III was in response to feedback from colleagues that the general population may contain a very small percentage of people that are affected by infrasound. Also, collaborators in the University of Minnesota Department of Otolaryngology have symptomatic patients who could be invited to testing.

Table 1: Overview of Human Response Testing Phases

	Number of Participants	Number of Stimulus Types	Repetitions	Individual Stimulus Duration (sec)	Measured Response Time (min)	Participants
Phase I	53	9, AM and Infrasound	3 (2 eyes closed and 1 eyes open)	30	13.5 min	Healthy, unaware
Phase II	15	2, Impulsive Infrasound (2 times for each repetition)	4 (2 eyes closed and 2 eyes open)	60	16 min	Healthy, unaware
Phase III	3	2, Impulsive Infrasound (2 times for each repetition)	4 (2 eyes closed and 2 eyes open)	60	16 min	Symptomatic, Aware

3.1 Stimulus

Phase I stimulus are presented in this section. No excessive amplitude modulation had been captured at the field site by the start of human response testing. To resolve this, audible range recordings from the field data were modified to create AM as indicated below and shown in Figure 2. Infrasound frequency recordings were also modified to exaggerate blade passing frequency and five harmonics by adding pure tones at these frequencies with random phases. The raw recordings from the field site were also played as part of the stimulus. "Flat" and "IRec" indicates the unaltered audible and infrasound recorded, respectively. Both recordings were measured 300 meters downwind of the Eolos single turbine site on June 6, 2016.

Background infrasound levels in the testing room were higher than turbine generated infrasound recorded at the field site as shown in Figure 5. To overcome this, the overall amplitude of the reproduced infrasound was increased so that the second (2.4 Hz) and third (3.2 Hz) harmonics were presented at ~95 dB. This level was ~ 10 dB higher than the noise floor which relates to ~40 dB higher than the infrasound recorded in the field in the frequency range of the first five harmonics.

Stimulus was recorded in the testing room during experimental design to compare actual infrasound levels in the testing room to target levels. This is shown in Figure 3.

- Audible Sound (Subwoofer)
    - "White"
        - (Gaussian) white noise.
        - Presentation Level: 40 dBA
    - "Flat"
        - Audible range recording used.
        - Band-pass filtered at 50 and 4000 Hz (3rd order Butterworth).
        - Presentation Level: 50 dBA
    - "AM"
        - "Flat" stimuli with 0.8-Hz modulation imposed on it with a depth of 10 dB (peak-to-trough dB difference).
        - Presentation Level: 50 dBA (measured at the modulation's peak)
- Infrasound (Rotary Woofer)
    - "IOff"
        - Rotary Woofer (fan) on, but no signal.
    - "IRec"
        - Infrasonic range recording used.
        - Low-pass filtered at 50 Hz (3rd order Butterworth).
    - "IEnh"
        - "IRec" stimuli's spectra smoothed with 100-(FFT)-point averaging window.
        - Random phases given to all spectral components.
        - Pure tones added to create a 0.8-Hz infrasound signal (with five harmonics).
            - Frequencies = [0.8, 1.6, 2.4, 3.2, 4.0, 4.8] Hz
        - Low-pass filtered at 50 Hz (3rd order Butterworth).

Table 2: Phase I Stimulus Conditions

IOff - White	IRec - White	IEnh - White
IOff - Flat	IRec - Flat	IEnh - Flat
IOff - AM	IRec - AM	IEnh - AM

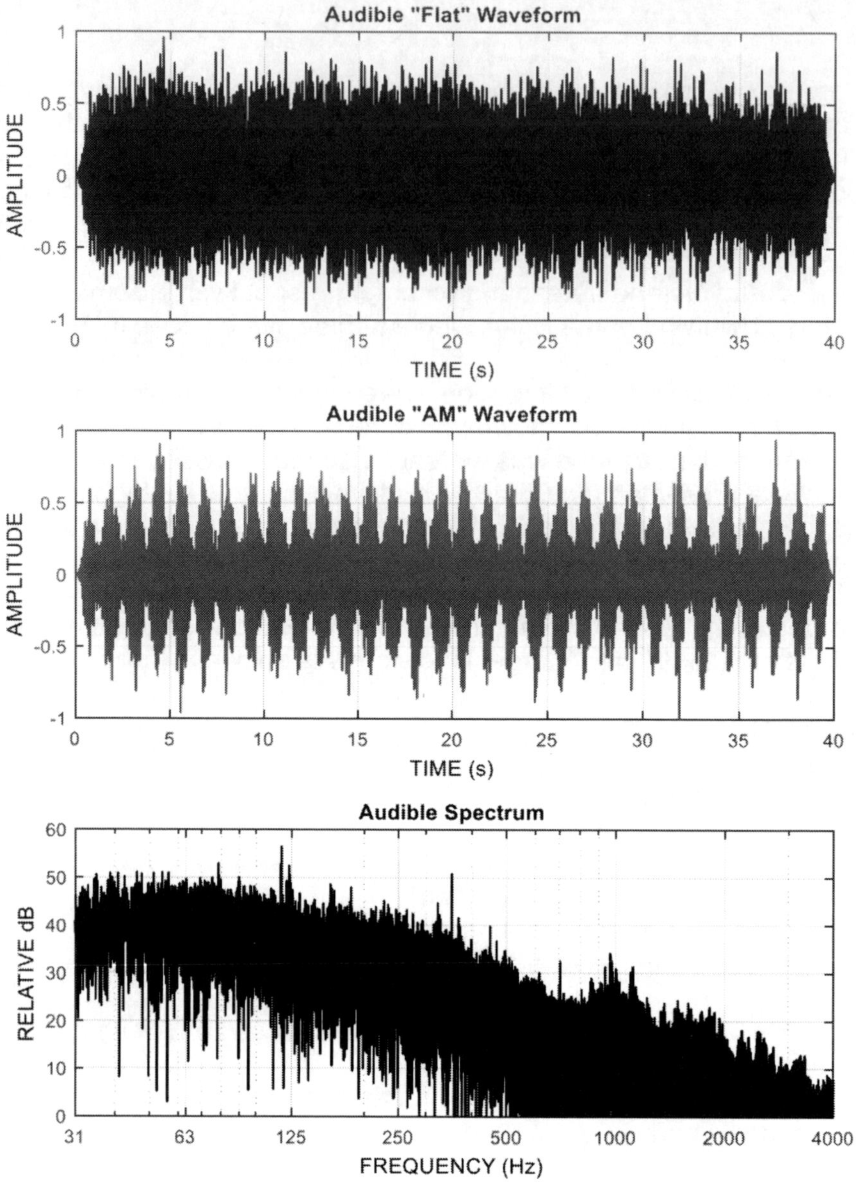

Figure 2: Phase I Field recorded noise audible stimulus and modified AM stimulus prior to amplification

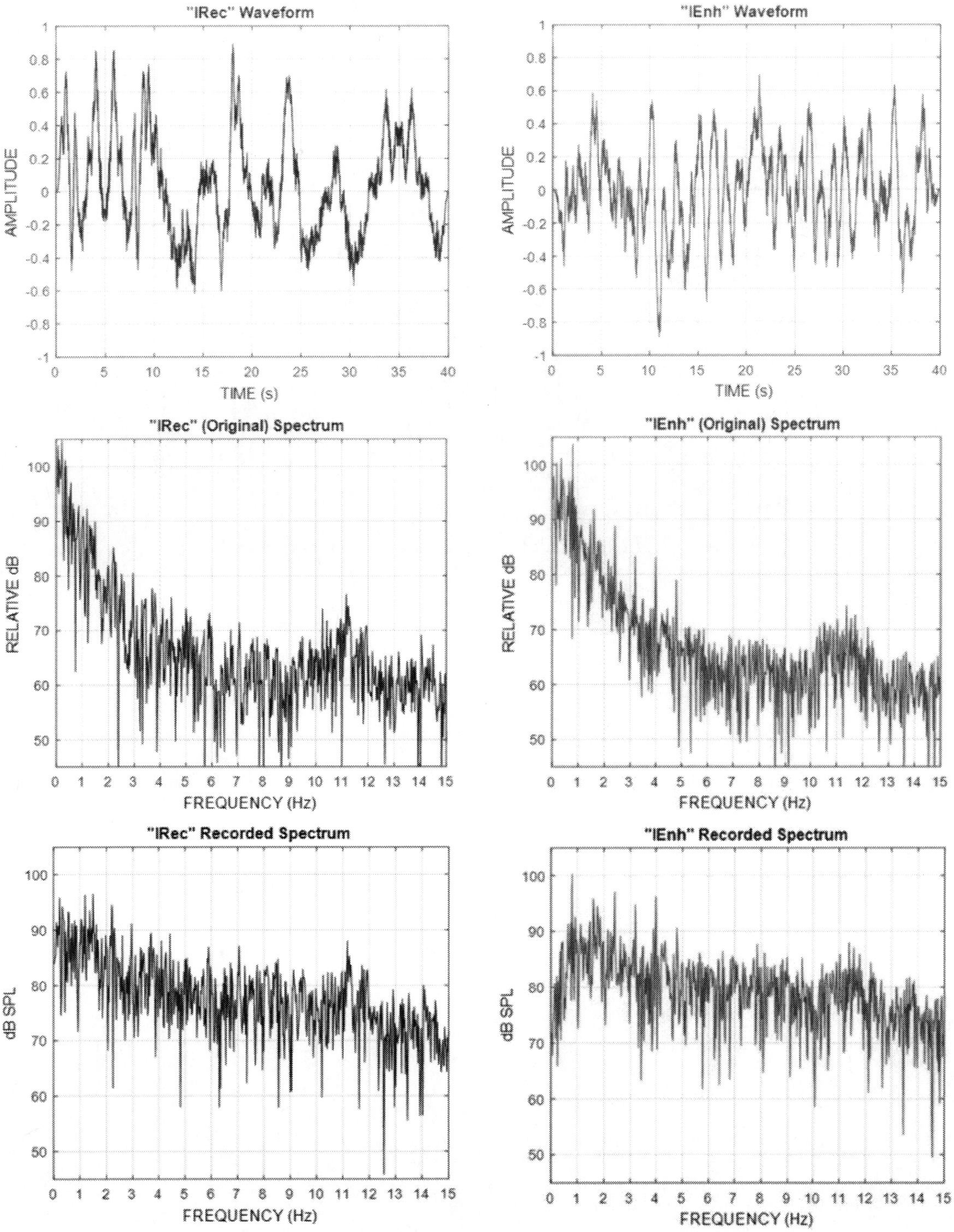

Figure 3:Phase I waveforms and spectra for the original (IRec) and enhanced infrasound stimulus (IEnh)

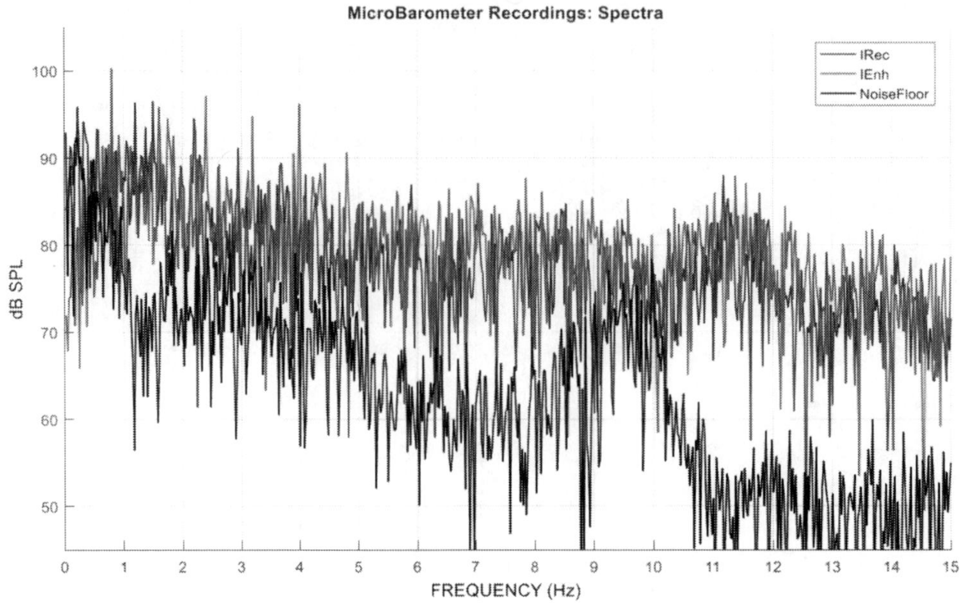

Figure 4: Phase I Spectra of testing room infrasound recordings

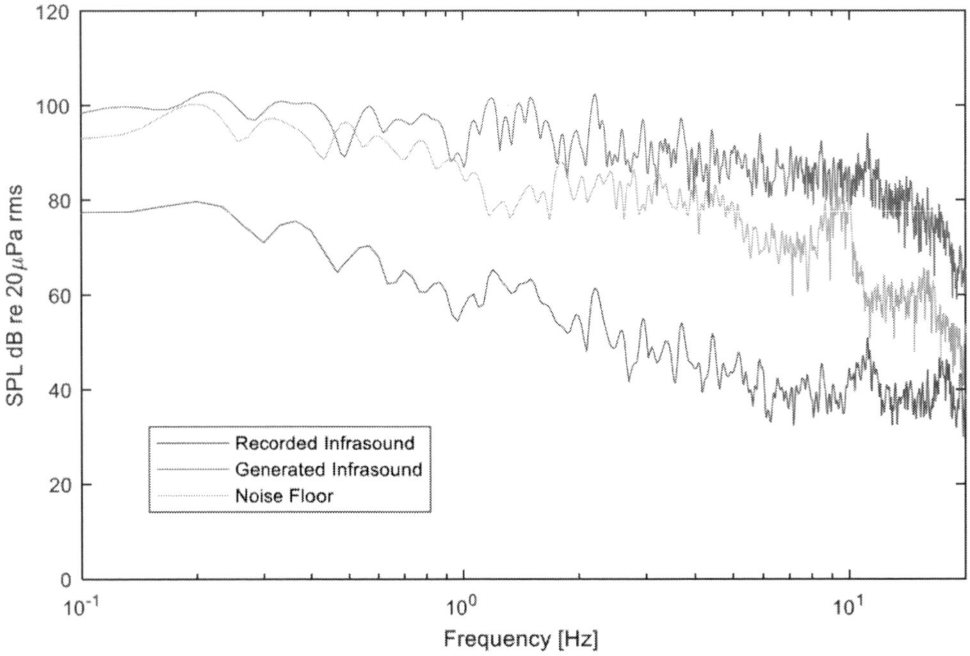

Figure 5: Spectra of infrasound for a 30 s segment of field measurement data (blue), the noise floor in the lab (yellow), and the reproduced infrasound (red)

After 53 participants were tested, the human response results were reviewed. There was a general lack of response to both AM and infrasound. Feedback from conferences indicated that the duration of stimulus may not be long enough. In addition, updated analysis of wind turbine infrasound gave a clear indication of the impulsive nature of wind turbine infrasound emissions as opposed to typical broadband background infrasound.

This updated analysis was based on infrasound collected from the Eolos Wind Research Station on June 10, 2012. For this dataset, turbine hub speed was measured independently and synced with the noise measurements via a laser tachometer. This allowed resampling of the infrasound to a fixed number of samples per rotation and ensemble averaging in order to remove destructive averaging caused by slight deviations from the design rotor speed of 15.5 rpm. This analysis only used blade rotations ranging from 15 to 16 rpm, or blade passing frequencies of 0.75 to 0.8 Hz. This filtering yielded a dataset of 1328 rotations to ensemble average. The time series and spectrum used for Phase II of human response testing is shown in Figure 7. Compared to the infrasound stimulus played in Phase I, which used random phases for the artificially added blade passing frequency and harmonic tones, the harmonics are much clearer in the spectrum although the presentation level of harmonics was same as in Phase I.

During Phase I testing, the rotary subwoofer was found to emit some extraneous audible noise in addition to the infrasound. To mitigate this artifact, a "masker" audio signal was created by recording audible noise during "IPulse" from the rotary woofer and adding white noise. This masker signal was used for all conditions. The duration of the stimulus was also extended to 60 seconds. There was no AM stimulus for Phase II testing.

- Audible Sound (Subwoofer)
  - "White"
    - (Gaussian) white noise.
    - Presentation Level: 50 dBA
    - Used only for the Pre- and Post-Test conditions.
  - "Masker"
    - Composite audio from:
      - 'RotaryWoofer.wav' (Microphone recording of the rotary woofer's fan while playing "IPulse", for fan pulsation.)
      - 60-dBA (Gaussian) white noise.
    - Band-pass filtered at 50 and 4000 Hz (3rd order Butterworth).
    - Presentation Level: 75 dBA
    - Used on all "Main Experiment" conditions.
- Infrasound (Rotary Woofer)
  - "IOff"
    - Rotary Woofer (fan) on, but no signal.
  - "IPulse"
    - Single rotation ensemble average of a 1328 resampled rotations [0.75<BPF<0.80]
    - The 3.9-s stimulus was resampled to 50 kHz and repeated to create a 70-s total duration.
    - Low-pass filtered at 50 Hz (3rd order Butterworth).

Table 3: Phase I and II Stimuli Conditions

IOff (+Masker)	IPulse (+Masker)

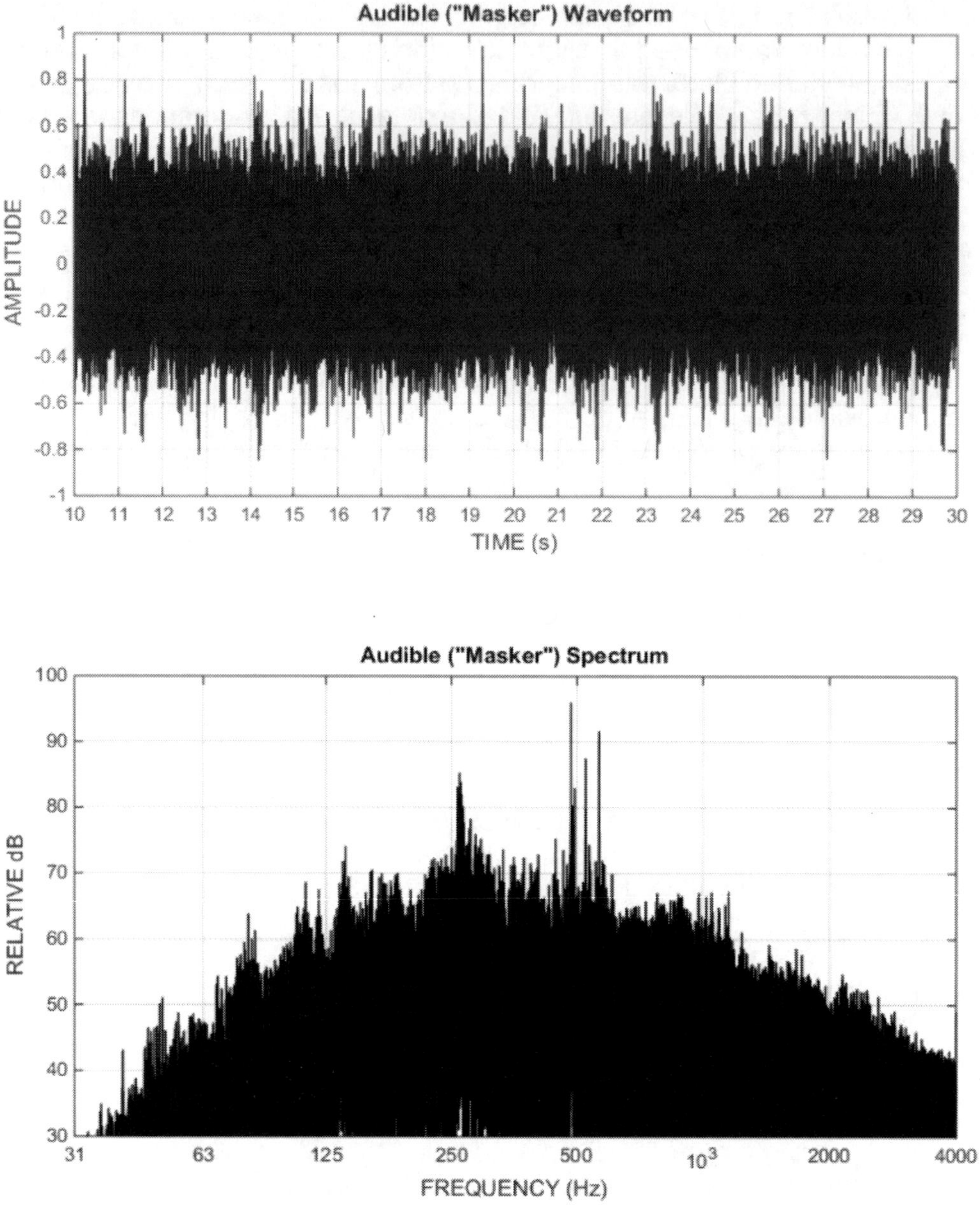

Figure 6: Audible "Masker" waveform and spectra prior to amplification

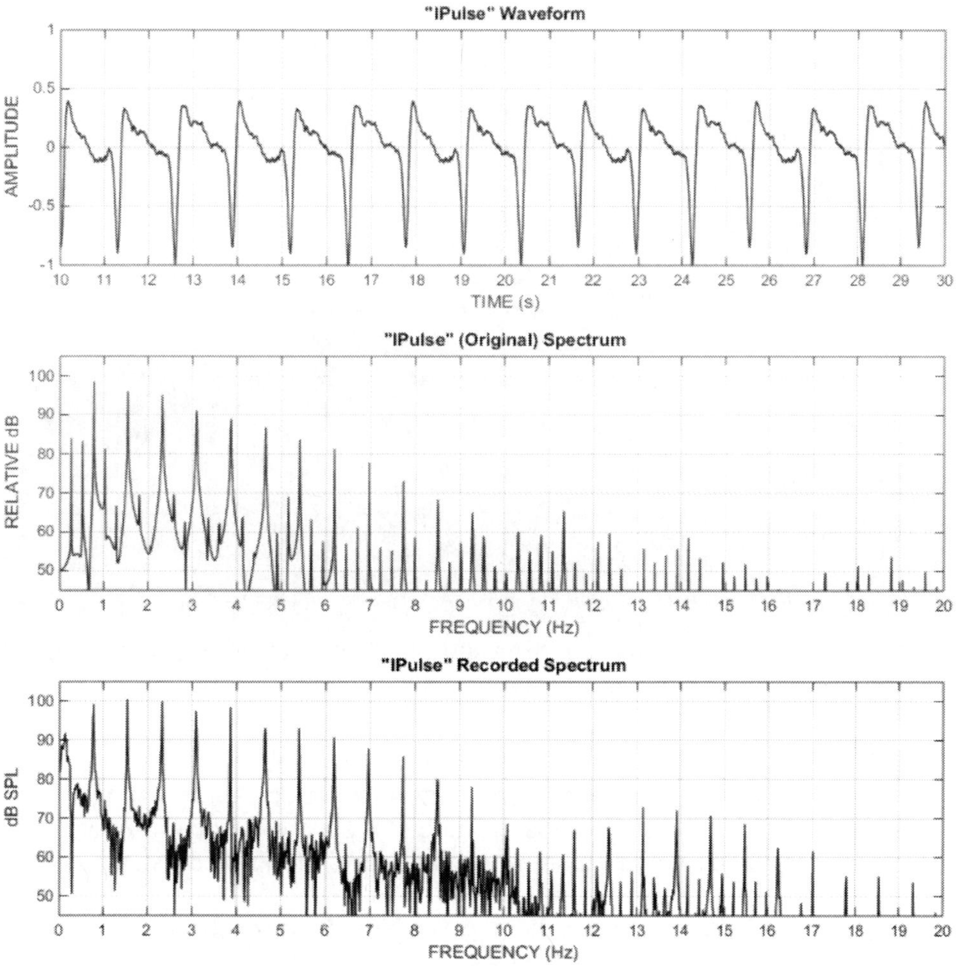

Figure 7: Time series and spectra of original and recorded "IPulse" signal from testing room. This signal was developed by resampling to a fixed number of samples per rotation and ensemble averaging.

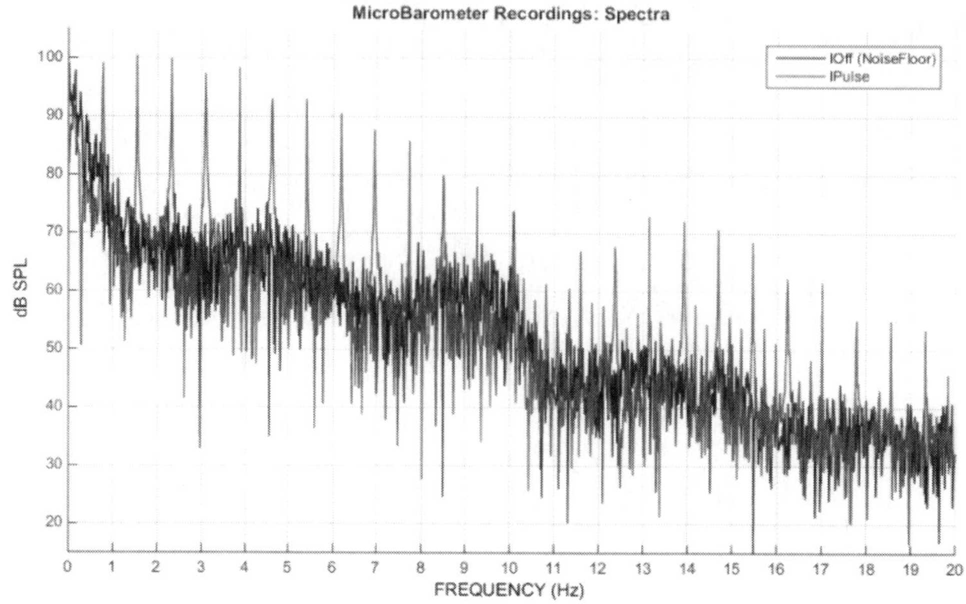

Figure 8: Phase II spectra of testing room infrasound recordings

Phase III testing used the same stimulus as Phase II, but with symptomatic participants who were aware research was related to wind turbine noise.

3.2 Response Analysis

The analysis was primarily performed on population statistics from each testing Phase. The primary metric was from the forceplate comparisons between pre and post testing for a baseline, AM stimulus on versus white noise, and infrasound stimulus on versus off.

Response to the question, *Did you sense anything besides the sound?*, after each individual stimulus was used to quantify the sensitivity index, d', with higher values indicating the signal could be more readily detected.

Response to the question, *How would you rate the condition?,* after each individual stimulus was plotted to give a rating value of the particular stimulus with 0.5 (neutral) being the default.

The post exposure symptoms questionnaire was also reviewed. This questionnaire was given after the testing was over.

4. Results

4.1 Phase I

The results of the sway for each stimulus/eye condition are summarized in Figure 9. It should be noted that there were twice as many eyes closed than eye open conditions which may affect error bars. Figure 10 collapses data by showing the percentage difference of sway during wind turbine noise stimulus (AM or infrasound) from associated references (white noise or infrasound off). The percentage difference combines all eye conditions and X/Y sway for the comparison. The largest percentage difference is post-test to pre-test. There was nothing remarkable about the other percentage differences.

Figure 11 shows the detection and rating for each individual stimulus. The Phase I population correctly detected infrasound for roughly half of the conditions. This may have been due to some an audible artifact of the rotary subwoofer operation. In general, rating values were neutral.

The post exposure questionnaire showed the most prevalent symptom was fullness of head/ears. There were no "severe" (3) ratings given. Fatigue symptoms were also noted but may have been related to the duration of the standing test.

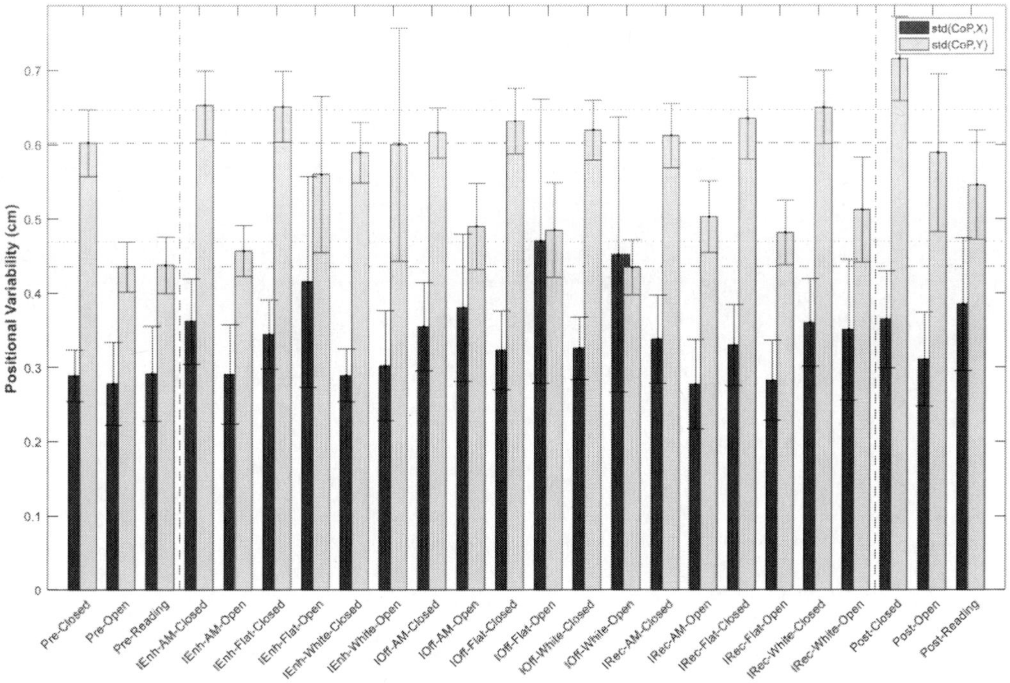

Figure 9: Phase I sway population statistics for each stimulus type/eye condition.

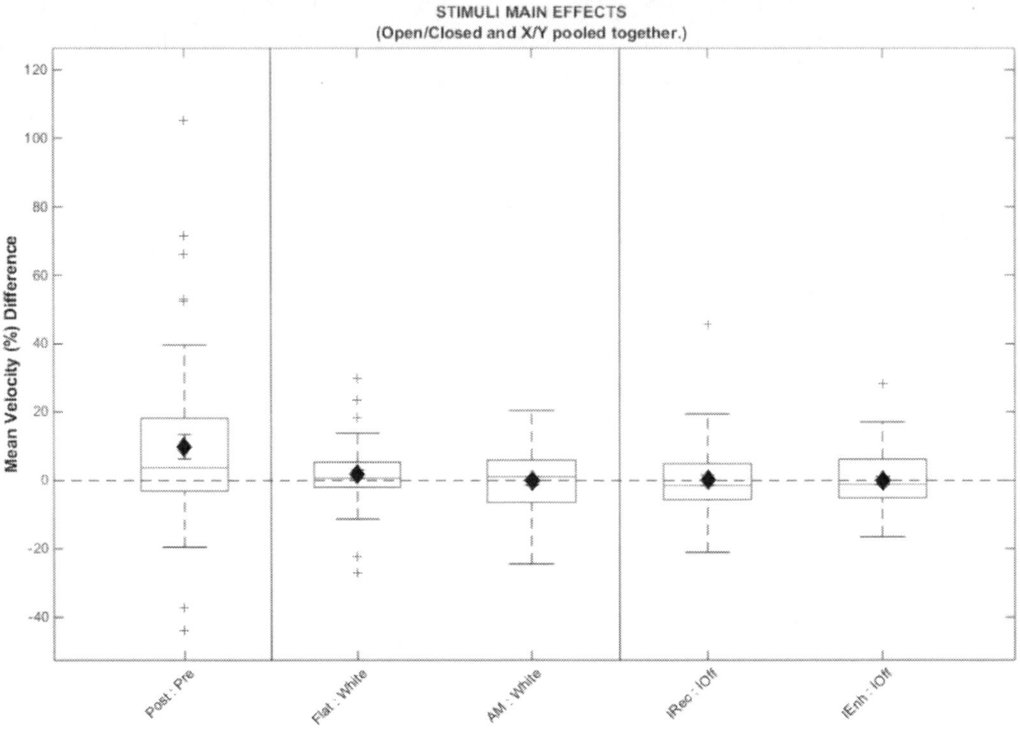

Figure 10: Phase I sway statistics of percentage differences from wind turbine noise stimulus to references

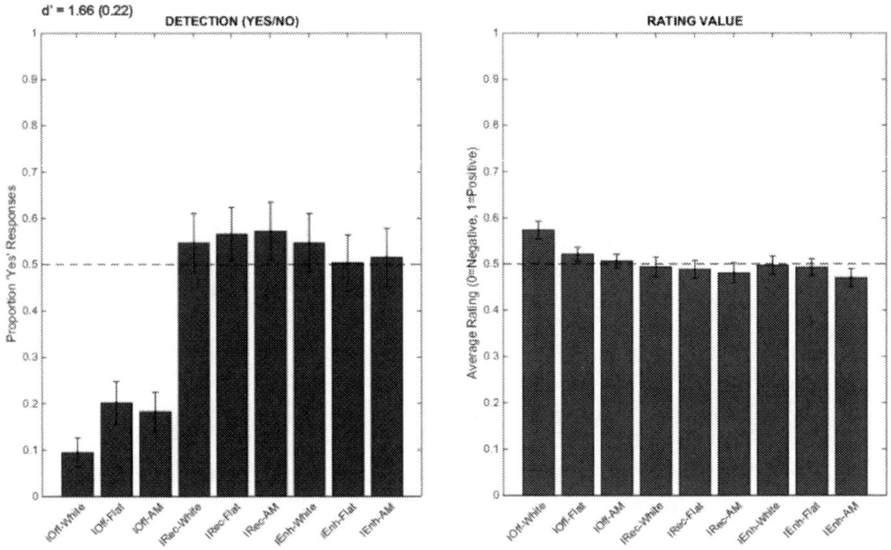

Figure 11: Phase I detection and rating for each stimulus

4.2 Phase II

The results of Phase II was even less remarkable than Phase I results with no significant difference in sway comparing the impulsive infrasound to the infrasound off condition. Figure 12 shows population statistics are nearly identical for both the eyes open and eyes closed conditions regardless of if the impulsive infrasound stimulus was on or off. There were an equal number of eyes closed and open for this Phase.

Figure 14 shows the detection and rating for each individual stimulus. The Phase II population correctly detected for less than half of the conditions. The addition of the "Masker" signal likely helped mitigate audible artifacts during rotary woofer operation. In general, rating values were neutral and nearly identical for both the infrasound off and on conditions.

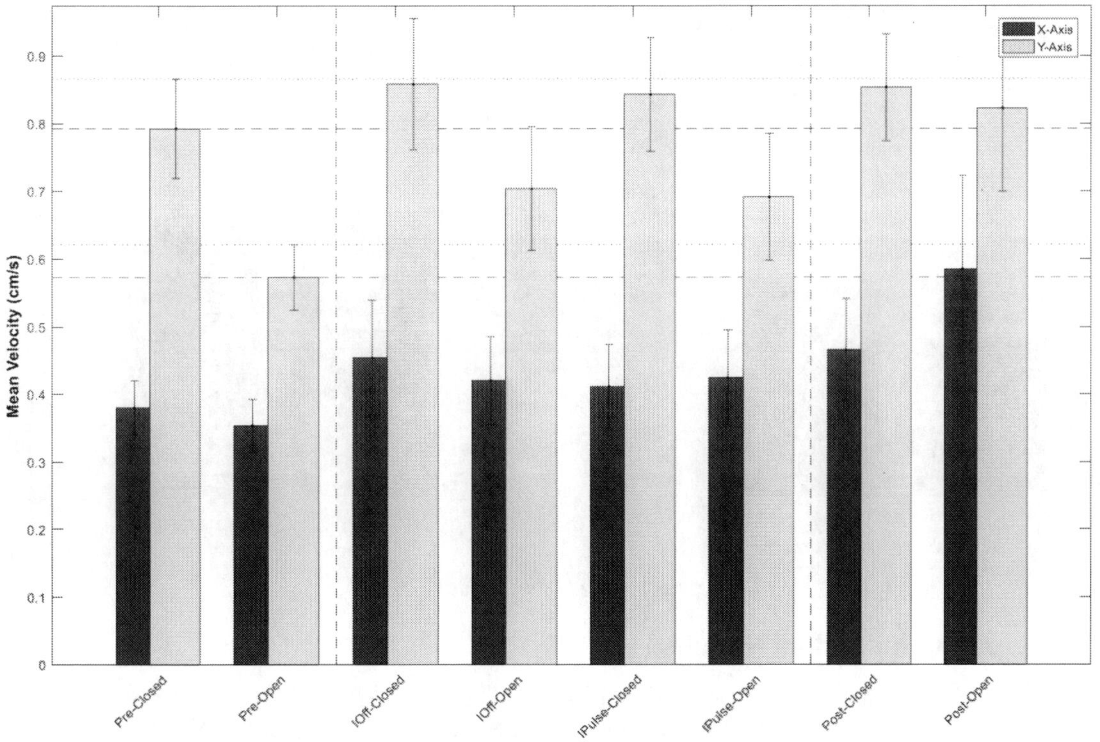

Figure 12: Phase II sway population statistics for each stimulus type/eye condition.

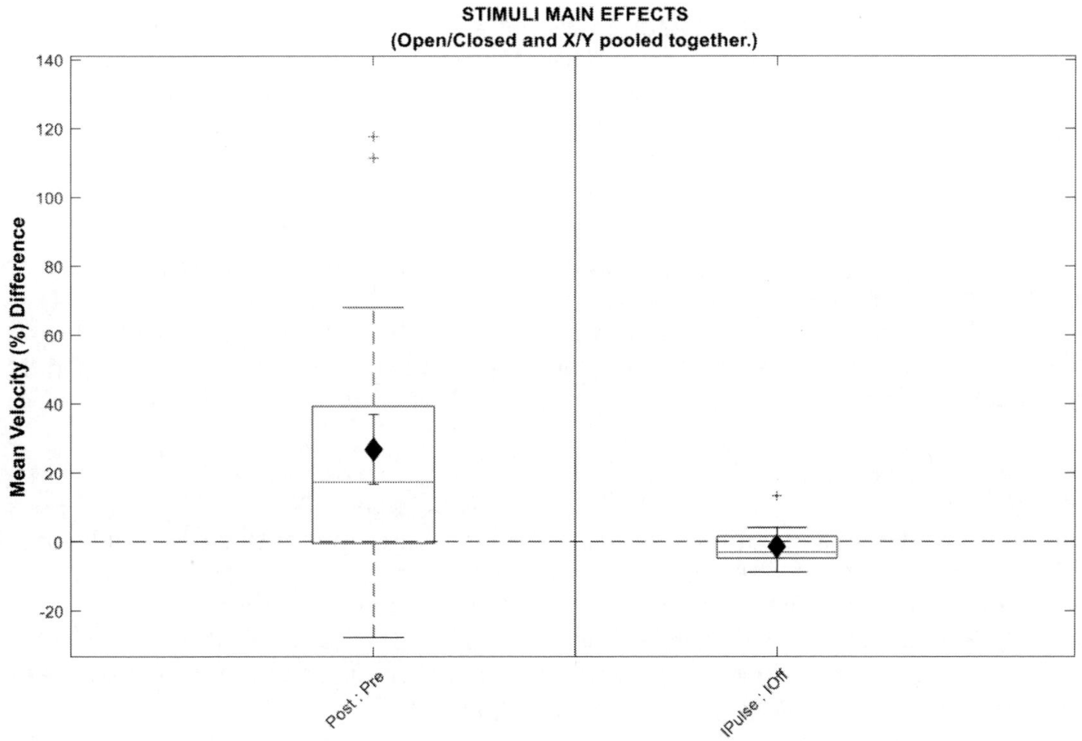

Figure 13: Phase II sway statistics of percentage differences from wind turbine noise stimulus to references

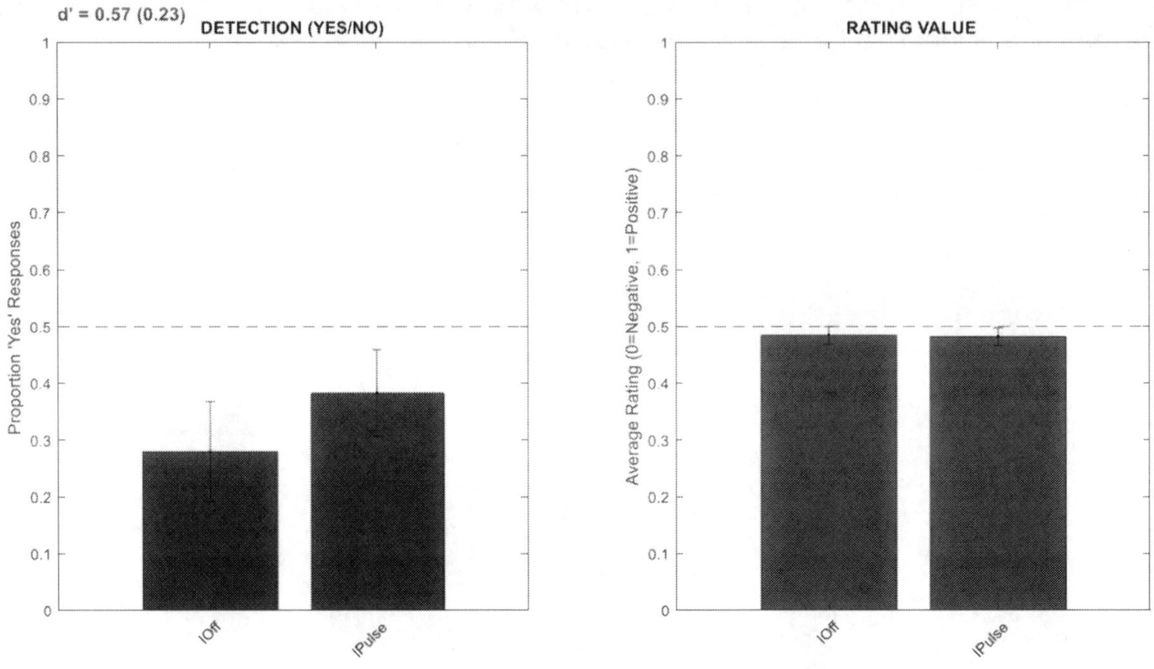

Figure 14: Phase 2 detection and rating for each stimulus

4.3 Outliers/Interesting Individual Results

The outliers in the two phases of testing have yet to be fully investigated. Correlations between sway, detection, rating value, the pre-test general heath survey and the post-test symptom survey should be performed. Some noticeable preliminary individual results are described here. There have only been three symptomatic people tested to date.

- A few participants could detect infrasound with $d' > 2$, but they reported no symptoms and no remarkable sway results. The audible "Masker" was used in Phase II to mitigate any other cues.
- Three out of 54 Phase I individual rated the stimuli with infrasound as slightly negative on sound quality ratings. One of those people was older, took the longest, and reported fatigue. The second reported not liking "airplane-like" sounds and had very high detection ability. The third person showed nothing notable.
- One younger individual with no symptoms and no detection swayed noticeably more on infrasound than other trials
- One symptomatic individual suffered from weakness in the knees and required a break from testing. When testing was restarted, weakness of knees was again reported in the absence of infrasound.

5. Conclusions

From the human response testing completed and analysed, there appear to be no significant relationship between sway or unpleasantness ratings for short stimulus durations of wind turbine noise involving AM or infrasound. After Phase 1, impulsive infrasound was used, duration was doubled, and an audible "Masker" signal was used. The result show even less response to that stimulus. The infrasound presented during human response testing was ~40 dB higher than that recorded in the field in the frequency range of 1 to 5 Hz.

Postural sway did not correlate well with reported symptoms and unpleasantness or detection ratings and may not be a good indicator of symptoms due to wind turbine noise. Longer exposure times with seated participants and relying on self-reporting may be a better approach.

Other general results are listed below.

- Post-exposure symptoms were rated as mild, with a predominance of discomfort, fatigue and head/ear fullness
- No one reported motion sickness
- No significant changes in postural sway in the presence of infrasound or AM sound; only one participant showed greater sway in infrasound conditions, but had no symptoms or negative feedback
- A minority of individuals reliably detected infrasound signals while simultaneously rating the sounds as neutral

Some potential reasons we did not encounter significant human response are listed below.

- Infrasound at levels presented does not cause illness/symptoms
- Illness due to infrasound may have very low prevalence in the healthy population and our sample may not be representative.
- Stimulus duration is short and not representative of the typical length of environmental exposures.
- Testing building and surrounding neighborhood contains background infrasound. Perhaps it is more notable in open areas.
- Our measures may be insensitive to subtle inner ear effects

Future work includes continued testing of patients who may be more susceptible to infrasound effects. Also, electrophysiologic testing including electrocochleography and vestibular evoked myogenic potentials may be utilized in future work to directly measure auditory system response.

References

Oerlemans, S., Sijtsma, P., & López, B. M. (2007). *Location and quantification of noise sources on a wind turbine*. Journal of sound and vibration, 299(4-5), 869-883.

Salt, A. N., & Hullar, T. E. (2010). *Responses of the ear to low frequency sounds, infrasound and wind turbines*. Hearing Research.

Moller, H., & Pedersen, C. S. (2004). Hearing at low and infrasonic frequencies. *Noise and Health, 6*(23), 37-57.

Salt, A., & Lichtenhan, J. (2014). How Does Wind Turbine Noise Affect People? *Acoustics Today*, pp. 20-28.

INCE*EUROPE*

**8th International Conference
on
Wind Turbine Noise
Lisbon – 12th to 14th June 2019**

Session I

Impact on People - 2

Organised by INCE-Europe
mail@inceeurope.org
www.inceeurope.org

Supported by Portuguese Acoustical Society

spa Sociedade Portuguesa de Acústica

INCE*EUROPE*

8th International Conference
on
Wind Turbine Noise
Lisbon – 12th to 14th June 2019

PIBE : a new French project for predicting the impact of wind turbine noise

David Ecotière, Cerema, Ifsttar, UMRAE, 11 rue Jean Mentelin, 67035 Strasbourg, France. david.ecotiere@cerema.fr
Benoit Gauvreau, Ifsttar Cerema UMRAE, Route de Bouaye, CS 4, 44344 Bouguenais Cedex, France. benoit.gauvreau@ifsttar.fr
Benjamin Cotté, ENSTA Paristech, IMSIA, Boulevard des Maréchaux, 91120 Palaiseau, France. Benjamin.cotte@ensta-paristech.fr
Michel Roger, Ecole Centrale de Lyon, LMFA, 36, avenue Guy de Collongue, 69134 Ecully Cedex, France. michel.roger@ec-lyon.fr
Isabelle Schmich-Yamane, EDF DTG, avenue de l'Europe, BP41, 38040 Grenoble Cedex 09, France. isabelle.schmich-yamane@edf.fr
Marie Cécile Nessi, EDF Renouvelables, Coeur Défense Tour B, 100, esplanade du général de Gaulle, 92932 Paris La Défense, France. marie-cecile.nessi@edf-en.com

Summary

The PIBE project is a new French research project that aims to improve wind turbine noise prediction methods and explore new solutions to reduce noise. The project brings together experts in aeroacoustics, sound propagation, experimental noise characterization and wind engineering. The research program is structured into 3 working groups (WP). The first aims to study amplitude modulation phenomena and focuses particularly on characterization and modelling of the dynamic stall of the flow around the blades, as well as on the conditions for amplitude modulation generation in the receiver. These phenomena are studied both in wind tunnels and in the vicinity of a wind farm. The second WP focuses on quantifying the uncertainties of noise prediction methods. To achieve this objective, the uncertainties and variabilities of the parameters influencing both the emission and propagation of noise are estimated; secondly, an uncertainty propagation model (combined with advanced and appropriate statistical methods) estimates the overall uncertainty. The last WP focuses on new noise reduction devices, using blades with modified leading and/or trailing edges. The effectiveness of the solutions will be characterized in the wind tunnel, both acoustically and aerodynamically. An estimate of their performance potential at a 1:1 scale is also expected during the project. The project is funded by the The French National Research Agency

1. Introduction

Wind energy is one of the promising energy sources to reach the objective set by the French regulation of increasing renewable energies to about one third of the final energy consumption by 2030. In spite of a strong growth of the wind energy sector these last 10 years, and in spite of a solid potential for development, France has fallen behind on this goal. This may be partly explained by the constraint framework in which wind energy is developing, as well as the opposition of wind farm neighbours who very often mention noise as a potential annoyance. In this context, first French collaborative research project on wind turbine noise, the PIBE project (2019-2023) aims to improve prediction methods for wind turbines noise and to explore new solutions for noise reduction.

2. Organisation of the project

The project is structured in three work packages (Figure 1). The first work package (WP) aims to study the amplitude modulation phenomena, which can be a major source of annoyance when they occur. This axis focuses particularly on understanding and characterizing the dynamic stall of the flow around the blades, as well as the conditions of amplitude modulation generation at the receiver. The second WP focuses on quantifying the variability of noise predictions. The last WP of the project aims to study and propose new noise reducing devices, using blades with modified leading and/or trailing edges.

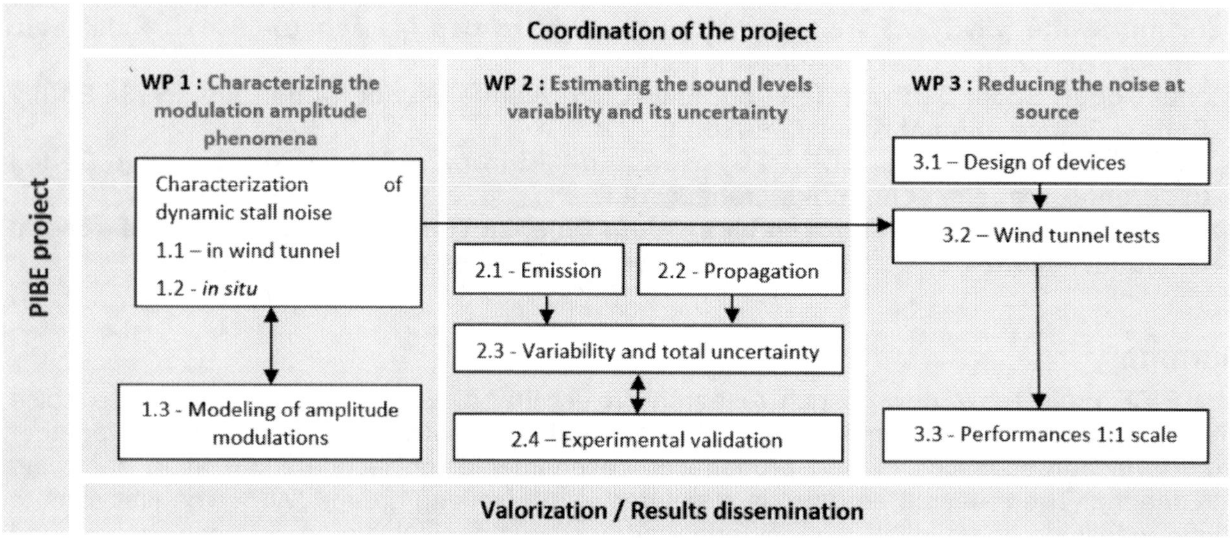

Figure 1: PIBE project organisation

The project is leaded by UMRAE. ENSTA and LMFA are involved in tasks dealing with aeroacoustics, UMRAE and EDF DTG in tasks dealing with experimental characterization of noise and with noise propagation, EDF Renouvelables provides its support in wind turbine engineering.

3. Scientific and technical objectives of the project

3.1 WP1 : Characterizing the amplitude modulation phenomena

Amplitude modulation phenomena, particularly those associated with dynamic stall on wind turbine blades, are a source of intense and intermittent noise that can cause noise annoyance (Lee *et al.*, 2011). WP 1 aims to characterize and understand the relationship between the occurrence of dynamic stalls and the meteorological and operational conditions of the wind turbine on a given site. Three approaches are explored in the project: Characterization of dynamic stall noise in a controlled environment (anechoic wind tunnel measurements), measurement of stall noise in situ and correlation with atmospheric conditions, modelling of wind turbine noise amplitude modulations.

The dynamic stall noise characterization is performed in a controlled environment using an original experimental device, consisting of a pitching airfoil with controlled rotational motion. Low frequency wall pressure measurements using a pressure scanner, and flow visualization around the profile by Particle Image Velocimetry (PIV) is performed. Coupled aerodynamic and acoustic studies is also investigated thanks to wall pressure measurements and acoustic measurements in the far field. In addition to the stall regime, the system is used to study the cyclic variations in trailing edge noise that contribute to amplitude modulation. Several inflow conditions are investigated (velocity, turbulence intensity). This experimental study is to our knowledge the first to study transient (cyclic) aspects such as dynamic stall noise, both aerodynamically and acoustically. These measurements are completed by flow calculations carried out using a CFD software dedicated to finite volume simulations using RANS (Reynolds-Averaged Navies-Stokes) or LES (Large-Eddy Simulation) models for incompressible flows (Archambeau et al., 2004).

In situ acoustic and meteorological measurements will be carried out next to a wind farm in order to identify the situations for which dynamic stall may occur and to characterize the sound levels generated in these situations. Sound spectra and audio signal recordings will be done at several distances, together with meteorological measurements (wind and temperature vertical gradient, turbulence) in order to characterize the acoustic propagation conditions, as well as the aeroacoustic emission conditions at blade level.

A modelling approach of amplitude modulations is also included in the project that consists in developing a time-domain model of wind turbine noise based on the acoustic analogy of Ffowcs Williams-Hawkings. This approach enables to better predict fluctuations in the amplitude of wind turbine noise at the receiver, and to produce sound synthesis that can be used for subsequent perceptual studies. Moreover, it accepts as input data expressions from precise profile theory, experimental results or numerical results. Time-frequency analyses will be conducted to evaluate the potential of a frequency approach with slowly time-varying parameters, based on the same Amiet theory. Predictions will be made from the wall pressure data and the turbulent velocity correlation functions measured or calculated. Indeed, as shown by (Moreau *et al.* 2009) the formulas used for trailing edge noise remain valid when the angle of attack is not too high (partial stall speed regime, noted as "light stall" in Figure 1). In the deep stall regime, however, the location where the pressure fluctuations are created moves away from the trailing edge and it is expected that the model is no longer valid. The simplified model proposed by (Moreau *et al.* 2009) fails in a dynamic regime. The feasibility of a more appropriate model will be studied in the project on the basis of experimental observations and flow simulations.

3.2 Estimating the variability of sound levels and associated uncertainties (WP2)

Wind turbine noise may present some time fluctuations at dwellings located in the vicinity of wind farms, even if the wind speed is stationary. This is often due to the variability of meteorology which may act on noise emission from the blades (see WP1) and on sound propagation between the source(s) and a local resident, as well to temporal evolution of ground acoustic properties. Currently, the acoustic impact studies of a wind farm project do not consider these phenomena, so there is no available information on the uncertainties of predicting the noise levels variability and a scientific and an industrial issue. Solving this problem would enable wind farm developers to estimate more precisely the risk of noise annoyance and to design optimally their wind farms.

To predict noise temporal level fluctuations, we need to estimate the variability related to the influence of the meteorology on the sound emission, in particular on the phenomenon of amplitude modulation. This task is therefore closely linked to WP1 and involve to a specific model that carry out a certain number of calculations, following advanced screening techniques, e.g. Morris, Monte Carlo, FAST, etc. (Saltelli *et al.*, 2008). If necessary, especially for reasons of calculation time, a multidimensional metamodel can be developped.

Predicting the variability of sound levels requires estimation of the variability associated with the influence of meteorology on sound emission, as well as that related to the influence of weather conditions and ground absorption variabilities on sound propagation. Total uncertainty is determined based on different classes of weather-type / ground / source-receiver positions. Today's scientific knowledge and numerical tools now make it possible to quantify the uncertainties coming from a large number of environmental parameters. The general approach in this project is to perform multiple numerical acoustic computations with a MWAPE model that takes into account the effects of ground and micrometeorology on propagation in a inhomogeneous medium (Lihoreau *et al.*, 2006; Cotté and Tian, 2015). The results will highlight the most influential sound propagation parameters, as well as quantify the uncertainties associated with these parameters on sound pressure level prediction.

Finally, an uncertainty propagation method of the input data on the output (sound level) is applied. For this, a sensitivity analysis of the global system to the different input parameters is carried out. This sensitivity analysis and estimate of the propagation of uncertainties throughout the prediction process makes it possible to quantify the relative influence of the parameters in relation to each other. Finally, a global uncertainty database on sound level predictions is obtained; these uncertainties will be related to both emission and propagation, depending on multiple parameters such as the state of the atmosphere (emission and propagation conditions), the nature of the ground (typology and acoustic impedance) and the source-receiver positions.

The same uncertainty propagation method will be adapted and applied to one or more engineering model(s). A parametric calculation tool will be developed within open source codes (www.code-tympan.org and www.noise-planet.org/noisemodelling.html). This will serve a dual purpose: On the one hand, it will improve knowledge of the global uncertainties inherent in engineering simulation methods, which represents a significant advance in the current state of the art; on the other hand, the developed tools will be made available to the open source community, thus contributing to the improvement of existing acoustic engineering practices and tools.

In order to validate the numerical results obtained, a large-scale *in situ* experimental campaign will be carried out on a wind farm. It will combine long-term acoustic measurements, measurements of ground absorption properties (Hess *et al*, 1990) and micrometeorological measurements (meteorological mast, 3D ultrasonic anemometers, wind LIDAR). The acoustic measurement campaign will last 1 year, at several distances from the wind farm. These points will be supplemented by other points during 2 intensive observation periods during the observation year. The database of these measures will be made available to the public *via* an Internet page. The purpose of the measurements will consist in particular in comparing the uncertainties of the sound pressure levels measured with those estimated by the reference model and which will have been previously fed by the uncertainties observed on the main influence parameters (ground, meteorology, etc.) during the measurement period.

3.3 WP3: Reducing noise at source

WP 3 focuses on research and design of systems that minimize the generation of aerodynamic noise. The systems considered in the project are retrofit type. In order to reduce trailing edge noise, different serration shapes have been proposed in the literature, with variable acoustic and aerodynamic performance and potentially high frequency noise regeneration. Alternative innovative solutions that reduce turbulence noise and also delay stall regime (static or dynamic) that has not been tested for wind turbine applications will be investigated in the project. Indicators based on acoustic performance, feasibility and aerodynamic performances of the devices are used in the project to rank the best devices.

The specific designs will be prototyped on a reduced scale in a laboratory and the devices thus manufactured will be integrated on defined airfoil. The conformity of the finished products will be ensured by rigorous metrological analysis. Measurements of parietal pressure and acoustic measurements with an array of 13 microphones will be done, and will enable to evaluate the aerodynamic performance of the devices, and in particular to identify the lift losses generated. The acoustic array will measure the radiated noise for different flow conditions and angle of incidence. A careful to ensure similarity conditions between the model of the trailing edge device and the 1:1 scale model of a complete wind turbine. We will evaluate the robustness of the solution(s) obtained under the various initial flow and device positioning conditions via an experimental design.

Transposing the acoustic performance of a scale model measured in a wind tunnel at a scale of 1:1 can be tricky if similarity rules between the two scales are not respected. Indeed, the same flow regime on a small and full-scale physical system will not generate the same aerodynamic effects as boundary layer separation or the generation of turbulent structures for example. Thus, extrapolation of an acoustic gain obtained in a wind tunnel could prove erroneous if similarity criteria based on dimensionless constants (Reynolds number and Mach number) are not taken into account when designing the scale models. The project proposes to analyze the scale effects of a trailing edge noise reduction device (serration type) between a scale model measured in a wind tunnel and a wind turbine at scale 1:1.

The possibility of relying on numerical simulations of relative boundary layer thicknesses at different scales on the same profile to justify transposition laws will be evaluated. The acoustic gain of the model and the wind turbine will then be deduced from the measurements taken at both scales. Finally, transposition rules between the scale model and the full-scale device will be defined.

4. Project outcomes

The main outcomes of the project will be:

- Development of a model for predicting amplitude modulation
- A database of experimental data on the wind tunnel characterization of noise due to dynamic stall at wind turbine blades.
- A database of experimental data on wind turbine noise propagation
- A database and a model for estimating the uncertainties of wind turbine noise predictions
- The evaluation and the development of new solutions for reducing wind turbine noise at source.

Acknowledgements

This research is funded by the French National Agency for Research within the convention N°ANR-18-CE04-0011.

References

Archambeau F., Mehitoua N., Sakiz M. (2004). *Code Saturne: a finite volume code for the computation of turbulent incompressible flows*. International Journal on Finite Volumes 1(1).
Cotté B., Tian T. (2015), *Prediction of variability in wind turbine noise calculations*, Wind Turbine Noise conf.

Hess H.M., K. Attenborough K., Heap N.W. (1990), *Ground characterization by short-range propagation measurements,* J. Acoust. Soc. Am., 87, 1975-1986.

Jones, A and McKenna T (2010) *A new method for statistical analysis: the two part reverse interrogation of sparse data* Jnl Reverse Statistics 55(2), 65-219

Lee S., Kima K., Lee S. (2011), *Annoyance caused by amplitude modulation of wind turbine noise*, Noise Control Eng. J. 59(1).

Lihoreau B., B. Gauvreau, M. Bérengier, Ph. Blanc-Benon, I. Calmet, (2006), *Outdoor sound propagation modeling in realistic environments: Application of coupled parabolic and atmospheric models*, J. Acoust. Soc. Amer. 120(1), 110-119

Moreau S., Roger M., Christophe J. (2009), *Flow features and self-noise of airfoils near stall or in stall*, AIAA paper 2009-3198.

Saltelli, A., M. Ratto, T. Andres, F. Campolongo, J. Cariboni, D. Gatelli, M. Saisana, and S. Tarantola. (2008). *Global Sensitivity Analysis: The Primer.* Wiley.

INCEEUROPE

8th International Conference

on

Wind Turbine Noise

Lisbon – 12th to 14th June 2019

I can still hear it and it's making me ill

Geoff Leventhall Consultant geoff@activenoise.co.uk

Summary

A few residents near wind farms may respond negatively when the turbines are audible …… or visible. They need both inaudibility and invisibility. However, if they have developed concerns about infrasound, inaudibility is not sufficient – I can't hear it, but I know it's the infrasound which is making me ill.

There is a range of responses amongst residents, from no effect up to interference with their normal quality of life, and it is the small group at the extreme end of this range who are severely affected and who, in general, become prominent as objectors to a wind farm. They are not content with compliance of the wind farm with its criteria. They are the windfarm neighbours whom we hear about.

This paper considers why a few people react strongly to what is generally a low level of noise. To say it is because they are "noise sensitives" is merely restating their problem, without advancing understanding. The paper considers influences and factors which promote adverse responses to noise. For example, how some residents may be motivated to develop negative "mindsets", and how such mindsets may be modified.

1 Introduction

A well-designed wind farm, which meets all the required criteria, is not necessarily free from complaints. This is because the criteria for control of noise from any source do not satisfy all those exposed to the source. Criteria are not designed to do so. Criteria, which are defined by legislators, based on information supplied by acousticians, balance competing needs and are typically set at levels for which 5%-10% of those exposed are annoyed.

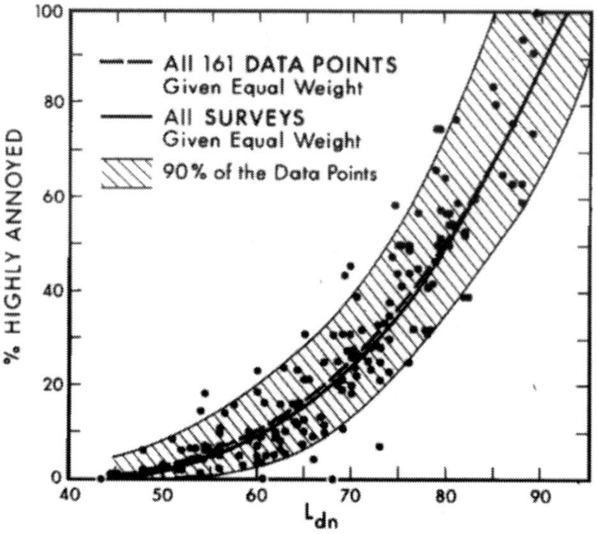

Fig 1 Shultz curve for annoyance by transportation noise

The well-known sigmoid curve, or Shultz curve , shows how annoyance varies with level, as in Fig. 1 (Schultz, 1978). This curve, which is used here for illustration, is for transportation noises, not wind turbines, but shows how there is a spread of about ±10dB of survey points around the averaged variation. There are two difficulties in use of the averaged variation.

1. If the criterion is set at, say, 10% highly annoyed, given by a sound level of about 60dB Ldn in Fig1, the spread above the average leads to the more affected people belonging to a more highly exposed average group. 2. Those who are exposed below the 10% criterion level include a small percentage of subjects who continue to be highly annoyed at very low levels and appear to be influenced by non-acoustic factors.

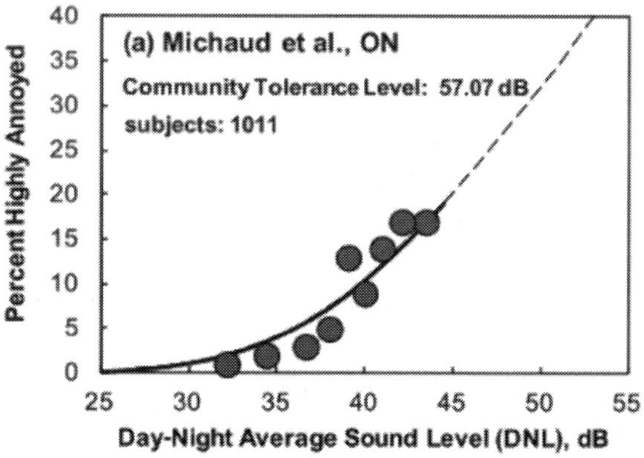

Fig 2 Variation of Wind Turbine annoyance with level

The equivalent curves for wind turbines are not as highly developed as for transportation noise (Fig.1). An example from Michaud is shown in Fig. 2 (Michaud *et al.*, 2016). Here the 10% highly annoyed level is at about 40dB DNL, compared with 60dB DNL for transportation noises.

2. Interaction of residents and wind turbines

The distress and ill health which is reported in the vicinity of some wind turbines is primarily attributed to noise, but visual effects and electromagnetic radiation are other potential contributors. Affected persons are not passive recipients and any reaction to the turbine can be interpreted as sending a response back to the turbine. This response will vary, depending on how residents relate to the wind turbine. If they believe the turbine is the source of their problems, affected persons may send a distress response. Others may feel that the turbine is a beautiful construction, which they enjoy looking at. The physical input, which comes from the turbine, produces an emotional output, such as annoyance, pleasure or indifference. Provision of a telephone complaint line may enable the emotional response to be converted into action.

Fig 3 illustrates these effects, where in addition to interactions with the wind turbine, residents may also be subject to ongoing influences, which act to modify their response. These influences include interventions from objector groups or individuals, negative press reports e.g. on infrasound and general "fright factors".(Deignan *et al.*, 2013), which may have influenced the listener prior to installation of the turbine.

Residents who experience extreme responses are outside standard noise dose concepts, reacting intensely to very low levels of noise.

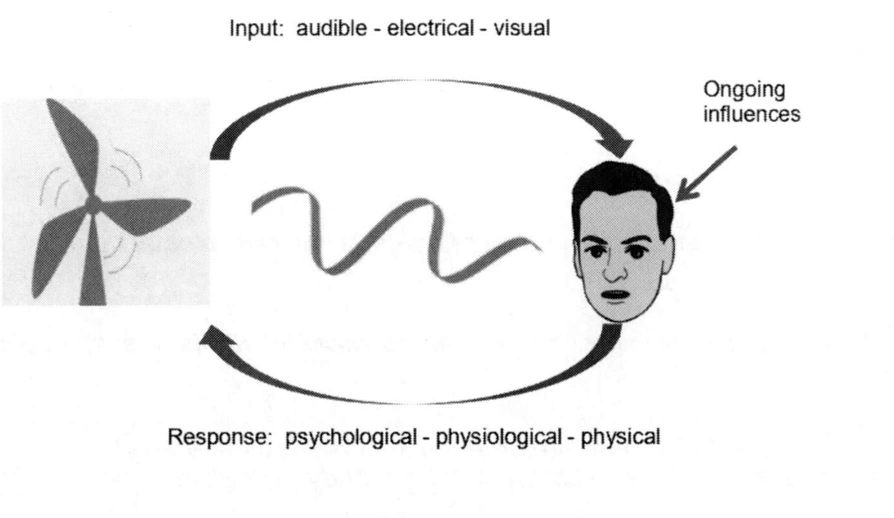

Fig 3 Wind turbine and listener interaction

3. Wind Turbines, Annoyance and Health

A number of reviews of effects of wind turbines on health were published round about 2010 but, as these largely used the same information, they came to similar conclusions. However, recent reviews, using wider and updated sources, have come to comparable conclusions to the earlier reviews. For example

> "This joint statement.........summarizes the results of the best research available and concludes that there is little scientific evidence that sound from wind turbines represents a risk to human health among neighboring residents." (Thorne *et al.*, 2019)

Summaries of annoyance from wind turbines and effects on sleep and health, which have been published in recent years include:

Personal and situational variables associated with wind turbine noise annoyance (Michaud et al., 2016)

Before–after field study of effects of wind turbine noise on polysomnographic sleep parameters
(Jalali et al., 2016)

A review of the possible perceptual and physiological effects of wind turbine noise (Carlile et al., 2018)

Pregnancy exposure to wind turbine noise and adverse birth outcomes: a nationwide cohort study (Poulsen et al., 2018)

Response to Noise Emitted by Wind Farms in People Living in Nearby Areas (Pawlaczyk-Luszczynska et al., 2018)

Variables associated with wind turbine noise annoyance and sleep disturbance (Radun et al., 2019)

Wind turbines and health (Thorne et al., 2019)

Health effects of wind turbines on humans in residential settings: Results of a scoping review (Freiberg et al., 2019a)

The influence of wind turbine visibility on the health of local residents: a systematic review (Freiberg et al., 2019b)

Impact of long term exposure to wind turbine noise on redemption of sleep medication and antidepressants. a nationwide cohort study (Poulsen et al., 2019a)

Long term exposure to wind turbine noise and risk for myocardial infarction and stroke: a nationwide cohort study (Poulsen et al., 2019b)

This selection from papers in the current literature illustrates the interest in the topic of wind farms and health.

4. Extreme responses

The widespread use of "annoyance" to describe an effect of wind turbines is unsatisfactory, as a single word does not express the multiple reactions which commence with awareness of the (impending) presence of the turbines, through auditory and visual stimuli, and potential progression to some highly stressed residents who claim that their life is disrupted by the turbines. Only a small number exhibit this extreme response, becoming highly annoyed by low noise levels. These are the small number of residents at the lower end of the response curve (Figs 1 and 2). Their response may be contributed to by other stimuli, in addition to turbine noise. (Blanes-Vidal and Schwartz, 2016) Extreme responders may not respond according to a noise dose relation, but are distressed by any perception of the turbines.

5. Managing Change

There is a parallel between the reactions of some people to wind turbines and the stages of grief, as experienced by a person who has been given a limited time to live, or a person who has lost a close family member. These stages were described by Kübler-Ross. (Kubler-Ross, 1973) , The Stages concept has been supported by later work (Holland and Neimeyer, 2010) and has been successfully applied to a range of areas in which it has become necessary to manage change in one's personal or professional life (Dyer, 1994). In this respect, Kübler-Ross applies to the personal adjustments required to adapt to a change in environment caused by introduction of wind turbines.

Kübler-Ross (Kubler-Ross, 1973) identified five stages of adaptation response, as in Fig 4:

1. Denial
2. Anger
3. Bargaining (which we will change to Negotiation)
4. Depression
5. Acceptance

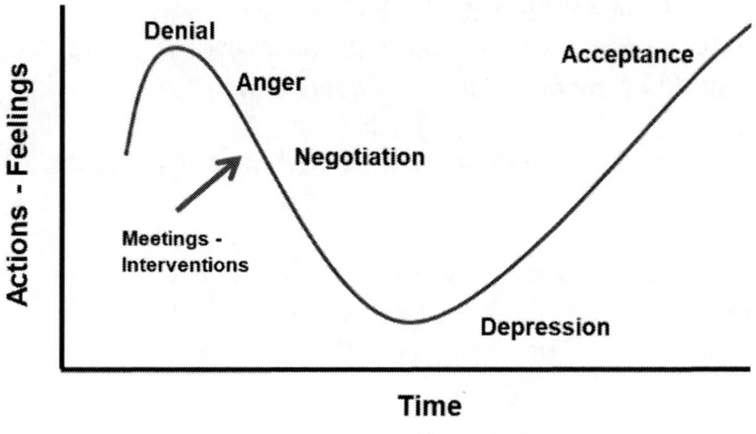

Fig.4 Kübler-Ross stages of adaptation

Consider a situation in which residents first hear that a wind farm is proposed in their locality. Some may welcome the proposal, others may be indifferent, but some may declare strong opposition. The first reaction of opponents is often **Denial** – it can't be true, they can't do this to us. Confirmation of the proposal leads to **Anger** and resentment, whilst plans are developed to challenge the decision. This leads to the **Negotiation** phase, during which there are public meetings, discussions with the wind farm developer, local authorities etc. The public meetings are often very heated and used by objectors to spread misinformation about turbines, whilst developers try to be reassuring.. Following negotiation, some of the residents who oppose the wind farm may suffer from **Depression**, feeling that there is no solution to the problem. Eventually they may come to **Acceptance** when they understand that they cannot stop the windfarm.

Those who exhibit high annoyance at low sound levels are the most likely to remain in the **Depression** stage.

Not everybody follows these five stages in sequence and some may move both forward and back within them, but they are a useful guide to understanding and managing an unwelcome change in our lives.

During the Anger and Negotiation stages, there will be local meetings, with interventions by people from outside the district, who wish to share their experiences. This may be a time of misinformation when illusory truths, for example on infrasound, proliferate (Leventhall, 2017) and susceptible residents are pushed further into depression, from which they are unable to escape. Some of these long-term depressed residents may exhibit physical symptoms, such as tension and poor sleep quality. They respond negatively to the presence of the turbines, and any indication of turbine operation is a trigger for distress. "I can still hear it and it's making me ill"

Alternatively, Negotiation can lead to a resolution of problems and enable the residents to feel that they have been consulted and listened to, that they are part of the process. Exclusion of residents from critical decision-making fosters opposition, whilst inclusion may lead to acceptance. The residents want to genuinely feel that they have been able to influence the outcome of the Negotiation phase. (Maris *et al.*, 2007) (Walker and Baxter, 2017) (Firestone *et al.*, 2018)

The most affected residents can be compared to "HUM Sufferers", who are distressed by sounds which cannot be measured or traced (Leventhall *et al.*, 2003) (Bommer *et al.*, 2016) (Frosch, 2016). A recent example is described in

https://www.leicestermercury.co.uk/news/uk-world-news/pensioner-says-neighbours-noise-driving-2668121

Wind turbine complainants and Hum complainants show much commonality. In both cases the symptoms relate to the listener's high stress levels, rather than direct physiological effects of a low level of noise. The symptoms of Wind Turbine

Syndrome (Pierpont, 2009) and Noise Stress (Leventhall 2002) (Nagai *et al.*, 1989; Møller and Lydolf, 2002) are compared in Table 1. The symptoms are seen to be very similar, demonstrating the Wind Turbine Syndrome as an example of stress from exposure to an unwanted noise. It is necessary to consider the paths by which the stress may develop. A controlling factor is a listener's attitude to the noise source – their mindset. Attitudes are malleable and may be conditioned by external influences, leading to an Illusory Truth[1] (Leventhall, 2017), whilst resentment is a very corrosive influence.

WTSyndrome (Pierpont)	NOISE STRESS (e.g. the HUM)
sleep disturbance headache ear pressure tinnitus dizziness vertigo nausea visual blurring tachycardia irritability problems with concentration and memory - panic episodes associated with sensations of internal pulsation or quivering "which arise while awake or asleep"	Insomnia headache pressure in ears or head dizziness nausea eye strain fatigue distraction nose bleed feels vibration muscle spasms palpitations skin burning stress tension etc.

Table 1 Comparison of noise stress from Wind Turbine Syndrome and other Sources

6. Misophonia and noise sensitivity

It may be difficult to distinguish between two persons, exposed to the same noise, when one is believed to be highly noise-sensitive and the other is believed to suffer from misophonia, which is a strong, acquired aversion to certain trigger sounds.

Noise sensitive people generally have a latency in their response, but are upset by continuous sounds of the type they find irritating.

The misophonic person is averse to specific trigger sounds, and has developed a learned, fast response to these. That is, there was an earlier time when the person responded normally to the sound, but the sound subsequently developed a specific trigger meaning, leading to misophonia. A characteristic of misophonia is that the misophonic response is fast and associated with both an emotional and physical

[1] Illusory Truths arise, for example, following repetition of a false statement until it becomes accepted through familiarity

response, such as uncontrolled muscle tightening, at the start of the exposure. (Edelstein *et al.*, 2013; Cavanna and Seri, 2015; Kumar *et al.*, 2017) . Whilst anger is the main resoponse associated with misophonia, the related phonophobia elicits fear of the sound.(Palumbo *et al.*, 2018) Both exhibit physical responses.

A low level of noise trigger is sufficient to stimulate a misophonic response. The trigger may develop to include other sounds, if these become associated with the original trigger. The misophonic person must avoid their trigger sound and take steps to prevent its association with additional sounds, so that these do not become independent triggers.

Application of misphonia concepts to wind turbine noise indicates a possibility that a small number of extreme responders may be exhibiting a learned misophonic reaction. Onset of misophonia is mainly amongst younger people, but the older are not immune and there is the potential of misophonia developing, particularly in some who were initially antagonistic to wind turbines. Residents with misophonic reactions to wind turbine noise are in a difficult situation, as they cannot easily remove themselves from the trigger noise, other than by changing location. Some do this.

7. Infrasound

Infrasound has a special place in discussions on wind turbine noise. Its adoption by objectors led to presention and misrepresentation as a "fright factor" and its supposed effects have become a firmly established illusory truth (Leventhall, 2017). Much of the evidence that has been put forward by objectors to support their claims that infrasound is harmful is taken from work at considerably higher infrasound levels than those from wind turbines. For example, Punch and James describe work at high levels as relevant, by implication, to the low levels from wind turbines.(Punch and James, 2016).

Current attitudes to infrasound were driven by the Wind Turbine Syndrome (WTS) (Pierpont, 2009), such that those who claim adverse affects from turbines are described as Wind Turbine Syndrome sufferers. The Wind Turbine Syndrome is said to be caused by an affect of infrasound from wind turbines on the vestibular and related systems in the body, but is not supported by scientific evidence. One scientific paper claims support for WTS, but the paper (Schomer *et al.*, 2015), although published in a well-known refereed journal, is clearly flawed, as can be seen as follows.

Schomer et al assume that the effects on the otolith, which is part of the balance system within the inner ear, are similar for both whole body vibration and for infrasound exposure and, after some calculations on the mechanics of the otolith, reach the conclusion that a 0.7Hz tone at 54 dB (0.01Pa) produces about the same to three times the force on the otolith as is caused by a $5m/s^2$ vertical, whole body acceleration at the same frequency. In this, the 0.7Hz tone is assumed to be a wind turbine blade pass frequency, whilst the $5m/s^2$ is related to US Navy criteria for nauseogenic effects. (Kennedy *et al.*, 1987).

Schomer et al are proposing that greater potential for nauseogenic effects occur from airborne 0.7Hz at 54dB (0.01Pa) than from a $5m/s^2$ whole body vertical acceleration at the same frequency. Developing this a step further ($d=a/\omega^2$) shows that the vertical displacement produced by the vibration is about 0.25m. A vertical displacement of 0.25m corresponds to a pressure change of 3Pa (hρg) or 104dB, which is considerably greater than that from the wind turbine. It is unlikely that the infrasound from vertical displacement contributes to nausea, since nausea also occurs from horizontal vibration, where there is no change in vertical height. (Golding, 2001). Thus, it appears that the nausea from vertical vibration is due to the effects of whole body movement, not infrasound. Consequently, Schomer et al have not shown that infrasound, at the levels from wind turbines, is a cause of nausea.

Recent work has exposed the ear to high sound levels, including infrasound, detecting vestibular responses by VEMPs (vestibular-evoked myogenic potentials), which indicate excitation of the saccule.(Jurado and Marquardt, 2019) Pure tone pressures were applied to the ear via a tube sealed into the ear canal, with increasing level as the frequency reduced, reaching 132dB at 4Hz. The conclusion was that the saccule "seems to be rather insensitive to airborne infrasound", even though the levels used were around 60dB higher than those from wind turbines.

The web site Stop These Things (STT) was set up in late 2012 as an antiwind communication and publishes a daily blog of anecdotes and pseudoscience, reprinting and commenting on press articles which attack wind energy. Infrasound is a regular topic. STT, which is published anonymously, is noted for its misrepresentations and exagerations

Some recent STT posts featuring infrasound include:

> *Home Wreckers: Finnish Study Finds Wind Turbine Infrasound Unsafe For Residents Living Within 15 Km*
> ***February 1, 2019***
>
> *Pulsing Punishment: Wind Turbine Infrasound Delivers Perpetual Torment for Neighbours*
> *January 24, 2019*
>
> *'Green' Energy Guinea Pigs: Wind Industry's American Victims Monitored for Infrasound Effects on Heart & Health*
> *December 10, 2018*
>
> *Silent Killer: Why Wind Turbine Infrasound Causes Serious Health Problems for Wind Farm Neighbours*
> *December 6, 2018*

Heart-stopping: German Research Finds Low-Frequency Wind Turbine Noise & Infrasound Cardiac Health Risk
November 10, 2018

And many more like these, illustrating the illusory truths with which the public has to contend.

A commonly used vehicle for spreading fears about infrasound is letters to local newspapers. A recent publication states:

> *The wind industry ignores the infrasound generated by its turbines, focusing only on audible sound. New York state goes along with this deception, overlooking the life-threatening affects of infrasound when its dangers are so extensively documented. The state's action (or inaction) is tantamount to condoning the use of residents as guinea pigs, people deprived of informed consent.*
>
> https://www.wind-watch.org/news/2019/03/25/tuning-out-infrasound-dangers/

8. Nocebo and Placebo effects

The Nocebo/Placebo effect was originally described in the medical context, where symptoms and treatments are influenced by expectations and conditioning. Recognition of the effect goes back to at least the early 1960s (Kennedy, 1961), but modern brain imaging has shown its neurological basis. (Dodd *et al.*, 2017) The outcome of communication to patients, either directly or by implication, illustrates the crucial importance of information transfer in creating expectations (Benedetti *et al.*, 2007; Bensing and Verheul, 2010; van Laarhoven *et al.*, 2011; Reicherts *et al.*, 2016; Chavarria *et al.*, 2017).

These papers make clear that the Nocebo/Placebo responses are well established in general clinical work and are powerful in their operation, an operation which is largely based on expectations and conditioning. It is a short step to consider nocebo/placebo as an element in health related responses to wind turbines, within a "background noise" of assertions that wind turbines are harmful to health.

The first direct application of the nocebo effect to wind turbines was by Chapman et al and has been supported by follow-up work (Chapman *et al.*, 2014; Crichton *et al.*, 2014a; Crichton *et al.*, 2014b; Crichton and Petrie, 2015; Tonin *et al.*, 2016; Chapman and Crichton, 2017). However, the importance of expectations was investigated earlier (Crichton *et al.*, 2013), with the following results:

> *During exposure to audible windfarm sound and infrasound, symptoms and mood were strongly influenced by the type of expectations. Negative expectation participants experienced a significant increase in symptoms and a significant deterioration in mood, while positive expectation participants*

reported a significant decrease in symptoms and a significant improvement in mood.

Chapman at al 2014 introduced the nocebo to wind turbine studies and considered the effect in a community, using submissions and publicity on a pending wind farm hearing as sources of information. Some of the media material was described as "frightening".

In addition to reviewing the literature, Crichton carried out subject-based experiments with the general outcome that subjects who have been influenced by negative expectations present more symptoms than those who have been exposed to positive expectations, as was confirmed independently by Tonin (Tonin *et al.*, 2016)

The Nocebo/Placebo are established and well documented effects in the conditioning of responses to wind turbines. Objectors are not comfortable with them because of the light they shine on objector activities. Lacking facts to counter the use of Nocebo, objectors turn to ridicule and personal attacks. For example:

> **Pierpont** *They are not fabricating these symptoms. Their symptoms are not Simon Chapman's silly "nocebo effect." The symptoms are — real! Really and truly caused by IWT infrasound.* (Pierpont, 2017)

Statements like this indicate a (deliberate?) misunderastanding of the Nocebo/Placebo effect, impying that symptoms induced by the effect are, in some way, unreal. This is not what the literature says.

Negative attitudes to wind turbines, developed through illusory truths, which are rife in the area of wind turbines, are highly likely to influence the residents' attitudes, so feeding the Nocebo effect. There is a conflict with leading objectors, who have persistently claimed a direct physiological action from wind turbines.

8. Is Help is available?

Is it possible to help severely affected residents, those in the Depression stage of Kübler-Ross, Fig.4, to live happier lives, which are not blighted by the presence of wind turbines. It has been shown that the "talking therapies", particularly Cognitive Behaviour Therapy (CBT) have promise in helping people to desensitise to a troubling noise.(Leventhall *et al.*, 2008; Leventhall *et al.*, 2012). Introductory information on CBT is given on https://www.rcpsych.ac.uk/mental-health/treatments-and-wellbeing/cognitive-behavioural-therapy-(cbt)

CBT is a form of "relaxation therapy". This is a broad term that describes a range of different therapeutic techniques. Many of these are simple, well-developed, procedures which generate a number of positive physiological and psychological benefits. The overall concept of a relaxation therapy is now well-accepted within mainstream medicine as a means of stress control and therapy is available on the

UK National Health Service. https://www.nhs.uk/conditions/stress-anxiety-depression/self-help-therapies/

CBT recognises that our thoughts, feelings and actions are connected and aims to reduce negative thoughts, so that problems are dealt with in a positive manner. CBT analyses our current problems and enables these to be dealt with positively, so reducing stress.

Relaxation is an essential part of therapy. For example, in their advice book on stress the British Medical Association, say that systematic relaxation, "Improves sleep, increases mental and physical performance, combats tiredness, decreases anxiety and tension".(Wilkinson, 2004) Essentially, it leads to a physiological and psychological state which is opposite to the state of stress, including the stress from exposure to noise.

In the application of CBT to helping noise sufferers, the first phase of the work included a group of nine participants who attended sessions over 10 weeks and were introduced to CBT concepts and techniques. Recognition of the difficulties of attending group sessions led to development as an internet-based distance learning project (Leventhall *et al.*, 2012).

Assessment, both before and after therapy, included a 25 point questionnaire (Noise Reaction Questionnaire NRQ) which explored:

- How the noise made sufferers feel (emotions)
- How it affected them physically (health)
- How it affected their daily activities and interactions with others (social)

Scoring was on a five point semantic scale

Not at all = 0
A little of the time = 1
Some of the time = 2
A good deal of the time = 3
Most of the time = 4

The results were encouraging, as shown in Fig. 5, where the main problems (highest response score) are shown by vertical dotted lines. These problems are

5 I have a hard time adjusting to the noise
9 The noise interferes with my quality of sleep
13 The noise makes it hard for me to fall asleep at night
15 The noise makes me feel agitated or restless
17 The noise makes me feel anxious
22 The noise makes me feel tired and fatigued
25 The noise prevents me from being able to relax

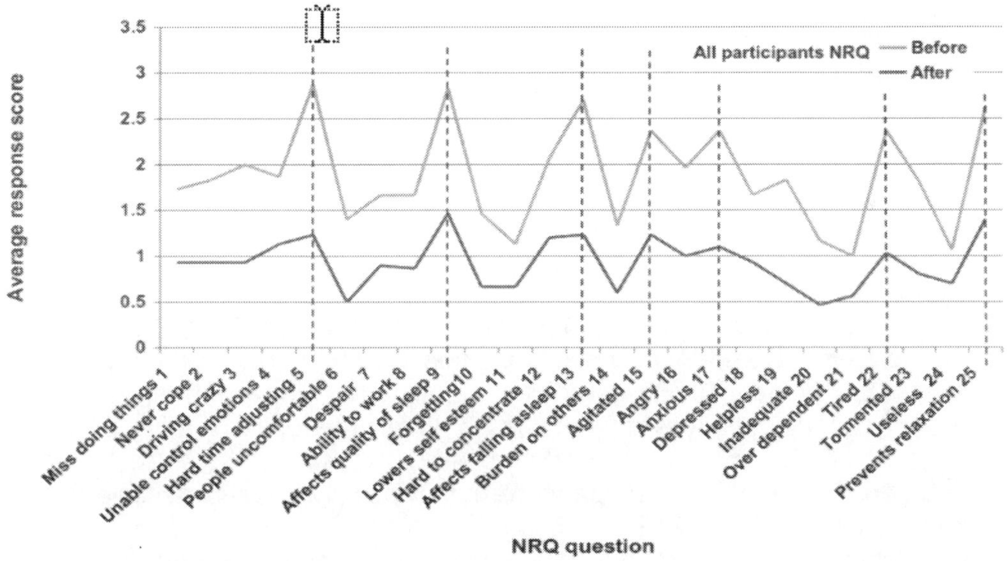

Fig 5 Before and after questionnaire responses. Averaged over subjects

The main problems are seen to be sleep/tiredness and anxiety/tension. The reduction in each of these was by 1 to 1.5 points on the five-point response scale, indicating a useful improvement in the subjects' responses.

This earlier work has shown the application of CBT to persons with noise problems and may be effective for those who have problems with wind turbine noise, but there are hurdles to be overcome. Experience with a large number of subjects showed that those who were most likely to benefit from CBT had been through all the environmental and medical processes and had failed to find the source of the noise. They were reconciled to living with their noise and were willing participants in the desensitisation process.

Another group, who believed they knew the source of the noise (often from a neighbour's property) and wanted it stopped. However, they were not always correct in their source assumption. Those who are affected by wind turbines know the source of their problems and may be reluctant to join a desensitisation program.

9. Summing up

The paper has considered those who respond acutely to nearby wind turbines, giving consideration to those who exhibit the most severe responses and have, through no fault of their own, been unable to adapt to the changes in their environment which accompany the introduction of wind turbines. Explanations for the severity of their responses are attempted, but this is a very difficult area. It is possible that severity of responses has been influenced by the strong anti-windfarm campaigns which raise fears of effects on health, but adverse effects can been rolled back by therapies such as CBT.

References

Benedetti, F., Lanotte, M., Lopiano, L., Colloca, L., 2007. When words are painful: unravelling the mechanism of the Nocebo effect. Neuroscience 147, 260-271.
Bensing, J.M., Verheul, W., 2010. The silent healer: The role of communication in placebo effects

Patient Education and Counseling 80, 293-299.
Blanes-Vidal, V., Schwartz, J., 2016. Wind turbines and idiopathic symptoms: The confounding effect of concurrent environmental exposures

Neurotoxicology and Teratology 55, 50-57.
Bommer, A., Young, A., Bruce, B., 2016. Mysterious and Annoying: Hums, Tones, and Other Hard-to-Detect Residential Noise Complaints. Proc Noise-Con 2016.
Carlile, S., Davy, J.L., Hillman, D., Burgemeister, K., 2018. A review of the possible perceptual and physiological effects of wind turbine noise. Trends in Hearing 22, 1-10.
Cavanna, A.E., Seri, S., 2015. Misophonia: current perspectives. Neuropsychiatric Disease and Treatment 11, 2117-2123.
Chapman, S., Crichton, F., 2017. Wind turbine syndrome A communicated disease. Sydney University Press.
Chapman, S., Joshi, K., Fry, L., 2014. Fomenting sickness: nocebo priming of residents about expected wind turbine health harms. Frontiers in Public Health, doi: 10.3389/fpubh.2014.00279.
Chavarria, V., Vian, J., Pereira, C., Data-Franco, J., Fernandes, B.S., 2017. The Placebo and Nocebo Phenomena: Their Clinical Management and Impact on Treatment Outcomes. Clinical Therapeutic 39, 2017.
Crichton, F., Chapman, S., Cundy, T., Petrie, K.J., 2014a. The link between health complaints and wind turbines: support for the nocebo expectations hypothesis. Frontiers in Public Health 2, doi: 10.3389/fpubh.2014.00220.
Crichton, F., Dodd, G., G, S., Gamble, G., Petrie, K.J., 2014b. Can Expectations Produce Symptoms From Infrasound Associated With Wind Turbines? Health Psychology 33, 360-364.
Crichton, F., Dodd, G., Schmid, G., Cundy, T., Petrie, K.J., 2013. The Power of Positive and Negative Expectations to Influence Reported Symptoms and Mood During Exposure to Wind Farm Sound. Health Psychology . doi: 10.1037/hea0000037.
Crichton, F., Petrie, K.J., 2015. Health complaints and wind turbines: The efficacy of explaining the nocebo response to reduce symptom reporting. Environmental Research 140, 449-455.
Deignan, B., Harvey, E., Hoffman-Goetza, L., 2013. Fright factors about wind turbines and health in Ontario newspapers before and after the Green Energy Act. Health, Risk and Society, 234-250
Dodd, S., Dean, O.M., Vian, J., Berk, M., 2017. A Review of the Theoretical and Biological Understanding of the Nocebo and Placebo Phenomena. Clinical Therapeutics 39, 469-476.
Dyer, W.G., 1994. Toward a theory of entrepreneurial carreers. Entrpreneurship Theory and Practice, 7-21.
Edelstein, M., Brang, D., Rouw, R., Ramachandran, V.S., 2013. Misophonia: physiological investigations and case

descriptions. Frontiers in human neuroscience 7, doi: 10.3389/fnhum.2013.00296.
Firestone, J., Hoen, B., Rand, J., Elliott, D., 2018. Reconsidering barriers to wind power projects: community engagement, developer transparency

and place. Journal of Environmental Policy & Planning 20, 370-386.
Freiberg, A., Schefter, C., Girbig, M., Murta, V.C., Seidler, A., 2019a. Health effects of wind turbines on humans in residential settings: Results of a scoping review. Environmental Research 169, 446-463.
Freiberg, A., Schefter, C., Hegewald, J., Seidler, A., 2019b. The influence of wind turbine visibility on the health of local residents: a systematic review. International Archives Occupational and Environmental Health, https://doi.org/10.1007/s00420-00019-01403-w.
Frosch, F.G., 2016. Manifestations of a low-frequency sound of unknown origin perceived worldwide, also known as "the Hum" or the "Taos Hum". International Tinnitus Journal. 20, 59-63.
Golding, J.F., 2001. A motion sickness maximum around the 0.2 Hz frequency range of horizontal translational oscillation. Aviation, Space and Environmental Medicine 72, 188-192.
Holland, J.M., Neimeyer, R.A., 2010. An Examination of Stage Theory of Grief among Individuals Bereaved by Natural and Violent Causes: A Meaning-Oriented Contribution. OMEGA 61, 103-120.
Jalali, L., Bigelow, P., Nezhad-Ahmadi, M.-R., Gohari, M., et al, 2016. Before–after field study of effects of wind turbine noise on polysomnographic sleep parameters. Noise and Health 18, 194-205.
Jurado, C., Marquardt, T., 2019. On the Effectiveness of airborne infrasound in eliciting vestibular-evoked myogenic responses. Jnl low freq noise vibn and active control, DOI: 10.1177/1461348419833868.
Kennedy, R.S., 1961. The nocebo reaction. . Med. World (Lond.) 95, 203-205.
Kennedy, R.S., Allgood, G.O., van Hoy, B.W., Lilienthall, M.G., 1987. Motion Sickness Symptons and Postural Changes Following Flights in Motion-Based Flight Trainers Jnl Low Freq Noise Vibn 6, 147-154.
Kubler-Ross, E., 1973. On death and dying - (Book). Routledge.
Kumar, S., Tansley-Hancock, O., Sedley, W., etal, 2017. The brain basis for misophonia. Cuirrent Biology 27, 527-533.
Leventhall , G., 2002. 35 years of low frequency noise - Stephens Medal Lecture. Proc IoA 24, Proceedings CD.
Leventhall, G., 2017. Why do some people believe that they are "made ill" by wind turbine noise. Proc 7th International Confernce Wind Turbine Noise.
Leventhall, G., Pelmear, P., Benton, S., 2003. A review of published research on low frequency noise and its effects - Report for Defra. http://www.defra.gov.uk/environment/noise/research/lowfrequency/pdf/lowfreqnoise.pdf.
Leventhall , G., Robertson, D., Benton, S., Leventhall, L., 2012. Helping sufferers to cope with noise using distance learning cognitive behaviour therapy. J. Low frequency Noise, Vibration and Active Control 31, 193-204.
Leventhall, H.G., Benton, S., Robertson, D., 2008. Coping strategies for low frequency noise. Jnl Low Freq Noise Vibn 27, 35 - 52.
Maris, E., Stallen, P.J., Vermunt, R., Steensma, H., 2007. Evaluating noise in social context: The effect of procedural unfairness on noise annoyance judgments. Jnl Ac Soc America 122(6), 3483–3494.

Michaud, D.S., Keith, S.E., Feder, K., Voisescu, S.A., et al, 2016. Personal and situational variables associated with wind turbine noise annoyance. J. Acoust Soc Am 139.
Møller, H., Lydolf, M., 2002. A questionnaire survey of complaints of infrasound and low frequency noise. Jnl Low Freq Noise Vibn 21, 53 - 65.
Nagai, N., Matsumoto, M., Yamsumi, Y., Shiraishi, T., Nishimura, K., Matsumoto, K., Myashita, K., Takeda, S., 1989. Process and emergence of the effects of infrasonic and low frequency noise on inhabitants. Jnl Low Freq Noise Vibn 8, 87-89.
Palumbo, D.B., Alsalman, O., de Ridder, D., Song, J.J., Vanneste, S., 2018. Misophonia and Potential Underlying Mechanisms: A Perspective. Frontiers in Psychology 9, doi: 10.3389/fpsyg.2018.00953.
Pawlaczyk-Luszczynska, M., Zaborowski, K., Dudarewicz, A., et al, 2018. Response to Noise Emitted byWind Farms in People Living in Nearby Areas. International Journal of Environmental Research and Public Health 15, doi:10.3390/ijerph15081575.
Pierpont, N., 2009. Wind Turbine Syndrome. K-Selected Books.
Pierpont, N., 2017. ISO acknowledges motion sickness from low-frequency oscillatory motion below 1 Hz. North American Platform Against Wind Power, http://www.na-paw.org/Pierpont-ISO-9996.php.
Poulsen, H.A., Raaschou-Nielsen, O., Pena, A., Hahmann, A.N., et al, 2018. Pregnancy exposure to wind turbine noise and adverse birth outcomes: a nationwide cohort study. Environmental Research 167, 770-775.
Poulsen, H.A., Raaschou-Nielsen, O., Pena, A., Hahmann, A.N., et al, 2019a. Impact of long term exposure to wind turbine noise on redemption of sleep medication and antidepressants. a nationwide cohort study. Environmental Health Persepctives 127 https://doi.org/10.1289/EHP3909.
Poulsen, H.A., Raaschou-Nielsen, O., Pena, A., Hahmann, A.N., et al, 2019b. Long tern exposure to wind turbine noise and risk for myocardial infarction and stroke: a nationwide cohort study. Environmental Health Persepctives, https://doi.org/10.1289/EHP3340.
Punch, J.L., James, R.R., 2016. Wind Turbine Noise and Human Health: A Four-Decade History of Evidence that Wind Turbines Pose Risks. Hearinghealthmatters.org.
Radun, J., Hongisto, V., Suokas, M., 2019. Variables associated with wind turbine noise annoyance and sleep disturbance. Building and Environment 150, 339-348.
Reicherts, P., Gerdes, A., Pauli, P., Wieser, M.J., 2016. Psychological Placebo and Nocebo Effects on Pain Rely on Expectation and Previous Experience. The Journal of Pain, 17, 203-214.
Schomer, P., Erdreich, J., Parmidighantam, P., Boyle, J., 2015. A theory to explain some physiological effects of the infrasonic emissions at some wind farm sites. J. Ac Soc America 137, 1356–1365.
Schultz, T.J., 1978. Synthesis of social surveys on noise annoyance. J. Acoust Soc Am 64, 377-405.
Thorne, P.S., Osterberg, D., Johannsen, K., 2019. Wind turbines and health. Iowa Environmental Council, https://www.iaenvironment.org/webres/File/IEC_Wind_Health_Paper_2019_FINAL.pdf.
Tonin, R., Brett, J., Colagioiri, B., 2016. The effect of infrasound and negative expectations to adverse pathological symptoms from wind farms. Jnl Low Freq Noise Vibn and Active Control 35, 77-90.

van Laarhoven, A.I.M., Voglaar, M., Wilder-Smith, O.H., van Riel, P.L.C.M., 2011. Induction of nocebo and placebo effects on itch and pain by verbal suggestions. PAIN 152, 1486-1494.
Walker, C., Baxter, J., 2017. Procedural justice in Canadian wind energy dev elopment: A comparison of community-based and technocratic siting. Energy Research and Social Science 29, 160-169.
Wilkinson, G., 2004. Understanding Stress. Dorset: Family Doctor Publications.

INCE*EUROPE*

8th International Conference on Wind Turbine Noise
Lisbon – 12th to 14th June 2019

Session K

Source Prediction - 1

Organised by INCE-Europe
mail@inceeurope.org
www.inceeurope.org

Supported by Portuguese Acoustical Society

Sociedade Portuguesa de Acústica

INCEEUROPE

8th International Conference
on
Wind Turbine Noise
Lisbon – 12th to 14th June 2019

A Brief Study on Noise Propagation of Airfoils from Wind Turbines Using the Lattice Boltzmann Method

Sara Rodriguez
Sao Paulo State University (Poli-USP): sara.rodriguez@alumni.usp.br

Joseph Youssif Saab Jr.
Maua Institute of Technology (IMT): saab@maua.br

Alexandre Martuscelli Faria
Sao Paulo State University (Poli-USP): martuscellifaria@usp.br

Marcos de Mattos Pimenta
Sao Paulo State University (Poli-USP): marcos.pimenta@poli.usp.br

Summary

The Lattice Boltzmann Method (LBM) is a very promising Computationsl Fluid Dynamics (CFD) model, which attracts many researchers and engineers because it allows concurrent solving of many physical phenomena and it has the advantage to be highly parallelizable, i.e., simultaneous performance of multiple operations, in order to divide the domain into small parts and assign each one to a processor. The LBM resolves all scales, macro and micro, simultaneously and is gaining momentum for solving multiphysics and aeroacoustic problems. The LBM may solve in a simplicity of programming and moderate computational cost the Navier-Stokes equation, with a restriction that the Mach number must be low. This restriction does not hinder the application of the method to the simulation of the noise propagation of airfoils from wind turbines. Therefore, it is important to compare the LBM with other CFD methods to select the best way to predict the noise generated by wind turbines and its propagation to the far field. This article intends to make this investigation in the context of potentially improving the Pnoise module, which is a Poli-USP & TU-Berlin collaboration project implemented in the QBlade software, available for download under a General Public License. The aim is to allow limited validation for high-Reynolds numbers self-noise and inflow noise models under development for full-size rotor noise prediction.

1. Introduction

Wind energy is one of the most important renewable energy sources today, and has been used increasingly in the world. The largest source of wind power generation comes from horizontal axis wind turbines (HAWT) both in onshore and offshore areas. However wind energy needs prerequisites for its implementation, because of technical dependency of strong winds places and in locations that are capable to support the installation of large turbines (100m high for example). Figure 1 shows the rapid increase of the wind energy in the last decades.

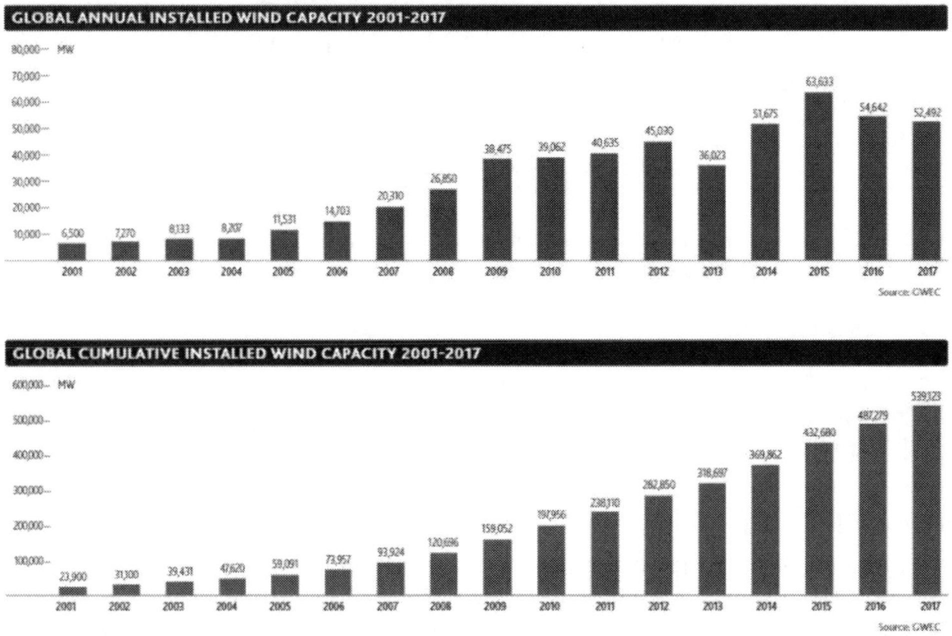

Fig. 1 Global annual installed and Global Cumulative Installed Wind Capacity [25].

Another important aspect is that the dimensions of wind turbines have grown over the years (see in Figure 2), specially the blades, this increase come with the multitude of benefits, in particular helping to drive down the cost of wind energy because wind speeds increase with altitude, thus the taller the turbine the faster it spins and the more energy it can generate. Furthermore, the wind is generally steadier higher up, easing power peaks and troughs and increasing reliability. The trend is to increase the size of the turbines and not the wind farms, reducing maintenance and operation costs.
One of the main obstacles to using wind turbines considering its large dimensions and the proximity in populated areas, increasingly recurrent situation, is the noise produced by them, which annoys residents and the local fauna nearby. Therefore, it is important predict properly the noise to define the best places for wind parks and enables the improve of the design of quieter turbine in order to avoid future problems.

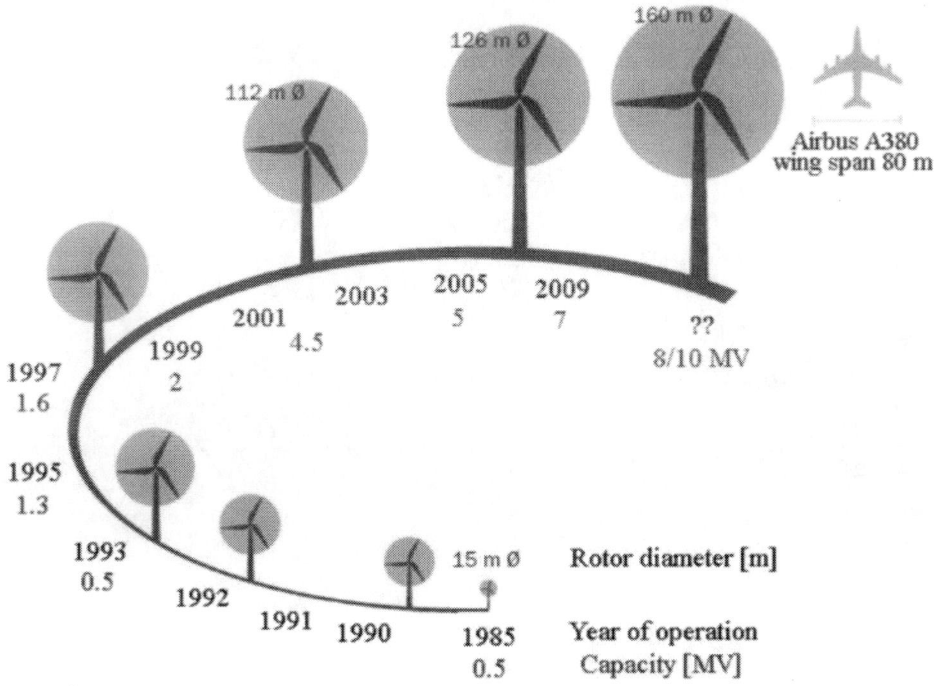

Fig. 2 Growth of wind turbines over time [8].

Other important fact is that an decrease of the sound pressure level by one decibel would increase the annual energy production by 20% [26] (see in figure 3).

The acoustic study of wind turbines may be performed with QBlade, which is an open-source calculation software, distributed under the GNU General Public License, made by TU Berlin, its purpose is the design and aerodynamic simulation of wind turbine blades. The *Polytechnic School of the University of São Paulo* joined the group to improve the software to include the noise module, called PNoise. The noise prediction of this application is currently implemented in two dimensional (2D) and in the near future in three dimensional (3D).

The main motivation of this work is to contribute to the enlargement of the use of wind turbines, making them quieter and more efficient and enabling them to be used in more places with the multiscale fluid dynamic base with using the Lattice Boltzmann Method (LBM).

2. The Lattice Boltzmann Method (LBM)

The Lattice Boltzmann method [1, 2, 11, 28] is a modern approach in Computational Fluid Dynamics and it is often used to solve the incompressible, time-dependent Navier-Stokes equations numerically. This method is easily capable to represent complex physical phenomena, ranging from multiphase flows to chemical interactions between the fluid and the surroundings.

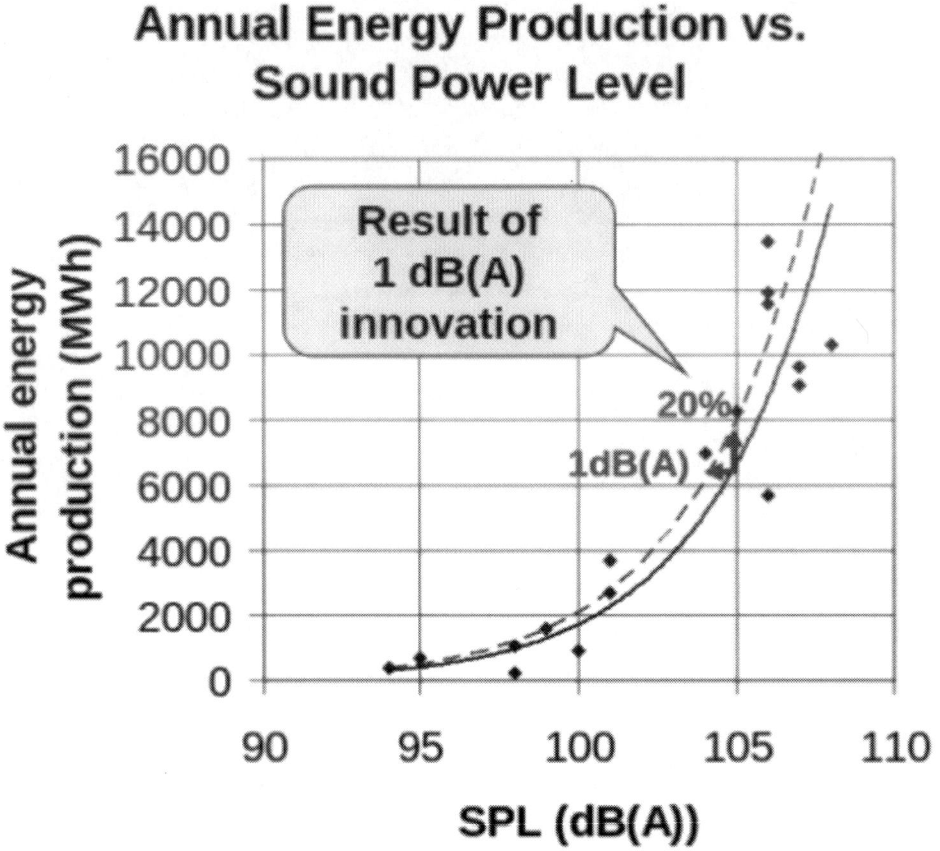

Fig. 3 Relation in increase in annual energy production by wind turbines and reduce the noise in 1dB(A) [26].

Based on a molecular description of a fluid, the LBM can directly incorporate physical terms stemming from a knowledge of the interaction between molecules. For this reason, it is an invaluable tool in fundamental research, as it keeps the cycle between the elaboration of a theory and the formulation of a simple corresponding numerical model.

Historically, LBM evolved from the Cellular Automata models, which is a computer model for the evolution of discrete states described by integer values, a different way of representation considering the lattice Boltzmann variables.

The Boltzmann equation is analogue of the Navier-Stokes equation at a molecular level and it describes the space-time dynamics of a statistical quantity called *probability distribution function*, which is defined in 6-dimensional phase space. The molecular level brings benefits comparing with the Navier-Stokes equation, because of no dependency of a separation of time scales and the ability to describe fluids in non-hydrodynamic regimes with large molecular mean free paths. Furthermore, the molecular model is able to capture transport phenomena such as friction, diffusion and temperature transport and derive the corresponding transport coefficients.

The LBM represents the position and velocity of interacting pseudo-molecular, "mesoscopic" particles of a fluid and acts on real-valued quantities, but it describes some dynamics in a discrete phase space, which is displayed like a lattice. The lattice Boltzmann equation can be derived directly from the continuous Ludwig Boltzmann's kinetic theory of gases equation, that considers gases/fluids like a large number of small particles with a randomly movement. The particle streaming and elastic particle collision are the base for calculation of exchange of momentum and energy. This process can be modelled by the Boltzmann transport equation, which is given by:

$$\frac{\partial f}{\partial t} + \vec{u} \cdot \nabla f = \Omega \tag{1}$$

where $f(\vec{x}, t)$ is the particle distribution function, $\vec{u}$ is the particle velocity, and Ω is the collision operator.

The LBM simplifies the Boltzmann's original idea of gas dynamics, because reduces the number of particles and keeps them arranged in the nodes of a lattice, similar to simplification of a football game like a Foosball table. For a 2D model one particle can stream in 9 directions, that includes it stay at rest. The model is commonly known the denominated $D2Q9$ (see in Figure 4), because is 2D and has 9 velocity vector, the microscopic velocities are denoted by $\vec{e}_i$, for $i = 0, ..., 8$, defined by:

$$\vec{e}_i = \begin{cases} (0,0) & i = 0 \\ (1,0), (0,1), (-1,0), (0,-1) & i = 1,2,3,4 \\ (1,1), (-1,1), (-1,-1), (1,-1) & i = 5,6,7,8 \end{cases} \tag{2}$$

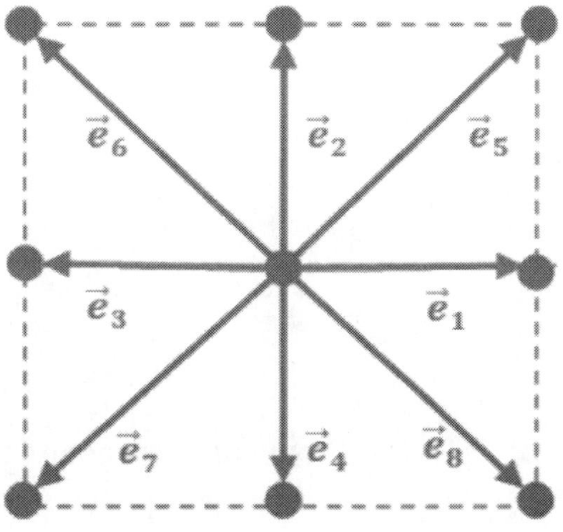

Fig. 4 Illustration of a lattice node of the $D2Q9$ model [1].

The definition of the macroscopic fluid density is given by the sum of microscopic particle distribution:

$$\rho(\vec{x}, t) = \sum_{i=0}^{8} f_i(\vec{x}, t) \tag{3}$$

The macroscopic velocity $\vec{u}(\vec{x}, t)$, consequently, is defined by an average of microscopic velocities $\vec{e}_i$ weighted by the distribution functions f_i:

$$\vec{u}(\vec{x}, t) = \frac{1}{\rho} \sum_{i=0}^{8} c f_i \vec{e}_i \tag{4}$$

The streaming and collision processes, which consists the main phases in LBM are gives by:

$$\overbrace{f_i(\vec{x} + c\vec{e}_i \Delta t, t + \Delta t - f_i(\vec{x}, t))}^{\text{Streaming}} = -\overbrace{\frac{|f_i(\vec{x}, t) - f_i^{eq}(\vec{x}, t)|}{\tau}}^{\text{Collision}} \tag{5}$$

where f_i^{eq} which is the equilibrium distribution, and τ is the relaxation time towards local equilibrium.

The model is implemented with the collision and streaming calculations performed separately, an special attention is given to the boundary nodes of the lattice. The Figure 5 presents the phases of the streaming progressively inside the lattice nodes.

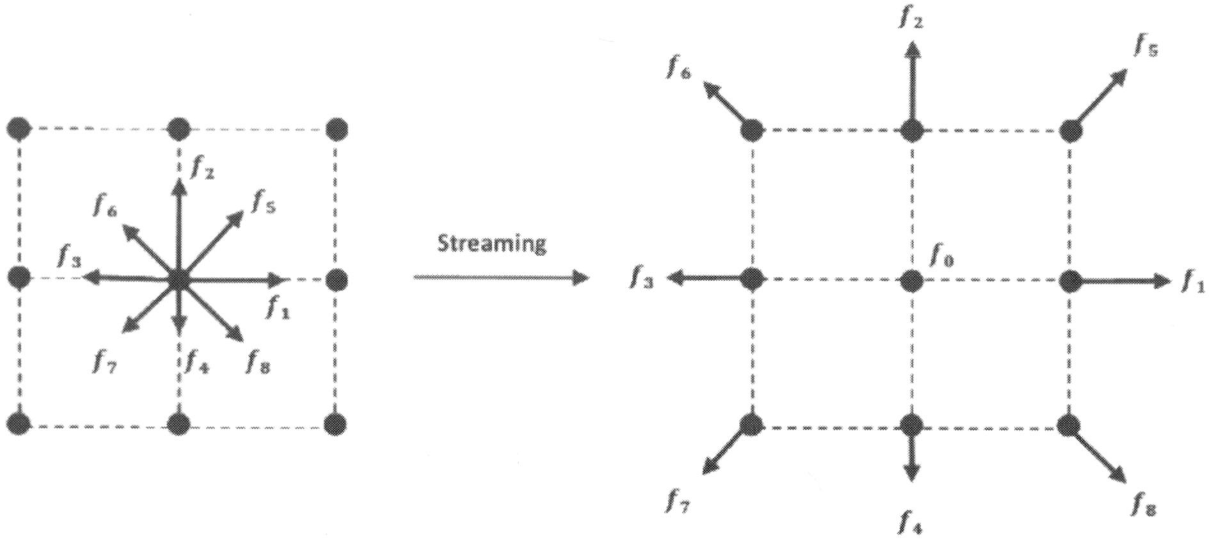

Fig. 5 LBM Streaming process [1].

For single-phase simulations is adequate use Bhatnagar-Gross-Krook (BGK) collision, where equilibrium distribution f_i^{eq} is defined by:

$$f_i^{eq}(\vec{x}, t) = \omega_i \rho + \rho s_i(\vec{u}(\vec{x}, t)) \tag{6}$$

and $s_i(\vec{u})$ is defined by:

$$s_i(\vec{u}) = \omega_i \left[3\frac{\vec{e_i} \cdot \vec{u}}{c} + \frac{9}{2}\frac{(\vec{e_i} \cdot \vec{u})^2}{c^2} - \frac{3}{2}\frac{\vec{u} \cdot \vec{u}}{c^2} \right] \tag{7}$$

where the weights designed by ω_i are:

$$\omega_i = \begin{cases} 4/9 & i = 0 \\ 1/9 & i = 1, 2, 3, 4 \\ 1/36 & i = 5, 6, 7, 8 \end{cases} \tag{8}$$

and the lattice speed is given by $c = \frac{\Delta x}{\Delta t}$. The fluid kinematic viscosity ν is related to the relaxation time τ, it is used in the $D2Q9$ model and defined by:

$$\nu = \frac{2\tau - 1}{6}\frac{(\Delta x)^2}{\Delta t} \tag{9}$$

The LBM algorithm can be sketched by the following steps:
1) Definitions of ρ, $\vec{u}$, f_i, and f_i^{eq};
2) Stream phase, move $f_i \rightarrow f_i^*$ in the direction of $\vec{e}_i$;
3) Calculation of macroscopic ρ and $\vec{u}$ from f_i^* using Equation 3 and 4
4) Obtain f_i^{eq} using Equation 6;
5) Collision phase, compute the updated distribution function $f_i = f_i^* - \frac{1}{\tau}(f_i^* - f_i^{eq})$ using Equation 5
6) Repeat steps 2 to 5.

When $\tau \rightarrow 1/2$ numerical issues can occur, so, during the phase of streaming and collision are required special treatments to the boundary nodes on the distributions in order to satisfy the necessary macroscopic boundary conditions (BCs).

3. Boundary Conditions

One of the most important concerns of creating the correct flow with the LBM is to accurately impose the desired boundary conditions. Although specifying boundary conditions on the Navier-Stokes equation is straightforward, imposing boundary conditions within a Lattice Boltzmann simulation can be very hard, especially on the Graphics Processing Unit (GPU). The most common situations for BCs are:

- Bounce Back: No-Slip Boundary Conditions;
- Boundary Conditions with Known Velocity;
- Periodic Boundary Conditions;
- Imposed Pressure Difference Boundary Conditions.

The BCs are difficult to apply because macroscopic boundary conditions from the Navier Stokes equation must be translated to boundary conditions on f and f_{eq}. Therefore is necessary determine appropriate equations for calculating these variables along boundaries given a macroscopic boundary condition and this is very important to the accuracy and stability of the LBM numerical solution.

Some of the popular BCs are the Bounce-back [27] and Zou-He [34].

In order to obtain a no-slip boundary condition for a specified boundary, we use the "Bounce Back Method". The most simple scheme for this method is to place a wall halfway between a wall and a fluid grid points and then "bounce-back" particles that stream into the wall. The Figure 6 presents f_4, f_7, and f_8 stream into the wall, and are bounced back by setting $f_5 = f_7$, $f_2 = f_4$, and $f_6 = f_8$.

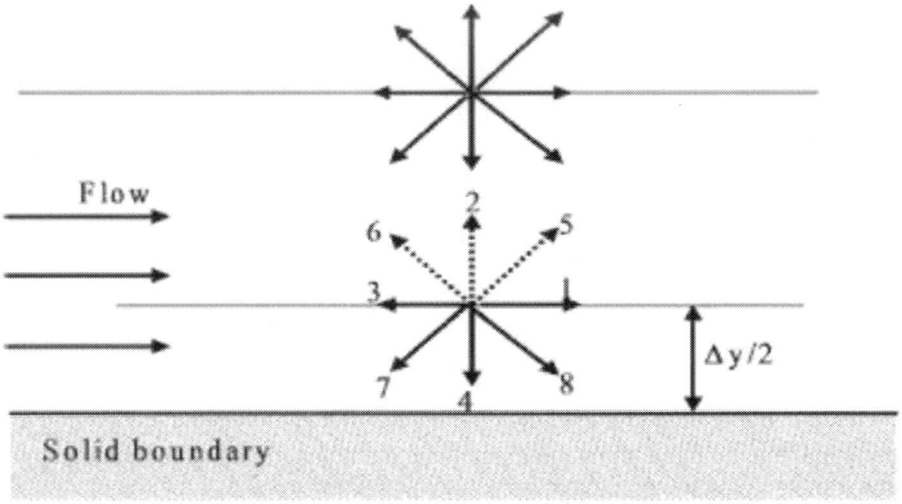

Fig. 6 Scheme of the Bounce Back Method [21].

Another common situation on simulations is to impose a constant velocity at an inlet and outlet. Considering the case presented by the Figure 7, in the west inlet there are particles f_6, f_3, f_7, f_2, and f_4 that will stream onto a point on the wall and the condition is to solve for f_5, f_1, and f_8 based on the constant velocity assumption. usually one should apply bounce back to the non equilibrium part of f on the boundaries, i.e. $f_5 - f_{5,eq} = f_7 - f_{7,eq}$ [34]. The f_{eq} depends on the aerodynamic variables, so the constant-velocity boundary condition will enter through $f_e q$.

The periodic boundary condition is more simple because one can directly relate the knows of one boundary to the unknowns of the other boundary. In the Figure 8 in (a) the unknowns are, after streaming, f_6, f_2, and f_5. However, along boundary (b) these three distributions are known. So it is possible to solve all of the unknowns by relating $f_{6,a} = f_{6,b}$, $f_{2,a} = f_{2,b}$, and $f_{5,a} = f_{5,b}$, and likewise for the other boundaries and for the unknowns of (b) relating them to the known variables of (a).

The pressure boundary condition is a particular case, because as pressure enters Lattice Boltzmann simulations in a subtle way, an example is relate the pressure to density by $\Delta P \propto \rho c_s^2$. To set f at each boundary one way is to follow a procedure similar to the BCs with known velocity but in the present case, setting ρ to a constant.

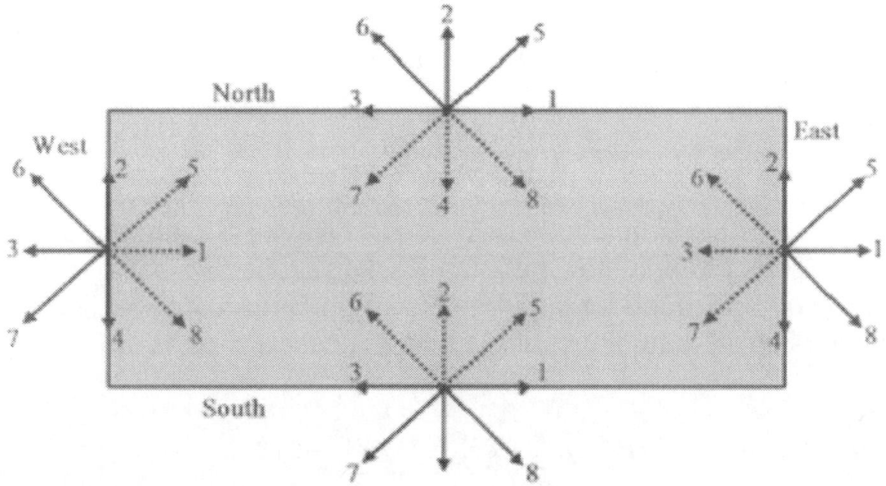

Fig. 7 Schematic showing the known and unknown distribution functions at respective boundaries of the domain [21].

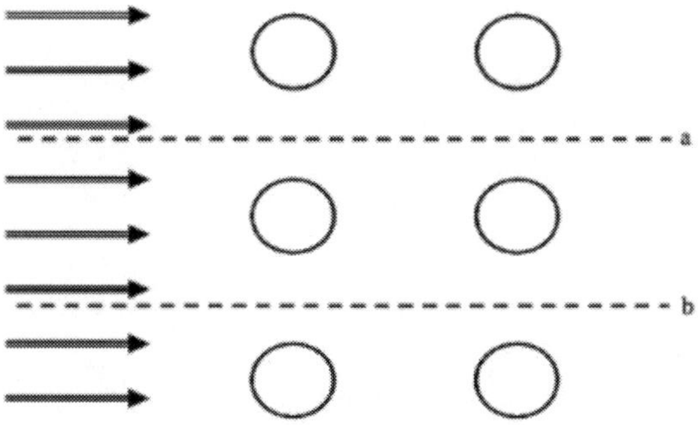

Fig. 8 Schematic illustrating the periodic boundary conditions and how they relate to each other and the flow [21].

The Lattice Boltzmann simulations become unstable when $\Delta\rho$ is large, so it must to adjust the resolution of the simulation until $\Delta\rho \sim 0.01$. For this method it is also important to make sure that the initial velocity and density matches that of the imposed flow.

4. Relations of the Lattice Boltzmann Method and Navier-Stokes Equations

The LBM is considered a special finite difference method [7] for solving the Boltzmann transport equation arranged like a lattice. Describing the Boltzmann transport equation in terms of the discrete distribution function:

$$\frac{\partial f_i}{\partial t} + \vec{e}_i \cdot \nabla f_i = \Omega_i \tag{10}$$

Discretizing the differential and collision operators:

$$\frac{f_i(\vec{x}, t + \Delta t) - f_i(\vec{x}, t)}{\Delta t} + \frac{f_i(\vec{x} + \vec{e}_i \Delta x, t + \Delta t) - f_i(\vec{x}, t + \Delta t)}{\Delta x} = -\frac{f_i(\vec{x}, t) - f_i^{eq}(\vec{x}, t)}{\tau} \tag{11}$$

Defining that $\Delta x = \Delta t = 1$, the Equation 5 is recovered.

The LBM may be used for fluid simulations because the Navier-Stokes equations can be recovered from the discrete equations through the Chapman-Enskog procedure using a multi-scaling expansion technique. The $ID2Q9$ model used for simulate Rayleigh-Bénard convection is able to recover the incompressible Navier-Stokes equations, using a multi-scaling expansion:

$$g_i = g_i^{(0)} + \epsilon g_i^{(1)} + \epsilon^2 g_i^{(2)} + \ldots \tag{12}$$

$$\frac{\partial}{\partial t} = \epsilon \frac{\partial}{\partial t_1} + \epsilon^2 \frac{\partial}{\partial t_2} + \ldots \tag{13}$$

$$\frac{\partial}{\partial x} = \epsilon \frac{\partial}{\partial x_1} \tag{14}$$

where $g_i^{(0)} = g_i^{eq}$ and $\epsilon = \Delta t$ is the expansion parameter. Is possible that up to $O(\epsilon)$ to derive the following continuity and momentum equations [11]:

$$\nabla \cdot \vec{u} = \epsilon \left(\tau - \frac{1}{2}\right) P + O(\epsilon^2) \tag{15}$$

$$\frac{\partial \vec{u}}{\partial t} + \nabla \cdot (\Pi^0) = \epsilon \left(\tau - \frac{1}{2}\right) Q + O(\epsilon^2) \tag{16}$$

$$\tag{17}$$

where $P \sim O(\epsilon)$ and $Q \sim O(\epsilon) + O(M^2) + \frac{c^2}{3}\nabla^2\vec{u}$ and M is the Mach number.

With the results of P and Q to Equations 16 and 17, the continuity equation is derived accurate to $O(\epsilon^2)$ and the momentum equation is derived to $O(\epsilon^2 + \epsilon M^2)$.

$$\nabla \cdot \vec{u} = 0 + O(\epsilon^2) \tag{18}$$

$$\frac{\partial \vec{u}}{\partial t} + \vec{u} \cdot \nabla \vec{u} = -\nabla p + \nu \nabla^2 \vec{u} + O(\epsilon^2 + \epsilon M^2) \tag{19}$$

5. Lattice Boltzmann Method and Acoustics

The LBM is commonly used to predict some wave behavior using the D2Q9 system with a delta function to excite in a defined node, see Figure 9, but it is not useful to the acoustic study because this solution vary with the system's viscosity and has no simple analytic solution.

The first studies of LBM acoustics [5, 6, 9, 12], the sound waves have been generated configuring an initial density and velocity distribution wave correspondent. These studies was focused in waves with a "infinite" dimension letting the initial wave distribution propagates through a periodic boundary.

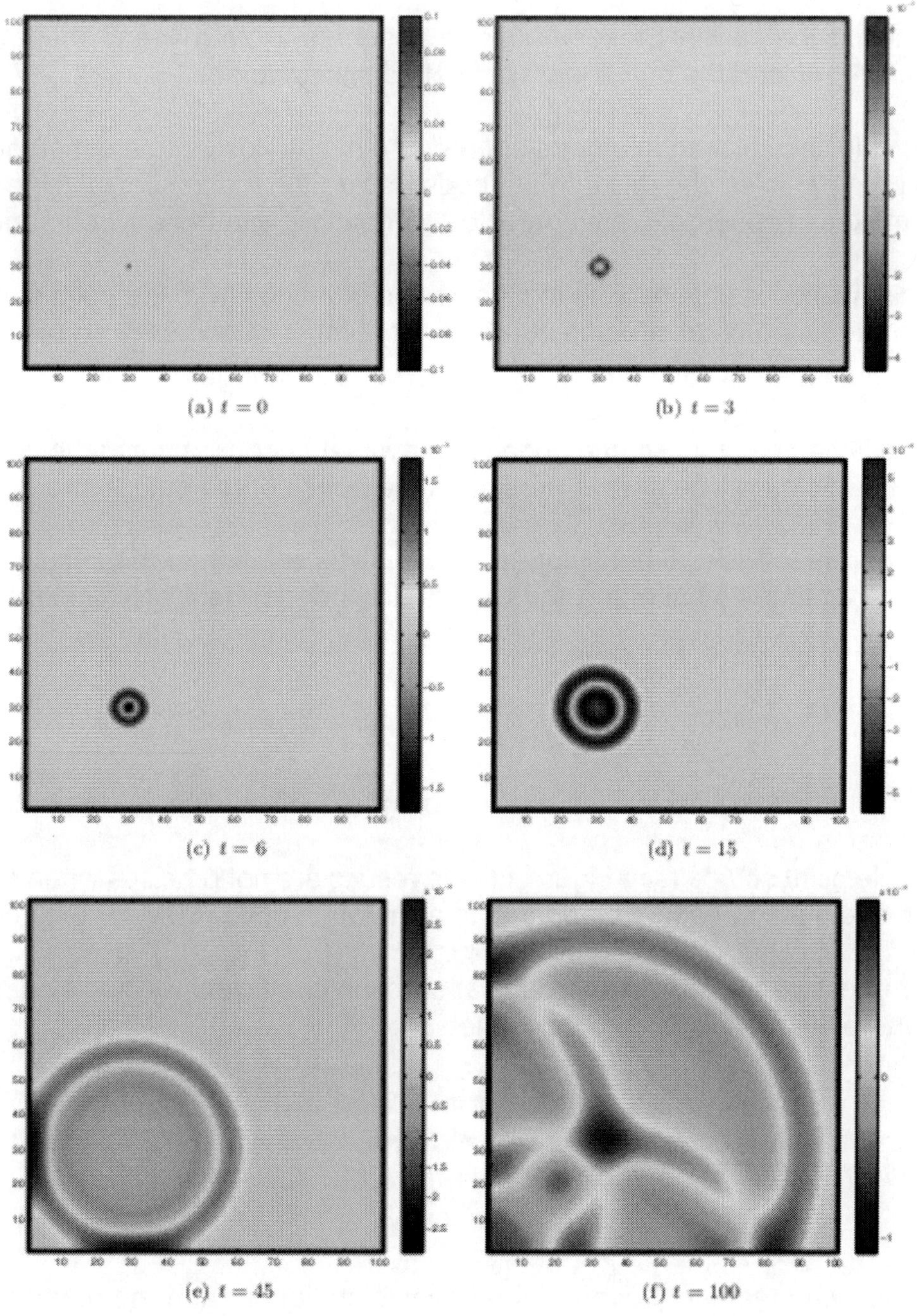

Fig. 9 Simulation of a walled system excited by a delta pulse [31].

For acoustic simulation of non-infinite waves the node of the acoustic source sends a sinusoidal signal. In the work of [31] is presented a simple way to define it, locking a node's density to a sine function around the equilibrium density:

$$\rho(\vec{x}, t) = \rho_0 + \rho_S \sin\left(\frac{2\pi}{T} t\right) \tag{20}$$

where ρ_S is the source amplitude point, and T the period of the oscillation in lattice units.

For wave propagation the Navier-Stokes equation assumes to be a linear phenomenon for small disturbances, so $\rho_S \ll \rho_0$ and the non-linear wave effects are avoided.

For macroscopic velocity $\vec{u}$ placed in a point source one may to lock it to a constant source velocity, like $\vec{u} = 0$ or to let the particles streams into the node decide the macroscopic velocity at the point of the source, that makes the source to emit particles compatible with the environment.

The equilibrium distribution is obtained setting the values of ρ and $\vec{u}$ with the point source node locked using the equation 20 to the node streams out the right number of particles, so the constraints of the equilibrium distribution are not valid $\left(\text{i.e. } \sum_i f_i^{(0)} = \sum_i f_i\right)$ as ρ changes.

So there is essential that the point source node is always set to equilibrium in the collision step to the correct number of particles will be always present in the point source before streaming.

Is necessary to guarantee the real behaviour of the analytic solution of the simulated system, so in cases 2d the z coordinate is infinite and the results are independent of z, according the following equation:

$$\rho_0 = A H_0^{(2)}(kr) e^{jwt} \tag{21}$$

that represents an outgoing cylindrical wave with z being an infinite line, in cylindrical coordinates.

Considering a single point source (see Figure 11) the results are not precises when the interest points are measured next to the source node, and that is a limitation of the LBM for the acoustic case.

The results are impressive because the viscous absorption coefficient α_S in k is calculated using the sum $4\nu/3 + \nu'$, where this viscosities are calculated by:

$$\nu = c_s^2 \left(\tau - \frac{1}{2}\right) \tag{22}$$

$$\nu' = \frac{2}{3}\nu \tag{23}$$

the comparison between the attenuation of the analytic stationary and the numerical solutions are remarkable (see Figure 10), validated, thus, the expected results for the lattice viscosity as presented

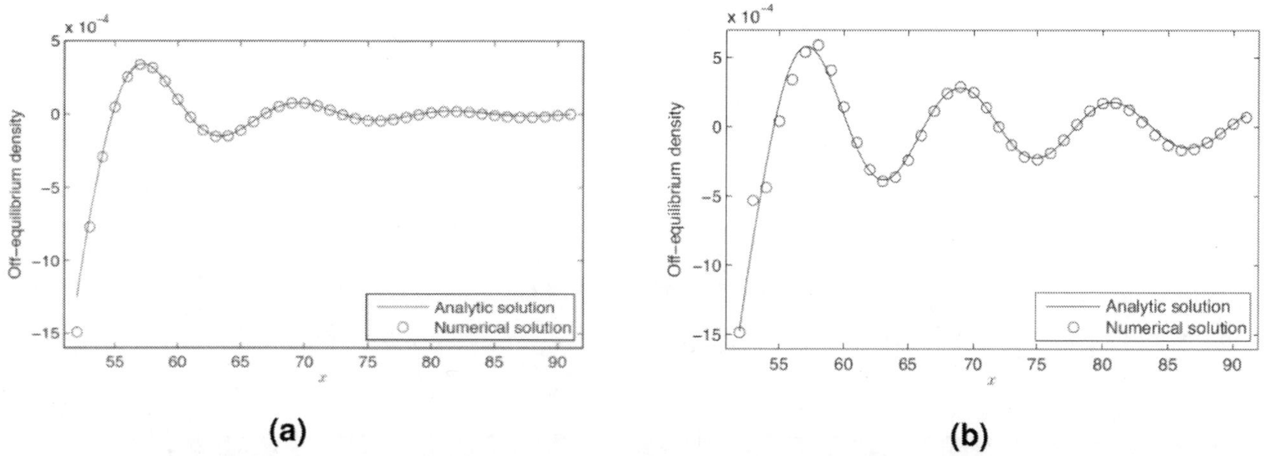

Fig. 10 Best comparison of transient numerical solution with the analytic stationary solution for $t = 75$ a) $\tau = 0.6$ and b) $\tau = 1$ [31].

in [31].

In the results obtained by [31] τ values were adjusted by the value of $|A|$, and an important observation is that for some values of τ the speed of sound measured are not correct and the wavelengths differ between the analytic and numerical methods, and this is an indication that the speed of sound depends on more than the viscosity.

Using a single source point to simulate an infinite line source in z direction results in a damped cylindrical wave, and simulating a continuous line of point sources along a given x in a periodic system results in a plate source that is infinite in y and z direction. The propagation are 2d and symmetrical in x and $-x$ directions of the plate, and this propagation is based by:

$$p_a = Ae^{-\alpha_s x} e^{j(\omega t - kx)} \tag{24}$$

and α_s is the *spatial absorption coefficient* and is given by:

$$\alpha_s = \frac{\omega}{c_s\sqrt{2}}\sqrt{\frac{\sqrt{1+(\omega\tau_s)^2}-1}{1+(\omega\tau_s)^2}} \tag{25}$$

The expected behavior is good but not adequate when the system approaches a steady state, see in Figure 12.

6. The Use of LBM Simulation of HAWT Noise

The LBM can predict far field noise according the studies [17, 33] and has also been used to simulate airfoil sections, like to investigate the TE noise [16, 29, 30].

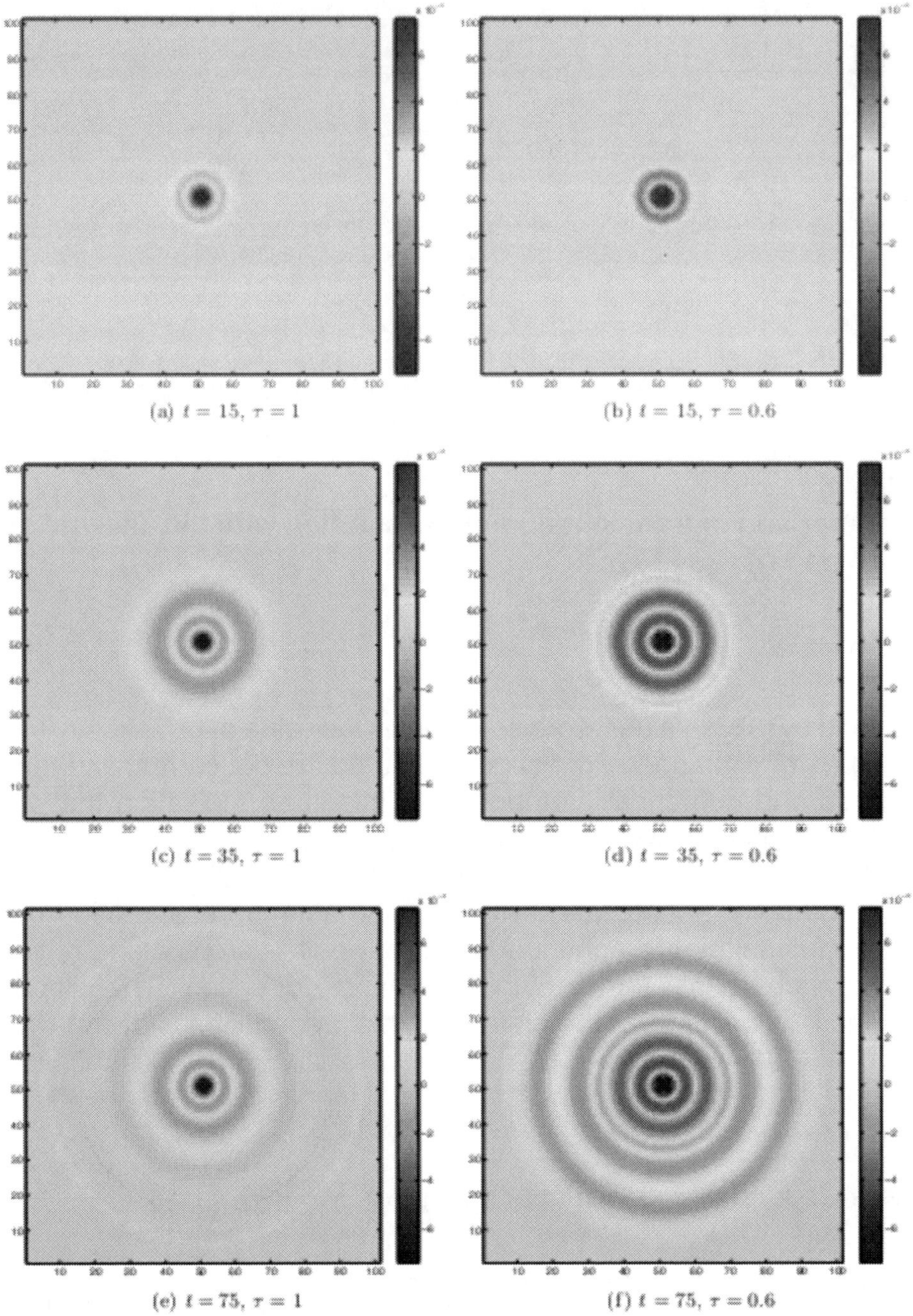

Fig. 11 Snapshots of off-equilibrium density in two different periodic systems by a point source, with for τ = 1 and τ = 0.6, considering the minimum pressure [31].

The main advantage of the LBM comparing with Computational Fluid Dynamics is that it can solve the compressible flow equations and obtain the hydrodynamic and acoustic pressure fluctuations.

The LBM segments the flow area in an unitary 3d elements called voxels and they may be grouped in variable resolution (VR) zones and this this implies in the refinement of the mesh. The open source

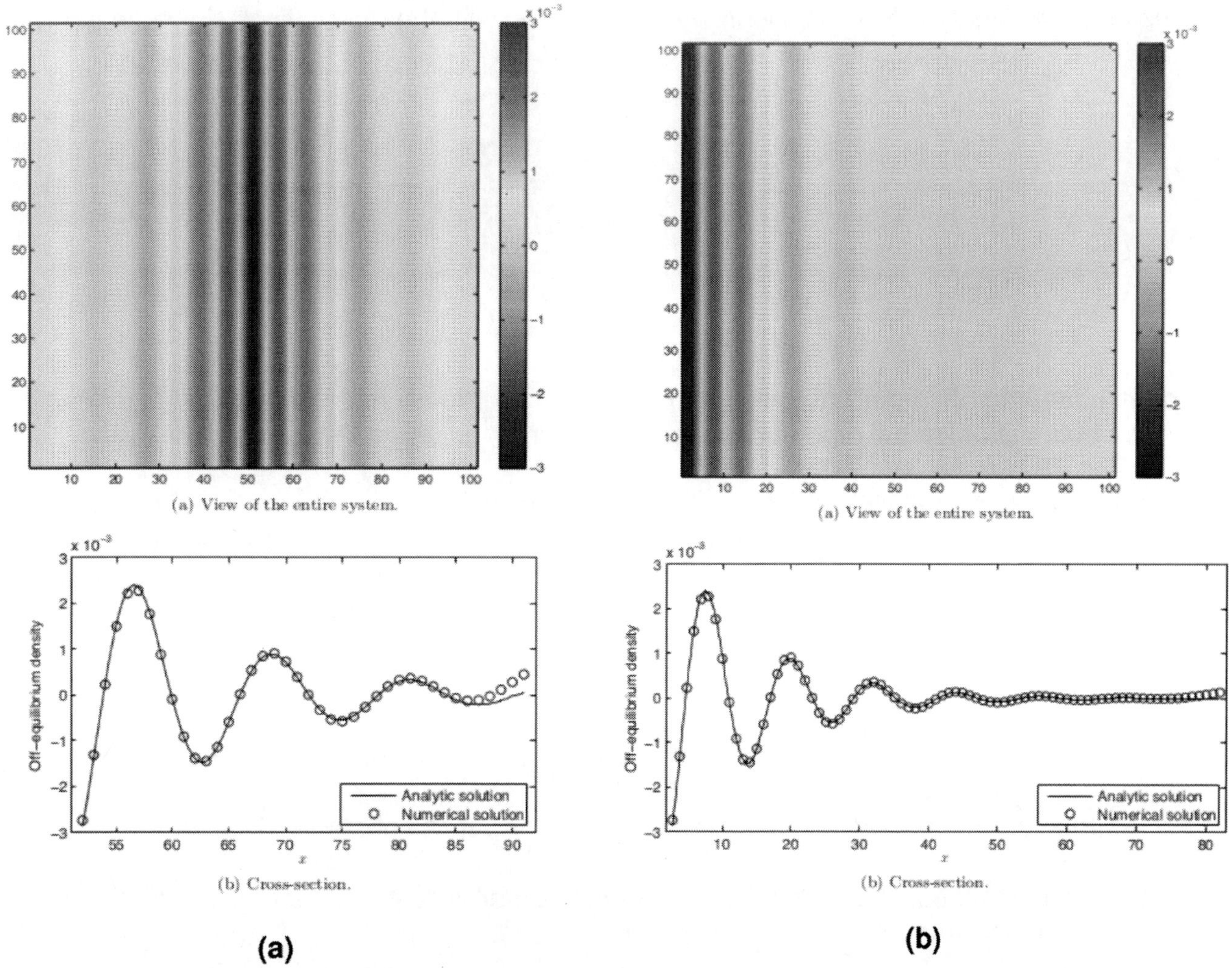

Fig. 12 The state of a system at $\tau = 1$ and $N = 101$ of point sources along a) $t = 75$ and $y = 51$, b) $t = 155$ and $y = 2$ [31].

Palabos.

The Palabos application consider the mesh refinement according to [14], which is based on the off-equilibrium and equilibrium parts of the distribution function are based by the Knudsen number:

$$f^{\text{neq}} \sim f^{\text{eq}} Kn \tag{26}$$

and the refinement criterion is determined by the measured Knudsen number (see equation 27), and this implies in the quality of the solutions when compared with the "theoretical" Knudsen number (see equation 28).

$$Kn = \frac{\lambda}{L} = \frac{\tau c_s}{L} \sim \frac{f_i^{(1)}}{f_i^{(0)}} \tag{27}$$

$$Kn = \frac{\lambda}{L} = \frac{\tau c_s}{L} = \frac{Ma}{Re} \tag{28}$$

To simulate an airfoil is important to set the boundary conditions, so the inlet is given a input velocity and in the outlet is defined by the atmospheric pressure.

The expected results for this simulation according [16] must be close to the obtained by other methods (f.e. LES) and the experimentally measured. It is also presumed that the far field will be more accurated then points measured next to the source nodes, because the microphones are placed in the simulation area, and it is possible to calculate a frequency dependent correction factor in 2d simulations:

$$\beta = 10 \log_{10} \left(\frac{f b^2}{c R} \right) \tag{29}$$

where b is the segmented span, c is the sound speed and R the observer distance.

The most significant source of noise is, according the literature [3, 29], located in the TE area from the airfoil, and it generates turbulent eddies from this site.

The most simple way to perform the simulation is with the *Exa PowerFLOW™*, and it gets a lot of computational cost (between two and three days). Using an open source software like *Palabos* the difficulty is greater given the need for great knowledge of computational programming.

Another use for the LBM is to obtain the acoustic directivity in complex shapes, like in [29] that investigated the fluid flow parameters in a clean and serrated wind turbine profile called DU95. The LBM also may simulate turbulent and transition structures and may help to predict the noise in the far field.

Fig. 13 Simulation of a WT airfoil in *Exa PowerFLOW™* [32]

7. QBlade and the PNoise Module

The QBlade is an open source made by the *Technische Universität Berlin* (T.U. Berlin) wind turbine calculation software, distributed under the General Public License (GPL).

This application allows users to rapidly design custom airfoils mainly for research purposes in a easy and intuitive way.

The noise module, called *Pnoise* was made and is still being performed in collaboration by the *Poli Wind* group, from the Poli-USP. Some of the works published by the group are [23] and [24].

This module has incorporated the trailing edge (TE) noise, that is is based on a modified Brooks-Pope-Marcolini (BPM) TE noise model [4] with turbulent boundary layer data provided by the analysis tool for airfoils nominated XFLR5 [10]. These two models were integrated inside the project of one wind turbine in the software QBlade [18–20, 22].

Some improvements are planned in the *PNoise* module with the inclusion of other self-noise sources, currently is being made the study of the leading edge (LE) noise for future implementation.

The "quasi-3D rotor" prediction tool is being made too and a challenge for this project is to project the calculated noise at the far-field position from the noise source, for this was considered the use of the LBM.

The LBM is able to gain in refinement and quality of simulation results and also to compare with other predictive methods, like using a parabolic equation [15] and statistical predictions [13].

Even if it is defined that all the particles are arranged at a point source, this positioning is updated according to their densities, as presented by equation 20, so, to create a one-sided plate source is made by setting the point sources next to a wall, and locking $\vec{u}$ to 0 the effects of the reflected particles are unconsidered. The behaviour would be similar in the D1Q3 lattice comparing with the D2Q9 lattice when the system is one-dimensional projected.

8. PNoise and LBM

The *PNoise* module in *QBlade* software is gradually increasing its noise analysis spectrum, like to include more noise sources as the leading edge noise and to implement the quasi-3d analysis.

The LBM is also being increasingly improved and it has brought us a new approach in the WTN area, mainly because of its great capacity to solve problems and to not depend on the Navier Stokes solutions, in effect almost all the studies in this area were performed using the commercial software called *Exa PowerFLOW™* and very little developed in open source applications, such as *Palabos*.

For the time the LBM may become another alternative for comparing the results from *PNoise* in order to validate the data obtained, however, to implement the LBM within *PNoise* there are certain difficulties. The parallelization (the ability to process in parallel a certain computational code to reduce simulation time) must be selected using the processor capacity from the computer that runs the software, the increase in simulation time, which would be a tremendous disadvantage in this implementation and the difficult to insert the LBM inside the *PNoise* code, among others.

The best way to introduce a LBM for WTN in the *PNoise* is using an existent open source software, created in the same computational language (like C based, the PNoise is created in "qt") and creating a specific model to be adjusted for each airfoil profile, and the best way to do this is using a geometry in "stl" file format what turns the simulation in LBM in 3d considering the *Palabos* software. If the geometry is included as a 2d, each node will need to have boundary conditions, that is hard to do.

Therefore, the LBM was demonstrated in the literature to be useful and may be used to enhance the results of the simulation of *PNoise* for validation, mainly in the far field noise but not integrated inside the *PNoise* because it requires great computational capacity and this is not the scope of *QBlade* software.

9. Conclusion

This work presented the LBM, its characteristics, their use for the WTN propagation study considering the literature data and its application considering the results obtained with the module *PNoise* from *QBlade* software.

The LBM has been increasingly used in the WTN literature because of the several advantages for fluid simulations over traditional CFD methods in the following cases:

- To deal with Complex boundaries using on-grid bounce-back;
- LBM may be applied to simulate flows with complex geometries such as porous media flows;

- LBM can be easily parallelized and thus can be applied to the Large eddy simulation (LES).

One fault expected is come from the derivation of the Navier-Stokes equations from LBM, because the compressible effects can produce a source of errors when solving the incompressible Navier-Stokes equation, but this is not a problem for the WTN due to consider an incompressible flow.

Some of the main characteristics of the LBM are the efficiency and flexibility, it was designed to solve complex physics and sophisticated algorithms, and run in high performance hardware. With the LBM some problems that requires great accuracy could finally be solved. The LBM is based on particle mesh and allows the creation of hybrid systems that may be coupled in particle phenomena.

Compared to classical CFDs, the LBM is the largest resource consumer because the discrete probability distribution functions used in this model requires more storage memory than the Navier-Stokes equation variables, it may be compensated by the computational efficiency and the use of modern computers that suit parallel architecture or using General Purpose Graphics Processing Units (GPGPUs) and this is a disvantage for simulation WTN.

The *PNoise* is a module that calculates WTN from the trailing edge, has been increasingly improved like it is planned to include the leading edge noise source and the quasi 3d modelling. One thinked point in this study is how to use the LBM with *PNoise*.

Therefore, the LBM was demonstrated in the literature to be useful and may enhance the results of the simulation of *PNoise* but it requires great computational capacity and requires much simulation time, and the simulation speed is one of the main advantages of the *Qblade* software.

The use to validate the results from *PNoise* are interesting, mainly to comparison in the far field region.

References

[1] Bao, Y. and Meskas, J. (2014). Lattice boltzmann method for fluid simulations.

[2] Begum, R. and Abdul Basit, M. (2008). Lattice boltzmann method and its applications to fluid flow problems. *European Journal of Scientific Research*, 22:216–231.

[3] Bertagnolio, F. (2008). Trailing edge noise model applied to wind turbine airfoils.

[4] Brooks F., T., Pope, D., and Marcolini A., M. (1989). Airfoil self-noise and prediction. 1218.

[5] Buick, J., Greated, C., and Campbell, D. (2007). Lattice bgk simulation of sound wave. *EPL (Europhysics Letters)*, 43:235.

[6] Buick, J., L Buckley, C., Greated, C., and Gilbert, J. (2000). Lattice boltzmann bgk simulation of nonlinear sound waves: The development of a shock front. *Journal of Physics A: Mathematical and General*, 33:3917.

[7] Chen, S., Martinez, D., and Mei, R. (1996). On boundary conditions in lattice–boltzmann methods. *Physics of Fluids - PHYS FLUIDS*, 8:2527–2536.

[8] Cortés E., Sánchez F., O. A. M. B. H. M. and M., Y. T. (2017). On the material characterisation of wind turbine blade coatings: The effect of interphase coating – laminate adhesion on rain erosion performance. *Materials (Basel).*

[9] Cosgrove, J., Buick, J., Campbell, D., and Greated, C. (2004). Numerical simulation of particle motion in an ultrasound field using the lattice boltzmann model. *Ultrasonics*, 43(1):21 – 25.

[10] Drela M., Y. H. and A., D. (2009). Xfoil/xflr5. *Cambridge, Ma.*

[11] Guo, Z., Shi, B., and Wang, N. (2000). Lattice bgk model for incompressible navier–stokes equation. *Journal of Computational Physics*, 165:288–306.

[12] Haydock, D. and Yeomans, J. (2001). Lattice boltzmann simulations of acoustic streaming. *Journal of Physics A: Mathematical and General*, 34:5201.

[13] Kelly, M., Barlas, E., and Sogachev, A. (2018). Statistical prediction of far-field wind-turbine noise, with probabilistic characterization of atmospheric stability. *Journal of Renewable and Sustainable Energy*, 10:013302.

[14] Lagrava, D., Malaspinas, O., Latt, J., and Chopard, B. (2012). Advances in multi-domain lattice Boltzmann grid refinement. *Journal of Computational Physics*, 231(14):4808–4822.

[15] Lee, S. (2015). Prediction of far-field wind turbine noise propagation with parabolic equation.

[16] Manegar, F., Carolus, T., and Erbsloeh, S. (2017). High fidelity airfoil trailing edge noise predictions via lattice-boltzmann simulations.

[17] Mann, A., Perot, F., Kim, M.-S., Casalino, D., and Fares, E. (2012). Advanced noise control fan direct aeroacoustics predictions using a lattice-boltzmann method.

[18] Marten, D. (2010). Extension of an aerodynamic simulator for wind turbine blade design and performance analysis. *TU Berlin,.*

[19] Marten, D. (2014). Qblade short manual v0.8.

[20] Marten, D. and Peukert, J. (2013). Qblade guidelines v0.6.

[21] Mohamad, A. (2011). *Lattice Boltzmann Method Fundamentals and Engineering Applications with Computer Codes.*

[22] Pechlivanoglou, G. e. a. (2009). Qblade. *Berlin.*

[23] Saab Jr, J. (2015). Displacement thickness evaluation for semi-empirical airfoil trailing-edge noise prediction model. *Journal of the Brazilian Society of Mechanical Sciences and Engineering*, 38.

[24] Saab Jr, J. (2016). Trailing-edge noise: development and application of a noise prediction tool for the assessment and design of wind turbine airfoils.

[25] Sawyer, S. and Dyrholm, M. (2017). Annual market update 2017. *Global Wind Report from Global Wind Energy Council (GWEC).*

[26] Stefan Oerlemans, P. F. (2012). Low-noise wind turbine design.

[27] Succi, S. (2003). *The Lattice Boltzmann Equation – For Fluid Dynamics and Beyond*, volume 22, page 368.

[28] Sukop, M. and T. Thorne Jr, D. (2006). Lattice boltzmann modeling: An introduction for geoscientists and engineers. *Lattice Boltzmann Modeling: An Introduction for Geoscientists and Engineers*.

[29] van der Velden, W., H Van Zuijlen, A., Jong, A., and Bijl, H. (2015). On the noise prediction of a serrated du96 airfoil using the lattice boltzmann method.

[30] van der Velden, W. C., van Zuijlen, A., and Ragni, D. (2016). Flow topology and noise emission around straight, serrated and slitted trailing edges using the Lattice Boltzmann methodology.

[31] Viggen, E. M. (2009). *The Lattice Boltzmann Method with Applications in Acoustics*. PhD thesis.

[32] Wouter, V. (2018). Wind turbine noise prediction.

[33] Zhu, T., Sturm, M., H Carolus, T., Neuhierl, B., and Perot, F. (2014). Experimental and numerical investigation of tip clearance noise of an axial fan using a lattice boltzmann method. volume 2.

[34] Zou, Q. and He, X. (1996). On pressure and velocity boundary conditions for the lattice boltzmann bgk model. *Physics of Fluids*, 9.

INCE*EUROPE*

8th International Conference
on
Wind Turbine Noise
Lisbon – 12th to 14th June 2019

Towards digital wind turbine noise certification

Wouter van der Velden and Damiano Casalino
Dassault Systemes Deutschland GmbH, Curiestraße 4, 70563, Stuttgart, Germany
wouter.vandervelden@3ds.com

Summary

Flow and acoustic simulations are performed on the NM80 full-scale 3-bladed wind-turbine, tested during the DanAero campaign. SIMULIA PowerFLOW's Lattice Boltzmann Method is used to predict the turbulent-resolved unsteady flow around the full wind-turbine, yielding good results with reference data. The noise at certification position for several rotor revolutions is computed using an auralization technique from simulated flow data stored along a portion of the rotor revolution. The auralization process is based on a frequency-domain Ffowcs-Williams and Hawking noise spectrum computation from blade wall-pressure pressure fluctuations, taking into account ground reflection. Noise spectra at several azimuthal copies of the microphones are finally corrected to take into account blade-motion Doppler effects end atmospheric absorption and smoothly appended to recover the full rotor revolution. Significant cost reduction in simulation time is found with the new post-processing process, making the entire process suitable for industrial turn-around-times.

1. Introduction

Wind energy; the solution to deal with an increasingly recognized and critical challenge for mankind. In order to reduce greenhouse gases, the demand for renewable energy, such as wind energy, is growing and growing. A study from van den Berg [13] showed that noise, and in particular the swishing character of the noise, is one of the most annoying aspects for onshore wind turbines. To protect public health, governments apply strict noise regulations for both maximum, average and modulated noise levels. Hence, noise constitutes as an important barrier for the widespread application of wind energy as many wind turbines must operate at reduced power, especially during the night when background noise levels are lower. This leads to a lower power output from the turbines, which will lead to an overall reduction of the annual energy production, finally resulting in a loss of revenue.

Therefore, when designing a wind-turbine, an important target to optimize the wind turbine efficiency is the blades shape. As the wind power is directly related to the exposed surface area, one market trend is to propose increased rotor size which has two consequences, among others. One is that prototyping becomes more expensive and more challenging to carry out. The other is that the flow-induced noise contribution keeps increasing and can be a significant source of annoyance for people living in a nearby community. Aerodynamic noise is related to the rotational speed, the presence of tower, the inflow properties, the blade tip design and the blade geometry. The latter is increasing in complexity, with the airfoil profile, chord, and twist distributions depending on the span location in order to trade-off aerodynamic efficiency and noise. The blade-tower interactions mostly impact the low frequencies, whereas the overall blade self-noise is responsible of the mid and high frequency content caused by the vortical flow past the blade tip, flow separation in the inner part of the blade, vortex shedding from a blunt trailing edge, and boundary-layer turbulent fluctuations passing by the trailing edge [2, 10]. Tip noise and blunt-trailing-edge noise can be prevented by a proper blade design. Flow separation noise may be significant in off-nominal conditions and can be also reduced by a proper aerodynamic design and by using vortex generators. Whereas, mitigating turbulent-boundary-layer-trailing edge noise (subsequently denoted as trailing edge noise), which is widely recognized as the dominant source of wind-turbine noise, is way more challenging.

The goal of this paper is to apply the knowledge gained airfoil trailing-edge noise prediction in wind-tunnel quasi two-dimensional conditions to a full three-dimensional rotating turbine. The full-scale 2.3 MW NM80 turbine used in the DanAero [9] measurement campaign is considered. One of the three blades of the turbine was specifically manufactured for this experiment and equipped with multiple sensors. The blade is a LM 38.8 m blade attached to the hub at a radius of 1.24 m from the rotor center. Therefore, the rotor radius at the blade tip is 40.04 m, which includes the pre-bend of the blades. A met mast located near the turbine was used to monitor atmospheric conditions.

In this paper, SIMULIA's PowerFLOW time-explicit, compressible and transient solver based on Lattice-Boltzmann Method (LBM) is used to predict the unsteady flow and acoustic near field around the NM80 wind turbine. When performing the numerical simulation, the full-scale geometry is truly rotated using a local rotating body-fixed reference frame method. This type of simulation implemented in LBM has been already used and validated across various rotating machinery applications.

The present study provides a detailed analysis of the aerodynamic quantities, such as surface pressure coefficient, force and momentum at a single wind speed. The noise at IEC 61400-11 certification position for several rotor revolutions is computed using an auralization technique from simulated flow data stored along a portion of the rotor revolution. The auralization process is based on a frequency-domain Ffowcs-Williams and Hawking noise spectrum computation from blade wall-pressure pressure fluctuations, taking into account ground reflection. Noise spectra at several azimuthal copies of the microphones are finally corrected to take into account blade-motion Doppler effects end atmospheric absorption and smoothly appended to recover the full rotor revolution.

2. Methodology

2.1. Lattice Boltzmann Method

The numerical solver, which is based on the Lattice Boltzmann Method (LBM) with a wall-modeled Very Large Eddy Simulation (VLES) model for turbulence [5, 6, 14, 15], is extensively used for a wide range of industrial applications. As relevant to the present study, the solver has been also used to simulate the noise of rotating machines like wind-turbines[11], helicopter rotors[12], installed aircraft propellers[1, 7] and turbofans[3, 4, 8].

Lattice-based methods are by nature explicit, transient and compressible, and are an alternative to traditional CFD methods based on the discretization of the Navier-Stokes equations and derived variations. The basic idea of LBM is to track the advection and collisions of fluid particles. Since the average number of particles in a representative volume of fluid far exceeds the compute power required to track them individually, the particles are grouped into a set of discrete i-directions. The computation follows the particle distribution function f_i which represents the number of particles in a unit of volume at a specific time and location moving with velocity c_i. As in statistical physics, the flow variables such as density and velocity are determined by taking the appropriate moments, i.e. summations over the set of discrete directions, of the particle distribution function.

For simulations of flows with arbitrary geometry rotating around a fixed axis, the 3-D dimensional computational domain is divided into an inner and an outer domain. The inner domain has a grid fixed with the rotating geometry so that the geometry does not have a relative motion with respect to the grid. This forms a "body-fixed" Local Reference Frame (LRF) domain with the rotating geometry. The grid in the outer domain is fixed with the ground and forms a "ground-fixed" reference frame domain. The inner and outer domains are connected by a closed, zero-thickness, transparent interface.

Due to the unique properties of PowerFLOW, which is intrinsically unsteady and compressible, the acoustic pressure field can be extracted directly from the computational domain. For the far-field noise prediction, an additional set of tools is used, as described in next sub-section.

2.2. Acoustic certification prediction model

The main source of wind-turbine noise is the blade trailing edge, and in particular the outer part of it due to the higher velocity. The characteristic frequency ranges of trailing-edge noise scales as the flow velocity divided by the integral length of the turbulent flow fluctuations in the boundary layer. A more rigorous dimensional analysis is not need to argue that a large scale separation between noise frequencies and rotational frequency exists. Therefore, the noise computed in a blade-rotating reference system reaches a statistical convergence in a portion of the rotor revolution. This argument is used to develop an analysis process that allows to recover accurate noise signals over several rotor revolutions by using a CFD simulation covering only a small portion of the rotor revolution. The simple trick used in the process is to reproduce the different noise levels measured by a ground-fixed microphone during one rotor revolution by computing the noise levels at several microphones distributed along a ring coaxial with the wind-turbine. Since no relative motion between microphones and blade occurs, an efficient implementation of the Ffowcs-Williams and Hawkings (FW-H) formulation in the frequency domain can be used. Ground reflection is taken into account in exact way through a mirroring technique directly implemented in the FW-H integration process.

A reflection coefficient can be prescribed in order to take into account acoustic absorption without phase variations.

Since the different blades generate the same noise levels at the same relative positions, another way to reduce the computational cost of the full wind-turbine noise computation is to resolve the boundary-layer turbulent fluctuations with a fine mesh only in the outer part of one blade. The FW-H surface integration is applied to this blade portion and finally the noise from the full turbine is computed by adding signals properly shifted in time.

The FW-H computation generates spectra at different azimuthal locations. The amplitude variation over time for every single narrow band is then used to generate a synthetic noise signals using a standard auralization technique. Since the phase of the signal for every band and blade is generated randomly, several rotor revolutions can be reproduced. Prior auralization, a blade-rotation Doppler correction is applied by considering a source located at $0.8R_t$ (80% outboard). The same source location is used to compute the atmospheric absorption as a function of source/microphone distance and frequency.

For the sake of the present analysis, a full ground reflection is considered (no absorption). A number of 72 azimuthal microphones are used to recover a full rotor revolution. The physical ground-fixed microphones are located at 100 m downwind position, as indicated by the IEC 61400-11 process. A schematic of the microphones is added to Figure 1. An atmospheric humidity of 70% is considered. Noise signals are auralized along 5 rotor revolutions. The final noise spectra are computed by using about 1500 Welch windows with 50% overlap applied to the whole signals. The resulting frequency range is approximately $30 - 8,000$ Hz.

3. Wind turbine geometry and computational setup

The horizontal axis wind turbine used in current study is extracted (by best means) from the DanAero experimental campaign. The NM80 turbine is a 2.3 MW, full scale three bladed turbine, with LM Wind Power 38.8 m mounted blades. Detailed tower, hub and nacelle geometries are not available in the benchmark definition, and are replaced, for the sake of the present study, by realistic/arbitrary wind turbine geometries.

In the computations, the full turbine system is modeled without any geometrical simplification. The blades are pre-bended and attached to a default hub with a 1.24 m offset from the rotor center, resulting in a total length of $R = 40.04$ m. Furthermore, they were pitched 5 degrees towards lower angles of attack and tilted 2.5 degrees towards the pressure side to have the correct default pitch and rotor cone angle respectively. The rotor center is 59.89 m above ground, and the nacelle encapsulates a box with dimensions of $12 \times 4 \times 4$ meter. The tower is modeled as a tapered cylinder with 2.1 m at the base and 1.25 m at the top. An overview can be found in Figure 2, including the ground-fixed coordinate system.

The full geometry is embedded and centered into a large simulation domain of $100R \times 100R \times 100R$ to mimic free-field conditions. The floor is modeled as a standard rigid wall, while the ceiling is modeled as a frictionless wall to avoid the development of a boundary layer. All other boundary conditions are appropriately set as inlet/outlet conditions. Acoustic sponge zones in outer regions provide an anechoic condition to prevent spurious reflections of acoustic waves propagating from the wind turbine to the far field.

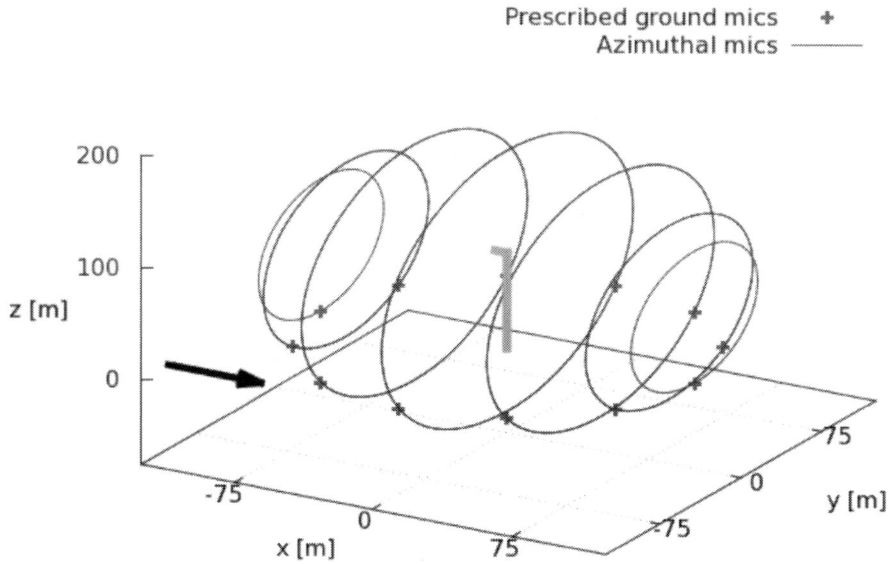

Fig. 1 Sketch of the ground microphone positions (red) and their azimuthal wind turbine axis rotating microphones (blue). Green indicates the turbine tower and nacelle, the arrow the wind direction

The grid generation process is fully automated and follows a user defined grid resolution scheme in regions of interest. The initial aerodynamic simulation is for converging and settling the rotor slipstream and the correct inflow around the blade. Highest resolution is placed around the outer 20 m part of a blade, ensuring local $y+$ values at the trailing edge do not exceed 80. The converged aerodynamic solution is then seeded onto an aeroacoustic simulation, which has an extra resolution to cover the transition strip, the resulting turbulent boundary layer, trailing edge and serrations. This results in a $y+$ value of 15 at inboard and 20 at outboard sections, or 0.45 mm, or 3200 voxels/grid cells along the average chord (1.5 m for the outer 20 m section). Details of the current resolution strategy are depicted in Fig 2. Transition location is variable along the blade span and extracted using the laminar-to-turbulent transition (LTT) model in aerodynamic modality in PowerFLOW, and cross-checked with XFOIL computations.

The simulation initial and boundary conditions are tried to be set as prescribed by the DanAero Task 39 1.1 case. This yields rigid construction (no deformation of the blades), no tilt, no tower shadow, no wind shear and no yaw angle error. The rotor speed is 12.3 RPM and the blades are pitched 0.15 degrees to the feather. A wind speed of $u_\infty = 6.1$ m/s is considered under standard atmospheric conditions ($\rho_\infty = 1.231$ kg/m^3 and $T_\infty = 19°$ C). As the prescribed atmospheric turbulence parameters in Task 39 are unrealistic for full scale CFD computations (long settling time and domain needed), standard turbulence parameters of $TI = 5$ % and $TL = 0.01$ m are used. This will have negligible effect on the aeroacoustic prediction, as the main goal of the present study is to evaluate trailing edge,

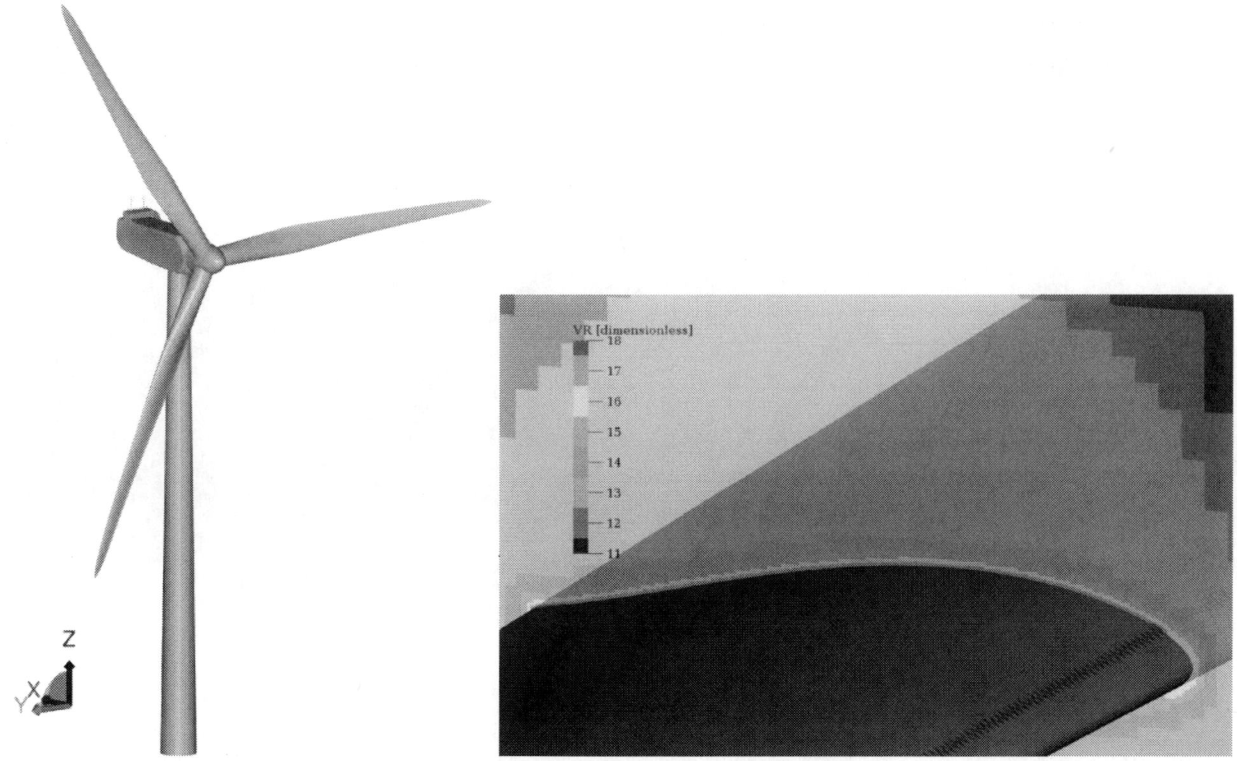

Fig. 2 Overview of the NM80 wind turbine geometry and close-up of the different refinement regions close to the blade

tip and stall noise. With these conditions, transition strips are placed around 6 % and 40 % of the suction and pressure side respectively.

As the aeroacoustic post-processing is optimally designed to handle large rotors under low RPM's, it is not required to run the aeroacoustic simulations for multiple revolutions to collect a 10 second bandwidth as stated by the IEC certification. Instead, only $1/12$th of the full revolution is modeled, i.e. 0.45 seconds. Together with an optimized setup of less than 300 million fine equivalent voxels (700 million overall), the user only need 85 kCPUh (i.e. < 3 days computation time on SIMULIA cloud) to obtain enough unsteady surface pressure data for the acoustic computation.

4. Results

The results are spitted into two section. Firstly, an overview of the aerodynamic performance is given, both qualitatively and quantitatively. Details of the unsteady velocity and pressure field give indication of the quality of the simulation, and of the quality of the certification noise prediction. The latter is discussed in the final section, detailing step by step the new certification noise process.

4.1. Aerodynamic information

To highlight the unsteadiness of the simulation, it is useful to observe and identify local separation at the root section of the blade. Therefore, in Figure 3, iso-λ_2 surfaces, colored by velocity magnitude are plotted. It can be observed that there is 1) a clear tip vortex present and 2) locally the flow separates at $1/4$th of the inboard section of the blade. Fully attached flow is therefore expected at the outboard sections of this turbine, under the current simulated conditions. The fact that no turbulence seems to convect at the trailing edges of the outboard section of the blade, is due to the fact that this results originates from the aerodynamic settling part of the simulation. No transition strip was modeled here, and $y+$ values are relatively coarse.

Fig. 3 Instantaneous snapshot of isosurfaces of λ_2, colored by velocity magnitude, extracted from the aerodynamic simulation

However, the simulation requirements of the aerodynamic simulation are sufficient enough to extract the mean quantities of the flow, such as the pressure distribution along different radial sections. In Figure 4, three spanwise locations are shown for which pressure data is extracted. A correlation with XFOIL 2-D results is made to cross-check the simulation results, showing good agreement. As expected, the largest pressure forces are obtained at midboard sections of the wing, where the flow is fully attached.

The aerodynamic generated noise studied in this paper is mainly affected by the turbulent boundary layer, convected past the trailing edge. Therefore, to have an idea of the type of turbulent flow, at

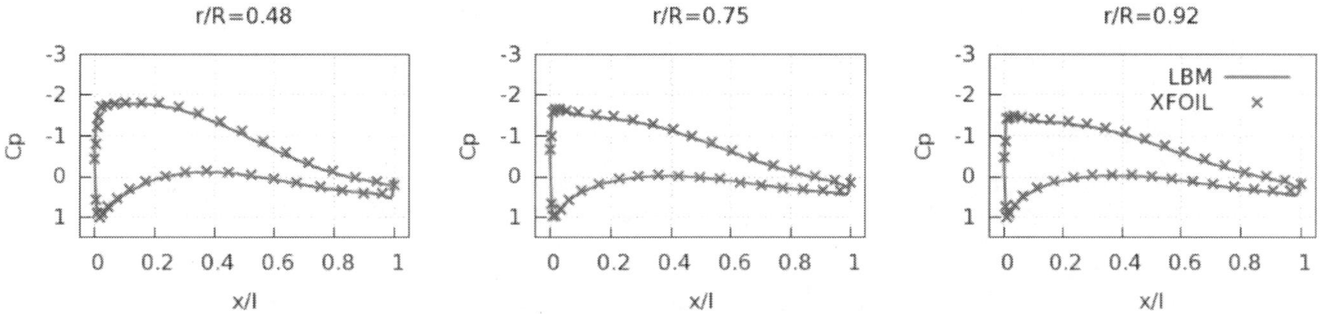

Fig. 4 Mean pressure distribution along different radial sections, integrated over multiple blades

different radial sections along the blade, the velocity magnitude in the rotating reference frame is plotted in Figure 5. It can be seen, first of all, that at all the three slices (locations according to Cp cuts from previous paragraph), a fully developed turbulent boundary has been formed thanks to the inclusion of the transition strip. At the inboard slice, at $r/R = 0.48$, the trailing edge is relatively thick and vortex shedding occurs. Traveling more outboard, the flow accelerates, the shedding diminishes, and the boundary layer becomes thinner.

Fig. 5 Snapshot of the boundary layer development along the blade span

The indication of turbulence convecting over the trailing edge can be visualized by means of a dilation field, which is proportional to the time derivative of pressure. Snapshots of the suction (top) and pressure side (bottom) are shown in Figure 6. The result shows that, at outboard positions, a turbulent boundary layer is built up from the transition strip. Along the entire outboard trailing edge, strong levels of pressure variations are found, indicating the primary acoustic source location on this blade.

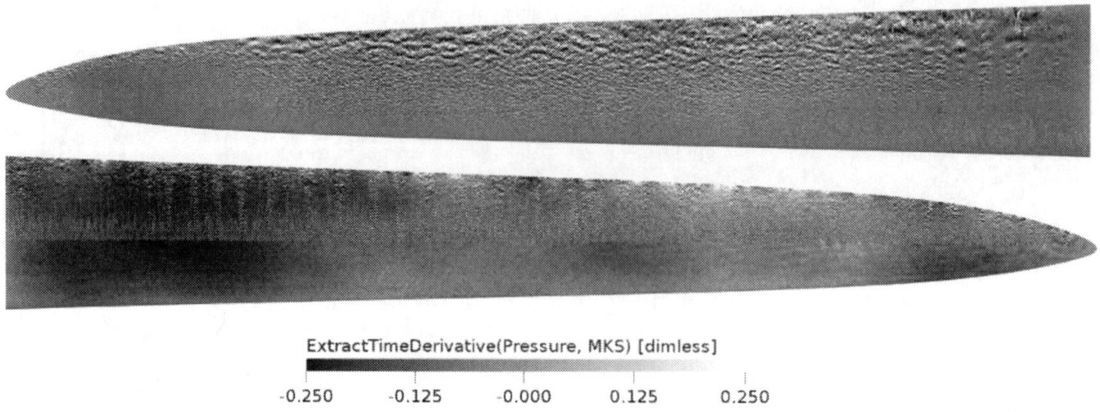

Fig. 6 Snapshot of time derivative of the pressure on the suction (top) and pressure (bottom) side

4.2. Certification noise analysis

The output of the acoustic certification prediction model involves various steps, which are described by their results in detail in this section. The unsteady surface pressure data, extracted from PowerFLOW, is the input into the model. Twelve physical ground microphones are used, distributed along a circle of 100 m radius around the turbine. In addition, a probe just 10 m away from the tower is used as another reference point. Each ground microphone is 72 times rotated along a coaxial ring, resulting in a total of 936 microphones at different positions.

The frequency-domain FW-H solver handles all these microphones, and calculate the resulting acoustic power spectral density using multiple Hanning windows. Ground reflection is taken into account already in this stage of the calculation. To indicate the spread obtained from different azimuthal probes (i.e. modulation due to blade rotation), the results of the downwind microphone is plotted in Figure 7, together with its Overall Sound Pressure Level (OASPL). Notice that these results come directly from the FW-H solver, and therefore correspond to the noise generated by one blade only, without Doppler effects. The large blue area is bounded by the minimum and maximum PSD obtained from all microphones. It indicates that, when calculating noise levels from slowly rotating blades, it is important to take into account the blade rotation with respect to the observer position, as the signal itself is very modulated over time. This is confirmed by observing the OASPL in Figure 7, as differences up to 15 dB are observed depending on microphone (and thus blade) angular position. Plotting in A-weighting scale yields similar variances between low and high level sound. This suggests that the swirling motion of the blades, at distances of 100 meter downwind from the tower, are still heavily been perceived by the observer.

When investigating a location closer to the tower, a stronger modulation of the source is expected. Indeed, when looking at Figure 8 (right), a variation between both locations is observed. Closer to the tower, the slanted distance is $\sqrt{100^2 + 60^2}/\sqrt{10^2 + 60^2} = 1.9$ times closer, and in generally slightly higher values in part of the revolution are observed. However, taking into account the distance variation, one would expect $20 \cdot \log_{10}(1.9) \approx 5.7$ dB, which is not observable for each single microphone on the right side of Figure 8. Instead, higher levels are observed for many of the 100 meter probes compared against the 10 meter azimuthal ground microphones. This means that, closer to the tower, the directivity effects of the isolated aerodynamic sources (e.g. trailing edge noise and tip noise) are

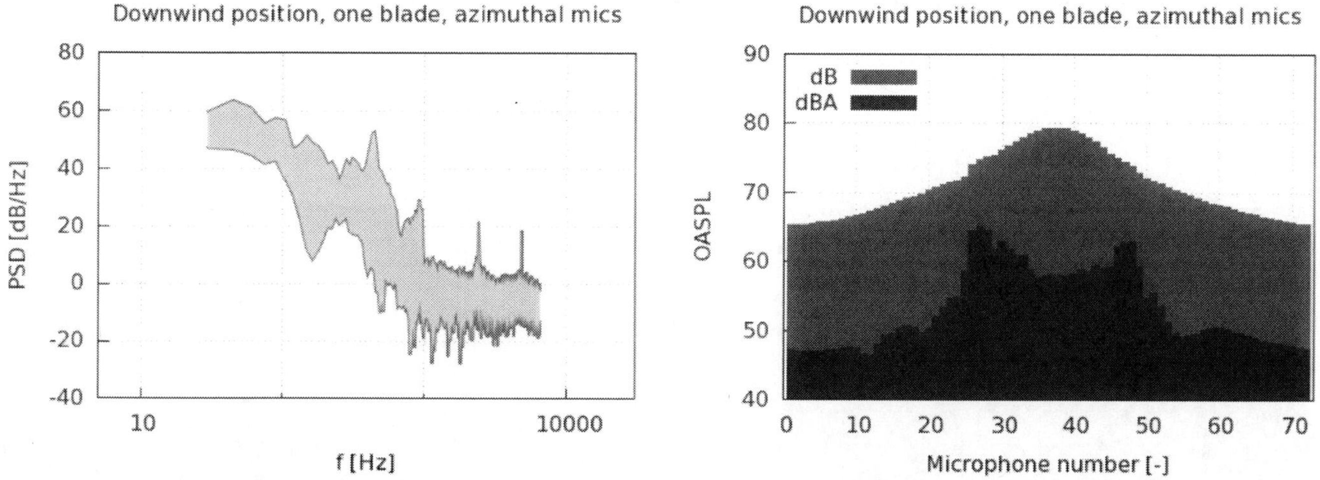

Fig. 7 FW-H PSD spectra (left) and OASPL (right) at all azimuthal locations for one downwind location

more pronounced. At a distance ten times further from the tower (the reference downwind position), the approximation of a point source can be made, and levels vary less over a revolution. This makes the variation of sound, and thus the swirling effect of a wind turbine, much more pronounced at locations close to the turbine, as shadow zones are more active. It can be seen that the variation at 100 meter is approximately 10 dB, while at 10 meter this increased to 35 dB.

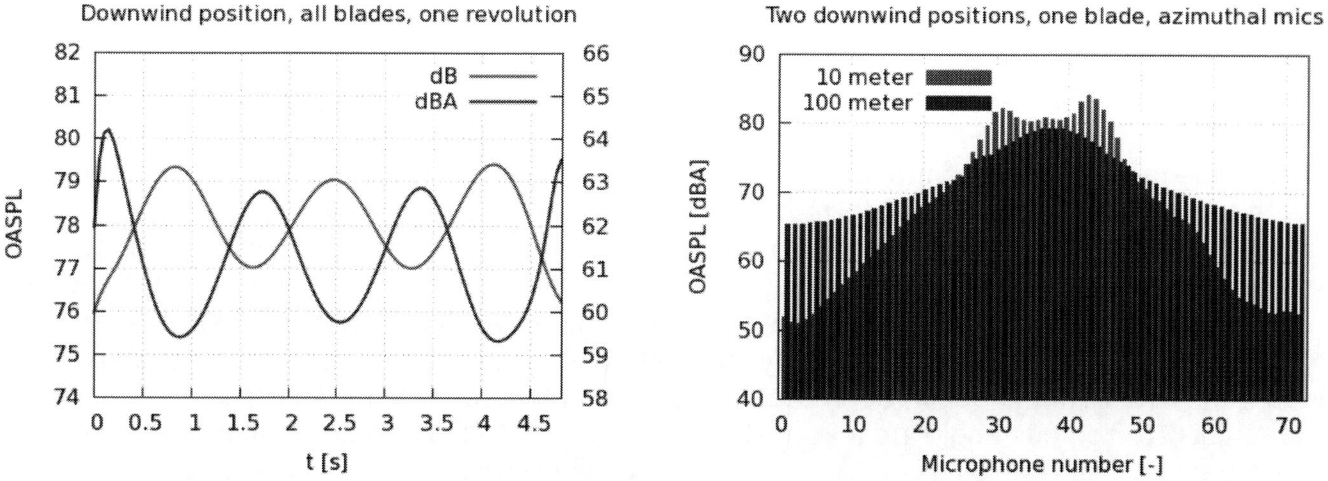

Fig. 8 Downwind position perceived OASPL in time (left) and for two downwind locations

Next step is to consider the full motion of the turbine, taking into account the three blades. Phase information is added to the computation, and one full revolution OASPL is depicted in Figure 8 (left). There is a shift of peak amplitude due to the A-weighting.

The final step in obtaining the certification apparent acoustic power (L_{WA}) and its spectrum, is to generate a synthetic signal based on auralization, thereby also taking into account atmospheric absorption and Doppler effects. A total of 5 revolutions is reconstructed, and its L_{WA} is calculated using:

$$L_{WA} = L_V - 6 + 10 \cdot \log_{10}\left[\frac{4\pi \cdot R_1^2}{S_0}\right], \tag{1}$$

where L_V is the background corrected A-weighted sound pressure level, R_1 the slant distance in meters from the rotor center to the microphone and S_0 a reference area, for simplicity here set to unity. The result is depicted in Figure 9 (left) and can be considered to be the official downwind noise estimation according to IEC certification.

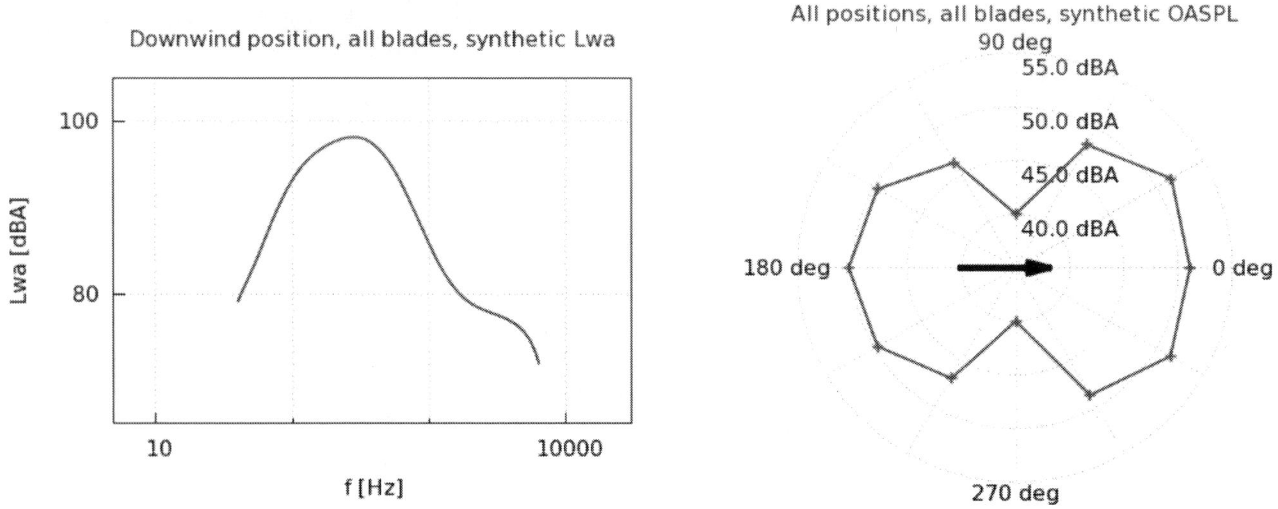

Fig. 9 Auralized apparant acoustic power Lwa (left) for the downwind position and its OASPL directivity (right)

Finally, the full process can be repeated for the other 11 physical ground microphones, oriented in a circle of 100 m radius around the turbine. The auralized OASPL is therefore plotted in a directivity pattern in Figure 9 (right). The wind direction is depicted by the arrow, and follows the positive x-axis along the default coordinate system. Zero degrees indicates the reference downwind position. The blade rotation is also oriented along the positive x-axis, meaning the down-stroke-going-blade is at the 270 degrees direction. From previous literature it is known that, when the observer is on the ground, the blade which convects its aerodynamic self-noise most efficiently, is the down-stroke-going blade [10]. The same is observed in the current computation, where OASPL levels at locations towards the 270 degrees angle are a few dB higher than at the 90 degrees location. A large shadow zone is present when standing in a side-wind position, while largest levels are seen in the downwind position.

5. Conclusion

Flow and acoustic simulations were performed on the NM80 full-scale 3-bladed wind-turbine, tested during the DanAero campaign. SIMULIA PowerFLOW's Lattice Boltzmann Method was used to predict the turbulent-resolved unsteady flow around the full wind-turbine, yielding good results with reference XFOIL data. A fully turbulent boundary layer was observed at the outboard sections of the wing, making it possible in this study to digitally analyze turbulent boundary layer noise at full wind-turbine level.

The noise at certification position for five rotor revolutions was computed using an auralization technique from simulated flow data stored along a portion of the rotor revolution. The auralization process is based on a frequency-domain Ffowcs-Williams and Hawking noise spectrum computation from blade wall-pressure pressure fluctuations, taking into account ground reflection. Noise spectra at several azimuthal copies of the microphones were finally corrected to take into account blade-motion Doppler effects end atmospheric absorption and smoothly appended to recover the full rotor revolution.

Significant cost reduction in simulation time was found with the new post-processing process of determining certification noise levels according to IEC standards, making the entire process suitable for industrial turn-around-times. Furthermore, the new process allowed to investigate the physical insight of the noise propagation due to different microphone placements and frequency domain analysis, where similar conclusions as in other, experimental reference, papers were found.

Acknowledgments

The DanAero projects were funded partly by the Danish Energy Authorities, (EFP2007. Journal nr.: 33033-0074 and EUDP 2009-II. Journal nr.: 64009-0258) and partly by eigenfunding from the project partners Vestas, Siemens, LM, Dong Energy and DTU.

References

[1] Avallone, F., Casalino, D., and Ragni, D. (2018). Impingement of a propeller slipstream on a leading edge with a flow-permeable insert: A computational aeroacoustic study. *International Journal of Aeroacoustics*, 17(6-8).

[2] Brooks, T., Pope, D., and Marcolini, M. (1989). Airfoil self-noise and prediction. *NASA Technical Report, Reference Publication 1218*.

[3] Casalino, D., Avallone, F., Gonzalez-Martino, I., and Ragni, D. (2019). Aeroacoustic study of a wavy stator leading edge in a realistic fan/ogv stage. *Journal of Sound and Vibration*, 442:138–154.

[4] Casalino, D., Hazir, A., and Mann, A. (2018). Turbofan broadband noise prediction using the lattice boltzmann method. *AIAA Journal*, 56(2).

[5] Chen, H., Kandasamy, S., Orszag, S. A., Succi, S., and Yakhot, V. (2003). Extended boltzmann kinetic equation for turbulent flows. *Science*, 301(5633):633–636.

[6] Chen, H., Orszag, S., Staroselsky, I., and Succi, S. (2004). Expanded analogy between boltzmann kinetic theory of fluid and turbulence. *Journal of Fluid Mechanics*, 519:301–314.

[7] den Ende, L. V., Avallone, F., Casalino, D., Eitelberg, G., and L. L. M. V. (2018). Aerodynamic and aeroacoustic investigation of swirl recovery vanes for propeller propulsion systems. *31st Congress of the International Council of the Aeronautical Sciences, Belo Horizonte, Brazil,*, pages 1–19.

[8] Gonzalez-Martino, I. and Casalino, D. (2018). Fan tonal and broadband noise simulations at transonic operating conditions using lattice-boltzmann methods. *AIAA Paper 2018-3919*.

[9] Madsen, H., Bak, C., Paulsen, U., Gaunaa, M., Fuglsang, P., Romblad, J., Olesen, N., Enevoldsen, P., Laursen, J., and Jensen, L. (2010). The danaero experiments final report. Technical report, RISO-R-1726 (EN), Technical University of Denmark.

[10] Oerlemans, S., Sijtsma, P., and Lopez, B. M. (2006). Local and quantification of noise sources on a wind turbine. *Journal of Sound and Vibration*, 299:869–883.

[11] Pérot, F., Kim, M.-S., Meskine, M., and Freed, D. (2012). Nrel wind turbine aerodynamic validation and noise predictions using lbm. *AIAA Paper 2012-2287.*

[12] Romani, G. and Casalino, D. (2017). Application of lattice-boltzmann method to rotorcraft aerodynamics and aeroacoustics. In *Proceedings of the 43rd European Rotorcraft Forum, Milan, Italy.*

[13] van den Berg, F., Pedersen, F., Bouma, J., and Bakker, R. (2008). Windfarm perception: Visual and acoustic impact of wind turbine farms on residents. *University of Groningen.*

[14] Yakhot, V. and Orszag, S. A. (1986). Renormalization group analysis of turbulence. i. basic theory. *Journal of Scientific Computing*, 1(1):3–51.

[15] Yakhot, V., Orszag, S. A., Thangam, S., Gatski, T. B., and Speziale, C. G. (1992). Development of turbulence models for shear flows by a double expansion technique. *Physics of Fluids A*, 4(7):1510–1520.

8th International Conference
on
Wind Turbine Noise
Lisbon – 12th to 14th June 2019

The Quasi-3D TE Rotor Noise Prediction Tool of the PNoise Code

Joseph Youssif Saab Junior, Maua Institute of Technology (IMT).
Poli-Wind Research Group: saab@maua.br

Sara Rodriguez, Polytechnic School, Sao Paulo State University (Poli-USP).
Poli-Wind Research Group: sara.rodriguez@alumni.usp.br

Alexandre Martuscelli Faria, Polytechnic School, Sao Paulo State University (Poli-USP).
Poli-Wind Research Group: martuschellifaria@usp.br

Marcos de Mattos Pimenta, Polytechnic School, Sao Paulo State University (Poli-USP).
Poli-Wind Research Group: marcos.pimenta@poli.usp.br

Summary

A 2D airfoil trailing-edge (TE) noise prediction module, *PNoise*, was developed in 2016 under a Poli-USP - TU-Berlin collaboration project and integrated into the *QBlade* wind turbine (WT) design platform, available under General Public License (GPL). Owing to the continued user interest and a large number of downloads of the host platform, a quasi-3D method was planned as follow-on for implementation which should return to the user the approximate, full blade and full rotor noise prediction. Also, the PNoise code is gradually receiving upgrades for inflow noise and other self-noise sources that will gradually benefit the quasi-3D method. Because the quasi-3D variation resorts iteratively to local 2D solutions, it also depends on the same underlying boundary layer thickness hybrid calculation and induced flow calculation methods of the 2D algorithm, with resulting accuracy affected by the errors inherent in each method plus the error deriving from the use of a finite number of blade sections in the blade and a discrete number of angular positions of the blade along the azimuthal plane to model the effect of the wind field, distance to observer and directivity function. Although the Poli-Wind group also research higher-order, longer-response time noise prediction methods (theoretical, CFD and Lattice-Boltzmann formulations) the PNoise code was developed with the early wind turbine (WT) rotor design task in mind and remains true to its original purpose with the addition of a quick turn-around-time quasi-3D tool. Provisions are made in the expanded blade noise method for the calculation of the overall sound pressure level (SPL) perceived by an observer (i) at a fixed location behind one single rotating blade or (ii) fixed in space, at any user-specified location in relation to the full rotor. Atmospheric and terrain propagation effects are not added at this time, only the distance between the observer and the many aeroacoustic sources, along with the respective directivity function suitable for each type of source. The opportunities for further developments include the addition of minor self-noise sources, transient wind field inlet condition, blade aeroelastic deformation effect calculations and noise propagation models to the farfield.

LIST OF ABBREVIATIONS AND ACRONYMS

Abbreviation	Meaning
AOA	Angle of Attack
BEM	The Blade Element Momentum method
BL	Boundary Layer
BPM	Brooks, Pope, Marcolini, Airfoil Self Noise Prediction Model
CFD	Computational Fluid Dynamics
FP	Flat Plate
HAWT	Horizontal Axis Wind Turbine
LBL	Laminar Boundary Layer
LE	Leading Edge
OASPL	Overall Sound Pressure Level (dB)
Poli – USP	Politechnic School - São Paulo State University - Brasil
PNoise	Poli-USP TE noise prediction code.
QBlade	TU Berlin-developed, public domain, Wind Turbine Performance and Structural analysis software.
Quasi-3D	Quasi Three Dimensional (method)
SPL	Sound Pressure Level (dB)
$SPL_{1/3}$	Sound Pressure Level for each 1/3 octave band (dB)
SPL_W	Weighted Sound Pressure Level (dB)
TBL	Turbulent Boundary Layer
TBL-FP	Turbulent Boundary Layer, Flat Plate model
TBL-TE	Turbulent Boundary Layer at the Trailing Edge
TE	Trailing Edge
TSR	Tip-Speed Ratio
VS	Vortex Shedding (Frequency) (Hz)
WT	Wind Turbine
WTN	Wind Turbine Noise
XFLR5	The XFOIL software with graphic user interface.
XFOIL	Hybrid, Potential Flow and Integral Boundary Layer Solver.
2D	Two-Dimensional (method)

LIST OF SYMBOLS

Letter	Meaning	Disambiguation, (section)	units
a	Axial induction factor		
a	Angle between $Y_B Z_B$ plane and local segment mid-section chord line.		^{0}
a'	Tangential induction factor		
AOA	Angle of Attack		^{0}
b	Geometric angle	Table 3, 20	^{0}
c, c_i	Airfoil chord		m
$c_{r,i+1}$	Cord at the outboard section of the blade segment considered		m
$c_{r,i}$	Cord at the inboard section of the blade segment considered		m
csv	Comma-separated value (text file format)		

$D*$	Displacement thickness (notation inside XFLR5 code)	m
HR	Hub radius	m
N/A	Not Available	
N/S	Not Selected	
N_s	Number of spanwise blade segments	
N_β	Number of blade angular positions in the azimuthal plane	
N_f	Number of frequencies in the spectrum	
n_b	Number of blades	
n	Number of blade segments	
Re	Reynolds Number	
Re_c	Reynolds Number, chord-based	
r_o	Outer radius for each blade segment	m
r_i	Inner radius for each blade segment	m
r	Rotor, blade radial position	m
r_{i+1}	Blade radius at the outboard section of the blade segment considered	m
r_i	Blade radius at the inboard section of the blade segment considered	m
T	Air temperature	$°C$
V_0	Velocity, freestream	m/s
V_{rel}	Velocity, wind relative	m/s
(X_e, Y_e, Z_e)	Observer coordinates at the far field (retarded coordinate system)	m
(X_B, Y_B, Z_B)	Blade Coordinate System	m
(X_{RT}, Y_{RT}, Z_{RT})	Rotated Trailing-Edge Coordinate System	m

GREEK LETTERS

Letter	Meaning	Disambiguation, (section)	units
α	Local angle of attack (AOA)		^{0}
β	Twist (angle) of the blade or local segment	Table 3, step 5	^{0}
β	Blade step angle on the azimuthal plane	Table 2	^{0}
δ^*	Boundary Layer Displacement thickness		m
θ_e	Directivity angle. Angle between the observer position and the chordwise airfoil plane, retarded coordination system.		
θ_p	Variable pitch angle of the blade	Table 3, steps 5, 20	^{0}
ϕ	Angle between the plane of rotation and the relative velocity.	Table 3, step 12	^{0}
ϕ	Angle, directivity, between the observer and the edge plane, projected in the plane perpendicular to the edge.	See Fig 1	^{0}
$\boldsymbol{\phi_e}$	Angle, directivity, between the observer and the edge plane, projected in the plane perpendicular to the edge, retarded coordinate system.	See Fig 1	
ω	Frequency, angular or rotational speed		$1/s$

1. Introduction

Since most airfoil self-noise prediction methods based on semi-empirical or simplified theoretical models are applicable to an airfoil section interacting with the flow, their application to a WT rotor must follow a procedure generally referred to as a *quasi-3D* method. This procedure has been applied, with some variations, in many researches (Glegg, 1987), (Brooks, Pope, & Marcolini, 1989), (Moriarty & Migliore, 2003), (Fuglsang & Bak, 2004), (Zhu, 2004), (Vargas, 2008), and (Kamruzzaman, Lutz, Nübler, & Krämer, 2011), among others.

In the current implementation scheme, the quasi-3D extension envisioned for the PNoise code (Saab Jr & Pimenta, Displacement Thickness Evaluation for BPM-Type Airfoil-TE Noise Prediction Model, 2015), (Saab Jr J. Y., Trailing-Edge Noise – Development and Application of a Noise Prediction Tool for the Assessment and design of Wind Turbine Airfoils, 2016), (Saab Jr & Pimenta, 2016), (Saab Jr., et al., 2016), (Saab Jr & Pimenta, Verification and Validation of the "PNoise" Airfoil Trailing-Edge Noise Prediction Module inside "QBlade"., 2017), (Saab Jr J. Y., Pimenta, Faria, & Rodriguez, 2018) intends to offer the user, whether a WT manufacturing enterprise or a researcher, with a unique set of features:

- Overall and spectral sound pressure level quick estimates, suitable for comparison and refinement of design.
- Flexibility for any airfoil, blade and rotor geometry.
- Flexibility for flow conditions specification and calculation, within noise models validity range.
- Flexibility on the source of the vertical turbulent scale to be employed in some of the noise models.
- Selection of airfoils self-noise sources to be modeled.
- Inflow noise source selection.
- A table for warning the user on all model validity limitations.
- User-defined number of blade segments and blade azimuthal positions for the quasi-3D noise estimation.
- Arbitrary, user-defined observer position at the farfield (for rotor noise estimation).
- Open-source, free software download, under GPL.

2. Methodology Overview

In the quasi-3D procedure, the generic steps listed on Table 1 are followed.

Table 1 – The steps of a generic quasi-3D noise calculation procedure for wind turbine rotors.

Step	*Description*
1	The blade is divided into N_S spanwise segments.
2	By determining iteratively, the axial and radial induction factors with the aid of a Blade Element Momentum (BEM) method (Hansen, 2008) the relative speed, Reynolds number, Mach number and AOA are calculated. The tip correction method and other corrections elected are applied.
3	One TBL 2-D calculation process (e.g. TBL-FP correlation, XFOIL, CFD, Experiment) provides the data necessary to calculate the source parameters of the model.

4 The TE source spectrum in 1/3 octave bands is obtained for each radial blade segment, at a fixed observer position with respect to the blade segment (i.e. an observer rotating with the blade and positioned at $(\theta = 90°, \phi = 90°)$, according to the reference frame of Figure 1.

5 A more physical situation of a rotating blade but with the observer fixed relative to the ground is then introduced with the aid of a coordinate systems transformation and use of a directivity function.

6 The total spectrum and OASPL can then be calculated from the logarithmic sum of all contributions at the observer position, fixed relative to the ground.

7 Propagation models may or may not be employed to the farfield

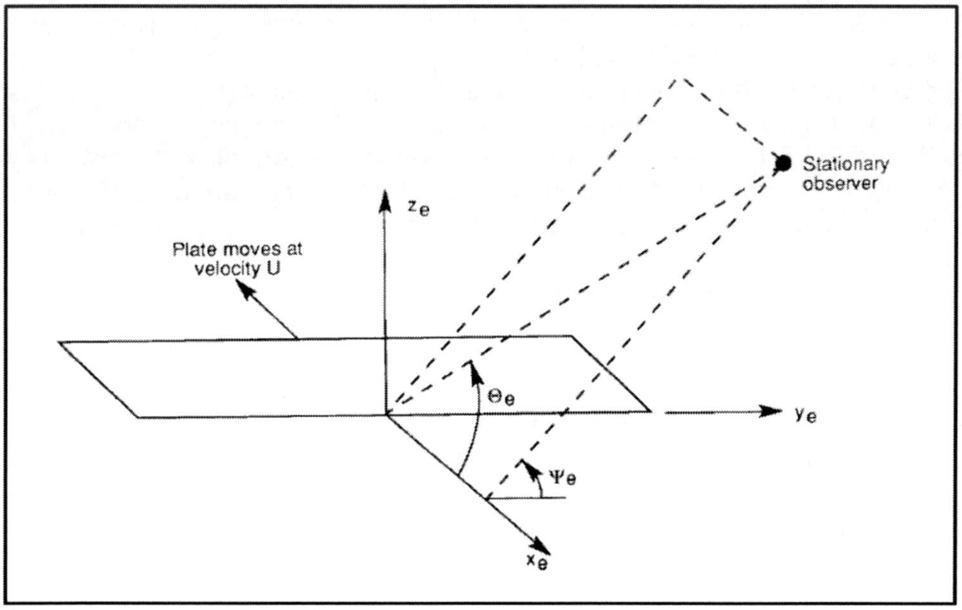

Figure 1 - The 3D retarded coordinate system with origin at the TE of a thin plate representing the airfoil moving in rectilinear motion of velocity U in negative x_e axis direction. $\Theta \to \theta$; $\Psi \to \phi$. Source: (Brooks, Pope, & Marcolini, 1989)

In section 3, the most important definitions for the PNoise quasi-3D extension are briefly discussed. In section 4 the specific flow diagram developed for the blade noise prediction method only is presented.

3. PNoise quasi-3D Method definitions

Noise calculation for a wide range of wind speeds, rotational speeds and pitch angles does not seem practical. A combination of 10 values for each of those parameters would result in a thousand different flow and geometry input combinations for analysis. Now suppose each blade of a 3-blade rotor would be split in 50 sections for the BEM calculations and each noise source would be calculated for each blade segment, for 8 different angular positions along the azimuthal plane and for 33 center frequencies. If only the TE noise model is selected for the calculation (which comprises 3 different sub sources as in the BPM model (Brooks, Pope, & Marcolini, 1989)), the algorithm would produce almost 12 million noise contributions, from a single source, to be log-summed at the observer location! Apart from complexity and output screen and file clutter, the computation time would become impractical for the purpose of preliminary design analysis.

This simple exercise suggests that the noise calculation for a specific wind condition and one (at a time) tip speed ratio (TSR) selected by the user would be more plausible, at first. This section discusses the minimum number of noise sources, i.e., blade segments and blade angular positions along the azimuthal plane necessary to render OASPL and noise spectrum representative at the observer position in the farfield.

3.1 Number of blade segments.

The blade discretization is accomplished in spanwise segments with specific chord length, twist angle and airfoil geometry, all considered in the calculations of the PNoise quasi-3D extension.
The discretization step is considered by some authors (Kamruzzaman, Lutz, Nübler, & Krämer, 2011) to be the most sensitive and difficult part to adjust in a quasi-3D extension of a blade noise method.
Glegg (Glegg, 1987), employed 5 radial positions, for each blade, in his application of the method, while Zhu (Zhu, 2004) employed 8 segments for the blades, like Kamruzzaman et al. (Kamruzzaman, Lutz, Nübler, & Krämer, 2011).
However, Vargas (Vargas, 2008) accomplished a sensitivity study (see Table 2) that indicates 13 segments for each blade as a number below which SPL variation is still significant at the farfield. Thus, in the PNoise quasi-3D method, the user is warned if less than 13 segments are adopted for the blade discretization process. As the QBlade embedded BEM method uses 40 blade segments as a default minimum, for high precision, the user might select the same discretization number used in the BEM method, in order to avoid a second discretization/interpolation process with associated error propagation.

3.2 Number of azimuthal positions for each blade.

For practical application purposes, Glegg (Glegg, 1987) suggested the average of 20 azimuthal angles for each blade, in the application of the quasi-3D method. However, Vargas (Vargas, 2008) also accomplished a sensitivity study on the number of optimal positions of the blade in the azimuthal plane N_β, versus the OASPL predicted. For this purpose, the blade was split in $N_S = 13$ segments and the frequency spectrum was divided in $N_f = 29$ centre frequencies. The sensitivity analysis result is displayed in Table 2. It shows that OASPL evaluation for more than 8 blade positions (equivalent to a step of 45°) on the azimuthal plane is not necessary. Also, Vargas found that this number of positions also kept the calculation time very low.
For this reason, the default number of steps for each blade along the azimuthal plane is kept in 8 for the current model.

Table 2 - Sensitivity studies of SPL_W (A-Weighted SPL) as a function of the number of positions of the blade in the azimuthal plane. The total number of computed pressure values is $N_p = N_\beta \cdot N_S \cdot N_f$. Source: (Vargas, 2008).

β step (°):	360	320	250	130	80	60	45	20	1
N_β:	1	2	2	3	5	6	8	18	360
N_p:	377	754	754	1131	1885	2262	3016	6786	135720
processing time (s):	0	0	0	0	1	1	1	3	59
SPL_W (dBA):	94.31	93.82	94.72	98.43	97.93	98.30	98.29	98.30	98.29

3.3 Definition of the reference system.

After analyzing the reference systems adopted by some authors (Zhu, 2004), (Broux, 2008), (Vargas, 2008), it was considered that all adopted coordinate systems had aspects that could be improved upon. After further conference with some Industry experts in the field[1] a decision was made to adapt the local reference frames suggested in the German Guideline for Certification of Wind Turbines, (Germanisher Lloyd and German Wind Energy Committee, 2010).

In the selected base-system, the WT geometry representation is simplified by the absence of rotor axis tilt and cone angles, but the main advantage of this system is that the blade and chord coordinate frames have one axis aligned with the pitch axis of the blade, while for the other frames studied some unusual spanwise axis were adopted, passing through "the middle of the blade" or "the middle of the chord", for instance, which are neither precisely described nor relevant geometrical points for calculations.

However, the improvement of the representation obtained by taking the pitch axis as part of reference system also brings about another simplifying assumption, i.e., the pitch axis is assumed to intersect the chord line of all sections at 25% of its length, from the TE. With this assumption it is the authors´ goal to make the pitch axis intersect the thickest section of each airfoil (or close to it), the region where the center of pressure and structural spar should be preferably located and often the axis through which different airfoil sections are stacked in order to assemble a blade.

The set of reference systems adopted do not affect the aerodynamic or structural routines of QBlade (Pechlivanoglou, Marten, G., N., & Wendler, 2009), (Marten D. , Extension of an Aerodynamic Simulator for Wind Turbine Blade Design and Performance Analysis, 2010), (Marten D. , Qblade Short Manual V0.8, 2014), (Marten D. , 2016).

For the quasi-3D rotor noise prediction calculation, five reference frames were adopted, among inertial and non-inertial systems. For details on all the reference systems and final coordinate transformation matrix, please refer to (Saab Jr J. Y., Trailing-Edge Noise – Development and Application of a Noise Prediction Tool for the Assessment and design of Wind Turbine Airfoils, 2016).

4. The resulting implementation plans.

Figure 2 displays one of the flow diagrams developed for integration of the PNoise quasi-3D code into the QBlade code, more specifically the flow diagram for the estimation of the full-blade noise, with the observer in a solidary, rotational movement at constant specified distance and directivity angles from the TE mid-point.

[1] Dr. George Pechlivanoglou, from SmartBlade and Dr. Erik Sloth, from Vestas Wind Systems.

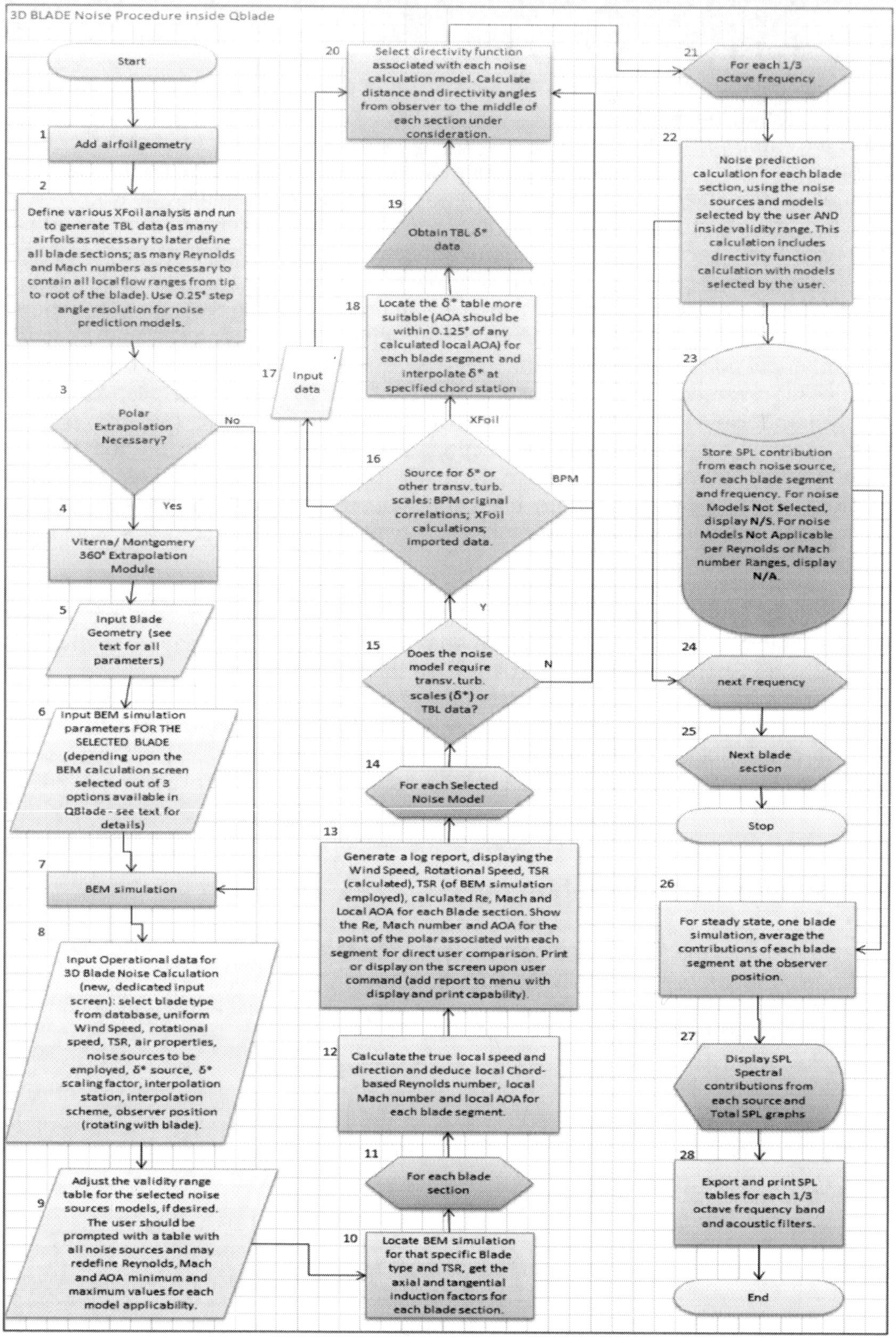

Figure 2 – Flow diagram for the PNoise quasi-3D extension for a single, full blade noise calculation.

4.1 Step by step description of the calculation procedure.

The description in Table 3 follows the same flow-chart element numbers shown in Figure 2.

Table 3 – Step by step description of the quasi 3D blade noise estimation procedure in PNoise code.

STEP	DESCRIPTION
1	• Airfoil geometry imported into XFoil/XFLR5 (Drela, X-FOIL, 2000), (Drela, Youngren, & Deperrois, XFOIL/XFLR5, 2009) module. All the airfoils to be employed later in the blade definition screen must be either generated or imported at this time.
2	• XFoil analysis for all airfoils to be employed in the blade construction shall be defined and run for Reynolds, Mach and AOA ranges that may occur during the blade operation. The outputs are necessary in the form of Polar Objects (for BEM method calculations) and D* tables (for noise calculations). One Polar Object shall be associated with each blade section later in the blade geometry definition routine.
3	• As of the development of this code, there is no noise model (TE or other source) applicable to angles of attack beyond stall, although there has been steady progress toward it (Bertagnolio, Madsen, Fischer, & Bak, 2015). However, WT blades are subject to this kind of flow during transient operation. Since there are some ongoing efforts pursuing modeling of the noise beyond the stall, there should be provisions for the Polar object calculated by the XFoil to be replaced by the 360° Polar object calculated by the QBlade 360° Polar module, which employs either the Viterna (Viterna & Janetzke, 1982) or Montgomery (Montgomerie, 2004) models.
4	• 360° Polar generation, if necessary, as a substitute for the XFoil polar.
5	• Input blade geometry @ HAWT ROTORBLADE DESIGN Module: o Rotor name. o Number of blades, *nb*. o Hub diameter (m) (divide by 2 to obtain HR, Hub radius). o Index number for each segment, *i*. o Outer Radius for each segment (m), r_o. o Inner Radius for each segment (m), r_i. o Chord (m) for each segment, c_i. o Twist angle (β°) for each segment, β_i. o Airfoil Object attributed to each segment. o Polar Object attributed to each segment. o Pitch angle ($\theta_p{}^{\circ}$) for the blade (when entered, will add a constant value to the twist at every segment of the blade). • When "Blade coordinates" checkbox is checked, the distance to each segment is measured from the blade root, i.e. the reference is at the hub radius (HR) obtained from the hub diameter informed. If it is not checked, the distance to each segment is measured in absolute coordinates, with reference at the hub axis. The finished geometry can be exported in the ".stl" format. • In case the user has made use of the blade optimization and/or blade scaling features of the QBlade code, the optimized/scaled blade geometry should replace the original entered geometry for all future calculations, including noise prediction modules. • The modifications that can be made to the blade with the Advanced Design menu options are made in the export drawing only and do not affect the BEM calculations. Also, the modifications implemented with this optimization/scaling tool is not the geometrical baseline used by the noise calculation routines.

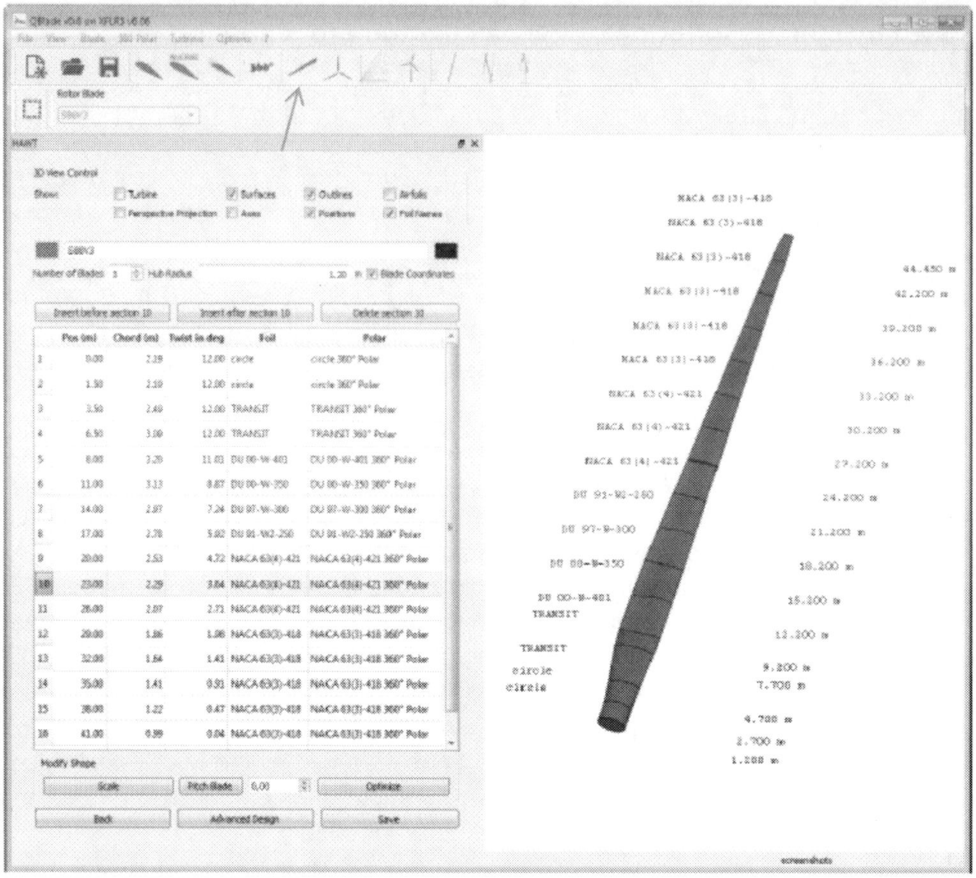

Screen shot from the (Marten & Wendler, Qblade Guidelines v0.6, 2013) manual.

6 • The next step it to input the BEM information in one of the modules that perform that calculation (Rotor BEM Simulation, Multiparameter BEM Simulation and Turbine BEM simulation).

7 • Calculation of the aerodynamic local features with the aid of the BEM method.

8 In this step, many groups of data are loaded for the 3D steady Blade noise calculation routine:

a) Geometry selection and flow conditions.

- Choose blade type in the database (with pitch angle).
- Wind Speed [m/s]
- Rotational Speed [RPM]
- TSR [ ]

Notice: only two of the 3 last parameters above should be informed, since their calculation is interdependent.

b) Air properties

- Air Density (default = 1.225 kg/m^3)
- Air Dynamic Viscosity (default = 1.78 E-5 Pa.s)

c) Selection/activation of noise sources (both available and planned sources shown).

- Inflow noise
    - Lowson´s model

- Amiet´s model
- Poli-Wind Modified Lowson model

- Self-noise sources
  - BPM TE Noise Model (select all here or individual sub-sources below)
  - Suction side of airfoil (attached flow).
  - Pressure side of airfoil (attached flow).
  - Separated flow on suction side (high Reynolds flow).

  OR
  - TNO-Blake TE Noise Model.

  OR
  - Amiet´s TE Noise Model.

- Separation-Stall Model
  - BPM Separation-Stall Model

- Laminar Boundary Layer Vortex Shedding Noise Model
  - BPM LBL-VS Model

- Tip Vortex Formation Noise Model
  - BPM Tip Vortex formation Model

  Trailing-Edge Bluntness Vortex Shedding Noise
  - BPM TE Bluntness VS Noise Model

d) δ^* (displacement thickness) source
  - XFoil calculation
  - Station for reading or interpolating δ^* value (default = 0.98c).
  - Original BPM correlations
  - Import data
  - δ^* scaling factor (default = 1)
  - Eddy Convection velocity as a fraction of the Mach number (default = 0.8*M)

e) Select directivity functions to be employed.

- Same options for high and low frequency sources of the 2D input screen apply here.

f) Define the observer position.

- In this option (single blade noise calculation routine), the observer rotates with the blade, at the fixed distance and angles specified by the user, the angular speed will be the same for the blade and the observer, both in non-inertial frames. The observer position should be determined in relation to the Blade Coordinate System (Germanisher Lloyd and German Wind Energy Committee, 2010). Its origin is at the Blade root (at hub radius) and rotates with the rotor. Its orientation to the hub is fixed. X_B in direction of rotor axis; Z_B radially; Y_B is set so that the system rotates clockwise:

  - X_B = +10 [m] (default)
  - Y_B = +10 [m] (default)
  - Z_B = blade radius/2 [m] (default value)

- The default position is 10 m downstream of the rotor plane, 10 m downstream of the blade in the rotation plane and in the mid-span of the blade. The following figure is provided in the input menu in order to help the user define the observer location:

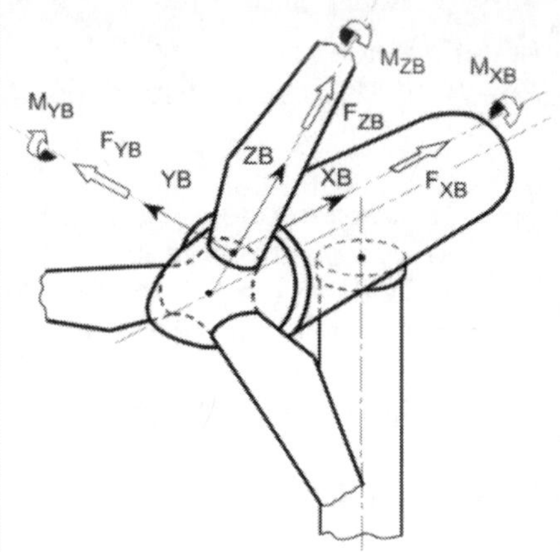

XB in direction of the rotor axis
ZB radially
YB so that XB, YB, ZB rotate clockwise

- The following warning notice is to be included in the code and displayed for the user at this screen: "Notice: The BPM directivity expression for High Frequency noise is not suitable for shallow angles ($\Theta \rightarrow 180°$). For details, see page 105 of (BPM, 1989)."

9
- Display for acceptance or adjustment, the table with validity range known for each selected noise source models. If the informed limits are deleted, the models will be applied regardless of validity. If the user inputs new validity ranges based in his/her experience, the values are stored and displayed as future default values, unless the INITIAL SYSTEM DEFAULT VALUES box is checked. For each noise source and model combination, the user will be able to set upper and lower limits for Reynolds Number, Mach Number and Angle of attack.

10
- For blade segments 1 to n, locate the BEM simulation results associated with the selected blade, for the same or closer TSR for which calculation is available. Save the Calculated TSR and BEM TSR values on the log.

11
- For blade segments 1 to n (root to tip; n was specified in the Blade input screen):

12
- Obtain the axial (a') and tangential (a) induction factors for each blade segment.
- Calculate the local tangential and local axial speed for the middle of each blade segment, employing the following relations:

$$tangential\ speed = \omega . r(1 + a')$$

Where:

$\omega = 2.\pi.\frac{n}{60}\ (\frac{rad}{s})$, where

n=rotational speed in rpm
r = radius of the segment mid-section, from the rotation axis (if absolute coordinates are in use) or from the hub radius (if root coordinates are in use), in meters.

$$axial\ speed = V_0.(1 - a)$$

Where:

$V_0 = wind\ speed\ (\frac{m}{s})$.

- Obtain the relative wind speed direction, ϕ, experienced by the rotating blade segment:

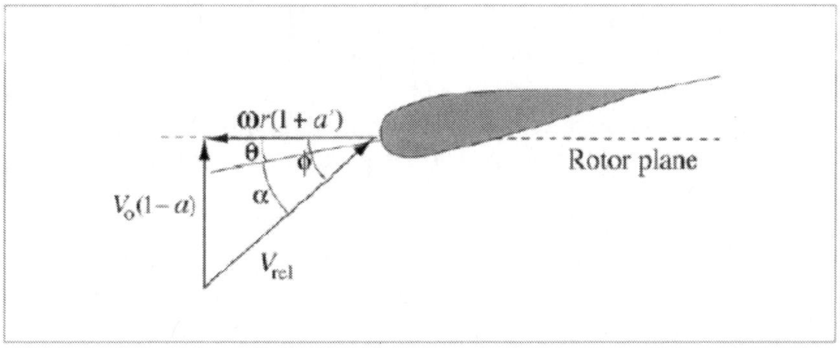

- Obtain the local blade AOA, α, by subtracting from the angle φ, the twist angle for that section:

$$\alpha = \phi - \theta$$

- Calculate the local Reynolds and Mach numbers, based on the relative wind speed $V_{rel.}$ and air properties supplied in the selected BEM module input:

Local Reynolds number = Air Density [kg/m^3] * Vrel [m/s] * Chord at mid span of the segment [m] / Air Dynamic Viscosity [Pa.s]

Local Mach number = Vrel [m/s]/ SQRT (1.4*286.9* (T+273.15)),

Where T = atmospheric air temperature, [˚C]. Default = 15 ˚C

13 Generate a log report, displaying:

- Calculated x used TSR data for the blade.

For each blade segment:

- Calculated x Re number data on the Polar Object associated with that blade segment.
- Calculated x used Mach number data on the Polar Object associated with that blade segment.
- Calculated local AOA x closer AOA angle available in the Polar Object associated with that blade segment.

The log is to be printed or displayed onscreen depending upon user selection on the menu (add report to menu with display and print capability).

14 • For each Selected Noise Model:

15 • Check if the noise model requires transversal turbulence length scales (δ*) or other Turbulent Boundary Layer data.

16 • Check user source option for δ* or other transversal turbulence length scales: BPM original correlations; XFoil calculations; imported data.

17 • Input data (in .csv format), if applicable.

18 • If the information is to be provided by the XFoil TBL calculation, locate the δ* table more suitable (AOA should be within 0.125° of any calculated local AOA) for each blade segment....

19 ....and interpolate δ* at the chord-station specified by the user (or default station of 0.98C, if not specified). Interpolate δ* data for one or both airfoil sides (pressure and suction sides), depending upon TE noise source selection in the input dialog.

20 • Locate the directivity function associated with each noise calculation model selected.

- Calculate distance and directivity angles from observer to the mid-span of the segment under consideration using the reference frame transformation matrices provided (different set for 3D blade and 3D rotor calculations).
- Save information for the noise calculation routine.
- Store the Mach and Reynolds information for the calculation.
- If original BPM correlation is to be employed, identification of the closest polar is not necessary.
- Get the δ* data for each applicable airfoil side.
- Transform the observer position in relation to all intermediate reference systems, until the observer position is obtained in relation to the TE Coordinate System and obtain the values for the BPM coordinates (r, θ, ϕ)

  - Input X_e, Y_e, Z_e
  - Attribute their respective values to X_B, Y_B, Z_B
  - For the blade, get the Pitch angle, θ_p
  - For each blade segment, get: $c_{r_{i+1}}, c_{r_i}, r_{i+1}, r_i, \beta$
  - Calculate $angle\ a, angle\ b, Z_B - (r_{i+1} - r_i)/2$
  - Calculate $\begin{bmatrix} X_{RS} \\ Y_{RS} \\ Z_{RS} \end{bmatrix}$
  - Calculate $Y_{RS} - 0.75 * (c_{r_{i+1}} - c_{r_i})/2$
  - Calculate $\begin{bmatrix} X_{RT} \\ Y_{RT} \\ Z_{RT} \end{bmatrix}$
  - Calculate (r_e, θ_e, ϕ_e)

The whole process may be summarized in the following matrix operations:

$$\begin{bmatrix} X_{RT} \\ Y_{RT} \\ Z_{RT} \end{bmatrix} = \begin{bmatrix} 1 & 0 & 0 \\ 0 & \cos b & \sin b \\ 0 & -\sin b & \cos b \end{bmatrix} \cdot \begin{bmatrix} X_{RS} \\ Y_{RS} - 0.75 * (c_{r_{i+1}} - c_{r_i})/2 \\ Z_{RS} \end{bmatrix}$$

$$\begin{bmatrix} X_{RS} \\ Y_{RS} \\ Z_{RS} \end{bmatrix} = \begin{bmatrix} \cos a & \sin a & 0 \\ -\sin a & \cos a & 0 \\ 0 & 0 & 1 \end{bmatrix} \cdot \begin{bmatrix} X_B \\ Y_B \\ Z_B - (r_{i+1} - r_i)/2 \end{bmatrix}$$

Where,

the angle *a* is the total angle between the Y_BZ_B blade reference system plane and the local midsection chord line. If the Y_BZ_B plane is set equal to the rotation plane, then the angle *a* is the equivalent to the θ angle in QBlade manual (Marten D. , Qblade Short Manual V0.8, 2014), p.34:

$$\theta = \theta_p + \beta$$

θ_p = pitch angle; β = the local twist angle.

$$b = \operatorname{atan}\left(\frac{c_{r_{i+1}} - c_{r_i}}{r_{i+1} - r_i}\right)$$

$$r_e \equiv r_{RT} = \sqrt{X_{RT}^2 + Y_{RT}^2 + Z_{RT}^2}$$

$$\theta_e = \pm \operatorname{atan}\left(\frac{Z_{RT}}{Y_{RT}}\right)$$

$$\phi_e = \pm \operatorname{atan}\left(\frac{X_{RT}}{Z_{RT}}\right)$$

21 - For each 1/3 octave frequency, using δ* information interpolated from XFoil or from original BPM correlations, or read from file, as selected by the user.

22 • Call the noise prediction calculation routine for each blade segment, for each noise source selected by the user AND inside validity range. This calculation includes directivity function calculation with models associated with each source.

23 • Store SPL contribution from each noise source, for each blade segment and frequency. For noise Models **N**ot **S**elected, display **N/S**. For noise Models **N**ot **A**pplicable per Reynolds, Mach number or AOA Ranges, display **N/A**.

24 • Next Frequency.

25 • Next blade Section.

26 • Since this is a steady-state, one blade simulation, average the contributions of each blade segment at the observer position. For *n* blade segments, calculate the average SPL as follows:

$$SPL_{Blade\_Total} = 10.log\left(\frac{1}{n}\left(\sum_{i=1}^{n} 10^{SPL_i/10}\right)\right)$$

27 • Display SPL Spectral contributions from each source and Total SPL graphs (output analog to the existing 2D noise module).

28 • Export and print SPL tables for each 1/3 octave frequency band, per source, and acoustic filters (A, B, C), as selected by the user on the menu.

5. Conclusions

A quasi-3D method was devised for rotor and blade noise prediction and the latter is described in detail in this paper. The method is intended to extend the current PNoise 2D TE noise prediction capabilities inside QBlade, due to significant user acceptance of that open-source, GPL code. The method is not new in essence, but combines the best practices recommended by many authors for various parameters with a more straightforward coordinate system and is optimized for the environment, the main goal of which is delivering a quick turn-around-time user experience for the preliminary, relative noise performance evaluation at the early design phase of WT equipment. Although conceived for preliminary design, the tool is also formal in a sense that it communicates clearly all model validity conditions and generate a detailed log of all calculations accomplished. The 3D extension was designed to draw as much synergy as possible from existing and tested features, like the boundary-layer code, the BEM code and 2D noise method previously implemented in QBlade or PNoise. The implementation and validation will be the objects of a future report, since coding is undergoing progress and a new model for inflow noise will be also included (Faria, Pimenta, Saab Jr, & Rodriguez, 2018), which is receiving the final analysis. The implementation has provisions for other noise sources (both in the 2D and 3D algorithms) and it is the authors' intent to add a complete menu of self-noise and inflow noise prediction sources and models to the QBlade platform, in order to complement its already comprehensive aerodynamic, structural and aeroelastic features.

References

Bertagnolio, F., Madsen, H. A., Fischer, A., & Bak, C. (2015). Experimental characterization of stall noise toward its modelling. *Wind Turbine Noise 2015* (pp. 1-12). Glasgow: INCE Europe.

Brooks, T. F., Pope, S., & Marcolini, M. A. (1989). *Airfoil Self-Noise and Prediction.* NASA. Langley: NASA.

Broux, F. (2008). *3D Extension of Wind Turbine Noise Prediction Tool IAGNOISE.* Stuttgart: IAG.

Drela, M. (2000). X-FOIL. http://web.mit.edu/drela/Public/web/xfoil/: MIT.

Drela, M., Youngren, H., & Deperrois, A. (2009). *XFOIL/XFLR5.* MIT, Cambridge, Ma.

Faria, A., Pimenta, M. M., Saab Jr, J. Y., & Rodriguez, S. (2018). Selection for a Leading-Edge Noise Model for PNoise. *Terceiro Congresso de Engenharia e Ciências das Três Fronteiras*, (pp. 1-15). Foz do Iguaçu, BR.

Fuglsang, P., & Bak, C. (2004). *Development of the Risø Wind Turbine Airfoils.* Risø, Roskilde.

Germanisher Lloyd and German Wind Energy Committee. (2010). *Guideline for the Certification of Wind Turbines.* Hamburg: Germanisher Lloyd.

Glegg, S. (1987, June). Significance of Unsteady Thickness Noise Sources. *AIAA Journal V. 25 (6)*, pp. 839-844.

Hansen, M. (2008). *Aerodynamics of Wind Turbines* (Second Edition ed.). London: Earthscan.

Kamruzzaman, M., Lutz, T., Nübler, K., & Krämer, E. (2011). Implementation and Verification of an Aeroacoustc Wind Turbine Blade Analysis Tool. *WTN 2011 Proceedings* (pp. 1-16). Rome: INCE Europe.

Marten, D. (2010). *Extension of an Aerodynamic Simulator for Wind Turbine Blade Design and Performance Analysis.* Technischen Universität Berlin. Berlin: TU Berlin.

Marten, D. (2014). *Qblade Short Manual V0.8.* Berlin: TU Berlin.

Marten, D. (2016). *QBlade v0.95 - Guidelines for Lifting Line Free Vortex Wake Simulations.* Berlin: TU Berlin.

Marten, D., & Wendler, J. (2013). *Qblade Guidelines v0.6.* Berlin: TU Berlin.

Montgomerie, B. (2004). *Methods for Root Effects, Tip Effects and Extending the Angle of Attack to 360 degrees with Application to Aerodynamics for Blades on Wind Turbines and Propellers.* Kista: Swedish Defense Research Agency Scientific Report FOI R 1305 SE.

Moriarty, P., & Migliore, P. (2003). *Semi-Empirical Aeroacoustic Noise Prediction Code for Wind Turbines.* NREL, Golden, CO.

Pechlivanoglou, G., Marten, D., G., W., N., M., & Wendler, J. (2009). *QBlade.* Technische Universität Berlin, Berlin.

Saab Jr, J. Y. (2016). *Trailing-Edge Noise – Development and Application of a Noise Prediction Tool for the Assessment and design of Wind Turbine Airfoils.* Retrieved 07 19, 2017, from USP Digital Library - Thesis and Dissertations: http://www.teses.usp.br/teses/disponiveis/3/3150/tde-14032017-140101/pt-br.php

Saab Jr, J. Y., & Pimenta, M. M. (2015). Displacement Thickness Evaluation for BPM-Type Airfoil-TE Noise Prediction Model. *6th International Conference on Wind Turbine Noise* (pp. 1-15). Glasgow: INCE Europe.

Saab Jr, J. Y., & Pimenta, M. M. (2017). Verification and Validation of the "PNoise" Airfoil Trailing-Edge Noise Prediction Module inside "QBlade". *7th International Conference on Wind Turbine Noise* (pp. 1-11). Rotterdam: INCE Europe.

Saab Jr, J. Y., Pimenta, M. M., Faria, A. M., & Rodriguez, S. (2018, Sept 28). Determination of local fow conditions and preliminary investigation on the validity of the PNoise code for large-size wind turbine. *Journal of the Brazilian Society of Mechanical Sciences and Engineering*, pp. 493-504.

Saab Jr, J., & Pimenta, M. (2016, Feb). Displacement Thickness Evaluation for Semi-Empirical Airfol Trailing-Edge Noise Prediction Model. *Journal of the Brazilian Society of Mechanical Sciences and Engineering, V.38, N.2*, pp. 385-394.

Saab Jr., J. Y., Pimenta, M. M., Piqueira, J. R., Marten, D., Pechlivanoglou, G., Nayeri, C. N., & Paschereit, C. O. (2016). *Verification and Validation of the Airfoil Trailing-Edge Noise Prediction Module (PNoise) inside QBlade v0.95.* São Paulo: POLI-USP.

Vargas, L. d. (2008). *Wind Turbine Noise Prediction.* Instituto Superior Técnico, Lisboa.

Viterna, L. A., & Janetzke, D. C. (1982). *Theoretical and Experimental Power From Large Horizontal-Axis Wind Turbines.* Cleveland, OH: NASA Lewis Research Center Technical Report TM-82944.

Zhu, W. (2004). *Modelling of Noise From Wind Turbines.* DTU, Lyngby, DK.

INCE*EUROPE*

8th International Conference
on
Wind Turbine Noise
Lisbon – 12th to 14th June 2019

Analysis of a high fidelity aero-servo-elastic process chain to assess low-frequency emissions from wind turbines

Florian Wenz, IAG, University of Stuttgart, florian.wenz@iag.uni-stuttgart.de
Levin Klein, IAG, University of Stuttgart, levin.klein@iag.uni-stuttgart.de
Thorsten Lutz, IAG, University of Stuttgart, thorsten.lutz@iag.uni-stuttgart.de
Philipp Rettler, SWE, University of Stuttgart, rettler@ifb.uni-stuttgart.de

Summary

The increase in computational resources enables high fidelity simulations of wind turbines which allow to capture evermore phenomena including the acoustic low-frequency emissions. One of the remaining challenges is to verify these numerical results in the present study. A high fidelity process chain is used to numerically investigate the low-frequency emissions from a generic wind turbine. It consists of a computational fluid dynamics (CFD) solver coupled to a multi-body simulation (MBS) solver to take aeroelasticity into account and includes a controller for a realistic load behaviour. The sound pressure at discrete observer positions surrounding the turbine is calculated by means of a Ffowcs-Williams-Hawkings (FW-H) acoustic solver. A comparison with measurements requires an appropriate post-processing and visualisation of the sound pressure signal. Statistical analyses are often well-suited since the operational conditions and wind turbine geometries are in many cases not accurately reproduced in the simulation. However, discrete events like the blade-tower interaction can sometimes be better described by time series. A few different evaluation approaches are applied on the numerical results to show their suitability and significance. Especially, a variable rotational speed hampers the analysis in frequency domain and needs to be considered. The computational results of a generic wind turbine are qualitatively compared to acoustic low-frequency measurements of a commercial wind turbine to qualitatively verify the numerical process chain.

1. Introduction

There are a lot of publications about low-frequency emissions from wind turbines. Hubbard and Shepherd [1] are among the first to mention acoustic pulses due to the blade-tower interaction at blade-passing frequency (BPF) and its harmonics. Nowadays this interaction is considered to be the major cause of the low-frequency emissions from wind turbines based on measurements [2] [3] as well as simulations [4] [5]. Therefore, the major part of the low-frequency noise has the same origin and it is adequate to focus on the coherent part of the acoustic signal. Hessler et al. [6] give an overview over the latest findings in the field of low-frequency noise of wind turbines. The importance of correctly capturing the blade passing pulse of the blade-tower passage is shown by Richarz et al. [7] who state that due to random phase shifts during propagation the inaudible low-frequency pulse can get audible. The local velocity and temperature gradients in a turbulent flow alter the local acoustic wave speed, when the length scale of the eddy and the acoustic wavelength are similar. This is mainly the case for higher harmonics which are otherwise masked by broadband noise but are enhanced in this way at an observer position.

1.1 Scope and Objectives

This paper addresses the blade-tower passage emission of wind turbines and the quality of its prediction with a numerical high-fidelity process chain. A CFD-MBS coupling [4] is used to run a complex wind turbine simulation including as many occurring effects as currently possible. More precisely, a generic 5 MW turbine with six flexible structures, namely, three blades, tower, foundation and drive train, is simulated. The included controller is forced to change pitch and rotation speed since a sheared turbulent inflow close to rated conditions is used. The acoustic pressure at defined observer positions is calculated based on the surface pressure using the FW-H equation. Afterwards the sound pressure time series is post-processed with different approaches. Depending on the operational conditions and the mechanism which are evaluated, their suitability and significance is shown with the focus on enabling comparisons to measurements.

2. Numerical Details

The low-frequency emissions from a generic 5 MW wind turbine are investigated numerically. The used high-fidelity process chain allows fully coupled CFD simulations under realistic turbulent inflow conditions including a pitch and rotational speed controller.

2.1 Numerical Process Chain

The simulation results were obtained using the high-fidelity process chain established by Klein et al. [4]. As described in this reference, the basis for the numerical simulations is the CFD solver FLOWer [8]. It is a compressible, dual time-stepping, block structured Reynolds-averaged Navier–Stokes (RANS) solver. The implemented CHIMERA technique allows the use of multiple grids for the structures and the background. The solver is continuously extended at the Institute of Aerodynamic and Gas Dynamics (IAG) [9], [10] improving its suitability for wind turbine simulations [11], [12]. The structural deformations are calculated with the commercial MBS solver SIMPACK [13] applying nonlinear beam theory.

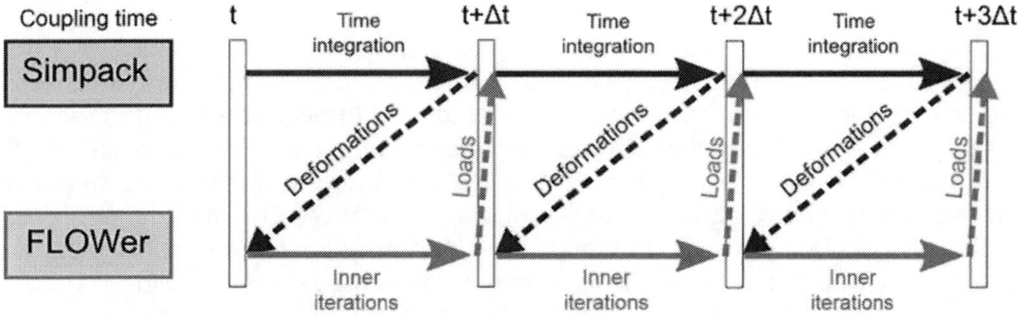

Figure 1: Explicit coupling scheme of the FLOWer–SIMPACK coupling *[4]*.

The coupling between FLOWer and SIMPACK enables combined calculation of rotating and non-rotating parts and allows to consider flexible deformations and rigid-body motions simultaneously. Hence, pitch motions and changes in rotational speed of the rotor caused by a variable speed collective pitch controller can be transferred from the MBS solver to the CFD solver. The employed coupling is a partitioned approach which uses an explicit coupling scheme depicted in

Figure 1. The communication is realized with 160 markers distributed over the turbine structures at which integrated loads and deformations are transferred. This fluid-structure interaction (FSI) takes the influence of unsteady structural deformation on the aerodynamics into account by deforming the CFD mesh.
The coupled FLOWer-SIMPACK simulations generate time series of the surface pressure distribution on the turbine. These are used to calculate the aeroacoustic signal at distant, predefined observer positions by means of the in-house FW-H solver ACCO. Volume sources generated by free-flow turbulence, ground reflections and nonlinear propagation due to atmospheric layering and turbulence are neglected. Hence, an undisturbed propagation and observers located in the acoustic far field are assumed.
For more information on the process chain in general and the implementation and functionality in detail please refer to Klein et al. [4].

2.2 Simulation Cases

The simulated turbine is derived from the generic 5 MW turbine developed by the National Renewable Energy Laboratory (NREL) [14], which was slightly modified in the OFFWINDTECH project [15]. The hub height is 90 m and the rotor diameter is 126 m. The rated conditions of the modified onshore configuration of the turbine are given with a rotational speed of 11.7 RPM at a hub wind speed of 11.3 ms^{-1}. The simulations without controller use rated rotational speed and the inflow conditions aim to meet rated wind speed at hub height. The atmospheric boundary layer (ABL) is represented by the power law with an exponent of 0.19. The generic atmospheric turbulence is generated using Mann's model [16] resulting a turbulence intensity at the turbine position of 16 %. Hence, a turbulent ABL inflow involves shear as well as turbulence.
The described numerical setup was used to investigate the impact of model complexity on the calculated low-frequency emissions by Klein et al. [4]. On top of these already presented cases the present paper shows results of a case with included controller and an additional degree of freedom, namely the drive train torsion. Overall, two different load cases (LC) and three different levels of fluid structure coupling (FSC) are considered. Table 1 summarises all simulation cases regarded hereafter.

Case name	Inflow	Flexible structures	Controller
LC2 [4]	uniform	none	no
LC2_FSC3 [4]	uniform	blades, tower, foundation	no
LC4_FSC3 [4]	turbulent ABL	blades, tower, foundation	no
LC4_FSC5	turbulent ABL	blades, tower, foundation, drive train	yes

Table 1: Definition of simulation cases.

2.3 Computational Setup

The CFD model is taken from Klein et al. [4]. It consists of 10 independent body meshes, namely three blades, tower, hub, nacelle, hub-nacelle connector and three blade-hub connectors, as well as a background mesh all connected via the CHIMERA technique. The resolution of the body meshes allows to fully resolve the boundary layer ($y^+ \leq 1$) and the background mesh is refined upstream of the turbine to propagate the atmospheric turbulence. Overall, the mesh consists of 86 million cells. The unsteady RANS simulations are performed using a second-order dual time-stepping scheme according to Jameson [17]. A physical time step corresponding to 0.75° azimuth (≈0.01068s) with 100 inner iterations is applied for the evaluated part of the simulations. The spatial discretization in body meshes is based on second-order central discretization with Jameson–Schmidt–Turkel (JST) artificial dissipation term [18]. The fifth-order WENO scheme is applied in the background mesh in order to reduce the dissipation of vortices. All simulations are fully turbulent and use the k-ω-SST model by Menter [19] for turbulence modelling.
The structural model in SIMPACK was built by Matha et al. [20]. The blades and the tower are each represented by multiple flexible bodies so that they are modelled nonlinearly, whereas hub, nacelle and foundation are kept rigid. The foundation is connected to the ground with a spring-

damper system [4]. In simulations with controller the drive train is also flexible with respect to torsion.

3. Evaluation of Simulation Results

The sampling rate of the sound pressure is equal to the physical time step of the simulations Δt=0.01068 s. Thus, the highest captured frequency is 46.8 Hz. However, the spatial resolution at the source, i.e. the surface mesh of the turbine, also limits the highest resolved frequency. The shortest wavelength properly simulated is limited by the largest surface cells. Assuming it takes ten cells to properly capture a pressure wave, the highest frequency to be well resolved is approximately 25Hz. All depicted sound pressure levels are unweighted and the time signals are cut to multiples of blade passages before evaluation.

3.1 Methods for Evaluation of Acoustic Data

Merchant et al. [21] give a good overview on approaches and techniques to evaluate acoustic signals statistically as well as in the time domain. In general the sound pressure is measured in Pascals (Pa). However, it is usually expressed as sound pressure level (SPL) in decibels (dB) due to the huge pressure range which is audible

$$SPL_{rms} = 20 \cdot \log_{10}\left(\frac{p_{rms}}{p_0}\right) dB.$$

The standard reference sound pressure for airborne sound p_0=20 µPa is used in this paper. The sound pressure p used for decibel calculation is expressed in root-mean-square values. This means the effective sound pressure is evaluated as it is defined in IEC 103-02-03. A single value results from this evaluation. It represents the broadband sound pressure level within a given time interval including all captured frequencies. This is the most often used average acoustic metric. However, this approach gives no information on the characteristic of the sound like peak sound level or tonality and does not allow to specifically evaluate the low-frequency emission. Therefore, more substantial evaluation methods need to be applied.

A typical approach to analyse acoustical data is a transformation into the frequency domain to gain insight into the frequency characteristics. This is normally done using Fast-Fourier-Transformation (FFT). The spectral sound pressure level is calculated using

$$SPL(f) = 20 \cdot \log_{10}\left(\frac{\frac{2}{\sqrt{2}} \cdot \left|\frac{FFT[p(t)]}{L}\right|}{p_0}\right) dB,$$

where L is the number of samples in the pressure signal p(t). The frequencies above the Nyquist frequency $F_s/2$ are discarded and hence the remaining frequency bins are doubled to get the single-sided FFT. Since the pressure signal is considered to be a superposition of sine waves it is divided by $\sqrt{2}$ to get the effective sound pressure level.

However, for not entirely periodic signals the interpretation of the resulting spectra with regard to the coherent component is difficult. Vanderkooy and Mann [2] introduce an approach to regularize quasi-periodic acoustic signals of wind turbines with varying rotational speed and obtain periodic data which are better suited for a FFT evaluation. The necessary information on the blade passage period is extracted from the original sound pressure signal by detecting local peaks. The duration of the original signal is than divided by the number of blade passages captured in the signal. The obtained mean blade passage period is discretised equidistantly so that the original sampling rate is roughly obtained. Each of the detected blade passage periods

is interpolated on one of the normalised mean blade passage periods. The string of these segments is the resampled signal with a warped time line.
For strongly time-dependent sound pressure signals the short-time Fourier transform (STFT) is a method of Fourier analysis to also detect the temporal change in the frequency spectrum. The time signal is divided into multiple segments which are individually transferred into the frequency domain using FFT. The changing spectra are visualised as a function of time in spectrograms. This method combines evaluation in the frequency domain and time domain and hence a trade-off between frequency resolution and time resolution is necessary. Overlapping time windows can be used to smooth the data in the time domain.
If the sound has a rather impulsive character evaluating the pressure waveform directly in the time domain is a good option. Vanderkooy and Mann [2] investigated the shape of the coherent acoustic blade passing pulse of multiple turbines which can be used as reference for simulation results. The shape of the acoustic pulse is often masked by background noise of different sources. Therefore, an approach is to reconstruct the signal by inverse transforming the first couple of BPF harmonics extracted from the spectrum.

3.2 Analysis of Average Values

The broadband root-mean-square sound pressure level SPL_{rms} is evaluated using the FW-H solver at a carpet of 3600 observers on the ground surrounding the turbine. The highest included frequency is, as already mentioned, 46.8 Hz and the evaluated signals have a duration of seven (LC2) and nineteen revolutions (LC4_FSC5), respectively. The outermost circle of observers has a radius of 2000 m. In Figure 2 and Figure 3 the noise carpets for the cases LC2 and LC4_FSC5 (compare Table 1) are shown. These plots are well suited to get information on the directivity of the acoustic low-frequency emission and its total level.

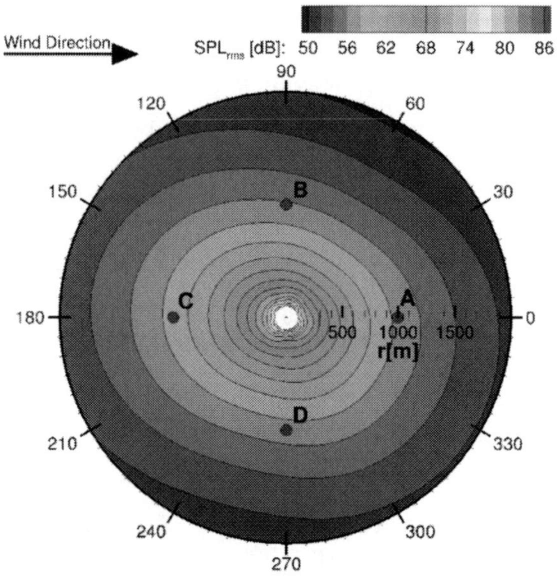

Figure 2: Unweighted broadband SPL (f<46.8 Hz) on the ground for case LC2.

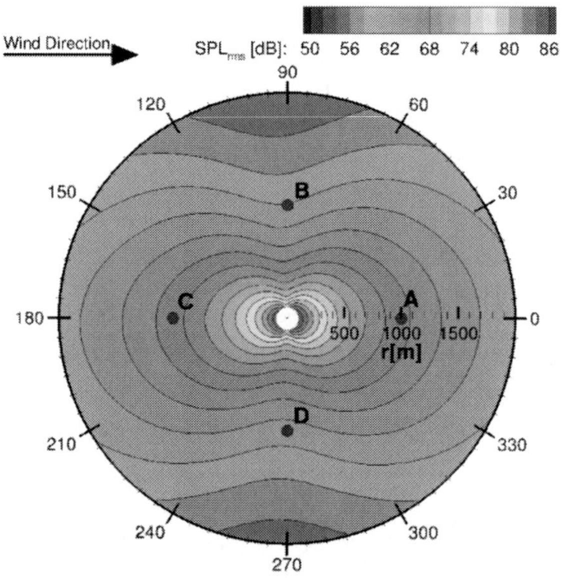

Figure 3: Unweighted broadband SPL (f<46.8 Hz) on the ground for case LC4_FSC5.

The sound pressure level for case LC4_FSC5 is significantly higher and the directivity is also slightly different. Since there are multiple levels of complexity between these two cases, looking at the intermediate cases helps to identify the causes. The values at four discrete observers A-D, each at a distance of 1000 m from the turbine, of all four cases are compared in Table 2.
As expected and already found by Klein et al. [4] the SPL is increased when deformation is considered. The main reason for that was identified as the decreased blade-tower distance [22]. Adding turbulence to the inflow further increases the SPL but only in upstream and downstream

direction. The impact of the controller is rather small. Certainly, the resulting strong directivity for such low-frequencies is an interesting finding. However, looking at these plots and values, it cannot be decided whether the increase in magnitude is due to more broadband noise or to higher tonal peaks. Generally, it is not possible to gain information about the character of the sound. Moreover, these values include also the energy of frequencies which might not be correctly resolved in the simulation because of the above mentioned mesh limitations. To get more insight an evaluation in the frequency domain is necessary.

Observer	LC2	LC2_FSC3	LC4_FSC3	LC4_FSC5
A	57.5	64.2	66.5	66.6
B	56.3	61.7	61.3	60.2
C	58.8	65.2	67.2	67.2
D	57.0	62.2	62.0	61.1

Table 2: Unweighted SPL_{rms} in dB (f<46.8 Hz) at discrete observers 1000 m from the turbine for all cases.

3.3 Analysis in Frequency Domain

Performing a Fast-Fourier-Transform of the sound pressure signal at discrete observer positions allows to analyse the frequency characteristic. The signal is cut into multiples of the blade passage to obtain an almost periodic input to the FFT. This approach was already used by Klein et al. [4] to evaluate the cases LC2, LC2_FSC3 and LC4_FSC3. The key observation was that the acoustic emission occurs mainly at the BPF and its harmonics. Moreover, it was found that the reduced blade-tower distance when considering flexibility increases the peak levels and turbulent inflow increases the broadband level. All these observations could be made by directly transforming the pressure signal with a FFT to the frequency domain.

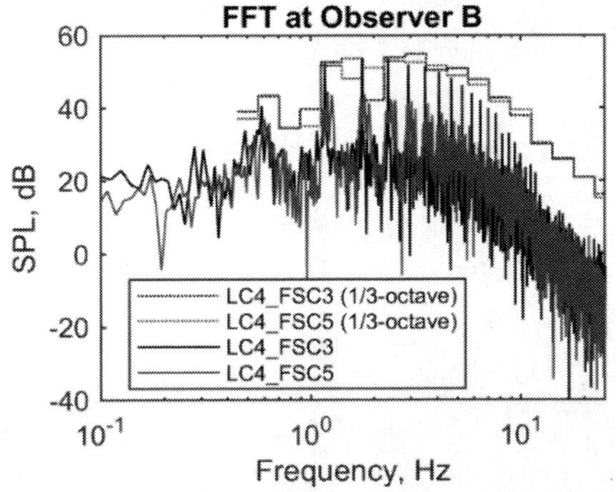

Figure 4: Spectra of unweighted SPL for cases LC4_FSC3 and LC4_FSC5 at observer B.

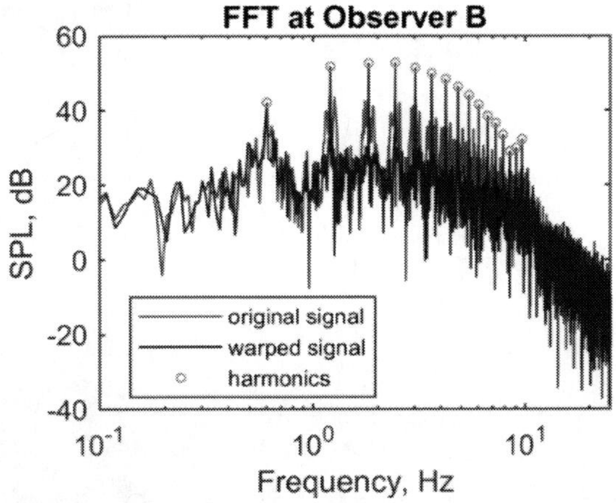

Figure 5: Spectra of unweighted SPL for case LC4_FSC5 at observer B of original and warped signal.

This approach yields meaningful results if the input signal is periodic which is given for all cases with constant rotational speed. However, the spectrum of the case LC4_FSC5, with a variable rotational speed, shown in Figure 4, has a significantly different shape. The issue is whether the difference has a real physical cause or is the consequence of an inappropriate evaluation? An indication that the low frequency emissions of the cases are actually very similar are the almost identical one-third octave band SPLs (see Figure 4). Since the requirement for a FFT of a periodic input signal is violated the differences are most likely caused by the evaluation approach. To confirm this, the time warp, described in section 3.1, is applied. The results are plotted in Figure 5 in comparison to the FFT of the unaltered signal.

The spectrum of the reconstructed periodic signal shows, as expected, very clearly the fundamental frequency and its harmonics. Now the spectrum is very similar to, for example, the one of case LC4_FSC3 in Figure 4. Therefore, it can be concluded that also in the case with controller the blade-tower interaction is the dominant acoustic source in the regarded frequency range. To verify that the controller did not alter the SPL caused by the blade passage, the SPL of the first 15 harmonics (exemplarily marked in Figure 5) are added up in Table 3.

	LC2	LC2_FSC3	LC4_FSC3	LC4_FSC5
SPL_{15harm} [dB]	56.1	61.7	60.7	59.8
% of SPL_{rms}	99.6	99.9	99.0	99.3

Table 3: Unweighted SPL of first 15 harmonics in dB at observer B and its ratio of SPL_{rms} (f<46.8 Hz) for all cases.

For all cases more than 99 % of the SPL_{rms} (f<46.8 Hz) is caused by the first 15 harmonics of the BPF. This evaluation is not possible for cases with varying rotational speed without reconstructing a periodic signal since otherwise the harmonic series is not clearly extractable (see Figure 5). It can also be seen that adding turbulence to the inflow reduces the ratio slightly compared to uniform inflow. This corresponds well with the observation of an increase in broadband noise.
Despite the advantages of the signal warping for frequency analyses, this evaluation method distorts the real character of the acoustic signal, especially if the blade passage period varies strongly. This is because the method is a kind of averaging and the time information gets lost. The mentioned STFT and a visualisation via spectrograms preserves the time information, namely the variations over time (see Figure 6).

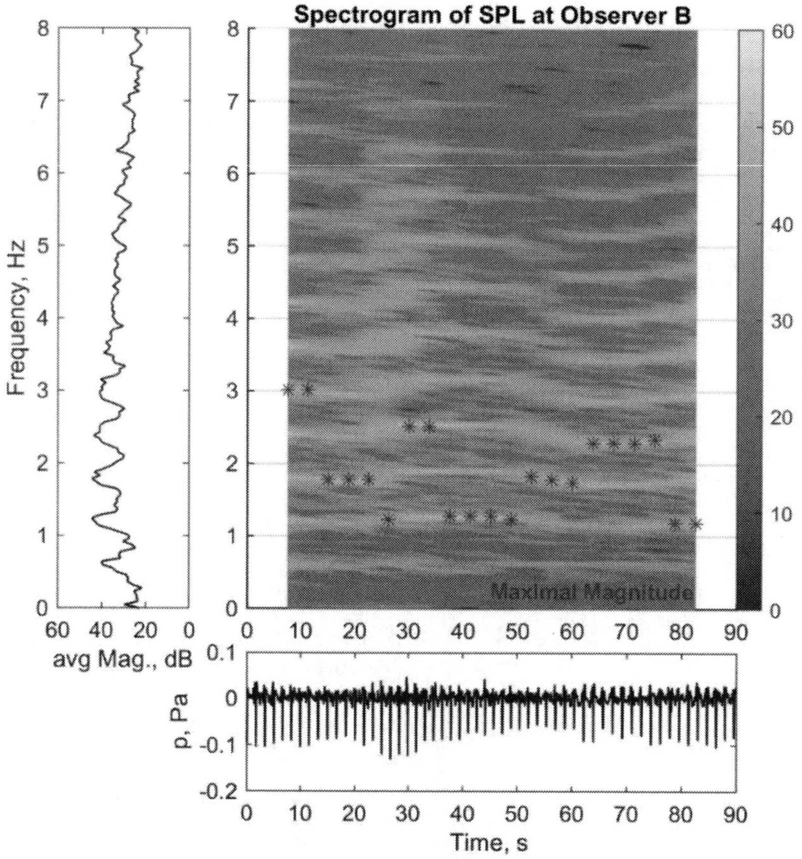

Figure 6: Spectrogram of case LC4_FSC5 at observer B.

The shown spectrogram was created with a window size of 15 s and a hop size of 3.75 s, resulting in 75 % overlap of the frames. A rectangular window was applied. The plot on the left of the figure is created by averaging the magnitude of the spectrogram for each frequency. The BPF and its harmonics are well visible and the magnitudes correspond to the findings in Figure 5 for the warped signal. Even though the BPF itself has rather low amplitudes, looking at its harmonics the increase of rotational speed at about 30 s as well as the decrease at approximately 60 s is obvious. The included red markers depict the maximal magnitude for each window. The maximal SPL value varies between 49.3 dB and 53.3 dB what is in accordance with the peaks in the FFT plots. Moreover it can be seen that the maximum jumps between the second and the fifth harmonic of the BPF.

3.4 Analysis in Time Domain

All the evaluations shown so far indicate that the blade-tower passing is the dominant event for acoustic low-frequency emissions. Although this is a periodically occurring event and can, as shown, be evaluated in the frequency domain, the passage itself is a fairly impulsive sound. Therefore, looking at the pressure waveform allows some interesting observations as well as enable another way to compare simulation and measurement. In Figure 7 the warped pressure signal of case LC4_FSC5 is plotted for observers B and C. At the side of the turbine, namely observer B, the blade passing is very prominent (upper graph). Downstream of the turbine (observer C), however, the blade passing is almost not visible since many fluctuations are superposed (lower graph). This is because the turbulent inflow interacts with the blades which emit mainly in upstream and downstream direction (see Klein et al. [4]). Hence to visualise the shape of the blade passing pulse at observer C the pressure signal is reconstructed using the first 15 harmonics (marked in Figure 5). Now a clear pulse is visible, even though it is significantly different than at observer B. Comparable differences in the shape of coherent blade passing pulses extracted from measurements of different wind turbines were also observed by Vanderkooy and Mann [2].

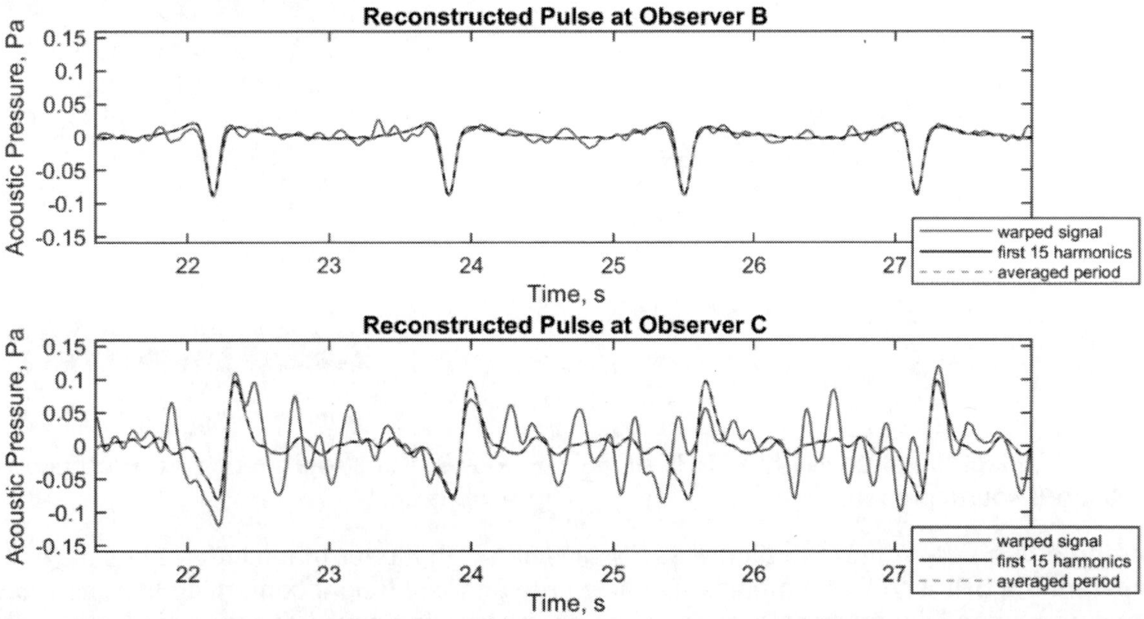

Figure 7: Pressure signal of case LC4_FSC5, reconstructed signal using 15 harmonics and averaged periods.

Another interesting finding made by Vanderkooy and Mann [2] is that averaging all periods removes most of the noise and only the blade passing pulse remains. This is also true for the simulation results of case LC4_FSC5 and confirms that the broadband noise is of stochastic nature and has no tonal components, leaving the blade tower interaction as only source of low-frequency tonality.

4. Qualitative Comparison with Measurements

The simulation results shown so far match the expectations and reasonable explanations could be found for the observations. Still at least a qualitative comparison to measurements is essential to verify the numerical process chain and its results. Obviously, there are no measurements for a generic wind turbine. Therefore, absolute values cannot be verified, but the relative values of peaks in FFTs as well as the shape of the spectrogram and the blade passing pulse can be compared. The measurements shown in this paper were carried out by Stuttgart Wind Energy (SWE) of the University of Stuttgart by using the procedure explained in [23]. A GRAS 47AC free-field infrasound microphone with a frequency range from 0.09 Hz to 20 kHz (±3 dB) was used. It was placed in a distance of about 140 m southeast (125°) of a single multimegawatt wind turbine. During the evaluated period of the measurement the wind came from a direction of 145°-180° with a velocity of about 5 ms^{-1} at hub height. Hence, the microphone was placed more or less upstream of the turbine. Since the numerical results indicate that the acoustic emission is similar in upstream and downstream direction a qualitative comparison with observer C is reasonable. Figure 8 shows the spectrum of a 400 s interval of the measured acoustic pressure. Before the FFT was performed the sampling rate was reduced to 100 Hz resulting in 50 Hz to be the highest resolved frequency. A period with relatively constant wind speed and hence constant rotational speed was chosen. As for the simulation results, discrete peaks occur which can be linked to the BPF and its harmonics. Again not the BPF itself but a higher harmonic has the highest amplitude. The first 15 harmonics contain 84 % of the SPL_{rms} (f<50 Hz). This ratio is considerably lower than for the simulations, but the measured signal also contains background noise sources not related to the wind turbine.

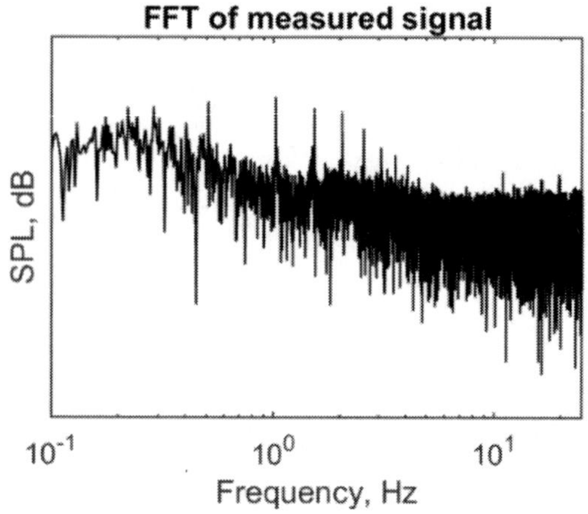

Figure 8: Spectrum of unweighted SPL of measured sound pressure.

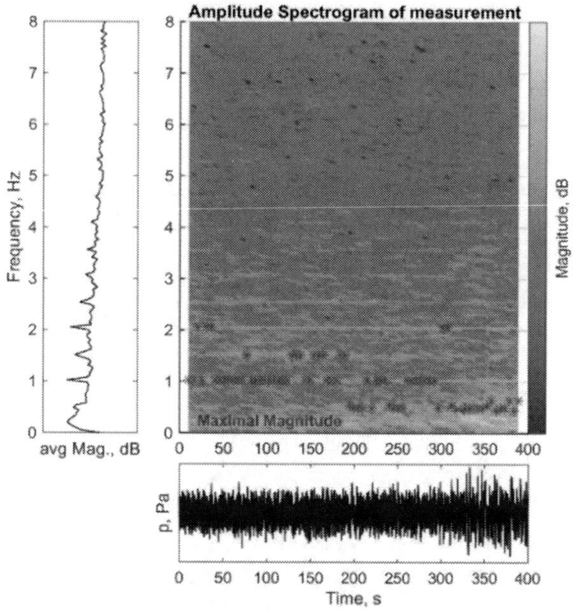

Figure 9: Spectrogram of measured acoustic pressure.

Looking at the spectrogram in Figure 9, narrow lines with higher magnitudes are visible. These represent the BPF and its harmonics. The lines are almost straight confirming that the rotational speed was indeed almost constant. As for the simulation, the maximal amplitude jumps between the harmonics. Overall, the first seven harmonics are visible in the plot on the left in Figure 9 showing the time-averaged SPL per frequency. This is less than in the simulation, but it should be noted that the dimensions of the simulated and measured wind turbine are significantly different and the wind speed was much lower in the measurement. In addition, an increase in amplitude is evident below 1 Hz starting after about 300 s whereas the higher harmonics continue unaltered. This suggests the presence of not turbine-related sound sources in the signal.

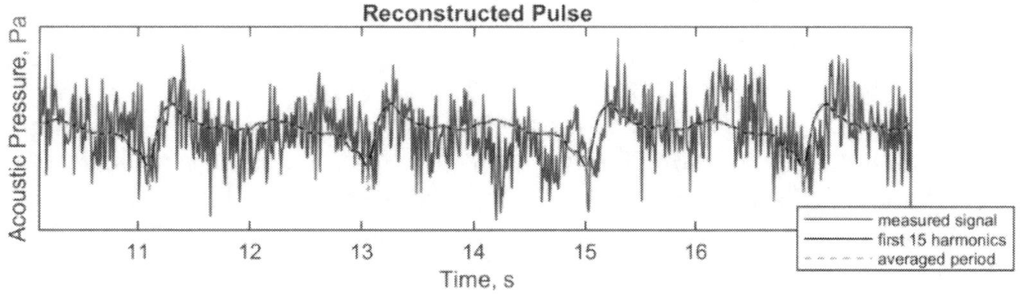

Figure 10: Measured pressure signal, reconstructed signal using 15 harmonics and averaged periods.

The reconstructed coherent pulse is depicted in Figure 10 in comparison to the original signal. Again in the signal itself the blade passing is hard to detect. The reconstructed waveform has an evident similarity with the pulse of the simulation at observer C (Figure 7 bottom). This is an indication that the simulation captures the blade-tower passage accurately.
Overall, the evaluation of the measured acoustic pressure shows similar phenomena as the simulation results and hence the process chain is able to capture the main effects contributing to the low-frequency emission from wind turbines.

5. Conclusions

The results of simulations with differing complexity performed by Klein et al. [4] are compared by means of different evaluation methods. The general directivity of the emission can be represented using the root-mean-square SPL evaluated at multiple observer positions. As long as the rotational speed is constant the character of the low-frequency emissions can be well visualised with a FFT. The tonality at the BPF and its harmonics is the dominant characteristic. When the rotational speed varies during the evaluation interval the FFT no longer reveals the tonality that clearly. A time warping must be applied or a spectrogram must be used to visualise the prominence of discrete frequencies. The reconstructed coherent pulse shows the general waveform of the blade-tower passage. The shape changes according to the observer position.
Some simulation results could be qualitatively confirmed by applying the same evaluation methods on a measured sound pressure. However, a final verification of the numerical process chain is only possible by simulating the commercial wind turbine at the same operational conditions as during the measurement. This would also allow to quantitatively validate the simulation results.

Acknowledgment

The studies were conducted as part of the joint research project "Objective Criteria for Seismic and Acoustic Emission of Inland Wind Turbines (TremAc), FKZ 0325839A", funded by the German Federal Ministry for Economic Affairs and Energy (BMWi). The authors are grateful for the financial support. The authors gratefully acknowledge the High-Performance Computing Center Stuttgart (HLRS) for providing computational resources within the project WEALoads.

References

[1] H. H. Hubbard and K. P. Shepherd, "Aeroacoustics of large wind turbines," *The Journal of the Acoustical Society of America,* vol. 89, no. 6, pp. 2495-2508, 1991.

[2] J. Vanderkooy and R. Mann, "Measuring wind turbine coherent infrasound," in *International Conference on Wind Turbine Noise*, Glasgow, 2015.

[3] K. L. Hansen, B. Zajamsek and C. H. Hansen, "The Occurrence of Nocturnal Wind Farm Rumbling Noise," in *7th International Conference on Wind Turbine Noise*, Rotterdam, 2017.

[4] L. Klein, J. Gude, F. Wenz, T. Lutz and E. Krämer, "Advanced computational fluid dynamics (CFD)–multi-body simulation (MBS) coupling to assess low-frequency emissions from wind turbines," *Wind Energy Science,* no. 3, pp. 713-728, 2018.

[5] Y. Yauwenas, B. Zajamšek, J. Reizes, V. Timchenko and C. J. Doolan, "Numerical simulation of blade-passage noise," *The Journal of the Acoustical Society of America,* vol. 142, p. 1575–1586, 2017.

[6] G. Hessler, G. Leventhall, P. Schomer and B. Walker, "Health Effects from Wind Turbine Low Frequency Noise & Infrasound Do Wind Turbines Make People Sick? That is the Issue," *Sound and Vibration,* vol. 51, no. 1, pp. 34-44, 2017.

[7] W. Richarz and H. Richarz, "Propagation Through A Turbulent Atmosphere Makes Blade Passage Harmonics Audible," in *7th International Conference on Wind Turbine Noise*, Rotterdam, 2017.

[8] N. Kroll, C.-C. Rossow, K. Becker and F. Thiele, "The MEGAFLOW project," *Aerospace Science and Technology,* vol. 4, no. 4, pp. 223-237, 2000.

[9] U. Kowarsch, M. Keßler and E. Krämer, "High order CFD-simulation of the rotor-fuselage interaction," in *39th European Rotorcraft Forum*, Moscow, 2013.

[10] P. Weihing, J. Letzgus, G. Bangga, T. Lutz and E. Krämer, "Hybrid RANS/LES capabilities of the flow solver FLOWer-application to flow around wind turbines," in *The 6th Symposium on Hybrid RANS-LES Methods*, Strassbourg, 2016.

[11] K. Boorsma, J. Schepers, S. Gomez-Iradi, I. Herraez, T. Lutz, P. Weihing, L. Oggiano, G. Pirrung, H. Madsen, W. Shen, H. Rahimi and P. Schaffarczyk, "Final report of IEA Task 29, Mexnext (Phase 3)," IEA, ECN, Petten, the Netherlands, 2018.

[12] N. Sørensen, N. Garca, S. Voutsinas, E. Jost and T. Lutz, "Aerodynamics of Large Rotors WP2 Deliverable 2.6 Effects of complex inflow for the AVATAR reference rotor and NM80 rotors," Technical Report, Petten, the Netherlands, 2017.

[13] SIMULIA, "Simpack Multi-Body Simulation Software," Dassault Systèmes, [Online]. Available: http://www.simpack.com/.

[14] J. Jonkman, S. Butterfield, W. Musial and G. Scott, "Definition of a 5-MW reference wind turbine for offshore system development," National Renew. Energ. Lab.(NREL), Golden, CO, United States, 2009.

[15] D. Bekiropoulos, T. Lutz, J. Baltazar, O. Lehmkuhl and N. Glodic, "D2013-3.1: Comparison of benchmark results from CFD-Simulation," Deliverable report, KIC-OFFWINDTECH, 2013.

[16] J. Mann, "The spatial structure of neutral atmospheric," *J. Fluid Mech.,* vol. 273, p. 141–168, 1994.

[17] A. Jameson, "Time dependent calculations using multigrid, with applications to unsteady flows past airfoils and wings," in *10th Computational Fluid Dynamics Conference*, 1991.

[18] A. Jameson, W. Schmidt and E. Turkel, "Numerical solution of the Euler equations by finite volume methods using Runge Kutta time stepping schemes," in *14th fluid and plasma dynamics conference*, 1981.

[19] F. R. Menter, "Two-equation eddy-viscosity turbulence models for engineering applications," vol. 32, no. 8, pp. 1598-1605, 1994.

[20] D. Matha, S. Hauptmann, T. Hecquet and M. Kühn, "Methodology and results of loads analysis of wind turbines with advanced aeroelastic multi-body simulation," in *DEWEK*, Bremen, 2010.

[21] N. D. Merchant, K. M. Fristrup, M. P. Johnson, P. L. Tyack, M. J. Witt, P. Blondel and S. E. Parks, "Measuring acoustic habitats," *Methods in Ecology and Evolution,* vol. 6, no. 3, pp. 257-265, 2015.

[22] C. Hansen, B. Zajamšek and K. Hansen, "Infrasound and Low-Frequency Noise from Wind Turbines," in *3rd Symposium on Fluid-Structure-Sound Interactions and Control*, Perth, 2015.

[23] F. Calarco, P. W. Cheng, T. Zieger and J. Ritter, "Acoustic and seismic emissions from wind turbines," in *German Wind Energy Conference DEWEK*, Bremen, 2017.

INCE*EUROPE*

8th International Conference on Wind Turbine Noise Lisbon – 12th to 14th June 2019

Session L

Source Prediction - 2

Organised by INCE-Europe
mail@inceeurope.org
www.inceeurope.org

Supported by Portuguese Acoustical Society

Sociedade Portuguesa de Acústica

INCE*EUROPE*

8th International Conference
on
Wind Turbine Noise
Lisbon – 12th to 14th June 2019

Simulation of sound radiation of wind turbines using large-scale finite element models

Thomas Grätsch, Marc Zarnekow, Frank Ihlenburg
Hamburg University of Applied Sciences
thomas.graetsch@haw-hamburg.de

Summary

In this paper the development of a vibro-acoustic finite element model at large scale is presented that can be used for structural-borne and air-borne path analysis in order to identify the critical path of sound propagation within a wind turbine. The investigation is focused on the transfer of mechanical vibrations from the gear box through the drivetrain. It is shown that a simplified numerical-analytical model of gear vibrations leads to mechanical resonance peaks that correspond to tonalities in measured far-field sound spectra.

1. Introduction

The control of noise emissions from wind turbines has become an issue of growing attention and strict regulation over the last decades where tonalities in the far-field spectrum are of particular importance. These far-field tonalities can be traced back to mechanical vibration sources within the drivetrain like the gear box or the generator. Measurements show that tonalities in the far field sound (pressure) spectra correspond with meshing frequencies of the gear box high speed stage.

The focus in this research is the simulation of transfer paths from vibrational sources (e.g. the gear box or generator) to radiating components of the turbines (e.g. blades or nacelle cover) with the goal to identify critical transfer paths within the turbine and further to find efficient design measures to reduce the radiation of tonal noise. In our study a 2.5 MW wind turbine of our project partner at operational state is considered.

The paper is organized as follows: In Section 2 a hybrid numerical-analytical approach is presented for modeling the excitation sources at the gears which combines a series of detailed finite element analyses for different meshing configurations of helical gears with analytical models for the transient gear dynamics. In Section 3 a large-scale finite element model of the whole wind turbine is developed which includes all major parts like the drivetrain, the mainframe, the nacelle, the blades and the tower in sufficient fine mesh resolution. In this model the calculated excitations as described in Section 2 are used as input data to represent the operating load spectra. The focus in this section is on the analysis of structure-borne transfer paths. Further, some brief considerations regarding accuracy, convergence and solution efficiency of the finite element model are made. Based on the dynamic response of structural parts, in Section 4 the model is also used to assess the radiation of air-borne sound where analytical methods are employed for calculating the frequency spectrum in the far-field in order to identify the critical tonalities. Section 5 shows a first example of application of the new developed model.

2. Numerical-Analytical Analysis of Gear Excitation

Gear vibrations due to varying mesh stiffness are a major source of tonal noise radiated from wind turbines. In order to obtain realistic excitations at operational state, in this section a hybrid numerical-analytical model of the gear dynamics is presented that combines a series of detailed finite element analysis of meshing configurations with analytical models for the transient dynamics of helical gears.

2.1 Finite Element Modelling of Meshing Stiffness

In a first step, the angle-dependent mesh stiffness of the gears is obtained from a sequence of static FE analyses that cover a complete meshing cycle. In this approach, the geometry of meshing gears can be modelled in a desired level of detailing including arbitrary geometric properties of interest such as helical angles, profile shift, or shape errors. The finite element model uses contact analysis in the area of potential contact between the gear flanks. A prerequisite for obtaining a stable and accurate solution is that the finite element resolution is sufficiently fine. Fig. 1 shows a section of a typical finite element model of a helical gearing.

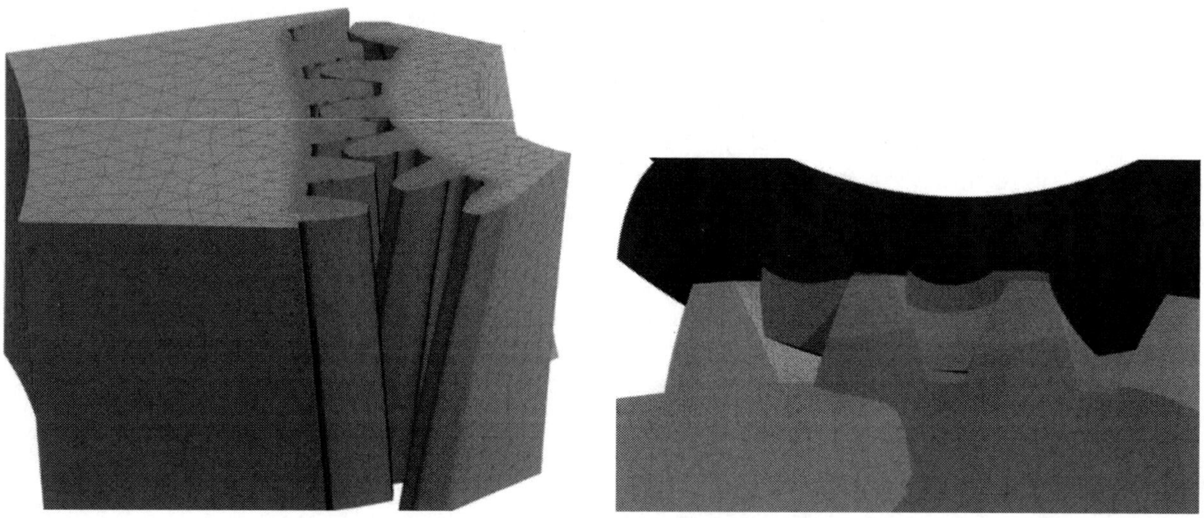

Fig. 1: Finite element model of a helical gearing

Our approach is validated with results from the literature, which shows excellent agreement for spur gears [1], see Fig. 2, while helical gears have not yet been extensively studied in the literature. For the considered spur gear connection, our results seem even more realistic than reported since the meshing stiffness changes rapidly as it should be in a realistic meshing configuration when a double tooth contact suddenly changes into a single tooth contact.

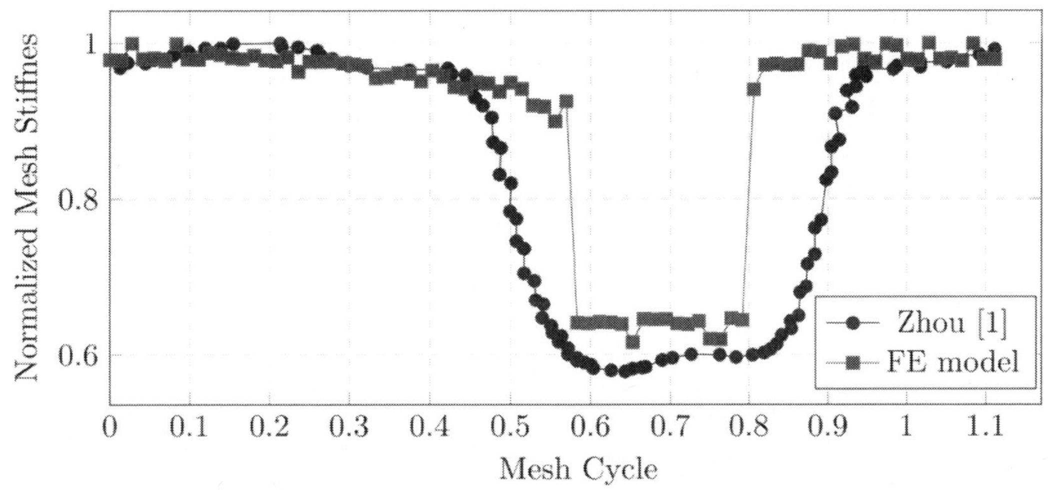

Fig. 2: Validation of mesh stiffness for spur connection, [1]

After having validated our approach, the procedure is applied to our actually considered wind turbine. Fig. 3 shows the results of a meshing stiffness analysis for a typical helical gearing.

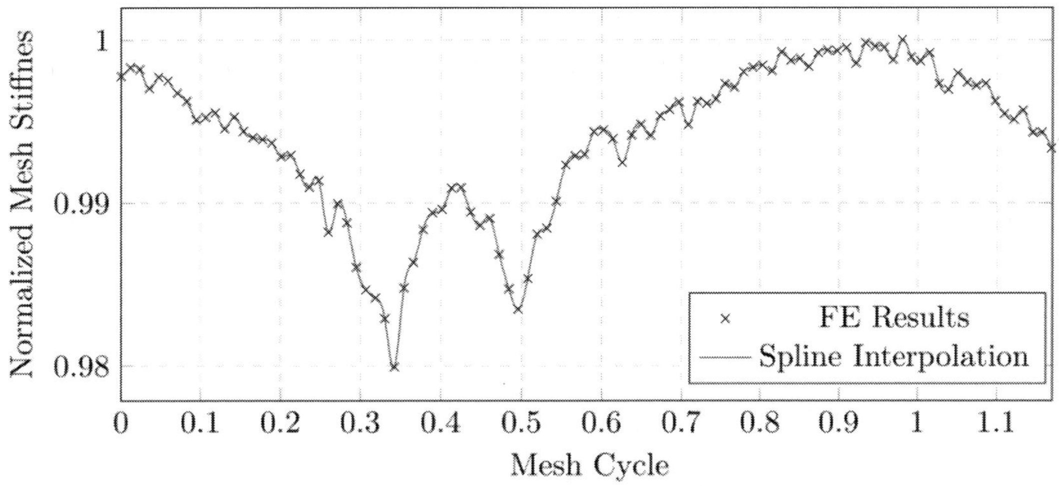

Fig. 3: Varying mesh stiffness as for a helical gear connection

2.2 Analytical Model of Gear Dynamics

The dynamic interaction within a gear pairing is simulated using a two-body model, see [2], with four degrees of freedom as shown in Fig. 4. The bodies are coupled via a linear spring-damper with a time-dependent spring constant $k_m(t)$ that is obtained from Fig. 3. The model also includes the stiffness and the damping of the gear support structure (e.g. bearing, shaft etc.).

Replacing the rotational degree of freedom by the transmission error (TE), the analytical model consists of three degrees of freedom which are the transmission error and the radial gear displacements along the line-of-action. The system is subject to parametric excitation by the time-varying stiffness $k_m(t)$ and to forced excitation by an error function $e(t)$ that can be used to include the effect of fabrication errors or wear.

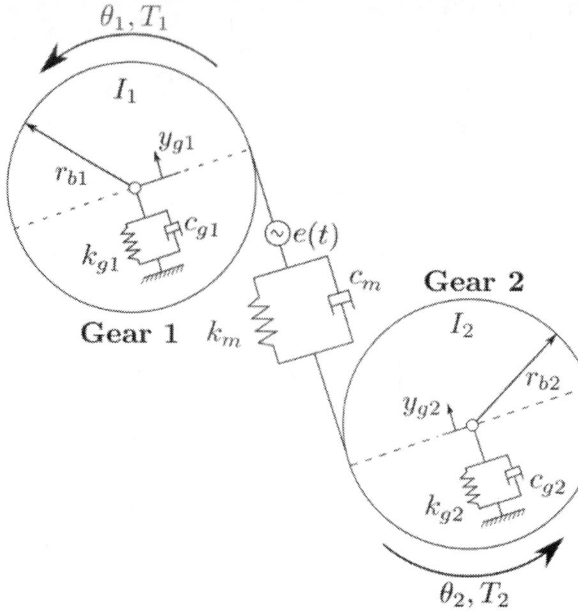

Fig. 4: Analytical model of gear dynamics

The resulting analytical model is a system of ordinary differential equations of second order which can be numerically solved in time with standard methods such as the Runge-Kutta scheme:

$$m_1\ddot{y}_{g1} + c_{g1}\dot{y}_{g1} - c_m\left(\dot{x} - \dot{y}_{g1} + \dot{y}_{g2}\right) + k_{g1}y_{g1} - k_m\left(x - y_{g1} +_{g2}\right) = -c_m\dot{e} - k_m e$$

$$m_2\ddot{y}_{g2} + c_{g2}\dot{y}_{g2} + c_m\left(\dot{x} - \dot{y}_{g1} + \dot{y}_{g2}\right) + k_{g2}y_{g2} + k_m\left(x - y_{g1} +_{g2}\right) = c_m\dot{e} + k_m e$$

$$m_e\ddot{x} + c_m\left(\dot{x} - \dot{y}_{g1} + \dot{y}_{g2}\right) + k_m\left(x - y_{g1} +_{g2}\right) = c_m\dot{e} + k_m e - m_e W_0\left(\frac{r_{b1}^2}{I_{g1}} - \frac{r_{b2}^2}{I_g 2}\right)$$

The stiffness parameter $k_m(t)$ is obtained from the above mentioned finite element analysis of the gear meshing. The solution of the system of differential equations is obtained in the time domain and can be transformed into the frequency domain by a Fast Fourier Transformation, see Fig. 5.

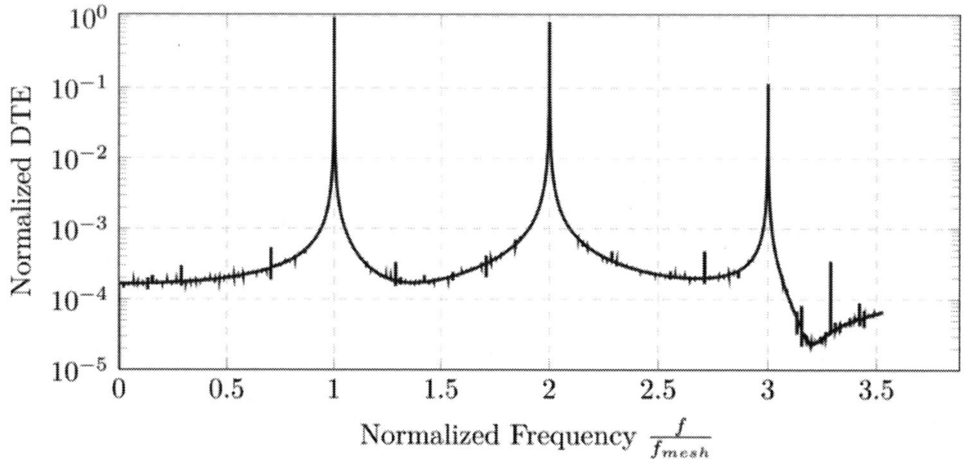

Fig. 5: Frequency spectrum of the transmission error

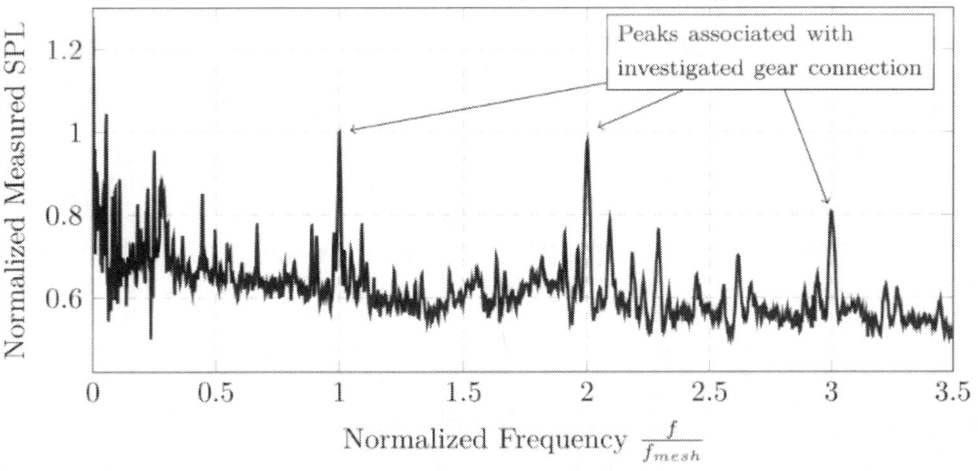

Fig. 6: Measured sound frequency spectrum of wind turbine gear box on a test-rig

To evaluate the results of the gear excitation spare modelling, the calculated excitation spectrum is compared with a sound pressure level spectrum measured on a gearbox test-rig. As can be seen from Fig. 6, the calculated transmission error frequency spectrum is in sound qualitative agreement with frequency response graphs of sound radiation, obtained from acoustic measurements on the gearbox test rig. Over an analysis spectrum of several KHz, the gearbox associated peak frequencies in both spectra agree very good and the relation of the peaks in the calculated excitation spectrum correspond well with the relation of the measured ones.

Critical tonalities in the sound spectrum of wind turbines are mainly caused by mechanical components in the drive train, for example the meshing gears. In Fig. 7 it can be seen that in the far-field sound spectrum of the wind turbine characteristic peaks occur at the same frequencies as calculated with the gear excitation model or as seen in the gear test rig measurement. Therefore, there is a need to investigate the transfer path of sound propagation within the turbine.

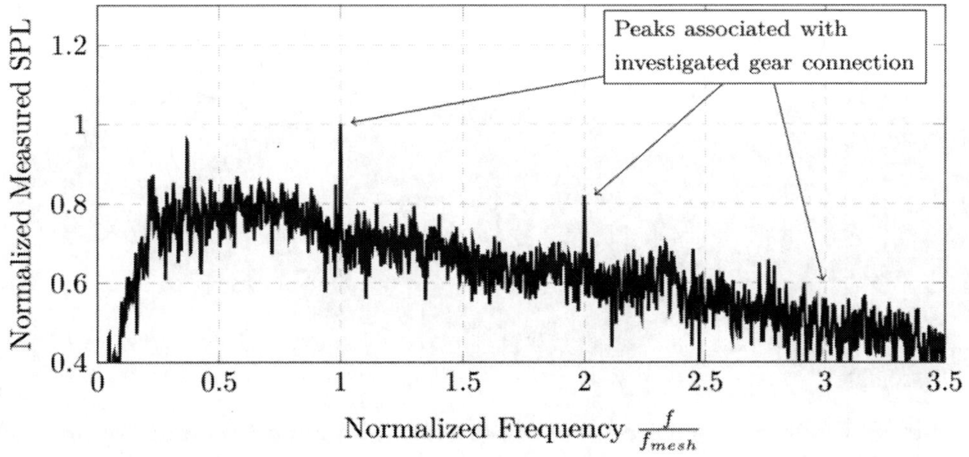

Fig. 7: Measured far-field sound spectrum of wind turbine gear box on a test-rig

The model for large-scale transfer path analysis is discussed in the next section, where either the transmission error as a displacement excitation or a force excitation at the bearings of the gear box can be applied as input, where both input data are directly obtained as a solution of the above given system of differential equations.

3. Development of Large-Scale Finite Element Model

For identifying the structure-borne transfer path of sound propagation a large scale finite element model of the whole wind turbine is developed. The finite element model consists of more than 10 million degrees of freedom and is solved in a broad-frequency range.

3.1 Finite Element Modelling

The finite element model includes all major parts like the drivetrain, the mainframe, the nacelle cover, the blades and the tower. The overall wind turbine model is modeled bottom-up, that means, at first, each component is modeled separately. The main steps for this component-wise modeling include the CAD generation and/ or import, the identification of relevant parts for the transfer analysis, defeaturing of local disturbances that are not necessary to resolve in the finite element model, and meshing of each component.

The large-scale finite element model is eventually built using the single components. As a part of the full model Fig. 8 shows the finite element model of the drivetrain when mounted on the mainframe.

The main modeling steps of this assembling step are the adequate definition of the interfaces between the components. Interfaces, where negligibly small deformations are assumed, are modeled as rigid connections, as for example between the main shaft and the shrink disc of the planet carrier or at the massive connection between main bearing and main frame. For other interfaces between components, such as bearings, mount or similar connections, stiffness and damping matrices are employed at the interfaces, which account for the related stiffness and damping of the connection. To obtain the 6x6 stiffness matrices in these cases, the nominal load at the operating point is applied in all 6 degrees of freedom in space to obtain the corresponding stiffness as the derivation of the force-displacement curve of the respective bearing or other connecting elements.

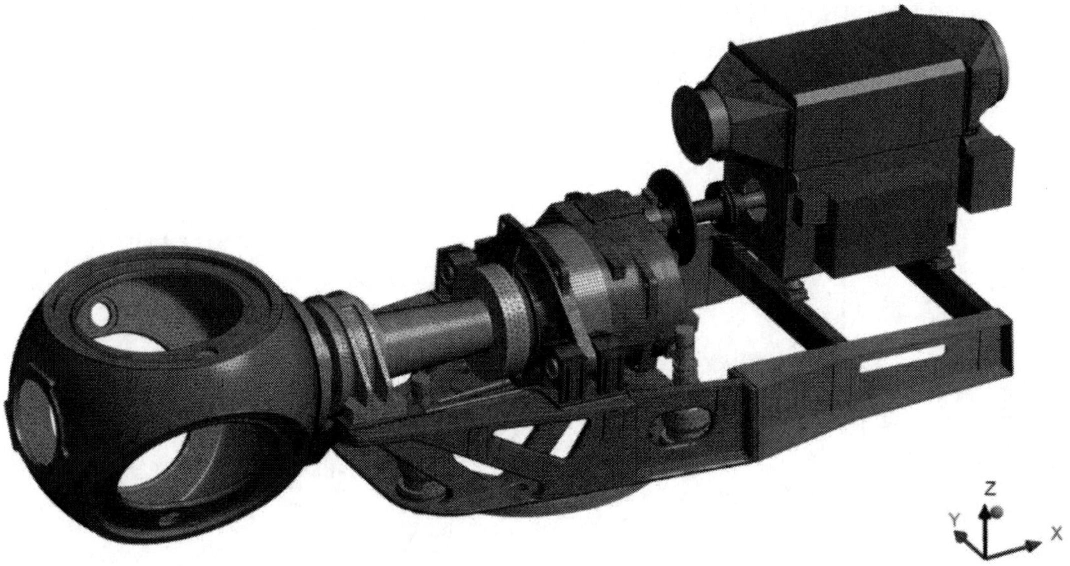

Fig. 8: Finite element model of the drivetrain mounted on mainframe

3.2 Mesh Density Evaluation

A crucial question in finite element analysis is whether the results are accurate and reliable. In order to assure that the discretization error at the component level is sufficiently small - hence that the solution quality of the computation is high - some considerations have been made:

For each single component several mesh density analyses are performed to evaluate the mesh quality. First of all, convergence curves of major local output quantities are calculated. Next, the model assurance criterion (MAC-value) is used to compare different finite element meshes with each other, where the reference solution is a so-called overkill solution with an extreme fine mesh resolution. All modes up to 1.5 times the highest frequency of interest for the vibro-acoustic simulation are compared. Further, in the same manner, the frequency response assurance criterion (FRAC-value) is employed to ensure the convergence in the frequency response up to the highest frequency of interest. Fig. 9 and 10 show the MAC- respective FRAC-values of the main frame before and after having found the sufficient fine mesh resolution.

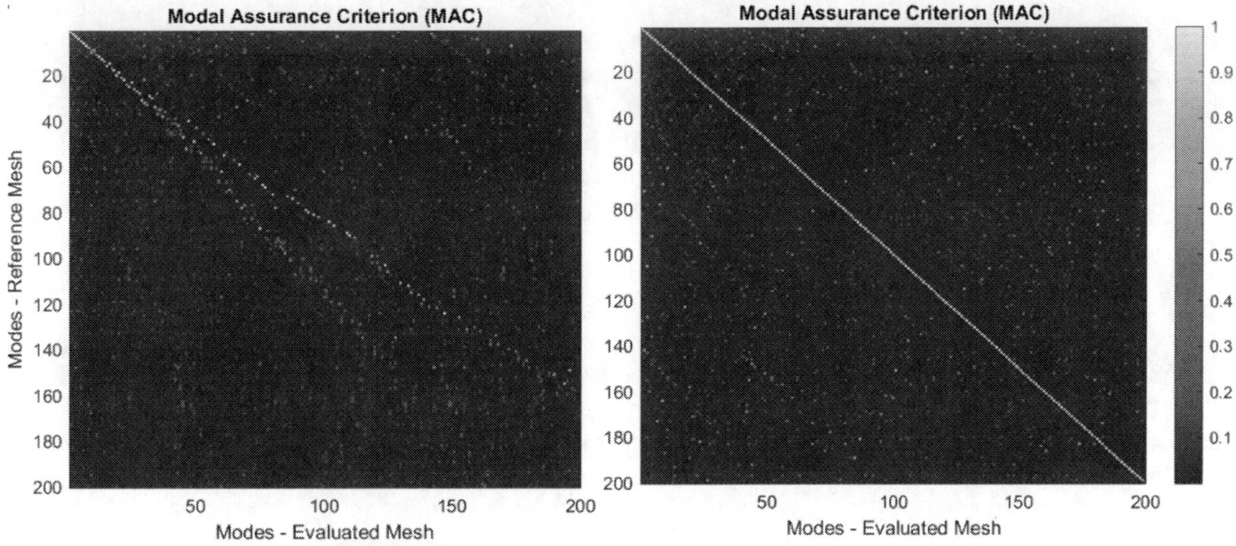

Fig. 9: MAC-values of a coarse mesh (left) and a sufficiently fine mesh (right)

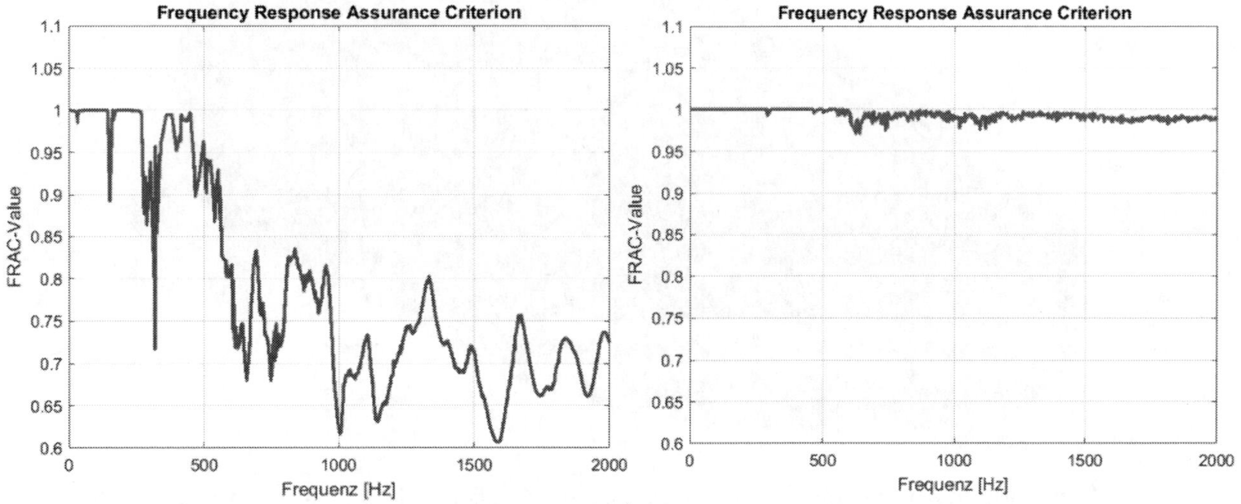

Fig. 10: FRAQ-values of a coarse mesh (left) and a sufficiently fine mesh (right)

3.3 Calculating Dynamic Transfer Functions

In order to find the critical transfer path, a frequency response analysis is conducted where a unit excitation is applied at the potential excitation source that accounts for the excitations during operation. In doing so, the transfer function is obtained which is equal to the frequency response at every desired location of the full model for every result quantity of interest. In order to obtain the response under operating conditions the linear behaviour of a frequency response analysis is exploited according to the linear equation

$$x(f) = T(f) \cdot TE(f)$$

where $x(f)$ denotes the frequency response at operation conditions, $T(f)$ denotes the transfer function obtained as described above by applying a unit excitation and $TE(f)$ denotes the operational excitation such as the transmission error at the gear box bearing according to Fig. 5. This procedure can be applied at the same time for all relevant excitation sources, and due to the linearity of the analysis, a superposition can be carried out to obtain the overall response or to evaluate the different contributions of the single excitation sources.

In addition to the displacement frequency response some advanced structural output quantities can be calculated in this approach such as velocities, energies or the ERP level of certain parts of the structure. As an example, Fig. 11 shows the displacement response of the nacelle cover at a critical frequency of 566.5 Hz when excited by the transmission error at the gear box bearing. The corresponding ERP level is shown in Fig. 12 for certain panels of interest. This results can now be used to identify critical panels and to suggest design changes to reduce the peaks.

To deal with the computational effort of the full model, standard model reduction methods of structural dynamics like the component mode synthesis method are employed. When applying these methods it can be shown for the considered wind turbine that the computational effort for solving the full system can be reduced by 90% in both computing time and memory allocation while the resulting error is negligibly small that is less than 1% in the considered frequencies for all modes up to 2000 Hz.

Fig. 11: Frequency response of the nacelle cover at 566.5 Hz due to gear excitation

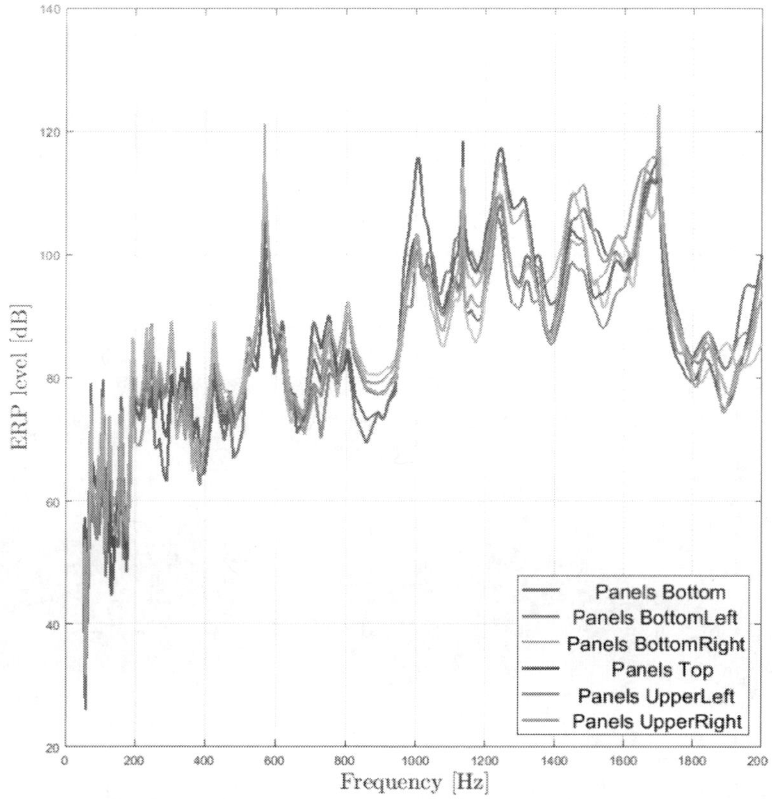

Fig. 12: ERP level of different panels of the nacelle cover from 0 to 2000 Hz due to gear excitation

4. Air-Borne Transfer Path Analysis

Dynamic structural results like the ERP level can provide useful results in the evaluation of the sound radiation behaviour of structures. But since simplifications and assumptions are made here, e.g. an ideal radiation efficiency is assumed and dynamic properties of cavities are not considered, their application can be limited under certain circumstances. Therefore, some additional air-borne transfer path analysis is necessary.

To investigate the air-borne transfer path, the full-scale finite element model can also be used for acoustic simulations. In this regard, the acoustic domain within the nacelle is modelled with acoustic finite elements. In this approach, the acoustic fluid field is coupled with the structural field, which is commonly referred to as fluid-structure interaction (FSI). FSI can generally be modelled via two different approaches, either as fully coupled (two-way coupling) or weakly coupled (one-way coupling).

The fully coupled formulation defines a mutual interaction of structure and fluid fields over both kinematic and kinetic coupling conditions at the interface and is commonly described by the following formulation, see [3]:

$$\left(-\omega^2 \begin{bmatrix} M_s & 0 \\ \rho_0 R^T & M_f \end{bmatrix} + j\omega \begin{bmatrix} C_s & 0 \\ 0 & C_f \end{bmatrix} + \begin{bmatrix} K_s & -R \\ 0 & K_f \end{bmatrix}\right) \begin{Bmatrix} u_s \\ p_f \end{Bmatrix} = \begin{Bmatrix} f_s \\ f_f \end{Bmatrix}$$

In this coupled problem the acoustic domain is interacting with the structural part, which is the nacelle cover. Spacial displacement and pressure degrees of freedom are employed in the coupled analysis, where the number per element depends on the actual type of the finite element used.

The weak FSI-coupling defines only a kinematic coupling condition at the interface by assuming that the acoustic fluid field does not influence the structure. Therefore, the structure part is solved independently, and subsequently the so-obtained structural vibrations are applied to the acoustic fluid as boundary conditions. The weak coupling is computationally more efficient, but it is, of course, a more simplified approach than the full coupling.

The described acoustic simulations can be used to investigate the acoustic transfer inside the nacelle, see Fig. 13 (left), where the pressure distribution due to vibrating nacelle cover panels is shown. Based on the results of the acoustic simulation, further acoustic finite element models are used to evaluate the sound radiation into the far field due to the vibrating panels, see Fig. 13 (right).

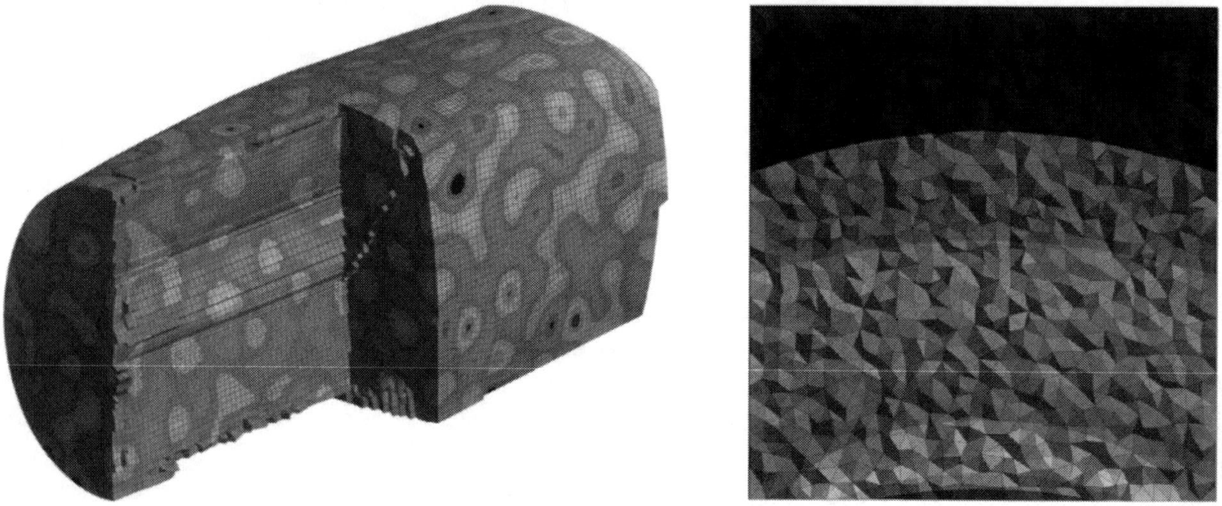

Fig. 13: Pressure distribution within nacelle due to weakly coupling to vibrating panels (left) and a section of an acoustic simulation of radiation into the far-field (right)

5. Computational Example

As an application of the presented approach a computational example is considered. The model consists of the whole drivetrain as shown in Fig. 8 including the nacelle cover. The finite element model consists of 2.5 Million nodes yielding 7.5 Million degrees of freedom. The CMS method is used for model reduction.

The goal in this principal scenario is to identify the critical transfer path between a known (vibrational) source and a receiver. As investigated excitation source in this case, the transmission error of the high speed gear stage is applied at the meshing gears. The aim is to evaluate how much and via which paths this gear vibration are radiated by the nacelle cover. Therefore the received quantity is described by the equivalent radiated power of the nacelle cover panels. That means for simplification reasons the acoustic radiation is approximated by the ERP.

There are basically three potential transfer paths to transport the sound from the vibration source (high speed gear stage) to the receiver (nacelle cover). The three possible transfer paths are through the torque arm (TA), through the main bearing (MB), or via the generator (GEN), see the three coloured lines in Fig. 14. As an additional aspect of this theoretical computational example, the transfer path via the generator is bridged by a rigid connection directly to the main frame. It is expected, that by applying this modification and its direct and stiff connection to the receiver, an artificially dominant and critical transfer path is established.

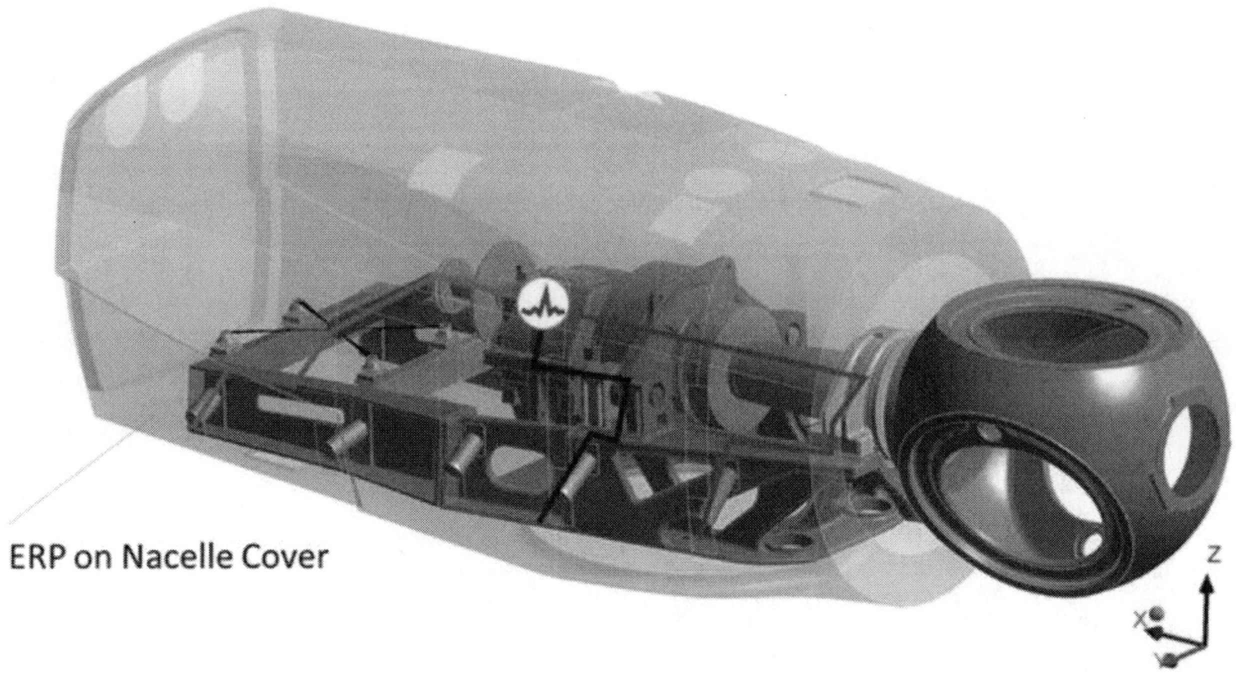

Fig. 14: Three possible transfer paths from gear excitation (source) to nacelle cover (receiver)

In order to find the critical path, the idea is to isolate one of the paths from the other by cutting the other two paths in the finite element model, respectively. For this, the assumption is made that the influence and the interaction between the three paths on the global dynamic behavior is negligibly small. This assumption can be confirmed by superposing the ERP response of the three individual paths and comparing this curve with the ERP response obtained by a full model analysis including all the three paths. Fig. 15 shows the calculated ERP level of the nacelle cover and compares the results of the reference model with all paths combined and the superposition of the ERP levels of the isolated contribution of the single transfer paths. Therefore it can be confirmed, that the modeling for this approach has an negligible small influence on the ERP levels.

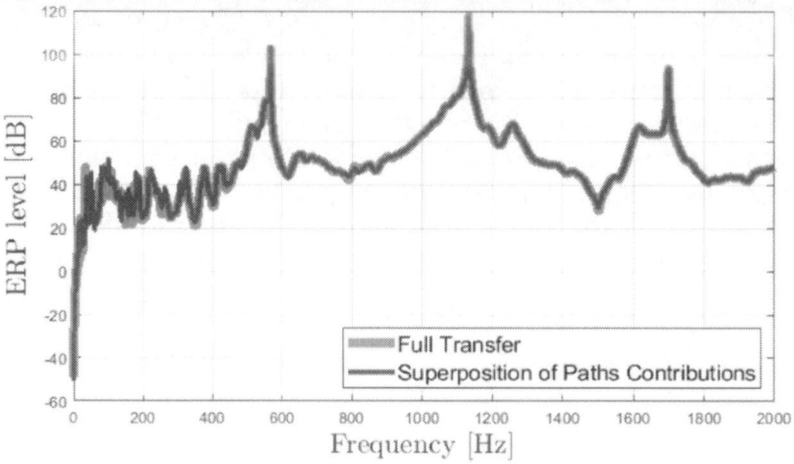

Fig. 15: Validation of modeling approach of isolating the path contributions

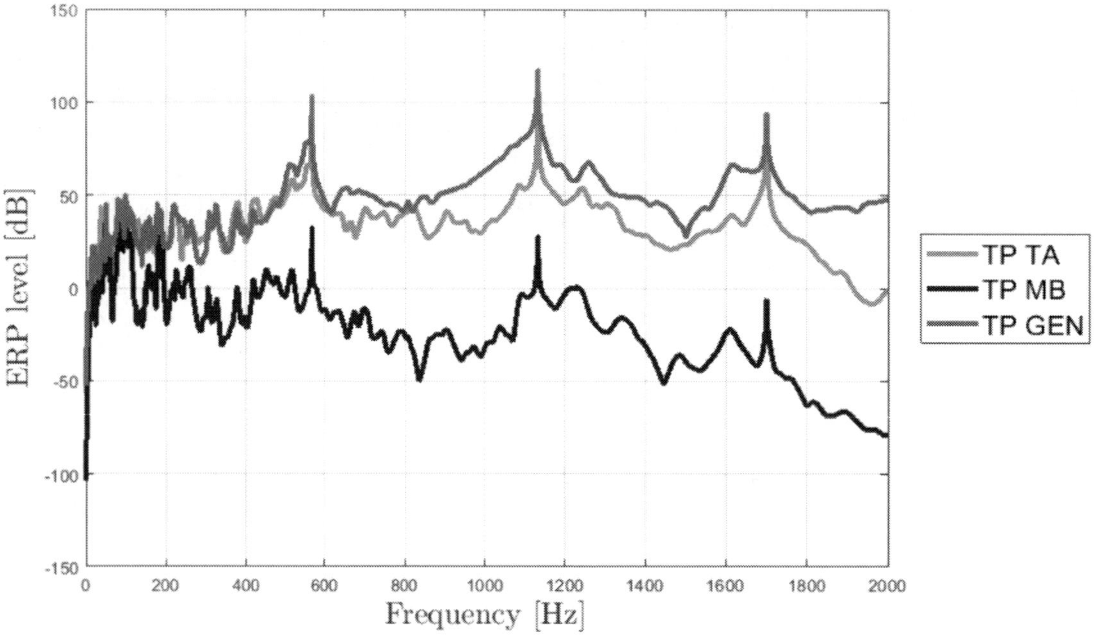

Fig. 16: ERP level of nacelle cover depending on transfer paths via torque arm (TA), main bearing (MB) or generator (GEN)

Fig. 16 shows the contributions of the three different transfer paths. As it can be seen the transfer path along the generator dominates nearly the complete frequency range while the transfer path via the torque arm is less dominant and the contributions by transfer path via the main bearing are negligible small at frequencies above 200 Hz.

Therefore the obtained results confirm the initial assumption that the modified bridged generator path represents a dominating and critical transfer path in this example scenario. The results show that the modelling approach provides the ability to evaluate the contributions of different transfer paths and therefore to identify critical vibration and sound transmission.

The identification of such critical transfer paths through the described modelling approach can be used to identify constructive problems and improvement measures. However, it must be added that absolute statements for a real turbine are not possible at this point, because the generator is not fully resolved. Hence, when using a more detailed generator model, this transfer path might be less dominant. However, this computational example shows how the presented approach can be employed in general in order to identify the critical transfer path of sound propagation.

6. Conclusions

In this paper, a finite element modelling approach to simulate the vibro-acoustic transfer path behaviour of a wind turbine under operating conditions has been presented.

The excitation spectrum of a gear stage is calculated by an hybrid analytical-numerical approach combining detailed static finite element with analytical gear dynamics. This approach is verified on a spur gear connection and then extended to a helical gear connection of a wind turbine gearbox. The gear dynamic modelling provides the frequency spectrum of the gear excitation under operating conditions, which has been compared to an measured sound pressure spectrum.

The structure-borne transfer paths of the considered wind turbine are investigated using a dynamic large-scale finite element model. Further, the aspects of mesh density evaluation and model reduction techniques are discussed. The dynamic large-scale finite element model is used to calculate transfer functions for all relevant quantities of interest. Subsequently, the transfer functions are combined with the previously determined load spectra to obtain the behaviour under operating conditions.

In addition to the structure-borne transfer path, the air-borne transfer path is addressed with different acoustic finite element modelling approaches that are used to investigate the acoustic behaviour of cavities (e.g. inside the nacelle) or to investigate the radiation of vibrating panels (e.g. nacelle cover, blades, or tower) .

A computational example of large scale finite element analysis shows how the contributions from several transfer paths can be evaluated separately in order to show how the presented approach can be employed in general in order to identify the critical transfer path of propagation.

References

[1] Zhou, J and Wenlei, S (2014) *Vibration and Noise Radiation Characteristics of Gear Transmission System* Jnl Low Frequency Noise, Vibration and Active Control 30(4), 485-502

[2] Özgüven, H and Houser, D (1988) *Dynamic analysis of high speed gears by using loaded static transmission error* Jnl of Sound and Vibration 125 (1), 71-83

[3] Sandberg, G, Davidsson, P and Wernberg, P (2009) *Fundamentals of Fluid-Structure Interaction* In: Computational Aspects of Structural Acoustics and Vibration, 23-101

8th International Conference
on
Wind Turbine Noise
Lisbon – June, 12-14th, 2019

Airfoil LE Noise prediction supplement for *PNoise* Code

Alexandre Martuscelli Faria, Polytechnic School, Sao Paulo State University (Poli-USP): martuscellifaria@usp.br

Joseph Youssif Saab Junior, Maua Institute of Technology (IMT): saab@maua.br

Sara Rodriguez, Polytechnic School, Sao Paulo State University (Poli-USP): sara.rodriguez@alumni.usp.br

Marcos de Mattos Pimenta, Polytechnic School, Sao Paulo State University (Poli-USP): marcos.pimenta@poli.usp.br

Summary

This work is part of an ongoing collaboration project between Poli-USP and TU-Berlin and aims to further improve the airfoil noise prediction module, *PNoise*, which is integrated into the wind turbine design environment, *QBlade*, available under General Public License. This paper discusses some of the aspects around selection, implementation and testing of an airfoil leading-edge (LE) noise prediction model to supplement the TE noise model of the *PNoise* code. The interaction of the LE with the turbulent inflow is an important source of wind turbine (WT) noise, especially in the case of small scale WT which behaves as non-compact noise sources. LE noise is usually confined to low frequencies (<1 kHz) and is strongly dependant on turbulence parameters, such as turbulence intensity and the turbulence integral length scale. Amiet's broadband noise theory and Lowson's semi-empirical method are discussed in the context of the model setup, along with the appropriate turbulence spectrum modelling. Results of the implemented model are then compared to experimental data from recent literature, in order to provide initial verification of the model. A discussion on compatibility with the current *PNoise* code is also presented.

1 Introduction

An airfoil in a turbulent flow experiences a fluctuating lift which radiates noise to the far-field. This fluctuating lift is a result of the unsteady pressure field produced by the airfoil in response to turbulence (Staubs, 2008). The turbulent flow field can be either produced upstream the airfoil, by the presence of inflow distortions and other aerodynamic elements, or it can be also consequence of the development of a turbulent boundary layer over the airfoil surface, in case of a steady inflow. The upstream mechanism is linked with the noise produced close to the airfoil leading edge, while the mechanism related to the turbulent boundary layer is a self-noise mechanism, discussed in details by Saab (Saab, 2016) and Saab and Pimenta (Saab and Pimenta, 2016).

The two noise generation mechanisms coexist and are responsible for the overall noise spectra. Normally, for WT applications, airfoil self-noise constitutes the dominant noise source. For certain flow conditions, however, i.e. when the incoming turbulence intensity and the integral length scale of the inflow eddies are large enough, the pressure fluctuations caused by the boundary layer eddies is smaller compared to the pressure fluctuations due the turbulent inflow, and the turbulent inflow noise mechanism is predominant over the self-noise.

Characterized as an interaction noise source, and perceived as a swishing noise, turbulent inflow noise is caused by the flow-surface interaction, when the atmospheric turbulence encounters the rotor blades. Since turbulence is not a uniform phenomenon, its characteristics depend on local atmospheric parameters, such as eddy size and turbulence intensity.

The eddy size is the most important parameter for determining the LE noise (Zhu, 2004). Due to the turbulence structure and the atmospheric stability, eddies of a wide range of sizes interact with the blade.

Studies conducted by Paterson and Amiet, Oerlemans and Migliore and Moreau, Roger and Jurdic have shown that the turbulent inflow noise is usually confined to the lower frequencies (<1 kHz), where the turbulent structures responsible for the LE noise generation are the larger structures (Paterson and Amiet, 1976) (Oerlemans and Migliore, 2004) (Moreau, Roger and Jurdic 2005). Therefore, the blade can be simplified as an acoustic dipole, which source strength is equal to the total fluctuating lift on the blade surface.

Estimation methods for quantifying LE noise should take into account parameters such as turbulence intensity, the longitudinal integral length scale, the largest turbulent structure, and the horizontal axis wind turbine (HAWT) geometric data, since turbulence depends on atmospheric conditions for specific height values (Staubs, 2008).

The main objective of this paper is to discuss the application of different methodologies for LE noise prediction and the respective turbulence spectrum modeling. First, a qualitative study of a full scale HAWT noise prediction is presented. Directly after that, a single airfoil LE noise prediction is compared to experimental measurements conducted on acoustic tunnel by Juknevicius and Chong (Juknevicius and Chong, 2018).

This work gives an overview of airfoil LE noise prediction, focusing on Amiet's theory (Amiet, 1976) and its extensions, such as inclusion of directivity proposed by Lowson (Lowson, 1992) and the more adequate turbulence spectrum model, proposed firstly by Simonich *et al.* (Simonich *et al.*, 1986) and further discussed by Santana (Santana, 2015).

Being this work part of *PNoise* code improvement, by Poli-Wind Research Group, compatibility with the current version is also discussed in this paper.

2 Methodologies for airfoil LE noise prediction

2.1 Amiet's broadband noise theory

A theoretical formulation and a semi-empirical methodology, for predicting the turbulent inflow noise, were introduced by Amiet with agreement to Kirchhoff and Curle theories, in order to ensure more reliability to the prediction against measurements. This methodology evaluates the far-field acoustic power spectral density produced by an airfoil in a subsonic turbulent stream, given in terms of characteristic quantities of the turbulence (Amiet, 1975). The theoretical approach, illustrated by Figure 1, corresponds to the compute the acoustic response of an airfoil of $2b$ chord and $2d$ span subjected to a turbulent flow with mean velocity U in the x direction. The noise source S is placed at the center of the airfoil, at the (x_0, y_0, z_0) coordinate system and the observer O is placed at the far-field, represented by the (x, y, z) coordinate system. This is a more general case, since it considers the observer placed at an arbitrary position of the far-field, with the free stream extending to infinity, what suggests suitability for WT noise prediction.

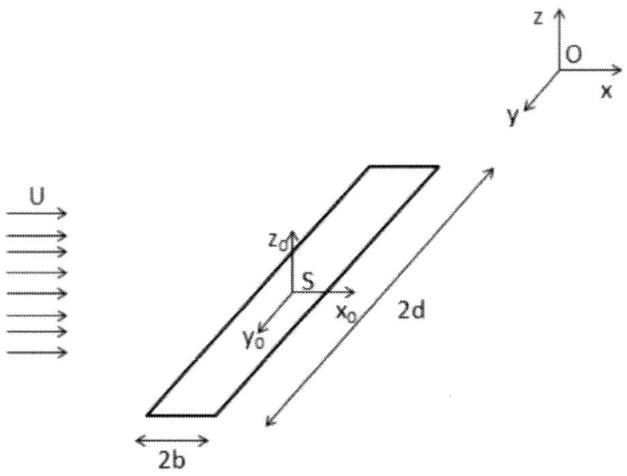

Figure 1: Amiet problem representation

The far-field is described by Amiet's formulation as:

$$S_{pp}(x, y, z, \omega) = \left(\frac{\omega z}{4\pi c_0 \sigma^2}\right)^2 U d\pi \Phi_{ww}(K_x, K_y) \left|\mathcal{L}(x, K_x, K_y)\right|^2 \tag{1}$$

where $\Phi_{ww}(K_x, K_y)$ is the turbulent velocity energy spectrum and $\mathcal{L}(x, K_x, K_y)$ corresponds to the acoustic lift response function.

Derivation of $\mathcal{L}(x, K_x, K_y)$ is discussed in details by Santana (Santana, 2015), where the flow field is described as a partial differential equation (PDE) problem, which consists of a canonical Helmholtz equation subjected to the boundary conditions of zero velocity potential upstream the airfoil leading edge; zero airfoil surface normal velocity (non-penetration condition) and zero pressure jump at the airfoil trailing edge (Kutta condition) and downstream.

$$\frac{\partial^2\varphi}{\partial\bar{x}^2}+\frac{\partial^2\varphi}{\partial\bar{z}^2}+\kappa^2\varphi=0 \tag{2}$$

$$\varphi(\bar{x},0)=0, \qquad \bar{x}\leq 0 \tag{3}$$

$$\frac{\partial\varphi}{\partial\bar{z}}(\bar{x},0)=\frac{-w_0 b}{\beta}e^{\overline{k_x}^*\bar{x}}, \qquad 0\leq\bar{x}\leq 2 \tag{4}$$

$$\left(i\overline{k_x}^*+\frac{\partial}{\partial\bar{x}}\right)\varphi(\bar{x},0)=0, \qquad \bar{x}>2 \tag{5}$$

The turbulent velocity energy spectrum, $\Phi_{ww}(K_x,K_y)$, on the other hand, can be modeled after von Kármán isotropic turbulence spectrum (Amiet, 1975) (Sinayoko and Hurault, 2015):

$$\Phi_{ww}(k_x,k_y)=\frac{4}{9\pi}\frac{\overline{u^2}}{k_e^2}\frac{(k_x/k_e)^2+(k_y/k_e)^2}{\left(1+(k_x/k_e)^2+(k_y/k_e)^2\right)^{7/3}} \tag{6}$$

where $\overline{u^2}=(IMc_0)^2$ and $k_e=\frac{\sqrt{\pi}}{L}\left(\frac{\Gamma(5/6)}{\Gamma(1/3)}\right)$.

Simonitch *et al.*, however, suggests that the turbulence spectrum should be modeled after Batchelor's Rapid Distortion Theory (RDT), because Turbulence rapid distortion takes place when a variation in the mean velocity field occurs due to change in the boundary conditions, e.g. turbulent flow approaching an airfoil (Simonitch *et al.*, 1986). Experimental evidences of rapid distortion were also reported by Santana (Santana, 2015). It is also necessary that the turbulence distortion occurs so rapidly that the contribution to the change in relative positions of the fluid particles from the turbulence is negligible (Batchelor and Proudman, 1954). In this case, the turbulence spectrum can be written as:

$$\Phi_{ww}(k_x,k_y)=\frac{91}{36\pi}\frac{\overline{u^2}}{k_e^2}\frac{(k_x/k_e)^2+(k_y/k_e)^2}{\left(1+(k_x/k_e)^2+(k_y/k_e)^2\right)^{19/6}} \tag{7}$$

2.2 Amiet's semi-empirical method

A second formulation presented by Amiet is a semi-empirical method, based on the acoustic tunnel experiment. An airfoil of $2b$ chord and $2d$ span is placed in a turbulent flow with mean velocity U in the x direction, as Figure 2 illustrates. The y coordinate extends in the spanwise direction and the origin of the coordinate system is placed at the center of the airfoil. The observer is located at the far-field, directly overhead the airfoil, represented as a microphone. This procedure is used to neglect the retarded time differences, what allows one to formulate the far-field sound in terms of the total fluctuating lift of the airfoil.

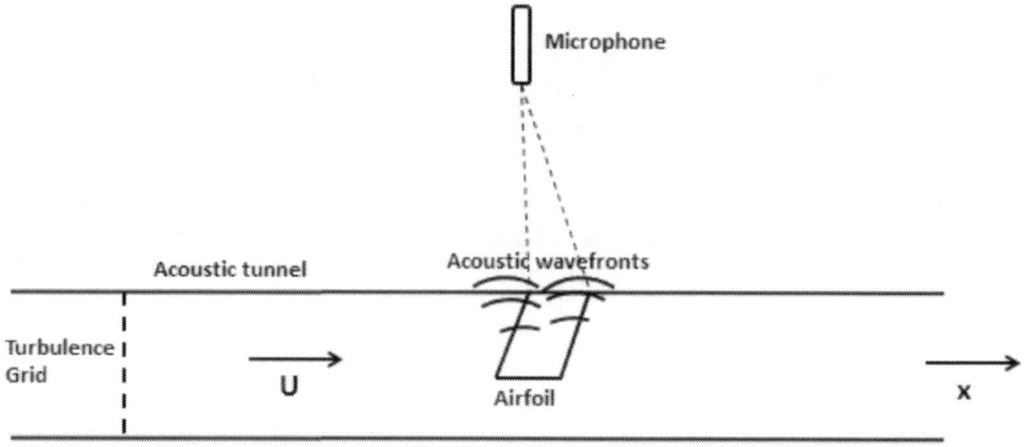

Figure 2: Airfoil in the free stream of an acoustic tunnel (adapted from R.K. Amiet, 1975).

In Amiet's experimental setup, an airfoil with a chord of 18 inches and a span of 21 inches was mounted between sideplates at zero angle of attack (AOA) in the UARL acoustic tunnel, and a turbulence generating grid was placed upstream the airfoil. The turbulence measurements indicated at the test section that turbulence properties were approximated by an isotropic homogeneous turbulence model. The integral length scale, L, of the turbulence was 1.25 inches. The streamwise turbulence intensity, I, was set to 4.4% for $U = 103$ ft/s.

Amiet have conducted one-third octave sound measurements with a microphone placed at 7 feet directly above the airfoil. Under the assumption that the mean flow is an one-dimensional flow, this simplified semi-empirical relation is obtained, for the one third octave sound pressure level, $SPL_{1/3}$, in dB relative to a pressure of $2 \cdot 10^{-4} \mu bar$:

$$SPL_{1/3} = 10 \log_{10}\left[\frac{Ld}{z^2} M^5 \frac{\overline{u^2}}{U^2} \frac{K^3}{(1+K^2)^{7/3}} \rho_0^2 c_0^4\right] + 181.3 \qquad (8)$$

It is important to draw attention here for the units of measure, that may influence the total SPL, because of the 181.3 constant, which is not calculated in S.I. Paterson and Amiet have presented the same formulation but for the CGS unit system (Paterson and Amiet, 1976):

$$SPL_{1/3} = 10 \log_{10}\left[\frac{Ld}{z^2} M^5 \frac{\overline{u^2}}{U^2} \frac{K^3}{(1+K^2)^{7/3}} \rho_0^2 c_0^4\right] + 58.4 \qquad (9)$$

Another question that one could point out is the influence of the airfoil AOA on LE noise. Measurements from Juknevicius and Chong evidence that, for symmetrical airfoils, only self-noise mechanism is influenced by AOA, while it does not play a significant role on turbulent inflow noise (Juknevicius and Chong, 2018).

2.3 Lowson's semi-empirical method

An alternative semi-empirical method for LE noise prediction was introduced by Lowson (Lowson, 1992), by modifying Amiet's semi-empirical method. Intended to be more suitable for WT applications, it presents modifications in order to provide a correction for the lower frequencies of the spectrum, and has introduced the concept of spherical directivity to turbulent inflow noise prediction, as already seen in the contemporary BPM TBL-TE (Brooks, Pope and Marcolini Turbulent Boundary Layer Trailing-Edge) noise prediction method (Brooks, Pope and Marcolini, 1989) (Moriarty and Migliore, 2003).

In Lowson's formulation, the total $SPL_{1/3}$ is firstly decomposed in terms of the high frequencies sound pressure level and the low frequency correction factor, LFC:

$$SPL_{1/3} = SPL_H + 10log_{10}\left[\frac{LFC}{1+LFC}\right] \tag{10}$$

For the high frequency domain, the evaluation of the sound pressure level follows:

$$SPL_H = 10log_{10}\left[\frac{\rho_0^2 c_0^2 sL}{2r_e^2} M^3 U^2 I^2 \frac{K^3}{(1+K^2)^{7/3}} \overline{D}_L\right] + 58.4 \tag{11}$$

where r_e is the total distance between source and observer, and $\overline{D}_L$ is the spherical directivity factor.

The directivity can be obtained by:

$$\overline{D}_L = \frac{\sin^2\theta \cos^2\varphi}{(1+M\cos\theta)^4} \tag{12}$$

where θ and φ are the directivity angles.

The low frequency correction factor is approximated by the following expression:

$$LFC = 10S^2MK^2\beta^{-2} \tag{13}$$

where S is the compressible Sears function that relates K and β^2 which can be written in the following form:

$$S^2 = \left(\frac{2\pi K}{\beta^2} + \left(1 + 2.4\frac{K}{\beta^2}\right)^{-1}\right)^{-1} \tag{14}$$

Units of measure are not provided by Lowson, what can lead to misunderstandings. However, as showed at the last sub-section, Paterson and Amiet method has already presented a constant value of 58.4 utilizing the CGS system of units (Paterson and Amiet, 1976), it can be assumed that Lowson also adopted that same system.

2.4 Modified RDT-Lowson's semi-empirical method

In face of Simonitch *et al.* and Santana's considerations on turbulence velocity energy spectrum modeling (Simonitch *et al.*, 1986) (Santana, 2015), one can follow a similar procedure to derive a modified Lowson's method, in order to model turbulence after the RDT spectrum.

This happens to change the constant value to 65.95 and the turbulence spectrum exponent to $19/6$. The result is an expression similar to Equation 11:

$$SPL_H = 10log_{10}\left(\frac{\rho_0^2 c_0^2 sL}{2r_e^2} M^3 U^2 I^2 \frac{K^3}{(1+K^2)^{19/6}} \overline{D}_L\right) + 65.95 \tag{15}$$

3 Methodology

Given that this work discusses LE noise prediction methods to be implemented in *PNoise*, and that the current version presents a modified BPM TBL-TE noise prediction method, let us consider Lowson's semi-empirical method the more suitable, because of taking into account spherical directivity. This reduces the comparison to terms of turbulence spectrum modeling, by considering turbulence being modeled after von Kármán isotropic turbulence or by Batchelor's rapid distortion theory.

First, as a qualitative example, the influence of turbulence modeling on Lowson's method is analyzed. The blade geometry chosen is the one presented in the DAN-AERO report (Madsen *et al.*, 2010) and flow conditions are the ones provided by Sinayoko and Hurault (2015). In this case, the blade has 0.9 m chord, total length of 38.8 m. The observer is placed at 104.5 m downstream of the turbine and 60 m below the hub. Turbulence integral length scale is set to 40 m and the incoming turbulence intensity is set to 12.5%. The flow Mach number is set to $M = 0.11$. The chord-based Reynolds number is $Re = 3.08 \cdot 10^6$. The respective TBL-TE noise prediction is also obtained from the *PNoise* noise prediction module, inside *QBlade* environment (Marten, 2016), at zero angle of attack.

The second part consists of validating the methods against experimental data. Juknevicius and Chong have conducted LE noise measurements for a NACA 0008 airfoil with 0.15 m chord and 0.498 m span subjected to a turbulent flow with mean velocity 50 m/s ($Re = 5 \cdot 10^5$), turbulence intensity 3.7% and turbulence integral length scale of 6.5 mm. Two other measurements were conducted with flow mean velocity of 30 m/s ($Re = 3 \cdot 10^5$) and 60 m/s ($Re = 6 \cdot 10^5$). The microphone was positioned 1 m directly overhead the airfoil LE (Juknevicius and Chong, 2018).

4 Results and discussion

4.1 Qualitative analysis – DAN-AERO WT

Figure 3 shows the predicted LE noise source by implementation of the modified Lowson-RDT-LE noise prediction method for the DAN-AERO WT, as well as the predicted TBL-TE noise source. The modified Lowson-RDT-LE appears to attend the low-frequency noise nature of the LE noise. For this prediction, TE noise becomes and is kept the dominant source for frequencies above 63 Hz, which is expected for a large HAWT.

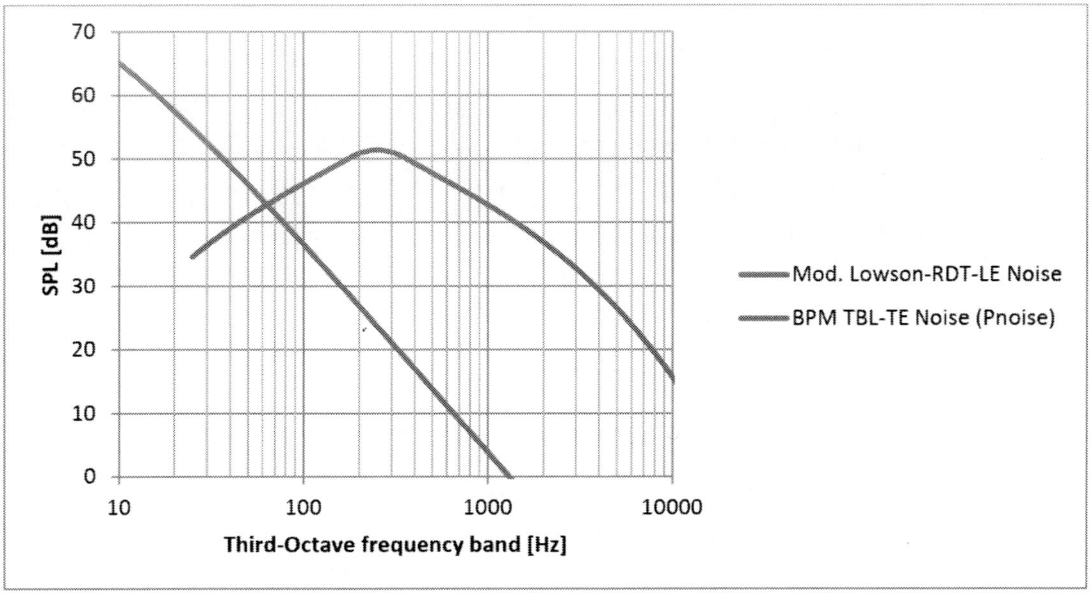

Figure 3: Predicted sound pressure levels for TE and LE noise by implementation of the PNoise BPM TBL-TE noise prediction method (red line) and the Modified Lowson-RDT LE noise prediction method (blue line) for the DAN-AERO wind turbine.

Figure 4 shows the predicted LE noise source by implementation of the standard Lowson LE noise prediction method for the DAN-AERO WT, as well as the predicted TBL-TE noise source. In this case, on the other hand, LE noise plays a significant role on the overall sound pressure level for the whole frequency spectrum, being also predicted as the dominant source for frequencies above 5 kHz, which does not represent the low-frequency noise nature of the LE source of a large scale WT.

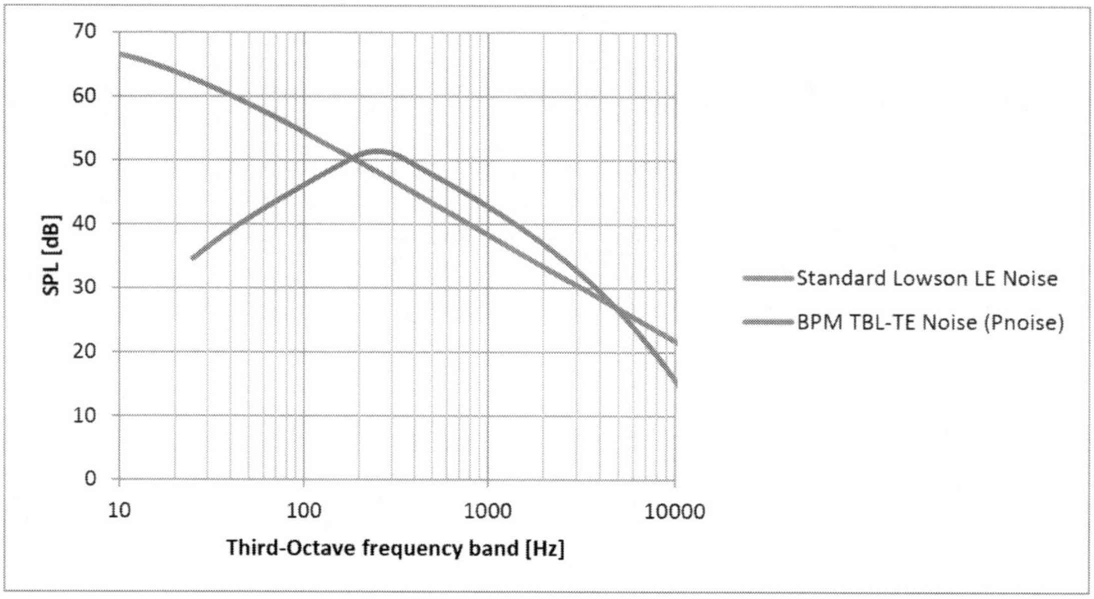

Figure 4: Predicted sound pressure levels for TE and LE noise by implementation of the PNoise BPM TBL-TE noise prediction method (red line) and the Standard Lowson LE noise prediction method (blue line) for the DAN-AERO wind turbine.

4.2 Quantitative analysis – validation against experimental data

As it can be seen in Figure 5, Figure 6 and Figure 7, the modified Lowson-RDT-LE noise prediction method provides a great approximation for the measured data, at the three different flow regimes, while the standard Lowson LE noise prediction method produces overestimations up to 33 dB at frequencies higher than 1 kHz. These overestimations were already noticed at the qualitative analysis, and yet confirmed in this validation. One should notice, however, that the experiment conducted by Juknevicius and Chong has only provided data for a NACA 0008 airfoil profile under low Re numbers ($< 10^6$).

Another noticeable fact is that all models based on Amiet's broadband noise theory begin assume the airfoil to be thin, so it does not take into account the real airfoil thickness. As already shown by Roger and Moreau and Devenport *et al.*, LE noise is strongly dependent on the airfoil thickness, and it decreases for thicker airfoils. An empirical expression for the airfoil thickness correction is proposed by Tian and Cotté, referenced from a NACA 0012 airfoil. This correction, however, does not contain the effects of camber and the nose radius (Tian and Cotté, 2016).

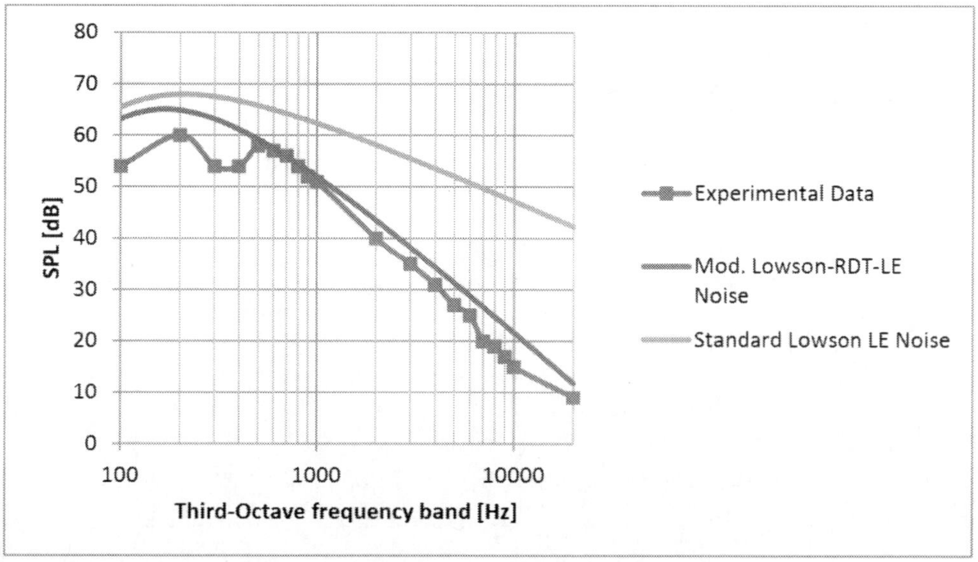

Figure 5: Measured LE noise of a NACA 0008 airfoil under a $Re = 5 \cdot 10^5$ flow (blue squares) against the predicted sound pressure levels: Standard Lowson LE noise prediction method (green line) and the Modified Lowson-RDT LE noise prediction method (red line) for the NACA 0008 airfoil.

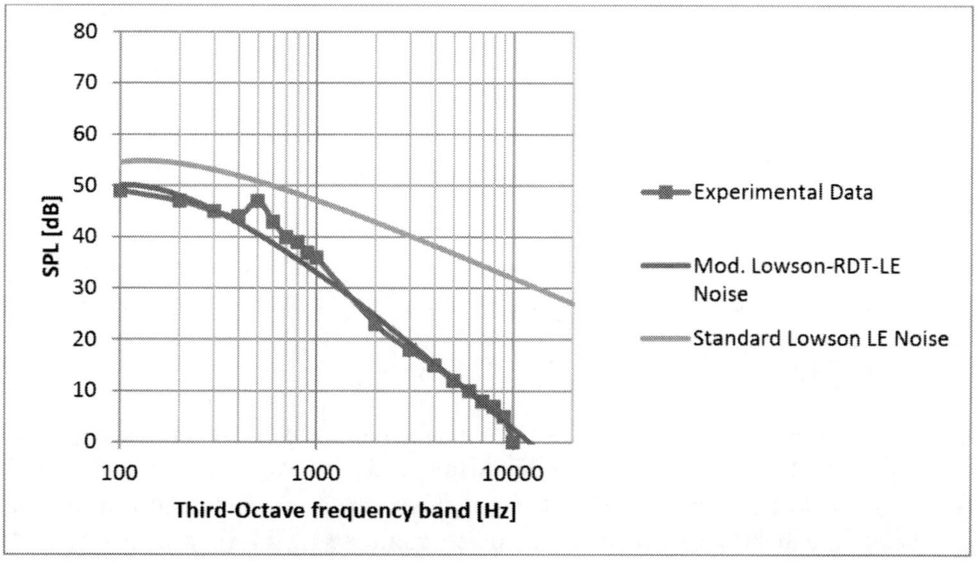

Figure 6: Measured LE noise of a NACA 0008 airfoil under a $Re = 3 \cdot 10^5$ flow (blue squares) against the predicted sound pressure levels: Standard Lowson LE noise prediction method (green line) and the Modified Lowson-RDT LE noise prediction method (red line) for the NACA 0008 airfoil.

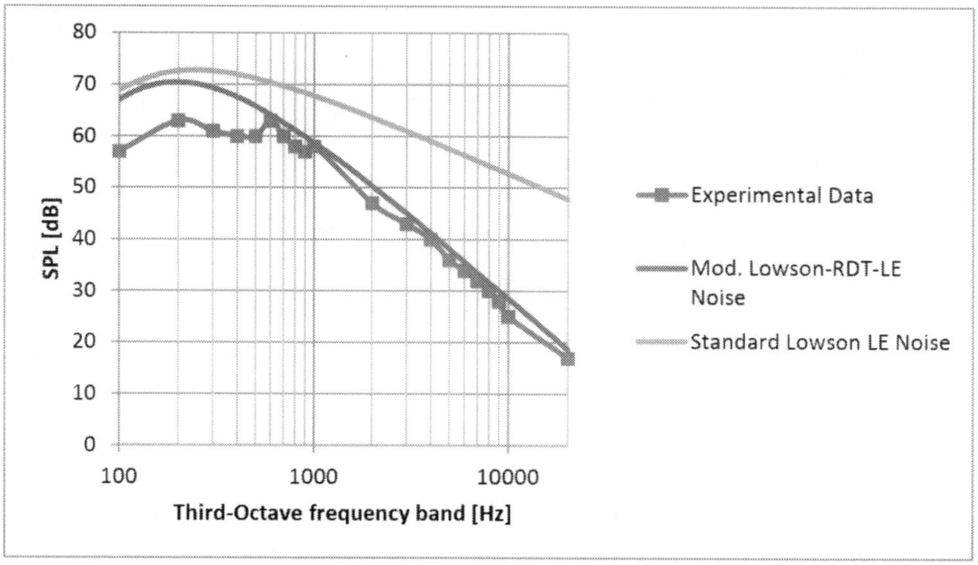

Figure 7: Measured LE noise of a NACA 0008 airfoil under a $Re = 6 \cdot 10^5$ flow (blue squares) against the predicted sound pressure levels: Standard Lowson LE noise prediction method (green line) and the Modified Lowson-RDT LE noise prediction method (red line) for the NACA 0008 airfoil.

5 Conclusions

This paper has presented an overview on turbulent inflow noise and its respective theoretical and semi-empirical prediction methods. As Lowson's method appeared to be more suitable for implementation on *PNoise* due to similarities with the BPM TBL-TE noise prediction method. The model was then discussed along with the turbulence spectrum modeling.

A modified Lowson's method was introduced with the turbulence spectrum modeled after Batchelor's rapid distortion theory. It was derived after Amiet's broadband noise theory analogously to the standard Lowson's method, with the von Kármán isotropic turbulence model.

Qualitative and quantitative analysis were then performed, respectively for a large scale WT ($Re = 2 \cdot 10^6$) and also for a NACA 0008 airfoil under a lower Re flow ($< 10^6$). The qualitative analysis has shown that RDT turbulence spectrum produces on the SPL curves a behavior closer to expectations, with the LE noise confined to lower frequencies (< 1 kHz), and with limited influence on the overall SPL. The quantitative analysis confirmed the qualitative analysis, displaying a curve fitting for the modified Lowson's method much closer to measured datathan the standard Lowson's method, and avoiding overestimations of up to 35 dB at higher frequencies. However, it demands further verification and validation for higher Reynolds numbers ($> 10^6$) and for airfoil geometries other than the NACA 0008.

Another future improvement is to include airfoil thickness effects along with the modified Lowson's method.

Acknowledgements

The authors would like to acknowledge National Council for the Improvement of Higher Education (CAPES) for the financial support.

References

Amiet RK. (1975). Acoustic radiation from an airfoil in a turbulent stream. Journal of Sound and Vibration: Vol. 41(Pt. 4):407-20.

Batchelor GK, Proudman, I. (1954). The effect of rapid distortion of a fluid in turbulent motion. The Quarterly Journal of Mechanics and Applied Mathematics; Vol. 7(Pt. 1):83-103.

Brooks TF., Pope DS., Marcolini MA. (1989). Airfoil Self-Noise and Prediction. NASA Reference Publication 1218.

Devenport W, Staubs J, Glegg S. (2010). Sound radiation from real airfoils in turbulence. Journal of Sound and Vibration 329 (2010) 3470–3483.

Juknevicius A., Chong TP. (2018). On the leading edge noise and aerodynamics of thin airfoil subjected to straight and curved serrations. Journal of Sound and Vibration: Vol. 425:324-43.

Lowson MV. (1992). Assessment and prediction of wind turbine noise. Flow Solutions Report. Bristol (England). Report No.: ETSU W/13/00284/REP.

Madsen HA, Bak C, Paulsen US, Gaunaa M, Dtu R, Fuglsang P, Glasfiber LM. (2010) The DAN-AERO MW Experiments Final Report Risø-R-Report. Vol. 1726.

Marten, D. (2016). PNoise Guidelines, Berlin: TU Berlin.

Moreau S, Roger M, Jurdic V. (2005). Effect of angle of attack and airfoil shape on turbulence interaction noise. Proceedings of the 11th AIAA/CEAS Aeroacoustics Conference Meeting and Exhibit; May 23-25, Monterey (CA). NASA.

Moriarty P, Migliore P. (2003). Semi-empirical aeroacoustic noise prediction code for wind turbines. NREL Technical Report. Golden, Colorado (US), Report No.: NREL/TP- 500-34478.

Oerlemans S, Migliore P. (2004). Wind tunnel aeroacoustic tests of six airfoils for use on small wind turbines. NREL Scientific Report. Golden, Colorado (US), Report No.: NREL/SR-500-35339.

Paterson RW, Amiet RK. (1976). Acoustic radiation and surface pressure characteristics of an airfoil due to incident turbulence. Proceedings of the 3rd AIAA Aero-Acoustics Conference; Palo Alto (CA). NASA.

Roger M, Moreau S. (2010). Extensions and limitations of analytical airfoil broadband noise models. International Journal of Acoustics 9 (2010) 273–305.

Saab Jr. JY. (2016). Trailing-Edge Noise - Development and Application of a Noise Prediction Tool for the Assessment and Design of Wind Turbine Airfoils [dissertation]. Escola Politécnica, Universidade de Sao Paulo (Brazil).

Saab Jr JY, Pimenta, MM. (2016). Displacement Thickness Evaluation for Semi-Empirical Airfol Trailing-Edge Noise Prediction Model. Journal of the Brazilian Society of Mechanical Sciences and Engineering, V.38, N.2, pp. 385-394.

Santana LD. (2015). Semi-Analytical Methodologies for Airfoil Noise Prediction [dissertation]. Katholieke Universiteit Leuven (Belgium).

Sinayoko S, Hurault J. (2015). On predicting wind turbine noise and amplitude modulation using Amiet's theory. Proceedings of the 6th Wind Turbine Noise Conference; 2015 Apr 20-23; Glasgow, Scotland. New Brighton: INCE Europe.

Simonich JC, Amiet RK, Schlinker RH, Greitzer EM. (1986). Helicopter Rotor Noise Due to Ingestion of Atmospheric Turbulence. NASA Contractor Report 3973, Langley Research Center, Hampton, Virginia.

Staubs JK. (2008). Real Airfoil Effects on Leading Edge Noise [dissertation]. Virginia Polytechnic Institute and State University, Blacksburg (VA).

Tian Y, Cotté B. (2016). Wind Turbine Noise Modeling Based on Amiet's Theory: Effects of Wind Shear and Atmospheric Turbulence. Acta Acustica united with Acustica vol. 102, 626-639.

Zhu WJ, Heilskov N, Shen WZ, and Sörensen JN. (2005). Modeling of aerodynamically generated noise from wind turbines. Journal of Solar Energy Engineering, vol. 127, no. 4, pp. 517-528.

INCEEUROPE

8th International Conference
on
Wind Turbine Noise
Lisbon – 12th to 14th June 2019

Numerical study of aerodynamic radiated noise of a Coflow-Jet Vertical Axis Wind Turbine

Dan Radulescu INCDT-COMOTI: dan.radulescu@comoti.ro
Marius Deaconu INCDT-COMOTI: marius.deaconu@comoti.ro
Georgel Vizitiu INCDT-COMOTI: georgel.vizitiu@comoti.ro
Narcisa Burtea INCDT-COMOTI: narcisa.burtea@comoti.ro

Summary
Vertical Axis Wind Turbines (VAWT) represent a suitable solution for urban areas since they are insensitive to wind direction and they produce less noise. One of the shortcomings of VAWT is represented by a lower efficiency comparing with Horizontal Axis Wind Turbine (HAWT). Coflow-Jet (CFJ) represents a new method to improve the aerodynamic performance of the airfoils due to increased circulation on the suction side. Using the CFJ improves the VAWT performance but a noise evaluation should be considered in this case. Operating wind turbines generate tonal and broadband noises affecting the living environment adversely; especially small wind turbines located in the vicinity of human living places. Therefore, it is important to determine the level of noise pollution of such type of wind turbine installation. The current study carries out numerical prediction for aerodynamic noise radiated from an H-Darrieus Vertical Axis Wind Turbine which uses CFJ. Incompressible transient simulation is conducted to obtain the instantaneous turbulent flow field. Noise prediction was performed by the Ffowcs Williams and Hawkings (FW–H) acoustic analogy formulation. Simulations were performed for three different tip-speed ratios (TSR). First, the mean torque coefficient is compared with the basic VAWT which does not use CFJ. Then, the study focuses on the broadband noises of the turbulent boundary layers and the tonal noises due to blade passing frequency. Results indicate a direct relation between the strength of the radiated noise and the rotational speed. Furthermore, noise directivity is investigated. It is concluded that the OASPL (Overall sound pressure level) is strongly reduced both tonal and broadband by the use of CFJ for a VAWT and further optimisation is possible.

1. Introduction
There are two principal types of VAWT rotor: lift-type and drag-type. Drag type rotors work by having greater drag on one side of the rotor axis than the other. Drag-type rotors are very simple, but the high drag on the half of the rotor which is travelling upstream limits their efficiency. In addition, the rotors experience a high level of thrust in the wind direction, which limits their maximum size.

Lift-type rotors use aerofoil shaped blades to generate lift, a component of which is in the direction of rotation, thereby producing torque. These rotors do not have the problems of drag-type rotors and have been shown to be able to achieve power coefficients similar to HAWTs [Paraschivoiu, 2002]. Unlike HAWT rotors, the angle of attack experienced by the blades on a VAWT is a function of azimuth, with the peak in angle of attack occurring near the most upstream point of the rotation. The fluctuation of the angle of attack is a function of the tip speed ratio (TSR), which is defined as the ratio of the blade tip speed to the wind speed. At lower TSR the angle of attack can exceed the static stall angle, which has a significant impact on rotor performance due to the occurrence of dynamic stall on the blades.

A second important feature of VAWT performance is the influence of the wake from the upstream blades. As the lift on the blades fluctuates, they shed vorticity, and this is exacerbated by dynamic stall. At the same time the spanwise variation in the pressure distribution results in a strong tip vortex being formed. The vertical wake is then advected through the rotor volume and subsequently interacts with the blades in the downstream half of the rotation, reducing the blade performance in this region. In the field of wind turbine noise, it is necessary to have an appreciation of the concept of noise annoyance.

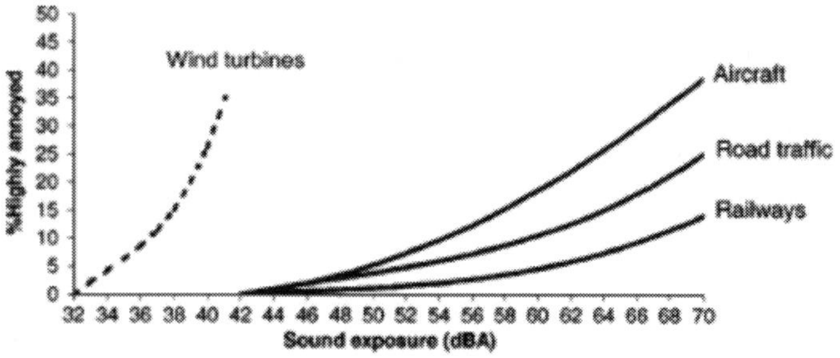

Fig. 1 - Typical annoyance curves

Measurements of wind turbine noise have often indicated that the sound levels radiated by wind turbines are lower than other sources yet still result in greater annoyance (see Figure 1, Pederson and Waye [2004]).

2. Aeroacoustic Noise sources for VAWT

Aeroacoustic noise sources of a VAWT include several categories which produce both tonal and broadband noise. The most important are considered: dynamic stall, blade- wake interaction, incoming turbulence –blade interaction.

2.1 Dynamic stall noise

Both experimental and computational studies have indicated that during dynamic stall an aerofoil can experience significantly unsteady loading. Unsteady forces radiate sound, and the rapid loading fluctuations observed during dynamic stall are likely to be a particularly intense sound source. During operation at lower TSR the blade angles of attack will exceed the static stall angle. Since the blade angle of attack is constantly changing, the blades experience dynamic stall, which results in a significantly different lift and drag behaviour relative to a static aerofoil: Leishman [2006] found that the dynamic loads can exceed the static values by up to 100%. The process of dynamic stall has been found to be dependent on a large number of factors including Mach number, Reynolds number, pitching rate and aerofoil profile, but a general description of the stages of dynamic stall has been outlined by Leishman [2006]. The stages of dynamic stall were first described by Carr et al. [1977] based on data from hot-wire measurements. The sequence below summarises the key points, such that they correspond to the five stages.

1. The aerofoil initially passes the static stall angle without any detectable change in the flow over the aerofoil and the boundary layer remains thin with no evidence of flow reversal. As the

aerofoil incidence increases further, the flow in the boundary layer starts to reverse at the surface.

2. Larger eddies start to form and full flow reversal moves from the trailing-edge to the leading-edge. When the flow reversal reaches the leading-edge, a strong leading-edge vortex is formed.

3. As the angle of attack continues to increase the separated leading-edge vortex grows and advects along the blade chord resulting in a loss of lift at the leading edge and a shift of the centre of pressure to the aft of the blade. This is the cause of the abrupt change in the pitching moment coefficient which is typically described as 'moment stall'. Carr et al. [1977] observed that the peak in the pitching moment curve occurred just before the core of the leading-edge vortex passes over the trailing edge of the blade. It is this large, intense vortex that maintains flow curvature over the aerofoil and prevents lift stall. As the vortex moves towards the trailing edge however, the flow curvature starts to reduce and the lift coefficient starts to drop.

4. Once the leading-edge vortex has passed off the trailing edge the aerofoil can no longer produce the flow curvature required for lift generation, and an abrupt drop in the lift coefficient occurs. After this point the flow over the aerofoil is fully separated as seen in full static stall hence the lift and drag coefficients tend towards their static values. Other smaller vortices may continue to form and be shed from the trailing edge resulting in further unsteadiness in the blade loading.

5. As the blade pitches down the flow attaches from the leading edge. The reattachment point moves along the blade chord at a velocity of approximately 25-35% of the free stream velocity. Depending on the pitching rate the process of reattachment may not be complete until the aerofoil is well below the static stall angle, so the static values of lift and drag will not be achieved until a lower angle of attack. Carr et al. [1977] noted that even once the flow is fully reattached the potential flow does not appear to return to unstalled conditions until the blade has passed its minimum angle and is already on the upstroke.

2.2 Blade wake interaction

The vorticity transport model simulations indicated that there are potentially significant interactions between the wake from the upstream half of the rotor and the downstream blades. In particular, the intense vortices shed from the upstream blade tips are seen to cause strong blade vortex interactions. These interactions with the tip vortices were shown to be more intense at higher TSR [Scheurich, 2011]. Another possible region of blade wake interaction was highlighted by Castelli et al. [2011] who suggested that the wake shed by the rotor hub could result in a strong blade wake interaction noise. As a VAWT rotates, the lift, and therefore the circulation, on the blades fluctuates, meaning that they will shed vorticity. At lower tip speed ratios, the blades may also experience dynamic stall, during which strong vortices are shed. The downstream blades will interact with the vertical wake from the upstream blades and this may generate impulsive blade loads by blade-vortex interaction. As the blade passes a vortex generated by a previous blade passage it experiences a change in angle of attack due to the vertical velocity component induced by the vortex. This in turn causes a change in lift on the blade and hence according to acoustic analogies produces a dipole type tonal noise.

2.3 Incoming turbulence noise

The loads on an aerofoil are strongly dependent on the flow conditions. If the oncoming flow is turbulent this will generate fluctuating loads on the blades, which will result in noise being radiated (see Amiet [1975]). As with the aerofoil self-noise, the radiated sound is highly dependent on the exact nature of the turbulent inflow. Unfortunately, it is very challenging to accurately characterise the turbulence in the wind, since it is governed by a wide range of parameters such as the surface roughness and the atmospheric temperature gradient. Detailed descriptions of some of the relationships that govern the levels of large-scale atmospheric turbulence are presented in the thesis by McIntosh [2009]. There is very little published information about the small scales of atmospheric turbulence and this is for two reasons.

Firstly, the vast majority of the energy in the wind is contained in the larger structures, commonly known as gusts, so from the perspective of power generation the small scales are irrelevant. Secondly, the local surface roughness will affect the turbulence spectra [McIntosh, 2009], meaning the small scales are likely to be strongly site specific. A scaling relationship for turbulent inflow noise was derived by Amiet [1975] and subsequently adapted by Glegg et al. [1986] for use in wind turbine noise simulations.

2.4 Other sources of aeroacoustic noise

Aerofoil self-noise is the noise generated by the interaction between an aerofoil and the flow disturbances produced in its own boundary layer and near wake. Brooks, Pope, and Marcolini [1989] wrote the definitive work on empirical self-noise modelling where they grouped these interactions into five categories and developed correlations for predicting the sound radiated by each source. The five source terms are:

1. Turbulent boundary layer trailing edge noise.
2. Separation/stall noise.
3. Laminar boundary layer vortex shedding noise.
4. Trailing edge bluntness vortex shedding noise.
5. Tip vortex formation noise.

The relative magnitude of these sources is dependent on the exact flow conditions over the aerofoil. In many practical applications, where the boundary layer is turbulent, the laminar boundary layer vortex shedding noise can be ignored since it is much quieter than the turbulent trailing edge noise. Fink and Bailey [1980] also noted that during stall the noise level can be up to 10dB greater than the noise due to TBL-TE noise at low angles of attack. At moderate angles of attack, the noise was still found to be due to the passage of turbulent, separated flow over the trailing edge, while at higher angles of attack the noise was radiated from the chord as a whole [Paterson et al., 1973].

3. Dynamics of VAWT

The aerodynamics of a lift-type VAWT are fundamentally unsteady. As the blade rotates around the hub the angle between the blade velocity and the wind velocity varies significantly. This creates a variation in angle of attack and hence a variation in the aerodynamic forces exerted on the blades. The magnitude of this variation is governed by the tip speed ratio (TSR) which is defined as the ratio of the blade speed to the wind speed. An approximation of the variation in blade angle of attack can be calculated using simple geometry by assuming that the free stream flow is uniform in both magnitude and direction over the whole rotor. This 'geometric' variation in angle of attack is shown in figure 2 for a range of TSR. It is important to note that the geometric angle of attack varies from the true angle of attack due to the deflection of the incoming flow by the rotor.

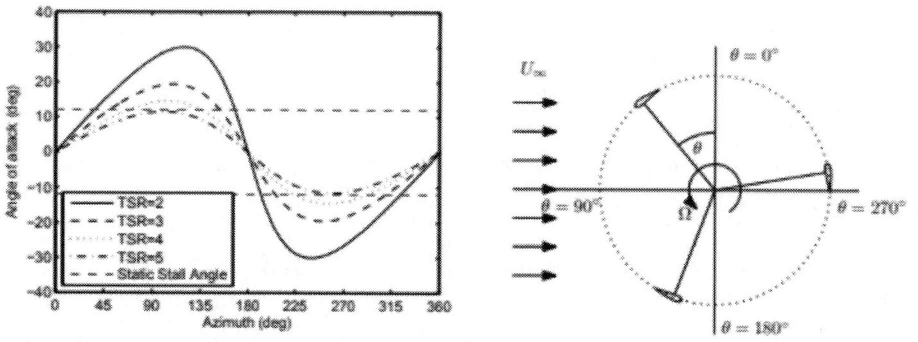

Fig 2. - VAWT blade angle of attack during one rotation at different TSR

From figure 2 it is clear that at low TSR the blades pass the static stall angle as they approach the most upstream point of the rotation. Since the angle of attack is constantly changing, the blades will experience dynamic stall at low TSR. As the TSR increases, the extent of the stall in the upstream half of the rotor decreases, which increases the amount of torque that the blades generate since the lift to drag ratio remains higher. The higher blade speed however, increases the level of parasitic drag on the rotor which simultaneously reduces the torque generated by the rotor. As such the design of a VAWT rotor is a trade-off between minimising both stall and drag in order to achieve maximum performance (figure 3).

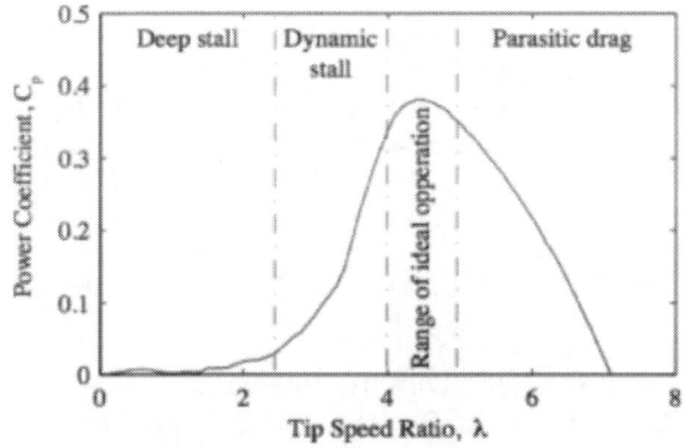

Fig 3. - Power coefficient optimisation vs TSR

The second key feature of VAWT operation is the fact that the blades in the upstream half of the rotation shed a wake that passes through the rotor and interacts with the blades in the downstream half of the rotation. This effect is exaggerated by the influence of the tip vortices shed by the upstream blades since their direction of rotation tends to cause the wake at the top and bottom of the rotor to be deflected towards the mid-span of the downstream blades.

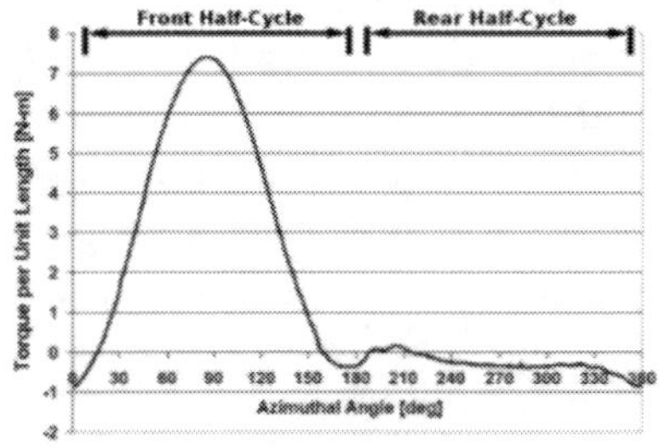

Fig 4. - Typical torque for one blade during complete rotation

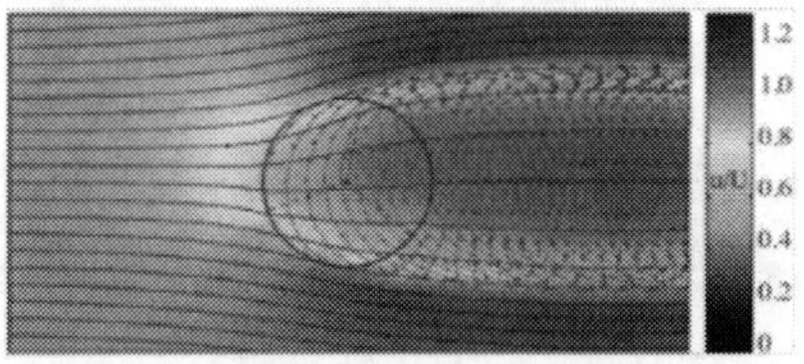

Fig. 5 - Velocity field after flow passing the VAWT

This potentially introduces further unsteady blade loading that could affect the noise radiated by a VAWT rotor. A typical torque graph produced by one blade during a full 360 ^{0} cycle is presented in figure 4.

It can be seen that most of the power is produced within from half cycle where the kinetic energy extraction is maximum. The remaining flow has a smaller average velocity due to this blockage effect as can be seen in figure 5.

4. Aerodynamic noise mitigation and performance enhancement using Coflow-jet principle

CFJ method consists in adding injection slots near the leading edge and suction slots near the trailing edge of a basic airfoil, as in figure 6. Air is injected near the LE and extracted near the TE using a recirculation pump, so we have the case of zero net mass flux since the mass of injected air is equal with the extracted mass.

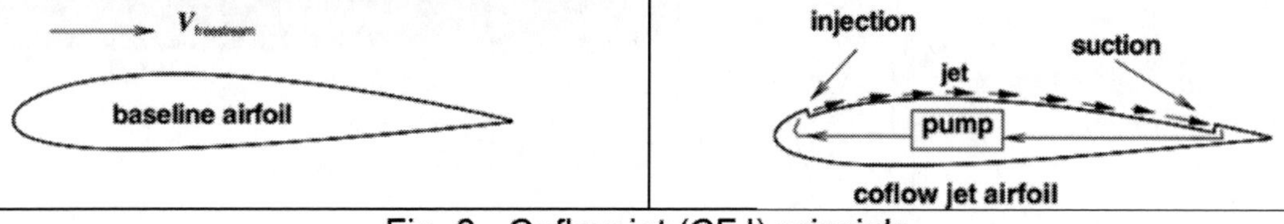

Fig. 6 - Coflow jet (CFJ) principle

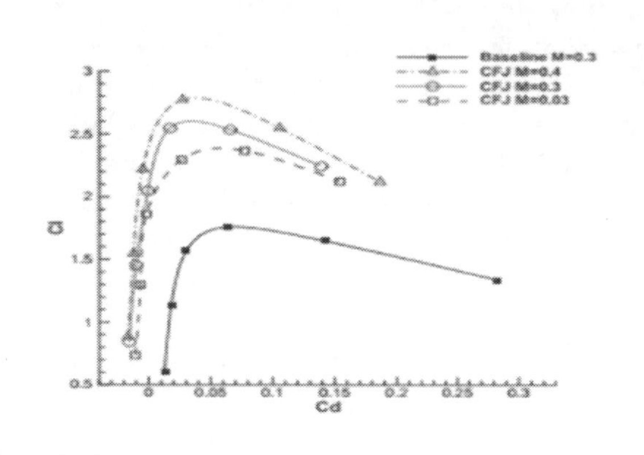

Fig.7 - Aerodynamic characteristics for base and CFJ airfoils

The aerodynamic results of such circulation are: increased boundary limit momentum, much higher stall AoA, decreased wake, much higher lift and a reduced induced drag. It must be mentioned that for some configurations it is possible to achieve even a negative induced drag which makes the solution highly attractive for new airplanes design. A typical performance graph showing Lift and Drag coefficients of CJF and baseline airfoils are presented in figure 7.

From an aeroacoustic point of view, the delayed stall produces less wake behind the profiles so it was expected that in case of VAWT to find a decrease of the blade-wake interaction noise.

5. CFD and Aeroacoustic simulation

5.1 Geometry and mesh

Numerical 2 D study was conducted on 3 blades, H-type VAWT with the following characteristics: R= 1.2 m; Basic blade profile NACA0021 with C=0.3 m. For CJF modified profile we used: $h_{injection\ slot}$ = 0.002 m; $h_{succion\ slot}$ =0.004 m. Injection and suction slots are placed at 6% chord and 80% chord distance from the leading edge as can be seen in figure 8.

Fig. 8 - CFJ airfoil for VAWT

Injection and suction slots were used on both sides of the airfoils due to the fact that during a complete rotation the angle of attack oscillates between negative and positive values so the suction side becomes pressure side. It should be mentioned that it is possible to use active control of the jets in order to use them alternatively only for the suction side.

The domain was divided in 3 circular parts, two static and one rotating containing the turbine blades (as seen in figure 9). Mesh was structured for the static domains and partially structured for the rotating domain. We used a number of 845000 cells and 853000 nodes. For the boundary layer we used an y^+ = max. 2.2 which was calculated after the simulation. Total domain D1 diameter was 6 times the VAWT diameter in order to keep the acoustic sources information inside.

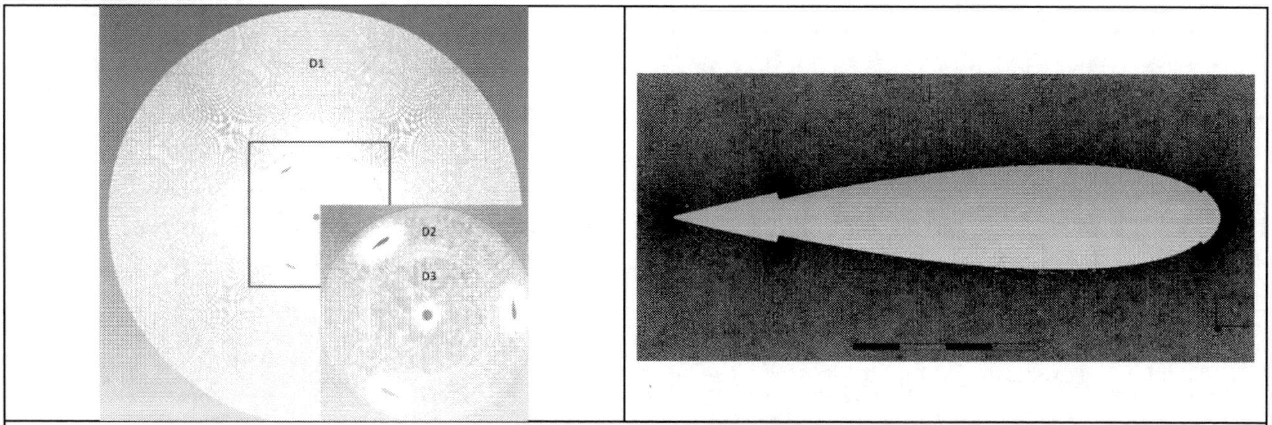

Fig. 9 - Domains and mesh

5.2 Numerical simulation

Numerical algorithm

Numerical simulation for aerodynamic and acoustic field was conducted with Ansys Fluent software. For the aerodynamic analysis, we used URANS and SST turbulence model. Transitory analysis was made using an implicit scheme using a time step which correspond to 1 degree of rotation. Every time step was iterated by 20 times. In the current paper, the discretization algorithm was conducted by SIMPLE with second-order upwind scheme by under-relaxation factors for ensuring convergence of iterative computations. The transient formulation was performed by second order implicit. Simulations were performed for Basic and CFJ airfoils at TSR 1.5; 2; 2.5. Injection and suction mass flow were established for each TSR in order to have no stall at maximum angle of attack for that configuration. This resulted from separated steady flow RANS simulations using CFJ airfoils at different attack angles and flow speed. This resulted in the minimum injection mass flow which was later used as a boundary condition for the VAWT simulation.

Boundary conditions

Parameter	Value
Incoming far field velocity	10 m/s
Turbulent intensity far-field	5%
Flow density	1.225 kg/m^3
Dynamic viscosity 1.789x10^{-5}	1.789 x 10^{-5} Ns/m^2
Turbulent intensity injection –suction slots	5%
Injection speed	90 m/s
Suction speed	45 m/s
Walls	Non-slip condition

6. Results

Aerodynamic and acoustic results show some interesting information by using the Coflow jet for a VAWT. Even that the main purpose of this study was the acoustic field evaluation, aerodynamic performance as Momentum coefficient are presented.

6.1 Aerodynamic performance

The use of CFJ for the VAWT increases the output power by reducing or completely eliminating the stall of the airfoils and due to the reactive force which appears due to the jets. However, this performance comes to a cost, which is the power consumption of the jets which must be evaluated and subtract it from the overall generated power by the VAWT. Since the study was dedicated to the acoustic field, this evaluation was not conducted here. Data and results regarding the Momentum coefficient (CM) and the comparison between Baseline (no jet) and CFJ solution for TSR 1.5; 2; 2.5 are presented below.

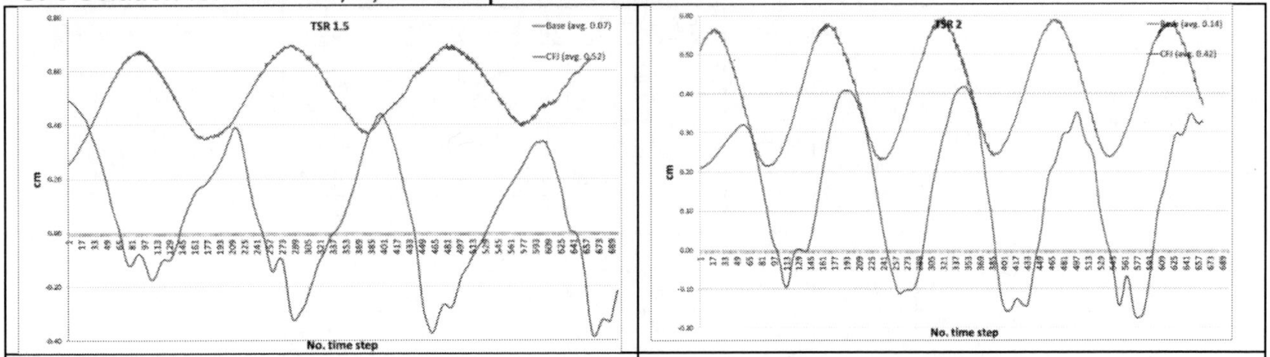

a) b)

Fig. 10 - Momentum coefficient (cm) Base vs CFJ a) TSR1.5; b) TSR2; c) TSR 2.5.

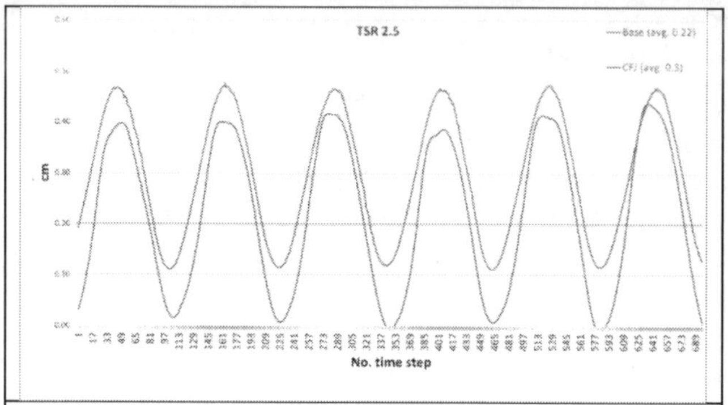

c)

Aerodynamic performance is dramatically improved by CFJ solution especially for low TSR.
It is worth to mention that this performance is achieved by using a supplementary power to inject the air. However, it has been previously demonstrated that it is possible to achieve a net gain using the CFJ as a super circulation method.

6.2 Aeroacoustic performance

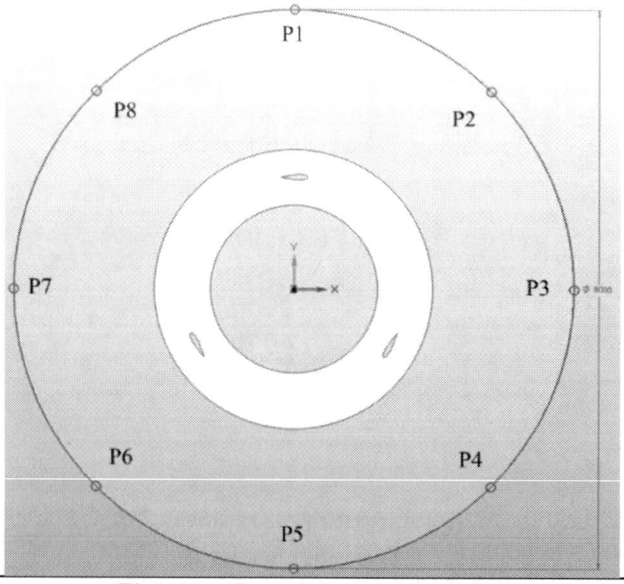

Fig. 11 - Receivers distribution

The acoustic field around the wind was calculated using the Ffowcs Williams-Hawkings analogy. We distributed 8 receivers around the turbine at 3 m distance from the rotating axis as can be seen from figure 11.
OASPL values for each receiver were calculated for TSR 1.5 ;2; 2.5 and are presented within Table 1. The values were calculated for overall frequency domain (0-600 Hz), infrasound (0-20Hz) and sound (20- 600 Hz) comparing Base with CFJ solution.

Table1 - OASPL for different receivers

Case	Virtual mic	Frequency domain dB					
		0-600Hz		0-20Hz		20-600Hz	
		Base	CFJ	Base	CFJ	Base	CFJ
TSR1.5	P1	99.4	97.9	99.2	97.8	85.7	82.6
	P2	96.6	95.7	96.2	95.4	85.2	82.9
	P3	100.4	99.2	100.1	99.0	87.2	85.6
	P4	104.7	103.2	104.5	103.1	90.4	86.7
	P5	106.9	106.0	106.8	105.9	87.9	86.4
	P6	106.3	105.4	106.3	105.3	86.4	85.9
	P7	106.7	106.1	106.6	106.0	91.4	89.5
	P8	105.0	104.6	104.8	104.4	90.4	89.5
	Lp_avg	**104.5**	**102.7**	**104.3**	**102.6**	**88.6**	**86.3**
	Diff.	-	**1.8**	-	**1.8**	-	**2.3**
TSR2	P1	98.6	97.0	98.4	96.8	85.91	82.70

	P2	96.2	93.7	95.9	93.5	84.38	79.91
	P3	98.7	98.0	98.4	97.9	87.02	83.59
	P4	103.9	102.3	103.7	102.2	89.68	84.26
	P5	106.8	105.0	106.7	104.9	90.08	84.65
	P6	107.5	106.5	107.5	106.5	85.44	83.97
	P7	107.7	106.4	107.6	106.4	89.30	84.75
	P8	105.0	104.1	104.9	104.0	90.48	88.45
	Lp_avg	**104.7**	**103.4**	**104.6**	**103.4**	**88.3**	**84.6**
	Diff.	-	**1.3**	-	**1.3**	-	**3.7**
TSR2.5	P1	102.1	93.9	101.5	93.8	93.11	77.20
	P2	97.9	88.4	97.6	88.2	86.88	74.96
	P3	98.6	93.1	98.4	93.0	85.64	77.81
	P4	102.1	96.9	101.9	96.8	89.46	81.82
	P5	105.2	99.6	104.9	99.6	93.22	79.55
	P6	107.5	102.1	107.4	102.0	88.43	82.75
	P7	107.9	102.8	107.9	102.7	87.62	85.20
	P8	106.5	100.8	106.4	100.8	90.06	81.46
	Lp_avg	**104.8**	**99.2**	**104.7**	**99.1**	**90.1**	**81.2**
	Diff.	-	**5.6**	-	**5.5**	-	**8.9**

It can be seen that it is possible to achieve an important noise reduction on all frequencies and even more on audible domain. The max average reduction of 8.9 dB was obtained for TSR 2.5 case. The explanation might be that for this TSR the jets speed was optimised in order to be the minimum speed at which stall is not present.

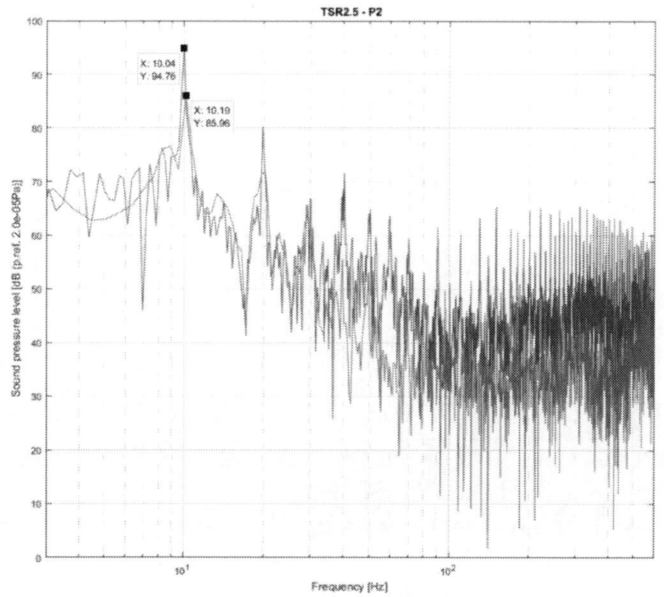

Fig. 12 - Noise spectrum analysis (Base - blue; CFJ - orange)

For smaller TSR it is possible that the jets were to strong and can add some broadband jet noise.
An interesting finding is that CFJ is capable to strongly reduce the tonal noise as can be seen from the FFT analysis at receiver P2 and presented in figure 12.
This was an expected result since the wake reduction due to CFJ reduces also the blade-wake interaction which is the main tonal noise source.
Another interesting finding is that VAWT noise has a distinct directivity and can be altered by the use of CFJ as can be seen from the figure 13.

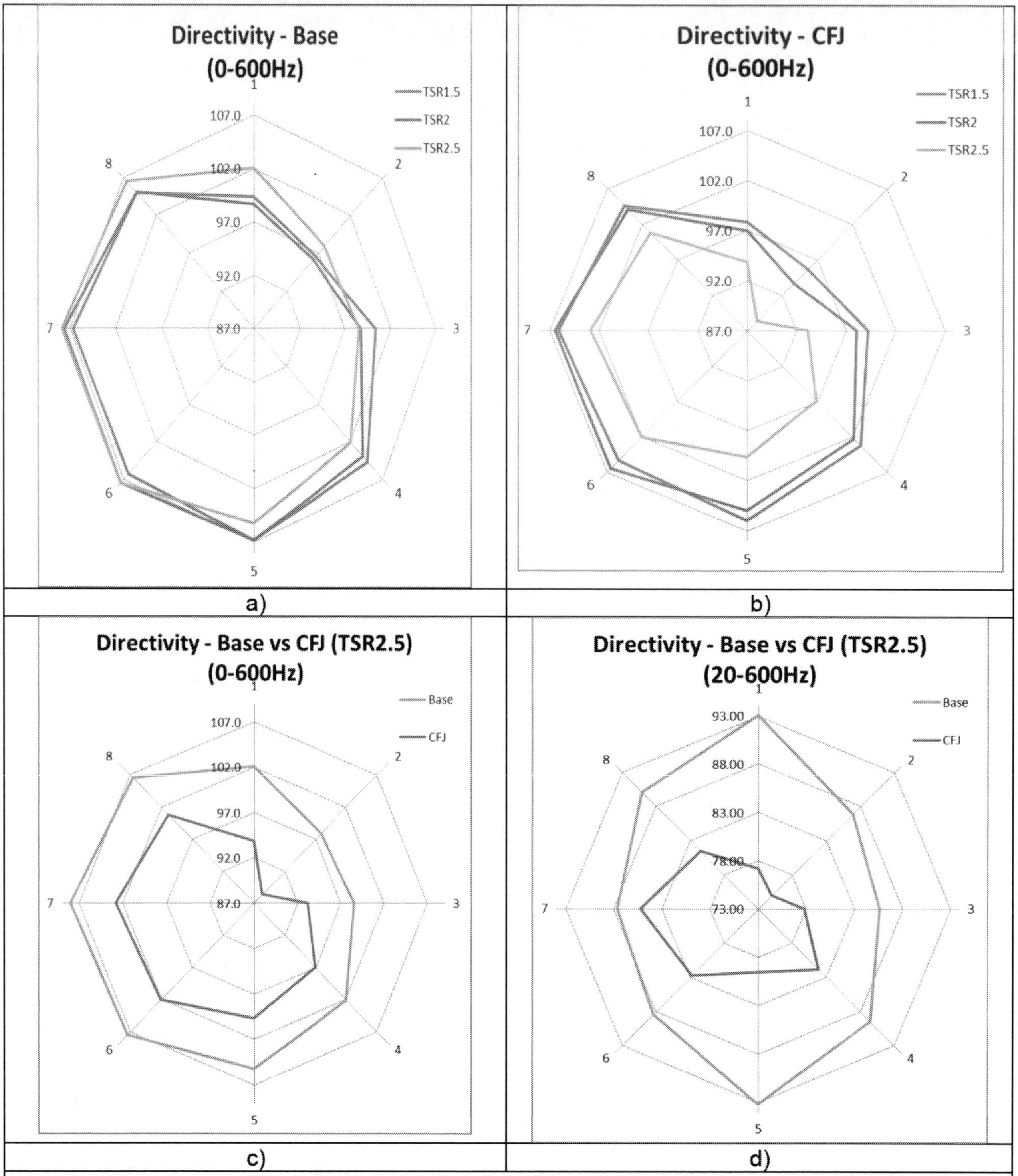

Fig. 13 - a) noise directivity base b) directivity CFJ c) directivity base vs CFJ TSR2.5 d) directivity base vs CFJ TSR 2.5 (20-600Hz)

It can be seen that CFJ can also modify VAWT noise directivity as can be seen in figure 13 d).

7. Conclusions

CFJ VAWT represents an interesting option for wind energy harvesting since it is efficient and quieter. Efficiency is much higher than the base but requires additional input power to recirculate the jet. However, it has been demonstrated by other authors that it is possible to achieve an important net gain in power output. Several interesting findings about noise field have been revealed by this study. The most important is that it is possible to get a strong OASPL reduction both broadband and tonal. This is the effect of wake reduction and thus the blade-noise interaction. These findings require a future investigation for the possibility of optimization the CFJ VAWT in terms of efficiency and noise production.

References

Paraschivoiu I.,Saeed F., Desobry V.(2002) *"Prediction capabilities in vertical axis wind turbine aerodynamics"* Berlin, Germany

Castelli M.R.(2012) *„Effect of blade number on a straight – bladed vertical axis Darrieus wind turbine"*, World Academy of Science, 61

Ferreira S., Bijl H., G. Van Bussel,G van Kuik(2007) *„Simulating dynamic stall in a 2D VAWT"*, Journal of Physics, Conference Series 75

RK Amiet (1976) "*Noise due to turbulent flow past a trailing edge*" Journal of Sound and Vibration, 47(3):387{393}. 16, 79

T Brooks (1993) *"Studies of blade-vortex interaction noise reduction by rotor blade Modification"* NASA STI/Recon Technical Report A, 959:90093. ix, 29

M Castelli, A Villa, and E Bernini (2011) "*CFD Analysis of the Influence of Central Shaft on Vertical Axis Wind Turbine Noise Emission"* In International Wind Turbine Noise Conference, Rome. x, 52, 56

C Ferreira (2009) "*The near wake of the VAWT: 2D and 3D views of the VAWT aerodynamics*" PhD thesis, Technische Universiteit Delft, . 6, 36, 40, 51, 53

S McIntosh (2009) *"Wind Energy for the Built Environment"* PhD thesis, University of Cambridge. x, xiii, 3, 20, 21, 36, 38, 39, 42, 45, 46, 49, 53, 88, 142, 144, 152

Zha, G.-C. and Paxton C. (2004) *"A Novel Airfoil Circulation Augment Flow Control Method Using Co-Flow Jet"* AIAA Paper 2004-2208, 2nd AIAA Flow Control Conference, Portland, Oregon.

INCEEUROPE

8th International Conference
on
Wind Turbine Noise
Lisbon – 12th to 14th June 2019

A Wind Turbine Noise Code Benchmark - Round 1

Franck Bertagnolio
DTU Wind Energy, Roskilde, DK-4000, Denmark

Second C. Author
Institution, City, Province, Zip Code, Country

Summary

The aim of this study is to compare and validate wind turbine noise prediction codes from various institutes and companies. This effort is part of the IEA Wind TCP Task 29 (Wind Turbine Aerodynamics) and IEA Wind TCP Task 39 (Quiet Wind Turbine Technology). The benchmark is divided into 3 rounds which will be dealt with incrementally in time, and the focus of the present article is on the first round. Note that this study concentrates on aerodynamic noise generation, therefore mechanical noise and long-range atmospheric propagation effects are not considered.

1. Introduction

Wind turbine noise emissions are commonly measured in the field according to the IEC-64100-11 standard [4]. Microphones are placed on the ground downstream of the turbine at a specified distance from the tower (equal to the height of the tower plus half of the rotor diameter). The wind turbine is considered as a monopole noise source and is thereby supposed to emit the same noise levels in all directions. Thus, the mesured sound pressure levels can be related to the sound power levels of the turbine. Immission levels at dwellings can be evaluated using a variety of methods predicting the propagation losses, from simple semi-empirical formulas to advanced simulation methods such as Parabolic Equations or Computational Aero-Acoustics. Furthermore, the results of the above standardized measurements are binned according to the wind speed to reflect the actual variations of the noise emission levels with respect to the wind conditions. Modern turbines produce typically less noise at low wind speed because of the lower rotational speed, and nearly constant noise above rated power when the rotational speed is normally kept fixed and power is regulated by the controller through the blade pitch.

However, as far as the actual noise emissions are concerned (i.e. when considering the wind turbine as the source of noise), the physics of noise generation mechanisms is more complicated than the above conceptualization. The assumption of a monopole noise source is certainly a first order approximation. Nevertheless, note that for wind turbine certification it is considered as satisfying to measure according to the IEC-64100-11 standard since microphones should be located directly downstream of the turbine, where the maximum noise levels are expected. In this sense, the IEC standard is a worst case situation. Yet, the noise generation mechanisms (e.g. trailing edge noise considered as one of the most potent source of noise in the audible range) present strong directivity features [1, 3]. These effects are not accounted for in the measurements according to the above standard.

Another aspect of wind turbine noise is the fact that multiple noise generation mechanisms are simultaneously at play. These can be segregated in two main categories: mechanical and aerodynamic noise. Mechanical noise results from structural vibrations of the turbine components (tower, nacelle and blades) as well as gear noise. Usually, it is identified as tonal noise with frequencies associated to the eigenfrequencies of these components or to the rotational speed of the shaft or the gear components. Although potentially annoying and subject to regulations related to tonal noise, mechanical noise is not considered herein. The present study concentrates on aerodynamic noise, i.e. noise generated by the interaction of turbulent vortices with the blade surfaces. These turbulent features can be self-generated, as it is the case for the turbulent boundary layer flow developing along the blade airfoil sections producing noise when passing by the trailing edge. Alternatively, they can originate from the turbulent atmospheric flow impacting the blades or some other external source of turbulence (e.g. wake from an upstream turbine).

Aerodynamic noise generation involves complex phenomena and their mutual interactions. Firstly, fluid flow turbulence is a difficult topic with highly non-linear processes which are not trivial to predict. Most of the theory of turbulence is based on statistical averaging which somehow facilitates some aspects of flow prediction in contrast to the deterministic prediction of the chaotic behavior of the turbulent structures. This is quite relevant for wind turbine noise generation since the time-scales of the turbulent motions generating noise are quite small compared to the time-scales of the parameters influencing the wind turbine operation (e.g. changing mean wind speed over the rotor disk or rotational speed). Secondly, turbulence interaction with the hard surfaces generating noise are also rather complicated phenomena, e.g. noise scattering at the trailing edge. As a consequence, accurately predicting noise from wind turbines can be challenging.

Wind turbine manufacturers have a long practical experience of noise emissions from their wind turbines and have access to a lot of noise measurement data. To predict the noise emission from a turbine, in many cases they rely on semi-empirical models which can be accurately tuned thanks to the above practical experience and know-how. The next step in developing modeling tools capable of predicting wind turbine noise is to introduce more physics in the models and try to describe more faithfully the actual processes involved. This may become a critical asset as wind turbine technology develops with the use of advanced aerodynamic features such as flaps, morphing blades, winglets, etc.

To the best authors' knowledge, there does not exist a commercial simulation tool that is dedicated to the prediction of wind turbine noise as a source, although several of these codes can be used in this context []. As a result, wind turbine manufacturers and research institutions alike separately are developing their own prediction tools. When wind turbine manufacturers can rely on their expertise and historical data, it is sometimes difficult for the researchers to find reliable data to validate their models. Contrastingly, the manufacturers do sometimes have limited resources to develop more advanced simulation tools and usually rely on their existing more empirical tools. In both cases, exchanging experience and comparing results may benefit the two communities.

The aim of the present study is to define a comparison benchmark for wind turbine noise simulation codes. This initiative was taken as part of IEA Wind TCP Task 39 (Quiet Wind Turbine Technology) in collaboration with IEA Wind TCP Task 29 (Analysis of Aerodynamic Measurements). The comparisons are based on an existing wind turbine which was extensively measured during the DANAERO project [5, 6]. These measurements were conducted on is a 2.3 MW NM80 wind turbine located in Tjæreborg, Denmark. One of the blades was specifically manufactured for this project and extensively equipped with aerodynamic sensors, as well as surface pressure high-frequency microphones flush-mounted in the outer part of the blade which are relevant for studying aeroacoustic emissions. Some noise measurements according to the IEC-64100-11 standard also exists and may become part of the present benchmark once this has been negotiated and agreed with the current owner of these data.

The first objective of this comparison benchmark is to make sure that the noise predictions from the different codes are based on (nearly) identical aerodynamic data. Indeed, aerodynamic noise is by essence driven by aerodynamic quantities. In this respect, this benchmark is actively connected to the IEA Wind TCP Task 29 Phase IV which currently focuses on using the DANAERO database to validate aeroelastic codes for wind turbines. Therefore, the present study also includes the validation of a restricted set of aerodynamic data considered as crucial for accurately evaluating the noise emissions. The validation of the aerodynamic noise predictions is the core objective of the present benchmark. This is dealt with in two steps. Firstly, the surface pressure fluctuations measured with high-frequency microphones on the blades can be used to partly validate noise prediction codes, particularly for trailing edge noise but possibly turbulent inflow noise as well. Secondly, the noise emissions of the whole turbine in different configurations are investigated.

2. General description of the benchmark

2.1. Generalities

The present benchmark is a combination of a series of code validation through simplified test cases and actual measurement data. In the simplified cases, the analysis of the results should reveal some shortcomings of the actual numerical models and their implementation. In particular, some of these test cases will assume axial symmetry of the flow around the rotor axis and this symmetry should be reflected in the computed results, both aerodynamic and acoustic ones. Furthermore, comparisons between results from the various codes should bring some light on the various methods, assuming that high-fidelity models such as CFD can be taken as reference. Concerning the use of actual measurement data, those collected during the DANAERO experiment will be considered.

Briefly, the DANAERO experiment is a series of measurement campaigns conducted in year 2009

Fig. 1 The NM80 wind turbine in Tjæreborg

on a 2.3MW NM80 wind turbine with a hub height of 60 m [5] (see Fig. 1). One of the three blades of the turbine was specifically manufactured for this experiment and equipped with multiple sensors. Aerodynamic sensors (pressure taps, Pitot tubes, microphones) are distributed along the blade span at several given radii (see Fig. 2). The blade is a LM 38.8 m attached to the hub at a radius of 1.24 m from the rotor centre. Therefore, the rotor radius at the blade tip is 40.04 m. A met mast located near the turbine was used to monitor atmospheric conditions. The project also included a series of wind tunnel experiments for which 2D reproductions of given blade airfoil sections were measured. These are not considered as part of the present benchmark so far. Note that surface pressure microphones were flush-mounted on the blade near the outer most instrumented radial section. These measurements are quite relevant for the validation of noise emission models (i.e. trailing edge noise, and possibly turbulent inflow noise).

In the present benchmark, we are interested in validating:

- The aerodynamic part for the wind turbine noise codes using pressure tap sensors and Pitot tubes. These validations are a subset of those conducted as part of Task 29 Phase IV Case IV.1 and therefore mainly orientated toward participants to the present benchmark who do not participate to Task 29.

- The estimation of turbulent boundary layer quantities near the trailing edge relevant for trailing edge noise modeling. These quantities were not measured during the measurement campaigns, but validation will be based on cross- checking the results obtained by the various participants (and existing experience on this type of data, e.g. BANC benchmark [2]).

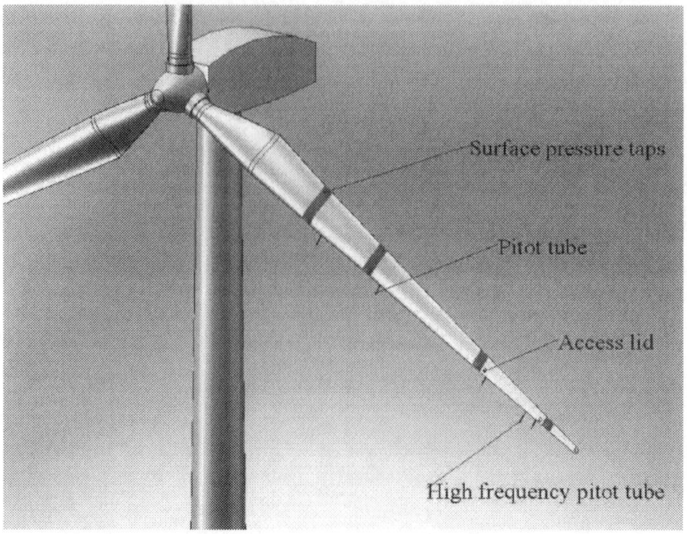

Fig. 2 Equipment on the LM 38.8 m blade

• The estimation of the surface pressure fluctuations (more specifically their spectra) for simulation codes using an approach that do provide such quantity. This validation is mainly relevant for turbulent inflow and trailing edge noise prediction models. The validation can be performed using measurement data from surface pressure microphones on the test blade.

• The prediction of the acoustic noise immission in the far-field (in fact at the distance recommended by the IEC 61400-11 standard). The participants are expected to provide their turbulent inflow and trailing edge noise predictions, but may also include other noise sources (e.g. separation noise if separation is detected, tip noise) if they wish so.

Furthermore, according to the previous description of the benchmark, the benchmark is divided into 3 rounds:

- Round #1 is concerned with the validation of the codes for four idealized cases including symmetrical cases, as well as rigid and flexible rotors. Two parameters are varied in order to quantify their influence on the acoustic results: the rotor rotational speed and wind shear.

- Round #2 is concerned with actual cases from the DANAERO experiment. The specific aim of this round is to compare numerical results with existing measurement data.

- Round #3 is concerned with comparisons of the wind turbine noise codes by calculating noise characteristics as a function of wind speed. Noise was not measured as part of the DANAERO experiment, but noise measurements were conducted as part of the certification procedure for this turbine and may possibly be used in this round for validation. For the time-being, the description of Round #3 is tentative and will be refined at a later stage.

Note that the two first rounds are integrated parts of IEA Wind TCP Tasks 29 & 39, while the third round is more orientated toward participants of Task 39. In this paper, we are only interested in the first round of this benchmark.

2.2. Test cases

As mentioned above, the benchmark is divided into 3 rounds. Each round contains a number of test cases to be simulated. The specific geometrical and aerodynamic inputs for each test case are shortly described in this section for Round #1. The results to be provided are specified in the following section.

For the **Round #1**, there are 4 test cases defined as follows:

Case 1.1
Same as Task 29 Case IV.1.1 and provide comparison results (i.e. aerodynamic and boundary layer quantities, surface pressure near leading and trailing edge, and immission noise) as specified in Section 3 below. The specifications of Case IV.1.1 amounts to an axi-symmetric configuration (no-rotor tilt or yaw, no tower shadow, no wind shear, no inflow turbulence, but the pre-bend is included) and a rigid rotor (i.e. no aeroelastic deformation of the blades).

Case 1.2
Same as Task 29 Case IV.1.2 and provide comparison results as specified in Section 3 below. The specifications of Case IV.1.2 are identical to Case IV.1.1, but for flexible blades. However, if the participant's wind turbine noise code cannot handle flexible blades, this case should be ignored and the participant should carry on with the following cases, assuming a rigid structure in Round #2 (see below).

Case 1.3
Same as Case 1.1, but with a different rotor speed.

Case 1.4
Same as Case 1.1, but with wind shear.

The main other parameters common to all calculations of Round #1 are reported in Table1. Note that the turbulence intensity and length scalse specified herein are only meant for the turbulent inflow noise modeling, not the atmospheric wind speed impacting the turbine for the aero-elastic calculations.

2.3. Results to deliver

The results that participants to the benchmark should deliver can be divided into 4 sets.

The first set is concerned with aerodynamic data at the 3 radius locations along the blades. The quantities of interest are mainly relative velocity (with and without induction), angle of attack, aero-dynamic forces (incl. lift and drag) and coefficients (C_L and C_D, respectively), as a function of time and/or blade azimuth angle. In addition, pressure coefficients around the blade are investigated.

The second set of data is related to trailing edge noise modeling. The results to deliver are quantities across the boundary layer near the trailing edge, both on pressure and suction sides, such as velocity profiles, turbulent kinetic energy, turbulent length scales, and integral quantities such as boundary layer thickness or displacement thickness.

The third set is concerned with surface pressure spectra near the trailing edge and the leading edge, which are of potential interest for trailing edge and leading edge noise, respectively.

The fourth and last set are the noise immission levels at pre-defined observer locations around the turbine on the ground, as well as one location at hub height directly downwind of the turbine. This

Table 1 Main computational input parameters for Round #1.

Quantity	Value
Tilt	0°
Coning	0°
Tower shadow	None
Air density	1.231 kg/m^3
Temperature	19°C
Wind speed	$V_H = 6.1$ m/s
Wind shear	None
Turbulence intensity	8.96%
Turbulence length scale	39 m
Rotor speed	12.3 rpm
Blade pitch angle	1.5° (>0 nose down)
Yaw error angle	0°
Transition location	$x/C = 0.065$ (Suction side)
Transition location	$x/C = 0.20$ (Pressure side)

last noise immission location is dedicated to check the sanity of the numerical models with respect to the axial symmetry defined in Round #1 (see above).

3. Preliminary results and comparison examples

As none of the participants have sent their results at the time of writing, some of the expected results are shown in this section in order to illustrate the specificities of the different test cases.

3.1. Aerodynamic results

In figure 3, some aerodynamic data on a given blade at the outer most radial position $r = 37$ m are plotted as a function of its azimuth angular position ψ for the 4 considered test cases. The quantities are the effective velocity, the angle of attack, the lift and drag coefficients. As expected, for test cases 1.1, 1.2 and 1.3 (i.e. without shear), the aerodynamic quantities do not vary with the azimuth position. Contrastingly, these quantities are a function of the azimuth for test case 1.4, which includes a wind shear, thereby breaking the symmetry with respect of the rotor axis as discussed earlier. Note that the angle of attack, as well as lift and drag, are highest when the blade points upwards (i.e. for $\psi = 0°$). Test case 1.3 exhibits a higher effective velocity because of the higher rotational speed. Furthermore, it can be noticed that the blade flexibility for test case 1.2 also modifies the aerodynamic properties with a higher angle of attack, at least at the considered radius.

3.2. Noise immission results

As the main goal of this benchmark is to validate wind turbine noise simulation code, a few examples of noise immission results in the vicinity of the turbine are shown here.

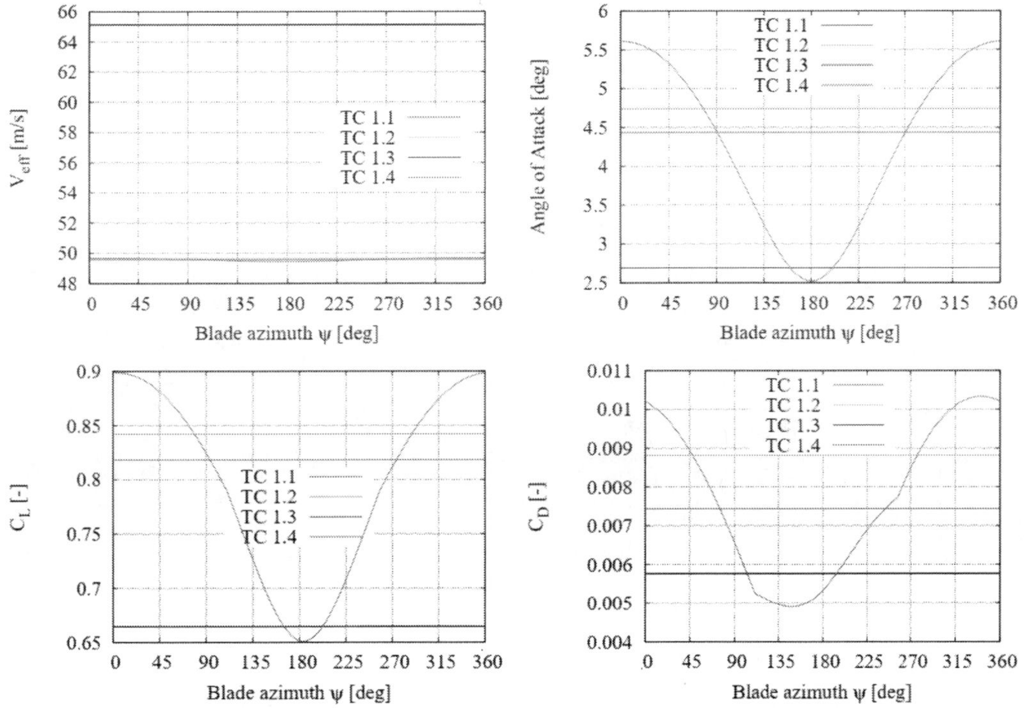

Fig. 3 Aerodynamic quantities as a function of blade azimuth location ($\psi = 0°$ when the blade points up)

The whole turbine noise immissions (i.e. the contributions of the noise emissions are integrated over the whole span of the 3 blades and over the whole frequency range) for the 4 test cases are plotted as a function of the azimuth of blade #1 during one of its revolution in Figs. 4, 5 and 6. In the first two figures, the immission location (denoted as P13) is directly downwind of the rotor at hub height. Therefore, this latter point is located on the rotor axis since the rotor is not tilted. In the third figure, the immission point (denoted as P7) is at the IEC standard position, directly downwind of the turbine on the ground. Furthermore, in these figures the noise levels are displayed by adding up all considered noise sources, i.e. here turbulent inflow noise, trailing edge noise and stall noise, as well as individually in the separate sub-figures.

As expected for the results at P13 in Fig. 4, the symmetry of the flow for test cases 1.1, 1.2 and 1.3 results in constant noise levels. However, for an asymmetical flow field (i.e with wind shear for test case 1.4), the noise levels exhibit temporal variations, in particular for the trailing edge and stall noise, whereas the turbulent inflow noise appears insensitive to the wind shear. The overall noise is slightly influenced by the shear. In the case of A-weighted noise as displayed in Fig. 5, the stall noise becomes dominant and the temporal variation of noise levels become apparent for the overall noise. It may be surprizing that stall noise is dominant in the present configuration. This is investigated below. Before that, it should be noted that the noise immissions at P7 in Fig. 6 illustrate the fact that this position breaks the symmetry of the noise emission as a result of the noise sources directivity.

In order to study the noise emissions in more details, map of the noise sources across the rotor disk are plotted, once again both for all added-up noise sources and individually. These maps represent the elementary contribution to the noise immission levels at given observer locations (i.e.

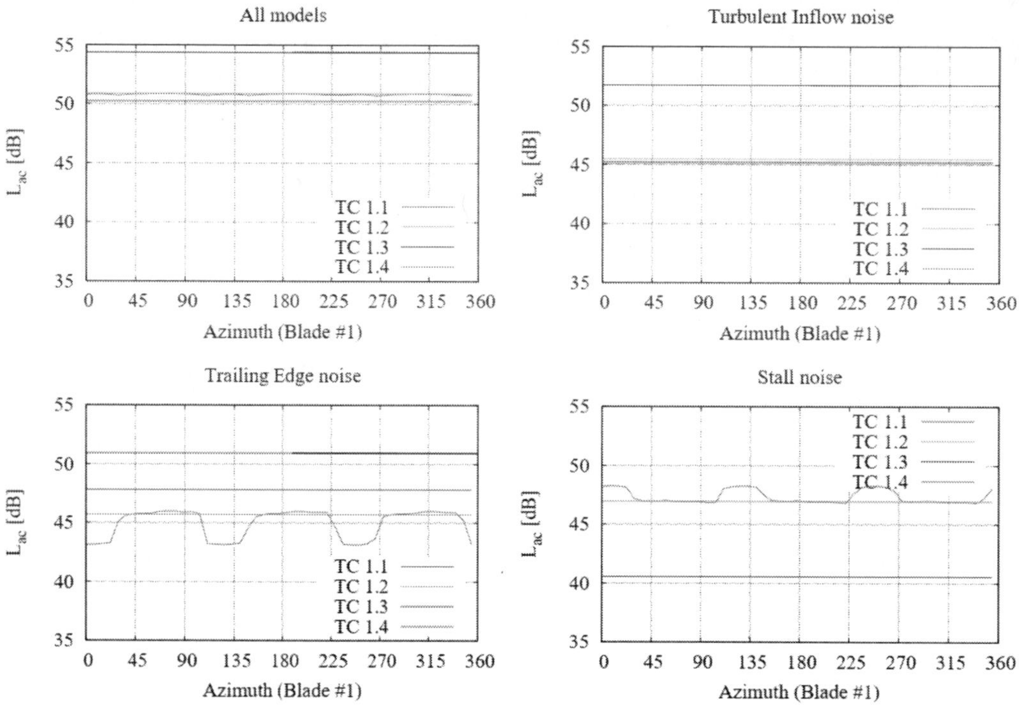

Fig. 4 Noise immission at P13 (downwind of rotor at hub height) as a function of blade #1 azimuth location

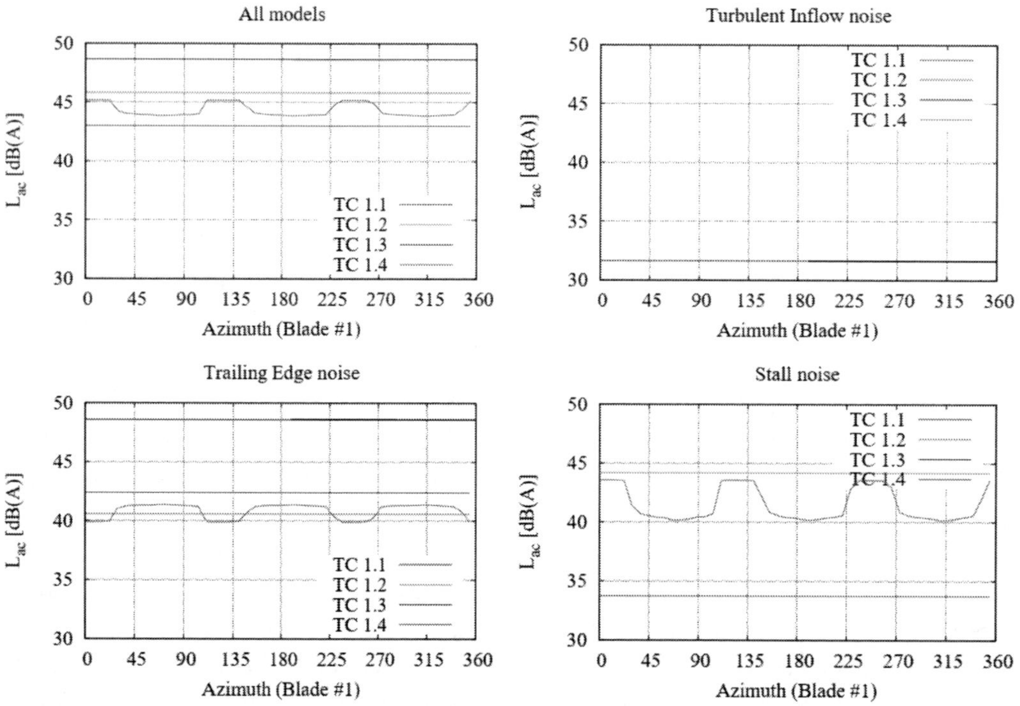

Fig. 5 A-weighted noise immission at P13 (downwind of rotor at hub height) as a function of blade #1 azimuth location

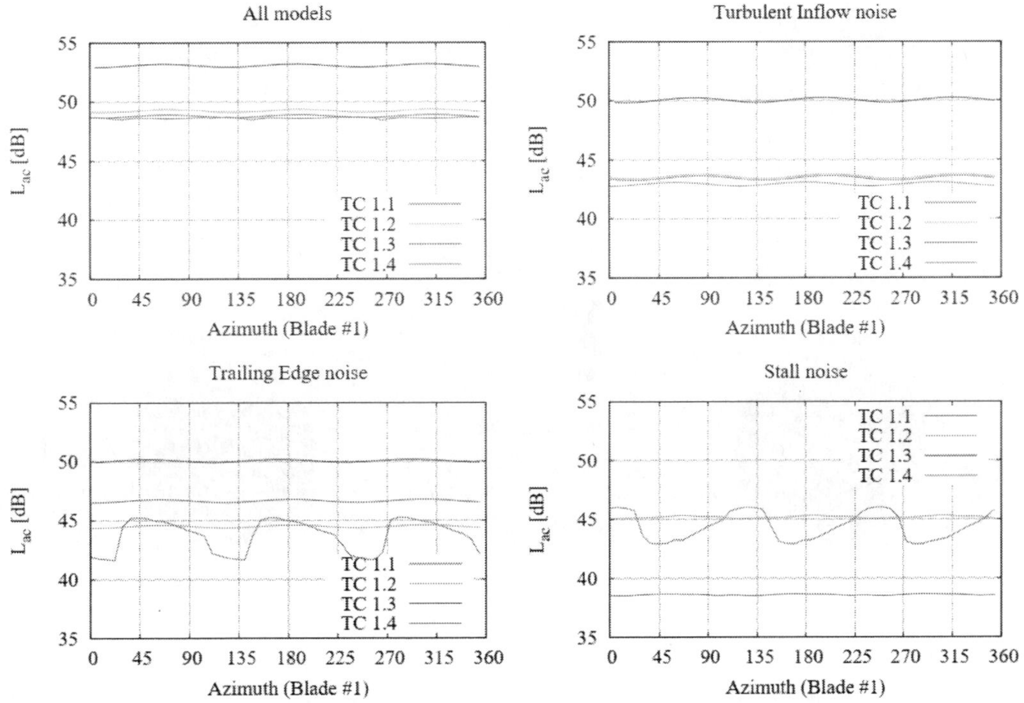

Fig. 6 Noise immission at P7 (downwind of rotor on the ground) as a function of blade #1 azimuth location

P7 or P13 here) from the local noise emissions across the rotor disk. Let us first consider test case 1.1. The map for position P13 is displayed in Fig. 7. The symmetry of the noise emissions is clear. Furthermore, it can be seen that stall noise is concentrated in the inner region of the rotor disk/blade where thick airfoil sections can be found, more easily triggering the occurrence of stall. The same map but for position P7 is displayed in Fig. 8. Once again, the asymmetry from the noise directivity patterns becomes apparent and it can be seen that higher noise levels are observed on the lower right part of the rotor disk, both for turbulent inflow and trailing edge noise. This effect is possibly a combination of directivity and the fact that the lower part of the rotor disk is closer to the observer at P7.

Finally, the noise map for test case 1.4 (with wind shear) for an oberver at position P13 is displayed in Fig. 9. It can be seen that stall noise is produced on a large upper part of the rotor disk. Indeed, it is where the wind speed is higher due to the wind shear and angles of attack are also larger (see Section 3.1). However, it was observed in the previous section the angles of attack remain relatively low and it is quite surprizing that stall is so widely spread. This indicates a potential problem in the simulation code which has to be investigated (or alternatively a misinterpretation of the results). As a matter of fact, one of the primary goal of the present benchmark is to detect such inconsistencies in the results and try to improve the prediction tools accordingly.

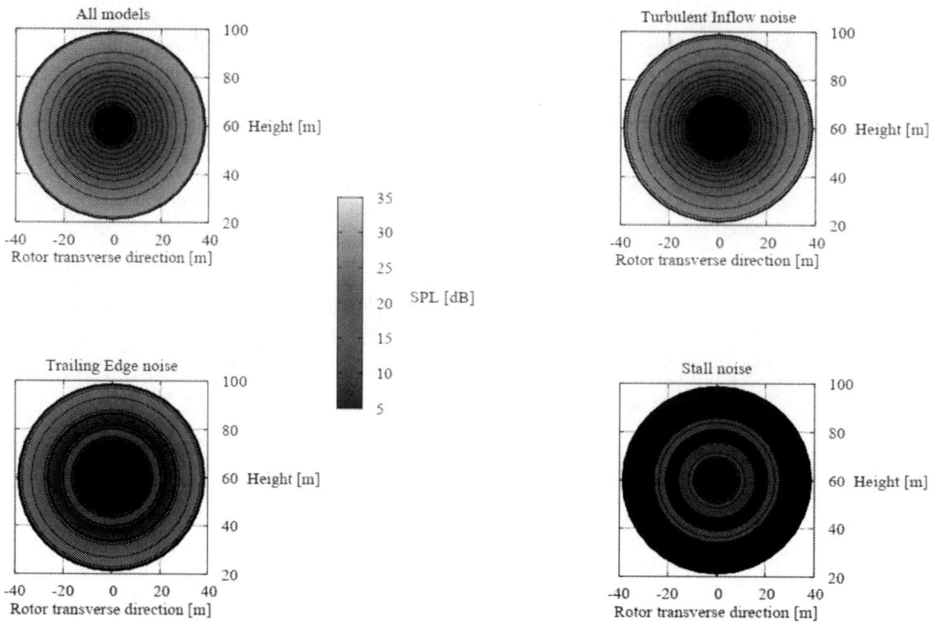

Fig. 7 Noise immission map at P13 (downwind of rotor at hub height) as seen from upwind the turbine - Test case 1.1.

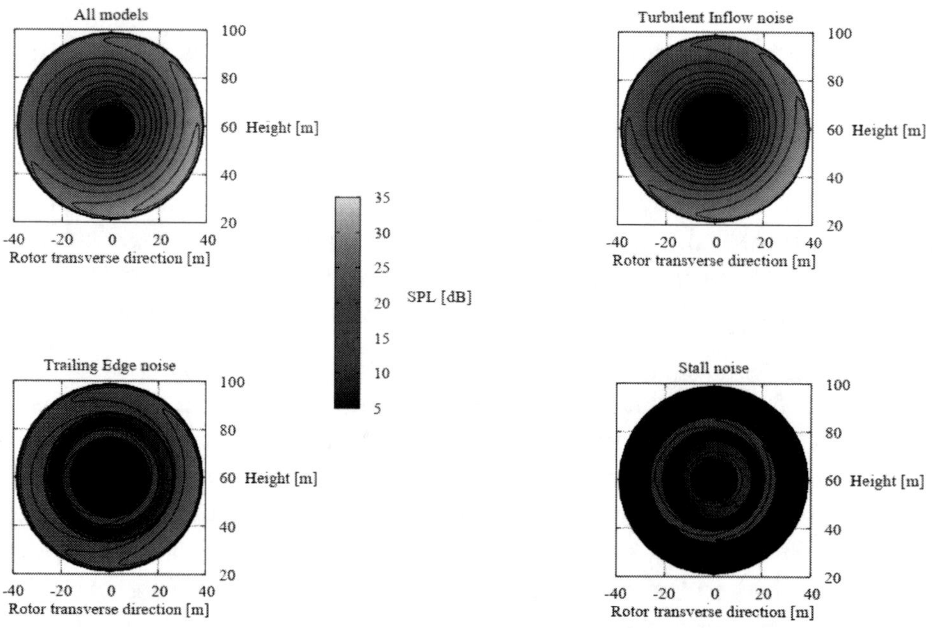

Fig. 8 Noise immission map at P7 (downwind of rotor on the ground) as seen from upwind the turbine - Test case 1.1.

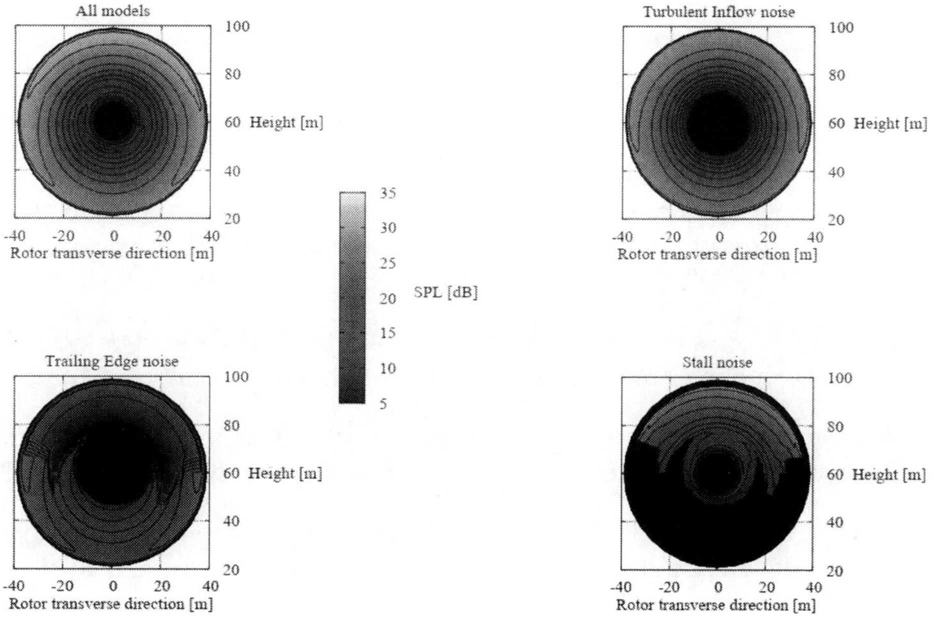

Fig. 9 Noise immission map at P13 (downwind of rotor on the ground) as seen from upwind the turbine - Test case 1.4.

4. Conclusion

A benchmark for wind turbine noise simulation codes comparison and validation is proposed. The main details of the numerical inputs to the various test cases have been presented (at least for the first round of this benchmark).

At the time of writing, none of the participants have had time to perform the required computations and send their results. However, it is expected that a number of participants will have conducted these before the start of the conference, and that comparisons and analysis of the results can be presented then.

A tentative timeline for the continuation of this benchmark follows. As mentioned above, it is expected that the results of the first round can be analyzed at the WTN 2019 conference, as well as during the next IEA Wind TCP Task 39 meeting which is planned as a side-event to the conference. Round #2 should be conducted during the second semester of year 2019, and Round #3 probably during the first semester of year 2020. However, conclusions from the initial analyses may alter this timeline. In particular, it may be necessary to come back on specific issues of the test cases if difficulties in understanding the results and their comparisons do arise.

Acknowledgments

This work was partly supported by the EUDP-2016 project Jr.nr. 64016-0056 funded by the Danish Energy Agency (Energistyrelsen). In addition, the present benchmark is conducted as a sub-task of the IEA Wind TCP research Task 39 (Quiet Wind Turbine Technology), as well as a part of IEA Wind

TCP research Task 29 Phase IV (Analysis of Aerodynamic Measurements).

References

[1] Arcondoulis, E. J. G., Doolan, C. J., Zander, A. C., and Brooks, L. A. (2010). A Review of Trailing Edge Noise Generated by Airfoils at Low to Moderate Reynolds Number. *Acoustics Australia*, 38(3):135–139.

[2] Herr, M., Ewert, R., Rautmann, C., Kamruzzaman, M., Bekiropoulos, D., Iob, A., Arina, R., Batten, P., Chakravarthy, S., and Bertagnolio, F. (2015). Broadband Trailing-Edge Noise Predictions - Overview of BANC-III Results. In 21th *AIAA/CEAS Aeroacoustics Conf. (Proc.)*, AIAA Paper 2015-2847, Dallas (TX).

[3] Hutcheson, F. V. and Brooks, T. F. (2006). Effects of Angle of Attack and Velocity on Trailing Edge Noise Determined Using Microphone Array Measurements. *International Journal of Aeroacoustics*, 5(1):39–66.

[4] IEC (2012). International Standard, Wind Turbines - Part 11: Acoustic Noise Measurement Techniques. IEC 61400-11, International Electrotechnical Commission, Geneva (CH). ISBN 978-2-83220-463-4.

[5] Madsen, H. A., Bak, C., Paulsen, U. S., Gaunaa, M., Fuglsang, P., Romblad, J., Olesen, N. A., Enevoldsen, P., Laursen, J., and Jensen, L. (2010a). The DAN-AERO MW Experiments. In 48th *AIAA Aerospace Sciences Meeting Including The New Horizons Forum and Aerospace Exposition (Proceedings)*, AIAA Paper 2010-645, Orlando (FL).

[6] Madsen, H. A., Bak, C., Paulsen, U. S., Gaunaa, M., Fuglsang, P., Romblad, J., Olesen, N. A., Enevoldsen, P., Laursen, J., and Jensen, L. (2010b). The DAN-AERO MW Experiments - Final report. Tech. Rep. Risø-R-1726(EN), Risø-DTU, Roskilde, Denmark.

INCE*EUROPE*

8th International Conference on Wind Turbine Noise
Lisbon – 12th to 14th June 2019

Session N

Aerofoil Design

Organised by INCE-Europe
mail@inceeurope.org
www.inceeurope.org

Supported by Portuguese Acoustical Society

spa Sociedade Portuguesa de Acústica

INCEEUROPE

8th International Conference
on
Wind Turbine Noise
Lisbon – 12th to 14th May 2019

Aeroacoustic Optimization of a Small Wind Turbine - Methodology and Experimental Validation

K. Volkmer, N. Kaufmann, T. Carolus
University of Siegen, 57076 Siegen, Germany

Corresponding author: kevin.volkmer@uni-siegen.de

Summary

Overall objective of the work presented is the aeroacoustic optimization of an existing small 3 m-diameter horizontal axis wind turbine, equipped with blades made of NREL's S83x airfoil sections. In this work, an acoustically improved airfoil shape was developed via a KAMRUZZAMAN/AMIET trailing edge noise prediction model that is integrated into an evolutionary algorithm. The acoustic and aerodynamic performance of both, the baseline and optimized airfoil segment, were validated experimentally in an aero-acoustic wind tunnel. In a second step new turbine blades were designed with the blade-element-momentum design method utilizing the new airfoil shape. A set of new blades was manufactured. The baseline and modified turbine were tested in a realistic free field environment, requiring statistical methods for the evaluation of power and sound data. The control of the turbine is such that in average it operates at design tip speed ratio (λ = 7.5). Off-design points are not considered in this study.

The aerodynamic noise of the wind turbine with the new optimized blades could be reduced by up to 4 dB in the frequency range from 1 to 6 kHz. The penalty is an increase of high frequency noise at around 7.5 kHz. This modification of the turbine's noise signature is clearly audible close to the turbine. In the distance of less than 100 m a listener does not perceive the high frequency component and experiences a substantial reduction of the perceived turbine noise. The overall aerodynamic performance of the new blades is equivalent or even slightly better as compared to the baseline.

Nomenclature

Symbols

$B_{1,2,3}$ = parameters in KAMRUZZAMAN model
C = chord length
C_d = section drag coefficient (for a two-dimensional airfoil)
C_f = skin friction coefficient (for a two-dimensional airfoil)
C_l = section lift coefficient
F = force per meter of airfoil section span
H = boundary layer shape factor
I = radiation integral
K = convective wavenumber
L = span
L_{Spp} = level of power spectral density of far field acoustic pressure
M = free stream Mach number
$OSPL$ = overall sound pressure level
$\overline{\overline{OSPL}}$ = spatial chordwise and angle of attack averaged overall far field sound pressure level
P = power
PT = penalty term
R_T = ratio of outer to inner boundary layer timescale
S_O = corrected observer distance
S_{pp} = power spectral density of far field acoustic pressure
St = Strouhal number
TI = turbulent intensity of ingested flow
U_c = convection velocity
c = wind speed
c_s = speed of sound
f = frequency
f_{obj} = objective function
f_{peak} = noise peak frequency
g = weighting factor
h = trailing edge thickness
k = acoustic wavenumber
l_y = spanwise correlation length
ls = length scale for BROOKS-POPE-MARCOLINI-model
m = number of design angles of attack
n = number of chordwise positions for wall pressure fluctuation predictions
n_{PT} = number of penalty terms
$ncrit$ = critical N-factor in XFoil
$nfft$ = number of discrete Fourier points
p = pressure
p,q = exponents in KAMRUZZAMAN model
u_τ = skin friction velocity
w = flow velocity
$x_{1,2,3}$ = Cartesian co-ordinates

Greek symbols

Φ_{pp} = power spectral density of surface pressure fluctuations

Π_c = COLE's wake strength parameter
α = effective angle of attack
β_c = CLAUSER's equilibrium parameter
δ^* = boundary layer displacement thickness
ε = lift-to-drag-ratio
θ = boundary layer momentum thickness
κ = frequency parameter
λ = tip speed ratio
ν = kinematic viscosity
ρ = air density
σ = standard deviation of wind speed
τ_w = wall shear stress
ω = angular frequency
ξ = chordwise airfoil coordinate starting at the leading edge

Subscripts
Spp = power spectral density of far field acoustic pressure
l = lift
d = drag
e = position at boundary layer edge
eff = effective
obj = objective
peak = noise peak
ref = reference
w = weighing
∞ = in free stream / at position far upstream

Abbreviations
BG = background
BPM = BROOKS-POPE-MARCOLINI
Exp. = experiment
Mic. = microphone
Pred. = prediction
TE = trailing edge

1. Introduction

Several targets are relevant for designing the twisted, tapered and carefully profiled blades of horizontal axis wind turbines. Naturally, the maximum energy output of a wind turbine is of primary concern and achieved by optimal aerodynamic design. Yet, mitigation of flow induced noise is on a par with efficiency, structural durability, cost, etc.

Overall objective of the work presented is the aeroacoustic optimization of an existing small 3 m-diameter horizontal axis wind turbine. This baseline wind turbine (design inflow velocity c_0 = 6 m/s, design tip speed ratio λ = 7.5) was designed via the classical GLAUERT/SCHMITZ blade-element-momentum method (see e.g. GERHARD et al. [1]). The blades were made of NREL's S83x airfoil sections which were designed as "a family of quiet, thick, natural-laminar-flow airfoils for 1 to 3 m-diameter turbines" (SOMERS [2]). In particular, the outward part of the blades is made from S834 airfoils with a thickness/chord ratio of 15 %. The baseline turbine has been operated for many years on the roof of a building at the University of Siegen, equipped with a large number of sensors, among those two stationary and turbine mounted weatherproof microphones and a phased microphone array, see appendix A. A pre-study (appendix B) confirmed that the main region of sound radiation is the outward part of the blade (but not the blade tip). This and the fact that the flow velocity around the blade unavoidably increases towards the tip led to the hypothesis that trailing edge (TE) noise is the dominant noise source. TE noise is caused by pressure fluctuations on the blade surfaces due to the turbulent boundary layer which develops in streamwise direction on the blade surfaces. The pressure fluctuations are scattered when convected past the trailing edge. Thus, details of the turbulent flow around each airfoil shaped blade section and - causally connected - the airfoil geometry itself becomes relevant.

Different techniques for trailing edge (TE) noise mitigation are currently under investigation or even state-of-the-art. State-of-the-art are rigid serrated trailing edges (HOWE [3] and GRUBER [4]), recently refined towards flexible serrations (KLEMME and CHRISTOPH [5]). Other research deals with blowing or sucking air through small slots at the blade surface for manipulating the boundary layer in the TE region (GERHARD [6] and WOLF et al. [7]). The design of per se low-noise airfoil shapes and their effect on the overall turbine noise emission has been tackled by wind turbine aerodynamicists but - to the knowledge of the authors - not much has been published (see e.g. LUTZ et al. [8]), OERLEMANS et al. [9]).

In this work, an acoustically improved airfoil shape is developed via a model-based optimization scheme. In a second step new turbine blades are designed utilizing the new airfoil shape while keeping all other design parameters of the turbine constant. A 2D airfoil segment and a set of new blades were manufactured. Eventually, both, an airfoil segment with the new profile and the complete turbine are investigated experimentally in an aero-acoustic wind tunnel and in situ for validation of the predicted noise reduction.

2. Methodology for airfoil shape optimization

Fig. 1 shows the structure of the airfoil optimization scheme developed. Core is a semi-analytical airfoil trailing edge sound prediction model. Boundary layer parameters, that are essential inputs into the acoustic model as well as the overall aerodynamic airfoil performance data (lift and drag as a function of angle of attack, i.e the polars), are determined by a panel method. Everything is integrated into an optimization loop, here an evolutionary algorithm. The result is an airfoil shape which should be optimal with respect to TE noise.

Advanced semi-analytical acoustic prediction schemes follow a two step procedure: Firstly, the prediction of the wall pressure fluctuations beneath the turbulent boundary layer and, secondly, the modeling of the scattering of these fluctuations at the trailing edge and the generation of sound. For the scattering at the trailing edge AMIET's far field noise model [10], extended by ROGER and MOREAU [11], is utilized in this work. The prediction of the wall pressure fluctuations is more challenging and much research has been undertaken in the recent years (see e.g. ROZENBERG et al. [12], CATLETT et al. [13, 14], HU and HERR [15], KAMRUZZAMAN et al. [16]).

Motivated by a comparative study by LEE and VILLAESCUSA [17] the model by KAMRUZZAMAN et al. [16] (in the following abbreviated as "KAMRUZZAMAN model") is employed.

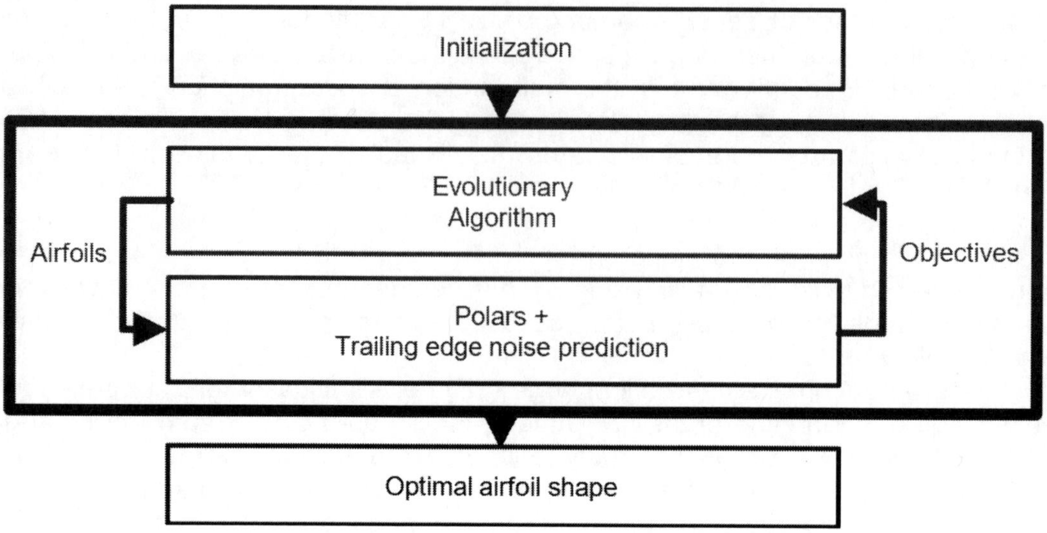

Fig. 1: Flow chart of airfoil shape optimization

2.1 KAMRUZZAMAN's wall pressure model

The KAMRUZZAMAN model [16] yields the wall pressure fluctuations Φ_{pp} in a single sided spectral formulation at a given chordwise position of the airfoil:

$$\frac{\Phi_{pp}(\omega) w_e}{\tau_w^2 \delta^*} = \frac{B_2 \cdot \left(\frac{\omega \delta^*}{w_e}\right)^2}{\left[\left(\frac{\omega \delta^*}{w_e}\right)^p + B_1\right]^q + \left[B_3 \left(\frac{\omega \delta^*}{w_e}\right)\right]^7} \tag{1}$$

w_e is the velocity at the edge of the boundary layer, τ_w is the wall shear stress, δ^* the displacement thickness, $\omega = 2\pi f$ the angular frequency, $B_1 = 0.27$, $B_3 = 1.15\ R_T^{-2/7}$, $p = 1.637$, $q = 2.47$,

$$R_T = \frac{\delta^* u_\tau^2}{w_e \nu} \tag{2}$$

with u_τ being the skin friction velocity and ν the kinematic viscosity, and

$$B_2 = 0.45\left[1.75\left(\Pi_c^2 \beta_c^2\right)^m + 15\right] \text{ with } m = 0.5\left(\frac{H}{1.31}\right)^{0.3}. \tag{3}$$

$H = \delta^*/\theta$ is the boundary layer shape factor with θ being the boundary layer momentum thickness. β_c is CLAUSER's parameter and in the KAMRUZZAMAN model calculated as

$$\beta_c = \left(\frac{1.7 + \sqrt{2/C_f}\,(1 - 1/H)}{6.1} \right)^2 - 1.81, \tag{4}$$

where C_f is the skin friction coefficient. The free stream pressure gradient is

$$\Pi_c = 0.277, \text{ if } \beta_c \leq -0.5,$$

and otherwise

$$\Pi_c = 0.8(\beta_c + 0.5)^{0.75}. \tag{5}$$

It is worth to note that, as in other recent wall pressure spectral models, inputs are local boundary layer parameters like the velocity at the boundary layer edge as well as parameters which take into account the upstream history of the boundary layer like the boundary layer displacement thickness (cp. ROZENBERG et al. [12]).

2.2 AMIET's trailing edge noise model

The computed wall pressure spectrum serves as an input into AMIET's trailing edge far field noise model [10]. The trailing edge is equivalent to the x_2-axis in spanwise direction, Fig. 2; x_1 is the chordwise coordinate and x_3 the direction perpendicular to the airfoil surface; the origin of the coordinate system is at mid span. The chord length of the airfoil is C, its span L, the flow angle of attack α and the free stream flow velocity w_∞ are also denoted. Additionally, a coordinate ξ is introduced as a chordwise airfoil coordinate starting at the leading edge.

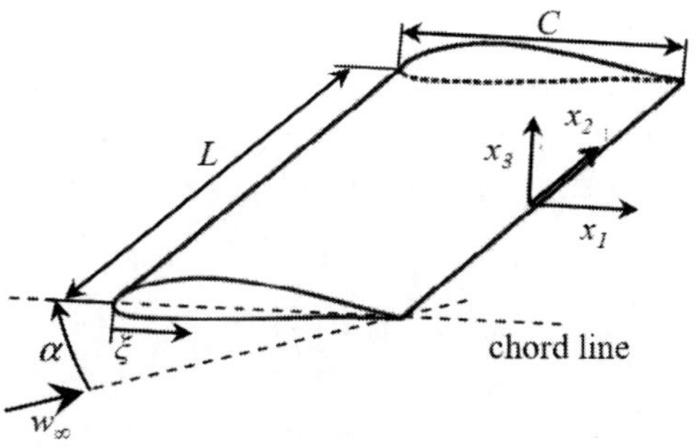

Fig. 2: Airfoil section and co-ordinate system; origin at mid span at the trailing edge

ROGER and MOREAU [11] added back scattering at the leading edge and a 3D extension to AMIET's original model [10]. The 3D extension is required for predictions at oblique observer positions with non-zero x_2-components. The first extension is utilized, whereas the second is implemented but not used as no oblique observer positions occur in the context of this work. The simplified model for large aspect ratios L/C is used. The spectral density of the acoustic far field sound pressure is

$$S_{pp}(\boldsymbol{x},\omega)=\left(\frac{x_3}{S_0^2}\right)^2\left(\frac{\omega C}{4\pi c_s}\right)^2 2\pi L\left|I\left(\frac{K}{C},\frac{kx_2}{CS_0},\kappa\right)\right|^2\frac{1}{\pi}\Phi_{pp}l_{x2} \tag{6}$$

with the speed of sound c_s, the observer distance, corrected for convection effects,

$$S_0=\sqrt{x_1^2+\left(1-M^2\right)\left(x_2^2+x_3^2\right)}, \tag{7}$$

the wall pressure spectrum Φ_{pp} and the spanwise correlation length l_y. The spanwise correlation length l_y is calculated as in TIAN et al. [18] with the help of the CORCOS model [19]. The airfoil response function I is - among others - a function of the acoustic wave number $k = \omega/c_s$ and the convective wave number $K = \omega/U_c$ with U_c being the convection velocity. Details about the airfoil response function can be found in [11]. Also super- and subcritical gusts are included which occur for $\kappa_2 > 0$ and $\kappa_2 < 0$, respectively, where

$$\kappa^2=\left(\frac{KM}{1-M^2}\right)^2-\frac{K_2^2}{1-M^2} \tag{8}$$

and the free stream Mach number $M = w_\infty/c_S$. As the large aspect ratio assumption is applied, the aerodynamic wave number in spanwise direction is $K_2 = kx_2/S_0$.

Note that a modification of the 2D airfoil shape (keeping all other dimensions of the airfoil section constant) will affect the wall pressure fluctuations only. Hence, minimizing trailing edge noise basically reduces to minimizing the near trailing edge wall pressure fluctuations. AMIET's acoustic model simply "transfers" these fluctuations into the acoustic far field.

2.3 Aerodynamic airfoil performance

A panel method as implemented in *XFOIL* by DRELA [20] is used to determine all overall airfoil performance parameters like pressure distribution, lift and drag, but also the boundary layer parameters. The section lift coefficient C_l is calculated by the lift force per unit length of airfoil section span F_l (for an infinite wing) and the section drag coefficient with the drag force per unit length of airfoil section span F_d (for an infinite wing) acting on a reference chord length C_{ref} = 1 m

$$C_l\equiv\frac{F_l}{0.5\rho w_\infty^2 C_{ref}} \tag{9}$$

$$C_d\equiv\frac{F_d}{0.5\rho w_\infty^2 C_{ref}} \tag{10}$$

with the free stream velocity w_∞ and the fluid density ρ. The lift to drag ratio is calculated as

$$\varepsilon\equiv\frac{C_l}{C_d}. \tag{11}$$

The boundary layer parameters are normalized with the chord length (as the boundary layer displacement thickness δ^* and momentum thicknesses θ) or with the inflow velocity (as the velocity at the boundary layer edge w_e).

2.4 Optimization scheme

The airfoil geometry, i.e. the shape of pressure and suction side, is parameterized via Bezier curves as in KAUFMANN [21].

An evolutionary algorithm is applied, since - in contrast to e.g. gradient based schemes - it is able to find the global optimum. It is described by BAMBERGER [22], who was inspired by the textbook from THÉVENIN and JANIGA [23].

The objective function is set as

$$f_{obj}\left(\boldsymbol{x}_{Bez}\right) = -\overline{\overline{OSPL_{obj}}} - \sum_{i}^{n_{PT}} g_i PT_i \,. \qquad \textbf{(12)}$$

$\overline{\overline{OSPL_{obj}}}$ is an overall sound pressure level averaged over a) the range of design angles of attack[1], b) several chordwise positions (for evaluation of wall pressure spectra) per angle of attack (see appendix D). This type of averaging ensures a non-local optimization. PT_i are n_{PT} penalty terms which are weighted by weighing factors g_i. The penalty terms are formulated such that

1. the thickness of the airfoil increases monotonically from leading edge to the maximum thickness and then decreases monotonically towards the trailing edge of the airfoil
2. extremely thin airfoils close to the trailing edge are avoided
3. separation is excluded all along the chord for the targeted angle of attack by sorting out airfoil shapes associated with boundary layer shape factors larger than 2.5 which is inspired by CLAUSER [25]
4. a decrease of lift below a set limit is avoided

Since the primary target of this work is a low noise airfoil, neither a restriction for the maximum airfoil thickness (often necessary for structural reasons) is implemented nor is the lift-to-drag ratio considered in the objective function.

The population size in the evolutionary algorithm is 150 individuals. Crossover between the parent parameters, i.e. Bezier points, followed by mutation is applied to produce the individuals of the next generation. The mutation range producing the following generation is confined to a percentage of the difference between the lowest and highest magnitude of the respective parameter in the previous generation. This percentage is 90% in the first 40 generations and 40% after. Only the best individual, the elite, will survive and be transferred to the next generation. Crossover is only performed within the first 40 generations. Convergence is reached if the Bezier points of the best individual vary less than 0.1% of the difference of the initial parameter limits within the last 40 generations.

---

[1] A turbine blade of a fixed pitch, variable speed turbine operates in a relatively small range of angles of attack (variations are mainly caused by atmospheric gusts and turbulences, GIGUERE and SELIG [24]). Hence, here the range of angles of attack is limited to $\alpha = 2° - 6°$.

3. Results

3.1 Airfoil

The typical value of the Reynolds number for this small-wind-turbine application is in the order of 200,000. Nevertheless, blade surface roughness due to operation in the real world and atmospheric free stream turbulence may cause a fully turbulent flow around the turbine blades and hence the airfoil segments (see eg. MCAULIFFE et al. [26]). Therefore, as the worst case scenario, the airfoil segment is optimized assuming of a fully turbulent boundary layer along pressure and suction side. To mimic a fully turbulent flow, tripping is applied at 2% and 5% of the airfoil section of chord length C on the suction and pressure side, respectively.

The geometries of the baseline airfoil and the obtained optimized airfoil KV200 are depicted in Tab. 1. It can be seen that the resulting maximum thickness of the KV200 is less than the one of the baseline profile, as the optimizer had the freedom to vary the thickness. The polars of lift, lift-to-drag-ratio as predicted from *XFOIL* and the far field sound from the combined KAMRUZZAMAN/AMIET/XFOIL trailing edge sound model are presented in Fig. 3. As compared to the S834 the lift polar of the new KV200 is similar whereas the lift-to-drag-ratio of the KV200 has even improved (although not included in the optimization objectives). One has to keep in mind that for the baseline and optimized airfoils - as stated earlier - tripping was applied very close to the leading edge in order to mimic a fully turbulent boundary layer. Of course, a S834 airfoil with natural transition would have a substantially better lift-to-drag-ratio. The far field sound pressure level *OSPL* was predicted based on the wall pressure fluctuations at ξ = 0.98 C and again the far field contributions of pressure and suction side are added. (This is different from the objective function that was based on the average of wall pressure spectra at different chordwise positions for a range of angle of attacks.) As the angle of attack is increased, the *OSPL* values of both airfoils increase as well. At the chosen chordwise position *XFOIL* predicts local flow separation on the suction side of the S834 close to an angle of attack of 6° and, hence, the KAMRUZZAMAN model [16] is not applicable anymore. This is already noticeable by the sharp drop of the *OSPL* values of the S834 in the lower diagram in Fig. 3 close to that angle. Surprisingly, the predicted *OSPL* of the optimized airfoil KV200 is up to 10 dB lower as compared to the baseline airfoil.

Tab. 1: Baseline and optimized airfoil

	Baseline airfoil
	Optimized for low far field noise

The baseline airfoil S834 and the optimized airfoil KV200 are also investigated experimentally in a small aeroacoustic wind tunnel as in appendix C. In order to perform a fair comparison both airfoils are compared at a same lift coefficient. Two different values of the lift coefficient, C_L = 0.38 and 0.64 are chosen for validation. In order to compensate for wind tunnel effects the geometrical angle of attack is set such that the static pressure distributions from the experiment and *XFOIL* agree.

The experimental results depicted in Fig. 4 deviate considerably from the predicted. Nevertheless, they demonstrate that the far field sound emitted by the optimized airfoil KV200 is reduced in the low frequency range of approx. 100 to 600 Hz by up to 4 dB as compared to the baseline S834. The BROOKS-POPE-MARCOLINI (BPM)-model [27] supports trailing edge noise to be expected at this frequency range (see appendix E). At medium frequencies a slight increase of noise is present for the optimized KV200 and the higher lift configuration. The experimental data in the high frequency range are irrelevant, since they are hidden in the empty wind tunnel background noise.

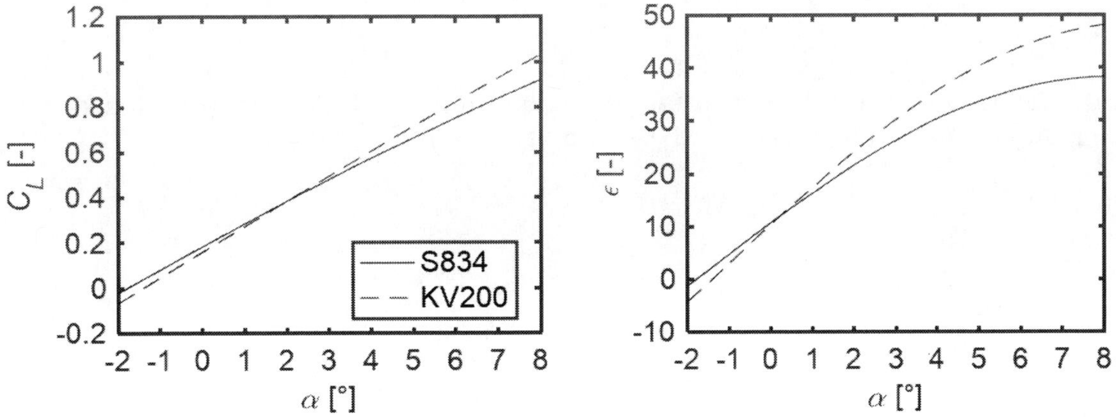

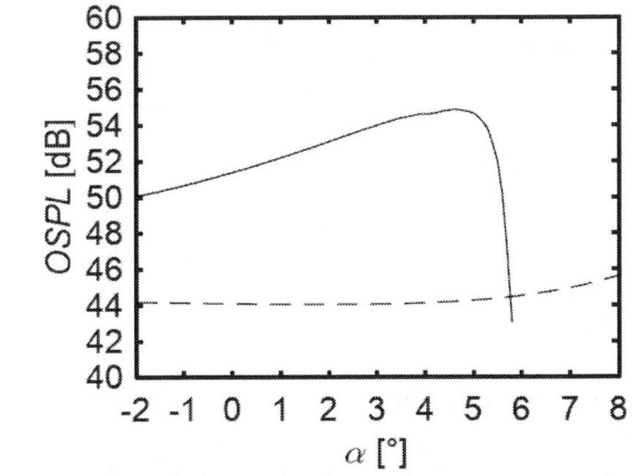

Fig. 3: Comparison of the S834 (baseline) and the optimized KV200 airfoil (predictions from *XFOIL* and the KAMRUZZAMAN/AMIET/XFOIL trailing edge sound model); Upper left: lift polar; upper right: polar of lift-to-drag-ratio; lower left: polar of overall sound pressure level at chosen far field position

3.2 Blade

Employing the acoustically optimized KV200 airfoil new blades B-KV200 for replacing the baseline blades are designed. Given the similar lift polars of the KV200 and the S834, the main blade parameters such as the spanwise distribution of chord length and pitch angle could have kept equal or similar. Fig. 5 shows the contour of the new blade and some section views.

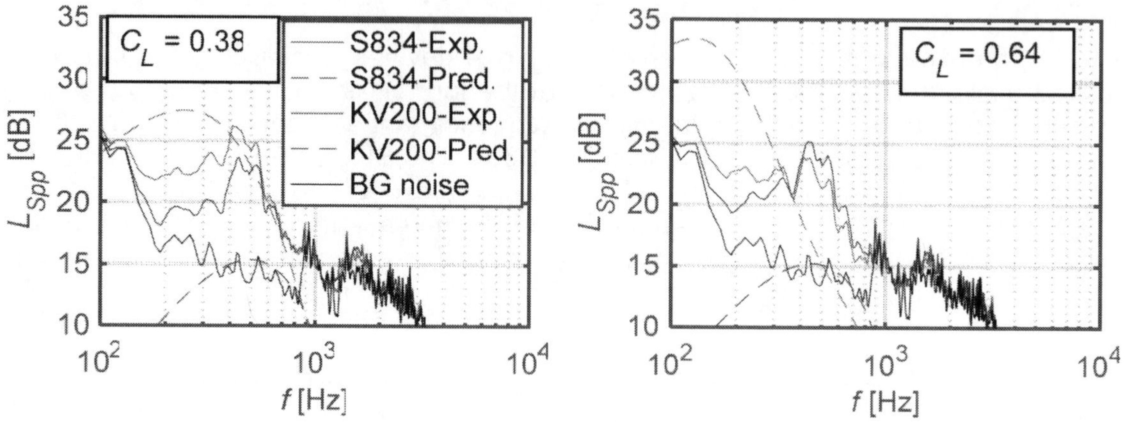

Fig. 4: Measured and predicted far field sound (the black lines indicate the empty wind tunnel measured background (BG) noise)

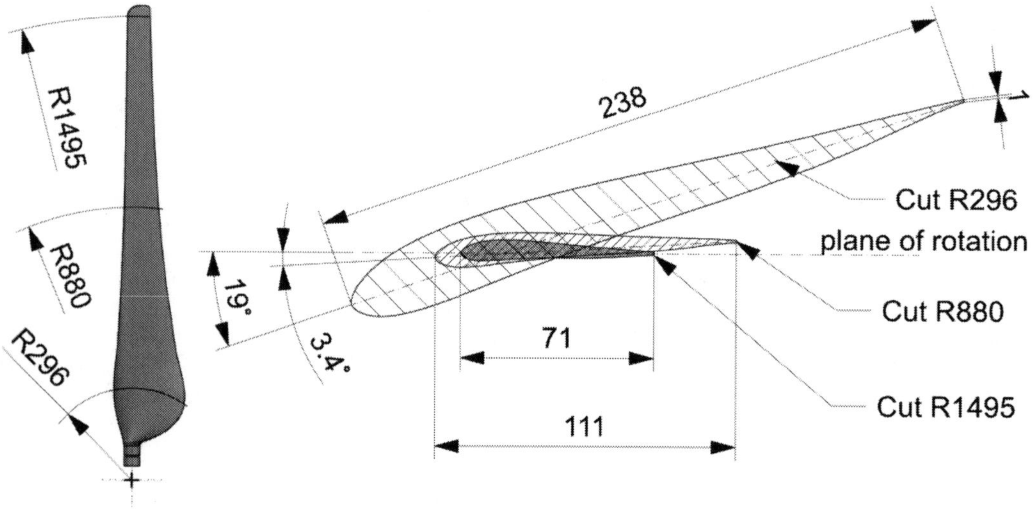

Fig. 5: Contour and section views of the new blade B-KV200

Two half shells were manufactured from carbon fibre reinforced raisin in a 3D milled mold, Fig. 6, and glued together. Special attention was given to thickness of the trailing edge. As compared to the baseline blade it was reduced from 1.5 mm ± 0.1 mm to 1 mm ± 0.1 mm. The purpose was to keep the ratio of trailing edge thickness to displacement thickness h/δ^* and hence the possible bluntness noise low (this is guided by the BPM-model [27], appendix E).

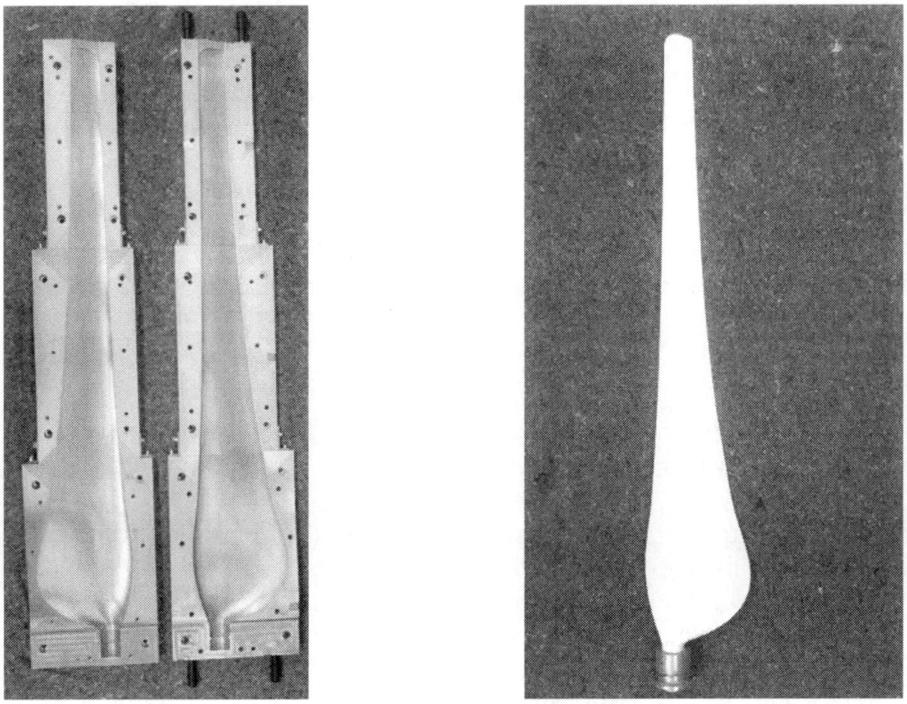

Fig. 6: The two halfs of the mold and the new blade B-KV200

3.3 Experimental turbine performance validation

As suspected from the description in appendix A the University of Siegen wind turbine test site is a low wind speed site in a non-ideal urban environment. The terrain is non-flat. The site is located at a large flat roof of the tallest building in the area, but, depending on the wind direction, with the potential of vortices developing on the roof edges. This requires measurements over several days or even months and a statistical analysis of all experimental data.

A prerequisite for a fair comparison of the baseline and the acoustically optimized turbine is that the power output of both turbines is similar. Fig. 7 (left) demonstrates that - although the measurements from both turbines were taken at different times - the probability distributions of the turbulent intensity are very similar. The turbulent intensity of the wind speed is

$$TI = \sigma_{ceff} / c_{eff} \tag{13}$$

with the standard deviation of the effective wind speed σ_{ceff}. Hence, TI is regarded as equivalent for both measurement campaigns. Fortunately, also the shaft power distribution of both turbines turned out to be very similar as seen in Fig. 7 (right). This means that the new blades B-KV200 are aerodynamically more or less equivalent.

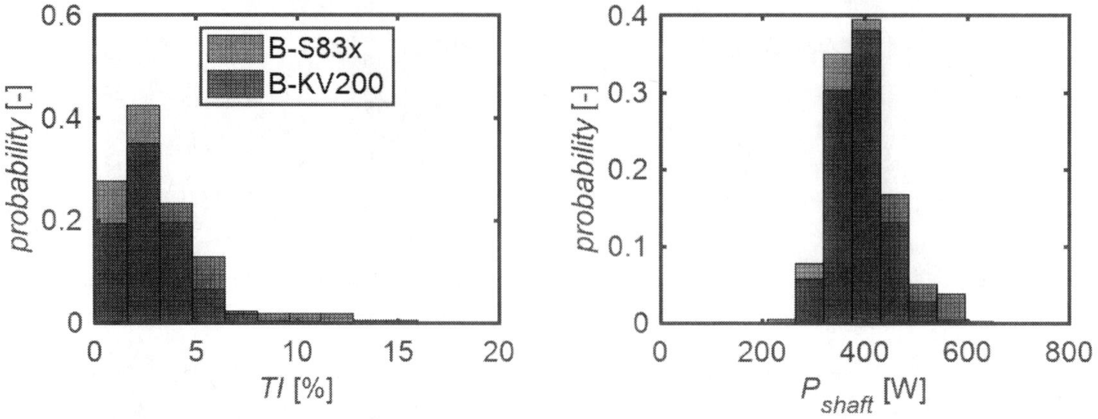

Fig. 7: Probability distributions of recorded 1 s time frames, left: turbulent intensity of upstream wind, right: turbine shaft power

The median sound pressure spectrum from both turbines from the off- and on-axis turbine mounted microphone is shown in Fig. 8. It should be noted that the spectra for the new blade appears less smooth as the number of available time windows utilized to build the median spectra was considerably smaller. The signal-to-noise ratio is superb. After all, the new blade design offers a noise reduction of up to 4 dB in the psychoacoustically important frequency band between 1000 and 4000 Hz. This is also clearly audible. Note that in this frequency range TE noise is expected considering the BPM-model (appendix E). The frequencies are higher compared to the wind tunnel measurements as the chord length of the blade is much smaller while the Reynolds number is the same. On the other hand a significant increase of noise is found towards higher frequencies. An explanation for this increase cannot be given yet.

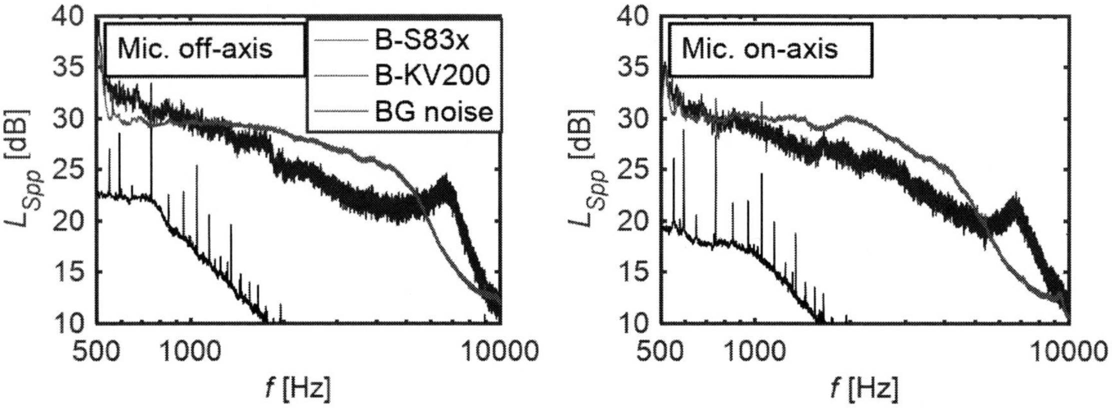

Fig. 8: Sound pressure at off- and on-axis turbine mounted microphone: Median power spectral density from baseline and new blade; BG = background noise

4. Summary and conclusions

The aerodynamic noise of a small wind turbine, originally equipped with NREL's S83x airfoil shaped blades, could be reduced by up to 4 dB in the frequency range from 1 to 6 kHz. The penalty is an increase of high frequency noise at around 7.5 kHz. An explanation for this increase cannot be given yet, in particular, since the wind tunnel measurements with the airfoil segment do not show this behaviour (even without tripping (not shown here)). Tip noise is not expected as the lift distribution is kept the same compared to the baseline turbine. Additionally, the frequency is too high for bluntness noise to be expected (appendix E). One has to keep in mind that the wind turbine blades - in contrast to the blade elements in the wind tunnel - are not tripped. Laminar effects could produce this excessive sound. The modification of the turbine's noise signature is clearly audible close to the turbine. In the distance of less than 100 m a listener does not perceive the high frequency component and experiences a substantial reduction of the turbine noise. The control of the turbine is such that in average it operates at design tip speed ratio (λ = 7.5). Off-design points are not considered in this study. The overall aerodynamic performance of the new blades is equivalent or even slightly better than the baseline.

This progress had been achieved by a new aeroacoustically optimized airfoil shape. By integrating a combined KAMRUZZAMAN/AMIET trailing edge noise prediction model into an evolutionary algorithm, a new airfoil shape was identified. Key idea was to "design" the turbulent boundary layer close to the trailing edge such that it promises lowest trailing edge sound. Despite the relatively poor performance of the semi-empirical (but computationally not costly) trailing edge sound prediction model, wind tunnel tests at a Reynolds number of 200,000 confirmed that the new airfoil produces less self-noise than the benchmark.

Acknowledgment

The authors thank Julien Christophe from the von Kármán Institute for Fluid Dynamics in Sint-Genesius-Rode, Belgium for discussions regarding AMIET's theory and supplying a code with AMIET's trailing edge noise model. Moreover, we gratefully appreciate the assistance by Dipl.-Ing. B. Homrighausen and Mr. J. Graf for designing the mold and manufacturing the blades.

Appendix A: The University of Siegen wind turbine test site

The University of Siegen wind turbine test site is located at a large flat roof of the tallest building in the area, but, depending on the wind direction, with the potential of vortices developing on the roof edges. The wind turbine to be tested is equipped with and surrounded by several sensors, Fig. A1:

- three wind speed cup anemometers
- three wind direction sensors
- rotational speed sensor
- voltage and current sensors
- rain sensor
- two permanent, weatherproof microphones (Microtech Gefell WME 952)

Fig. A1: Wind turbine and measurement stations M1, M2 and M3; each station is equipped with a cup anemometer and a wind vane type wind direction sensor

Cup anemometers and wind direction sensors are mounted on three stations positioned on a circle with a radius of 2.5 turbine diameters around the turbine (cp. IEC 61400-12-1 [29]) abreast of the nacelle. The effective wind speed c_{eff} relevant for the turbine is determined by spatially averaging all anemometer signals which are not in the wake of the turbine. The wake is assumed to cover a 74° cone (cp. IEC 61400-12-1 [29]) downstream of the turbine.

The positions of the two turbine mounted microphones are depicted in Fig. A2. The purpose of attaching the microphones to the permanently yawing turbine is to measure the sound always in well-defined positions relative to the turbine rotor. The possible draw back is that vibrations from the structure may contaminate the microphone signals. A separate study proved that this is irrelevant.

Within a special campaign, a microphone array consisting of 40 microphones irregularly located on an area of 0.75 m^2 (CAE Noise Inspector™) can be mounted on the wind turbine nacelle downstream of the rotor and even the wind vane, the center of the array being close to the rotational axis of the turbine. Standard beam forming is used for post-processing.

Data of all sensors - excluding the microphones - are sampled and saved to hard disc continuously 24 hours per day with a sampling rate of 10 Hz. Microphone data sampling with 40 kHz is only triggered when the wind speed is in an interesting range in order to save disc space. Data aquisition and storage is managed via an inhouse Labview™ based code.

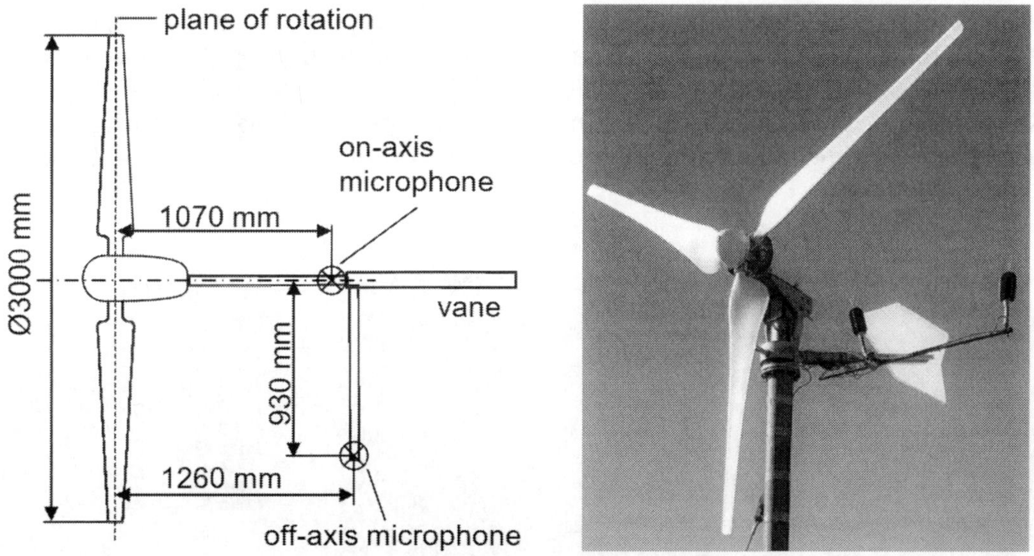

Fig. A2: Turbine mounted microphones (on-axis and one off-axis), left: schematic top-view diagram (not to scale)

As this roof-mounted wind turbine faces a rather dynamic wind speed it was decided to evaluate short time data windows of 1 s. To compensate for scatter of data through the dynamic wind speed and the time delay between measured wind speed and turbine speed, statistical methods are applied for evaluation. A 1 s window is as assigned "valid" if it satisfies following conditions:

1. The arithmetically averaged wind speed is in a range of ±1% of the design wind speed (6 m/s)
2. The arithmetically averaged tip speed ratio is in a range of ±1% of the chosen design tip speed ratio (7.5)
3. The electrical power is > 50 W in order to exclude recordings where the turbine was not yet feeding power into the grid (idling)
4. No precipitation (rain or snow)

Non-acoustic data are arithmetically averaged for each time window. The turbine shaft power P_{shaft} is derived from the calibrated electric generator power.

The acoustic data can be contaminated by background noise from the environment (road, train traffic etc.). Hence, for each valid time window a spectrum is evaluated. The resulting set of spectra are then reduced to one median spectrum. The median is insensitive against outliers (HANSEN et al. [30]). For evaluating the signal-to-noise ratio of the acoustic turbine signature, the background noise is evaluated in the same way but with the turbine stopped.

Appendix B: Prestudy: Detection of noise sources of the baseline wind turbine

Detection of noise sources via microphone array

Fig. B1 shows the acoustic map during operation of the turbine at around design wind speed. It can be seen that the outer span but not the tip of the turbine is the main noise source. As the microphone array (CAE Noise Inspector™) is mounted with its centre close to the axis, a rotationally symmetric acoustic picture would be expected, i.e. the acoustic picture should resemble a red donut. In contrast just the right side of the turbine on the picture in Fig. B1 is highlighted. A non-symmetrical distribution of the individual array microphones off the axis is thought to be the reason. However, the dominant noise source is still proved to be the outer blade region but not the tip.

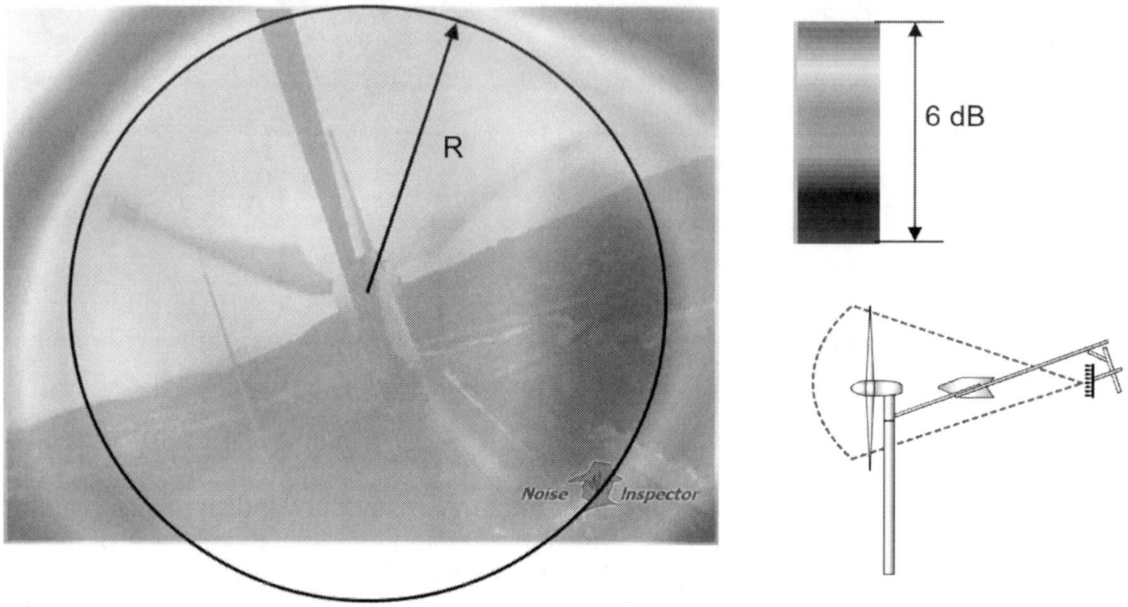

Fig. B1: Sound pressure map for the frequency range 630 - 10000 Hz. Lower right: Microphone array mounted downstream of the wind vane, the red frame shows the field of observation of the microphone array

Detection of noise sources via amplitude modulation

For our purpose the evidence of amplitude modulation can prove the blades to be the dominant noise source. Amplitude modulation is defined as the periodic varying noise levels a microphone receives under following conditions: a) the dominant noise source are the rotating blades and b) the microphone is located off-axis (rotational axis). The frequency of the variation is the blade passing frequency of the turbine. For the here considered small wind turbine it is around 15 Hz under design operating conditions. Those facts can be used to detect if blade noise is the dominant noise source in chosen frequency bands of the recorded noise spectrum. Here we slightly modify the method for evaluation of amplitude rating from the IOA Noise Working Group (BASS et al. [31]) to consider the higher blade passing frequency of the small baseline wind turbine:

A subset of 60 1 s time windows (instead of 100 10 s windows in the original IOA method)

1. Each time frame is band pass filtered to search for the occurrence of amplitude modulation in this frequency band (here: 1/3rd-octave bands instead of the fewer frequency bands in the original method)

2. Each band pass filtered time frame is subdivided into 1 ms intervals (instead of 100 ms in the original method)
3. Calculation of overall sound pressure level for each 1 ms time signal
4. Search for amplitude modulation occurring with blade passing frequency following the procedure of the IOA method
5. If in 50% or more of the 60 time frames occurrences of amplitude modulation are found in the noise records, amplitude modulation is proven according to the methodology.

Fig. B2 shows the found amplitude ratings for the whole spectrum in the recordings of the microphone off-axis. The fact that amplitude modulation was found for frequencies larger than 500 Hz proofs that the dominant noise source in the acoustic records has to be blade noise in these frequencies. For frequencies of 500 Hz or lower no amplitude modulation can be found indicating that other noise sources (e.g. vibrations or electrical generator) are dominant.

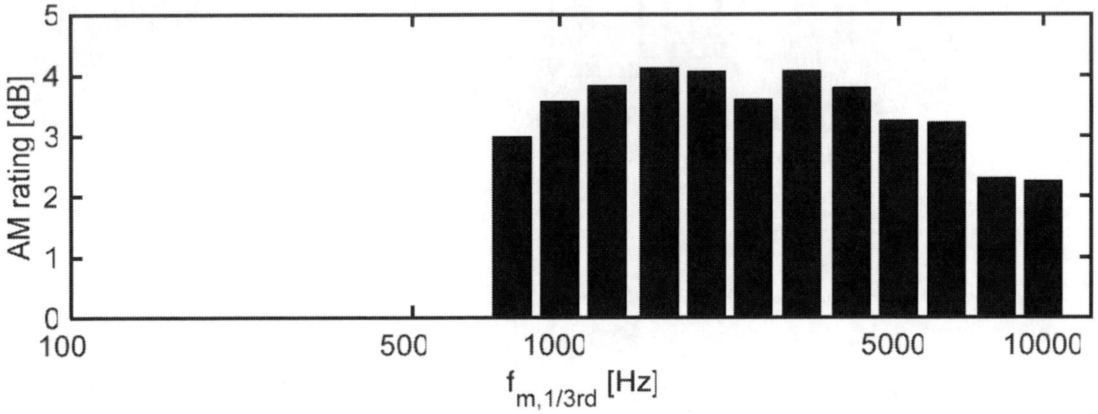

Fig. B2: Occurrence of amplitude modulation found in the noise spectrum of the microphone off-axis

Appendix C: Experimental wind tunnel setup

The aeroacoustic wind tunnel of the University of Siegen is schematically depicted in Fig. C1. It is designed to test airfoil sections with a chord length *C* of 200 mm and a span *L* of 266 mm. The airfoil sections are vertically mounted between side plates and 0.5 *C* in front of the nozzle exit.

Since the airfoil section is placed into a jet flow and not in a free flow, the geometric angle of attack used for mounting the airfoil section is not equal to the effective angle of attack α the airfoil experiences. Hence, for mounting the airfoil section an angle of attack correction by BROOKS [32] is utilized and checked via a comparison of the streamwise pressure distributions form measurements and a *XFOIL*-prediction [20].

The far field noise is measured with 12 microphones (1/2" Brüel & Kjaer™, type 4190) aligned on a sector of a circle with 0.5 m radius around the trailing edge (depicted in Fig. C2). Here we consider only the noise at microphone M10 on the pressure side is evaluated as it offers a high signal to background noise ratio. Jet shear layer diffraction errors were neglected because of the low flow velocities.

Time signals of far field microphones are captured for 30 s. The signal analysis is based on the power spectral density which was obtained by the function *pwelch* in MATLAB™ Vers. 9.1. The parameters chosen for *pwelch* are a *Hanning* window, the overlap between sections is 0.5 (MATLAB standard) and the number of discrete Fourier points (*nfft*) is equal to the section length. The spectra of each section is averaged to obtain the final spectrum with a frequency resolution of Δf = 10 Hz.

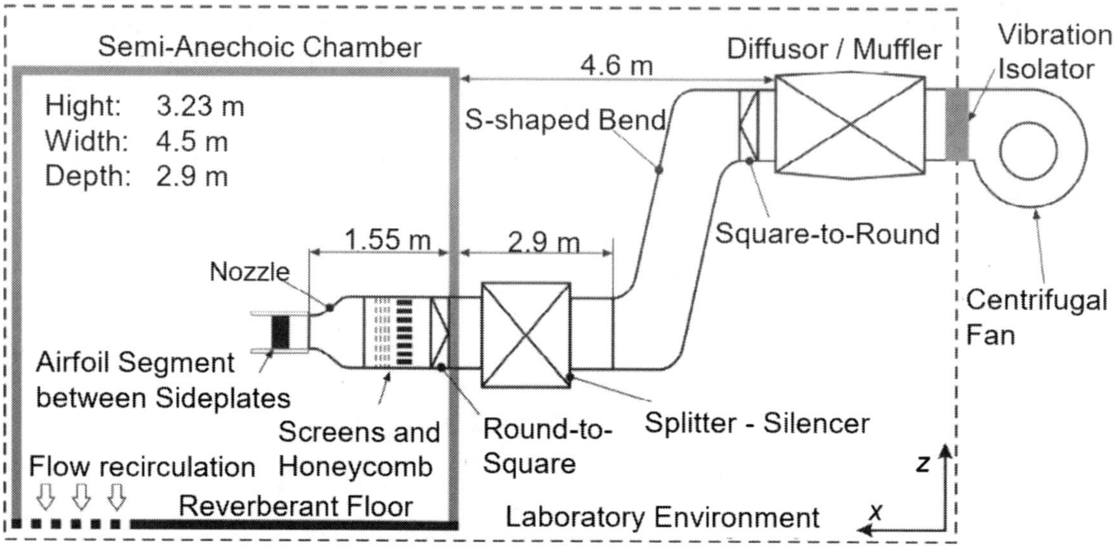

Fig. C1: Schematic layout of the aero-acoustic wind tunnel (not to scale)

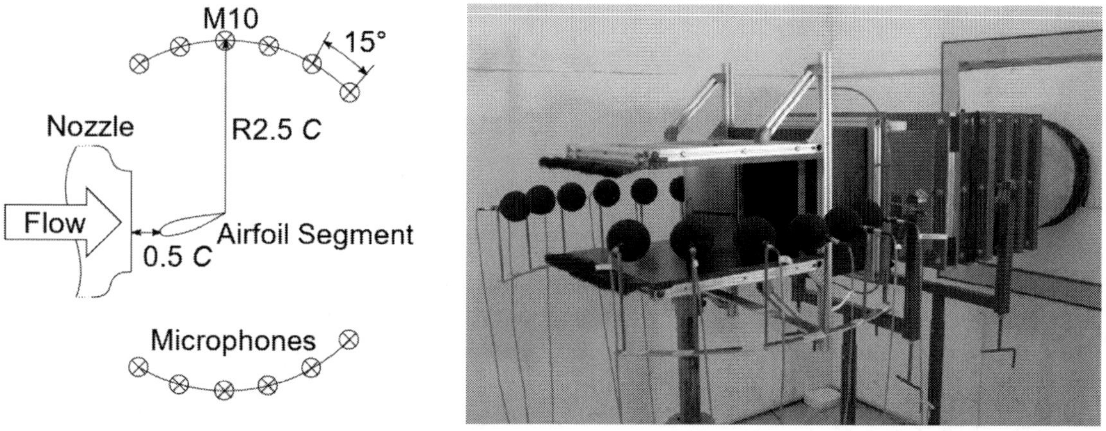

Fig. C2: Left: Schematic diagram of microphones around airfoil section (not to scale); right: Test rig and microphones in semi anechoic chamber

Appendix D: Definition of averages and levels

The power spectral density of the far field sound pressure S_{pp} calculated with eq. (6) can be written in terms of a level

$$L_{Spp} = 10\log_{10}\left(\frac{S_{pp}\cdot\Delta f_{ref}}{p_{ref}^2}\right) \quad \text{dB} \tag{D1}$$

with the reference sound pressure p_{ref} = $2{\cdot}10^{-5}$ Pa and the reference frequency bandwidth Δf_{ref} = 1 Hz. Integration of the power spectral density of the wall pressure fluctuations and far field sound pressure, here over the frequency range 100 Hz to 10 kHz, yields the overall far field sound pressure level *OSPL*

$$OSPL = 10\log_{10}\left(\frac{\int_{100\text{ Hz}}^{10000\text{ Hz}} S_{pp}(f)\,df}{p_{ref}^2}\right) \quad \text{dB.} \tag{D2}$$

The predicted power spectral density of the far field sound pressure S_{pp} is a function of frequency *f*, chordwise position ξ (that is used for the prediction of the wall pressure fluctuations) and the angle of attack α of the airfoil section. AMIET's model predicts the far field sound based on pressure fluctuations at one single chordwise position. Usually, this chordwise position is chosen to be *close to* the trailing edge but not exactly at the trailing edge. There is no accepted rule for choosing this position. Moreover, performing an optimization, based on one single chordwise position, would lead to a very local modification of the airfoil shape. Moreover, the sound pressure level for a complete range of angle of attacks shall be reduced. Therefore, a spatially (in stream or chordwise direction) and for a range of angle of attacks averaged overall far field sound pressure level is utilized. Averaging is done over *n* chordwise positions between 80% *C* to 98% *C* and *m effective* angles of attack between 2° and 6°:

$$\overline{\overline{OSPL}}_{obj} = 10\log_{10}\left(\frac{\frac{1}{m}\sum_{i=1}^{m}\left[\frac{1}{n}\sum_{j=1}^{n}\left[\int_{100\text{ Hz}}^{10000\text{ Hz}} S_{pp}(f,\xi_j,\alpha_i)\,df\right]\right]}{p_{ref}^2}\right) \tag{D3}$$

Appendix E: An estimate of trailing edge and trailing edge bluntness noise frequencies

A first estimate of the characteristic frequencies for trailing edge and trailing edge bluntness noise is obtained via the BROOKS-POPE-MARCOLINI (BPM) model [27] There, the spectral peak is at a Strouhal number value

$$St_{peak} = \frac{ls\cdot f_{peak}}{w_\infty} = 0.1, \tag{E1}$$

where f_{peak} is the peak frequency and *ls* a characteristic length scale. The length scale is either the boundary layer displacement thickness δ^* for trailing edge noise or the trailing edge thickness *h* for the trailing edge bluntness noise.

Within the wind turbine application the blade section at spanwise position r/R = 0.9 is considered to be the most dominant regarding noise emission. For a lift coefficient C_L = 0.64 (according to lift distribution on wind turbine) for both, the S834 and the KV200 airfoil shape at Re = 200,000, XFoil by DRELA [20] is utilized to predict the displacement thickness on the suction side. With these inputs the expected characteristic peak frequencies can be determined, Tab. E.

Tab. E: Predicted peak frequencies for blade sections according to the BPM-model

	Airfoil segment for wind tunnel test		Wind turbine	
	S834	KV200	S834	KV200
C	200 mm		80 mm	
h	0.4 mm		1.5 mm	1.0 mm
δ^* (XFoil predicted)	6.3 mm	2.4 mm	2.5 mm	1.0 mm
h/δ^*	0.06	0.16	0.6	1.0
f_{peak} (TE noise)	230 Hz	600 Hz	1600 Hz	4200 Hz
f_{peak} (TE bluntness noise)	3700 Hz		2700 Hz	4100 Hz

The predicted spectral peak of trailing edge noise is well within the dominant frequencies in wind tunnel and wind turbine measurements, and of trailing edge bluntness noise (if present) well within the dominant frequencies of the wind turbine measurements (while hidden in the background noise in the wind tunnel measurements (if present). According to the BPM-model bluntness noise is only relevant for a relative thickness of the trailing edge h/δ^* > 0.3. Hence, bluntness noise is not expected to be present at the wind tunnel tests but might be present at the wind turbine measurements. The trailing edge thickness of the new blade was reduced in order to keep h/δ^*. Indeed, there is no indication of bluntness noise in the turbine's acoustic signature (Fig. 8).

The airfoil optimization considered the worst-case scenario of fully turbulent boundary layers. However, the position of laminar to turbulent transition on the turbine blades is not fixed with tripping but natural. This has an effect on the predicted displacement thickness and, hence, both noise mechanisms. To evaluate this effect, the displacement thickness can also be predicted via XFoil when utilizing a *ncrit* (critical N-factor) of 1 instead of the here applied full-tripping. Essentially, this corresponds to natural transition. The effect on the predicted displacement thickness is insignificant (values not shown here) as compared to the effect of the geometrical airfoil shape.

References

[1] Gerhard, T., Sturm, M., Carolus, T. H.: Small Horizontal Axis Wind Turbine: Analytical Blade Design and Comparison With RANS-Prediction and First Experimental Data. Proceedings of the ASME Turbo Expo 2013

[2] Somers, D. M.: The S833, S834, and S835 Airfoils, National Renewable Energy Laboratory Report No. NREL/SR-500-36340, Golden, Colorado, USA 2005

[3] Howe, M. S.: Noise produced by a sawtooth trailing edge. The Journal of the Acoustical Society of America Vol. 90 (1991) No. 1, p. 482

[4] Gruber, M.: Airfoil noise reduction by edge treatments, Dissertation, Institute of Sound and Vibration Research, University of Southampton, 2012

[5] Klemme, T., Matthias, C. Rotorblatt, Windernergieanlage, Verfahren zum Herstellen eines Rotorblatts und zum Aufstellen einer Windenergieanlage, Offenlegungsschrift, DE102015012428A1, Senvion GmbH, 22297 Hamburg, DE, 25.09.2015, Deutschland

[6] Gerhard, T.: Umströmte Tragflügelsegmente: Auswirkungen des Hinterkantenausblasens auf Strömungsgrenzschicht und Schallemission, Ph.D. dissertation at the University of Siegen. Aachen, Shaker Verlag, 2016

[7] Wolf, A., Lutz, T., Würz, W., Krämer, E., Stalnov, O., Seifert, A.: Trailing Edge Noise Reduction of Wind Turbine Blades by Active Flow Control. Wind Energy Vol. 18 (2015) No. 5, p. 909–923

[8] Lutz, T., Herrig, A., Würz, W., Kamruzzaman, M., Krämer, E.: Design and Wind-Tunnel Verification of Low-Noise Airfoils for Wind Turbines. AIAA Journal Vol. 45 (2007) No. 4, p. 779–785

[9] Oerlemans, S., Schepers, J. G., Guidati, G., Wagner, S.: Experimental demonstration of wind turbine noise reduction through optimized airfoil shape and trailing-edge serrations - NLR-TP-2001-324, Nationaal Lucht- en Ruimtevaartlaboratorium (Ed.), National Aerospace Laboratory NLR, 2001

[10] Amiet, R. K.: Noise due to turbulent flow past a trailing edge. Journal of Sound and Vibration Vol. 47 (1976) No. 3, p. 387–393

[11] Roger, M., Moreau, S.: Back-scattering correction and further extensions of Amiet's trailing-edge noise model. Part 1 - Theory. Journal of Sound and Vibration Vol. 286 (2005) No. 3, p. 477–506

[12] Rozenberg, Y., Robert, G., Moreau, S.: Wall-Pressure Spectral Model Including the Adverse Pressure Gradient Effects. AIAA Journal Vol. 50 (2012) No. 10, p. 2168–2179

[13] Catlett, M. R., Forest, J., Anderson, J. M., Stewart, D.: Empirical Spectral Model of Surface Pressure Fluctuations beneath Adverse Pressure Gradients. 20th AIAA/CEAS Aeroacoustics Conference. 2014

[14] Catlett, M. R., Anderson, J. M., Forest, J. B., Stewart, D. O.: Empirical Modeling of Pressure Spectra in Adverse Pressure Gradient Turbulent Boundary Layers. AIAA Journal Vol. 54 (2016) No. 2, p. 569–587

[15] Hu, N., Herr, M.: Characteristics of Wall Pressure Fluctuations for a Flat Plate Turbulent Boundary Layer with Pressure Gradients. 22nd AIAA/CEAS Aeroacoustics Conference

[16] Kamruzzaman, M., Bekiropoulos, D., Lutz, T., Würz, W., Krämer, E.: A semi-empirical surface pressure spectrum model for airfoil trailing-edge noise prediction. International Journal of Aeroacoustics Vol. 14 (2015) No. 5-6, p. 833–882

[17] Lee, S., Villaescusa, A.: Comparison and Assessment of Recent Empirical Models for Turbulent Boundary Layer Wall Pressure Spectrum. 23rd AIAA/CEAS Aeroacoustics Conference. 2017
[18] Tian, Y., Cotté, B., Chaigne, A.: Wind Turbine Noise Modelling Based on Amiet's Theory. 5th International Meeting on Wind Turbine Noise. 2013
[19] Corcos, G. M.: Resolution of Pressure in Turbulence. The Journal of the Acoustical Society of America Vol. 35 (1963) No. 2, p. 192–199
[20] Drela, M.: XFOIL: An Analysis and Design System for Low Reynolds Number Airfoils. In: Mueller, T. J. (Ed.): Low Reynolds Number Aerodynamics. Proceedings of the Conference Notre Dame, Indiana, USA, 5-7 June 1989. Lecture Notes in Engineering, Vols.54. Berlin, Heidelberg: Springer 1989, p. 1–12
[21] Kaufmann, N.: Optimierung des hydrodynamischen Entwurfs einer Gezeitenströmungsturbine, Diplomarbeit, Lehrstuhl für Strömungsmechanik und Strömungsmaschinen, Technische Universität Kaiserslautern, 2015
[22] Bamberger, K.: Aerodynamic Optimization of Low-Pressure Axial Fans. Berichte aus der Strömungstechnik, Ph.D. dissertation at the University of Siegen. Herzogenrath, Shaker, 2015
[23] Thévenin, D., Janiga, G.: Optimization and Computational Fluid Dynamics. Heidelberg, Germany, Springer Verlag GmbH, 2008
[24] Giguere, P., Selig, M. S.: New Airfoils for Small Horizontal Axis Wind Turbines. Journal of Solar Energy Engineering (1998) No. 120(2), p. 108–114
[25] Clauser, F. H.: Turbulent Boundary Layers in Adverse Pressure Gradients. Journal of the Aeronautical Sciences Vol. 21 (1954) No. 2, p. 91–108
[26] McAuliffe, B. R., Yaras, M. I.: Transition Mechanisms in Separation Bubbles Under Low- and Elevated-Freestream Turbulence. Journal of Turbomachinery Vol. 132 (2010) No. 1, p. 11004
[27] Brooks, T. F., Pope, D. S., Marcolini, M. A.: Airfoil Self-Noise and Prediction, NASA Reference Publication 1218 (Ed.), NASA Langley Research Center, 1989
[28] Schepers, J. G., Wuerz, W., Curvers, A., Mantesanz, A., Oerlemans S., Garcillan, L., Braun, K., Fischer, M., Lutz, T., Koegler, K., Herrig, A., Maeder, T.: SIROCCO: Silent rotors by acoustic optimisation, ECN Energy research Centre of the Netherlands, 2007
[29] IEC 61400-12-1:2017. Wind energy generation systems - Part 12-1: Power performance measurements of electricity producing wind turbines. [online]: https://webstore.iec.ch/publication/26603
[30] Hansen, C. H., Doolan, C. J., Hansen, K. L.: Wind farm noise - Measurement, assessment and control. Wiley series in acoustics, noise and vibration. Hoboken, John Wiley & Sons Inc, 2017
[31] Bass, J., Cand, M., Coles, D., Irvine, G., Leventhall, G., Levet, T., Miller, S., Sexton, D., Shelton, J.: A Method for Rating Amplitude Modulation in Wind Turbine Noise - IOA Noise Working Group (Wind Turbine Noise), Amplitude Modulation Working Group, Final Report, INSTITUTE OF ACOUSTICS, 2016 (Version 1. Auflage)
[32] Brooks, T. F., Marcolini, M. A., Pope, D. S.: Airfoil Trailing-Edge Flow Measurements. AIAA Journal Vol. 24 (1986) No. 8, p. 1245–1251

INCEEUROPE

8th International Conference
on
Wind Turbine Noise
Lisbon – 12th to 14th June 2019

Scales of turbulence on a wind turbine leading–edge

Ronan Serré
Vestas Design Center, R. Lionesa 31, 4465–332 Leça do Balio, Portugal

Kristian Balschmidt Godsk
Vestas Wind Systems A/S, Hedeager 42, Aarhus, 8200, Denmark

Tomas Vronsky
Vestas Technology UK Ltd, Stag Lane, Newport, Isle of Wight, PO30 5TR, United–Kingdom

Summary

Turbulence intensity and integral length scale are two parameters driving leading–edge noise radiation. In the wind turbine industry, these quantities are generally estimated from empirical relations for atmospheric boundary layers but the turbulence actually experienced by the leading–edge during the blade rotation in wind turbine applications is believed to be more relevant to characterise leading–edge noise and to significantly differs from atmospheric boundary layer turbulence. This contribution investigates the influence of turbulence intensity and integral length scale estimations on wind turbine leading-edge noise prediction and compare the turbulence properties in the atmospheric boundary layer with what can be typically found on a wind turbine leading–edge. A more representative scaling of the inflow turbulence is proposed.

1. Introduction

Aerodynamic noise from a wind turbine is seen to originates predominantly from the trailing–edge as a result of the scattering of the boundary layer pressure waves and from the leading–edge as a response to the incoming turbulence [1]. Trailing–edge noise being gradually reduced, the contribution of the leading–edge noise is gaining importance and in addition, being in the low frequency region, survives on higher propagation distances, being less affected by air absorption [2]. Leading–edge noise modeling principally relies on the turbulence intensity and integral length scale of the incoming disturbances [3]. In the wind turbine industry, these quantities are generally estimated from empirical relations for atmospheric boundary layers [4]. Coupled with empirical models for leading-edge noise, the numerical predictions are generally found in close agreements with field test measurements. But such agreements may only be a consequence of the empirical models for leading–edge noise being in fact specifically calibrated for the empirical estimation of the turbulence scales in the boundary layer and this might not hold for analytical models for leading–edge noise as the turbulence that needs to be specified is relative to the chordwise component of the inflow velocity [3]. Tian and Cotté [1] used Amiet theory for leading–edge noise modeling coupled with empirical atmospheric boundary layer turbulence but with specific constants to adapt the estimated turbulence. And a wide variety is found in the turbulence intensity and integral length scale values generally adopted [1, 4, 5], indicating that no clear consensus exists in the community. Two remarks follow: *i)* if constants are freely modifiable to determine turbulence properties in atmospheric boundary layers might lead to a non–homogeneous use of the noise models and to inconsistant results among the community and *ii)* the turbulence characterics of the atmospheric boundary layer might not be the most relevant for leading-edge noise prediction in wind turbine application. The early work on wind turbine noise prediction of Glegg et al. [6] initiated a discussion on the relevance of the integral length scale in atmospheric boundary layer for leading–edge noise modeling of wind turbines. In this study, the orders of magnitude were judged too high and it was proposed to set the integral length scale as the blade chord. However, the relevance of these conclusions are also questionnable as the analysis was based on a wind turbine with a rotor diameter of 20 m, that is, quite different from the present generation of wind turbines. The pioneering theoretical work of Amiet [3] specificies that leading–edge noise radiation is directly dependant on the turbulence impinging the leading–edge, in the chordwise direction. It should then be preferable to avoid relying on empirical relations suited for atmospheric boundary layers and instead rely as far as possible on inflow aerodynamic data. This contribution presents an empirical and an anlytical model for leading–edge noise prediction and the impact of the turbulence properties through three empirical formulations for turbulence estimation compared with direct measurement of the turbulence intensity and integral length in the chordwise direction of a wind turbine leading–edge from field test measurements of the DanAero campaign [5, 7] and numerical modeling with an in–house numerical tool.

2. Numerical tool

2.1. Description

The numerical tool, hereinafter referred to as VTS, consists in one aeroelastic and one aeroacoustic tool. The aeroelastic tool models the dynamic behavior of horizontal axis wind turbines operating in specified wind conditions, such as simulated turbulent wind and wind shear [8]. The program runs in the time-domain producing time-series of loads and deflections. Aerodynamics is calculated using the BEM method with additional modeling to account for unsteadiness, dynamic wake, dynamic stall and yaw and/or tilt conditions. The aeroacoustic tool uses the aerodynamic inflow prediction to predict trailing–edge and leading–edge noise of isolated aerofoils or wind turbine blades from empirical [4, 9] or analytical models [10], according to the IEC standard [11]. The attenuation of the acoustic energy in atmospheric propagation by air absorption is modeled following the guidelines discussed by Salomons [2]. This contribution aims at addressing the leading–edge noise modeling but in field test measurements of wind turbine noise, the different contributions cannot be isolated. For this reason, when numerical predictions are compared to field test measurements of wind turbine noise, empirical modeling of the trailing–edge noise [9], referred to as "TBL–TE", is simply added for information. Only leading–edge noise empirical and analytical models are presented below.

2.2. Empirical modeling for leading–edge noise

The expression for the sound pressure level SPL in the empirical model reads as [4, 12]:

$$SPL = 10\ \log_{10}\left[\overline{D}_l \rho_0^2 c_0^2 \Lambda_{uu} \frac{L}{r_{obs}^2} M^3 T_i^2 \hat{k}^3 \left(1+\hat{k}^2\right)^{-7/3}\right] + 10\ \log_{10}\left(\frac{K_c}{1+K_c}\right) + 58.4 \tag{1}$$

with ρ_0 as the uniform medium density, c_0 the speed of sound, r_{obs} the observer distance, $M = U_0/c_0$ the Mach number, U_0 being the mean inflow velocity and L the aerofoil or blade element span length. The inflow turbulence parameters, T_i and Λ_{uu} are in this context defined as the turbulence intensity and integral length scale of the longitudinal velocity fluctuations in the atmospheric boundary layer, estimated by an empirical model described by Counihan [13], although it will hereinafter referred to as "Zhu" model. The model uses a constant defined as the surface roughness that is required to be ajusted by the user. Throughout this contribution, when referring to the Zhu model, the surface roughness is set as 0.01 m following Zhu et al. [4]. The function $\overline{D}_l$ is a low frequency directivity function as expressed by Brooks et al. [9]. with $\hat{k} = \pi f C/U_0$ a normalised wavenumber (with f the frequency and C the aerofoil or blade element chord length) and K_c a low frequency correction factor.

$$K_c = \frac{10M\hat{k}^2/\beta^2}{\left(2\pi\hat{k}/\beta^2\right) + \left(1+2.4\hat{k}/\beta^2\right)^{-1}} \tag{2}$$

with $\beta^2 = 1 - M^2$, the Prandtl–Glauert compressibility coefficient.

2.3. Analytical modeling for leading–edge noise

The expression for the instantaneous power spectrum density of the acoustic pressure S_{pp} in the analytical model, reads as [10]:

$$S_{pp}\left(\overrightarrow{x},\omega\right) = \left(\frac{\rho_0 \omega x_3}{2c_0 S_0^2}\right)^2 \pi U_0 \frac{L}{2} \phi_{ww}\left(k_1,k_2\right) \left|\mathcal{L}_{TI}\left(x_1,k_1,k_2\right)\right|^2 \tag{3}$$

with $\omega = 2\pi f$ the pulsation, $\overrightarrow{x}$ the observer coordinates in the frame of reference expressed in reference [9] (x_1 refers to the chordline, x_2 to the spanline and x_3 to the normal direction). The function S_0 accounts for the convective observer distance $S_0^2 = x_1^2 + \beta^2\left(x_2^2 + x_3^2\right)$. The function $\mathcal{L}_{TI}$ is an aeroacoustic transfer function described in details in reference [10]. The function ϕ_{ww} is the two–dimensionnal wavenumber spectrum of the upwash velocity fluctuations and can be estimated by Kármán model [3]:

$$\phi_{ww}\left(k_1,k_2\right) = \frac{3}{4\pi} T_i U_0 \Lambda_{uu}^2 \frac{\Lambda_{uu}^2\left(k_1^2+k_2^2\right)}{\left[1+\Lambda_{uu}^2\left(k_1^2+k_2^2\right)\right]^{5/2}} \tag{4}$$

with k_1 and k_2 the wavenumbers in the chordwise and spanwise directions, respectively. It relies on the turbulence intensity T_i and the integral length scale Λ_{uu}, here, specifically based on the chordwise velocity component [3].

2.4. Aerofoil noise assessment

Empirical and analytical modeling are first applied to estimate leading–edge noise on a NACA–0012 isolated aerofoil in a wind tunnel to replicate a study published by Paterson and Amiet [14] for a range of Reynolds number based on the chord length varying from 0.5×10^6 to 2.5×10^6 (corresponding to Mach numbers ranging from 0.12 to 0.5), at zero degree angle of attack. The turbulence intensity and integral length scale are imposed as 4 % and 0.03 m, respectively [14]. Figure 1 presents the sound pressure level spectra SPL, in dB. The empirical modeling (dashed lines) underestimates significantly the sound pressure levels at all frequencies. The analytical modeling (solid lines) is in better agreement with the measurements (symbols), some singularities notwithstanding. In the high frequency region however, the sound pressure levels tend to be overestimated.

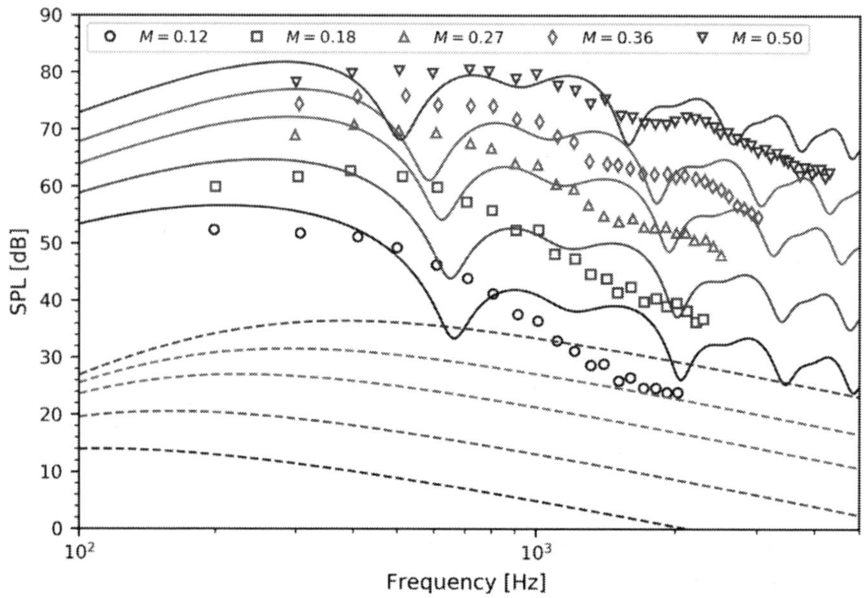

Fig. 1 Aerofoil noise prediction on a NACA–0012. Measurements from Paterson and Amiet [14] (symbols), empirical modeling (dashed lines) and analytical modeling (solid lines). The colours indicate the Mach number.

2.5. Wind turbine noise assessment

Empirical and analytical modeling are then applied to estimate the acoustic power radiated by a representative wind turbine and compare with field test measurements. The Vestas V80 is a 2 MW horizontal axis wind turbine with hub height of 60 m and a 40 m blade span. Here, the turbulence intensity and integral length scale are estimated as proposed in reference [4], in the empirical modeling framework, in the context of atmospheric boundary layer. The total acoustic power L_wA is depicted in figure 2 for a range of wind speed at hub height V_{hub} and the corresponding sound power level spectrum $L_w A f$ at peak noise ($V_{hub} = 11$ m/s) in figure 3. An estimation of trailing–edge noise from empirical modeling [9] is added to the figure, to account for the fact that the contributions from the field test measurement cannot be isolated. In figure 2, the empirical modeling gives a more realistic prediction, slightly lower than trailing–edge noise noise and measurements while the prediction from analytical modeling is clearly overestimated. The same observation is made in the frequency domain (figure 3) with the low frequency region being clearly overestimated by the analytical modeling.

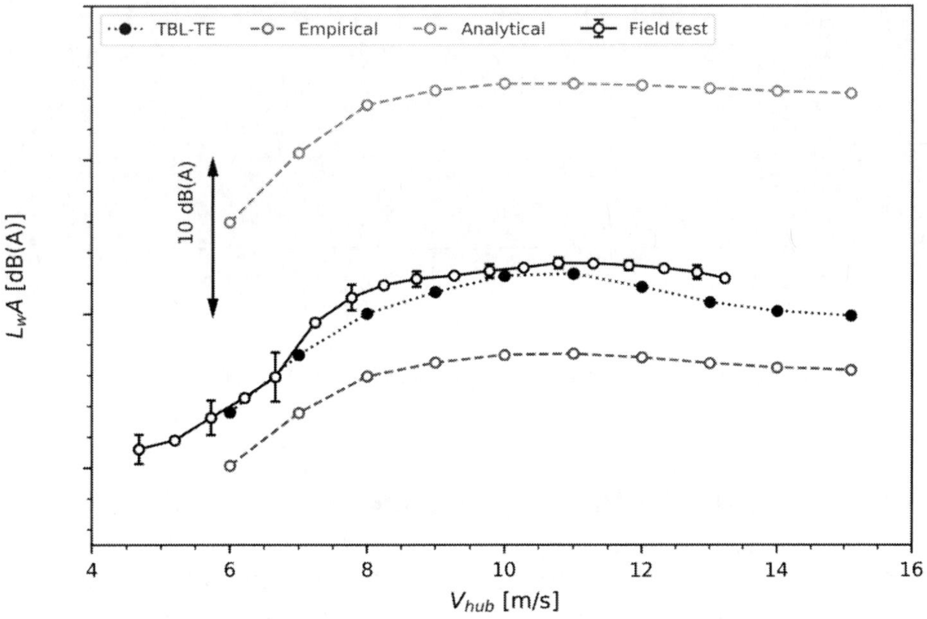

Fig. 2 Acoustic power. Atmospheric boundary layer turbulence. V80 wind turbine.

2.6. Preliminary discussion

For the leading–edge noise prediction of an isolated aerofoil, the analytical modeling is in better agreement with the measurements than the empirical modeling. However, for the leading–edge noise prediction of a wind turbine, the empirical modeling provides a better agreement. In the context of inflow turbulence, it is observed that the analytical modeling provides a realistic prediction when the inflow turbulence is prescribed but fails when the inflow turbulence is estimated by an atmospheric boundary layer model. In other words, the empirical modeling might give satisfaction only as a consequence of being calibrated to a specific inflow turbulence and this specific inflow turbulence might not be relevant to describe the actual inflow turbulence experienced by a wind turbine leading–edge, thus explaining the inaccuracy of the analytical modeling on wind turbine noise. Consequently, there is a need to assess the typical values for turbulence actually impinging a wind turbine leading and propose new inflow turbulence modeling in order to develop the use of analytical modeling for leading–edge noise prediction in wind turbine applications.

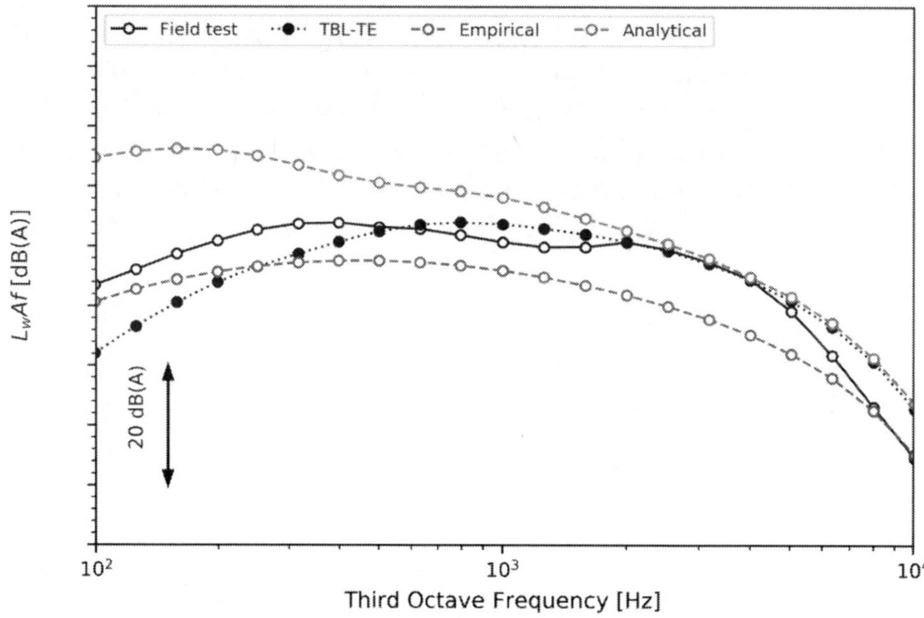

Fig. 3 Acoustic power spectra. Atmospheric boundary layer turbulence. V80 wind turbine. $V_{hub} = 11$ **m/s.**

3. Direct estimation of the inflow turbulence

Aerodynamic data from a field test measurement campaign, referred to as DanAero [7], and VTS are used to estimate the typical values for the turbulence actually impinging a wind turbine leading–edge. In the DanAero experiment, aerodynamic inflow conditions were measured for several spanwise locations on a 2.3 MW NM80 horizontal axis wind turbine with hub height of 57 m and a 40 m blade span, for a wind speed measured at hub height of approximately 6.2 m/s. The objective is to have a first estimation of the order of magnitude of the inflow turbulence characteristics from two different sources, through field test measurements and aerodynamic modeling to assess how far it might deviate from the inflow turbulence values estimated with an empirical atmospheric boundary layer approach. The Vestas wind turbine selected for VTS computations, the V80, is relatively similar to the wind turbine instrumentalised in the DanAero experiment. For VTS computations, the wind shear $V(z)$ is specified to fit the measured wind velocity profile, from IEC standard [8]:

$$V(z) = V_{hub} \times \left(\frac{z}{z_{hub}}\right)^{\alpha} \tag{5}$$

with z the vertical height above ground and z_{hub} the hub height. The exponent α determines the velocity profile and is imposed as $\alpha = 0.08$. The wind shear velocity profiles, as recommended by the IEC standard, measured from VTS and in the DanAero experiment are shown in figure 4. The atmospheric conditions are then consistent between the field test measurements and the numerical modeling. In the present contribution, one sensor location is used, at 80 % of the blade span ($r/R = 0.8$) to estimate the turbulence intensity and integral length scale of the chordwise velocity component and compared with the same estimation from VTS on the V80 (at the same relative spanwise location). The chordwise component of the inflow velocity u is retrieved from the DanAero

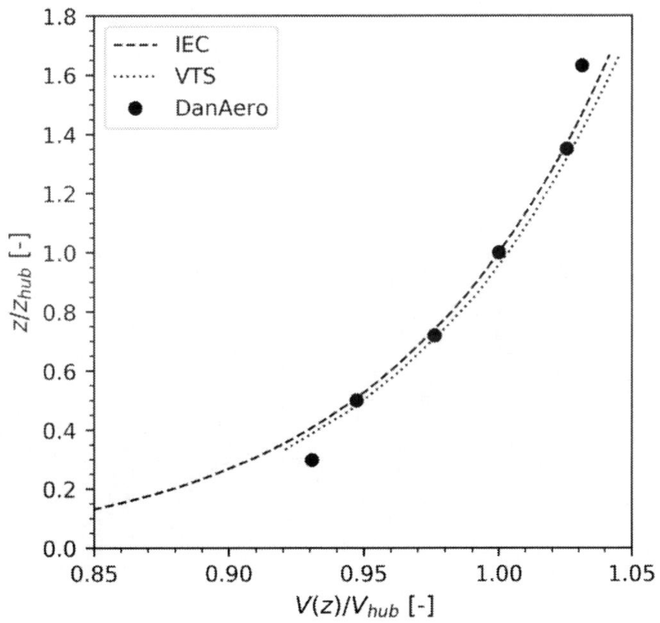

Fig. 4 Wind shear velocity profiles as specified by IEC standard, imposed in VTS and measured in the DanAero experiment.

experiment (with a sampling frequency of 35 Hz) and from VTS (with a sampling frequency of 25 Hz). In both cases, the acquisition time is 600 s (10 minutes). The velocity signals are considered made of a mean part (phase averaged) and a fluctuating part as $u = \overline{u} + u'$ with the overline denoting the mean part and the prime the fluctuating part. The turbulence intensity is defined as [15]

$$T_i = \frac{\sigma_{u'}}{\overline{u}} \tag{6}$$

with $\sigma_{u'}$ the standard deviation of the chordwise velocity fluctuations. The integral length scale is defined as the length scale over which the most energetic eddies are correlated. Following the estimation procedure from Thacker et al. [16], the chordwise velocity autocorrelation function is defined as

$$R_{uu}(\tau) = \frac{\overline{u'(t) \cdot u'(t+\tau)}}{\sqrt{\overline{u'(t)^2} \cdot \overline{u'(t+\tau)^2}}} \tag{7}$$

where τ is a variable time step in the signal history. The integral time scale follows as the time integration until the first zero–crossing of the autocorrelation:

$$t_{uu} = \int_0^\infty R_{uu}(\tau)\, \mathrm{d}\tau \tag{8}$$

then leading to the integral length scale definition as

$$\Lambda_{uu} = \overline{u} \times t_{uu} \tag{9}$$

The autocorrelation function and the corresponding integral length scale are estimated every 1 s from 100 s to 500 s in the available time signals. The autocorrelation functions estimated from DanAero and

VTS are presented in figure 5, averaged over the time blocks, with the variable time step τ normalised by the corresponding rotation period T. The standard deviation is relatively low, indicating that there might be no significant changes in the integral length scale with the blade azimuthal position in the rotation cycle.

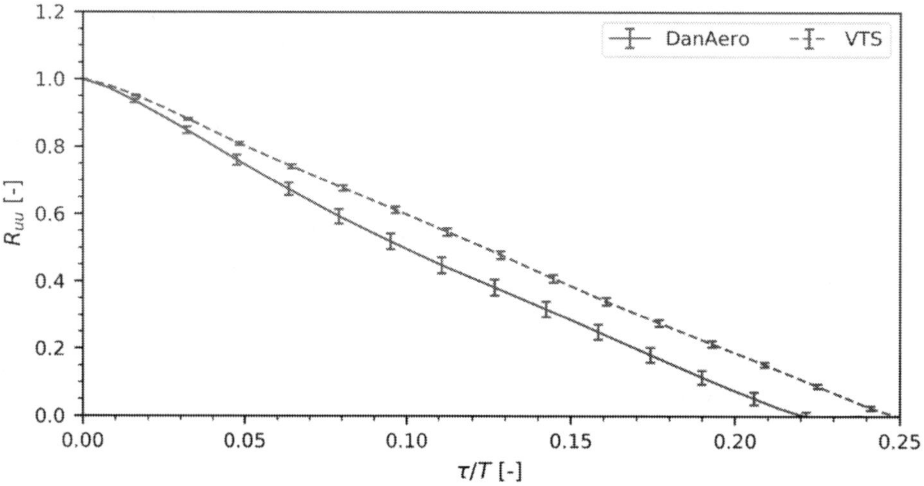

Fig. 5 Autocorrelation functions estimated from measurements in the DanAero experiment and VTS. The time is normalised by the rotation period.

4. Comparison with empirical estimations for turbulence

The values for the turbulence intensity and integral length scale directly estimated from the DanAero experiment and VTS are compared to values estimated from three empirical formulations for atmospheric boundary layers, applied with the same conditions used in VTS, on the V80 wind turbine. The three empirical atmospheric boundary layer models are: *i)* an aforementioned model referred to as "Zhu" from Zhu et al. [4], *ii)* a model referred to as "Tian" from Tian and Cotté [1] and *iii)* a model referred to as "Bertagnolio", following an empirical relation described in reference [5]. The same time discretisation by blocks of 1 s between 100 s and 500 s previously used is also applied to the empirical formulations in order to see the evolution of the inflow turbulence with time and in particular, to notice any effect of the blade azimuthal position in the rotation cycle. Figure 6 and figure 7 respectively present the turbulence intensity and the integral length scale from the DanAero experiment, VTS and the three empirical formulations along one rotation period. Regarding the turbulence intensity (figure 6), only the Zhu formulation predicts an evolution with the blade azimuthal position, although relatively small. The empirical formulations predict a turbulence intensity significantly higher than the direct estimation from the DanAero experiment and VTS (between 10 % and 15 % for the empirical estimations and around 2 % for the direct estimations). Regarding the integral length scale (figure 7), the three empirical formulations predict an evolution with the blade azimuthal position and here, relatively significant. However, from the direct estimations, the integral length scale does not seem to be affected by the blade azimuthal position. The striking fact is however the order of magnitude of the different length scale estimations: a few dozen meters from the direct estimations and a few hundred meters from the empirical estimations. The mean values (and the standard deviation associated with

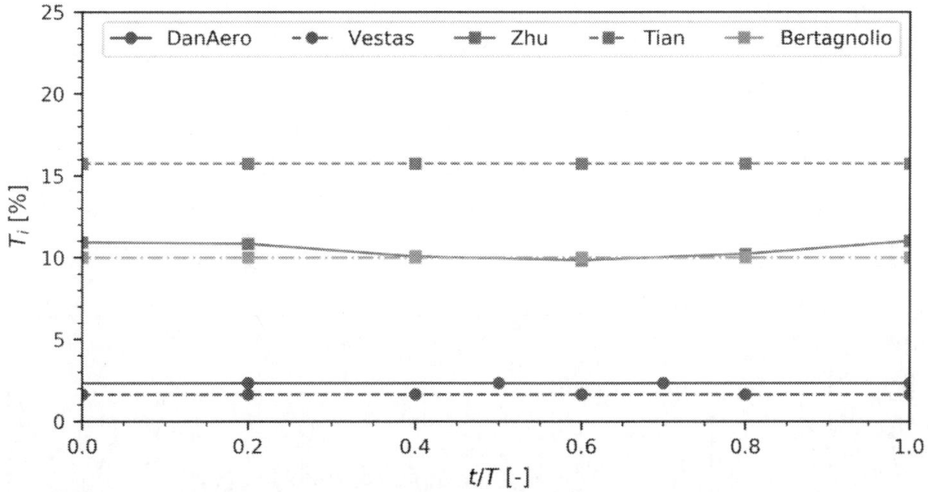

Fig. 6 Turbulence intensities. The time is normalised by the rotation period.

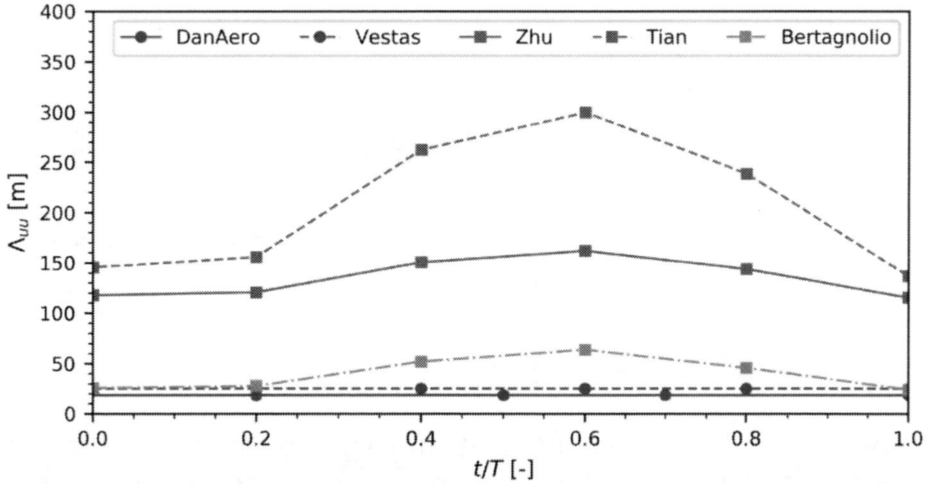

Fig. 7 Turbulence integral length scales. The time is normalised by the rotation period.

the integral length scale) are summarized in table 1.

	DanAero	VTS	Zhu	Tian	Bertagnolio
T_i [%]	2.3	1.6	10.4	15.7	10.0
Λ_{uu} [m]	18.1 ± 0.7	25.0 ± 0.3	139.3 ± 17.4	221.4 ± 61.1	43.3 ± 14.6

Table 1 Mean values with standard deviation for the turbulence intensity T_i and integral length scale Λ_{uu} estimations.

5. Influence of the turbulence inflow conditions on noise modeling

The impact of the different inflow turbulence estimations on the leading–edge noise prediction is now discussed. The total acoustic power is depicted in figure 8 and the corresponding sound power level spectrum at peak noise ($V_{hub} = 11$ m/s) in figure 9 from empirical modeling. An estimation of trailing–edge noise from empirical modeling [9] is added to the figure. In both figures it is observed that if the direct estimations for the inflow turbulence is used, the empirical leading–edge noise modeling underestimates the field test measurements but manages to yield relative agreements with empirical inflow turbulence estimations. In the context of analytical modeling, Roger and Moreau [10]

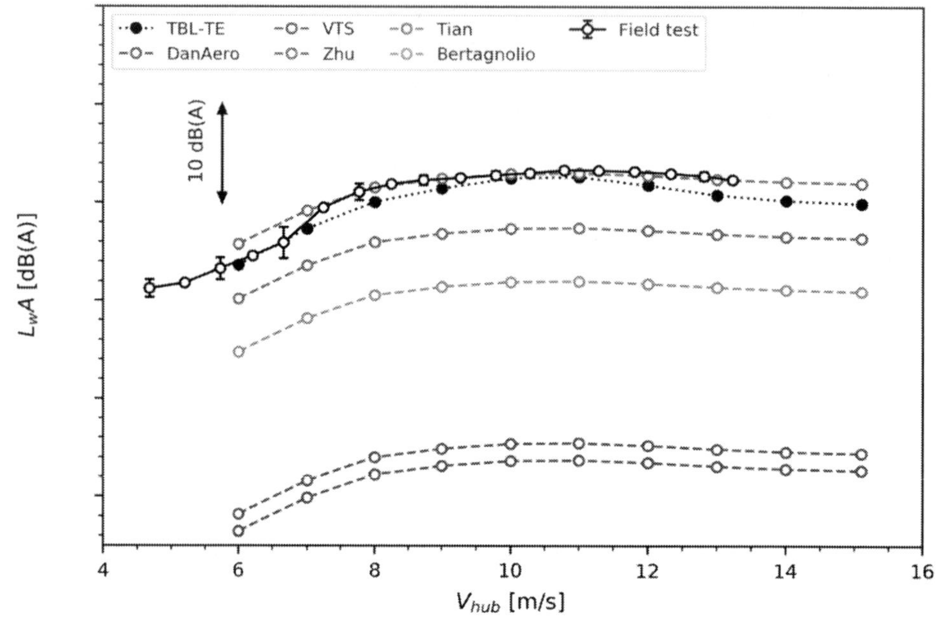

Fig. 8 Acoustic power. Empirical modeling. V80 wind turbine.

stated that the velocity fluctuation spectrum in the mid–span plane $\phi_{ww}(k_1, 0)$ scales as the upwash velocity power spectral density $S_{ww}(\omega)$. Figure 10 presents this scaling with the velocity fluctuation spectrum in the mid–span plane $\phi_{ww}(k_1, 0)$ estimated with the Kármán model for the different inflow turbulence estimations. The low frequency sampling notwithstanding, the velocity spectrum is better captured with the direct estimations of the inflow turbulence than with the empirical estimations and specially the peak frequency. The empirical estimations are expected to overpredict the leading–edge noise prediction. The total acoustic power is depicted in figure 11 and the corresponding sound power level spectrum at peak noise ($V_{hub} = 11$ m/s) in figure 12 from analytical modeling. An estimation of trailing–edge noise from empirical modeling [9] is added to the figure. Here, the opposite observation is made. Analytical modeling with empirical inflow turbulence overpredicts the leading–edge noise but gives relative agreement with the direct inflow turbulence estimations.

6. Towards a scaling of the inflow turbulence

Evidences suggest that the empirical estimation of the inflow turbulence from atmospheric boundary layer models do not account for the inflow turbulence actually experienced by a wind turbine

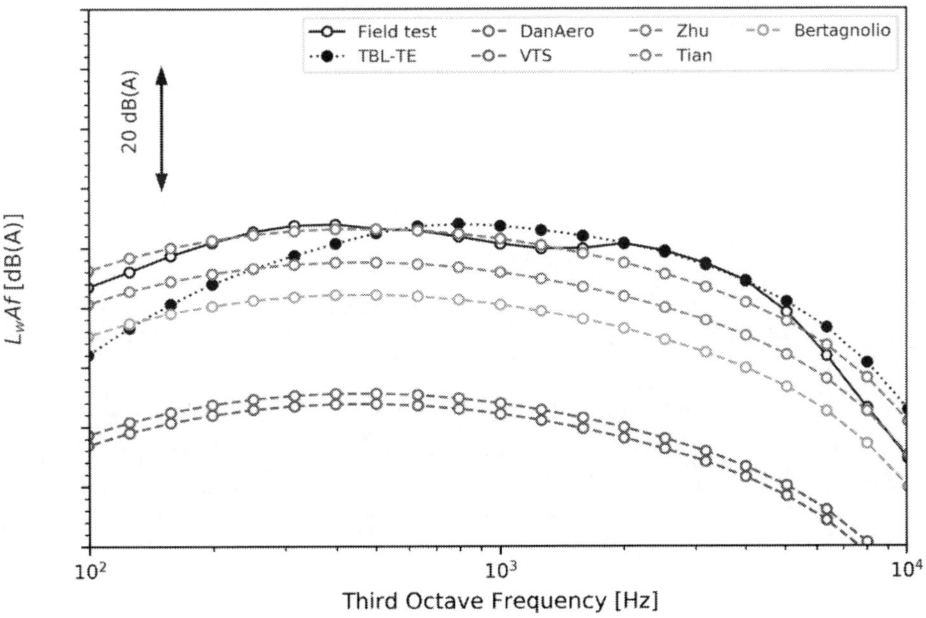

Fig. 9 Acoustic power spectra. Empirical modeling. V80 wind turbine. $V_{hub} = 11$ **m/s.**

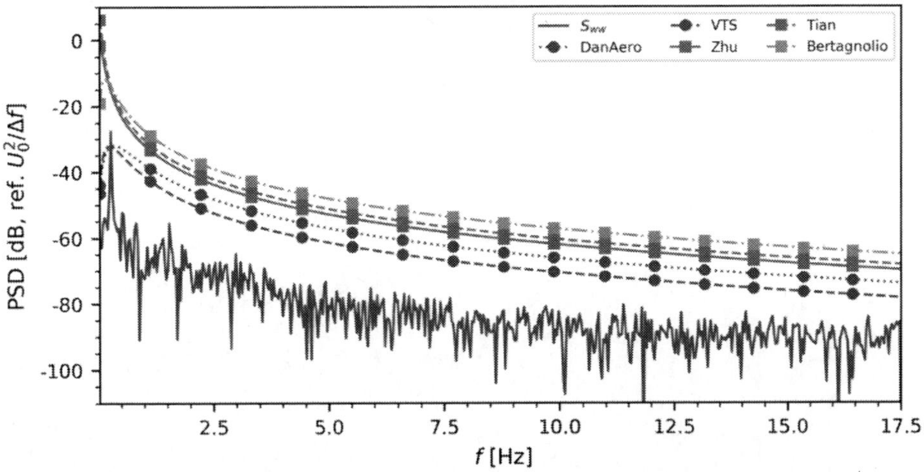

Fig. 10 Velocity spectrum scaling. The amplitudes are scaled with the mean inflow speed U_0 **and the frequency resolution** Δf**.**

leading–edge that is required for analytical leading–edge noise modeling. Instead, the turbulence properties estimated from the numerical tool are in fact similar to what can be estimated from field test measurements. From the same numerical tool VTS applied on a range of wind turbines, for several wind speeds at hub height, a more representative scaling can be proposed. Figure 13 presents the data reduction of turbulence intensities estimated from VTS for a range of wind turbines and wind speeds at hub height. A collapse of the data can be observed that leads to propose a scaling in the form of equation 10, with r as the radial location.

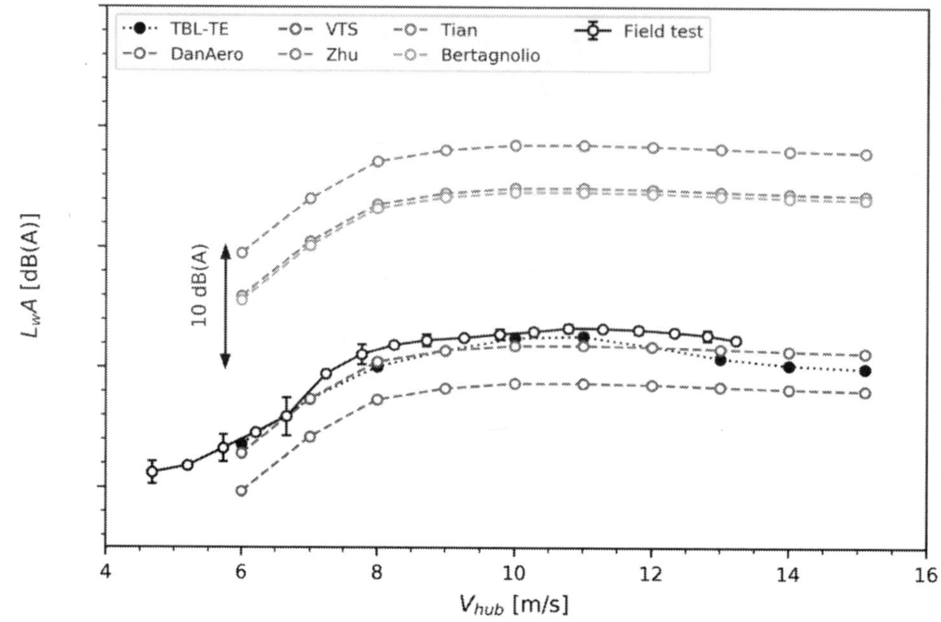

Fig. 11 Acoustic power. Analytical modeling. V80 wind turbine.

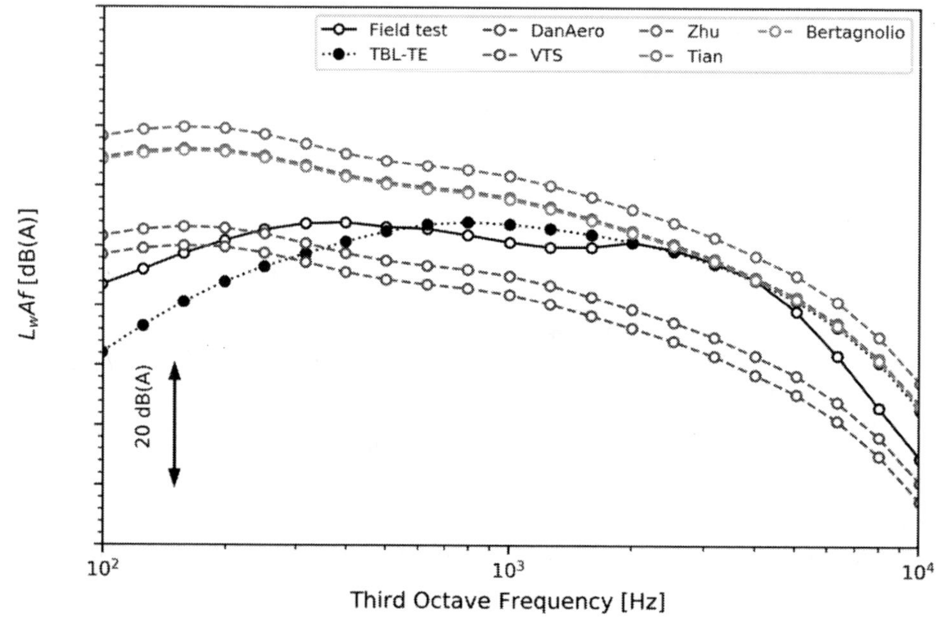

Fig. 12 Acoustic power spectra. Analytical modeling. V80 wind turbine. $V_{hub} = 11$ **m/s.**

$$T_i = A\left(\frac{V_{hub} \times T}{r}\right)^B + C \tag{10}$$

The turbulence intensity is expressed in percentage. Figure 13 presents the data reduction of integral length scales estimated from VTS for a range of wind turbines and wind speeds at hub height. A

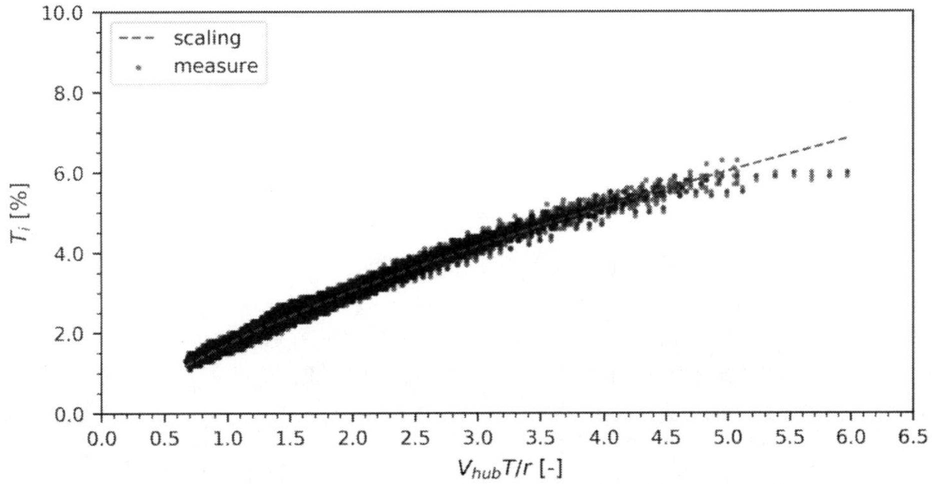

Fig. 13 Data reduction of turbulence intensities estimated from VTS.

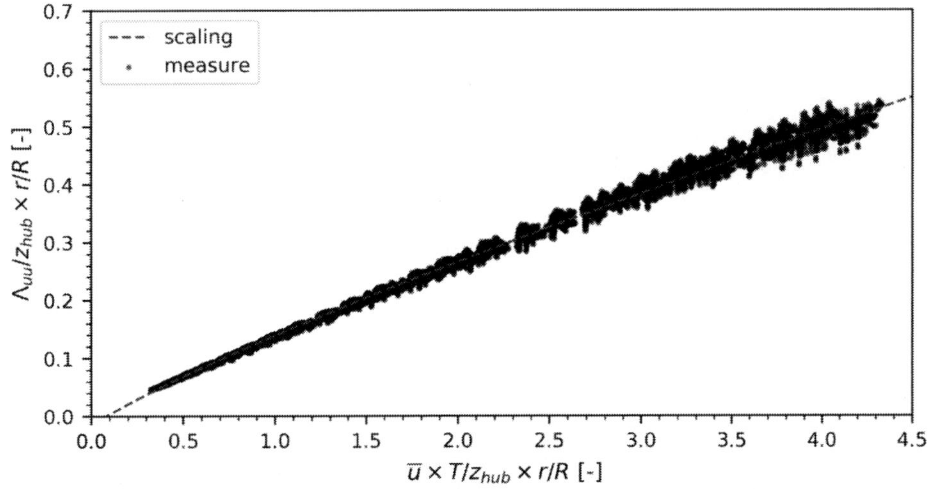

Fig. 14 Data reduction of integral length scales estimated from VTS.

relative collapse of the data can be observed that leads to propose a scaling in the form of equation 11, with R, the blade tip radius. The coefficients, found with a non–linear least squares method [17], are summarized in table 2.

$$\frac{\Lambda_{uu}}{z_{hub}}\frac{r}{R} = A\left(\frac{\overline{u}\times T}{Z_{hub}}\frac{r}{R}\right)^{B} + C \tag{11}$$

If the scaling provides a satisfying agreement with the direct estimation for the turbulence intensity (figure 13) as a consequence of a better collapse of the normalised data, the scaling of the integral length scale however (figure 14) tends to diverge towards the tip ($r/R \sim 1$). This suggests to view the proposed scalings as preliminary results. Figures 15 and 16 show the acoustic power prediction (of both trailing–edge and leading–edge noise combined) and the corresponding frequency content (at V_{hub} = 11 m/s) from the empirical modeling with empirical estimation of the inflow turbulence

	A	B	C
Scaling coefficients for T_i	2.376	0.643	−0.651
Scaling coefficients for Λ_{uu}	0.158	0.853	−0.020

Table 2 Scaling coefficients for the inflow turbulence.

and the analytical modeling with the scaling of the inflow turbulence. When the scaling is applied with the analytical leading–edge noise modeling, the agreement is satisfying at low wind speed but rapidly starts overpredicting the total acoustic power at higher wind speeds (figure 15). To assess the source of the discrepancies at higher wind speeds, a more thorough analysis needs to be carried with experimental measurements for several wind speeds at hub height. Then, the ability of the numerical tool to represent turbulence levels for a range of wind speeds at hub height could be validated.

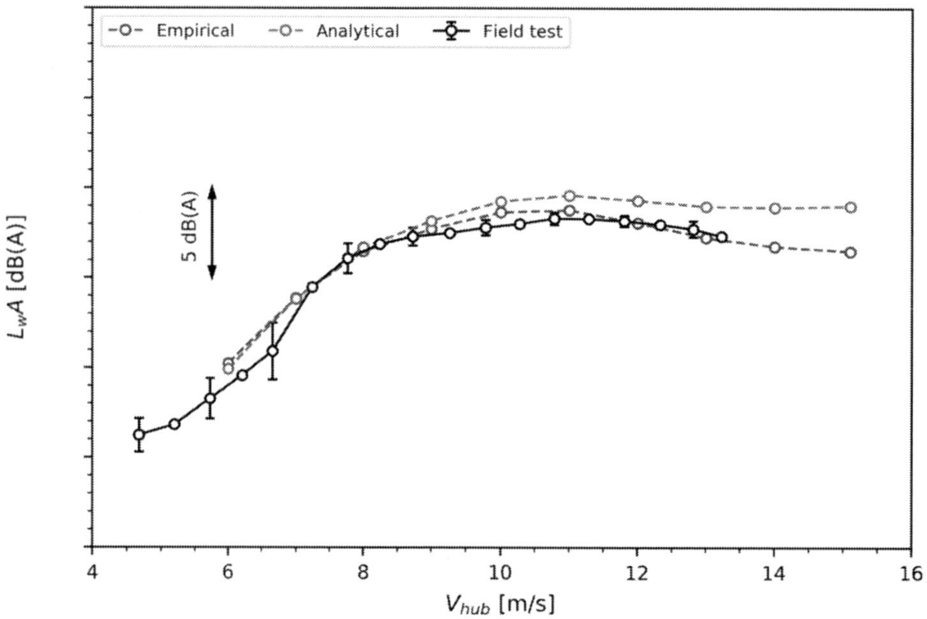

Fig. 15 Total acoustic power. Empirical modeling with atmospheric turbulence and analytical modeling with inflow turbulence scaling. V80 wind turbine.

7. Conclusion

The description of the inflow turbulence, in terms of turbulence intensity and integral length scale is crucial for leading–edge noise prediction. From the materials presented in this contribution, its estimation from empirical formulations of atmospheric boundary layer is believed to be valid only for empirical leading–edge noise modeling that are in fact specifically calibrated for these levels of turbulence. However, when using analytical leading–edge noise modeling, the empirical turbulence estimations fails and knowledge of the turbulence based on the chordwise velocity component at the leading–edge is needed. Turbulence intensity and integral length scale impinging a wind turbine

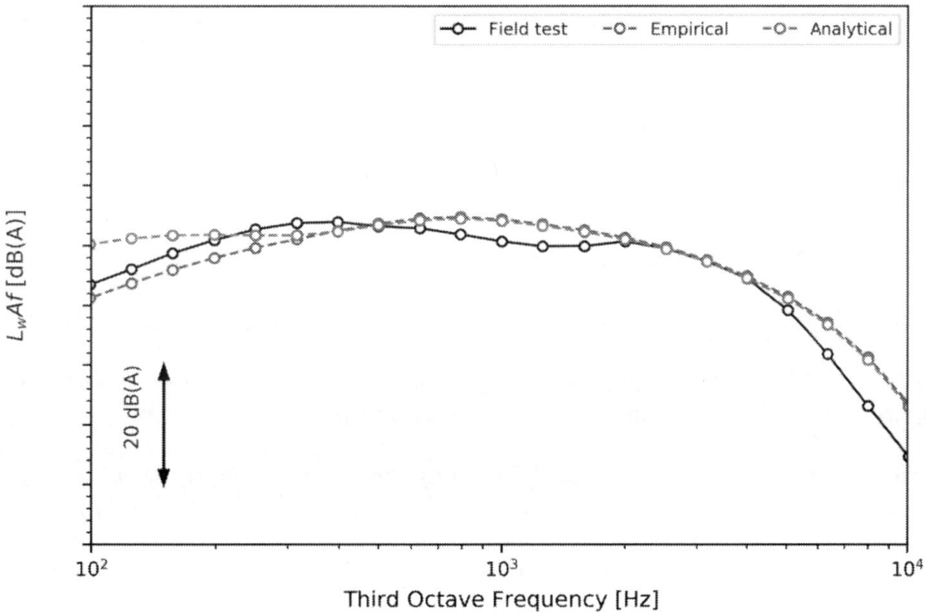

Fig. 16 Total acoustic power spectra. Empirical modeling with atmospheric turbulence and analytical modeling with inflow turbulence scaling. V80 wind turbine. $V_{hub} = 11$ **m/s.**

leading–edge have been measured and have been shown to differ significantly from the empirical estimations. Direct estimations show lower turbulence intensity and lower integral length scale than what can be estimated from the empirical formulations. The direct estimations also indicate relatively constant turbulence levels during a rotation period and that is also in contradiction with the empirical estimations. This contribution has shown that, with analytical leading–edge noise modeling, the leading–edge turbulence can be estimated from a low–order numerical tool to give representative noise predictions but that should not lead to disregard the importance of atmospheric boundary layer turbulence models but instead should lead to re-organise the methodolgy. The atmospheric boundary layer turbulence modeling should be used in the aerodynamic part of the numerical tool to yield estimation of the chordwise velocity at the leading–edge of the blades and once that is known, the turbulence intensity and integral length based on the chordwise component could be directly estimated and used as input for the analytical leading–edge noise modeling.

References

[1] Y. Tian and B. Cotté. Wind turbine nosie modeling based on amiet's theory: effects of wind shear and atmospheric turbulence. *Acta Acustica united with Acustica*, 102:626–639, 2016.

[2] E. M. Salomons. *Computational atmospheric acoustics.* Kluwer Academic Publishers, 2001.

[3] R. K. Amiet. Acoustic radiation from an airfoil in a turbulent stream. *Journal of Sound and Vibration*, 41(4): 407–420, 1975.

[4] W. Zhu, N. Heilskov, W. Shen, and J.: Sørensen. Modeling of aerodynamically generated noise from wind turbines. *Journal of Solar Energy Engineering*, 127:517–528, 2005.

[5] N. Troldborg, C. Bak, H. Aagaard Madsen, and W. R. Skrzypinski. DANAERO MW: Final report, 2013.

[6] S. A. L. Glegg, S. M. Baxter, and A. G. Glendinning. The prediction of broadband noise from wind turbines. *Journal of Sound and Vibration*, 118(2):217–239, 1987.

[7] H. A. Madsen, C. Bak, U. S. Paulsen, M. Gaunaa, N. N. Sørensen, P. Fuglsang, and L. Jensen. The DAN–AERO MW experiments. In *48th AIAA Aerospace Sciences Meeting*, number AIAA-2010-0645, 2010.

[8] International Standard. Wind turbines – part 1: Design requirements. IEC 61400–01, 2005.

[9] T. F. Brooks, D. S. Pope, and M. A. Marcolini. Airfoil self-noise and prediction. NASA, 1218, 1989.

[10] M. Roger and S. Moreau. Extensions and limitations of analytical airfoil broadband noise models. *International Journal of Aeroacoustics*, 9(3):273–305, 2010.

[11] International Standard. Wind turbines – part 11: Acoustic noise measurement techniques. IEC 61400–11, 2012.

[12] M. V. Lowson. Assessment and prediction of wind turbine noise. Flow Solutions Ltd, 1993. ETSU W/13/00284/REP.

[13] J. Counihan. Adiabatic atmospheric boundary layer: a reviewed and analysis of data from the period 1880-1972. *Atmospheric Environment*, 9:871–905, 1975.

[14] R. W. Paterson and R. K. Amiet. Acoustic pradiation and surface pressure characteristics of an airfoil due to incident turbulence. NASA, 1976.

[15] T. F. Geyer, E. Sarradj, and M. Hobracht. Noise generated by a leading edge in anisotropic turbulence. In *Inter–Noise*, 2016.

[16] A. Thacker, S. Loyer, and S. Aubrun. Comparison of turbulence length scales assessed with three measurement systems in increasingly complex turbulent flows. *Experimental Thermal and Fluid Science*, 34:638–645, 2010.

[17] R. L. Burden and J. D. Faires. *Numerical analysis*. Brooks / Cole, 2010.

8th International Conference
on
Wind Turbine Noise
Lisbon – 12th to 14th June 2019

Aeroacoustic Assessment of Wind Turbine Blade Tips

M. Herr, C.-H. Rohardt and B. Faßmann
German Aerospace Center (DLR), Braunschweig, Germany, D-38108

J. M. Pereira-Gomes
DNW-NWB, Braunschweig, Germany, D-38108

Summary

The trend to large wind turbines with increased blade diameters requires efficient noise reduction to increase public acceptance and to avoid further limitations in land use or shut-down times during night. Within the German national wind energy project BELARWEA improved methods to support the design of both, aerodynamically efficient and low-noise, wind turbine rotors are developed and validated. Aeroacoustically driven 2D profile design, 3D winglet design and 2D/3D CFD and CAA analysis are supplemented by the transfer of passive noise reduction technologies from aerospace applications to wind turbine blades. Experimental demonstration and tool validation is provided in systematic evaluation steps, (i) at 2D blade sections in the Acoustic Wind-Tunnel Braunschweig (AWB) of DLR, and (ii) at 3D blade tips in the larger acoustic facility DNW-NWB, operated by the German-Dutch Wind Tunnels foundation. The current paper provides a general overview of the DNW-NWB experiments and available validation data. Beyond this, the focus is set on the aeroacoustic evaluation of a new profile contour RoH-W-18%c37 (vs. NACA 64-618 reference profile) in the 2D as well as 3D test environments. The DNW-NWB tests were conducted on two 1:6 blade tip models in both the closed (3.25 m x 2.80 m) and ¾-open wind-tunnel test sections. Numerical results from a largely non-empirical hybrid CFD/CAA airfoil noise prediction approach, as applied during the profile design process, show excellent agreement with the finally measured 2–4 dB reduction of maximum sound pressure levels.

1. Introduction

According to current knowledge, trailing-edge noise (TEN), as generated in the outer 20–25% of the rotor radius, represents the most relevant noise contributor at modern large wind turbines. Important indicators to support this statement are i) measurements of noise radiation directivities corresponding to the typical TEN cardioid-type characteristics, ii) noise source localization results at wind turbine rotor blades acquired with phased microphone arrays, and finally, iii) the experimentally proven efficiency of trailing-edge modifications on overall wind turbine noise in field measurements [19–21]. TEN at rotor blades is generally modeled as 2D problem while separating the rotor in distinct blade elements. Its key mechanism is the fluctuating pressure induced by the boundary-layer turbulent eddies and its propagation to the farfield due to the scattering from the trailing edge (TE) [10, 16]. The overall blade noise is then estimated through summation of the respective radial contributions that are calculated based on the local and overall operational conditions. Typically, the contribution of blade tip noise, attributable to 3D tip vortex formation, is negligible. An acoustically driven blade design conventionally foresees a reduced loading, i. e. reduced power production in the very tip region. Since published field test data are still limited, the relevant literature does not provide clear statements concerning the importance of additional flow-induced noise sources, i. e. flow separation noise and turbulent-inflow leading-edge interaction noise. The noise ranking of the latter two contributions appears individually dependent on the turbine type and site quality.

2. General approach

The project BELARWEA* aims at an extension of today's aeroacoustic evaluation and design capabilities in wind energy; current analyses are predominantly based on a 2D assessment, whereas full-3D evaluations are generally not considered due to non-affordable computational expenses. The prediction of 2D-profile TEN has been successfully documented in the BANC† framework [14, 15]. Moreover, the noise reduction efficiency of TE modifications (porous TE add-ons like serrations, brushes, etc.) has been previously verified at 2D blade sections [28]. However, these forerunner 2D studies still lack a systematic clarification to which extent the derived results can be transposed to full-scale rotational rotor geometries. The current experimental study intends to cover the first step from 2D to 3D static conditions, whereas in a future - planned - second phase of the project also the particular effects of the rotation are sought to be identified. The major objective lies in the provision of high-quality validation data for both 2D-based and 3D CAA prediction methods. A fully 3D numerical evaluation methodology would offer the mid-term perspective to add more degrees of freedom in today's design processes.

2.1. 2D numerical assessment

Figure 1 gives a general overview of the hybrid CFD/CAA prediction approach applied to support the design of a new profile contour RoH-W-18%c37. Profile design was conducted in an iterative process using the panel method XFOIL [3] with successive aeroacoustic evaluations. The current paper restricts to the description of the latter 2D acoustic evaluation procedure, whereas ongoing full 3D simulation work will be published in the future.

*The German acronym BELARWEA ("Blattspitzen für Effiziente und LärmArme Rotoren von WindEnergie-Anlagen") stands for "blade tips for efficient and low-noise wind turbine rotors".

†AIAA/CEAS Workshops on Benchmark Problems for Airframe Noise Computations

In a first analysis step, a steady RANS computation is performed by use of the DLR TAU-code [11, 27]. The CFD provides a time-averaged turbulent flow around the airfoil and the related turbulence statistics. In a second step, the DLR CAA-solver PIANO [2] is applied. Time-dependent linear propagation equations are solved on structured multi-block (SMB) meshes to compute the sound field. The resulting acoustic quantities, i. e. the spatially and time-resolved sound pressure, the acoustic particle velocity, or the sound intensity can be evaluated at user-selected microphone positions. These are usually located circularly around the trailing edge at a certain distance larger than the chord length l_c (here: $r_{\text{mic}} = 2.5\, l_c$).

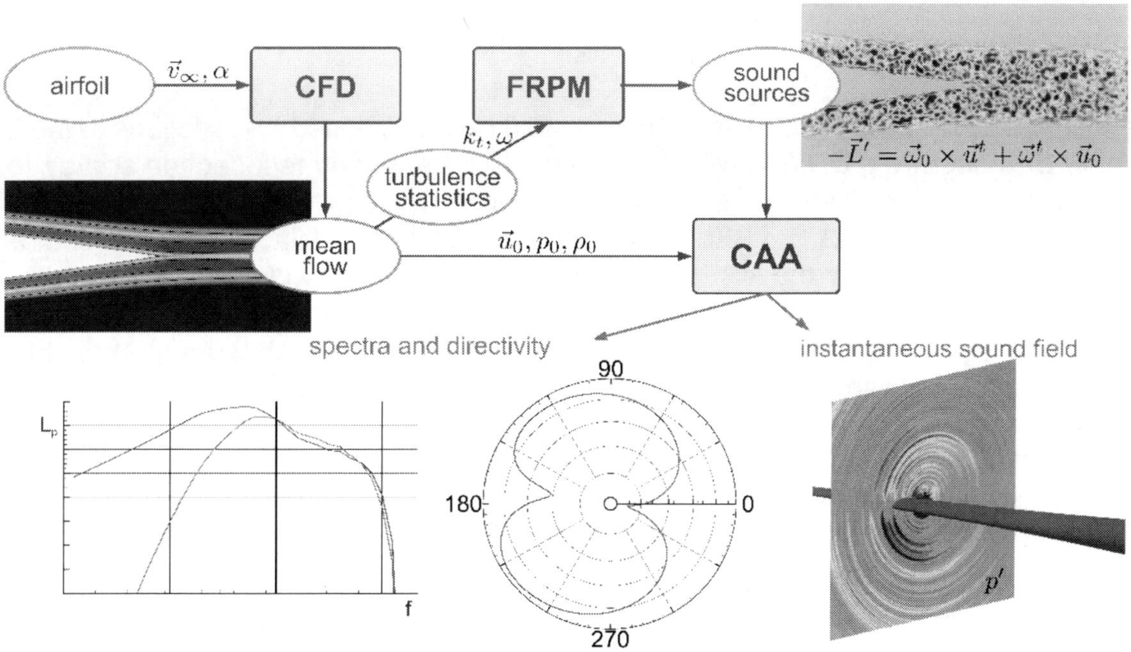

Fig. 1 Schematic of the hybrid RANS-based CAA prediction method.

On the right-hand side of the linear propagation equations, sound sources are explicitly imposed. A synthetic turbulence method provides fluctuating vorticity according to the turbulence statistics of the RANS solution. The Fast Random Particle-Mesh method (FRPM) [4, 5, 7, 8] realizes time-dependent fluctuations from time-averaged turbulence statistics. The standard method applied in this paper generates Gaussian correlated synthetic turbulence of local integral length scale and variance proportional to the turbulence kinetic energy distribution.

The steady time-averaged RANS flow provides the mean-flow over which the CAA simulation is carried out. Furthermore, the turbulence statistics provided by RANS is utilized to generate the unsteady vortex sound sources that drive the governing equations. In free field, this turbulence is coupled with the CAA solver, which is based on the 4th order accurate DRP scheme proposed by Tam & Webb [29]. The temporal discretization is realized by a low-dissipation, low-dispersion Runge-Kutta (LDDRK) algorithm by Hu [17]. Sound due to the interaction of vorticity with the trailing-edge is generated as part of the CAA simulation step. These vortex dynamics are dominated by the linear contributions to the source terms.

Farfield TEN data are scaled to represent 3D sound pressure level spectra at a span of s_{norm} = 1 m and a microphone distance of r_{norm} = 1 m. A 2D/3D scaling is applied as presented by Ewert et al. [6], taking into account the local Ma number and setting the model constant in Eq. (1) as $\zeta = 1.4$. Based on a definition of the spanwise coherence length scale Brooks & Hodgson [1] report $\zeta = 1.4 \ldots 1.6$.

$$\Delta L_{p,i}^{\text{norm}} = \underbrace{10 \log_{10} \left(\frac{\zeta}{2\pi} \frac{s_{\text{norm}}}{r_{\text{mic}}} \text{Ma} \right)}_{\text{2D to 3D correction}} + \underbrace{20 \log_{10} \left(\frac{r_{\text{norm}}}{r_{\text{mic}}} \right)}_{\text{distance normalization}} \tag{1}$$

The prediction quality of this approach was previously documented along with code-to-code and experimental data comparisons within BANC category 1 (TEN) [14, 22, 23] and continuing benchmark activity with wind energy industry [9].

2.2. Test description

Tests were conducted in both the closed (3.25 m × 2.80 m) and ¾-open test sections of the Low-speed Wind-Tunnel Braunschweig DNW-NWB. Experiments in the closed test section served to measure the aerodynamic coefficients and to characterize the tip vortex flow fields. An overview of the model variants is provided in Figure 2. The BELARWEA reference blade tip is a 1:6 down-scaled derivative of the outer 20% radius portion of the NREL-5-MW-reference rotor [18] which is based on NACA 64-618 profile sections. Corresponding local chord lengths l_c are listed in 1 along with the positioning of the sections with static pressure measurement instrumentation. 4 Kulite-sensors (pinhole arrangement with a 0.4-mm diameter) were placed in close vicinity of the TE in sections 1–4 on the suction side of each model. The model span (span-wise coordinate: z with origin at model root) amounts to 2.1 m.

Table 1 Definition of the instrumented blade sections 1–5 (c_p: distributions of static pressure coefficients) with reference to the blade element numbers '#' of the NREL 5 MW rotor blade according to Jonkman [18]. R, r; **rotor radius.**

#	1:1			NWB (1:6)			comment
	R	% R	c_p	r, m	z, m	l_c, mm	
-	50.40	80.00	-	8.40	0.000	356.4	root profile
14	52.75	83.73	1	8.79	0.392	352.6	
15	56.17	89.15	2	9.36	0.961	344.3	
16	58.90	93.49	3	9.82	1.417	322.7	
-	59.98	95.20	4	10.00	1.596	300.9	
-	60.48	96.00	-	10.08	1.680	287.5	partition line to winglets
17	61.63	97.83	5	10.27	1.872	231.3	configurations without winglets

The tip planform in the outer 2.2% of the rotor radius has been arbitrarily defined based on usual tip geometries‡. To account for the static wind-tunnel conditions the original blade twist was removed

‡Thanks are due to the advisory board of the BELARWEA project, particularly to S. Erbslöh, Senvion GmbH. Note that the NREL 5-MW model rotor [18] is defined in terms of blade elements only and lacks a detailed tip planform geometry definition.

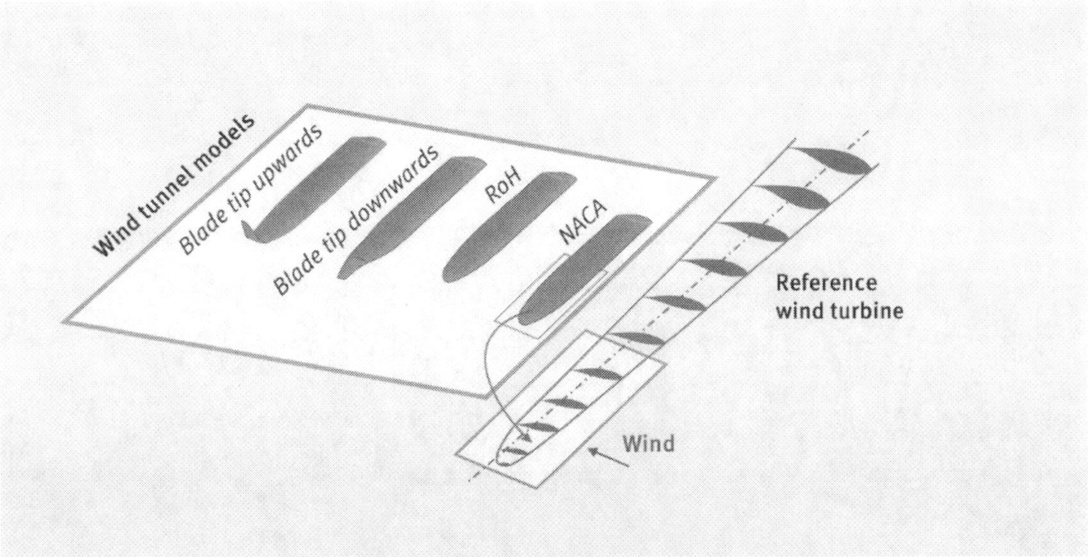

Fig. 2 DNW-NWB wing tip model variants.

and the chord distribution adjusted in radial (rotor) / spanwise (wind-tunnel) direction to consider the effect of varying local flow speed on local Reynolds numbers in the rotational case. Three additional blade tip variants are based on the new profile geometry RoH-W-18%c37; one with an identical planform and zero twist as the reference and the remaining two with exchangeable winglet geometries replacing the original tip. The two model variants without winglets were tested with brush retrofit devices based on a preselection made during the forerunner 2D AWB studies [28].

Tests in the closed test section served to measure the aerodynamic coefficients and to characterize the tip vortex flow fields. Acoustic measurements in the ¾-open test section comprised

- sound source identification with two phased microphone arrays, array 1 (~3 m diameter, 140 microphones) facing the suction side (SS) of the model and array 2 (~1 m diameter, 96 microphones) its pressure side (PS);
- TEN measurements in the approximate 2D-region of the model with a directional microphone system (elliptical mirror with 1.6 m diameter);
- comparative measurements with free field reference microphones;
- measurements of the hydrodynamic surface pressures close to the TE in different spanwise sections further characterized by static pressure distribution measurements, cf. Table 1.

Figure 3 gives an overview on the acoustic measurement setup. The obtained measurement database is applicable to (i) the validation of 2D-based TEN predictions and (ii) full 3D tip noise predictions. The validation concept for TEN predictions is shown in the bottom figure.

The model variants with RoH-W-18%c37 vs. NACA-64-618 profiles were tested for varying tripping conditions; the used denominations 'NAT' and 'FUL' accordingly refer to non-tripped turbulent boundary layers (TBL) with corresponding natural transition and tripped TBL, respectively. For configuration 'FUL' zigzag strips of 0.4 mm thickness were applied at 5% chord on the SS and 10% chord on the PS. Infrared thermography measurements confirmed the efficiency of the tripping devices. An additional configuration 'NATFIX' was introduced to ensure well-defined comparisons with the AWB measurements; in this configuration, boundary-layers were tripped at 42% chord on the SS and 52% chord on the PS to suppress the generation of a laminar separation bubble at the SS of the RoH-W-

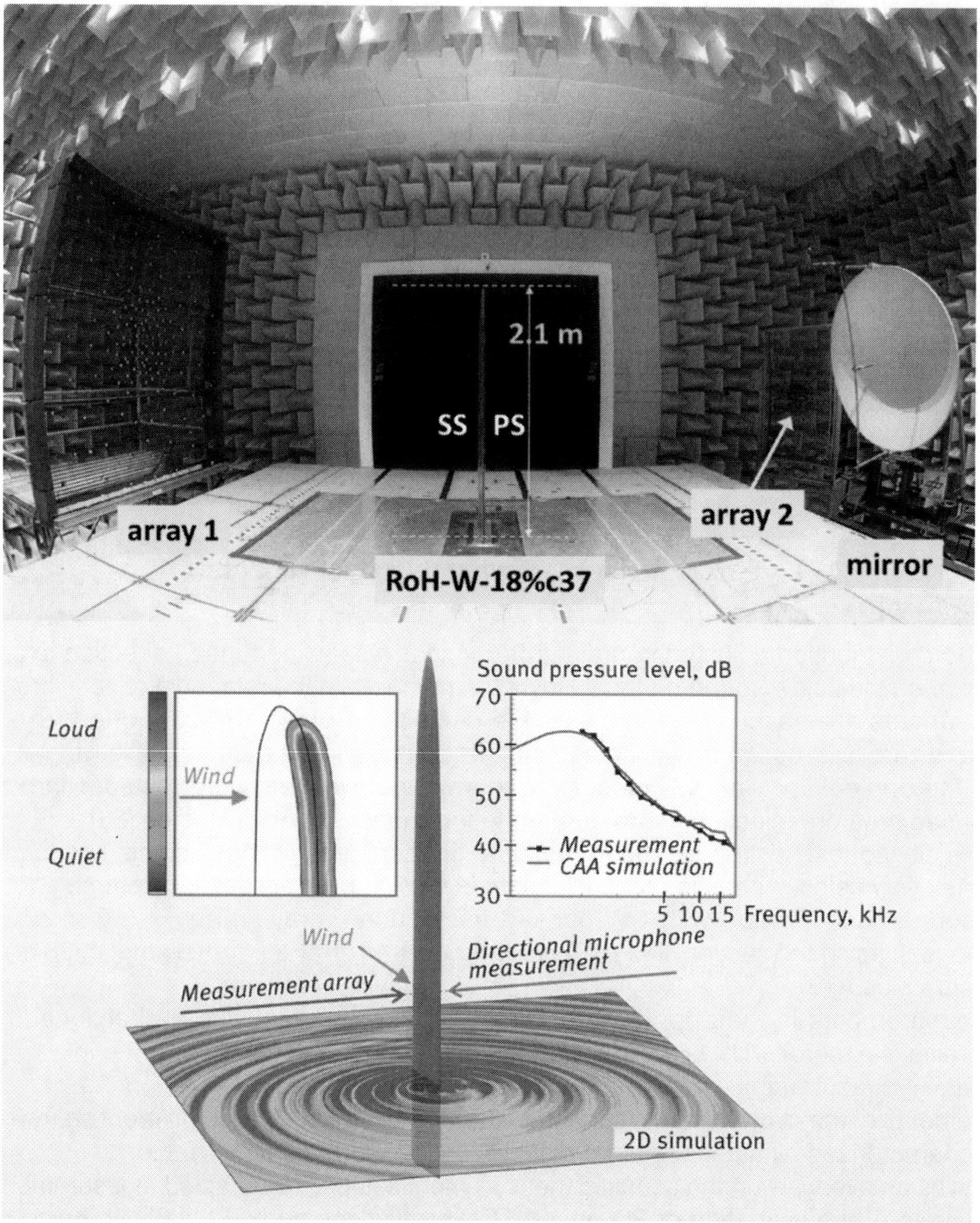

Fig. 3 DNW-NWB test setup and validation concept for 2D-based TEN prediction method.

18%c37 profile [28]. Note that the data post-processing is still ongoing; the following presentation of results will concentrate on the variants without winglets. The smaller array 2 was mainly dedicated to capture directivity effects induced by the winglets so that corresponding results will be retained for future communication on full 3D simulation and validation.

3. Results

3.1. 2D numerical results

The 2D numerical predictions obtained during the design process prior to the measurements are summarized in Figure 4; results refer to section 2 with l_c = 0.3443 m. Overall sound pressure levels OASPL are presented for stream velocities of u_∞ = 60 m/s and u_∞ = 80 m/s along with profile lift coefficients c_l, lift-over-drag ratios c_l/c_d and angles-of-attack α. Accordingly, an OASPL reduction of about 2–2.5 dB was expected for the design lift coefficient c_l = 1.15.

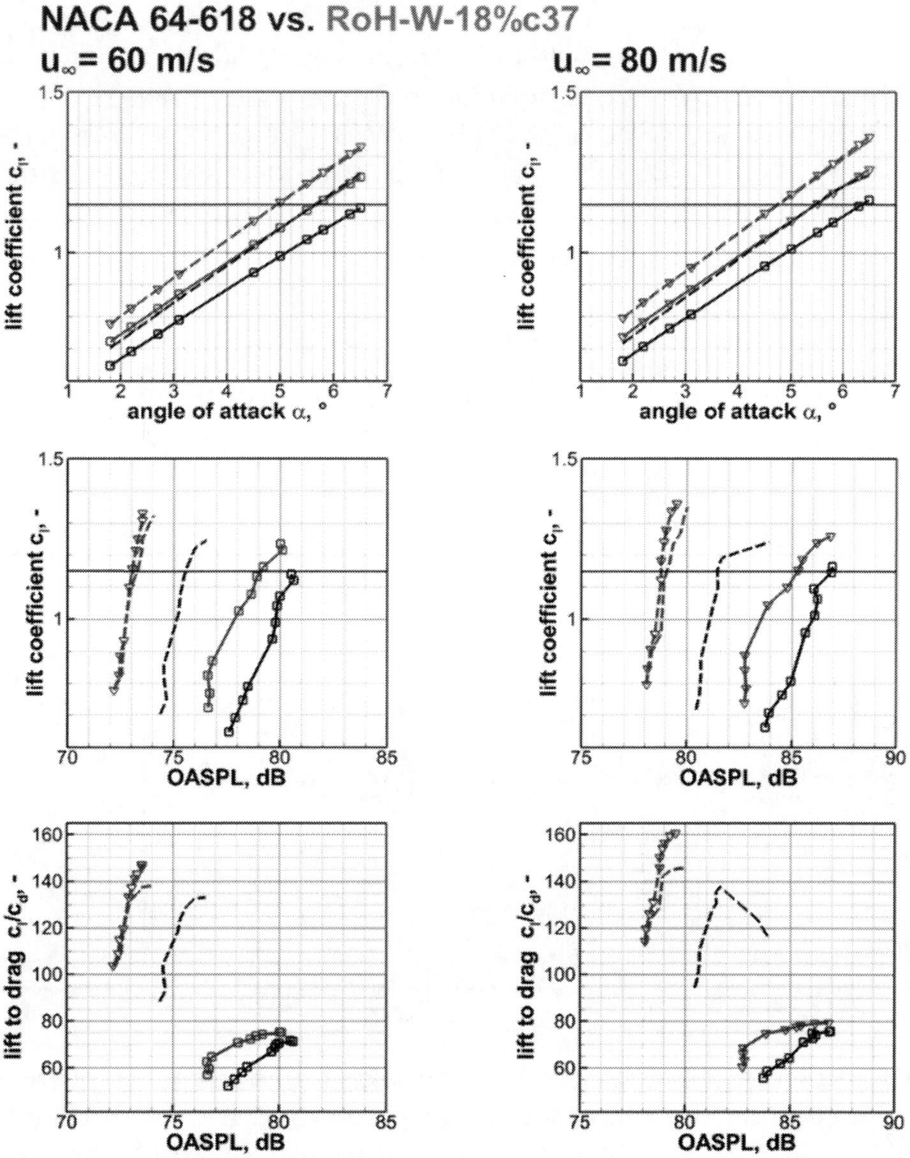

Fig. 4 2D CFD/CAA predictions for the NACA 64-618 (black color) vs. RoH-W-18%c37 (red color) profile contours. Symbols denote tripped configurations; i. e. FUL (solid lines) or NAT-FIX (dashed lines), dashed lines without symbols indicate natural transition NAT.

3.2. General overview on the DNW-NWB test data

The current setup in the DNW-NWB served with extended ranges of test parameters when compared to the preparatory studies at 2D blade sections in AWB with comparable chord (AWB: l_c = 0.3 m); i. e. tests could be performed at larger angles-of-attack (larger model aspect ratio of 6.7 compared to 2.7 in the AWB) and test velocities of up to 80 m/s (compared to 60 m/s in AWB). Moreover, due to the cantilever setup the background noise from the corner junctions (side-plates in AWB, test section floor in NWB) was reduced so that the usable frequency range could be extended below 1 kHz, the well-documented low-frequency limit in most of the published TEN experimental data. TEN peak frequencies are consequentially expressed in both Kulite (source) and farfield TEN data. TEN data in the spectral peak region appear still contaminated by the junction, however, to a lower extent when compared to AWB test conditions.

A first tentative cross-comparison of results from 2D blade section measurements in AWB vs. the blade tip model measurements in DNW-NWB is provided in Figure 5 along with 2D CFD/CAA predictions for configurations RoH-W-18%c37 FUL and NATFIX. AWB prediction and measurement results are scaled from u_∞ = 50 m/s to u_∞ = 80 m/s. One-third octave band sound pressure levels $SPL_{1/3}$ in

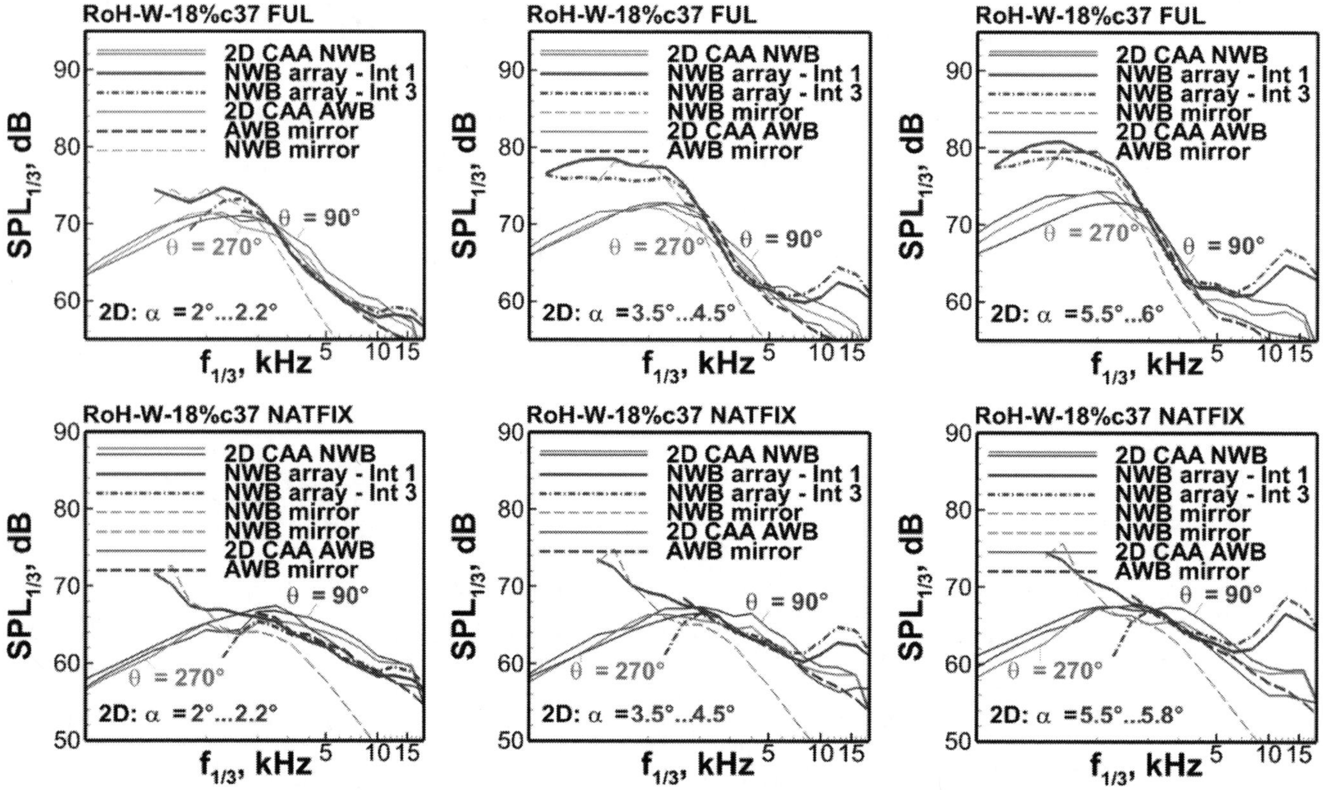

Fig. 5 Comparison of 2D CFD/CAA predictions with measurements at the RoH-W-18%c37 configuration, u_∞ = 80 m/s. The directivity angle θ refers to the wind-tunnel center line and to the TE position at α = 0°; θ = 90° (red color) corresponds to the position of the center microphone of array 1 facing the SS, θ = 270° (green color) to the convection-corrected focus position of the directional microphone system facing the PS (same definition, i. e. θ = 270°, holds for the AWB setup marked by blue color).

this representation format are arbitrarily normalized to a 1-m span and a 1-m distance. The shown microphone array data sets are based on a first assessment by means of standard beamforming, as provided by DNW-NWB directly after the test. Given the selected integration areas, it is not expected that spectral maxima will have to perfectly collapse with 2D simulations or AWB measurements, respectively. The integration areas are displayed in the Appendix, cf. Figure 13; Int 1 covers the whole model (2.1 m), whereas Int 2 (span: 1.4 m) excludes the floor region to initially suppress noise by the floor junction. Again, NWB conditions are selected based on local conditions at section 2, where also the elliptical mirror far focus was set. Aerodynamic angles-of-attack α were not exactly identical for both test environments (blue: AWB; red: NWB), but close enough to derive the following interim conclusions:

- Directional microphone and array measurements in DNW-NWB lead to consistent results; low-frequency peak levels which are dominated by floor-junction noise at least for the quieter model configurations (i. e. NATFIX and NAT) are of same order when comparing red solid lines (NWB array Int 1) with green dashed lines (NWB mirror). Due to the TEN directivity it is expected that for non-zero angles-of-attack levels beyond the TEN peak should be lower for the measurement systems facing the PS of the model, cf. the 2D simulation results for orders of magnitude. However, spectral decay slopes in the mirror spectra appear over-corrected by the herein applied theoretical gain and resolution assumptions according to Schlinker [12, 26]. Note that an in-situ calibration of the directional microphone system (as available for the smaller 1.4-m-diameter AWB directional microphone system) could not be afforded during the BELARWEA campaign. The latter is foreseen in the future to guarantee comparable data reliability as in the AWB.
- 2D simulation and AWB 2D blade section measurement results show almost perfect agreement, whereas discrepancies in spectral shape appear for the wing tip model spectra. Spectral decay slopes beyond the TEN peak are almost identical for the wing tip array data. This is an unexpected result if assuming the overall TEN of the wing tip model to be the result of an independent summation of slice-wise 2D TEN contributions.
- Deviations between 2D and wing tip model results are observable in the peak region and at frequencies around 10–15 kHz where tip noise contributions appear in the microphone array source maps, cf. Figure 6. The relative importance of the latter is dependent on the configuration. Exact reconstruction of absolute TEN peak levels and a clear identification of 3D effects on TEN spectra will require the application of more advanced array data post-processing methods to i) reliably correct for excess noise contributions and to ii) enabling a more detailed separation of overall noise into smaller section-wise contributions.

Some additional example results are shown below in Figures 6– 8 to illustrate the extent of the acoustic database. The noise reduction effect of brush TE devices was demonstrated, see Figure 6. However, due to the installation conditions at the wing tip model the achieved overall noise reduction (of up to 6 dB) was slightly reduced when compared to the results obtained in the AWB [28] under idealized 2D conditions.

For the same selected model configurations with and without brush the effects of test angle-of-attack (Figure 7) and velocity (Figure 8) are shown in terms of unsteady surface pressure narrowband spectra (figures left) and one-third-octave band spectra of farfield TEN (figures right), as measured with the directional microphone system. Note that the brush TE does not significantly affect the incoming hydrodynamic pressures in the source region.

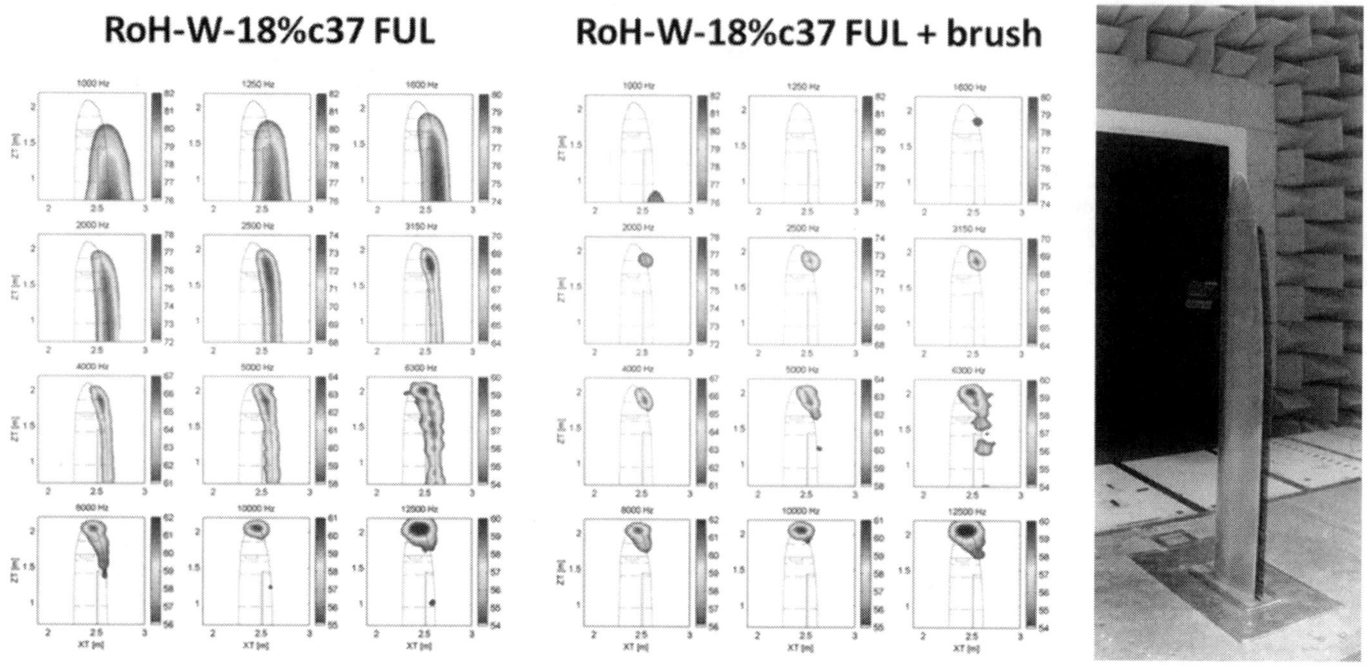

Fig. 6 Source maps from standard beamforming (microphone array 1, pointing to the SS, in 1/3-octave bands) for α = 3.1° and u_∞ = 80 m/s. Brush extends to z_T = 1.835 m.

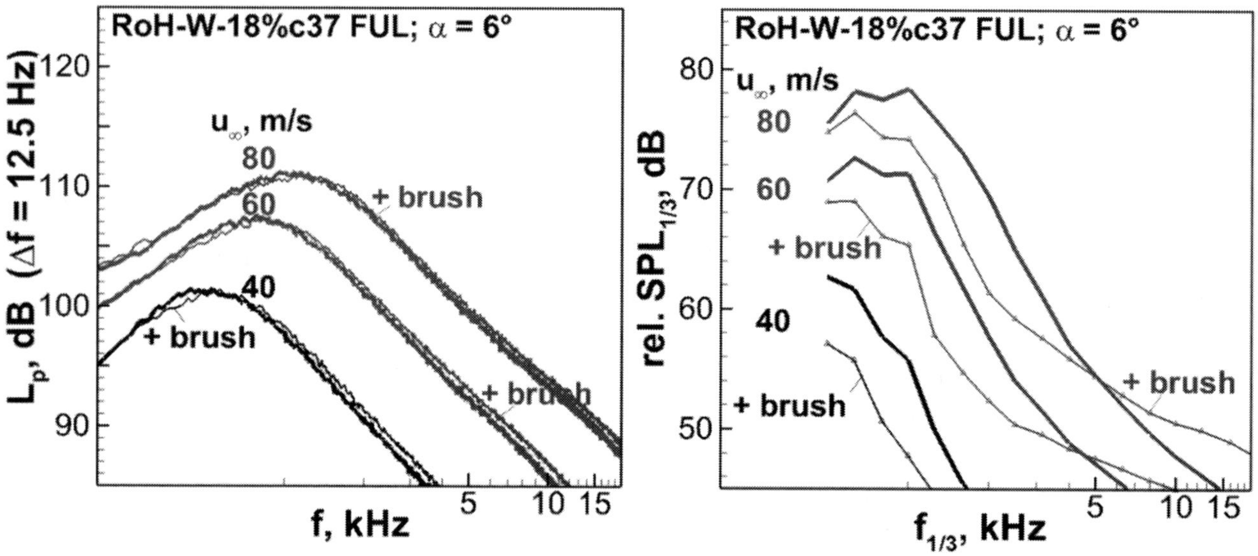

Fig. 7 Effect of flow speed on (left) surface pressure narrowband spectra (Kulite No. 2 at the height of the elliptic mirror focus, Δf = 12.5 Hz, re. 20 μPa) and TEN farfield noise (elliptic mirror data when focusing at the convection-corrected TE location at z = 0.9611 m), α = 6°.

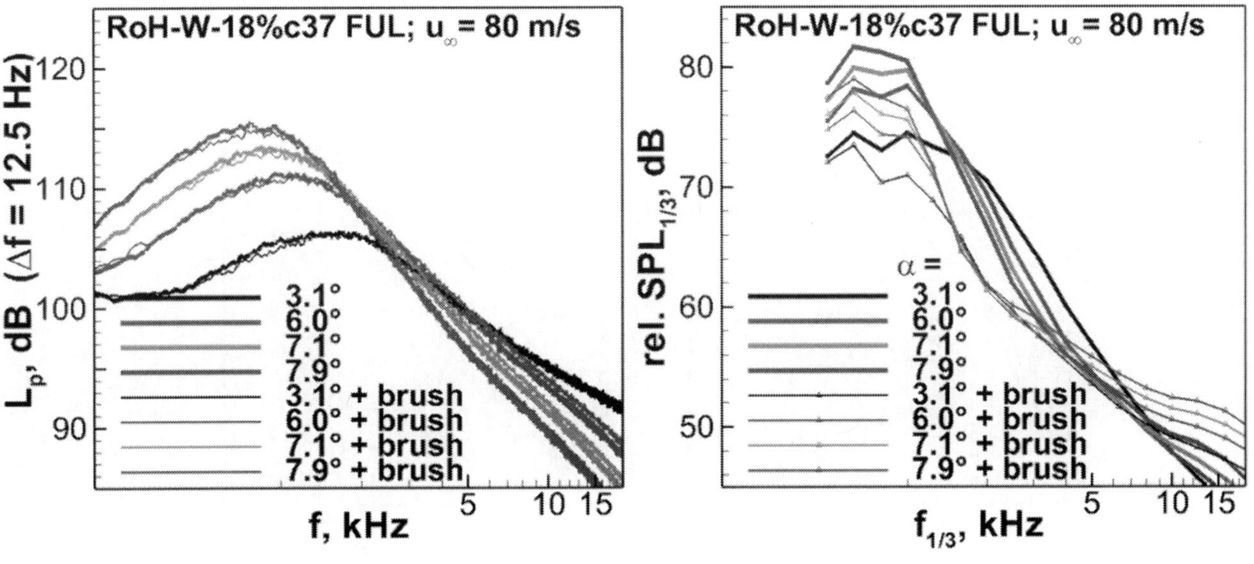

Fig. 8 Effect of test angle-of-attack on (left) surface pressure narrowband spectra (Kulite No. 2, Δf = 12.5 Hz, re. 20 μPa) and TEN farfield noise (elliptic mirror data when focusing at the convection-corrected TE location at z = 0.9611 m), u_∞ = 80 m/s.

3.3. Aeroacoustic evaluation of the RoH-W-18%c37 profile design

In the following, a selection of key figures are provided, documenting the noise reduction effect of the RoH-W-18%c37 profile contour, when compared to the NACA 64-618 reference. For better readability, the same color coding for the two model variants as in Figure 4 is kept in the remainder of this report; red color refers to the RoH-W-18%c37 contour and black color to the NACA 64-618 reference.
The aerodynamic conditions selected for the subsequent aeroacoustic evaluation are given in Figure 9 and Tables 2 to 3. Figures 10 and 11 summarize measured surface pressure spectra close to the TE for the TBL configurations NAT and FUL at four different spanwise positions. Static pressure distributions at the respective sections can be found in the Appendix, cf. Figures 14 and 15, indicating moderate flow separation for the configuration NACA 64-618 FUL already at α = 8.1 °.

Table 2 Comparison conditions with approximately equal lift (TBL configuration NAT); geometric vs. corresponding aerodynamical angles of attack (α_g/α) for c_L-values in Figure 9.

NAT	c_L = 0.75...0.76	c_L = 1.02...1.03	c_L = 1.12	c_L = 1.14...1.15
NACA 64-618	4.6°/3.8°	8.0°/7.0°	9.5°/8.4°	10.0°/8.8°
RoH-W-18%c37	3.8°/3.0°	7.0°/5.9°	8.2°/7.0°	8.5°/7.3°

Corresponding farfield sound measurement results are represented in Figure 12 for configurations NAT (top) and FUL (bottom). The figures also contain the predicted TEN spectra for local conditions at section 2. Measured and simulated c_p distributions are shown on the top of each set of

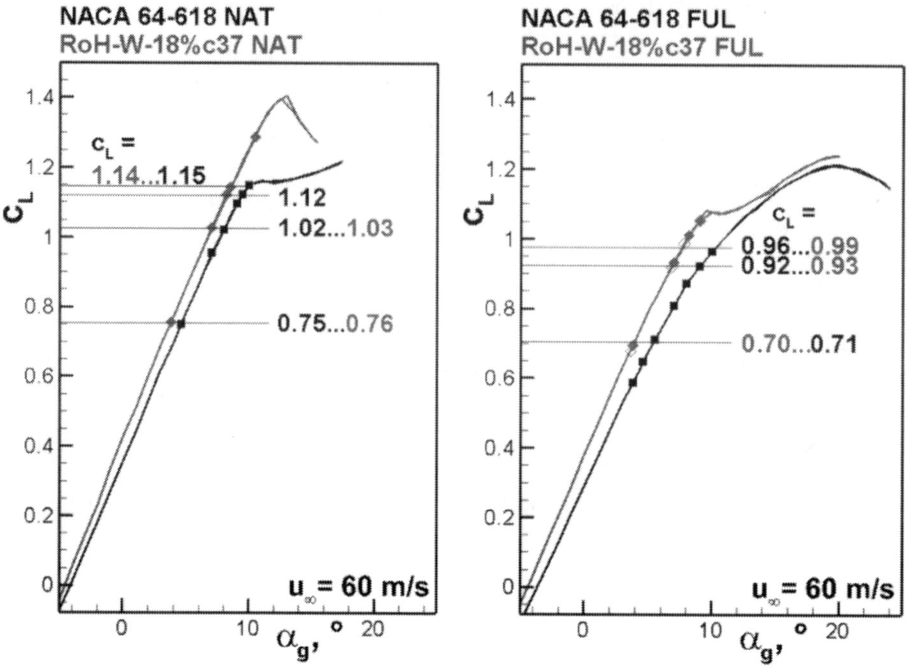

Fig. 9 Total lift coefficients c_L vs. geometric angles of attack α_g from force balance measurements (3/4-open test section, as measured). Filled symbols denote available acoustic measurements for these conditions at u_∞ = 40; 60; 70; 80 m/s, open symbols mark additional data points for u_∞ = 60 m/s only .

Table 3 Same as Table 2 but TBL configuration FUL.

FUL	$c_L = 0.70\ldots0.71$	$c_L = 0.92\ldots0.93$	$c_L = 0.96\ldots0.99$ (1.01)
NACA 64-618	5.5°/4.8°	9.0°/8.1°	10.0°/9.0°
RoH-W-18%c37	3.8°/3.1°	7.0°/6.0°	7.8°/6.8° (8.2°/7.1°)

figures. As expected from the design specifics and previous AWB measurements, configuration RoH-W-18%c37 NAT features a laminar separation bubble (with turbulent reattachment) at the SS that can be suppressed by tripping slightly upstream (configuration NATFIX). Equivalent available measurements at both the NACA 64-618 and the RoH-W-18%c37 variants in configuration NATFIX have shown consistently modulated spectra for both airfoils so that the received level differences are not significantly affected by selection of the respective tripping configuration. To reduce the number of graphs in this overview the presentation of results is limited to configuration NAT.

Overall, the observed trends in both Kulite measurement data and 2D CAA predictions are well-represented in the farfield TEN measurement data, i. e. significant reductions of the TEN peak levels are documented. The reduction in peak level increases with angle-of-attack. It is, however, accompanied by a corresponding noise increase at mid-to-high frequencies which would indeed lead to slightly reduced OASPL reductions, when A-weighting is applied. In particular for configuration NAT the measured frequency of the crossing location of the two spectra is perfectly captured by the

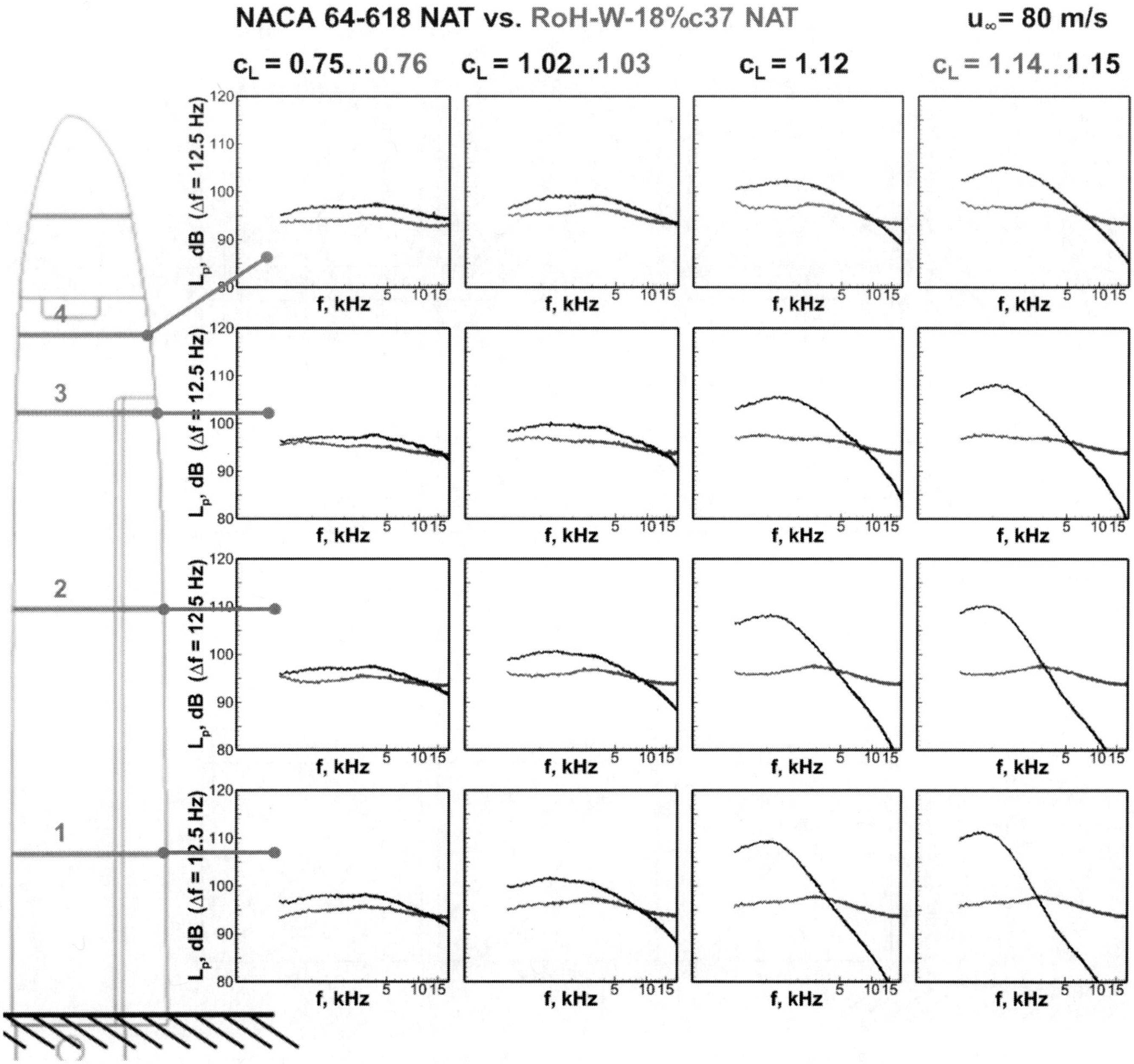

Fig. 10 Measured surface pressure narrowband spectra (re. 20 μPa) for configuration NAT.

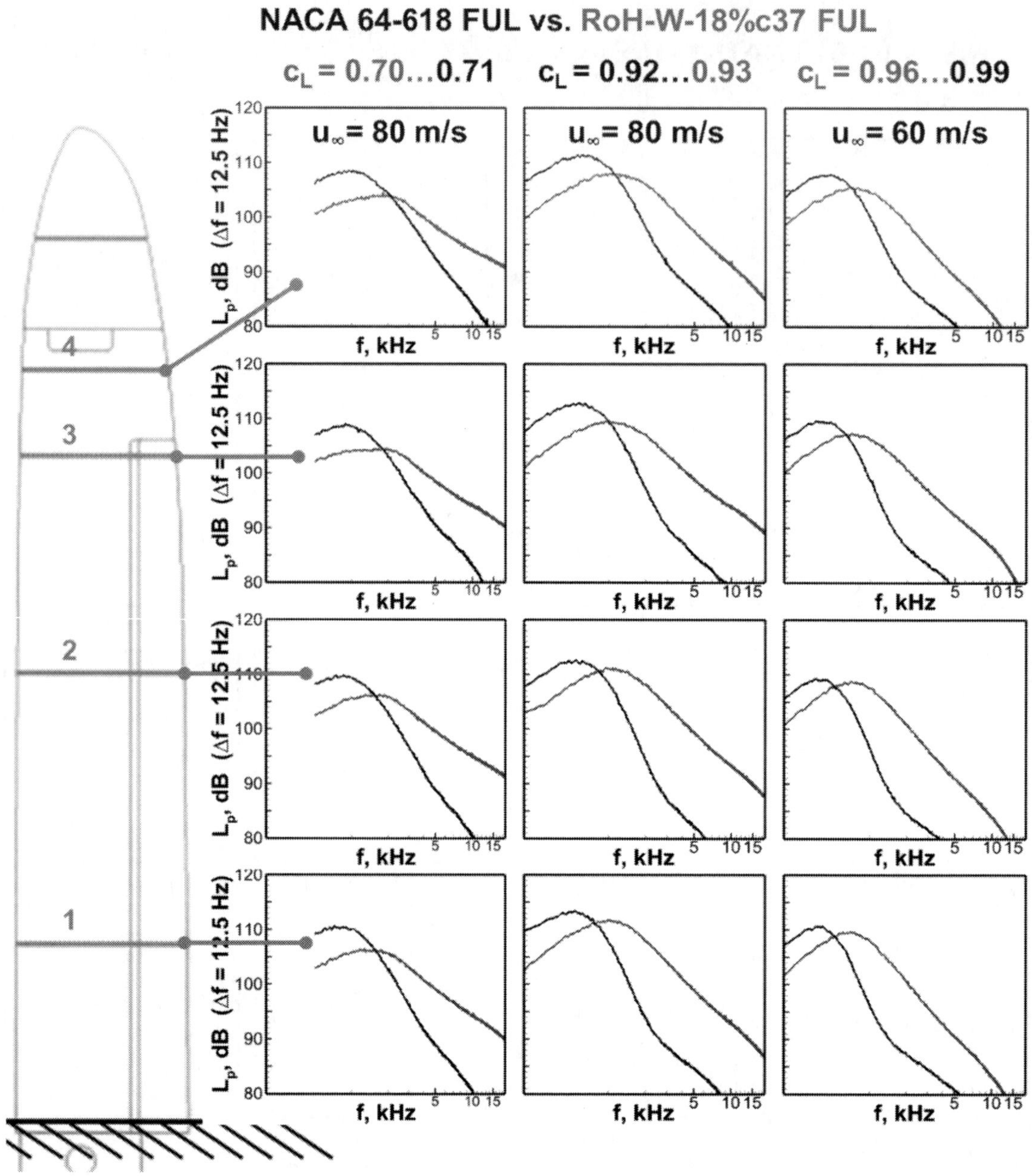

Fig. 11 Measured surface pressure narrowband spectra (re. 20 μPa) for configuration FUL.

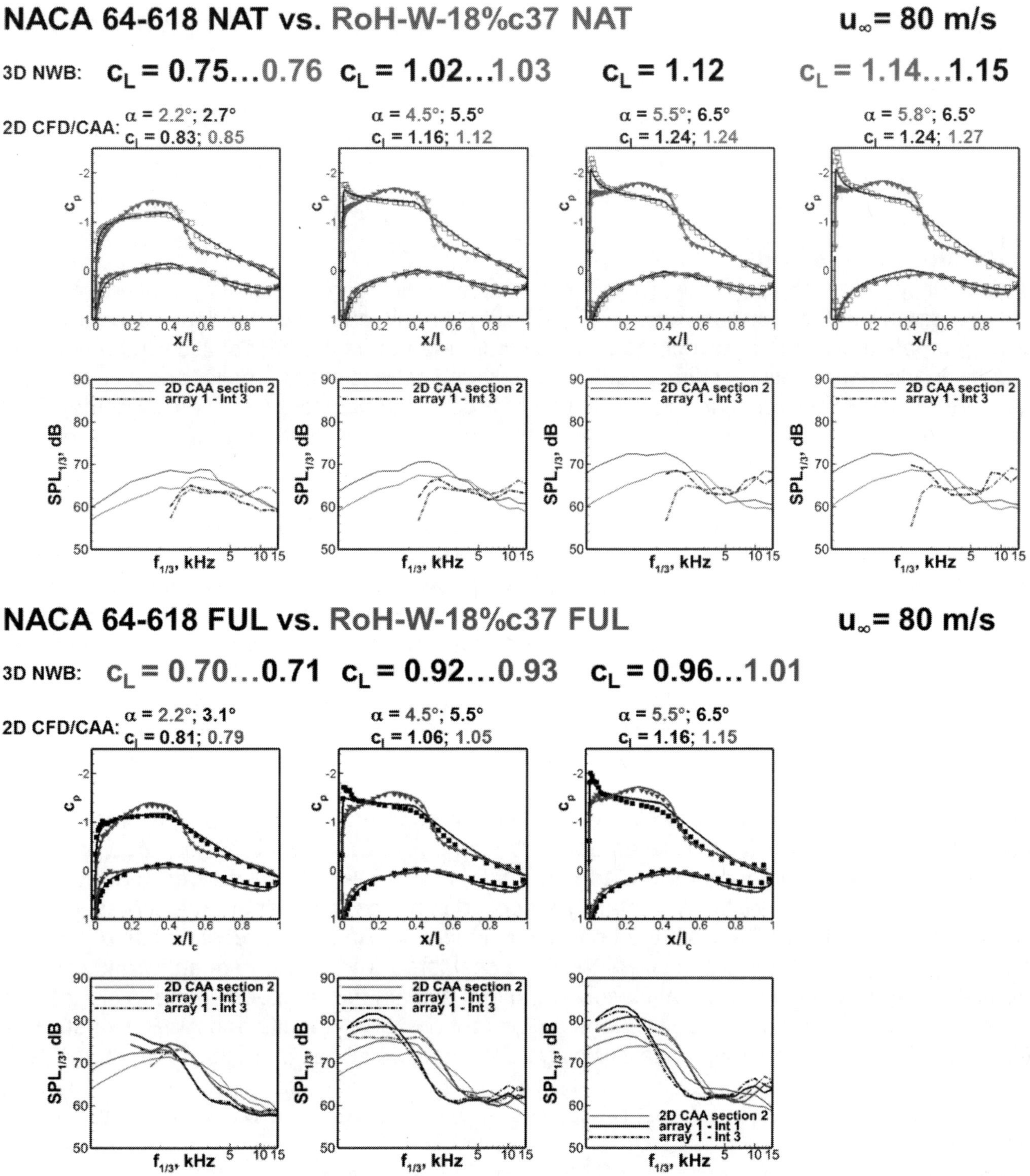

Fig. 12 1/3-octave band farfield TEN spectra (bottom line of figures), as measured with the microphone array 1, compared to 2D prediction results for conditions in section 2 and corresponding measured (symbols) and simulated (solid lines) c_p distributions (top line of figures). Filled symbols refer to tripped configurations FUL and NATFIX, open sy[illegible] tion NAT.

2D CAA predictions. This is lesser the case for configuration FUL because the flow separation at the NACA 64-618 is not reproduced in the 2D CFD results.

A-weighted representations of levels are omitted here because these are considered meaningful only for full scale conditions, being based on a previously specified (i. e. realistic) turbine application. The interested reader might refer to reference [13], where non-weighted and A-weighted results are shown for the original BELARWEA design conditions which where different from either AWB or NWB conditions. Design conditions were focusing at a projected experimental turbine and, therefore, represent a compromise between typical full scale and wind-tunnel conditions.

4. Conclusion and outlook

The present paper wraps-up first results from an extensive set of aeroacoustic experiments conducted in the framework of the German project BELARWEA. Small-scaled wind turbine blade tip models were systematically tested over a broad range of test parameters in the acoustic wind tunnel DNW-NWB, hence, supplementing previous tests at 2D blade sections in the smaller acoustic test facility AWB. Both experimental and numerical studies are reported aiming at the successful demonstration and validation of existent DLR noise prediction methodologies. A 2D-based, largely non-empirical prediction method was applied to support the design of a new airfoil contour RoH-W-18%c37. When compared to a NACA 64-618 reference contour, a 2–4 dB reduction of maximum sound pressure levels, corresponding to an OASPL reduction of 2–2.5 dB was predicted for the design lift coefficient c_l = 1.15. Measured noise reductions at the blade tip models at equivalent conditions perfectly confirm this design target. Application of brushes has been documented to achieve an additional noise reduction effect that can reach up to ~6 dB. However, the used validation/verification data base will have to be further analysed and improved prior to derive further conclusions on detailed 3D effects on noise generation. Apart from excess noise contributions from the model/floor junction, maximum noise levels are induced by trailing-edge noise (TEN) at low- to mid frequencies while noise contributions from the outer tip regions are located in the higher frequency range. The latter can reach significant levels comparable to the low-frequency peak, if a large laminar extent of the boundary layer can be realized.

Future steps will include an improved (slice-wise) post-processing of the available microphone array data for a better comparability of derived absolute sound pressure levels as measured on 2D wing sections in AWB and portions of the blade tip models in DNW-NWB. Moreover, more advanced deconvolution algorithms will be systematically applied in hopes to better suppress present excess noise contributions originating from the floor/model junctions. An on-site recalibration of the 1.6-m-diameter acoustic mirror system in DNW-NWB is considered worthwhile to quantify and correct for potential shear-layer-induced signal losses that might explain the unexpected steep slopes of the TEN spectra when compared to corresponding microphone array results and AWB acoustic mirror data. The full measurement data base covering free stream velocities of 40 m/s up to 80 m/s will be further analyzed with regard to commonly known TEN scaling laws. Based on the correspondingly normalized data sets, the applied post-processing and resulting data validity ranges can be further examined.

Ongoing work includes the full 3D CFD/CAA evaluation of the winglet configurations by means of a new hybrid DLR method, combining the RANS-based stochastic turbulence model FRPM with an efficient propagation solver by means of a fast-multipole boundary element method (DLR code FMCAS) [24, 25]. The collected database will serve for validation of the new methodology.

Appendix

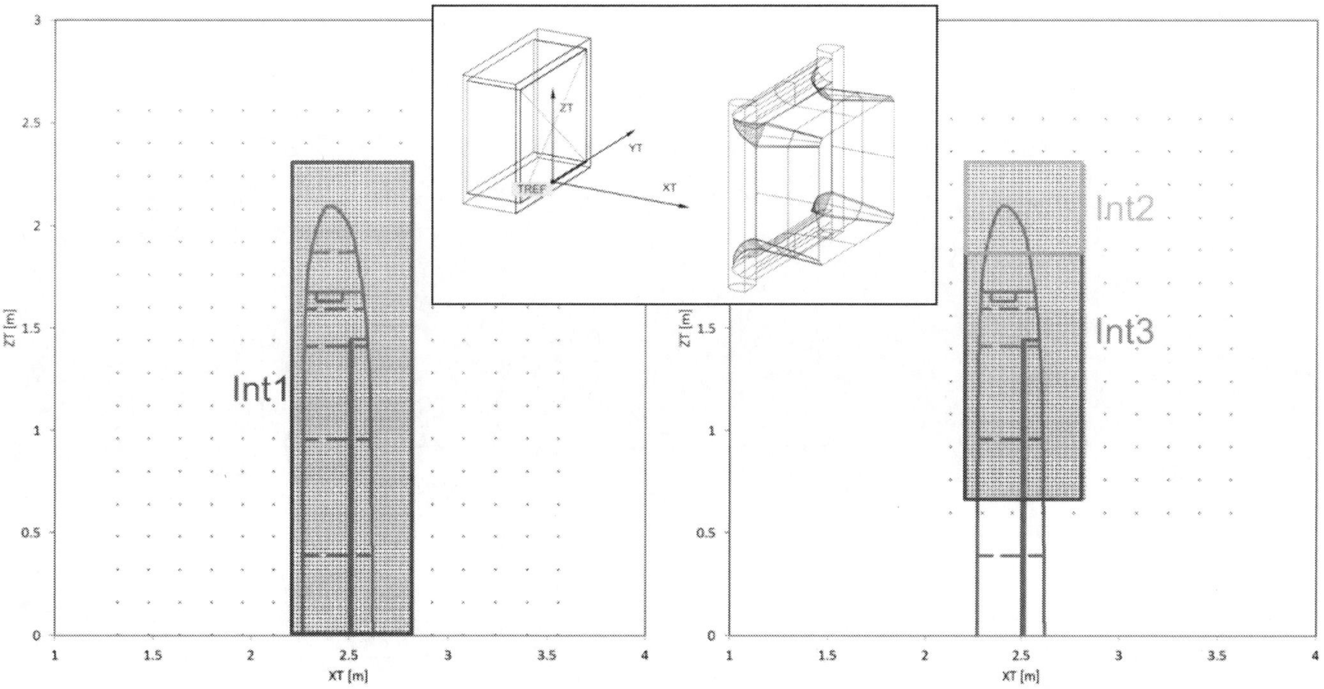

Fig. 13 Wind-tunnel coordinate system and integration areas used for the array post-processing.

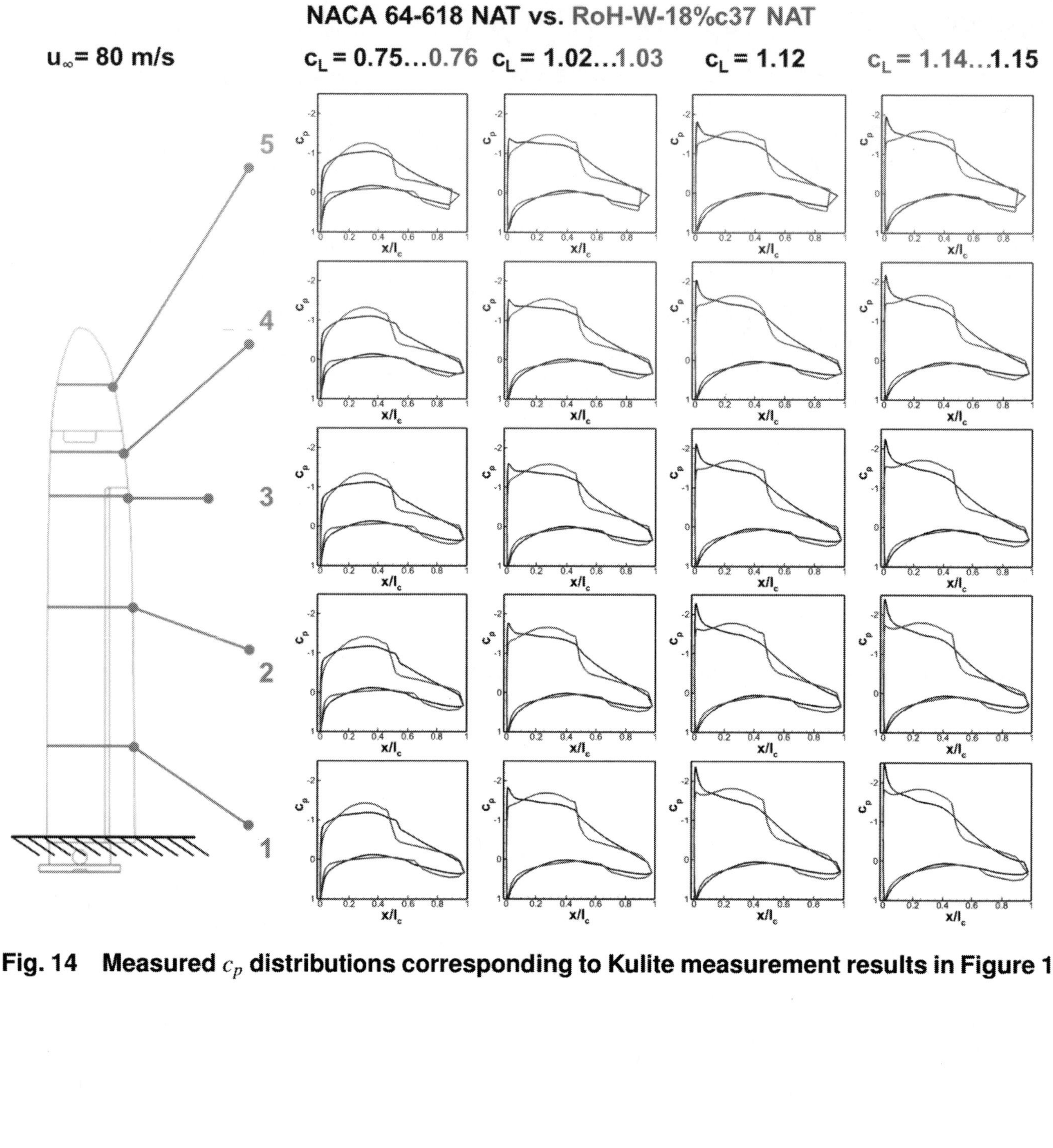

Fig. 14 Measured c_p distributions corresponding to Kulite measurement results in Figure 10.

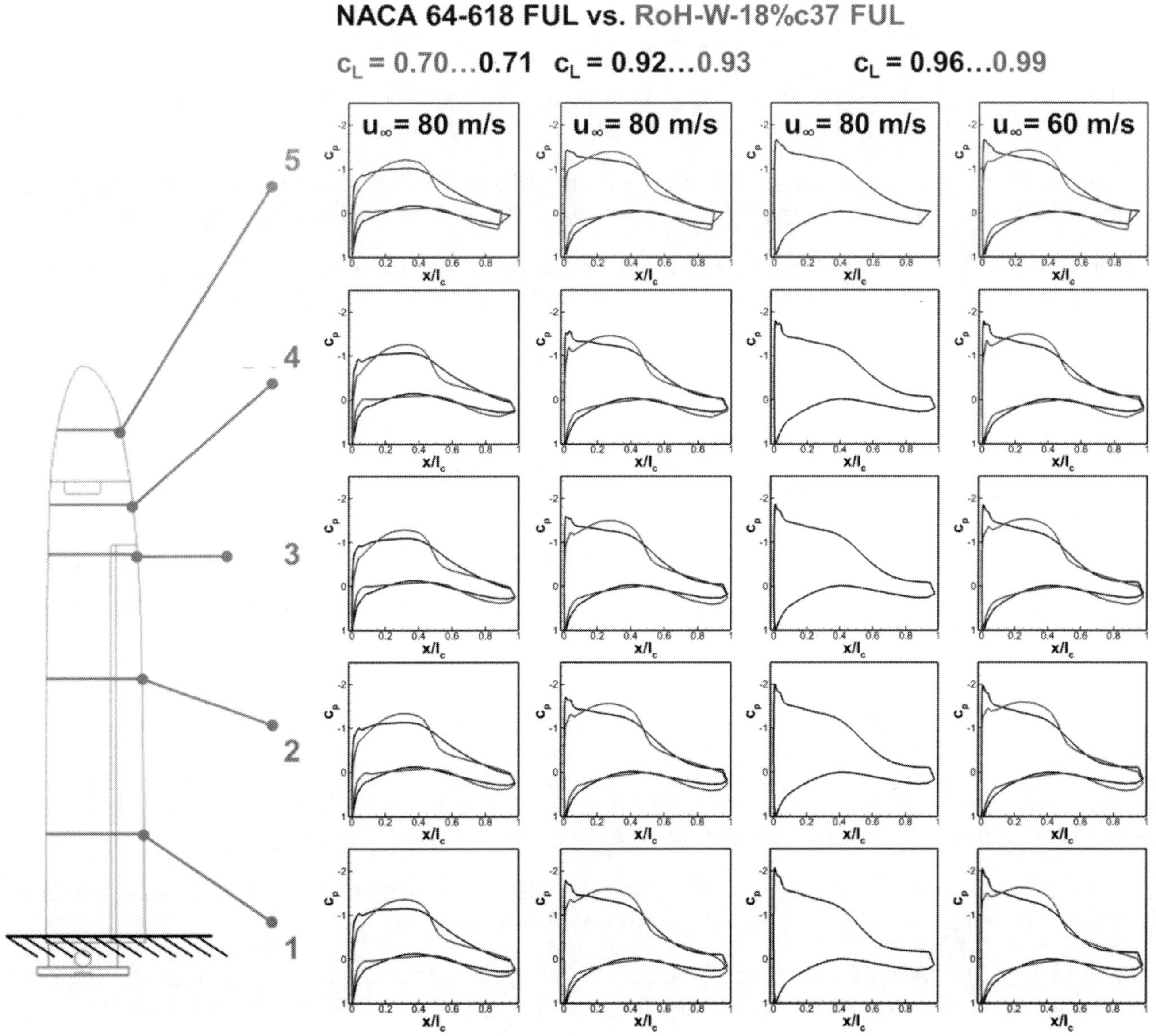

Fig. 15 Measured c_p distributions corresponding to Kulite measurement results in Figure 11.

Acknowledgments

This work has been conducted within the project BELARWEA (ref. 0325726) funded by the German Federal Ministry for Economic Affairs and Energy (BMWi). The authors are grateful to the BELARWEA consortium.

References

[1] Brooks, T. F. and Hodgson, T. H. (1981). Trailing Edge Noise Prediction from Measured Surface Pressures. *Journal of Sound and Vibration*, 78(1):69–117.

[2] Delfs, J. W., Bauer, M., Ewert, R., Grogger, H. A., Lummer, M., and Lauke, T. G. W. (2008). Numerical Simulation of Aerodynamic Noise with DLR's Aeroacoustic Code PIANO.

[3] Drela, M. (1989). An Analysis and Design System for Low Reynolds Number Airfoils. Conf. procs., Conference on Low Reynolds Number Airfoil Aerodynamics, University of Notre Dame, France.

[4] Ewert, R. (2007). RPM—the Fast Random Particle-Mesh Method to Realize Unsteady Turbulent Sound Sources and Velocity Fields for CAA Applications. AIAA Paper AIAA 2007-3506.

[5] Ewert, R. (2008). Broadband Slat Noise Prediction based on CAA and Stochastic Sound Sources from a Fast Random Particle-Mesh (RPM) Method. *Computers & Fluids*, 37:369–387.

[6] Ewert, R., Appel, C., Dierke, J., and Herr, M. (2009). RANS/CAA Based Prediction of NACA0012 Broadband Trailing Edge Noise and Experimental Validation. AIAA Paper AIAA 2009-3369.

[7] Ewert, R., Dierke, J., Siebert, J., Neifeld, A., Appel, C., Siefert, M., and Kornow, O. (2011). CAA Broadband Noise Prediction for Aeroacoustic Design. *Journal of Sound and Vibration*, 330(17):4139–4160.

[8] Ewert, R. and Edmunds, R. (2005). CAA Slat Noise Studies Applying Stochastic Sound Sources Based on Solenoidal Digital Filters. AIAA Paper AIAA 2005-2862.

[9] Ferret Gasch, O., Oerlemans, S., Faßmann, B. W., Herr, M., Bertagnolio, F., Fischer, A., Arnold, B., and Lutz, T. (2019). Trailing Edge Noise Prediction of Wind Turbine Airfoils: A Benchmark Exercise. AIAA Paper AIAA 2019-XXXX, number to be assigned.

[10] Ffowcs Williams, J. E. and Hall, L. H. (1970). Aerodynamic Sound Generation by Turbulent Flow in the Vicinity of a Scattering Half Plane. *Journal of Fluid Mechanics*, 40(4):657–670.

[11] Gerhold, T., Galle, M., Friedrich, O., and Evans, J. (1997). Calculation of Complex Three-Dimensional Configurations Employing the DLR-τ-Code. AIAA Paper AIAA 1997-0167.

[12] Herr, M. (2013). *Trailing-Edge Noise—Reduction Concepts and Scaling Laws.* Dissertation, Technische Universität Braunschweig, published as DLR Research Report 2013-32, ISSN 1434-8454, ISRN DLR-FB-2013-32.

[13] Herr, M., Ewert, R., Faßmann, B., Rautmann, C., Martens, S., Rohardt, C.-H., and Suryadi, A. (2018). Noise Reduction Technologies for Wind Turbines. *Notes on Numerical Fluid Mechanics and Multidisciplinary Design (NNFM)*, 136, Hrsg: A. Dillmann, G. Heller, E. Krämer, C. Wagner, S. Bansmer, R. Radespiel, R. Semaan, New Results in Numerical and Experimental Fluid Mechanics XI, Contributions to the 20th STAB/DGLR Symposium Braunschweig, Germany, 2016, Springer International Publishing AG:611–621.

[14] Herr, M., Ewert, R., Rautmann, C., Kamruzzaman, M., Bekiropoulos, D., Iob, A., Arina, R., Batten, P., Chakravarthy, S., and Bertagnolio, F. (2015). Broadband Trailing-Edge Noise Predictions—Overview of BANC-III Results. AIAA Paper AIAA 2015-2847.

[15] Herr, M. and Kamruzzaman, M. (2013). Benchmarking of Trailing-Edge Noise Computations—Outcome of the BANC-II Workshop. AIAA Paper AIAA 2013-2123.

[16] Howe, M. S. (1978). A Review of the Theory of Trailing Edge Noise. *Journal of Sound and Vibration*, 61(3):437–465.

[17] Hu, F. Q., Hussaini, M. Y., and Manthey, J. L. (1996). Low-Dissipation and Low-Dispersion Runge–Kutta Schemes for Computational Acoustics. *Journal of Computational Physics*, 124(1):177–191.

[18] Jonkman, J., Butterfield, S., Musial, W., and Scott, G. (2009). Definition of a 5-MW Reference Wind Turbine for Offshore System Development. NREL Technical Report NREL/TP-500-38060.

[19] Oerlemans, S. (2016). Reduction of Wind Turbine Noise using Blade Trailing Edge Devices. AIAA Paper 2016-3018.

[20] Oerlemans, S., Fisher, M., Maeder, T., and Kögler, K. (2009). Reduction of Wind Turbine Noise using Optimized Airfoils and Trailing-Edge Serrations. *AIAA Journal*, 47(6):1470–1481.

[21] Oerlemans, S., Sijtsma, P., and López, B. M. (2007). Location and Quantification of Noise on a Wind Turbine. *Journal of Sound and Vibration*, 299:869–883.

[22] Rautmann, C. (2017). *Numerical Simulation Concept for Low-Noise Wind Turbine Rotors*. Dissertation, Technische Universität Braunschweig, published as DLR Research Report 2017-35, ISSN 1434-8454, ISRN DLR-FB-2017-35.

[23] Rautmann, C., Dierke, J., Ewert, R., Hu, N., and Delfs, J. (2014). Generic Airfoil Trailing-Edge Noise Prediction using Stochastic Sound Sources from Synthetic Turbulence. AIAA Paper AIAA 2014-3298.

[24] Reiche, N., Ewert, R., and Delfs, J. (2016). Realization of Arbitrary Vorticity Spectra using Generic Stochastic Turbulence. AIAA Paper AIAA 2016-2964.

[25] Reiche, N., Lummer, M., Ewert, R., Delfs, J. W., and Alavi Moghadam, S. M. (2015). Towards High-Lift Noise from Fast Multipole BEM with Anisotropic Synthetic Turbulence Structures. AIAA Paper AIAA 2015-2672.

[26] Schlinker, R. H. (Oct. 1977). Airfoil Trailing Edge Noise Measurements with a Directional Microphone. AIAA Paper 77-1269.

[27] Schwamborn, D., Gerhold, T., and Heinrich, R. (2006). The DLR TAU-Code: Recent Applications in Research and Industry. In *ECCOMAS European Conference on Computational Fluid Dynamics 2006*.

[28] Suryadi, A., Martens, S., and Herr, M. (2017). Trailing-Edge Noise Reduction Technologies for Applications in Wind Energy. AIAA Paper AIAA 2017-3534.

[29] Tam, C. K. W. and Webb, J. C. (1993). Dispersion-Relation-Preserving Finite Difference Schemes for Computational Acoustics. *Journal of Computational Physics*, 107(2):262–281.

INCE*EUROPE*

8th International Conference
on
Wind Turbine Noise
Lisbon – 12th to 14th June 2019

Development of Design Guidelines for Low Noise But High Yield Wind Turbines

C. Hornung, Th. Lutz, E. Krämer
Institute of Aerodynamics and Gas Dynamics, University of Stuttgart, Germany

Summary

The reduction of wind turbine noise has been an ongoing effort over the past years, during which a variety of methods were developed. Today the reductions which are the easiest to achieve are already exploited, for example mechanical noise or stall noise is no longer an issue for modern wind turbines. In the audible range basically only trailing edge (TE) noise and inflow noise remain. However, reducing these further is a challenge which, most probably, only myriads of small steps during the whole design process of a turbine will conquer. In the perspective of a silent turbine design, the first steps take place during the design of the airfoils followed by those during the blade design. Subsequently adequate add-ons, for example serrations, and maybe even expensive active noise reduction methods as suction are selected. Finally the operation management must be suitably chosen. The focus of the present study is the aspect of loft design. It aims at the development of universal guidelines for blades with low trailing edge noise emissions but nevertheless high yield. To determine the TE noise, the in-house code IAGNoise+ is employed, which is based on a modified TNO-model. The power is calculated by a blade element method. In order to approach the subject in a structured way, a design of experiment method as developed by Taguchi was chosen. It is conducted for different sections, and thus different airfoils, in the outer part of a reference blade from the TUM reference turbine. The examined parameters are chord, pitch and twist. The analysis focuses on the aeroacoustics and aerodynamics. Restrictions due to loads are assumed to be equal to the reference blade. In can be shown that for each blade section a combination of angles and chord exist, for which an optimum between noise and power is achieved.

1. Introduction

The tonal components in the emitted aeroacoustic sound spectrum of wind turbines are usually amplified by feedback loops or periodic vortex separation. They are thus based on distinct processes, which can be manipulated relatively easy in order to reduce the noise emission. The broadband part of the spectrum, however, which is related to interactions of the blade with turbulence is significant more challenging to reduce. Positively influencing the complex mechanisms underlying the turbulence production and dissipation in order to reduce the noise sources is extremely challenging. Thus, the prevailing sound source in the audible range of the emitted spectra from wind turbines is still the turbulence induced trailing edge noise (short: TE noise). Over the years various methods have been proposed in literature to conquer TE noise.

Already during the design of the airfoils, attention should be paid to a noise-optimised design by keeping the turbulent length scale and the turbulent kinetic energy close to the TE as low as possible, in particular on the suction side. This is achieved by delaying the laminar to turbulent transition, but at the same time avoid separation bubbles. In addition it should be aimed for a soft pressure recovery by balancing the pressure regain between suction and pressure side [15]. Rautmann [18] suggest the use of a concave suction side close to the TE and a shift of load to the pressure side by using rear loading to compensate for the loss in lift. [18]. Different authors achieve noise reduced airfoils by utilizing optimization routines paired with semi-empirical or computational aeroacoustics (CAA) methods. Compared to the baseline airfoils, noise reductions by more than 2 dB are obtained. [2, 13, 21]

The blades of the rotor are then designed based on the set of (noise-optimised) airfoils. In all studies found, in which the noise reduction of the rotor is compared to the reduction of the airfoil itself, a lower or no noise reduction was measured for the rotor compared to the optimized airfoil only. This makes it clear that it is of great relevance how exactly the rotor is assembled from the blade sections. Some studies have already dealt with the topic and improved the geometry of individual rotors with the help of an optimization routine. Most of the optimized rotors had slightly larger chord and an increased twist angle than the baseline versions. [3, 5, 13, 14, 21].

A commonly chosen method to reduce noise emission in particular for already built turbines is to modify the operation mode. Since TE noise scales with the fifth power of the inflow Ma number, a reduction of rpm and thus the tip speed ratio immediately reduces the noise emission. However, as pointed out by [3] this results in higher loads on generator or gearbox and is thus only feasible within a small range without changing to a more expensive drive train.

Further noise reductions are achieved with so called add-ons. Serrations are already state-of the art. Almost all manufacturers now equip the blades with them as standard. The narrow triangles attached to the TE showed very promising results in wind tunnel test, where reductions of up to 6 dB were achieved. On the rotor, however, the reduction is less pronounced with 2 -3 dB [17]. Novel approaches include different forms of serrations, leaving the classical triangle behind, porous TE, combed serrations or even only combs. In particular the latter ones have proven significant effect in wind tunnel measurements. [4, 8, 17, 19, 20]

In addition, active methods are also investigated in research. Since they are usually associated with considerable costs for additional systems and it is difficult to design additional systems for the long operating life of wind turbines over several decades, so far, they have only been used in experimental studies. For example, the boundary layer on the suction side of the blade can be sucked in through a perforated plate to reduce the boundary layer thickness and the turbulent kinetic energy at the TE [1].

An alternative method is to influence the boundary layer by plasma actuators. [12]
To reach the goal of silent but at the same time power and cost efficient turbines, different described methods have to be combined. The least complex approaches are now used as standard and have already led to a significant reduction in noise emission. In order to further reduce trailing edge noise, it is inevitable to consider the noise mechanisms in the entire design process of the turbine.

The work presented here addresses the blade design process. Instead of optimizing a specific rotor, as is the case in the literature (see above), the goal is the development of general guidelines for blade design for low noise but nonetheless high yield. For this purpose two different turbines were examined, one of them is a commercial multi-megawatt turbine, the other is the TUM reference turbine designed in the framework of the IEA wind task 37 [3]. The results of the study are illustrated at the example of the reference turbine, but the conclusions drawn coincided for both turbines which supports the universality of the guidelines. To make the study more efficient a structured approach is chosen based the Taguchi method, a Design of Experiment (DoE) approach [7]. In addition, rather than examining the whole blade with multiple influencing parameters, it was focused on certain blade sections, similar to the blade element method (BEM) commonly used for power based turbine design.

The paper is structured as follows: Section 2 explains the general procedure followed, as well as the theory behind the noise and power prediction method employed. Afterwards the results of the approach are presented and the derivation of general guidelines is discussed in section 3. The conclusions drawn from the study for blade design are summarized at the end of the paper.

2. Methods

The TUM reference turbine is a model turbine with a rotor diameter of 130 m and a rated power of 3.6 MW at a tip speed ratio (TSR) of 8.869 [3]. With a cut in wind speed of 3 m/s the turbine is explicitly designed for use in low wind conditions, where noise emission plays an increased role due to low background noise. The blades are constructed on the basis of a DU airfoil series.
Instead of examining the whole blade, as conducted by the studies presented above, only sections of a blade are considered in the present work. The benefit of this quasi-2D approach is a faster evaluation and a significant reduction of the number of factors influencing the noise emission. Since the TE noise stemming from the outer part of the rotor prevails the overall spectrum the study is concentrated on two blade cuts with different relative thicknesses at 75 % and 98 % of the rotor diameter, which corresponds to a DU91-W2-250 and a DU08-W-210 airfoil, respectively. The highest elevation of emitted noise over background noise is found near the operating point for rated power. Below that point, rpm and thus tip speeds are smaller, leading to lower noise emission because of the Ma^5 scaling of TE noise and above the background noise increases, because of higher wind speeds. Thus, this point was selected for the here presented analysis. The reader has to keep in mind that the study is performed with the goal of a power and sound optimized turbine for low to rated power operating conditions. Restrictions due to structure or load as well as due to negative stall effects at very high wind speeds or fluttering at stand still are not taken into account. These effects of course determine the boundary conditions of the aerodynamically-acoustically optimized turbine in an industrial development process.

2.1. Procedure

A section at any position along the blade can be described with the parameters depicted in figure 1 and the airfoil's geometry. Keeping the airfoil as well as the position along the blade constant and implying constant TSR and wind speed, two independent parameters, chord c and angle of incidence ξ, remain. Note that the angle towards the undisturbed flow is not equal to the angle of attack experienced by the airfoil. The parametric study is thus conducted by varying angle of incidence and chord for the two different sections. In order to evaluate the influence of both parameters as well as

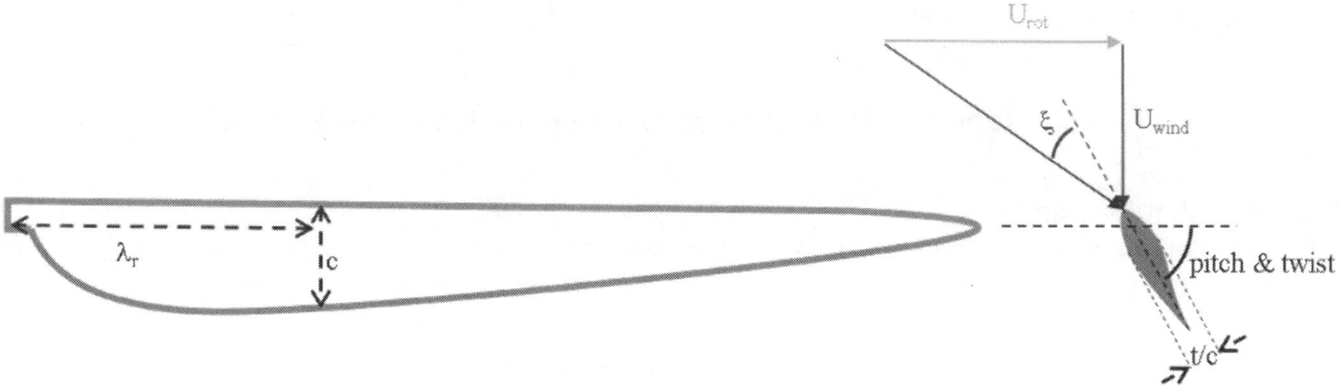

Fig. 1 Visualization of the quasi 2D approach with blade segments.

the interaction, the Taguchi method is employed. Different pre-desigend test matrices are available for this purpose, which satisfy the conditions of orthogonality and balance of the test design. Here the $L_{25}(5^6)$ matrix was used (cf. [7]). It comprises 25 different simulations, each with different settings. The two parameters, chord and angle of incidence were varied with 5 steps (levels) for each. Only the first two columns of the Taguchi test matrix were occupied, which makes it possible to evaluate the interactions between chord and angle of incidence. An overview over the range of the selected levels is given in table 1. The upper and lower boundaries for the chord were set to reasonable numbers around the design chord length but trying to explore the available space. The range for the angle of incidence were selected in such a way, that the lowest occurring lift coefficient was only just positive and the highest was close to stall (for the design Re number). The levels in between were equally spaced. After the evaluation of this first quite rough matrix, it was decided to define three further more refined matrices by selecting the most promising parameter ranges according to the matrix with the rough steps. For each combination of chord and angle of incidence the local power coefficient $C_{p,2D}$ and the local sound pressure level (SPL) are determined with respect to a spanwise width of the segment of 1 m. The power coefficient is calculated by means of the blade element method (BEM) (cf. section 2.4) and the SPL is predicted utilizing the in-house code IAGNoise+ based on RANS CFD simulations (cf. section 2.3). The inflow conditions for the RANS simulations are also gained from the BEM to consider the effects of induction on the noise emission. The results of the rough test matrices are then examined based on the Taguchi method. Afterwards the three refined matrices were defined for each section, in order to extract general design guideline.

	DU08-W-210 (rad pos.: 98 %)			
	coarse	fine 1	fine 2	fine 3
Chord [m]	0.1 - 1.7	1.3 - 1.7	0.1 - 0.9	0.1 - 0.9
Angle of incidence [°]	(-6.0) - 9.0	1.5 - 5.25	(-6.0) - 1.5	1.5 - 5.25

	DU91-W2-250 (rad pos.: 75 %)			
	coarse	fine 1	fine 2	fine 3
Chord [m]	0.5 - 2.6	1.75 - 2.375	0.5 - 1.75	0.5 - 1.75
Angle of incidence [°]	(-3.0) - 11.0	4.0 - 7.5	(-3.0) - 4.0	4.0 - 7.5

Table 1 Test matrices for the parametric study.

2.2. CFD simulations

CFD simulations were conducted to gain the polars for the BEM as well as for each test case to determine the noise. The simulations were conducted with the structured, compressible solver FLOWer, developed by the DLR and extended by the IAG. In order to use the same grid for all simulations of one airfoil a very fine mesh was designed consisting of approximately 0.15 Mio. cells in 18 Blocks. It was ensured that for all simulations the y^+ value in the cell closest to the wall is smaller than 1 and the cells expand with a ratio of 1.1 in wall normal direction. The far field was set to a distance greater than 80 chord lengths. To depict the effects of turbulence the two equation Menter SST Turbulence model was employed for the fully turbulent simulations.

2.3. Noise prediction

The noise is predicted using the in-house code IAGNoise+ [9]. Roughly sketched, the procedure can be divided in three steps: First boundary layer data close to the TE is extracted from a RANS CFD simulation. The data is then processed with statistical methods to determine the wall pressure fluctuations, employing a TNO type model. These type of models are based on a semi-analytical expression describing the power density of the frequency wave number spectrum of the wall pressure fluctuations (WPF) in dependence from turbulence characteristics. It is derived from a Poisson equation for pressure fluctuations via the use of a Greens-Function and Powel's reflection theorem. The turbulence characteristics are described with spectral model parameters. Within IAGNoise+ an enhanced version of the TNO model is implemented [9]. The governing equation for the power spectral density of the wall pressure fluctuations is:

$$P(k_1, k_3, \omega) = \\ 4\rho^2 \left(\frac{1}{k_1^2 + k_3^2}\right) \cdot \int_0^\infty \left(k_1^2 \left(\frac{\partial \overline{U_1}}{\partial x_2}\right)^2 + \frac{2\left(k_1^2 + k_3^2\right)}{15\nu}\varepsilon \right) \Lambda_2 \Phi_{22} \cdot \langle u_2'^2 \rangle \Phi_m e^{-2|\mathbf{k}|x_2} dx_2 \quad (1)$$

In the equation ρ represents the density, the wave-numbers in stream-wise, wall normal and span-wise direction are depicted by k_1, k_2, k_3 , respectively. The frequency is represented by ω. The vertical integral length scale Λ_2 is a measure for the size of the energy containing eddies in the boundary layer and the moving axis spectrum Φ_m describes the distortion of the spectrum when being convected in direction of the TE. The normalized spectrum of the vertical velocity fluctuations is given by Φ_{22} with the corresponding Reynolds stresses $\langle u_2'^2 \rangle$. Further model parameters are the dissipation ε and the viscosity ν. Compared to the original model it takes not only the interactions between the turbulence and the mean shear into account but also the interaction amongst the turbulence itself. Further, the anisotropy of the turbulent fluctuations is accounted for based on the approach by Kamruzzaman [11], but considering pressure gradients. Several smaller modifications include a different approach for the moving axis spectrum or the length scales. As a last step the results of the WPF model are fed into a far field model to determine the sound pressure level. All equations can be found in [9]. All in all, the different modifications improved the prediction quality of the model, in particular for high lift configurations. It has been shown, that with the current model it is possible to predict the noise of wind turbine rotors to a good degree of certainty. The CFD simulations were evaluated very close to the TE at 99,5 % of the chord. Predicting the noise in combination with FLOWer CFD SST simulations have been shown to gain good results for turbines as well as for airfoils [9, 10].

2.4. BEM

The airfoil polars, which are required for the BEM, are determined with the CFD code FLOWer. They were calculated for the design Reynolds number of 6 Mio. for the DU08-W-210 and 8 Mio. for the DU91-W2-250. Since the inflow conditions change in the course of the parameter study, the Reynolds number changes as well. However, the deviations due to differing Reynolds numbers for the different settings in the parametric study is neglected. The just-in-time calculation of polars with XFOIL with the exact required Reynolds number showed a larger difference to the final lift coefficient of the subsequent FLOWer simulation. Upon convergence of the CFD simulations the TE noise emission of each section is predicted with IAGNoise+. The contribution of each section to the overall power of the turbine is evaluated by a BEM formulation for the local power coefficient. For a whole turbine the power coefficient can be determined via (cf. e.g. [16]):

$$C_P = \frac{8}{\lambda^2} \int_{\lambda_h}^{\lambda} \lambda_r^3 a'(1-a) \left[1 - \frac{c_d}{c_l} tan\phi \right] d\lambda_r \tag{2}$$

The local power coefficient of each section with respect to 1 m $C_{P,2D}$ is calculated by breaking down the continuous integral for the overall rotor C_p in eq. (2) to discrete sections, resulting in:

$$\begin{aligned} C_{P,2D} &= \frac{8}{\lambda^2} \lambda_r^4 a'(1-a) \left[1 - \frac{c_d}{c_l} tan\phi \right] \Delta\lambda_r \\ &with \\ \Delta\lambda_r &= (\lambda_r(r+0.5m) - \lambda_r(r-0.5m)) \end{aligned} \tag{3}$$

In the equations a describes the axial induction whereas a' represents the tangential induction. The tip speed ratios at the hub, any radial position and the tip are denoted by λ_H, λ_r, λ, respectively. The lift- and drag coefficients are represented by c_l and c_d. In addition, the angle to the local inflow (including induction effects) φ is required to calculate the power coefficients.

3. Results and Discussion

In figure 2 the results of all test matrices are plotted. Shown are the local power coefficients over the axial induction colored by the total sound pressure level. Data points with an induction factor greater than 0.4 are excluded, since the classical BEM method is prone to errors for these induction factors (cf. also [6]). The symbols with lower $C_{P,2D}$ correspond to the DU91-W2-250 airfoil whereas the higher located points stem from the simulations with the DU08-W-210 airfoil. Besides few exceptions, the noise emission can be observed to increase with rising induction factor. The local power coefficient, however, shows a maximum at approximately 0.33, as expected from Betz's theory, and then decreases again. Therefore, over-induced turbines are associated with a higher noise emission for the same power coefficient. Thus, while an over-induction may be useful to maximize the energy yield for gusty inflow, this approach is not advisable for low-noise turbines.

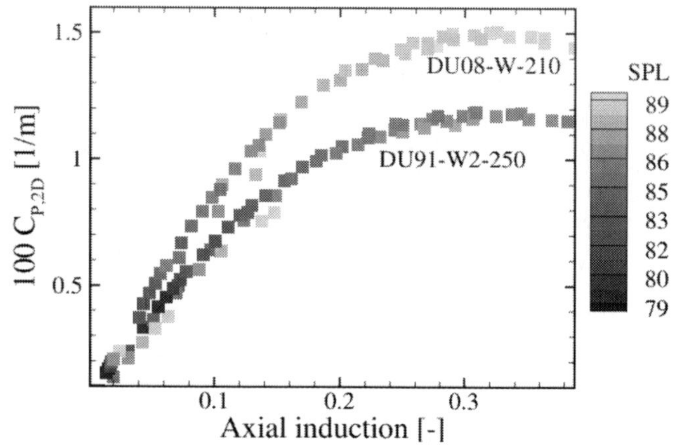

Fig. 2 Local power coefficient and SPL for both airfoils.

The strength of the Taguchi method lays in its statistical evaluation. Here the Analysis of Mean (ANOM) procedure is pursued. It was adapted by Taguchi to the orthogonal fields. It allows for an estimation of the tendency of the different parameters over the range of levels. In addition the influence of different factors can be compared. Thus, it gives an idea of the direction for optimization.[7] The analysis is shown in figure 3 for the coarse test cases only (cf. table 1). They comprise the widest range of both parameters and are thus more suitable to evaluate trends. Note that all points were included in the analysis, also those with induction factors larger than 0.4. However, since the trends are depicted by the BEM also for high induction factors, the overall tendencies shown in the plots should still be valid. On the left (fig. 3(a)) the results are depicted for the SPL. The grey dashed line shows the overall mean value. The green line focuses on the variation of the chord. Each point on the line represents the mean value over those simulations in which the chord length is set to the corresponding level. The conclusion can be drawn from the plot, that with higher chord length the SPL increases in average. This holds for both airfoils. The blue line shows the level averaged behaviour for the influence of the incidence angle. In contrast to the chord, here a local minimum can be observed. Overall the influence of the incidence angle is greater than that of the chord. The reason for the maximum being located at level five for the DU08-W-210 airfoil but at level one for the DU91-W2-250 airfoil lays in the greater contribution of the pressure side's noise emission at low angle of attack for the latter airfoil.

Turning the attention to the right handed figure (fig. 3(b)), shows the same analysis but now for the local power coefficient. The range of the parameters can be considered sufficient to find optimal points, since also here the maximum can be observed although the levels were chosen somewhat arbitrarily. For the DU08-W-210 airfoil it coincides for the factor of chord and angle at level four. For the thicker airfoil, DU91-W2-250, the maximum of the chord is located at level two, and the maximum for the angle at level four. As expected, the optimum levels in angle differ for SPL and $C_{P,2D}$, indicating

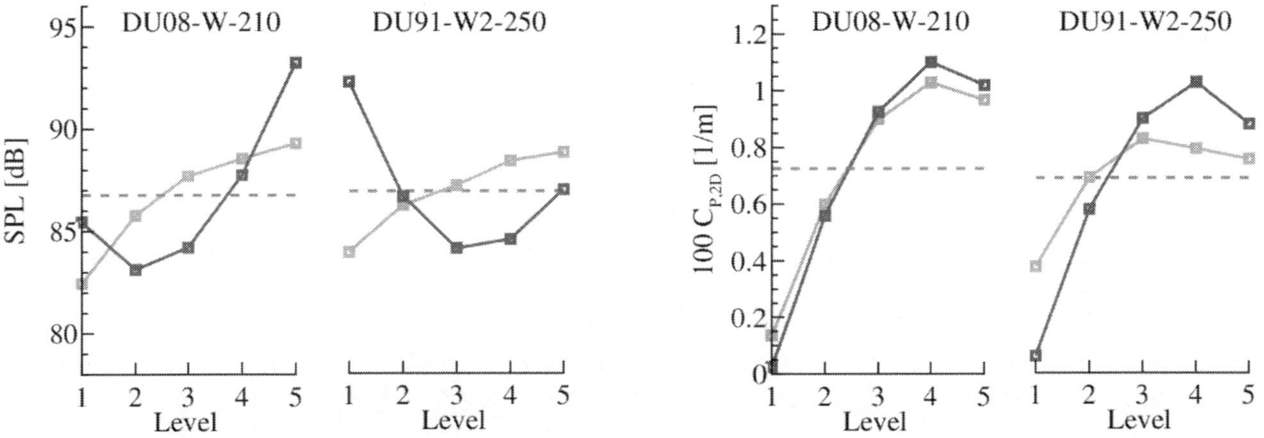

(a) ANOM SPL Caption

(b) ANOM Cp Caption

Fig. 3 ANOM for *SPL* and $C_{P,2D}$ with the overall mean value --, the influence of the chord -□- and of the angle of incidence -□-.

a Pareto front problem.
The ANOM separates the influence of different factors. The interaction between the two parameters can then be graphically evaluated as shown in figure 4.

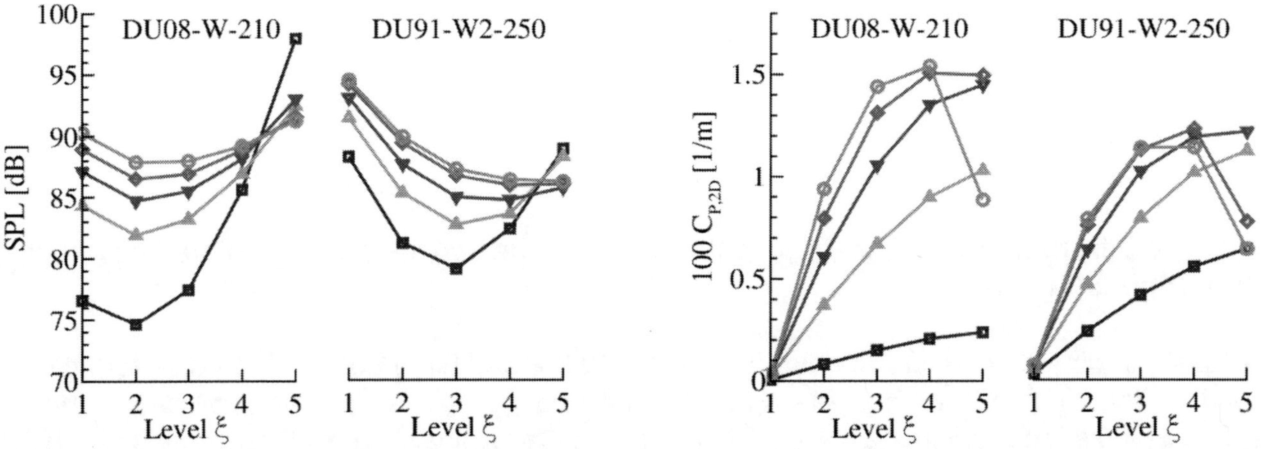

(a) SPL

(b) Cp

Fig. 4 Interaction diagram according to Taguchi with chord level 1 -□-, level 2 -△-, level 3 -▽-, level 4 -◇-, level 5 - ○ -.

Depicted are the results with the different levels of the angle of incidence on the abscissa. The different lines represent the levels of the chord length with dark blue being the smallest chord and bright blue representing the largest. Parallel curves would indicate that no interactions are present. Here, the directing of the curves stays the same over the lower range of levels for C_P as well as for

SPL. Thus, the parameters have synergetic interactions. An increase in chord leads to a similar result as an increase in angle of incidence. It can be concluded that both parameters have to be modified to achieve a low noise - high yield configuration.
In order to determine the best operating points for each airfoil, the results of all test matrices, including the refined ones are evaluated (cf. table 1). Points with induction factors higher than 0.4 are again excluded for the above mentioned reasons and because of their lack of relevance for real turbine design. In figure 5 the difference in local power coefficient and SPL between the original design and the simulations of the test matrices are depicted. They are defined as:

$$\Delta SPL = SPL_{test} - SPL_{design} \qquad \Delta C_{P,2D} = C_{P,2D,test} - C_{P,2D,design} \tag{4}$$

All points are colored by their corresponding lift-to-drag ratio, with the lift and drag coefficients taken from the polars of the design configuration. While the DU91-W2-250 airfoil is already operated in an

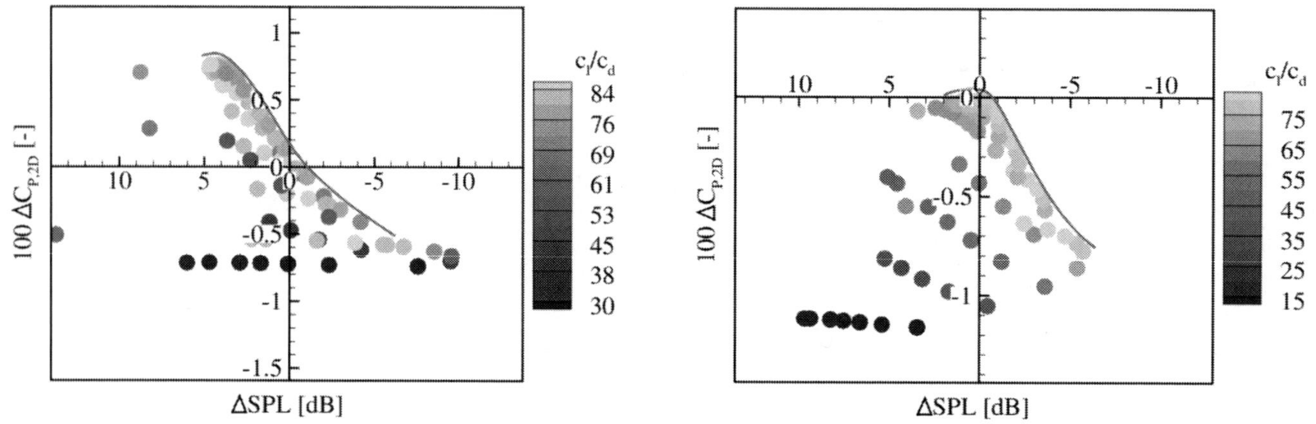

(a) DU08-W-210

(b) DU91-W2-250

Fig. 5 Pareto front for low noise -high yield design of the two sections, depicted as deviation from the original TUM design.

optimal design point in the reference turbine, there is still a little bit of potential in noise reduction for the DU08-W-210 airfoil. For both airfoils a point of maximum possible power can be determined. As typical for optimization problems a Pareto front forms out (grey line). Points located close to the line are characterized by the lowest possible noise emission for a certain local power coefficient for one distinct airfoil. It stands out that the airfoils are operated in their highest lift-to-drag ratio point when being close to the optimal line. The results for the DU08-W-210 airfoils are, however, influenced by Re number effects. Some data points with high lift-to-drag ratio are located slightly away from the Pareto front. They are, however, associated with the smallest tested chord length, which's Re number is only 20% of the design Re used to determine the polars.

4. Conclusion

A parametric study was conducted based on the blade and airfoil design from the TUM reference turbine to develop general guidelines for blades with low trailing edge noise emissions but nevertheless high yield. To facilitate the evaluation as well as computational time a quasi-2D approach was used. Two different sections, corresponding to two different radial positions and airfoil geometries were examined. Two different parameters were varied: The incidence angle (inclination towards the undisturbed inflow from pitch and twist) and the chord length. From the results several conclusions were drawn for the development of silent but powerful wind turbines. The most important being that for low noise wind turbines small effective angles of attack should be combined with high chord lengths aiming to operate the airfoils at their highest lift-to-drag-ratio point. Thus, for each radial position-airfoil combination a setting exists leading to the point of maximum power with minimal noise. These points are characterized by small effective angles of attack and high chord lengths. Over-induction should in general be avoided since it leads to higher noise only without increasing the power output. Restrictions due to structure or load as well as due to negative stall effects at very high wind speeds or fluttering at stand still were not taken into account.

Acknowledgments

Parts of this work were supported by Enercon GmbH Aurich. Further, we thank Pietro Bortolotti (previously University of Munich, now at NREL) for providing us with the geometry of the TUM reference turbine.

References

[1] Arnold, B., Lutz, T., and Krämer, E. (2018). Design of a Boundary-Layer Suction System for Turbulent Trailing-Edge Noise Reduction of Wind Turbines. *Renewable Energy Journal*, 123:249–262.

[2] Bertagnolio, F., Madsen, H. A., and Bak, C. (2009). Experimental Validation of TNO Trailing Edge Noise Model Validation and Application to Airfoil Optimization. In *EWEC Proceedings online.*

[3] Bortolotti, P., Sucameli, C., Croce, A., and Botasso, C. (2018). Integrated design optimization of wind turbines with noise emission constraints Integrated design optimization of wind turbines with noise emission constraints. In *IOP Conf. Series: Journal of Physics: Conf. Series 1037 (TORQUE).*

[4] Braun, K. A., Gordner, A., v.d. Borg, N., Dassen, A., Doorenspleet, F., and Parchen, R. (1998). Final Report: Serrated Trailing Edge Noise (STENO). Technical report.

[5] Fuglsang, P. and Madsen, H. A. (1998). Wind turbine design with numerical optimization and a semi-empirical noise prediction model.

[6] Guma, G., Bangga, G., Jost, E., Lutz, T., and Kräemer, E. (2018). Consistent 3D CFD and BEM simulations of a research turbine considering rotational augmentation. In *Torque Proceedings.*

[7] Henes, J. D. (2009). *Anwendung der Taguchi-Methode zur Quantifizierung von Fehler- und Unsicherheitsquellen der numerischen Aerodynamik.* Studienarbeit, Stuttgart.

[8] Herr, M., Rossignol, K.-S., Delfs, J., Lippitz, N., and Mößner, M. (2014). Specification of Porous Materials for Low-Noise Trailing-Edge Applications. *20th AIAA/CEAS Aeroacoustics Conference*, (June):1–19.

[9] Hornung, C., Lutz, T., and Krämer, E. (2019). A model to include turbulence-turbulence interaction in the prediction of trailing edge far fi eld noise for high angles of attack or slightly separated fl ow. 136:945–954.

[10] Hornung, C., Scheit, C., Napierala, C., Arnold, M., Bekiropoulos, D., Altmikus, A., and Lutz, T. (2017). Predicted and Measured Trailing-Edge Noise Emission for a 2.3 MW Wind Turbine. In *7th International Conference on Wind Turbine Noise*, Rotterdam.

[11] Kamruzzaman, M., Bekiropoulos, D., Wolf, A., Lutz, T., and Kramer, E. (2014). Rnoise: A RANS based airfoil trailing-edge noise prediction model. *AIAA conference*, pages 1–22.

[12] Koike, S., Kato, H., Kobayashi, H., Ura, H., Mitsuo, K., Atobe, T., Okubo, T., Nakakita, K., Watanabe, S., Tunnel, W., Researcher, A. S., Group, F. D., Researcher, S., Student, G., Tokyo, S.-k., and Leader, G. (2010). Time-Resolved PIV Applied to Trailing-Edge-Noise Reduction by DBD Plasma Actuator. *27th AIAA Aerodynamic Measurement Technology and Ground Testing Conference*, (July):AIAA 2010–4352.

[13] Lee, S., Lee, S., Ryi, J., and Choi, J.-s. (2013). Design optimization of wind turbine blades for reduction of airfoil self-noise. *Journal of Mechanical Science and Technology*, 27(2):413–420.

[14] Leloudas, G., Zhu, W. J., Sørensen, J. N., Shen, W. Z., and Hjort, S. (2007). Prediction and Reduction of Noise from a 2.3 MW Wind Turbine. *Journal of Physics: Conference Series*, 75(October 2005):012083.

[15] Lutz, T., Herrig, A., Würz, W., Kamruzzaman, M., and Krämer, E. (2007). Design and Wind-Tunnel Verification of Low-Noise Airfoils for Wind Turbines. *AIAA Journal*, 45(4):779–785.

[16] Moriarty, P. J. and Hansen, A. (2005). AeroDyn Theory Manual. Technical Report January, NREL - National Renewable Energy Laboratory.

[17] Oerlemans, S., Fisher, M., Maeder, T., and Kögler, K. (2009). Reduction of Wind Turbine Noise Using Optimized Airfoils and Trailing-Edge Serrations. *AIAA Journal*, 47(6):1470–1481.

[18] Rautmann, C. (2017). Numerical Simulation Concepts for Low-Noise Wind Turbine Rotors, Forschungsbericht 2017-35. Technical report, Deutsches Zentrum f"ur Luft und Raumfahrttechnik.

[19] Rubio Carpio, A., Avallone, F., and Ragni, D. (2018). On the Role of the Flow Permeability of Metal Foams on Trailing Edge Noise Reduction. *2018 AIAA/CEAS Aeroacoustics Conference*.

[20] van der Velden, W. C. and Oerlemans, S. (2017). Numerical study on combed teeth serrations for wind turbine noise reduction. *23rd AIAA/CEAS Aeroacoustics Conference*.

[21] Zhu, W. J., Shen, W. Z., and Sorensen, J. N. (2016). Low - Noise Airfoil and Wind Turbine Design. In *Wind Turbines - Design, Control and Applications*, pages 55–72.

8th International Conference
on
Wind Turbine Noise
Lisbon – 12th to 14th June 2019

Session O

Trailing Edge Noise Mitigation

Organised by INCE-Europe
mail@inceeurope.org
www.inceeurope.org

Supported by Portuguese Acoustical Society

INCE EUROPE

8th International Conference
on
Wind Turbine Noise
Lisbon – 12th to 14th June 2019

Numerical Investigation of the Porous Trailing Edge Noise Reduction Mechanism using the Lattice-Boltzmann Method

F. Manegar, University of Siegen, 57068 Siegen, Germany
E-Mail: farhan.manegar@uni-siegen.de

C. Teruna, Delft University of Technology, 2629HS Delft, The Netherlands
E-Mail: C.Teruna@tudelft.nl

F. Avallone, Delft University of Technology, 2629HS Delft, The Netherlands
E-Mail: F.Avallone@tudelft.nl

W. C. P. van der Velden, Dassault Systemes Deutschland GmbH, 70563 Stuttgart, Germany
E-Mail: wouter.vandervelden@3ds.com

D. Casalino, Delft University of Technology, 2629HS Delft, The Netherlands
E-Mail: D.Casalino@tudelft.nl

T. Carolus, University of Siegen, 57068 Siegen, Germany
E-Mail: thomas.carolus@uni-siegen.de

D. Ragni, Delft University of Technology, 2629HS Delft, The Netherlands
E-Mail: D.Ragni@tudelft.nl

A. Rubio Carpio, Delft University of Technology, 2629HS Delft, The Netherlands
E-Mail: A.RubioCarpio@tudelft.nl

ABSTRACT

Permeable materials can be installed at the trailing edge of an airfoil to mitigate turbulent boundary layer trailing edge noise in the low frequency range. Goal of this study is to investigate the physical mechanisms behind trailing edge noise reduction for a porous trailing edge. Numerical simulations are carried out and the results are validated against experimental data. The flow around a NACA 0018 airfoil at 0° angle of attack and at a chord-based Reynolds number equal to 2.8×10^5 is investigated. The transport phenomena within the porous region are reproduced with the Hazen-Dupuit-Darcy law, where the Darcian force coefficients are taken from experimental data. A comprehensive mesh study for both solid and porous trailing edges is conducted. The boundary layer integral parameters and the far-field sound pressure spectrum are in good agreement with the experiments. The noise reduction capability of the porous trailing edge is successfully predicted in the range of frequency of interest, however, simulations do not show the noise increase in the high frequency range, due to the absence of roughness in the computational setup adopted to model porosity. The spanwise correlation length of the pressure fluctuations close to the wall does not show any relevant difference between the solid and the porous trailing edge thus suggesting that the large scale structures are not affected by the porous insert.

1. INTRODUCTION

Airfoil self-noise is the major noise source in wind turbine applications [1]. In particular, turbulent boundary layer trailing edge (TBL-TE) noise, generated by the scattering of turbulent flow at the trailing edge [2] is the dominant noise mechanism [3]. The most common technique employed by wind turbine manufacturers is the installation of saw-tooth shaped serrations at the trailing edge to reduce TBL-TE noise. Other noise-reduction techniques are being developed to achieve noise reduction, such as, active boundary-layer suction system [4], porous trailing edges [5] and blade shape optimization [6].

Dedicated porous materials, typically applied to a certain percentage of the chord length, are made by a uniform and permeable structure with interconnected pores. The permeability typically reduces the impedance jump at the edge, with a consequent reduction of the radiated noise. Noise reduction capabilities of porous materials have been successfully demonstrated in several aerodynamic applications [7]. Porous materials are conventionally characterized with porosity and flow resistivity. Geyer *et al* [7] carried out an extensive set of measurements, using different materials and different streamwise extensions, but they did not find any exhaustive correlation between porous material characteristics and their noise levels. They found though that with a longer extension of the porous material, a larger measured noise reduction is achieved, together with a degradation of the aerodynamic properties (i.e., lift and drag).

More recently, Rubio-Carpio *et al* [8] investigated, using microphone array and time-resolved Particle Image Velocimetry (PIV) measurements, a NACA 0018 profile with a porous trailing edge made of metal foams. Several foams with different properties (i.e., porosity and flow resistivity) were investigated. Noise reduction was measured at lower to middle frequency (from 500 to 1600 Hz), while noise increase, attributed to the surface roughness, was found at higher frequencies. The effects of the flow connection through metal-foam on far-field acoustics was tested by adding a non-permeable layer [9]. For this configuration, no noise reduction was measured at lower to mid frequency, but noise increase in the higher frequencies was still present.

In this study, the experimental study of Rubio Carpio *et al* [8] is investigated numerically using the commercial software 3DS-Simulia™ PowerFLOW.

Since it is computationally expensive to simulate the entire geometry at the porous trailing edge, a porous media formulation is used.

2. COMPUTATIONAL METHOD

2.1 Flow solver

The lattice-Boltzmann (LB) commercial solver 3DS-Simulia™ PowerFLOW version 5.4b has been used because it has been proved to be reliable for trailing edge noise prediction [10, 11]. The LBM solves the Boltzmann equation on a Cartesian mesh, named as lattice. The discretization used for this particular application consists of 19 discrete velocities in three dimensions (D3Q19), involving a third-order truncation of the Chapman-Enskog expansion. The equations are solved with an explicit time integration approach and the stability of the solution is obtained by setting the Courant-Friedrichs-Lewly stability number to unity. To obtain the particle distribution, a collision term, based on a unique Galilean invariant [12], and an equilibrium distribution of Maxwell-Boltzmann [13] are adopted.

A Very Large Eddy Simulation (VLES) model is implemented to consider the effect of the sub-grid unresolved scales of turbulence. Following [14], a two-equations *k-epsilon* renormalization group is used to compute a turbulent relaxation time that is added to the viscous relaxation time. In order to reduce the computational cost, a pressure-gradient-extended wall-model is used to approximate the no-slip boundary condition on solid walls [15, 16]. The model is based on the extension of the generalized law-of-the-wall model [17] to consider the effect of pressure gradient.

A Darcy's porous media formulation is used to compute the flow within the foam. The parameters that characterize the macroscopic properties of transport phenomena within a porous material are: porosity, permeability and form coefficient. The porosity of the material is defined as:

$$\phi = 1 - \frac{\rho_p}{\rho_s} \tag{1}$$

where ρ_p and ρ_s are the density of the porous material and of the solid, respectively. In porous media, the Hazen-Dupuit-Darcy equation describes the pressure loss Δp when certain mass flow permeates through a porous material sample of thickness h:

$$\frac{\Delta p}{h} = \frac{\mu_\infty}{K} v_d + \rho_\infty C v_d^2 \tag{2}$$

where μ_∞ and ρ_∞ is the fluid's freestream dynamic viscosity and density respectively. K and C are the permeability and the form coefficient of the porous material. The coefficients of the Darcian velocity v_d (eq. 2) are defined as viscous ($R_V = \mu_\infty / \rho_\infty K$) and inertial ($R_I$ = C) resistance. The pressure gradient ∇p through a porous material is proportional to the local flow velocity u, as follows:

$$\nabla p = -R \cdot u \tag{3}$$

where R is the sum of viscous and inertial resistivity components (R_V and R_I respectively). This law is then applied in the LBM solver and readers are referred to [18] and [19] for a more detailed description.

The viscous and inertial resistances are taken from the experimental study conducted by Rubio Carpio *et al* [8].

2.2 Noise computation

The compressible and time-dependent nature of the transient CFD solution together with the low dissipation and dispersion properties of the LB scheme allow extracting the sound pressure field directly in the near-field up to a cut-off frequency corresponding to approximately 15 voxels per acoustic wavelength [20]. In the far field, noise is computed by using the Ffowcs Williams and Hawkings (FW-H) equation [21]. The formulation 1A, developed by Farassat [22] and extended to a convective wave equation is used in this study [23, 24]. The formulation is implemented in the time domain using a source-time dominant algorithm [25]. To compare computational results and simulations the scaling adopted in [11] is used:

$$L_{S_{pp},scaled} = L_{S_{pp}} + 10\log_{10}\left[\left(\frac{R}{R_c}\right)^2 \Bigg/ \left(\left(\frac{b}{b_c}\right)\left(\frac{M}{M_c}\right)^5\right)\right]. \tag{4}$$

where R_c = 1 m, M_c = 1 and b_c = 1 m, that are the reference listener location, Mach number and span of the airfoil.

3. NUMERICAL CASE SETUP

A two-dimensional NACA 0018 airfoil has been investigated. It has a chord length c = 200 mm and a span of b = 40%c (80 mm). Transition is forced at 20%c with a zig-zag trip with thickness of 0.6 mm, spanwise periodicity of 2.9 mm and length of 3.6 mm. The transition technique adopted here, though different from the actual experimental realization, has been found to be the one that best represents both the BL thickness and energy cascade.

The freestream velocity is U_∞ = 20 m/s, corresponding to a chord-based Reynolds number of 2.8×10^5. The effective angle of attack is α_{eff} = 0°, making it a zero-lift setup.

Fig. 1 shows a schematic of the simulation domain (not in scale). The reference coordinate system has its origin at the trailing edge of the airfoil (X/c = 0), where X and Y axes are aligned with the chord direction and its normal direction respectively.

The numerical grid consists of 10 zones of variable resolution (VR). Each VR zone consists of cubical volume cells called voxels. The size of voxels increases by a factor of two in adjacent VR zones. The VR zone containing the finest voxel is around the airfoil. The simulation domain is 100c long in the X and Y-axis. The airfoil is placed at the center of the simulation domain. The details regarding the number of voxels, time step corresponding to finest voxel resolution and computation time is provided in the next section, where different mesh resolutions are investigated.

Boundary conditions are provided in terms of velocity and pressure at all four boundaries (free-field boundary conditions) of the simulation domain in the X-Y plane, with periodic boundary conditions applied along the Z-axis. As seen in Fig. 1, for the porous configuration, a 1 mm thick outer layer of acoustics porous medium (APM) surrounding the porous medium (PM) is used at the trailing edge. Both medium applied in PowerFLOW describe the porous material as an equivalent fluid region with resistance governed by the Darcian forces. In addition to this, the APM also considers the acoustic absorption property of the material and governs the mass flow between the regular fluid region and the APM region through a porosity value. The simulated Mach number (M = 0.0588) is chosen to be the same as in the experiment. An anechoic outer layer is specified starting from a radius of 36c from the airfoil center to damp outward traveling acoustic waves and to prevent reflection by the domain boundaries.

All the mean and unsteady quantities shown in the mesh resolution study and results sections are evaluated with a sampling rate f_s = 20 kHz and are recorded for 10 flow passes (c/U_∞) (corresponding to a physical time of 0.1 s). They are evaluated at the red and blue dotted lines shown in Fig.2 for boundary layer and wake flow respectively. The far-field noise calculated from airfoil surface and FW-H permeable surface are evaluated with a sampling rate $f_{s,a}$ = 248 kHz and are recorded for 10 flow passes as well.

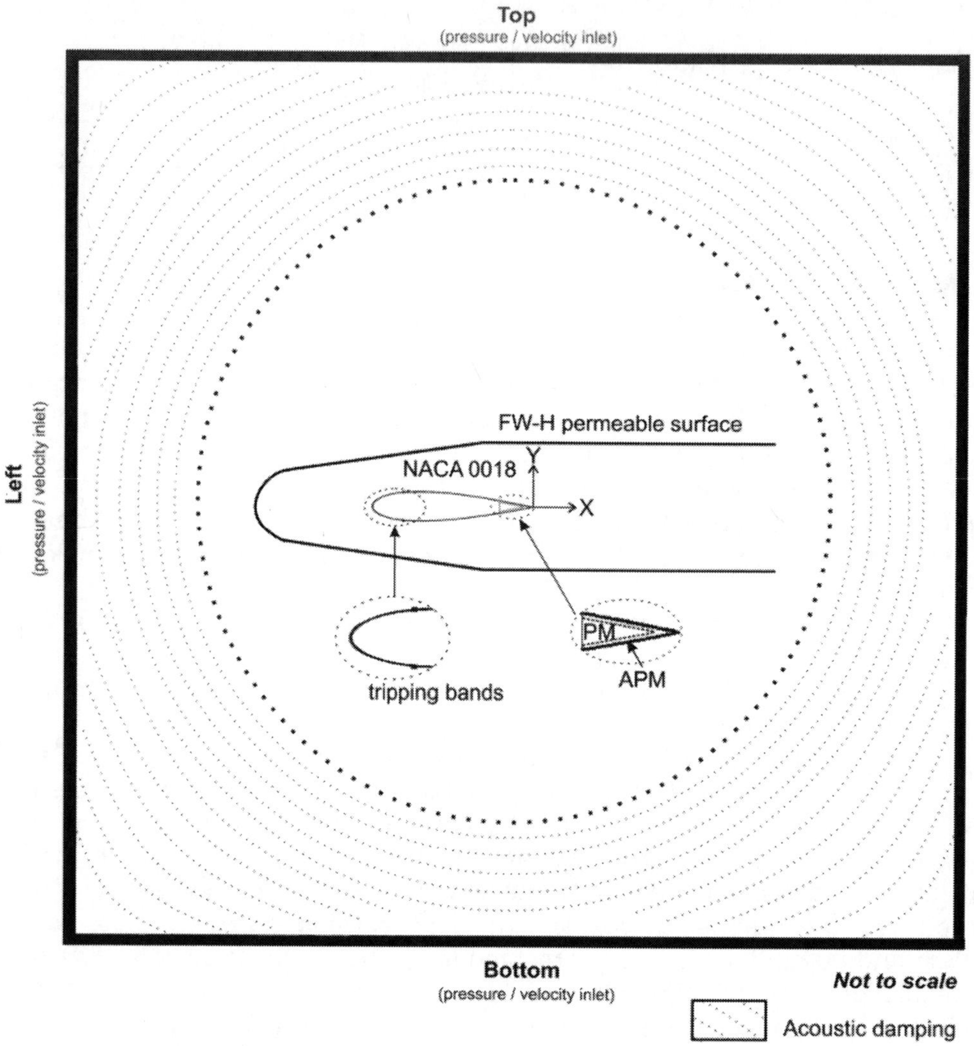

Figure 1. Schematic of airfoil segment and simulation domain in LBM (not to scale).

Mesh study

A mesh independence study is conducted aiming at evaluating the sensitivity to the voxel size and their distribution. The investigated configurations are named based on the surface y^+ value at the trailing edge. Table 1 reports the finest voxel size for each configuration, the corresponding time step, total number of voxels, total number of fine equivalent voxels (average number of voxels updated each time step) and the resulting computational time. The mesh configuration 'y^+ = 3 (only in the vicinity of the trip)' has the same distribution of VR zones as 'y^+ = 3', but with refinement applied only in the vicinity of the trip.

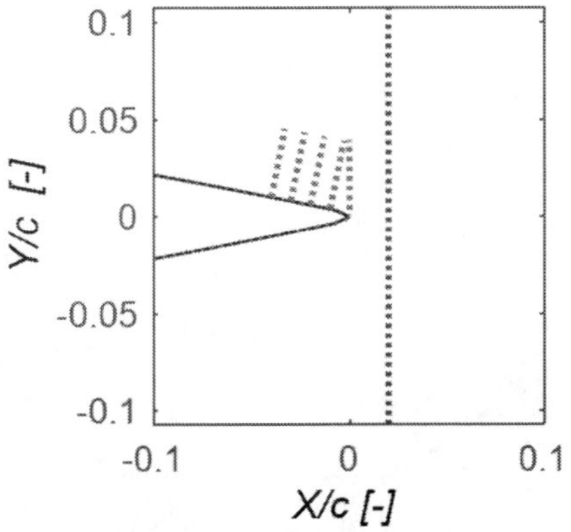

Figure 2. Locations at which velocity field data are sampled – red dotted lines: for the boundary layer, blue dotted lines: for the wake.

The configurations 'y^+ = 3' and 'y^+ = 2.25' differs only for the finest voxel size while keeping the same VR distribution. For the latter configuration, in order to reduce the computational costs, the span is decreased to 30%*c*, unlike the other two configurations where it is 40%*c*. Fig. 3 shows the mesh around trip region for the three configurations.

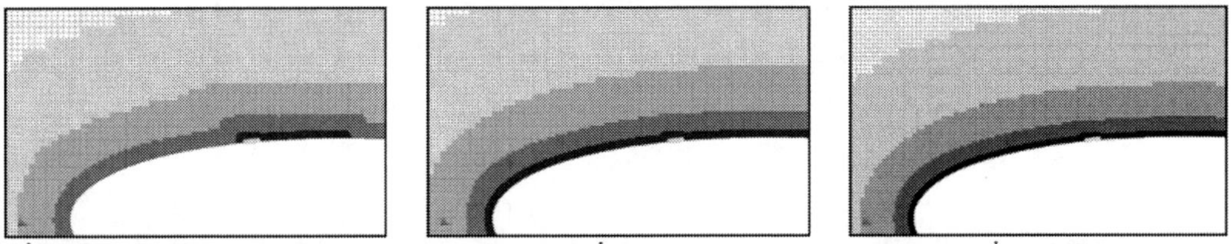

y^+ = 3 (only in the vicinity of the trip) y^+ = 3 y^+ = 2.25

Figure 3. Mesh generated around trip on suction side for the different mesh configurations.

Table 1. Mesh study details for solid trailing edge.

Parameters	**y^+ = 3 (only in the vicinity of the trip)**	**y^+ = 3**	**y^+ = 2.25 (span = 30%*c*)**
Finest voxel size (m)	7.81e-05	7.81e-05	5.86e-05
No. of voxels (million)	90	243	308
No. of fine equivalent voxels (million)	40	170	217
Smallest time step (s)	1.299e-07	1.299e-07	9.742e-08
Computational time (CPU hours) per flow pass on an Intel Xeon Haswell EP E5-2630v3 platform	1137	2880	4800

First the unsteady flow quantities are compared with experimental data for the two coarser mesh configurations, namely 'y^+ = 3 (only in the vicinity of the trip)' and 'y^+ = 3' to assess the influence of the grid on the flow development. This comparison is only performed for the solid trailing edge. Fig. 4 (left) shows the power spectral density of wall-parallel velocity fluctuations $|\Phi_{uu}|$ measured above the trailing edge at *Y/c* = 0.017 against the chord-based Strouhal number $St = fc/U_\infty$. Fig. 4 (right) shows the profile of root mean squared (R.M.S) values of the wall-parallel velocity (*u*) above the trailing edge, normalized with U_∞.

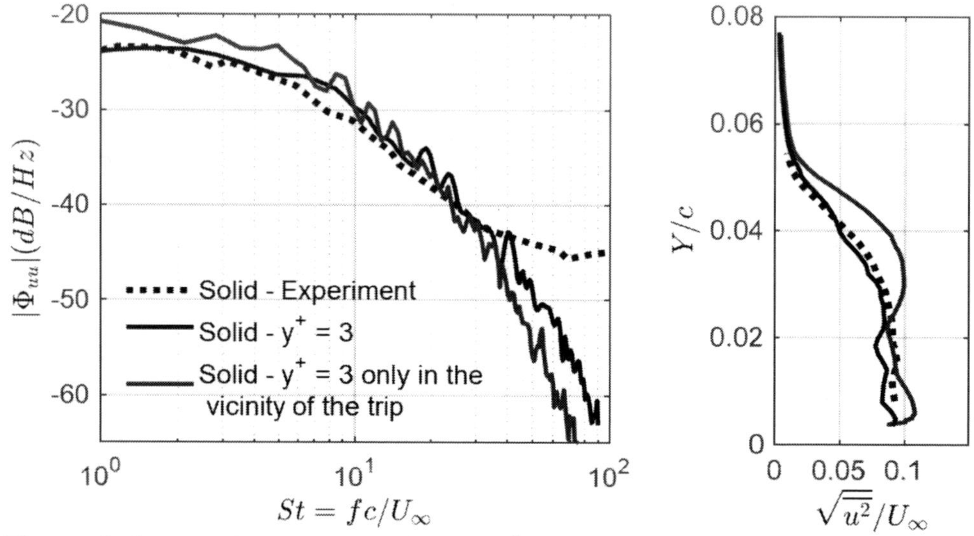

Figure 4. Comparison of 'y+ = 3' and 'y^+ = 3 (only in the vicinity of the trip)' with experiment (only for solid trailing edge). Left: Power spectral density of wall-parallel velocity fluctuations $|\Phi_{uu}|$ (X/c = 0, Y/c = 0.017). Right: Profile of R.M.S of wall-parallel velocity fluctuations above trailing edge.

From Fig. 4 (left) it is evident that the mesh configuration 'y^+ = 3 (only in the vicinity of the trip)' show more energy at lower frequencies (below *St* = 10) compared to experiment. The R.M.S of the streamwise velocity shows a broad peak between *Y/c* = 0.03 to 0.05 that is associated to the energy content discussed before. Refining the mesh uniformly around the airfoil solves the problem and a better agreement is found with the experiments. The peak for the 'y^+ = 3 (only in the vicinity of the trip)' is associated to turbulent structures generated by the zig-zag trip not sufficiently dissipated.

The convergence between mesh configuration 'y^+ = 3' and 'y^+ = 2.25' is further studied by comparing the unsteady wall-parallel and wall-normal velocity components as shown in Fig.5 and Fig.6, respectively. Spectra of the wall-parallel $|\Phi_{uu}|$ and wall-normal velocity fluctuations $|\Phi_{vv}|$ shown on the left side of the figures show grid independence. The R.M.S values shown on the right side of the figures show reasonable convergence in the case of solid trailing edge but some deviation in the porous trailing edge. It has to be mentioned that the simulation run for the 'y^+ = 2.25' mesh configuration only ran for a limited time instant due to excessive costs. From these plots, the mesh configuration 'y^+ = 3' is chosen for the remainder of this study.

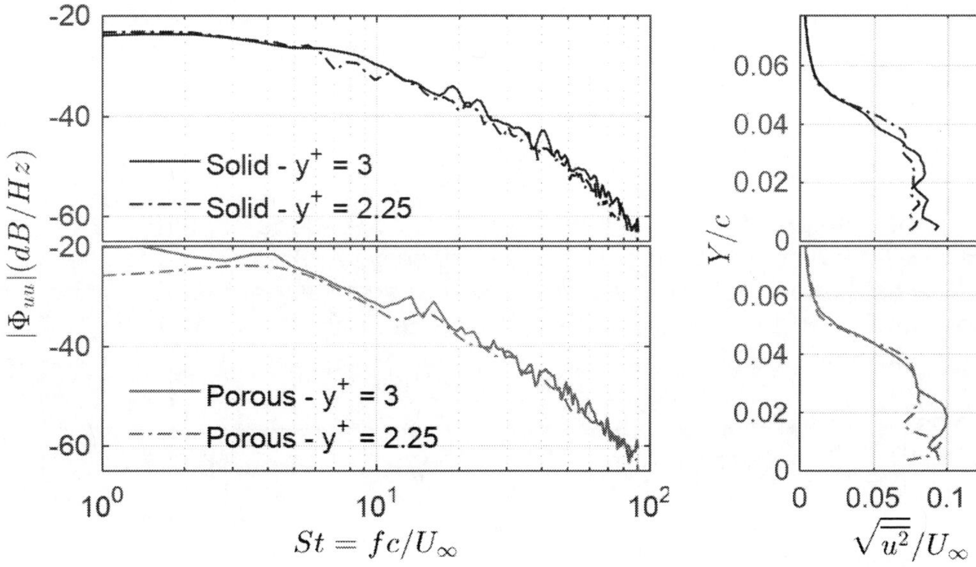

Figure 5. Mesh convergence study; left: Power spectral density of wall-parallel velocity fluctuations $|\Phi_{uu}|$ (X/c = 0, Y/c = 0.017); right: Profile of R.M.S of wall-parallel velocity fluctuations above trailing edge.

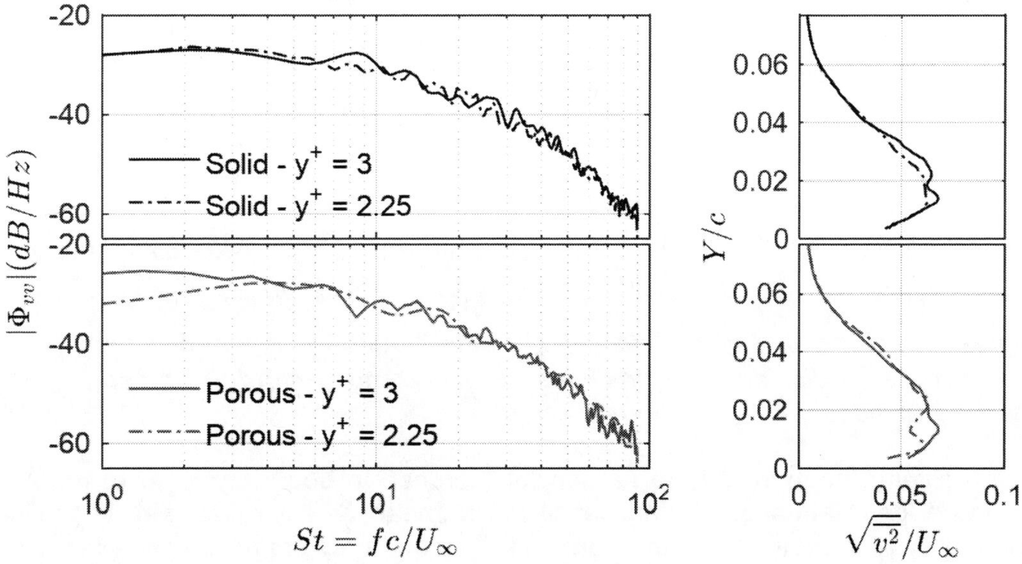

Figure 6. Mesh convergence study; left: Power spectral density of wall-normal velocity fluctuations $|\Phi_{vv}|$ (X/c = 0, Y/c = 0.017); right: Profile of R.M.S of wall-normal velocity fluctuations above trailing edge.

4. RESULTS

4.1 Comparison with the experimental results

Boundary layer statistics

Experiments have shown that a porous trailing edge leads to the development of a thicker boundary layer compared to a solid trailing edge [8, 26]. The reason for this phenomenon is attributed to roughness effects explained by Rubio Carpio *et al* [8] and to material resistivity explained by Geyer *et al* [26]. Simulation results show the same behavior as seen in the experiments. The boundary layer thickness (δ_{99}), displacement thickness (δ^*) and momentum thickness (θ^*) from simulations are validated against experimental results in Fig. 7. These integral quantities are then calculated as done in [8] (Fig. 2). δ_{99} is defined as the point where the wall-parallel velocity is 99% of the edge velocity (U_e). The agreement between the simulation and experiment is very good for the solid trailing edge case, while a slight disagreement is found for the porous trailing edge case.

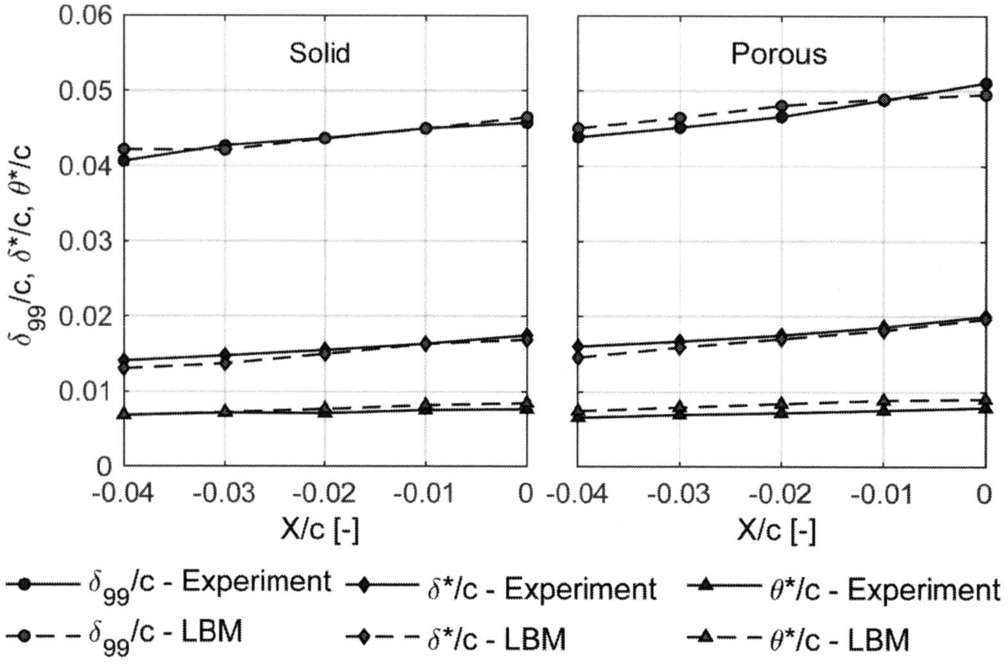

Figure 7. Boundary layer statistics showing boundary layer thickness (δ_{99}), boundary layer displacement thickness (δ^*) and momentum thickness (θ^*) on the suction side.

Even though in [8], the reason for increase in boundary layer of porous trailing edge is attributed to the deficit in the mean wall parallel velocity profile caused by roughness of the material, the simulation results here shows tendencies in favor of the explanations provided in [26]. As there is no roughness quantity modeled in the simulation, the increase in boundary layer for porous material could indeed be due to the resistivity properties of porous material.

Mean and turbulence quantities

In this section the mean and turbulence statistics obtained from simulations are compared against experiments. Simulation data are extracted and processed at the trailing edge (*X/c* = *0*), i.e. along the red dotted lines shown in Fig. 2. Fig. 8 shows the validation for a solid and porous trailing edge.

Profiles of the time-average wall-parallel velocity component *u* normalized with U_∞ are shown in Fig. 8 (left). A trend similar to the experiment is observed in the simulation. The porous trailing edge shows a slightly lower velocity compared to solid trailing edge for a given non-dimensional wall normal location. This difference was attributed to a higher surface drag.

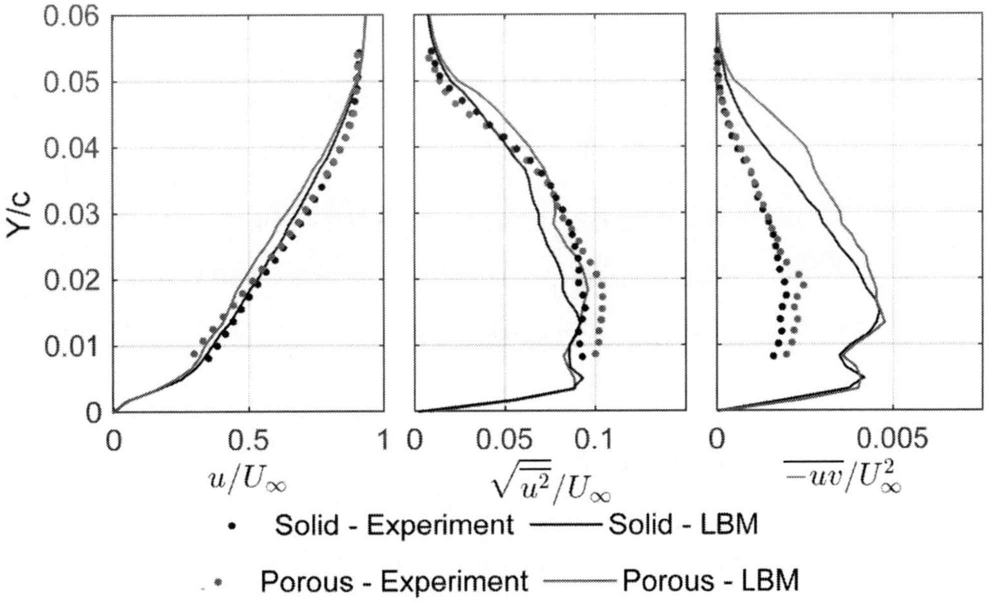

Figure 8. Mean and turbulence quantities at trailing edge (X/c = 0) on suction side.

The R.M.S profiles of wall-parallel velocity ($\sqrt{\overline{u^2}}$) normalized with U_∞ are shown in the middle plot of Fig. 8. For both solid and porous trailing edge, a reasonable good agreement is found with simulations capturing larger turbulent fluctuations with respect to the solid trailing edge.

The Reynolds stresses $\overline{-uv}$ normalized with U_∞ obtained from the simulations are larger than the experimental one. This disagreement, due to an over estimation of the fluctuations of the wall-normal velocity component (*v*), might be attributed to the different tripping used in the experiments. However, an experimental malfunction should also not be neglected, as this component is often difficult to be extracted and averaged from planar PIV data. Therefore, further research on this aspect is necessary. Nevertheless, the trends captured from simulations are similar to the experimental ones; the porous trailing edge generates larger fluctuations near to the wall compared to the solid trailing edge. This phenomenon is speculated in [8] to be due to the unsteady cross-flow through the porous material.

Computational data extracted in the wake (*X/c* = 0.02, at the blue dotted lines shown in Fig.2) are compared with the experiments in Fig. 9. As in the experiment, the porous trailing edge has a minor effect on the near wake flow showing only a small increase of the velocity fluctuations.

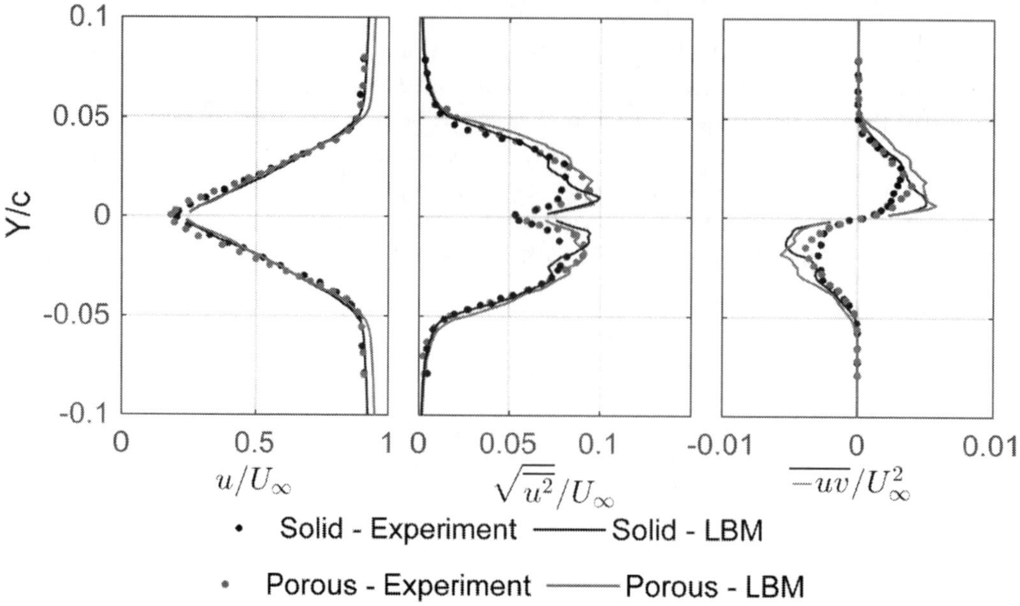

Figure 9. Mean and turbulence quantities in the near wake (X/c = 0.02).

Velocity fluctuation spectrum

The visualization of the instantaneous vorticity magnitude at the mid-span plane is shown for both configurations in Fig. 10. A qualitative comparison of the two figures show that the porous trailing edge weakly affect the large scale flow structures, in agreement with the previous results where only a minor difference in the amplitude of the turbulent fluctuations was found.

To verify if the porous materials affect the small scale structures, spectra of the wall-parallel velocity fluctuations at the trailing edge (X/c = 0, Y/δ_{99} = 0.3) are shown in Fig. 11 for the u component and in Fig. 12 for the v component. As for the previous case a good agreement is found with the experiments; the porous trailing edge shows higher energy associated at low St number (less than 10), in agreement with the flow visualization. The only difference with the experiments, still related to the roughness effect, is the St number at which the two curves cross each other. In the simulation no cross-over St number is found and the two spectra are similar for St > 10. In the experiments the curves cross each other at St = 15.

Fig. 12 shows the spectral content of wall-normal velocity (v) fluctuations at X/c = 0 and Y/δ_{99} = 0.3. As expected from the Reynolds stress profiles large energy content is present in the simulations with respect to the experiments. However, also in this case the energy distribution between the two configurations is similar.

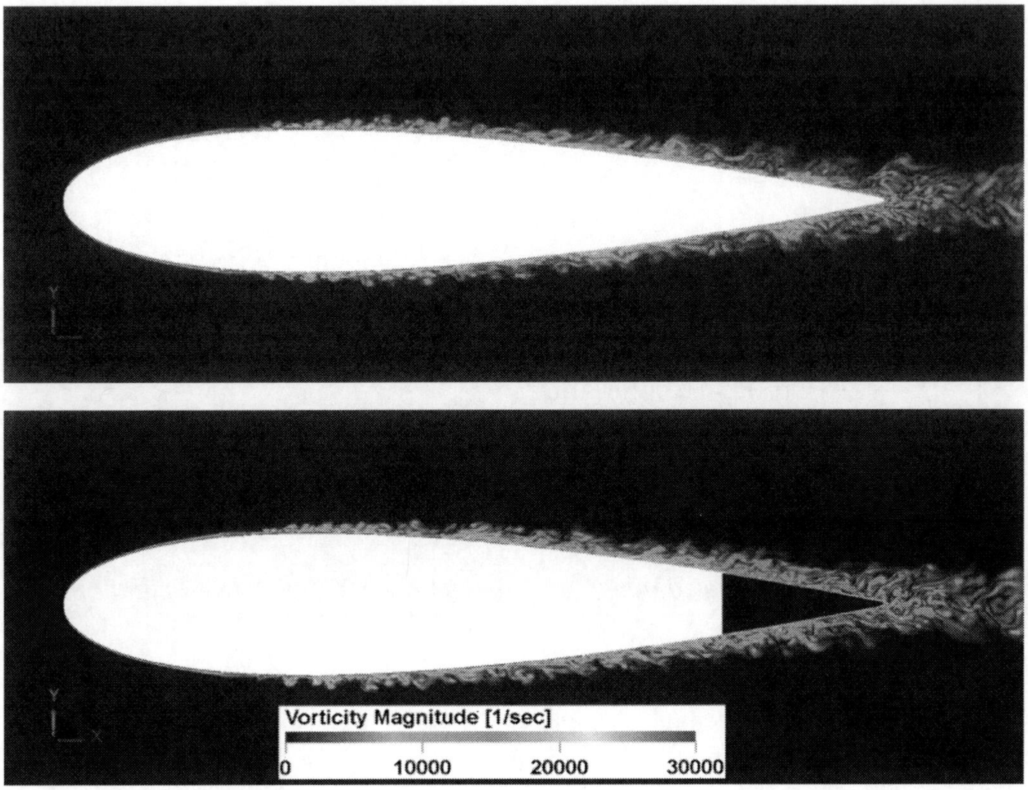

Figure. 10 Instantaneous vorticity magnitude fields at mid-span. Top – Solid trailing edge, bottom – porous trailing edge.

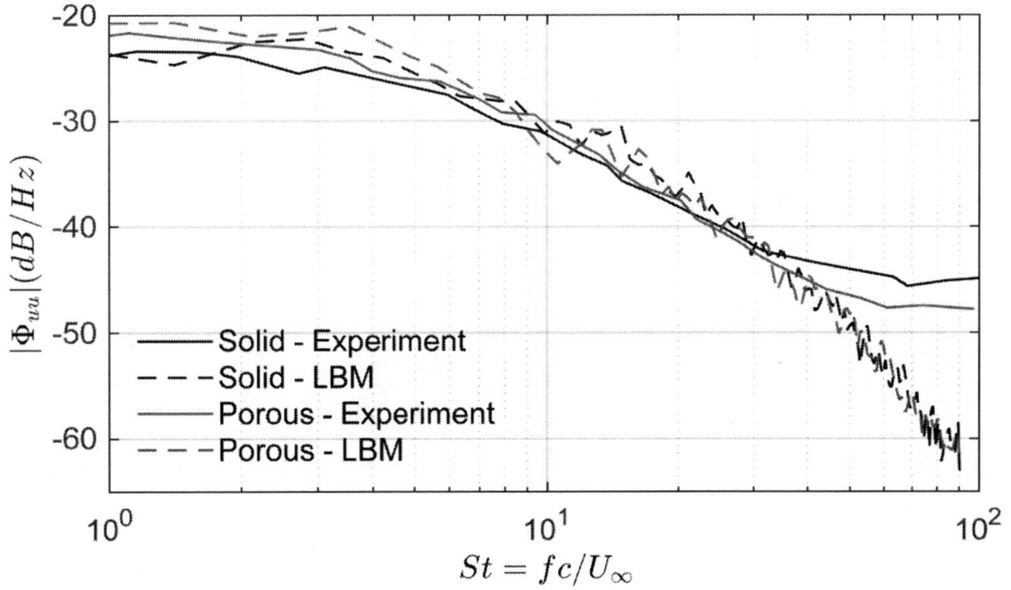

Figure. 11 Power spectral density of the wall-parallel velocity fluctuations at X/c = 0, Y/δ_{99} = 0.3.

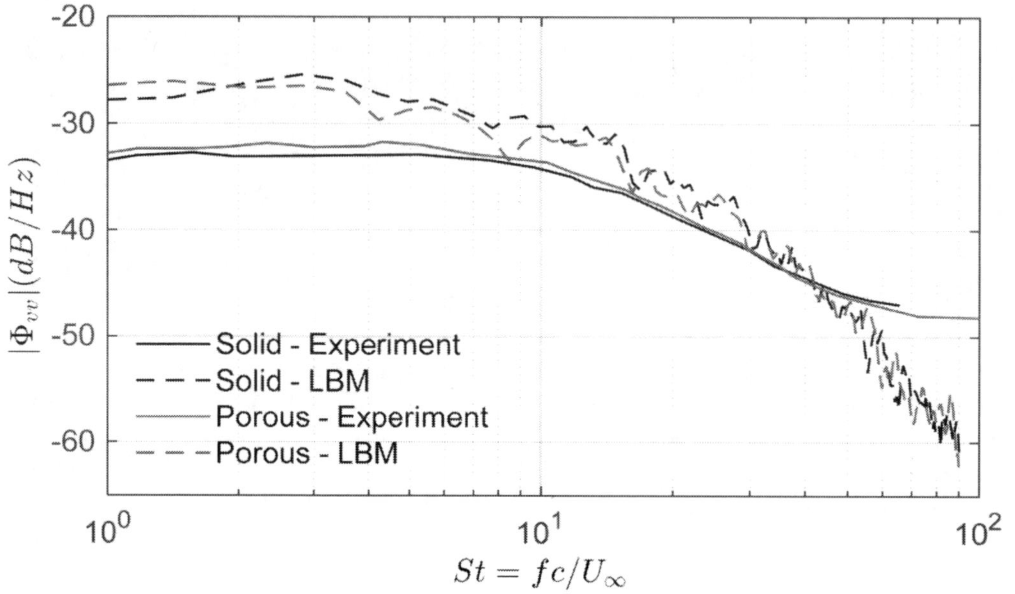

Figure. 12 Power spectral density of the wall-normal velocity fluctuations at X/c = 0, Y/δ_{99} = 0.3.

Acoustics

Far-field acoustic results are plotted in Fig. 13 where the far-field spectra are scaled according to eq. 4.

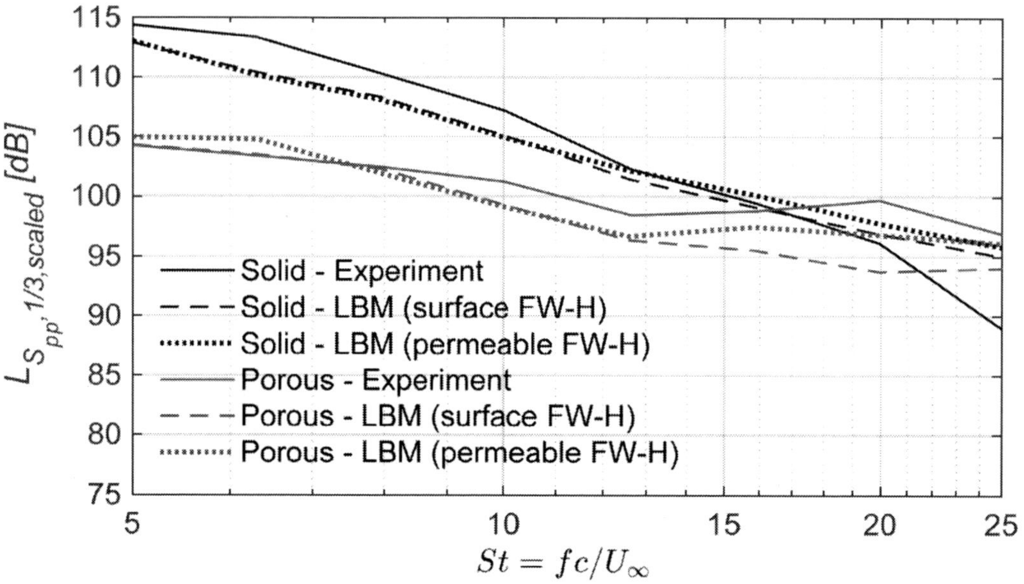

Figure. 13 Far-field sound pressure level between airfoils with solid and porous trailing edge.

The simulation predicts well the noise reduction capability of a porous trailing edge up to *St* = 16. The prediction for solid trailing edge is 2 dB less than the experiment, but on the whole the tendency is correct. Above *St* = 16 noise increase for the porous trailing edge is not found, whereas this is clearly present in the experiment. As stated in [8], this experimental noise increase is attributed to the roughness of porous material. In the simulation, only the properties of porous material are modeled while the surface roughness is not, which is most likely the case why the noise increase is not observed in the simulation results.

Directivity plots from simulations are shown in Fig. 14. They are obtained using 72 microphones equally spaced in a circle of radius equal to 7.4c at the airfoil mid-span. The far-field noise is further integrated over the non-dimensional frequency band reported in each plot. No experimental comparison is available. In the frequency range $4 < St < 16$ (where the porous trailing edge shows noise reduction) both configuration show a compact dipole. The porous trailing edge shows almost the same noise reduction tendency for all angles. However, in the frequency range $16 < St < 32$, there is almost no difference between the two configurations. The results show that the presence of a porous medium does not alter the directivity pattern, thus suggesting that the mechanisms behind the noise scattering are similar for the two configurations.

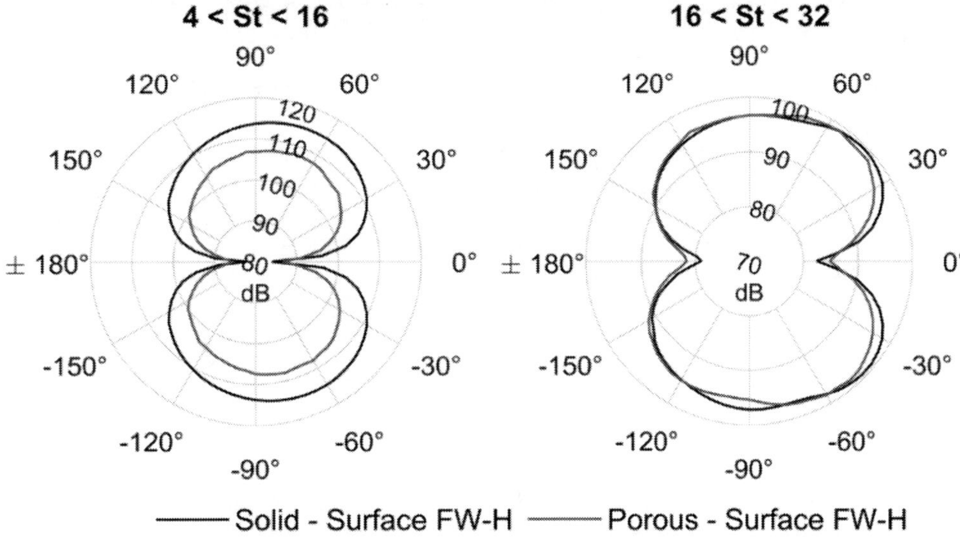

Figure. 14 Directivity plots of $L_{S_{pp},scaled}$ for the solid and porous trailing edge at two different non-dimensional frequency ranges: (left) $4 < St < 16$, (right) $16 < St < 32$.

4.2 Spanwise correlation length

Showkat Ali *et al* [27] reported that an internal hydrodynamic field within the porous material is present due to the penetration of the boundary layer flow within the porous material. This might alter the integral length scale of the turbulent structures. The spanwise correlation length (l_z) of the pressure fluctuations at 0.1 mm away from the wall, is calculated at various streamwise positions and plotted in Fig. 15. l_z is defined as:

$$l_z(x,f) = \int_{-\infty}^{\infty} \sqrt{\gamma^2(x,\Delta z,f)}dz \tag{5}$$

where γ^2 is the magnitude-squared coherence evaluated along the spanwise direction Z. Δz is the spanwise distance between the reference point and any other point along the span. γ^2 is computed with a periodogram approach using a Hanning window and 50 % overlap, and the resulting frequency resolution is 300 Hz.

An increase of l_z is found for configurations moving toward the trailing edge. More interestingly, only minor variations are found between the two configurations, thus suggesting that the flow is almost not affected by the presence of the porous medium.

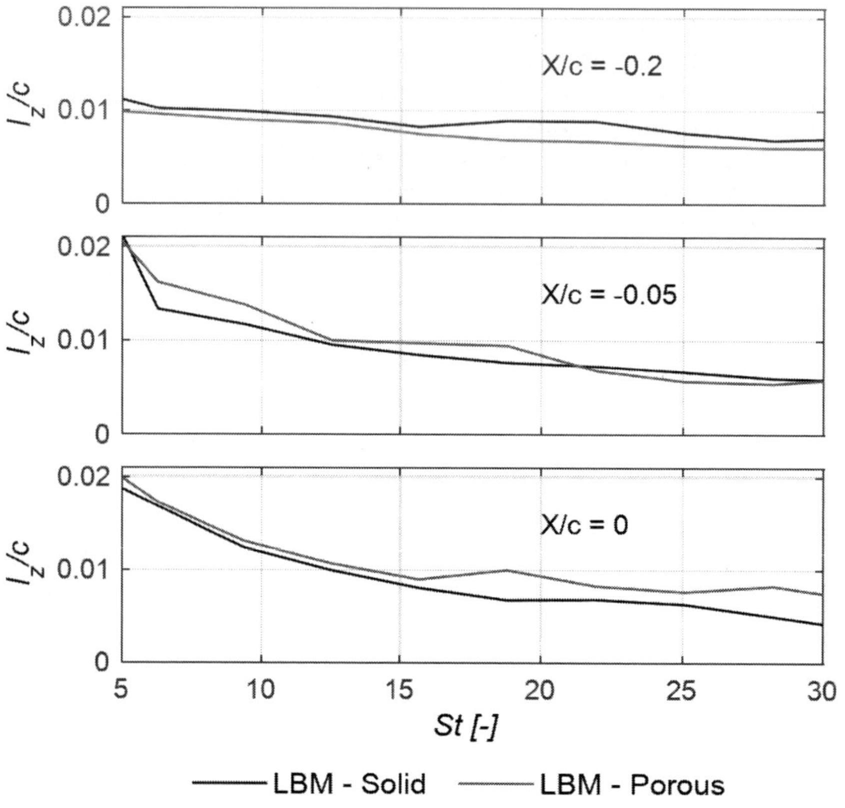

Figure. 15 Spanwise correlation length computed at 0.1 mm away in the wall-normal direction using pressure data.

In conventional trailing edge noise theory (Amiet [28]), l_z is one of the two most relevant parameters directly proportional to the noise radiation, where a larger l_z results in higher far-field noise. Since l_z is almost equal between the two configurations and the amplitude of the surface pressure fluctuations is larger for the porous trailing edge [8], confirmed here by the larger wall-normal velocity fluctuations, suggest that the Amiet's theory cannot be applied in this context. For this reason, the mitigation of the impedance mismatch between the two sides of the airfoil needs additional investigations. High fidelity simulations are therefore required and essential for this investigation.

5. CONCLUSIONS

Aeroacoustic Lattice-Boltzmann simulations of the flow around a NACA 0018 airfoil with solid and porous trailing edge are conducted successfully and validated against available experimental data. The transport phenomena within the porous region are reproduced with the Hazen-Dupuit-Darcy law, where the Darcian force coefficients are taken from existing experimental data. A comprehensive mesh study is conducted and a mesh resolution with resulting $y^+ = 3$ at the trailing edge is chosen and further analyzed throughout the study. A very good agreement is achieved between simulation and experiment for the boundary layer integral parameters and turbulence statistics, except for the wall-normal velocity fluctuations. The noise reduction capability of the porous trailing edge is successfully predicted in the simulation, but the phenomenon of noise increase observed at experiment is not captured by simulation - most likely owing to the roughness of porous material that is not modeled. Finally, the spanwise correlation length (l_z) of the pressure fluctuations in the wall-normal direction, is analyzed at various streamwise positions. It is observed that the flow structures are not strongly influenced by the porous medium.

6. ACKNOWLEDGEMENTS

Financial support for this work was partly provided by the Federal Ministry for Economic Affairs and Energy of Germany (BMWi) within the project RENEW (FKZ 0325838B).

We acknowledge gratefully Dassault Systemes Deutschland GmbH for providing the licenses and technical assistance to carry out the LBM simulations.

The first author thanks the TU Delft group for hosting him as visiting researcher in 2018/19 for eight weeks.

7. REFERENCES

1. T. F. Brooks, D. S. Pope and M. A. Marcolini, *"Airfoil Self-Noise and Prediction"*, NASA Reference Publication, NASA-RP-1218 (1989)
2. M. S. Howe, *"A review of the theory of trailing edge noise"*, Journal of Sound and Vibration, Volume 61, No. 3, pp. 437-465 (1978)
3. S. Oerlemans, P. Sijtsma and B. Méndez López, *"Location and quantification of noise sources on a wind turbine"*, Journal of Sound and Vibration, Volume 299, Issues 4–5 (2007)
4. B. Arnold, Th. Lutz and E. Krämer, *"Design of a boundary-layer suction system for turbulent trailing-edge noise reduction of wind turbines"*, Renewable Energy, Volume 123, Pages 249-262 (2018)
5. T. Geyer and E.Sarradj, *"Trailing Edge Noise of Partially Porous Airfoils"*, 20th AIAA/CEAS Aeroacoustics Conference Atlanta, GA (2014)
6. S. Oerlemans, M. Fisher, T. Maeder and K. Kögler, *"Reduction of Wind Turbine Noise Using Optimized Airfoils And Trailing-Edge Serrations"*, 14th AIAA/CEAS Aeroacoustics Conference (29th AIAA Aeroacoustics Conference), Vancouver, British Columbia, Canada (2008)
7. T. Geyer, E. Sarradj and C. Fritzsche, *"Measurement of the noise generation at the trailing edge of porous airfoils"*, Experiments in Fluids, Vol. 48, No. 2, pp 291-308 (2010)
8. A. Rubio Carpio, R. Merino Martinez, F. Avallone, D. Ragni, M. Snellen and S. van der Zwaag, *"Broadband trailing-edge noise reduction using permeable metal foams"*, INTER-NOISE and NOISE-CON Congress and Conference Pro-

ceedings, Vol. 255, Institute of Noise Control Engineering, pp. 2755-2765 (2017)
9. A. Rubio Carpio, F. Avallone and D. Ragni, *"On the role of the flow permeability of metal foams on trailing edge noise reduction"*, AIAA/CEAS Aeroacoustics Conference (2018)
10. F. Avallone, W. van der Velden and D. Ragni, *"Benefits of curved serrations on broadband trailing-edge noise reduction"*, Journal of Sound and Vibration, Vol. 400, pp. 167-177 (2017)
11. F. Avallone, W. van der Velden, D. Ragni and D. Casalino, *"Noise reduction mechanisms of sawtooth and combed-sawtooth trailing-edge serrations"*, Journal of Fluid Mechanics, Vol. 848, pp. 560–591 (2018)
12. H. Chen, R. Zhang and P. Gopalakrishnan, *"Lattice Boltzmann Collision Operators Enforcing Isotropy and Galilean Invariance"*, (aug 2015)
13. H. Chen, S. Chen and W. Matthaeus, *"Recovery of the Navier-Stokes equations using a lattice-gas Boltzmann method"*, Physical Review A, Vol. 45, No. 8, pp. R5339–R5342. doi:10.1103/PhysRevA.45.R5339 (1992)
14. V. Yakhot and S. Orszag, *"Renormalization group analysis of turbulence. I. basic theory"*, Jorunal of Scientific Computing, Vol. 1, No. 1, pp. 3–51. doi:10.1007/BF01061452 (1986)
15. C. Teixeira, *"Incorporating Turbulence Models into the Lattice-Boltzmann Method"*, International Journal of Modern Physics C, Vol. 09, No. 08, pp. 1159–1175. doi:10.1142/S0129183198001060 (1998)
16. D. Wilcox, *"Turbulence modelling for CFD (Third Edition)"*, DCW Industries, Incorporated (2006)
17. B. Launder and D. Spalding, *"The numerical computation of turbulent flows"*, Computer Methods in Applied Mechanics and Engineering, Vol. 3, No. 2, pp. 269–289. doi:10.1016/0045-7825(74)90029-2 (1974)
18. D. M. Freed, *"Lattice-Boltzmann method for macroscopic porous media modeling"*, International Journal of Modern Physics C, Vol. 9, No. 08, pp. 1491–1503 (1998)
19. C. Sun, F. Pérot, R. Zhang, P. T. Lew, A. Mann, V. Gupta, D. M. Freed, I. Staroselsky and H. Chen, *"Lattice Boltzmann formulation for flows with acoustic porous media"*, Comptes Rendus Mécanique, Vol. 343, No. 10-11, pp. 533–544 (2015)
20. G. Brès, F. Pérot and D. Freed, *"Properties of the Lattice–Boltzmann Method for Acoustics"*, 15th AIAA/CEAS Aeroacoustic Conference, Miami, Florida, pp. AIAA 2009–3395. doi:10.2514/6.2009-3395 (2009)
21. J. Ffowcs Williams and D. Hawkings, *"Sound Generation by Turbulence and Surfaces in Arbitrary Motion"*, Philosophical Transactions of the Royal Society of London A: Mathematical, Physical and Engineering Sciences, Vol. 264, No. 1151 (2010)
22. F. Farassat, and G. Succi, *"A review of propeller discrete frequency noise prediction technology with emphasis on two current methods for time domain calculations,"* Journal of Sound and Vibration, Vol. 71, No. 3, pp. 399–419 (1980)
23. G. Brès, F. Pérot and D. Freed, *"A Ffowcs Williams-Hawkings solver for Lattice-Boltzmann based computational aeroacoustics"*, 16th AIAA/CEAS Aeroacoustic Conference, Stockholm, Sweden, pp. AIAA 2010–3711. doi:10.2514/6.2010-3711 (2010)

24. A. Najafi-Yazdi, G. Brès and L. Mongeau, *"An acoustic analogy formulation for moving sources in uniformly moving media"*, Proceedings of the Royal Society A: Mathematical, Physical and Engineering Sciences, Vol. 467, No. 2125, pp. 144–165. doi:10.1098/rspa.2010.0172 (2010)
25. D. Casalino, *"An advanced time approach for acoustic analogy predictions"*, Journal of Sound and Vibration, Vol. 261, No. 4, pp. 583–612. doi:10.1016/S0022-460X(02)00986-0 (2003)
26. T. Geyer, E. Sarradj and C. Fritzsche, *"Porous Airfoils: Noise Reduction and Boundary Layer Effects"*, International Journal of Aeroacoustics, Vol. 9, No. 6, pp. 787–820. doi:10.1260/1475-472x.9.6.787 (2010)
27. S. A. Showkat Ali, M. Azarpeyvand, C. M. Szőke and C. R. Ilário da Silva, *"Boundary layer flow interaction with a permeable wall"*, Physics of Fluids 30(8):085111. doi: 10.1063/1.5043276 (2018)
28. R. K. Amiet, *"Noise due to turbulent flow past a trailing edge"*, Journal of Sound and Vibration, Volume 47, No. 3, pp. 387-393 (1976)

INCEEUROPE

8th International Conference
on
Wind Turbine Noise
Lisbon Portugal – 12th to 14th June 2019

Experimental Investigation of Self-Aligning Trailing Edge Serrations for Airfoil Noise Reduction

K. Stahl, F. Manegar, Th. Carolus
University of Siegen, 57078 Siegen, Germany

R. Binois
Senvion GmbH, Überseering 10, 22297 Hamburg, Germany

Corresponding author: kathrin.stahl@uni-siegen.de

Summary

Turbulent boundary layer trailing edge noise is considered to be the dominant noise mechanism in wind turbines. An effective noise attenuation add-on is saw-tooth serrations attached to the trailing edge. The serration amplitude and the wavelength are the main geometrical variables. This experimental study investigates another geometrical variable, the so-called serration flap angle. Two different serrations were attached to a two-dimensional DU93W210 airfoil segment and tested in the Acoustic Wind-Tunnel (AWB) of the German Aerospace Center (DLR) in Braunschweig at a Reynolds number of 1.2 mio. The experiment was carried out in two steps. Firstly, different *fixed* flap angles had been tested in order to find the acoustically most efficient flap angle. Secondly, *self-alignment* of the serrations was allowed, while the airfoil angle of attack was varied.

The initial hypothesis that self-aligning serrations are superior when the airfoil angle of attack is varying (e.g. due to gusts of pitch control) can partly be confirmed. More noise reduction is possible with steeper than self-aligning flap angles of the serration, adapted to each airfoil angle of attack. Then the serrations need to be fixed. Fixed serrations as an add-onto a blade, however, result in an increase of the flow induced force on the blade just due to the net increase of blade area - the streamwise pressure distribution is not affected by the existence of serration.

Nomenclature

Symbols	Units	Description
c	m	chord length
c_p	-	pressure coefficient
f	Hz	frequency
$L_{p(1/3)}$	dB	sound pressure level (1/3- octave band)
LE	-	leading edge
PS	-	pressure side
Re	-	Reynolds number
Ref	-	reference (i.e. without serrations)
SS	-	suction side
TE	-	trailing edge
u_∞	m/s	velocity
α_{eff}	°	effective angle of attack
α_{geom}	°	geometric angle of attack
δ	mm	boundary layer thickness
ϕ	°	flap angle (fixed serrations)
ϕ_{sa}	°	flap angle (self-aligning serrations)
λ	m	width of serrations (wave length)

1. Introduction

For modern large horizontal axis wind turbines, aerodynamic noise from the blades is generally considered to be the dominant noise source, provided that mechanical noise is adequately treated [1]. The noise generated by the blade can be tonal or broadband in character, and may be caused by several mechanisms, such as the turbulent boundary layers across the trailing edge (TE) (subsequently denoted as trailing edge noise), laminar boundary layer vortex shedding, vortices from blunt trailing edges, or the induced secondary flow in the blade tip region [2]. Several studies consider trailing edge noise to be the dominant noise of a wind turbine [3]. Saw-tooth shaped serrations as an add-on at the trailing edge of a blade are a common noise reduction technique for TE noise mitigation. The acoustically "optimal" "tooth length (or amplitude)" and "width (or wave length)" relative to the local airfoil boundary layer thickness δ have been determined in a previous study. The flap angle ϕ, i.e. the angle between the streamwise direction of the serrations and the tangent to the profile, has been studied by Arce in case of a NACA 0018 airfoil [4, 5]. It is felt, however, that more work is required concerning the effect of the flap angle. Furthermore, since add-ons like serrations in principle can increase the load on the blade, the effect of self-aligning flexible serrations is investigated. It is hypothesized that (i) self-aligning serrations are superior when the angle of attack to the blade is varying (e.g. due to gusts of pitch control), (ii) self-aligning serrations produce a noise reduction without increase of blade load.

In this work two different serrations were attached to a two-dimensional DU93W210 airfoil segment. A wide range of different fixed flap angles were tested and assessed in a wind tunnel experiment. Eventually, also self-alignment of the serrations was allowed utilizing a hinge.

2. Airfoil Segment and Experimental Setup

Airfoil segment

Experimental investigations were carried out with a DU93W210 as the reference airfoil segment, equipped with serrations, Fig. 1. The model for the investigations in this paper has a chord length *c* of 0.3 m (*x*/*c* = 1 always indicates the trailing edge (TE) of the airfoil without serrations) and a span of 0.8 m. The DU93W210 is an airfoil with a maximum thickness of 0.21*c*, dedicated to wind turbine application and developed at the Delft University of Technology [6]. The thickness of the trailing edge of the airfoil (not the serrations) is 0.005*c* and hence 1.5 mm. Boundary layer tripping was applied close to the leading edge using a 0.2 mm zigzag trip on the suction side and 0.4 mm zigzag trip strip on the pressure side, Fig. 2. For measuring the chordwise pressure distribution on the pressure (PS) and suction side (SS) at mid-span 39 and 15 static pressure taps, respectively, are used. Naturally, close to the TE and along the serrations pressure taps were not feasible.

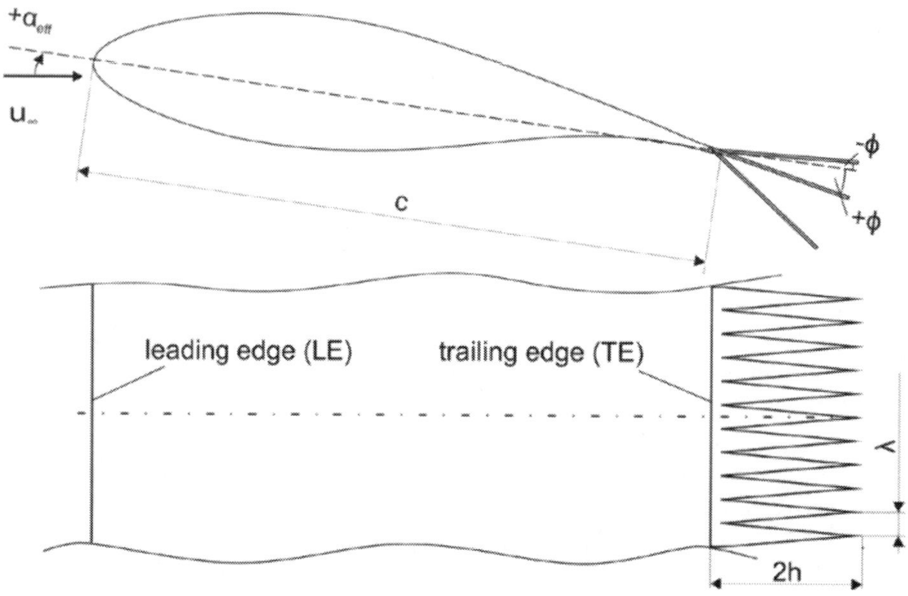

Figure 1: Airfoil and serrations.

The geometrical tooth and wave length of the serrations were selected according to a previous numerical Lattice-Boltzmann aeroacoustic simulation study.

Among eight different serrations with ratios of h/δ from 1.5 to 2.25[1] and λ/h from 0.2 to 0.6 the maximum efficiency in TE noise reduction was achieved for $h/\delta = 2$ and $\lambda/h = 0.35$.[2] Hence, these geometric dimensions are used for the trailing edge serrations named here S *I*. For practical reasons serrations S *II* with a somewhat shorter tooth as compiled in Tab. 1 and slightly less reduction capability were investigated as well.

Table 1: Geometry of serrations used in this study

	TE boundary layer thickness[1]	Absolute dimensions		Non-dimensional dimensions	
Denomination	δ [mm]	h [mm]	λ [mm]	h/δ [-]	λ/h [-]
S *I*	16	32	11.3	2	0.35
S *II*	16	24	7.5	1.5	0.31

All serrations were Laser cut from sheet metal of 0.5 mm thickness and attached to the TE via a hinge along the complete span. This allows self-aligning of the serrations in the flow, but various fixed flap angles could also realized with a set screw at the side plates. Fig. 3 shows the airfoil segment assembly ready for wind tunnel testing.

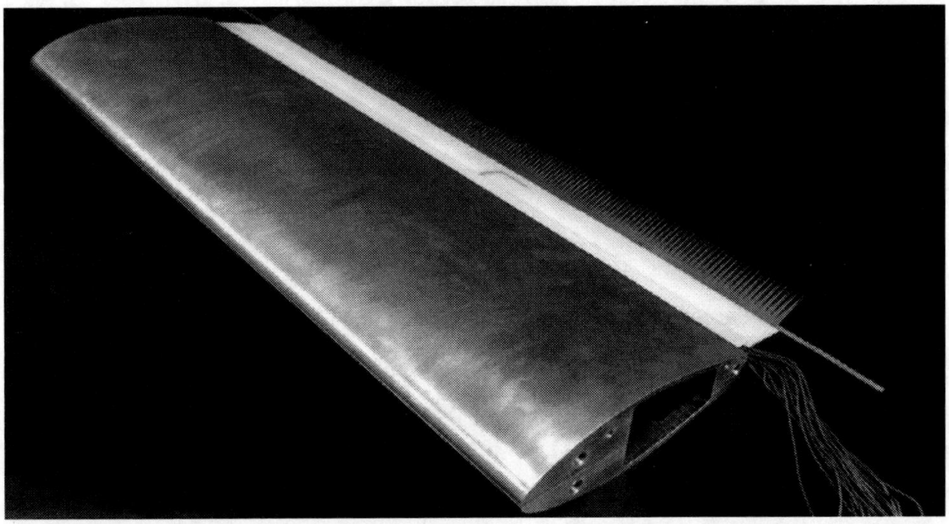

Figure 3: Airfoil assembly for experimental tests.

Experimental design

The experiments were carried out in two steps. Firstly, different *fixed* flap angles ranging from $\phi = -2.4°$ to $+33.6°$ in steps of 6° of the serrations *S I* were tested. Objective was to identify the acoustically most efficient flap angle. Throughout this test

[1] The boundary layer thickness $\delta = 16$ mm at the TE was derived from a XFOIL simulation for an effective angle of attack to the airfoil $\alpha_{eff} = 3.5°$ and a Reynolds number of 1.2 mio.

[2] Similar results can be seen by GRUBER et al. [7].

the airfoil angle of attack was set to $\alpha_{eff} = 3.5°$. Secondly, *self-alignment* of the serrations was allowed using serrations type *S II*, while the airfoil angle of attack was varied from $\alpha_{eff} = 2.5°$ to 3.5° and 4.5°. In addition, as soon as the self-aligning flap angle ϕ_{sa} was determined, fixed flap angles of $\Delta\phi = \pm 4°$ around each ϕ_{sa} were set. A summary of the parameter tested is given in Tab. 2. All measurements were done at a flow velocity of $u_\infty = 60$ m/s which corresponds to Re = 1.2 mio.

Table 2: Test parameters α_{eff} *and* ϕ for the parametric study, "*sa*" = self-aligning

Experiment	α_{eff}	Serrations	ϕ
Fixed flap angle study	3.5°	S *I*	0° (Ref), -2.4°, +3.6°, +9.6°, +15.6°, +21.6°, +27.6°, +33.6°
Self-aligning flap angle study	2.5°	S *II*	Ref, ϕ_{sa}, $\phi_{sa} + 4°$, $\phi_{sa} - 4°$
	3.5°		Ref, ϕ_{sa}, $\phi_{sa}+4°$, $\phi_{sa} - 4°$
	4.5°		Ref, ϕ_{sa}, $\phi_{sa} + 4°$, $\phi_{sa} - 4°$

Aero-acoustic wind tunnel

The experiment was carried out in the Acoustic Wind Tunnel Braunschweig (AWB) of the DLR. The AWB is an open-jet low noise facility with a rectangular nozzle exit of 0.8 m by 1.2 m and a maximum flow velocity of 65 m/s, as shown in Fig. 4. Sound absorbing linings at the chamber walls, floor and ceiling enable measurements at free field conditions for frequencies f > 250 Hz [8]. The dimensions of the anechoic chamber surrounding the free jet are 6.9 m x 6.9 m x 3.5 m with a test section length of 3.65 m [9]. The airfoil is mounted horizontally between two vertical side plates which are mounted to the nozzle. Integrated rotatable plates allowed the variation of the angle of attack.

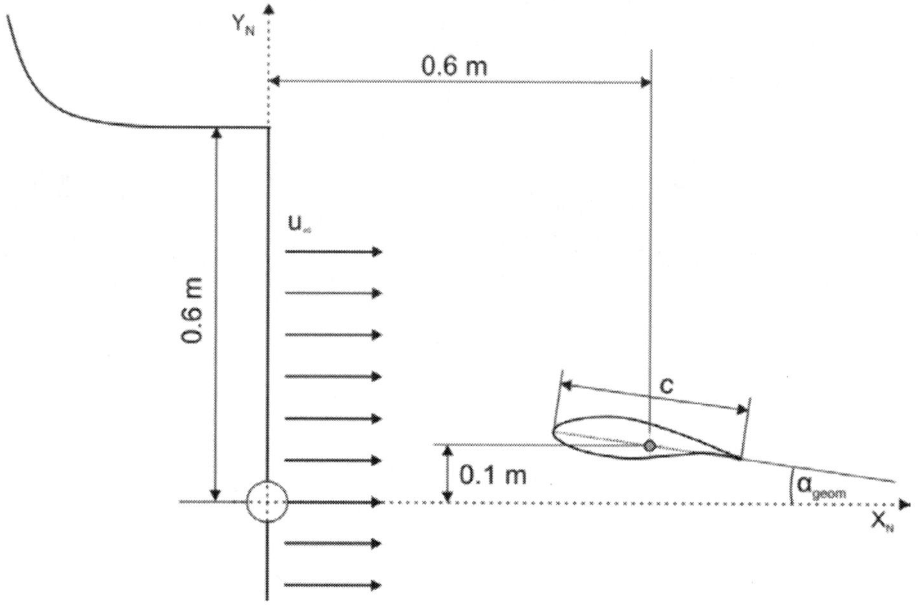

Figure 4: Experimental setup in the Acoustic Wind Tunnel Braunschweig (AWB).

Measurement techniques and data reduction

The static pressure distribution is obtained from the static pressure at each pressure taps in terms the pressure coefficients $c_p = (p-p_\infty)/0.5\rho u_\infty^2$. The measurements were made using a PSI 8400 acquisition system coupled with two 32 pressure ports scanner modules. The sampling rate was set to 50 Hz and averaged static pressure values were computed from 15 samples. The PSI 8400 system performs automatically the calibration and offset correction of the pressure modules [10].

Acoustic data were acquired through an elliptical mirror system pointing to the pressure side of the blade section and a microphone array (96 microphones pointing to the suction side) to determine noise radiated to either side, Fig. 5. Sound data were recorded at a sampling rate of 52 kHz. FFT signal processing resulted in a narrowband frequency bandwidth of Δf = 12.7 Hz. Other acquisition and processing parameters are: 16-bit dynamic range, 30 s measurement time, 30 kHz anti-aliasing low-pass filter, 500 Hz high-pass filter, Hanning window.

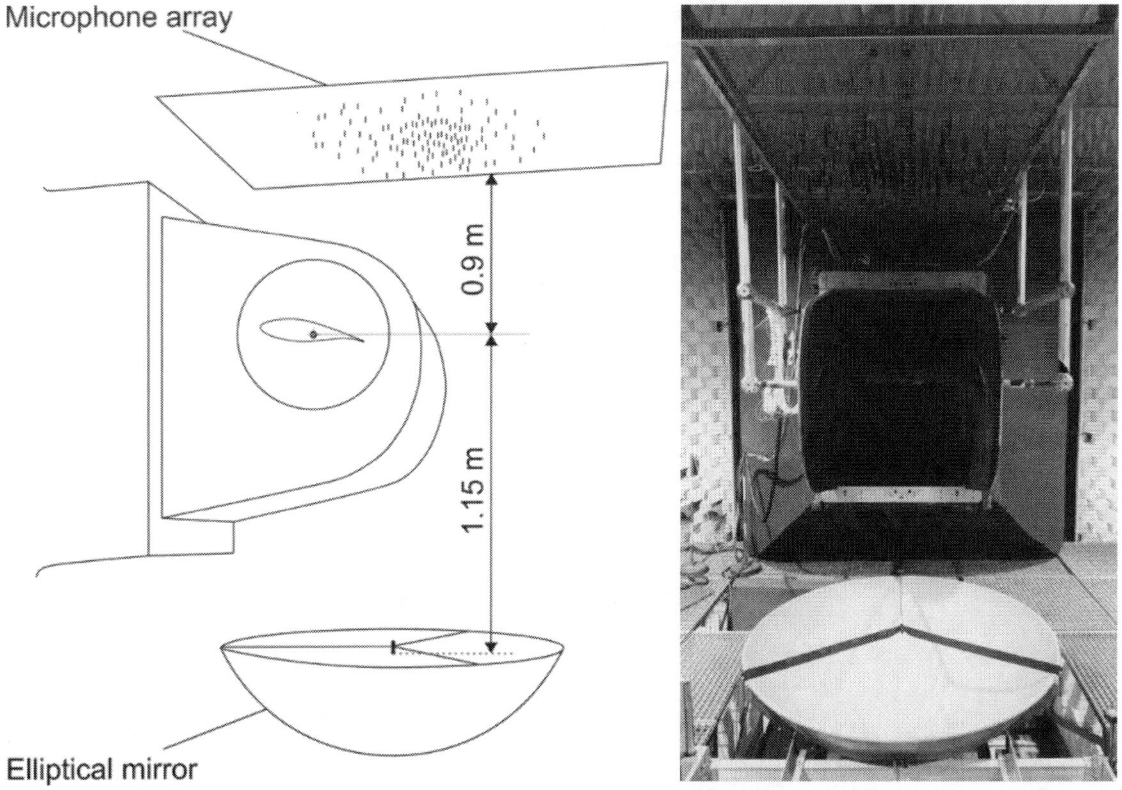

Figure 5: Measurement techniques in the experimental set up: Elliptical mirror and microphone array.

The mirror is mounted on a traverse. By moving the mirror in chordwise direction at mid-span below the airfoil section a localization of the noise source is possible. For all measurements mirror axis was oriented at 90° to the wind tunnel centre line. A calibrated 1/4-inch (6.35 mm) condenser microphone (Brüel & Kjær type 4136) was used. To provide absolute TE noise levels and to recover the "true" spectral shape, extensive corrections of the directional microphone data where performed. These account for: (i) Sound wave convection, (ii) extraneous wind-tunnel noise sources, (iii) system response function (effective spatial resolution and gain) and (iv) source distribution. Shear-layer refraction and scattering effects are negligible for frequencies $f \leq 10$ kHz [11] and were not corrected. Nevertheless, the lowest frequency to be resolved is 1 kHz. All results are presented in one-third octave sound pressure levels ($L_{p(1/3)}$).The microphone array has an aperture of 0.97 m and is located at a distance of 0.9 m above the blade's SS in the TE region. Except for directivity effects these measurements are expected to provide redundant TE sound information when compared to the mirror measurement data. The determination of absolute TE noise levels from the phased array data is done using diagonal removal and the CLEAN-SC method [12]. The microphone array measurements allow for cross checking the mirror results while also indicating the source distribution on the complete segment.

Effective angle of attack

For an airfoil in a confined jet, it is essential to identify the effective angle of attack for each set geometrical angle of the segment in the wind tunnel. This is due to the fact that the lift produced in a confined jet is not the same as the finiteness of the jet leads to significant flow deflection. A comparison of a calculated pressure distribution with the pressure distribution from the wind tunnel establishes a relationship between the two angles as in Table 3. The pressure distribution is calculated employing XFOIL by DRELA [13]. (As in the experiment, a turbulent flow is also forced in the simulation. The tripping positions are at 0.02c on the suction side and at 0.05c on the discharge side, Fig. 6 shows the distributions.)

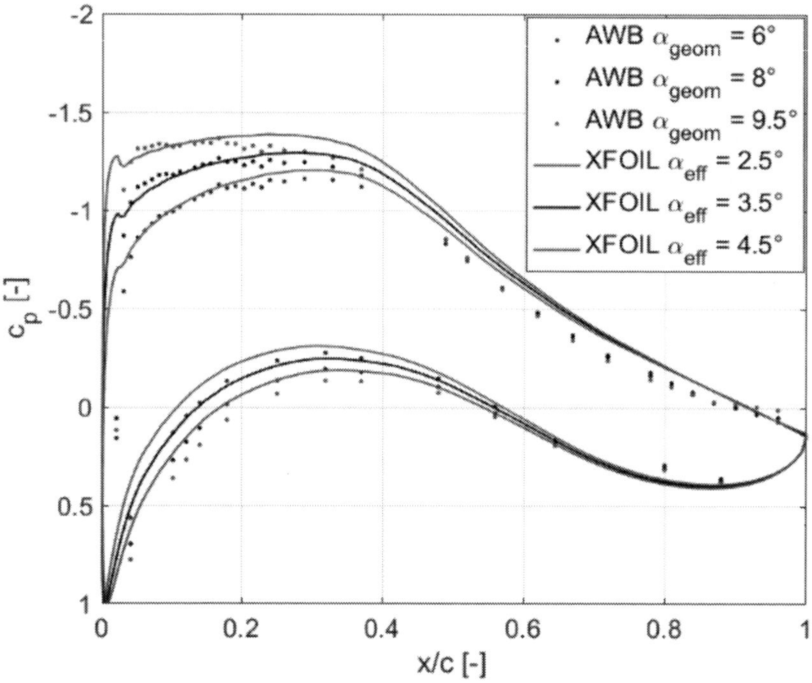

Figure 6: Measured and XFOIL-predicted static pressure distributions for corresponding geometrical and effective angles of attack.

Table 3: Corresponding geometrical and effective airfoil angles of attack

α_{geom}	6°	8°	9.5°
α_{eff}	2.5°	3.5°	4.5°

3. Results and Discussion

Fig. 7 depicts sound pressure maps for several 1/3-octave bands as obtained from the microphone array data. The reference airfoil and the airfoil with fixed flap angle (ϕ = 9.6°) at α_{eff} = 3.5° are compared. The effect of the serrations in the trailing edge region is clearly visible.

A more quantitative analysis is given in the next paragraphs. Fig. 8 shows the dependency of the spectra on ϕ for serrations S *I*. The four plots in Fig. 8 a) stem from the microphone array, in Fig. 8 b) from the mirror. In general, the results are independent of the measuring method. Since the signal-to-noise ratio is better with the mirror, only the mirror diagrams Fig. 8 a) are discussed further. A noise reduction can be observed irrespective of the flap angle up to the so-called cross-over frequency. The cross-over frequency is where serrations produce more sound than the airfoil without serration, here somewhere in the 1.6 or 2 kHz 1/3-octave band. As pointed out earlier, the lowest frequency resolvable by the mirror is 1 kHz. It is expected that the serrations are an effective mean for noise reduction far below this 1 kHz limit. As an example the noise reductions $\Delta L_{p,1/3}$ (i.e. sound pressure level with vs. w/o serrations) in the 1/3-octave band with a center frequency f_m = 1 kHz for a wide range of serration flap angles is plotted in the upper diagram of Fig. 9. Clearly, maximum noise reduction is achieved for fixed serration flap angles between ϕ = 10° and 15°. The self-aligning flap angle is lower than the optimum angle and provides smaller but still substantial noise reduction. The lower diagrams in Fig. 9 illustrate the effect of the airfoil angle of attack α_{eff} = 2.5°, 3.5° and 4.5° - here in case of serrations type *S II*. The self-aligning serration flap angles become smaller as the airfoil angle of attack is increased, and again: Noise reduction is existent but not maximum as for fixed flap angles.

Fig. 10 contains data for the case when the serrations are allowed to self-align in the flow. It is worth to note that in all experiments the self-aligning serrations stayed absolutely stable in the airfoil wake, no flow induced oscillations perpendicular to the mean flow were observed. Thus the self-aligning flap angle ϕ_{sa} measured is rather accurate. The torque to turn the serration out of their self-alignment position is remarkably large. The red dotted arrows in Fig. 9 indicate the data points ϕ_{sa} for self-aligning serrations. Irrespective of α_{eff} , the noise reduction is always approximately 5 dB in the 1 kHz frequency band. Clearly, ϕ_{sa} is acoustically not absolutely optimal but already shows a substantial noise reduction. And ϕ_{sa} varies with α_{eff} in the same sense as the optimal flap angles. Not shown here, the same tendencies are observed in other frequency bands up to the cross-over frequency. Hence, the initial first hypothesis that self-aligning serrations are superior when the airfoil angle of attack is varying (e.g. due to gusts of pitch control) can partly be confirmed. More noise reduction is possible with steeper than self-aligning flap angles of the serration, fixed for each α_{eff}. This points to the second initial hypothesis, that self-aligning serrations produce a noise reduction without increase of blade load, i.e. the flow induced forces on the blade. Fig. 11 shows the measured streamwise static pressure distribution for

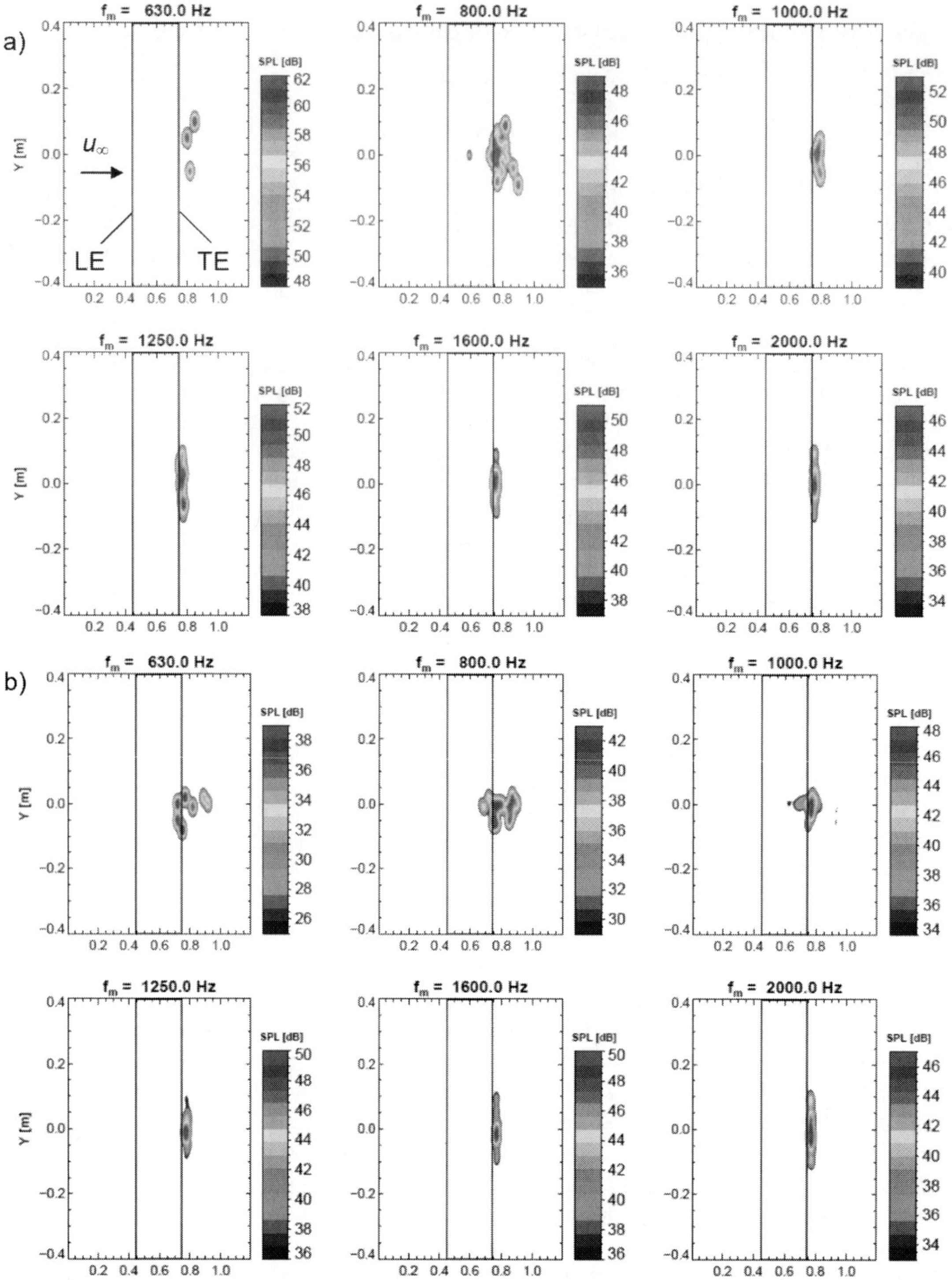

Figure 7: 1/3-octave band sound source map measured with the microphone array at α_{eff} = 3.5°, a) without serrations, b) with serrations S *I* at a fixed flap angle ϕ = 9.6°.

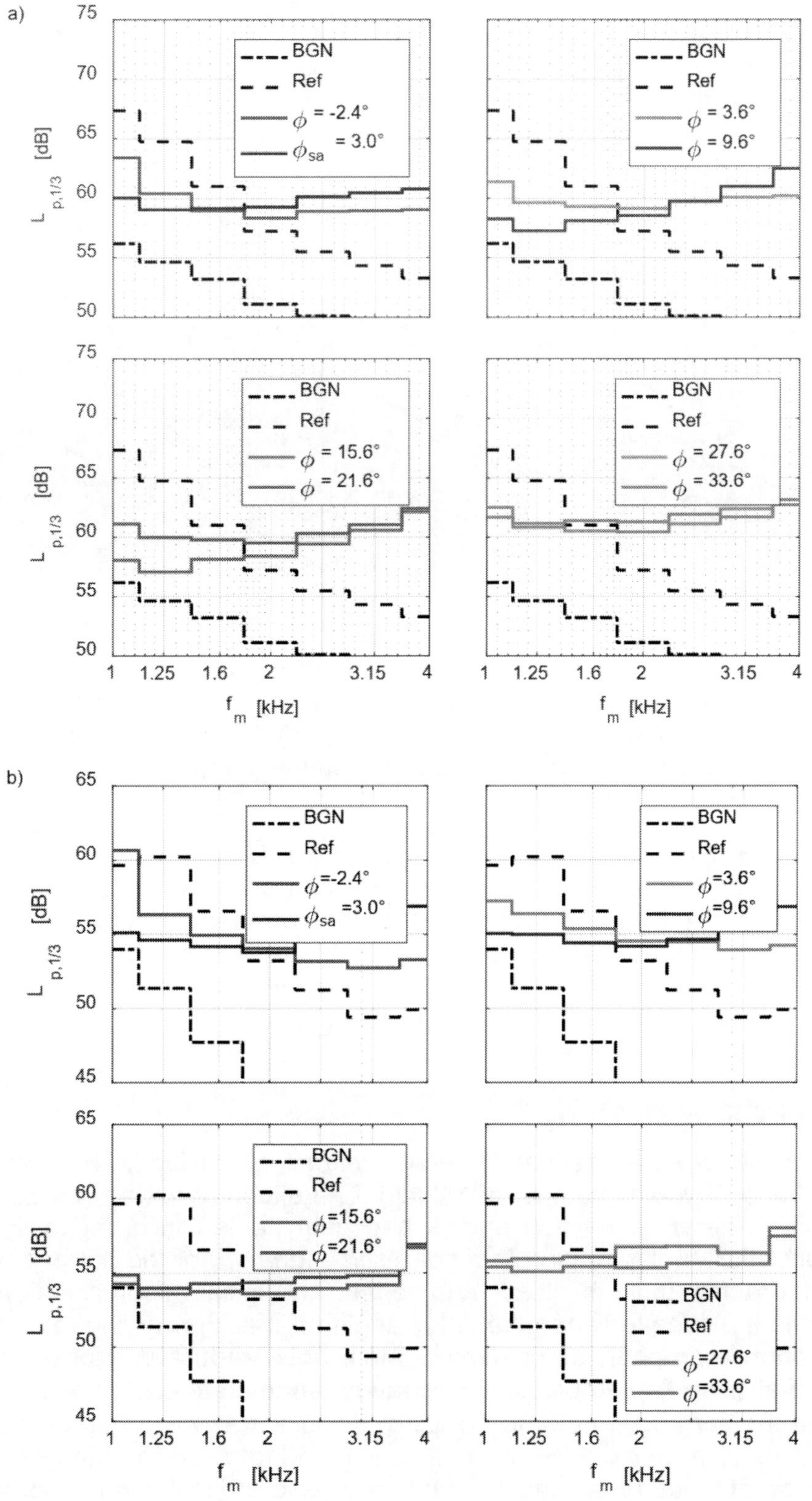

Figure 8: Far field 1/3-octave noise spectra for various fixed and self-aligning (sa) flap angles ϕ and α_{eff} = 3.5°, serrations S / a) from mirror, b) from microphone array.

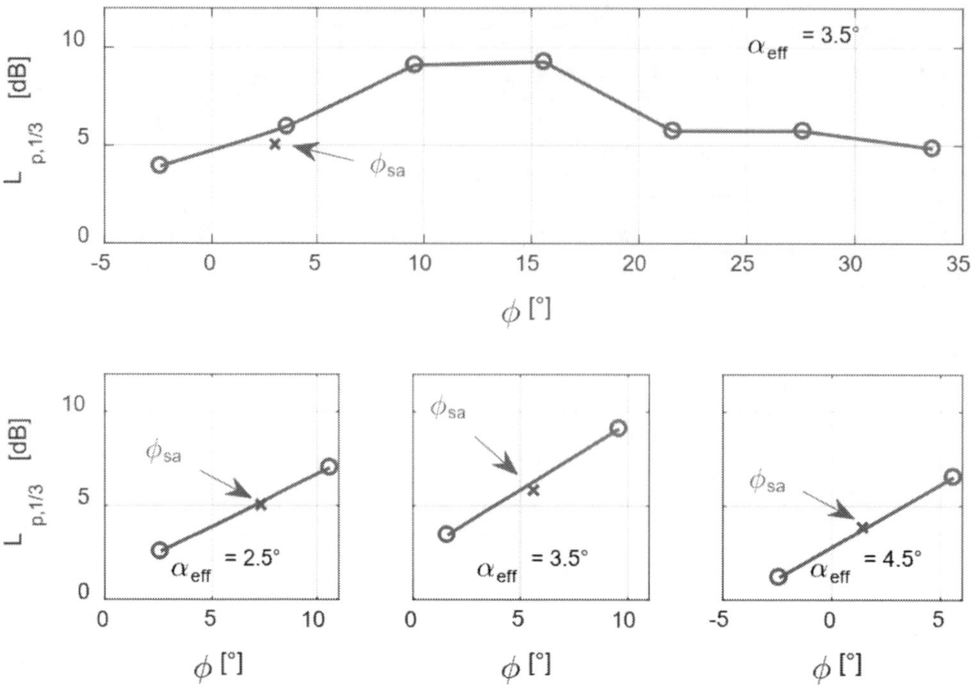

Figure 9: Noise reduction $\Delta L_{p,1/3}$ (with vs. w/o serrations) in the 1/3-octave band with a center frequency f_m = 1 kHz; upper: as a function of serration flap angle ϕ (serrations S *I*, taken from Fig. 8*)*; lower: as a function of airfoil angle of attack α_{eff} (S *II*, taken from Fig. 10*)*; the red crosses indicate the self-aligning flap angle, i.e. self-alignment of the serrations is permitted.

fixed and self-aligning serrations at α_{eff} = 3.5° and the reference without serrations. Note that x/c = 1 indicates the TE of the airfoil without serrations. For all practical reasons the pressure distribution is not affected by the serrations. Hence, fixed serrations as an add-on to a blade result in an increase of the flow induced force on the blade just due to the net increase of blade area - for the serrations tested here by appr. 8%. Self-alignment of the serrations will prevent this.

4. Summary and Conclusions

Two selected serrations were attached to a two-dimensional DU93W210 airfoil segment and tested at a Reynolds number of 1.2 mio. The experiment was carried out in two steps. Firstly, different *fixed* flap angles had been tested in order to find the acoustically most efficient flap angle. Secondly, *self-alignment* of the serrations was allowed, while the airfoil angle of attack was varied. The initial hypothesis that self-aligning serrations are superior when the airfoil angle of attack is varying (e.g. due to gusts of pitch control) can partly be confirmed. More noise reduction is possible with steeper than self-aligning flap angles of the serration, adapted to each airfoil angle of attack α_{eff}. Then the serrations need to be fixed. Fixed serrations as an add-onto a blade, however, result in an increase of the flow induced force on the blade just due to the net increase of blade area - the streamwise pressure distribution is not affected by the existence of serration. A work around is the integration in the initial blade design - not considered here.

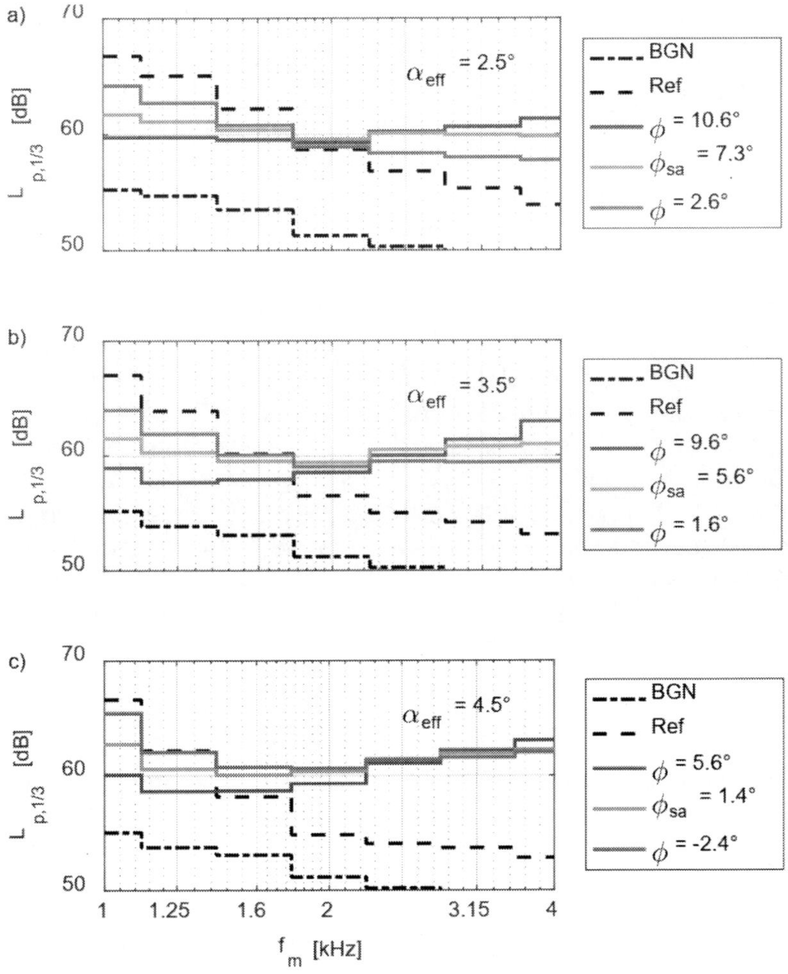

Figure 10: Far field 1/3-octave noise spectra: Self-adjusting serrations compared to fixed flap angles and the reference for different angles of attack (S *II*).

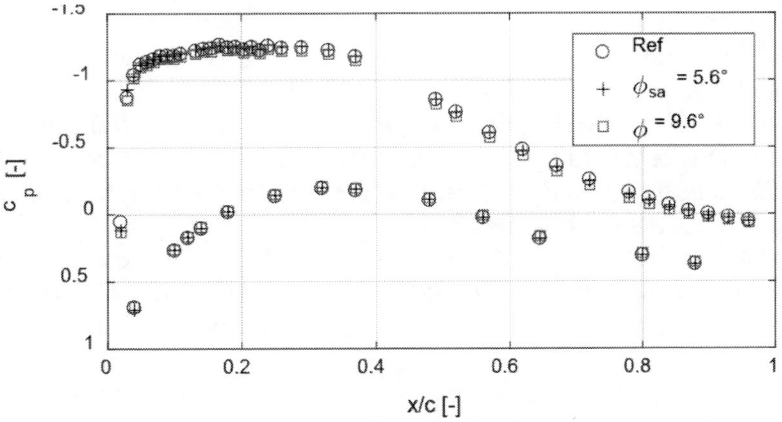

Figure 11: Streamwise static pressure distribution for fixed and self-aligning serrations (α_{eff} = 3.5°). "Ref" is the airfoil section without serrations. *x/c* = 1 indicates the TE of the airfoil without serrations.

5. Acknowledgement

Financial support for this work was partly provided by the Federal Ministry for Economic Affairs and Energy of Germany (BMWi) within the project RENEW (FKZ 0325838B). We also would like to thank B. Homrighausen and E. Bender from the University of Siegen for the design and manufacture of the instrumented model and for their support during the experimental phase in the wind tunnel. Finally we would like to thank the DLR Team in Braunschweig, especially M. Herr and K. Rossignol for the help and support during our measurement campaign.

6. References

[1] WAGNER, S., BAREISS, R., GUIDATI, G., *"Wind Turbine Noise"*, Springer, 1996.

[2] OERLEMANS, S., FISHER, M., MAEDER, T. AND KÖGLER, K., *"Reduction of Wind Turbine Noise using Optimized Airfoils and Trailing-Edge Serrations"*, 14th AIAA/CEAS Aeroacoustics Conference, Vancouver, Britisch Columbia Canada, 2008.

[3] OERLEMANS, S., SIJTSMA, P. AND MÉNDEZ LÓPEZ, B., *"Location and quantification of noise sources on a wind turbine"*, Jounal of Sound and Vibration, Volume 299, Issues 4-5, 2007.

[4] ARCE, C., RAGNI, D., PRÖBSTING, SE., SCARANO, F., *"Flow Field Around a Serrated Trailing Edge at Incidence"*, AIAA SciTech, 33rd Eind Energy Symposium, Kissimmee, Florida, 2015.

[5] ARCE, C., MERINO-MARTINEZ, R., RGNI, D., PRÖBSTING, S., AVALLONE, F., SINGH, A. AND MADSEN, J., *"Trailing Edge Serrations — Effect of Their Flap Angle on Flow and Acoustics"*, 7th International Conference on Wind Turbine Noise, Rotterdam, 2017.

[6] TIMMER,W.A., VAN ROOIJ,R.P.J.O.M., *"Summary of the Delft University Wind Turbine Dedicated Airfoils"*, 41st Aerospace Sciences Meeting and Exhibit, Reno, Nevada, 2003.

[7] GRUBER, M., JOSEPH, P.F., CHONG, T.P., *"On the mechanisms of serrated airfoil trailing edge noise reduction"*, 17th AIAA/CEAS Aeroacoustics Conference, Portland, Oregon, 2011.

[8] POTT-POLLENSKE, M. and DELFS, J., *"Enhanced Capabilities of the Aeroacoustic Wind Tunnel Braunschweig"*, 14th AIAA/CEAS Aeroacoustics Conference, Vancouver, British Columbia, Canada, 5-7 May 2008, AIAA Paper 2008-2910, 2008.

[9] HERR, M., "*Trailing-Edge Noise*", Dr.-Ing. Dissertation, Technische Universität Carolo-Wilhelmina zu Braunschweig, 2013.

[10] ROSSIGNOL, K. S., HERR, M., SURYADI, A., *"Senvion AWB Studie Part 2: Acoustic measurements at a wind tunnel model in AWB, in cooperation with Universität Siegen"*, Technical Report, Deutsches Zentrum für Luft und Raumfahrt e.V., Braunschweig, 2018.

[11] DOBRZYNSKI, W.; NAGAKURA, K., GEHLHAR, B. AND BUSCHBAUM, A., *"Airframe Noise Studies on Wings with Deployed High-Lift Devices"*, 4th AIAA/CEAS Aeroacoustics Conference, Toulouse, France, June 2-4 1998, AIAA Paper 98-2337, 1998.

[12] SIJTSMA, P., *"CLEAN based on spatial coherence"*, International Journal of Aeroacoustics", Vol. 6, No. 4, pp. 357-374, 2007.

[13] DRELA, M., YOUNGGREN, H., *"XFOIL 6.94 User Guide"*, Technical Report, Massachusetts Institute of Technology, Cambridge, 2001.

INCE EUROPE

8th International Conference
on
Wind Turbine Noise
Lisbon – 12th to 14th June 2019

Measurement of sound efficiency of trailing edge serrations (TES) on wind turbines in the Jura mountains.

Xavier Falourd, PRONA SA : x.falourd@prona.ch
Lukas Rohr, PRONA SA : l.rohr@prona.ch
Dominique Bollinger, HEIG-VD: dominique.bollinger@heig-vd.ch

Summary

ADEV Windkraft AG operates two wind turbines in St-Brais (Jura mountains in Switzerland). Given the particular topography and the proximity of the dwellings, the noise of wind turbines is clearly audible even if the immission limit values are respected. To overcome this effect, a retrofit of the blades has been done through the integration of trailing edge serrations (TES).

The main purpose of the study was to quantify the efficiency of the TES through the comparison on measurements done before and after the retrofit, measurements done at IEC61400-11: 2012 distances and at dwellings. The measurement protocol and all the data collected made it possible to extract reliable statistics for all the operating conditions of the wind turbines. The comparison of sound levels by wind speed class before and after installation of the trailing edge serrations (TES) shows that the reduction of the noise level of wind noise is between 2 dB and 4 dB at the emission for upper wind 6m/s at hub height. This paper presents the measurement campaign and the data analysis

1. Introduction

The ADEV company operates two wind turbines to St-Brais. Given the particular topography and the proximity of dwellings, the noise of the wind turbines is clearly audible even if the immission limits are respected. To compensate for this effect, a trailing edge serrations (TES) system was installed on each of the blades of the wind turbines at the end of the spring of 2017. The goal of this project is to control their acoustic efficiencies by measurements of control.

1.1 Principle of qualification of the wind turbine noise

Phases no.1 (spring 2017) and no.2 (fall 2017) were performed before and after the installation of TES and have permitted to collect acoustic, meteorological data and WT parameters of the two turbines so as to compare sound levels in bin classes at emission and at dwellings. Acoustic measurements were undergone near wind turbines (distance IEC61400-11:2012) and in two places of immission (dwellings at about 465 m and 565m of the nearest turbine) for two phases of about 25 days.
The comparison of noise levels by class of rpm before and after installation of the serrations of trailing (TES) at distances IEC61400-11 edge allows to quantify the reduction achieved near the wind turbines but also at dwellings (by extrapolation on the acoustic propagation and/or measurement).

Acoustic measurements at the immission locations also serve to qualify the audibility of the sound of the west wind turbine (WEA2) essentially during night time and weekend days.

2. Acoustic measurement protocol

Four sound level meters are used to perform simultaneous measurements of noise at the two emission locations (range IEC61400-11) and at two immission locations. Sound level meters are recoding continuous sound level LAeq, Leq in third octave bands (of 6.3 Hz to 20 kHz) as well as audio recordings. The positions of sound level meters for two measurement phases are similar with a precision of about two meters. A meteorological mast is located near the IEC61400-11 West point. The collected data, only for the second period, complete the wind turbine operational data sets.

The two sound level meters near WT are equipped with a secondary screen mounted on a rigid plate. Two sound level meters located at the immission points are placed on the ground and they are equipped with a secondary screen formed by a box open acoustic fabric-covered.

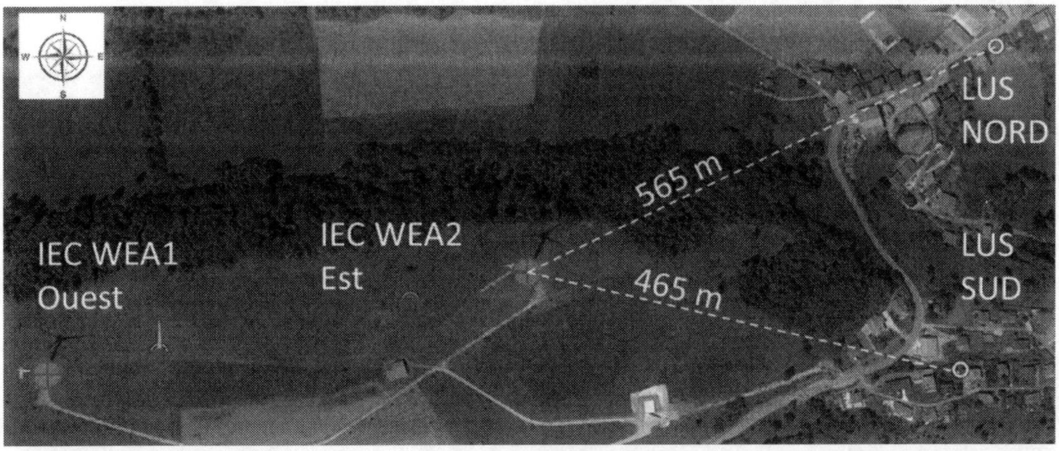

Figure 1: Locations of sound level meters and wind turbines WEA1 and WEA2

Figure 2: Pics of IEC61400-11 boards

Figure 3: Pics of microphone casing at immission points.

3. WT operational parameters

The two turbines operational data sets are documented for each phase of measurement through:

- Electric Power
- Rotation speed (rpm)
- Wind turbine orientation
- Wind speed at hub height

The temporal data of each parameter are delivered:

- phase no. 1 with a resolution of one minute,
- phase no. 2 with a resolution of 10 minutes.

3.1 Wind rose SCADA – Phase no.1

The figures below represent wind roses of each wind turbine for the first measurement period prior to installation of TES. It is remarkable that for this period the turbine WEA1 (West) is exposed to a wind stronger than the WEA2 (East) wind and the prevailing wind directions are somewhat different.

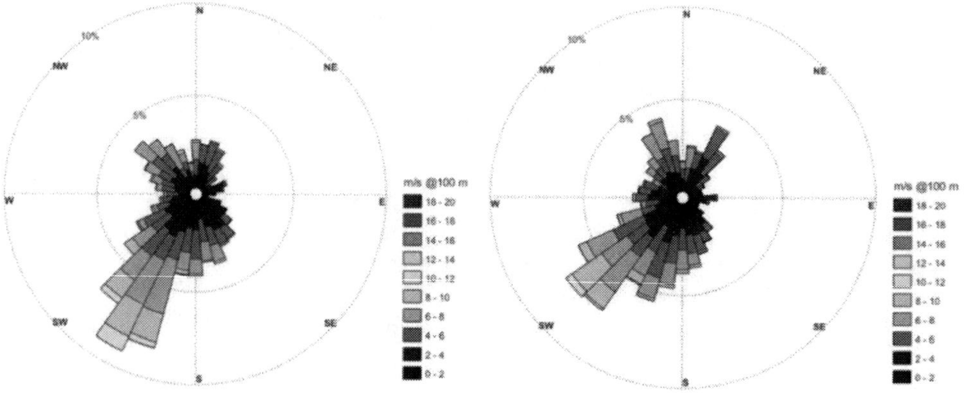

Figure 4: Wind roses phase no.1 (WT) - measurements of the wind at hub height. Left (West) WEA1, right WEA2 (East)

3.2 Wind rose SCADA – Phase no.2

The operational data sets of phase no.2 are delivered with a resolution of 10 minutes involving some interpolation of the data to have a resolution of one minute as for the first phase. The following figures present wind roses of each wind turbine for the second period after TES installation. It is remarkable that on this period the two wind roses are very similar with a predominance for winds from the Southwest.

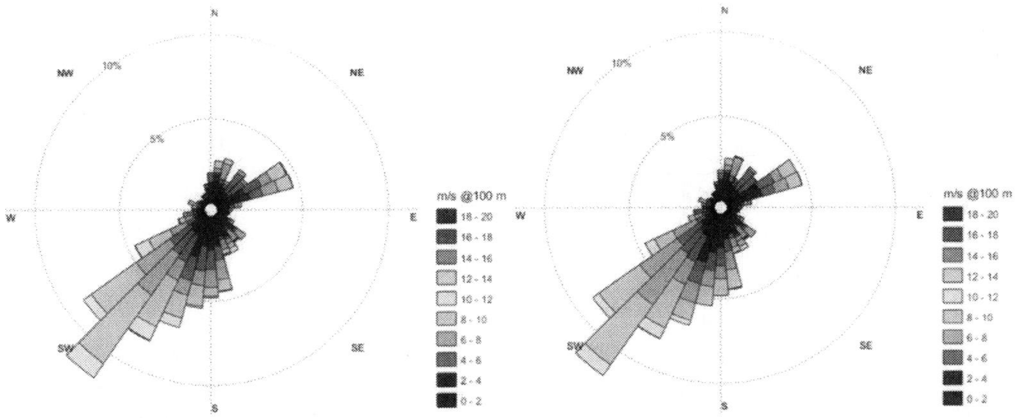

Figure 5: Wind roses phase no.2 (WT) - measurements of the wind at hub height (h = 78 m). Left (West) WEA1, right WEA2 (East)

3.3 WT statistics – Phase no.1

The following figure shows the two turbines operating statistics during the first phase of measurement. The night limitation performed on the wind turbine is (WEA2) is clearly distinguished when the wind speed is more than 8 m/s. There are two modes of power of the wind turbine is (WEA2) according to rotation speed (RPM).

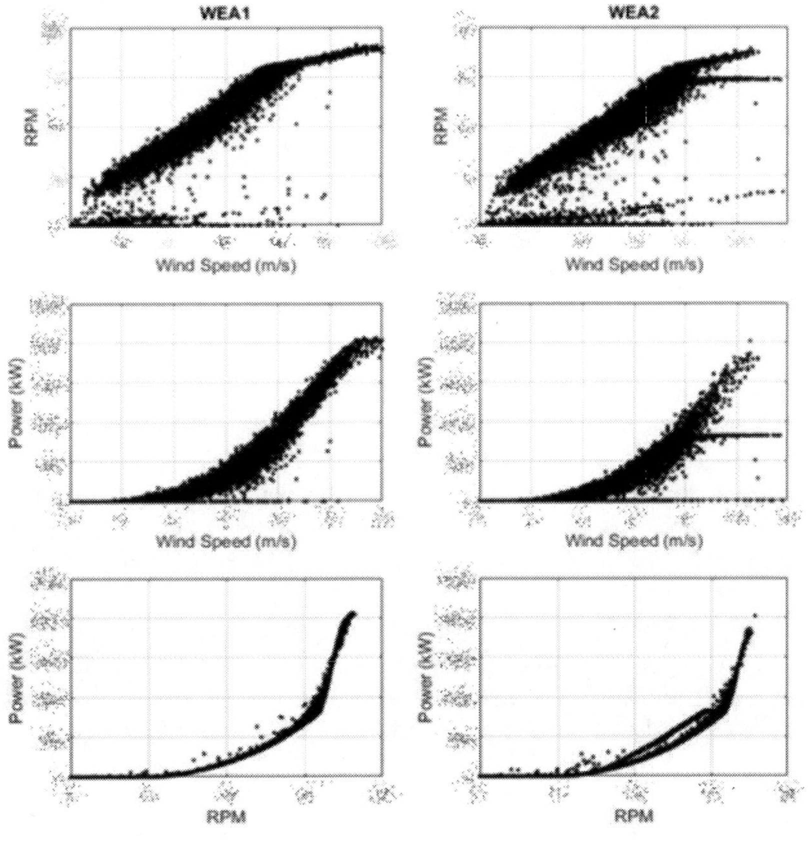

Figure 6: WT parameters - Phase no.1

3.4 WT statistics – Phase no.2

The following figure shows the two turbines operating statistics during the second phase of measurement. The night limitation being no more effective the statistics of the two turbines are very similar.

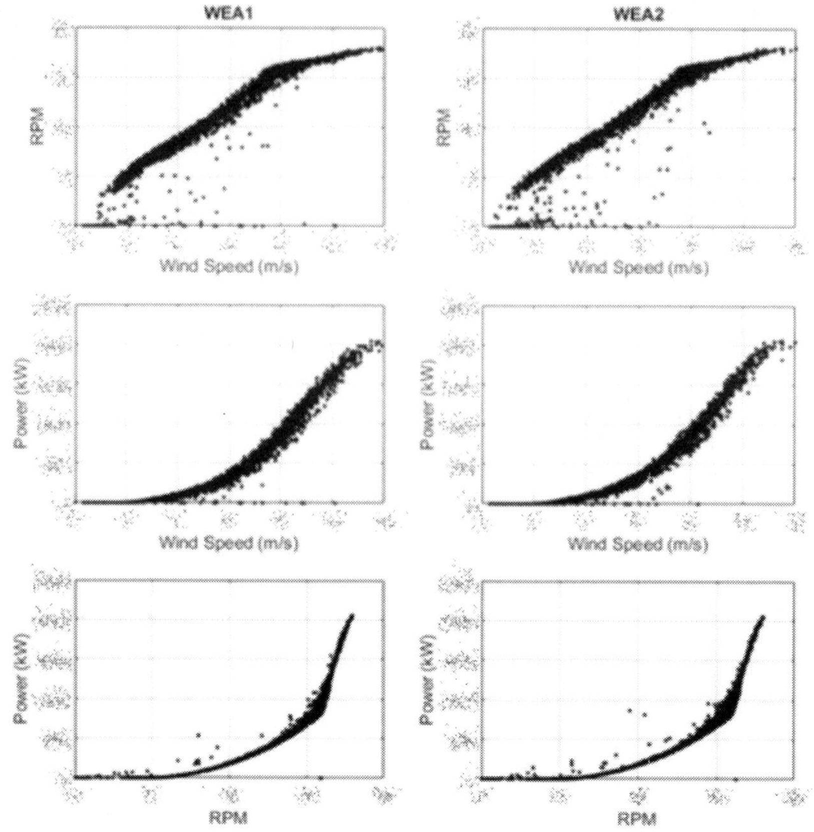

Figure 7: WT parameters - Phase no.2

4. Assessments by stop and go phase

The stop phases are designed to discriminate the wind noise of ambient noise at each point of measurement. According to the conditions of ambient noise, the sound of each turbine is quantifiable by differentiation of sound level LAeq or third octave bands.

Figure 8 is a report of a sequence alternating go-stop and go measurement of WEA1 (West) and WEA2 (East) wind turbines. This type of report indicates wind at hub height, RPM data for each wind turbine as well as noise data at the point of measurement.

All the reports of the stop and go tests indicate that the ambient noise on the wind farm site is low enough to allow significant acoustic measurements at distance IEC61400-11: 2012. On the set of measurements the emergence of wind noise on the ambient noise is located between 6dB approximately (wind 3 m/s, 6 rpm) and 25 about dB (wind 10 m/s, 18 rpm).

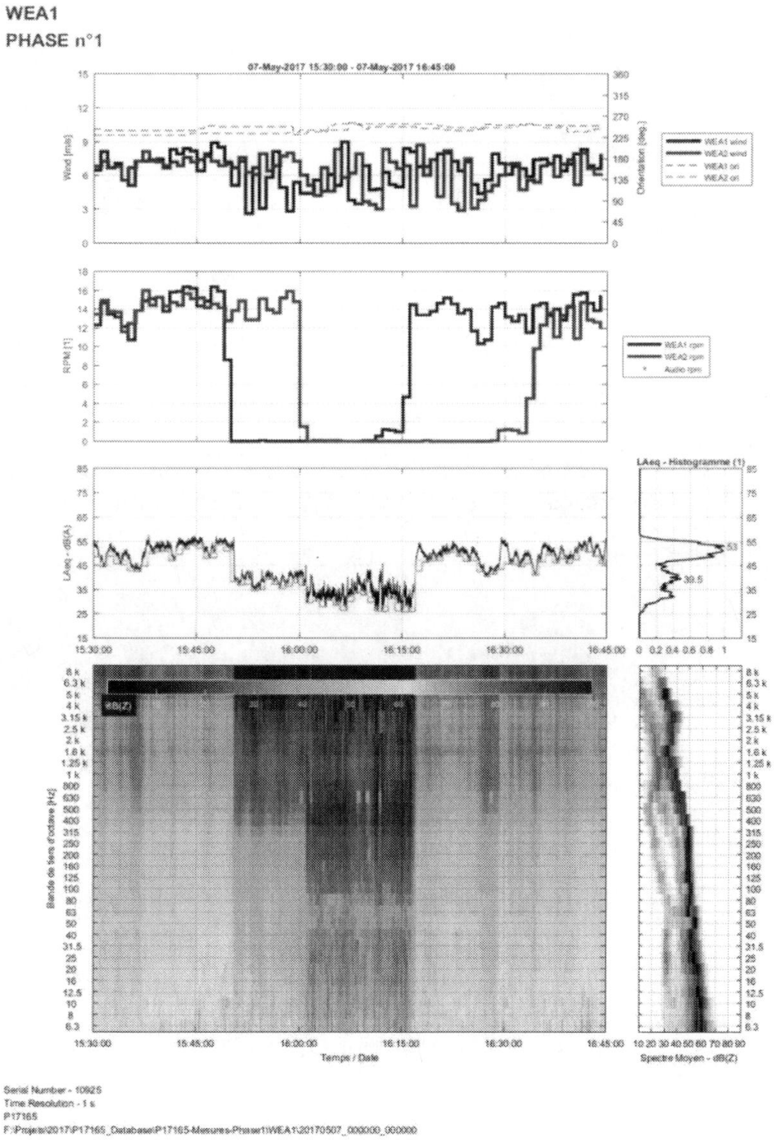

Figure 8: Report of measurement of wind noise at the IEC-WEA1 point to a sequence of stops (stops represented by the RPM curves to 0).

5. Noise statistics near the two wind turbines

Figure 9 and Figure 10 illustrate noise levels minimum LAeq (appointed Lmin in graphics) statistics for a period of observation of one minute depending on the speed of rotation (RPM) of the WEA1 (West) and WEA2 (East). The dispersion of the data is small, the average value per class of RPM allows to characterize the noise level by RPM class with good accuracy.

Note

Usually the sound levels are expressed relatively to wind at 10 m height. For the two phases (before and after TES) only operational data sets are available. Wind turbine noise being mainly composed of aerodynamic noise, the rotation speed of the rotor (RPM) is selected in place of the wind speed: wind turbine noise levels are almost a linear function of the rpm.

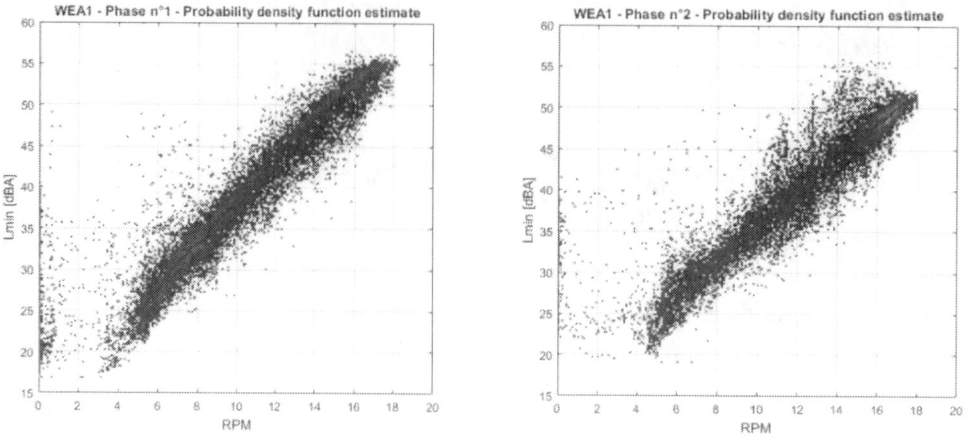

Figure 9: Raw statistics of the wind turbine WEA1 (West) - classified Acoustic measurements according to RPM data.

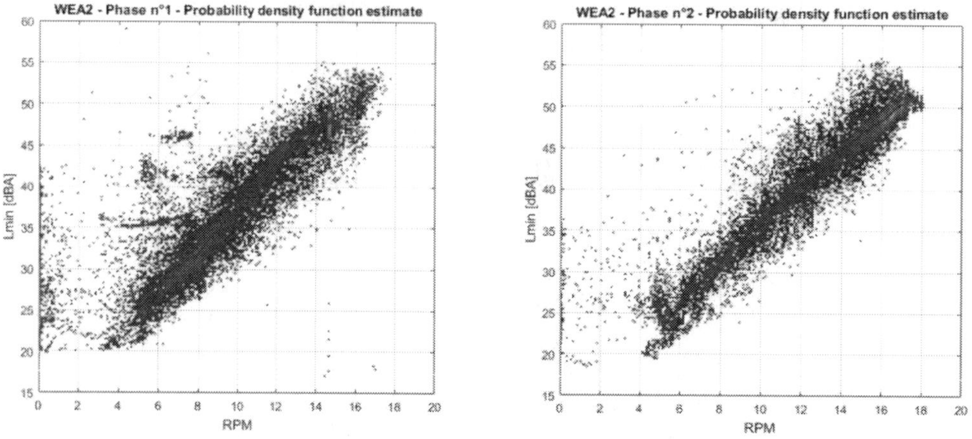

Figure 10: Raw statistics of the wind turbine WEA2 (East) - classified Acoustic measurements according to RPM data

The comparison of the mean values of the two turbines by phase noise is given in the Figure 11.

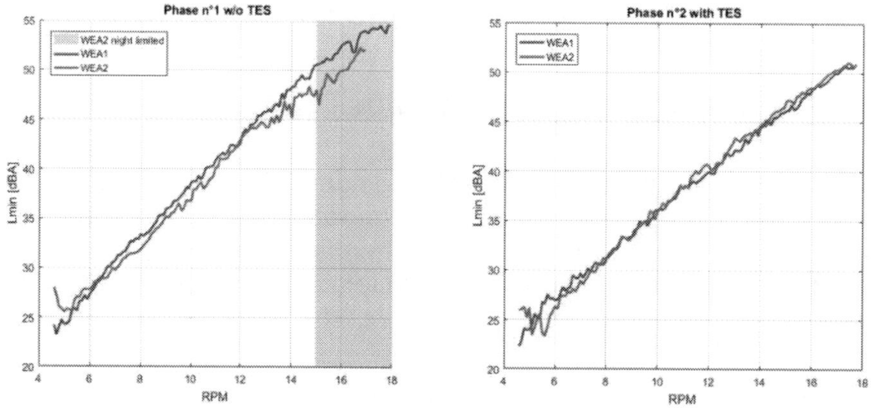

Figure 11: Comparison of noise from wind turbines WEA1 (West) and WEA2 (East) before and after TES installation.

For the phase no.1 despite a good similarity WEA1 and WEA2 curves are disjoint and can be explained by the fact that the wind rose is different for the two WT (speed and direction) and during periods of high wind speed, night limitations were applied to the turbine WEA2 influencing then the night statistics.

However for the phase no. 2, which offers roses of the same wind and without night limitation, the curves are considered the same. It validates the principle of measurement near the distances IEC61400-11 wind turbines: 2012.

6. Statistics on the efficiency of TES

The assessment of the efficiency of TES is performed by comparison of the average sound levels collected at distances IEC61400-11: 2012 the two phases of measurement. The following figures show the comparisons and the quantification of the efficiency of TES by speed of the rotor (RPM).

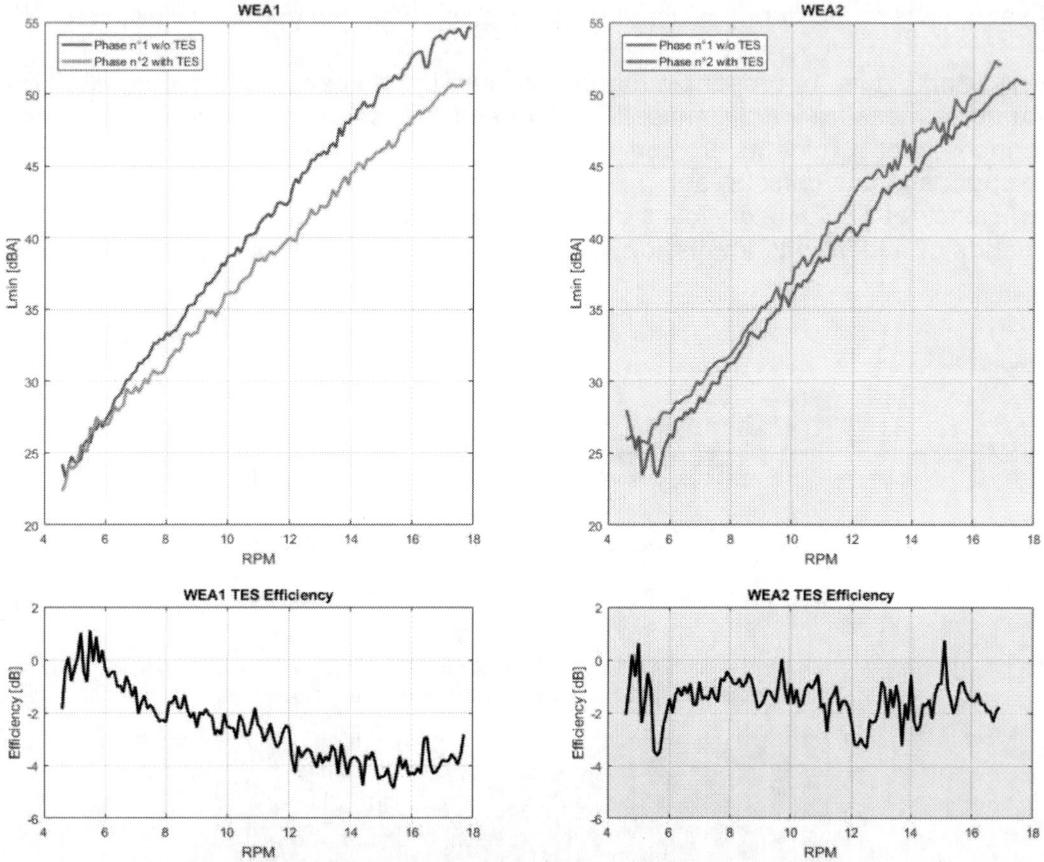

Figure 12: Average statistics of wind turbines WEA1 (West) and WEA2 (East) - top: comparison of noise levels before and after installation of TES: downstairs: statistical estimation of efficiencies. WEA1 is the reference for estimating turbine (no night clamping the first phase instead of WEA2).

For wind WEA1 (West), the measured efficiency offers a gradual reduction of the noise level between 0 dB and 4 dB according to the RPM. Under the assumption that the turbulence in exit of blades develop at high speed, efficiency of TES to reduce noise to airflow is greater at full speed than at low speeds.

For the wind WEA2 (East), measured efficiency offers a 2 dB of noise reduction on average but does not increase the efficiency according to the RPM. For the interpretation of these results, it is to consider that the wind turbine WEA2 (East) was subject to a night limitation during phase no.1 and the wind roses differ before and after the TES installation.

In conclusion, according to the terms of interpretation expressed previously, the two curves of assessment of efficiency of TES indicate a reduction in the noise level of at least 2 dB for the rotation speeds above 8 RPM. The results obtained near the wind turbine WEA1 (West) are the most reliable in terms of conditions of measurements (similarity of the wind roses).

7. Spectra and TES efficiency

In order to evaluate the influence of TES on the spectra of wind turbine noise, the comparison is carried out on a representative period during which the power of the wind turbine is maximum. In fact the database for statistical comparisons on the efficiency of TES to the immission by wind (speed, orientation) class is limited following the night limitation of the WEA2 turbine during phase no. 1. For each measurement phase, a period of 30 seconds is found during which the wind noise is audible to the immission locations.

Figure 13 and Figure 14 present in the third octave band spectra in the range 63 Hz - 4 kHz of wind noise measured at distances IEC61400-11 (emission) and immission locations. For both periods, measurements are corrupted by extraneous noise with significant influences on the spectral areas (pink areas).
The efficiency of TES is significant mostly in the frequency bands between 125 Hz and 1 kHz (bands of audibility to immission locations) and in the bands of frequencies above 1.6 kHz.

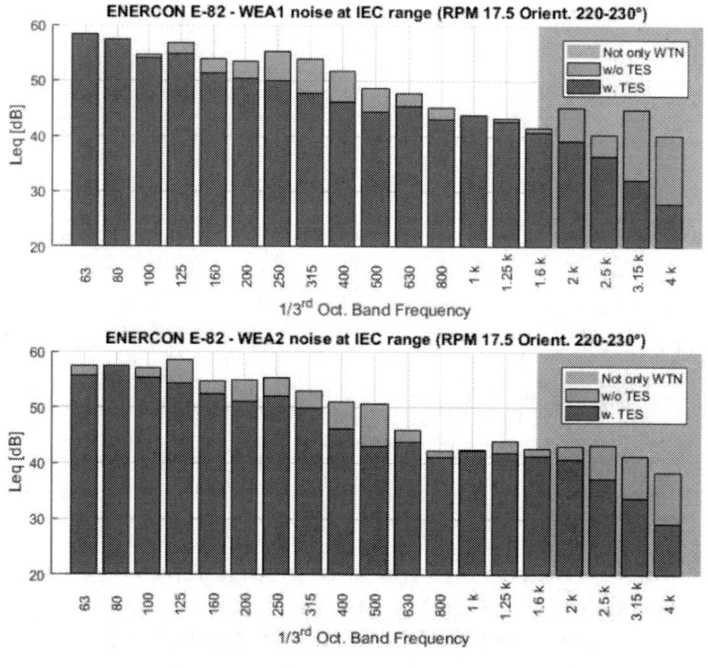

Figure 13: Spectra of IEC61400-11 distance before and after installation of TES

At dwellings the efficiency of TES is figured out for these same frequency bands indicating a reduction in noise levels.

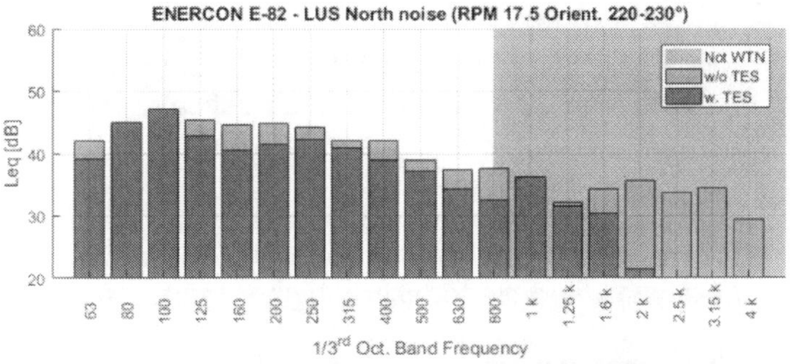

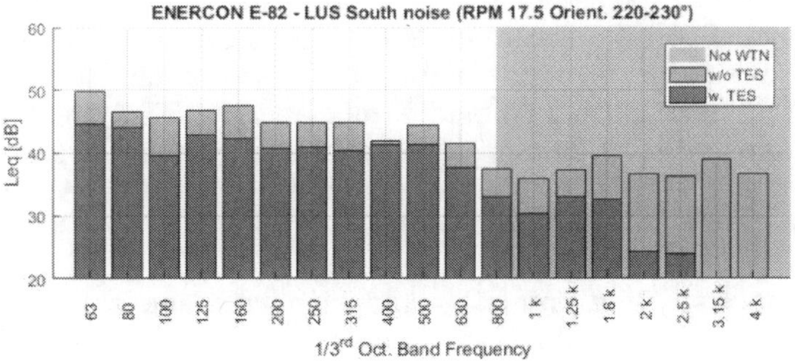

Figure 14: Spectra to read before and after installation of TES

The following table shows the noise levels LAeq (1 min) and L90 (L90 fractile index is the noise level LAeq met or exceeded 90% of the time, compared to the total duration of the measurement) corresponding to the emission and immission sites. For two wind measured at distances IEC61400-11, L90 levels indicate a reduction in the noise level of about 3 dB.

Sites / Sound Levels	**Phase n°1** RPM 17.5 Orientation 220-230°		**Phase n°2** RPM 17.5 Orientation 220-230°		**Différence**	
	LAeq dBA	L90 dBA	**LAeq** dBA	L90 dBA	**LAeq** dBA	L90 dBA
IEC WEA2	57.6	55.3	53.7	52.1	3.9	**3.2**
IEC WEA1	57.1	55.1	53.6	52.1	3.5	**3**

Table 1: LAeq (1 min) and L90 sound levels for the two periods and for each measurement site.

8. Noise levels statistics near WT and at dwellings

Sound levels at the IEC61400 locations (WEA1, WEA2) and immissions sites (LUS North, LUS South) for the two phases are presented in the Figure 15 in their statistical forms according to the wind speed at hub height of the wind turbine WEA2 (West).

The average statistical noises at distance IEC61400-11 are represented by the red lines for the wind turbines WEA1 and WEA2. These last curves are then transposed forward to the immission locations: the calculation of wind noise at the immission locations is based on extrapolation of the noise measured at distances IEC61400-11 decreased in terms of geometric divergence and losses by absorption according to the distance; the effects of soil and other spread effects due to field (relief) and barriers (buildings) are not factored by this modeling.

For every full curves (black and dotted) chart shows indicative values of wind turbine noise according to the data of the wind turbines. In the light of the statistical distributions of sound levels at the immission locations, it is remarkable that the observed minimum sound levels correspond to forecasts of sound levels calculated both statistically and by modeling.

The wind turbine WEA1 (West) participates only partially to the total wind turbine noise (less than 1 dB input) which is dominated by the nearest wind turbine WEA2 (East). Sound levels of wind turbine noise corresponding to the minimum values of the noise at immission locations; wind turbine noise is most likely still audible for winds higher than 6 m/s.

Although wind noise immission locations is mainly dominated by the WEA2 (East) wind turbine for which comparisons between the two phases are delicate due to the difference of the wind roses and the night limitations (only phase no.1), it can be concluded on the basis of a meticulous reading of these statistics that the efficiency of TES measured close to wind turbines can be transposed to the immission locations. Indeed the cross-reading of the statistics of the WEA1 and WEA2 turbines with modeling calculations allow to observe the reduction of wind noise at the immission locations.

As featured in noise reports at the immission (sound levels and spectra), wind noise is low emerging otherwise not present in the third octave bands above 800 Hz according to the ambient noise conditions. Thus, it would be interesting to express the same statistics for a sound level calculated in bands of frequencies from 20 Hz to 800 Hz.

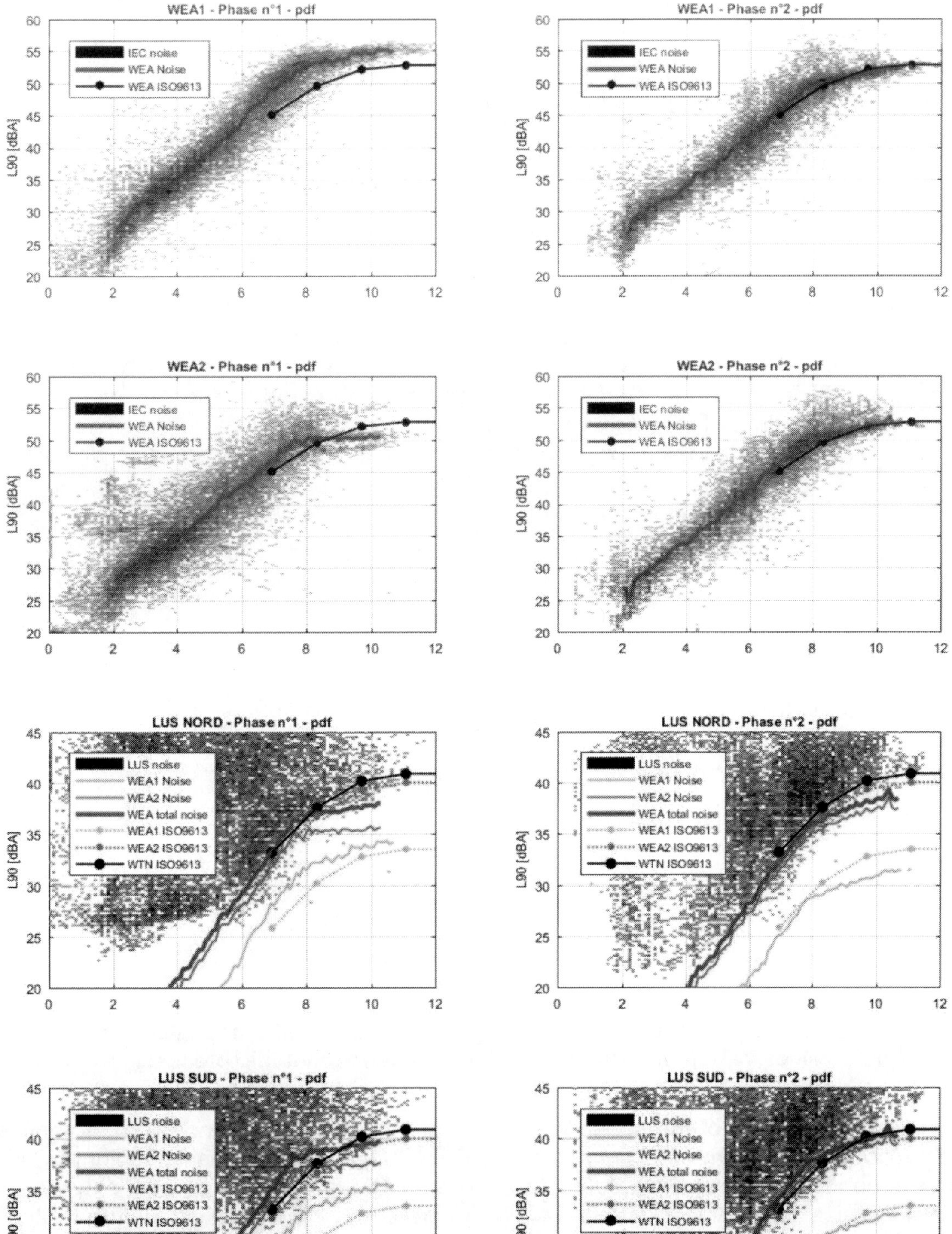

Figure 15: Statistics of sound levels L90 (1 minute). On the left, phase no. 1; right phase no. 2. Grey dots figures the raw noise data. Red curves are statistics of WTN. Black dot curves are approximation of the wind turbine noise according to ISO9613-2. The wind speed is measured at hub height

9. Remarks on the evaluation of audibility

The strong emergence of wind turbine noise on the ambient noise being permanent at distance IEC61400-11, the wind turbine noise (repeating audible pattern correlated with the rotation of the rotor) is clearly distinguished in sound recordings. The estimate of speed of rotation RPM of each wind turbine is achievable if based on a suitable data processing in strips of third octave with sufficient temporal resolution (here 125 ms). When operating the transformed of Fourier on specific bands of the sonogram in third octave bands, it is possible to express an instantaneous RPM with a resolution of about 3 seconds.

The following figure shows for a period of several days the extreme values of operational data (blue: maximum values, minimum values cyan) and the RPM values calculated from the audio analysis (in pink). For the speeds of the rotor at 3 rpm, the algorithm fails (lack of emergence of wind turbine noise). For speeds of rotation of the rotor higher at 3 rpm, estimates of the instantaneous rotation speed are therefore in good line as bounded by the extrema of the WT data. The characterization of the audibility of noise wind in a site, based on the hypothesis of the emergence of wind turbine noise on the ambiant noise is therefore achievable with this type of treatment.

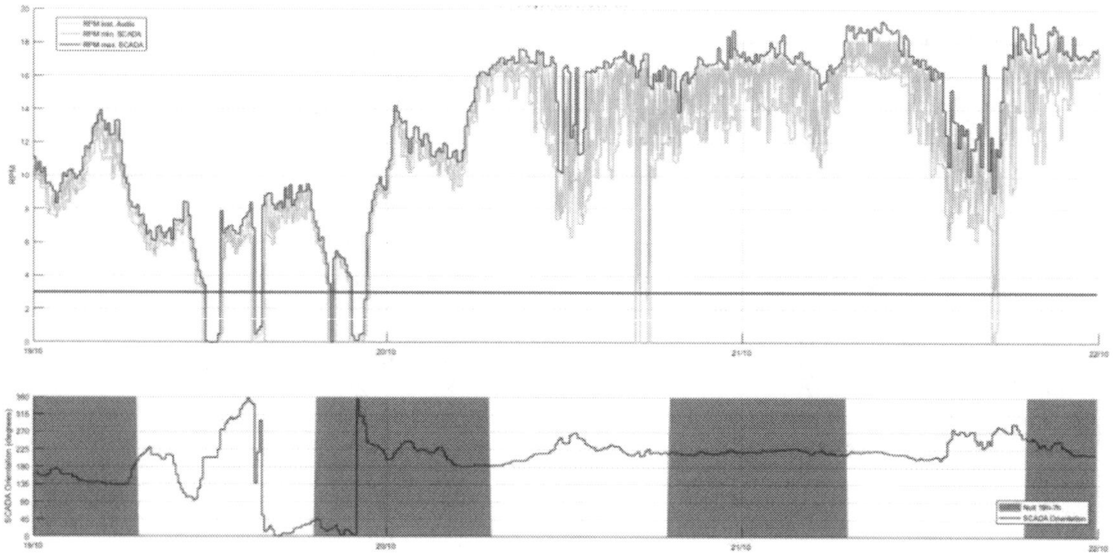

Figure 16: Assessment of the instant speed by audio analysis. In blue and cyan max and min of data values WT; in pink the instant RPM values from audio processing.

These same assessments of the speed of rotation of the rotor were conducted on immission sites. The following figure shows estimates of the RPM speed at the immission point LUS North of the phase no. 2. The calculated values of the instant RPM (in red) are present only when the wind turbine noise is detected. These "audibility" data are mainly found during night-time periods and weekend days i.e. the ambient noise is low.

This method, similar in principle to the one for the assessment of the amplitude modulations, could be studied more specifically for the qualification of the audibility phases of noise from wind turbines. The deepening of this method (robustness and precision) is not part of this project but remain of interest for the quantification of the audibility according to the swiss regulations.

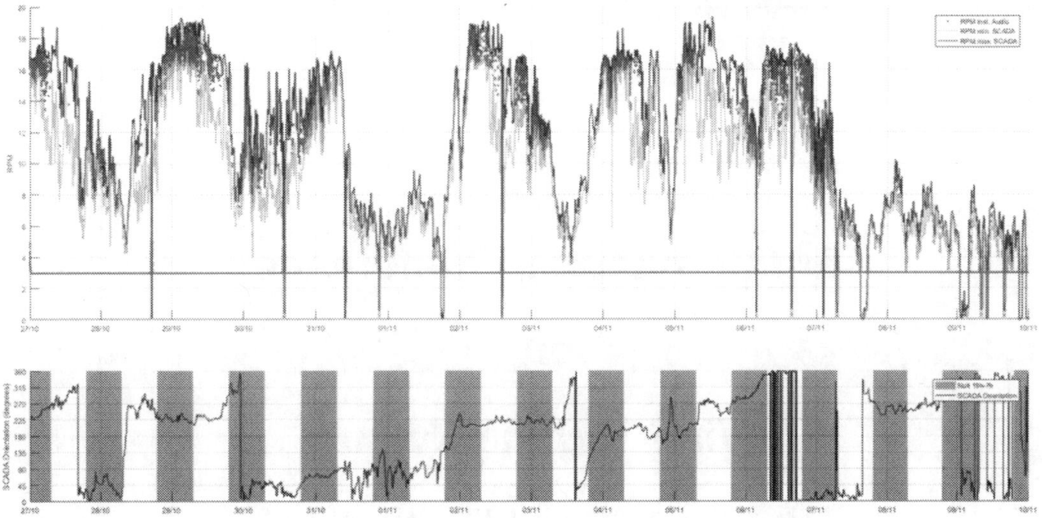

Figure 17: Evaluation of rotation speed RPM by audio analysis on Figueras North LUS - Phase data n ° 2

10. Conclusions

The main purpose of the study was to quantify the efficiency of TES, efficiency measured at distances IEC61400-11: 2012 near wind turbines. The measurement protocol and all of the data collected helped to extract reliable statistics to qualify wind turbine noise. The quantification of the efficiency of TES at immission locations has been estimated by extrapolation due to the ambient noise conditions and the night limitation (operated before the TES installation). Based on our statistical analyses and other listening to audio recordings, wind turbine noise was decreased by the TES which are effective in the audible band between 125 Hz and 1 kHz, frequency bands characteristic of wind turbine noise at mean and long distances (immission locations). Comparison of noise by rotation class (rpm) or wind speed class before and after installation of the trailing edge serrations (TES) shows that the reduction of the sound level of wind noise is between 2 dB and 4 dB near the wind turbine and about 2 to 3 dB at the immission for wind speed higher than 6 m/s or 12 rpm.

References

Proceedings Wind Turbine Noise 2015, 2017

IEC61400-11:2012 « Wind Turbines - Part 11: Acoustic noise measurement techniques »

« Wind Turbine Noise », D. Bowdler and G. Leventhall 2012

« A Method for Rating Amplitude Modulation in Wind Turbine Noise» IOA Noise Working Group (Wind Turbine Noise) 2016

8th International Conference

on

Wind Turbine Noise

Lisbon – 12th to 14th June 2019

Hydrodynamic Analysis of Trailing Edge Serrations with Blunt and Rounded Edges

Alper Celik, Bin Zang, Yannick D. Mayer, Xiao Liu and Mahdi Azarpeyvand*
University of Bristol, Bristol, United Kingdom, BS8 1TR

Summary

A comprehensive experimental study is performed to examine the effects of large serrations on the surface pressure fluctuations of a flat plate test rig. Two different configurations, consisting of a sharp edge profile and a rounded edge profile, are tested to understand the effect of serrations on the noise generated at the trailing edge and the associated flow field, compared to a straight trailing edge flat plate. The test rig is heavily instrumented with microphones and pressure taps. The tests are performed at serration root-to-tip distance based Reynolds number of $Re = 1.8x10^5$. The results are presented for a range of frequencies $50Hz < f < 6000Hz$, where a broadband noise increase is observed around $f \simeq 150 - 200Hz$ for both serrated configurations. It is shown that the structures generating the increase in the energy content in low frequencies evolve through the leading edge, then gain strength as they develop downstream and amalgamate at the tip of the serration. This was confirmed by both the sound pressure level results and the surface pressure coherence analysis.

*Corresponding Author, m.azarpeyvand@bristol.ac.uk

1. Introduction

Over the last few decades, the increased air traffic and rapidly developing wind turbine farms in urban and rural areas accelerated the concerns about noise emissions. Aerodynamic noise generated by the turbulent flow past a trailing edge over the airfoil is one of the most important sources of noise. Among all the mechanisms that Brooks *et al.* [1] have shown to contribute to airfoil self-noise, turbulent boundary layer interaction with the trailing edge is the main source generating the broadband noise. The increased awareness about noise pollution and foreseen noise pollution reduction targets by institutions (such as ACARE Flightpath 2050), researchers have focused on decreasing the noise levels of these inseparable instruments of modern life. Several passive control methods including brushes [2], leading and trailing edge serrations [3, 4, 5, 6] as well as the active control techniques such as wall-normal suction [7, 8] and blowing [9] or dielectric barrier discharge plasma actuators [10] have been investigated in the literature. Serrations, geometry which is inspired by the wing of an owl, is the most common method applied owing to its easy application and relatively low effects on the aerodynamic performances [11].

Analytical models initiated by Howe [12] to predict the noise reduction by the serrated trailing edges over predicts the noise reduction level. Although the models keep evolving with new studies such as Lyu *et al.* [3, 13, 14],the integration to calculate the far-field noise from wall pressure fluctuations requires the assumption of frozen turbulence. With this assumption the statistical properties of the convected eddies over the trailing edge is assumed to be not effected by the existence of the trailing edge which leads the pressure spectra to be formulated at the trailing edge in cartesian coordinate based wave numbers. These pressure spectra are then propagated to calculate far-field noise. Recent studies show that these assumptions might be considered to be the main source of error in predicted noise levels [15]. Having a non-straight trailing edge, the boundary layer and associated pressure spectra change through the leading edge of the serration. These changes in the pressure spectra on the serration should be taken into account to improve far-field noise predictions.

The necessity to obtain a sound knowledge on the flow structures and pressure spectra on the serration geometry accelerated both numerical [16, 17, 18] and experimental studies [2, 19, 20, 21]. In their DNS study [16], Jones and Sandberg showed that a highly three-dimensional flow dominates the tip of the serration, where stream-wise vortices from upstream smoothly develop into span-wise vortices. Recently conducted experimental studies [15, 20, 22, 21], support the outcomes of the numerical studies. However, there lacks a common view on how the serrations mitigate the broadband noise.

This investigation aims to map the wall pressure fluctuations and corresponding spectra in detail over a serration geometry. To achieve a highly instrumented serration geometry considering the size restrictions due to microphones, a set of relatively large serrations were manufactured. The serration root-to-tip length is five times the boundary layer height developed on the serration. Although the design may not appear to be optimal, the authors believe that most of the results are applicable to serrations of normal dimensions. The results may also shed light on serrated airfoils or wind turbine blades operating in sub-optimal conditions.

2. Experimental Set-Up

The experiments were performed in the University of Bristol Aeroacoustic Facility, which is a closed-circuit, open-jet anechoic wind tunnel. The anechoic chamber has a cut-off frequency of 160Hz with physical dimensions of 6.7m x 4.0m x 3.3m and the nozzle has an area of 500mm x 775mm which allows a steady operation from 5m/s to 45m/s. The experiments were performed for a range of flow velocities $10m/s \leq U_{\infty} \leq 40m/s$ with turbulence intensity levels below 0.2% [23]. For brevity, only the results at free stream velocity of $30m/s$ is presented.

To understand the flow physics over serrations, a highly instrumented flat plate test-rig was designed and manufactured at the University of Bristol. A total number of 84 Knowles FG-23329-P07 miniature pressure transducers were instrumented to obtain a very detailed near-field pressure map and allow possible future experiments the flat plate test. 75 microphones were installed on the serrations and 8 were installed on the flat plate. The microphones are 2.6mm in diameter with a circular sensing area of 0.8mm. The serrations and flat plate were machined to have 0.4mm pinhole mask to avoid pressure attenuation at high frequencies.

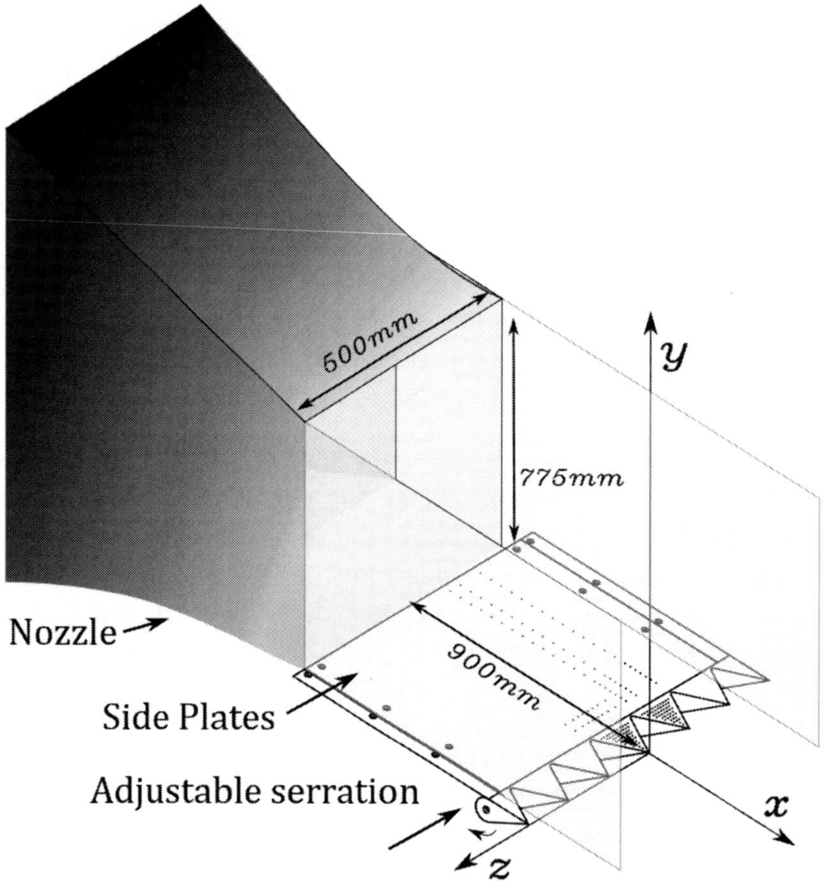

Figure 1 Schematics of the flat plate serration rig.

Velocity measurements were conducted using a 55P16 single-wire constant temperature hot-wire anemometry (CTA). Dantec StreamlinePro CTA 91C10 modules were used with 55P16 single-wire probes. The probes was calibrated by using the Dantec 54H10 type calibrator. The probe was controlled with a two-axis ThorLabs LTS300M traverse system.

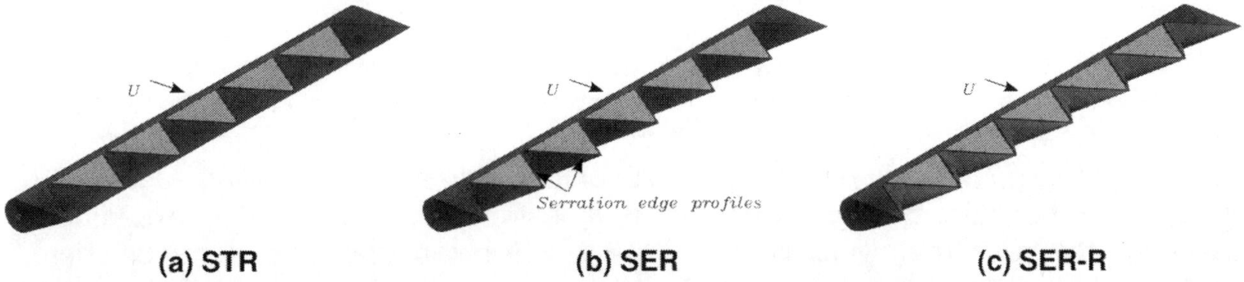

Figure 2 Schematics of the trailing edge models a) Straight trailing edge (STR) b)Serrated trailing edge (SER) and c) Serrated and rounded trailing edge (SER-R)

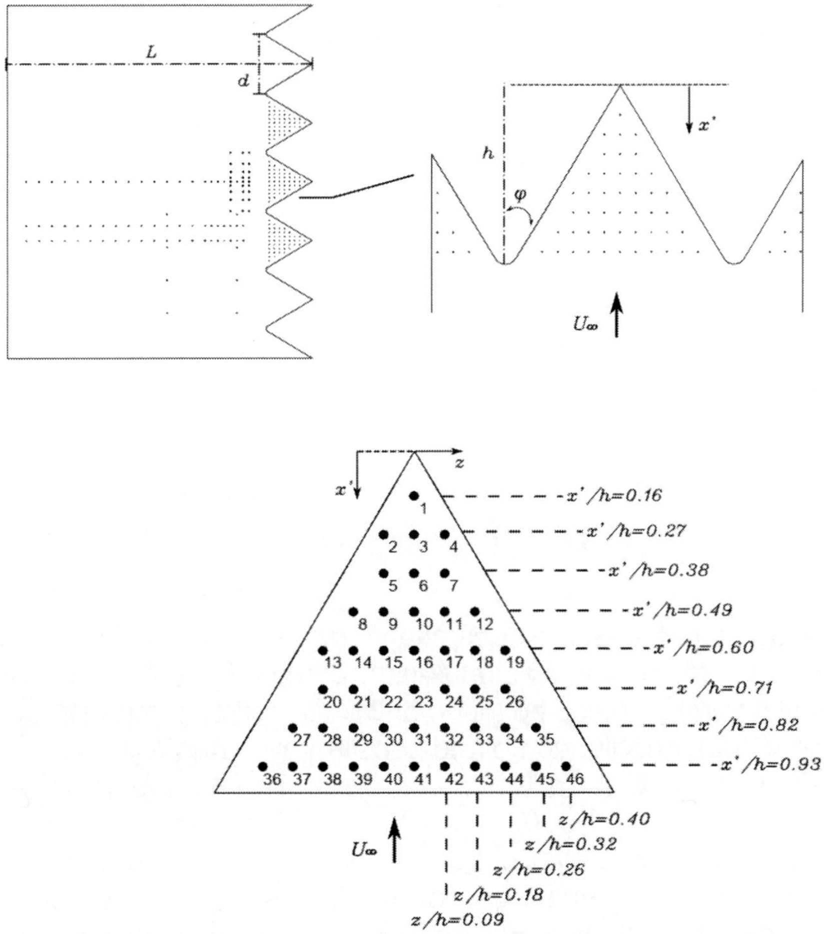

Figure 3 Schematics and geometric details of the flat plate model.

All near-field and hot-wire measurements were performed for 16 seconds at a sampling frequency of $2^{16}Hz$. The near-field and hot-wire data were acquired by using four National Instrument PXI-e4499 cards and were driven by a Matlab script. Pressure measurements were carried out using two 32 channel Chell MicroDAQ pressure scanners at a sampling frequency of 1000Hz for 16 seconds. The scanners have a full-scale accuracy of 0.05%. The averaged results were used to present non-dimensional pressure coefficient C_{prms}.

Figure 1 illustrates the flat plate rig mounted to the lower lip of the wind tunnel nozzle. Side plates were used to assure the two-dimensionality of the flow. The serration part can be tilted around z-axis to have an angle with respect to free stream flow. The serration can be adjusted for different angles from 0° to 30°. 3-D printed add-on fillings were used to obtain a straight trailing edge and rounded trailing edge. Schematics of these fillings are demonstrated in Figure 2. The experiments were conducted for the free-stream velocity of $U = 30m/s$, which corresponds to serration chord based Reynolds number of $Re \approx 1.8x10^5$.

The geometric details of the flat-plate rig are presented in Figure 3. The total length of the rig was $L = 0.9m$. Serration root-to-tip distance is $h = 0.091m$ and span-wise width is $d = 0.105m$. The serration angle is $\varphi = 32°$. A local coordinate x' is designated to clarify the microphone locations, as shown in Figure 3. The stream-wise and span-wise locations are labeled and presented at the bottom row of Figure 3. Taps used for the steady pressure measurements had the same spatial distribution as microphone locations.

3. Results & Discussion

3.1. Surface pressure fluctuations over the serration

The power spectral density (PSD) of the pressure fluctuations measured at each microphone is calculated as

$$\phi_{pp}(dB/Hz) = 20\log_{10}\left(\frac{\bar{p}'}{\bar{p}_{ref}}\right), \qquad (1)$$

where $\bar{p}'$ refers to the averaged pressure fluctuations and $p_{ref} = 20\mu Pa$ is the reference pressure. $\bar{p}'$ was obtained by converting the power spectral density of the measured signal at each microphone to pressure. Figure 4 demonstrates the comparison of the ϕ_{pp} values for the straight trailing edge (STR), serrated trailing edge (SER), and the combined serrated and rounded trailing edge (SER-R) cases for microphone 1 at the tip of the serration ($x'/h = 0.16$). The ϕ_{pp} results for the STR demonstrate a 6dB linear decay from 100Hz to 1000Hz, which could be interpreted as a sign of an attached turbulent boundary layer flow. The serrated trailing edge results demonstrate a completely different phenomenon with a broadband ϕ_{pp} hump between $50Hz < f < 600Hz$, which implies an increase in the energy content at low frequencies. SER-R results have a similar trend to that of SER. However, the broadband hump has a lower peak magnitude and elongates towards lower frequencies. These broadband humps may be the footprint of a large scale hydrodynamic field and will be investigated in the following sections.

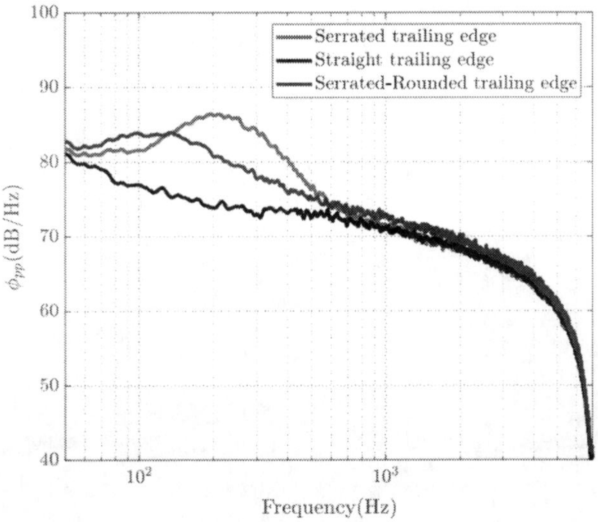

Figure 4 Comparison of the power spectral density of the pressure fluctuations for Straight trailing edge (STR), Serrated trailing edge (SER), and Serrated and rounded trailing edge (SER-R) at microphone location 1, $x'/h = 0.16$, $p_{ref} = 20\mu Pa$

Figure 5 further investigates the source of the energy content increase, where the ϕ_{pp} graphs for microphones at locations 3,6,10,16 (stream-wise), and 4,7,12,19 (along the edge profile) are shown on the left and right column of the figure, respectively. In the case of the STR trailing-edge, a similar trend is observed for all the measurement locations, as expected. The SER case, however, shows a distinct characteristic when all the figures are examined. Along the serration mid-plane on left hand side column, the broadband hump does not appear until the tip of the serration. On the contrary, following the edge of the serration, the broadband hump exists even at microphone location 19, and gradually increases in magnitude towards the tip for the SER case. This may be interpreted as the footprint of the physical structures evolving on the serration edge profile (on both sides) and amalgamating at the tip, which was also discussed in [21]. For the SER-R case, where the sharp edge profile was changed with rounded edge profiles, the broadband hump exists at a lower frequency range, peaked at around $f = 150Hz$. Moreover, the magnitude of the peak is lower compared to SER case above $\approx 200Hz$, but higher below that.

Contours of non-dimensional root-mean-square (RMS) coefficients for the pressure fluctuations p' are presented in Figure 6. The $C_{p_{rms}}$ values were calculated using Equation 2, where p_i is the ith data measured, $\bar{p'}$ refers to the averaged pressure over the entire measurement duration, and the dynamic pressure p_{dyn} was calculated based on free stream velocity. The very dense contours at the edges are due to interpolation between the zero valued cells and the measured region, hence should not be taken into consideration.

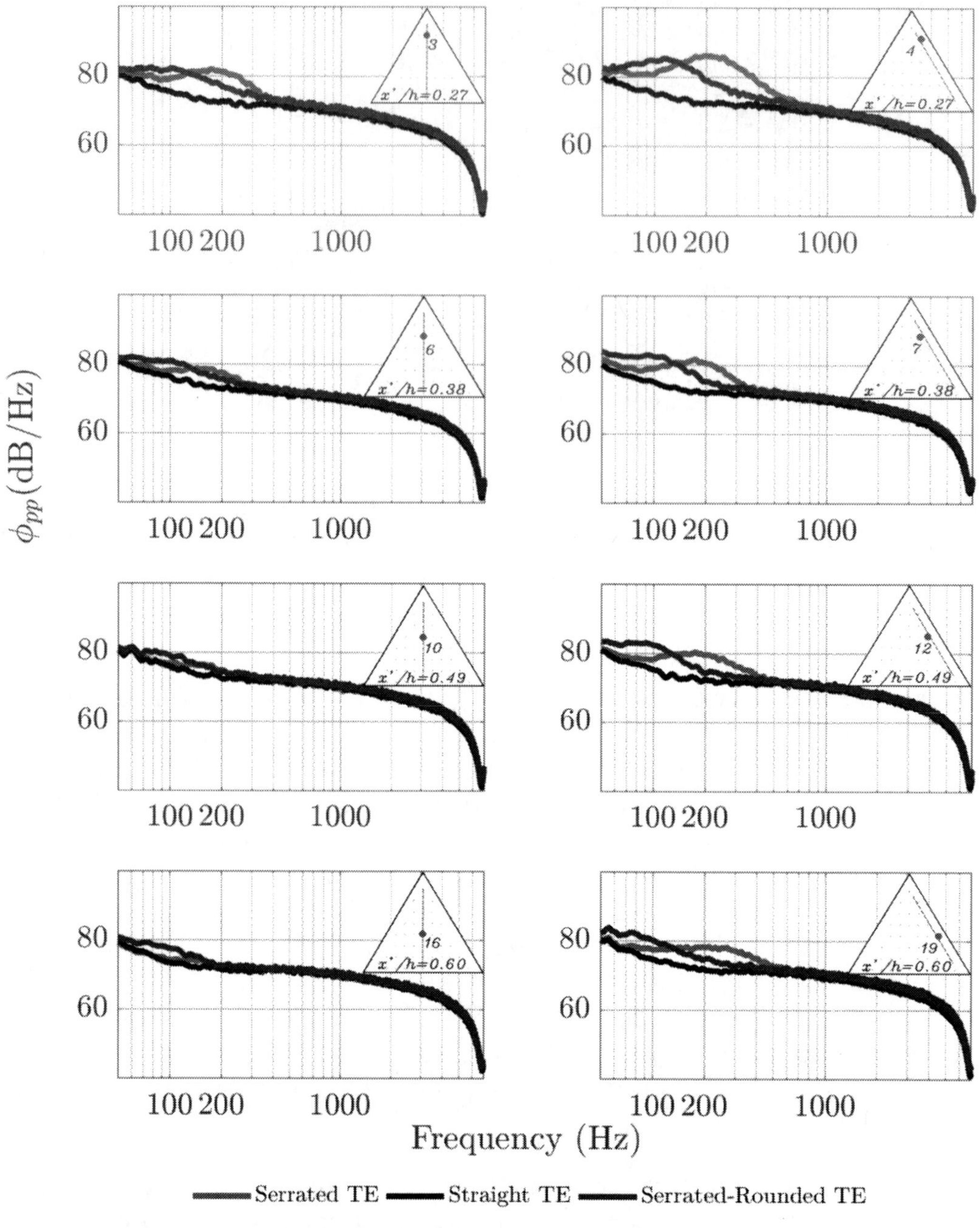

Figure 5 Comparison of the power spectral density of the pressure fluctuations measured at microphones 3,6,9,16 (stream-wise, on left column) and 4,7,12,19 (leading-edge, on right column) for STR, SER and SER-R, $p_{ref} = 20\mu Pa$

$$C_{p_{rms}} = \frac{\sqrt{\sum_{i=1}^{N}(p_i - \bar{p})^2/N}}{p_{dyn}}, \tag{2}$$

Considering the STR case, $C_{p_{rms}}$ value is mostly evenly distributed with a slight increase towards the trailing edge. For both the SER and SER-R cases the $C_{p_{rms}}$ values increase significantly compared to the STR case. For the SER case specifically, the increase is more pronounced at the tip, and the footprint of the leading edge structures can be seen on the left edge with very high $C_{p_{rms}}$ values. At $x'/h = 0.93$, where the connection with flat plate starts, a low $C_{p_{rms}}$ region is observed. The results of the SER-R are very similar to that of the SER, where the increased $C_{p_{rms}}$ region is more symmetrical with respect to the symmetry plane of the serration.

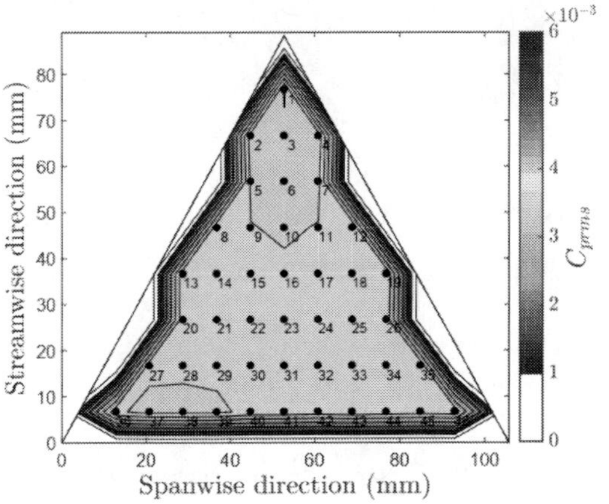

(a) Straight trailing edge

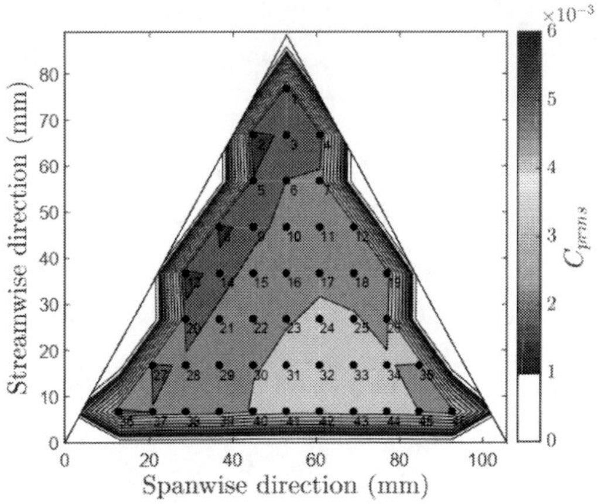

(b) Serrated trailing edge

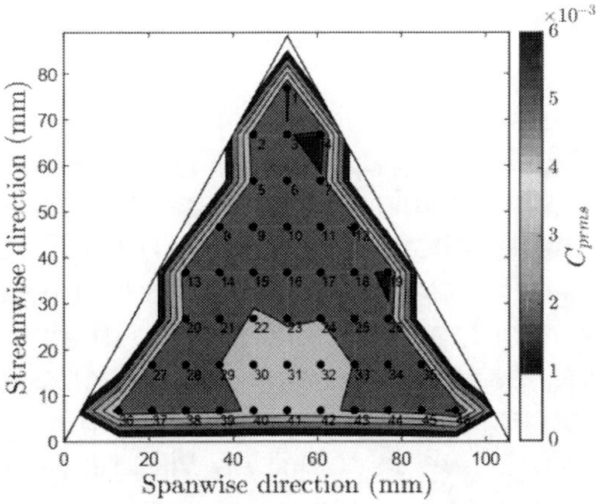

(c) Serrated and rounded trailing edge

Figure 6 Contours of non-dimensional pressure fluctuations C_{prms} for STR, SER and SER-R cases.

3.2. Surface pressure correlation analysis

3.2.1. Span-wise coherence and length scale

The span-wise coherence and the associated length scales are widely used tools to study the coherent structures in the flow. As briefly discussed in the introduction, the span-wise length scale of the structures have a crucial role in the prediction of the far-field noise from wall pressure fluctuations. The magnitude-squared span-wise coherence and associated length scales can be calculated from

$$\gamma^2_{p'_i p'_j} = \frac{\left|\phi_{p'_i p'_j}\right|^2}{\left|\phi_{p'_i p'_i}\right|\left|\phi_{p'_j p'_j}\right|}, \tag{3}$$

and

$$\Lambda_p = \int_0^{\infty} \gamma_{p'_i p'_j}(\xi_z)\, d\xi_z, \tag{4}$$

where $\gamma^2_{p'_i p'_j}$ stands for the span-wise coherence calculated between two microphones as i and j. $\phi_{p'_i p'_j}$ stands for the cross-power spectral density, and ξ_z represents the span-wise distance between two microphones.

Figure 7 presents the span-wise coherence results at five different stream-wise locations $x'/h = 0.16, 0.27, 0.38, 0.49, 0.6$ and $x'/h = 0.93$. The very left column represents serration and shows the microphones used to calculate the span-wise coherence. In Figure 7, the marker "x" stands for the reference microphone. The microphones, which were used to calculate the coherence with respect to the reference microphone, are colour-coded. The same colour-code is also implemented in the graphs to ease the interpretation. The rest of the columns from left to right represent the STR, SER and SER-R configurations.

Considering the results of the straight trailing edge case, i.e., STR, the span-wise coherence between the reference microphone and the closest microphones (mic3-4,mic6-7,mic10-11 and mic16-17) does not change significantly for all the locations examined. The broadband hump between $100Hz < f < 1000Hz$ peaks at $f = 400Hz$. The magnitude of the coherence decreases towards upstream locations,i.e, $x'/h = 0.93$. However, for the serrated case, the magnitude of the span-wise coherence at $x'/h = 0.27$ between microphones 3 and 4 significantly increases compared to the STR case, which may an indication of an increase in length scale of the structures. The peak frequency for the span-wise coherence occurs at $200Hz$, which coincides with the peak observed in ϕ_{pp} graphs for the SER. The use of the rounded serration edge profiles decrease the magnitude of the coherence, which means the structures created by this case are smaller compared to the SER.

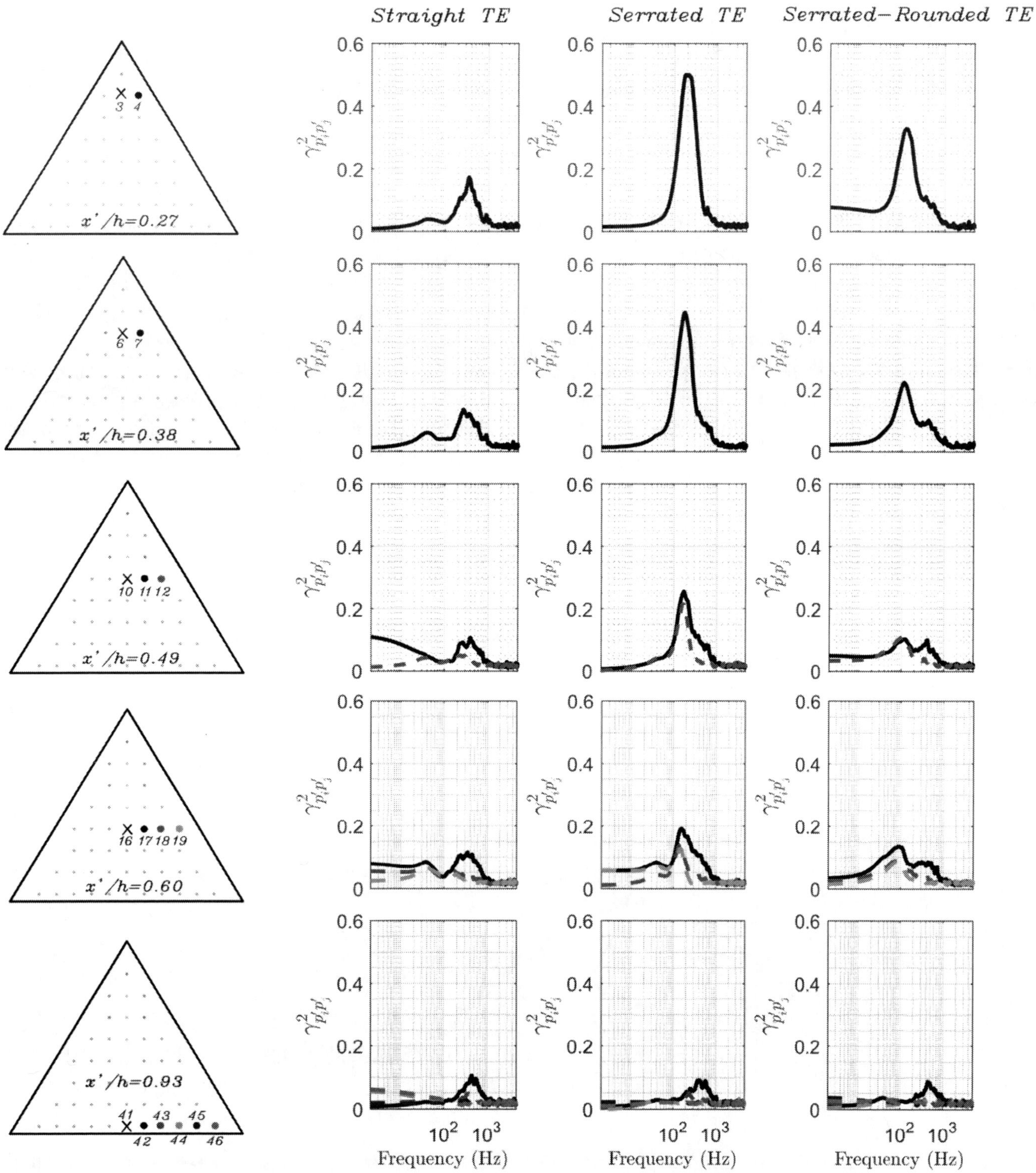

Figure 7 Comparison of span-wise $\gamma^2_{p'_i p'_j}$ among the STR, SER and SER-R at stream-wise stations $x'/h = 0.27, 0.38, 0.49, 0.60$ and $x'/h = 0.93$ from top to bottom.

Comparing all the coherence results over the entire serration, from $x'/h = 0.27$ to upstream $x'/h = 0.83$, from bottom row to top row, the use of STR is observed to cause a very slight increase in coherence values. Although all three cases have the same coherence behaviour at the very upstream station $x'/h = 0.93$, the SER and SER-R gradually increase their coherence value with the closest microphones at each station. This observation is in good agreement with those reported [20]. The results clearly show that trailing edge serrations have a significant effect on the span-wise coherence of the flow structures, and the intensity of the effect varies depending on the edge profile.

Figure 8 displays the effect of serration on the length scales of the coherent structures and the evolution of these structures from the root of the serration to the tip along the serration edge profile. The calculations were performed at $x'/h = 0.60$, among microphones 16,17,18 and 19 using Equation4. The figure shows that at high frequencies ($f > 1000Hz$) all three configurations converge, mostly due to limited spatial resolution. At lower frequencies, the three cases exhibit unique behaviour. The SER case has a peak at around $f = 160Hz$, which was previously observed in ϕ_{pp} and span-wise coherence results at $x'/h = 0.60$. The magnitude of the length scale nearly doubles the magnitude of straight trailing edge. Similarly the SER-R also exhibits a peak at around $f = 100Hz$ with a 50% magnitude increase compared to STR case. Moreover, this result supports the idea of relatively stronger turbulence activities occur on the trailing edge when it is serrated, compared to a straight one. Hence, far-field predictions based on the frozen turbulence assumption, which assumes there is no change in turbulence behaviour on the trailing edge, should consider length scale changes observed in serrated cases to improve the predictions.

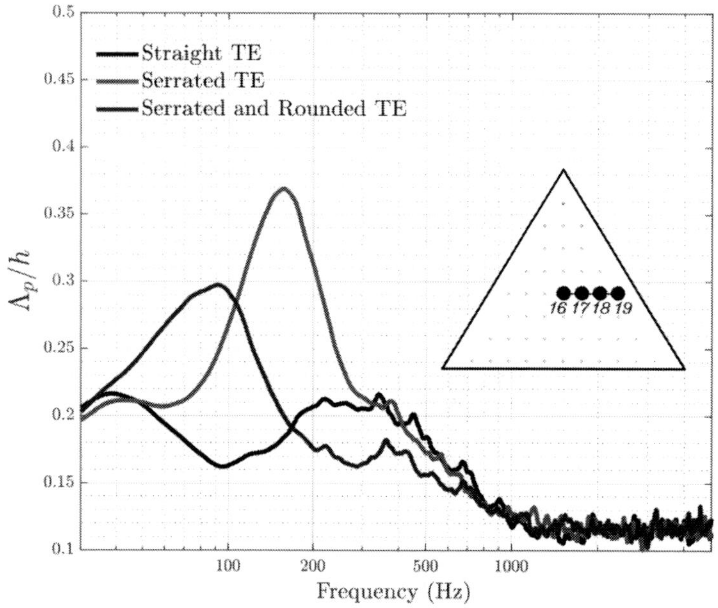

Figure 8 Span-wise length scales Λ_p based on the wall surface pressure measurement along the span at $x'/h = 60$.

3.2.2. Stream-wise coherence

Stream-wise coherence measured between the microphones are calculated using Equation3. Figure 9 presents the results of stream-wise coherence between reference microphone, shown by the marker "x", and upstream microphones at non-dimensional span-wise distances of $z/h = 0, 0.09, 0.18$ and $z/h = 0.26$. These locations are shown on the very left column of the figure. The microphones, which were used to calculate coherence with respect to reference microphone, are colour-coded. The same colour-code is also implemented in the graphs to ease the interpretation. The three columns from left to right represent the STE, SER and SER-R cases.

Considering the straight trailing edge case, STR, the trend of the stream-wise coherence does not change between the span-wise measurement locations ($z/h = 0$ to $z/h = 0.26$). A broadband hump occurs between $100Hz < f < 1200Hz$. Specifically, the coherence magnitude and frequency range is quite similar between coherence functions of mic3-mic6 at $z/h = 0$, mic4-mic7 $z/h = 0.09$, mic12-mic18 at $z/h = 0.18$ and mic19-mic26 at $z/h = 0.26$. This observation may indicate that the flow is well developed and the structures on the straight trailing edge do not change in the span-wise and stream-wise directions.

In contrast, the coherence behaviour dramatically changes with the introduction of serrations. On the symmetry line, $z/h = 0$, a very high magnitude broadband peak is present around $f = 200Hz$ between mic1 and mic3. Moreover, this peak exists for the coherence functions between the reference microphone "x" and upstream microphones until microphone 23. The broadband hump observed in STR still exists, but in lower magnitude. At $z/h = 0.09$, a similar trend with $z/h = 0$ is observed. Moving towards the serration edge profile, the magnitude of the peak at $f = 200Hz$ diminishes and a combined double hump occurs. This observation may be interpreted as follows: The strong turbulent structures are generated on the serration edge between microphone 4 ($x'/h = 0.27$ and $z/h = 0.09$) and microphone 12 ($x'/h = 0.49$ and $z/h = 0.18$) and develop towards downstream. Since these structures are not yet generated at upstream of microphone 12, the sensed effect is minimum. These results are well in-line with [21], where the vortical structures were claimed to be generated at around $x'/h = 0.5$, on the edge of the serrations. This also validates the initial assumption of applicability of the results of this study to normal sized serrations and the flow structures generated on them.

For the rounded-serrated case, SER-R, a similar double hump behaviour is observed. However, the magnitude of the peak is lower compared to the SER case and the peak frequency is at around $f = 150Hz$. An interesting result considering the measurement location $z/h = 0.27$ is the coherence magnitudes are lower than to that of STR. This unexpected result may be due to the rounded edges and requires further investigation.

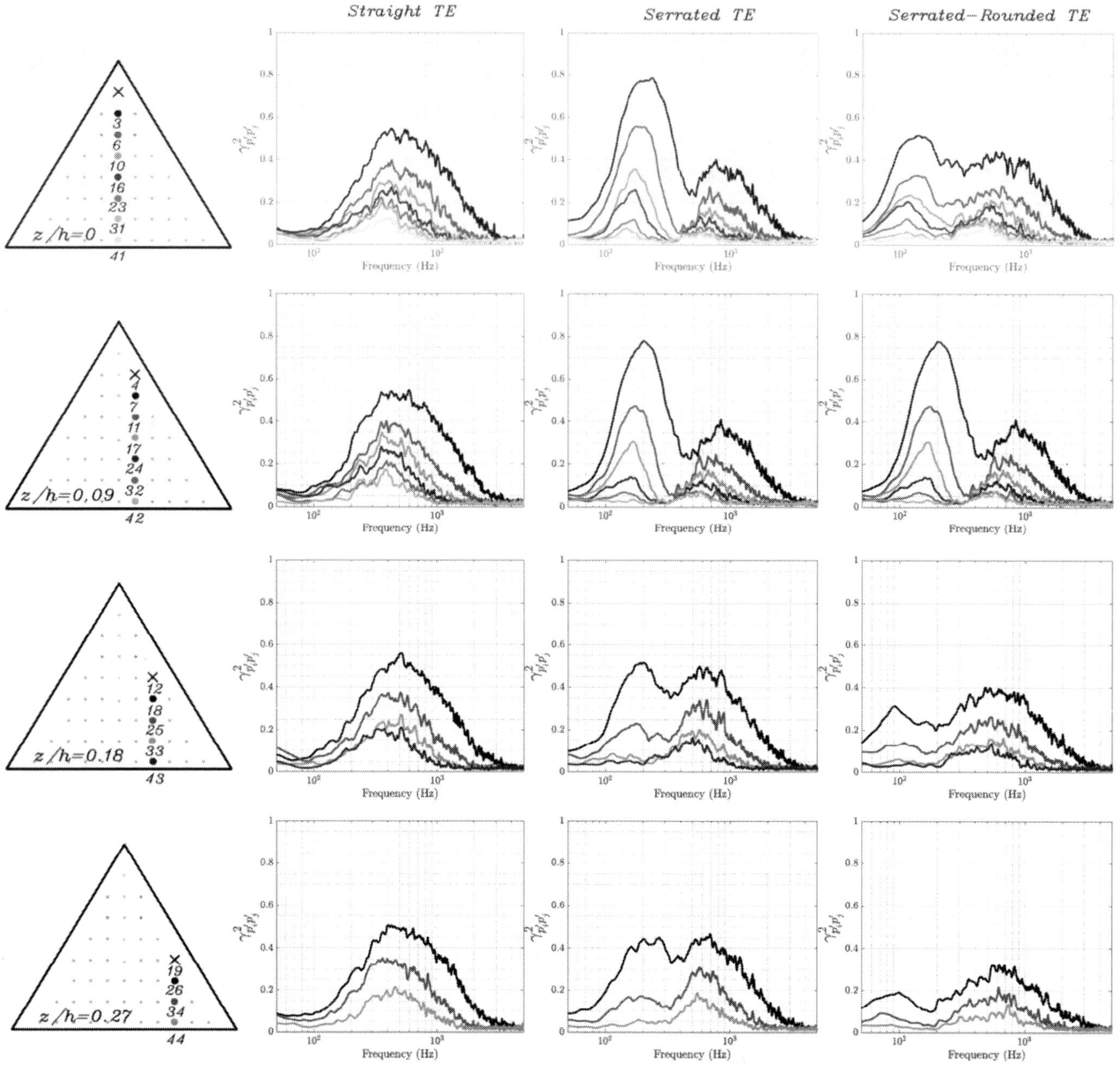

Figure 9 Comparison of stream-wise $\gamma^2_{p'_i p'_j}$ among STR, SER and SER-R span-wise stations $z/h = 0, 0.09, 0.18$ and $z/h = 0.26$ from top to bottom.

3.2.3. Temporal Correlation Analysis

To better understand the flow behaviour and assess the size of the turbulent structures, auto-correlation of the measured wall pressure is calculated using

$$R_{p'_i p'_j}(\tau) = \frac{\overline{p'_i(x_i, t+\tau) p'_i(x_i, t)}}{p'^2_{x_{i_{rms}}}}, \qquad (5)$$

where p'_i is the wall pressure signal from microphones located at x_i, $p'^2_{x_{i_{rms}}}$ is the root-mean-square of the pressure fluctuation p'_i and τ represents the time delay between the signals. Two figures, consisting of four graphs, are presented to aid a more extensive evaluation flow structures on serrations. Figure 10 includes the results for microphones 1, 3, 6 and 10 located at $x'/h = 0.16, 0.27, 0.38$ and $x'/h = 0.49$ respectively. Figure 11 is constructed to show the evolution of the structures on the serration edge profile for microphones 1,4,7 and 12 which are located at the same stream-wise distances as those shown in Figure 10. The auto-correlations are plotted against the non-dimensional time $\tau U/h$.

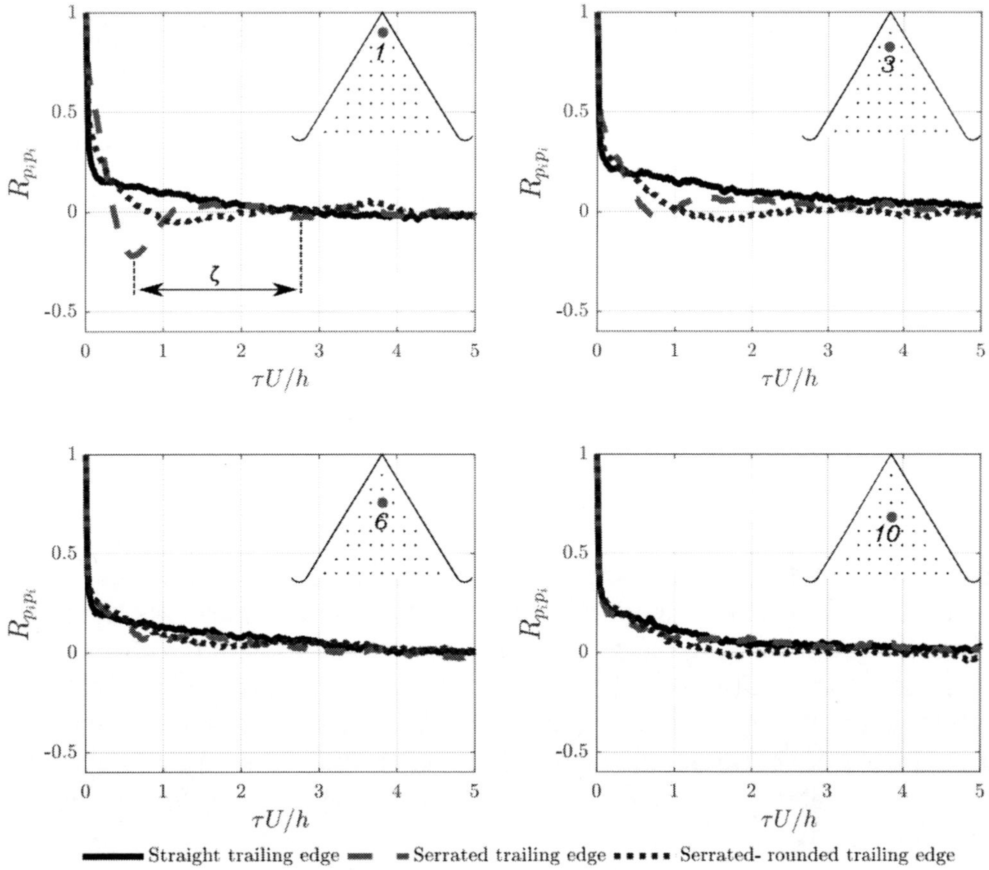

Figure 10 Autocorrelation of wall pressure fluctuations as a function of $\tau U/h$ for microphones 1,3,6 and 10 (stream-wise).

Considering the auto-correlation result for microphone 1 at $x'/h = 0.16$ in Figure 11, the STR case, shows a very fast decay at around $\tau U/h = 0$ and then slowly converges to a zero value, which is a

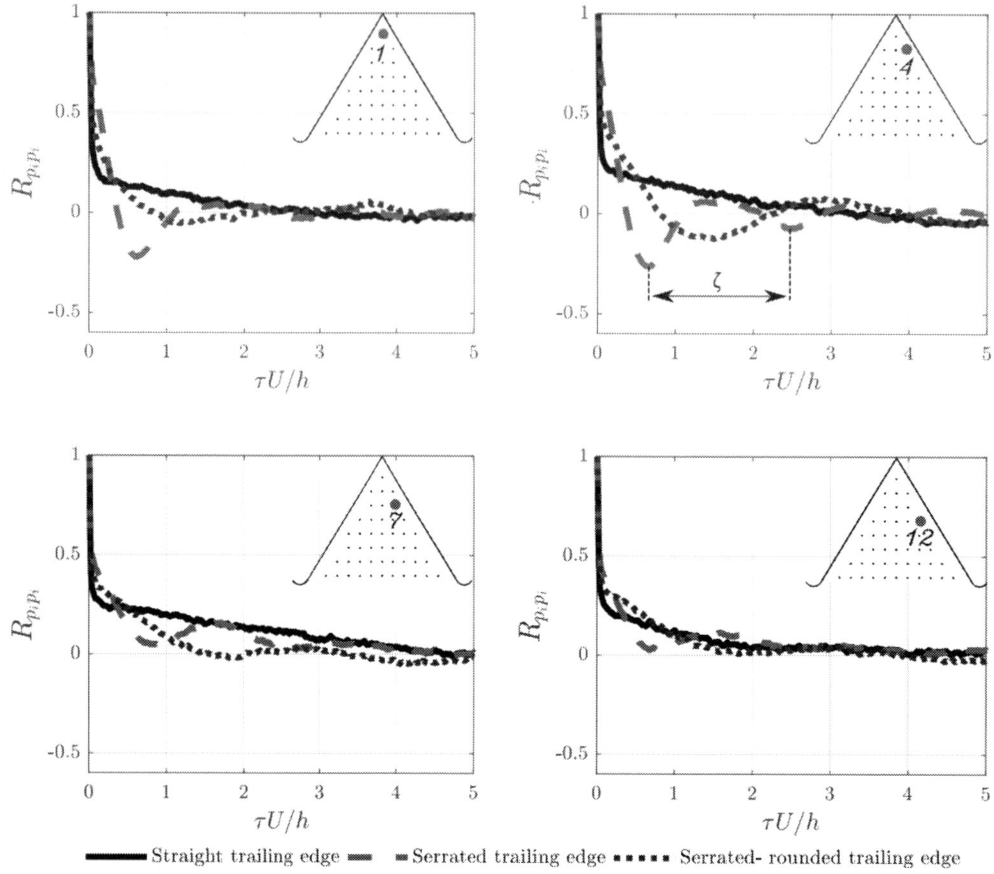

Figure 11 Autocorrelation of wall pressure fluctuations as a function of $\tau U/h$ for microphones 1,4,7 and 10 (edge-wise).

typical boundary layer auto-correlation profile. The fast decay is due to very fine scale structures in the turbulent boundary layer. The trailing edge serration, however, imposes a completely different behaviour. The slower decay and the large negative cusp are an indicator of the larger scale vortical motions in the boundary layer [22]. The oscillation periodicity ζ is an indicator of a quasi-periodic stream-wise vortical structure organization. The periodicity $\zeta = 2.13$ refers to an hydrodynamic event at $f = 153Hz$. For SER-R case the quasi-periodicity is hardly pronounced. At microphone location 3, although this behaviour still exist, it appears as if it is damped. Moving further upstream all cases collapse into a similar trend.

In Figure 11 the effect that the serration creates is more prominent and the oscillatory behaviour is observable over the entire upstream stations. At microphone location 4, the oscillation periodicity is similar, however the negative cusp is even deeper for both SER and SER-R case. At microphone locations 7 and 12, the quasi-periodicity is still observable but with lower correlation magnitude. This is a further validation of the importance of the edge profiles on generated turbulent structures.

4. Conclusion

The present study investigates the effect of large serrations on trailing-edge noise generated through the flow past a flat plate. Emphases are placed on the detailed mapping and analyses of near-wall pressure fluctuations, by means of steady and unsteady pressure measurements.The experiments are conducted in an open-jet anechoic wind tunnel at serration length h based $Re = 1.8x10^5$ for two different configuration for serrations. The power spectral density of the measured pressure (ϕ_{pp}) and fluctuating pressure distribution suggest that an elevated energy content exists for the serrated cases, which possibly contributes to the broadband hump observed along the surface ϕ_{pp} at relatively low frequencies between $50Hz < f < 600Hz$. Results from stream-wise and span-wise coherence analyses further corroborates these observations and indicate that a noticeable difference between the peak frequencies of the hump between straight ($200Hz$) and rounded serrations ($150Hz$). Last but not least, temporal auto-correlation on the surface microphones points to the presence of large scale vortical motions along the serrations, as compared to that of a developed profile for conventional straight trailing-edge flat plate. The present results affirm the notion that serrations play an important role in modifying the generation of trailing-edge noise and their different length scales should be accounted for when to predict the associated far-field noises.

Acknowledgments

The authors would like to thank Dr. Syamir Alihan Showkat Ali and Mr. Jonathon Luke Bowen for their valuable contributions.

References

[1] T.F. Brooks, D.S. Pope, and M.A. Marcolini. *Airfoil self-noise and prediction.* Tech. rep. National Aeronautics of Space Administration, Office of Management, 1989.

[2] M. Herr and W. Dobrzynski. "Experimental Investigations in Low-Noise Trailing Edge Design". In: *AIAA Journal* 43.6 (2013), pp. 1167–1175.

[3] B. Lyu, M. Azarpeyvand, and S. Sinayoko. "Prediction of noise from serrated trailing edges". In: *Journal of Fluid Mechanics* 793 (2016), pp. 556–588.

[4] X. Liu et al. "Aerodynamic Performance and Wake Development of Airfoils with Serrated Trailing-Edges". In: *AIAA Journal* 55.11 (2017), pp. 3669–3680.

[5] S.A. Showkat Ali, M. Azarpeyvand, and C.R. Ilário da Silva. "Trailing-edge flow and noise control using porous treatments". In: *Journal of Fluid Mechanics* 850 (2018), pp. 83–119.

[6] S.A. Showkat Ali et al. "Boundary layer flow interaction with a permeable wall". In: *Physics of Fluids* 30.8 (2018), p. 085111.

[7] A. Wolf et al. "Trailing edge noise reduction of wind turbine blades by active flow control". In: *Wind Energy* 18.5 (2015), pp. 909–923.

[8] M. Szoke and M. Azarpeyvand. "Active Flow Control Methods for the Reduction of Trailing Edge Noise". In: Proceedings of the 23rd AIAA/CEAS Aeroacoustics Conference, Denver, CO. AIAA 2017-3004.

[9] M. Szőke, D. Fiscaletti, and M. Azarpeyvand. "Effect of inclined transverse jets on trailing-edge noise generation". In: *Physics of Fluids* 30.8 (2018), p. 085110.

[10] I. Ayumur, N. Chiho, and A. Masahito. "Suppression of Tonal Trailing-Edge Noise From an Airfoil Using a Plasma Actuator". In: *AIAA Journal* 51.7 (2013), pp. 1695–1702.

[11] S. Oerlemans et al. "Reduction of Wind Turbine Noise Using Optimized Airfoils and Trailing-Edge Serrations". In: *AIAA Journal* 47.6 (2013), pp. 1470–1481.

[12] M.S. Howe. "Noise produced by a sawtooth trailing edge". In: *The Journal of the Acoustical Society of America* 90.1 (1991), pp. 482–487.

[13] B. Lyu and M. Azarpeyvand. "On the noise prediction for serrated leading edges". In: *Journal of Fluid Mechanics* 826 (2017), pp. 205–234.

[14] M. Azarpeyvand and Joseph P.F. Gruber M. "An analytical investigation of trailing edge noise reduction using novel serrations". In: Proceedings of 19th AIAA/CEAS Aeroacoustics Conference, Berlin, Germany. 2013.

[15] D. Ragni et al. "Measurements of near-wall pressure fluctuations for trailing-edge serrations and slits". In: *Experiments in Fluids* 60.1 (Nov. 2018), p. 6.

[16] L.E. Jones and R.D. Sandberg. "Acoustic and hydrodynamic analysis of the flow around an aerofoil with trailing-edge serrations". In: *Journal of Fluid Mechanics* 706 (2012), pp. 295–322.

[17] F. Avallone, W.C.P. van der Velden, and D. Ragni. "Benefits of curved serrations on broadband trailing-edge noise reduction". In: *Journal of Sound and Vibration* 400 (2017), pp. 167–177.

[18] R.D. Sandberg and L.E. Jones. "Direct numerical simulations of low Reynolds number flow over airfoils with trailing-edge serrations". In: *Journal of Sound and Vibration* 330.16 (2011), pp. 3818–3831.

[19] M. Gruber and Joseph P.F. Azarpeyvand M. "Airfoil trailing edge noise reduction by the introduction of sawtooth and slitted trailing edge geometries". In: Proceedings of 20th International Congress on Acoustics, ICA 2010, Sydney, Australia. 2010.

[20] F. Avallone, S. Pröbsting, and D. Ragni. "Three-dimensional flow field over a trailing-edge serration and implications on broadband noise". In: *Physics of Fluids* 28.11 (2016), p. 117101.

[21] T.P. Chong and A. Vathylakis. "On the aeroacoustic and flow structures developed on a flat plate with a serrated sawtooth trailing edge". In: *Journal of Sound and Vibration* 354 (2015), pp. 65–90.

[22] C.K.W. Tam et al. "The sources of jet noise: experimental evidence". In: *Journal of Fluid Mechanics* (2008), pp. 253–292.

[23] Y.D. Mayer et al. "Design of an Aeroacoustic Wind Tunnel Facility at the University of Bristol". In: Proceedings of the 24th AIAA/CEAS Aeroacoustics Conference , Atlanta, GA. AIAA 2018-3138.

INCEEUROPE

8th International Conference
on
Wind Turbine Noise
Lisbon – 12th to 14th June 2019

Session Q

Regulations

Organised by INCE-Europe
mail@inceeurope.org
www.inceeurope.org

Supported by Portuguese Acoustical Society

Sociedade
Portuguesa
de Acústica

INCEEUROPE

8th International Conference
on
Wind Turbine Noise
Lisbon – 12th to 14th June 2019

Current regulations for the protection against noise from wind turbines in Germany

Andrea Bauerdorff, German Environment Agency: andrea.bauerdorff@uba.de

Summary

The wind energy sector is growing steadily in Germany. With 29213 wind turbines and an installed capacity of around 52931 MW [1], Germany ranks first in Europe. For a successful realization of the transformation of the energy system it is necessary to increase the acceptance of wind turbines in the neighborhood. The basis for the approval of wind turbines in Germany is the German Federal Immission Control Act. According to this act installations subject to licensing shall be established and operated in such a way that this does not involve harmful effects on the environment. This requirement is specified in the "Technical Instructions on Noise Abatement, TI Noise" which contain noise immission values. The total noise exposure at a decisive immission point may not exceed these values. An important part of the noise assessment of wind turbines is a prediction of the noise immission. For this purpose, the TI Noise as well as the recommendations for the protection against noise exposure from wind turbines of the federal states are used in Germany. These recommendations have been fundamentally revised under consideration of the current state of knowledge in 2017. The procedure for the protection against noise from wind turbines in Germany will be explained in detail.

1. Introduction

Wind turbines are of great importance to meet the demand for electricity with renewable sources of energy. Therefore a successful realization of the transformation of the energy system in Germany is essential.

In Germany, wind turbines with a height of more than 50 m, are subject to licensing pursuant to the German Federal Immission Control Act [2]. Principally, installations subject to licensing shall be established and operated in such a way that this does not involve harmful effects on the environment or other hazards, considerable disadvantages and considerable nuisance to the general public and neighborhood. Moreover, precautions should be taken to prevent harmful effects on the environment. Within the scope of the licensing process not only environmental aspects are taken into account but also all other legal aspects in connection with the construction of a new wind turbine or the repowering of existing ones. For instance, building regulations as well as nature conservation requirements are also to be considered. One important element of the licensing process deals with the protection against noise. The regulations are laid down in the Sixth General Administrative Provision to the Federal Immission Control Act, which is called "Technical Instructions on Noise Abatement, TI Noise" [3]. It is applied to various kinds of industrial plants as well as commercial installations. This wide range of applicability includes wind turbines. The regulations for wind turbines also comprise the "recommendations on the protection from noise of wind turbines" of the Federal States of Germany [4]

2. Determination of noise immissions

The determination of noise immissions of wind turbines is carried out according to the TI Noise and the additional "recommendations on the protection from noise of wind turbines" of the Federal States of Germany [4]. The main steps of the noise assessment procedure for wind turbines are shown in figure 1.

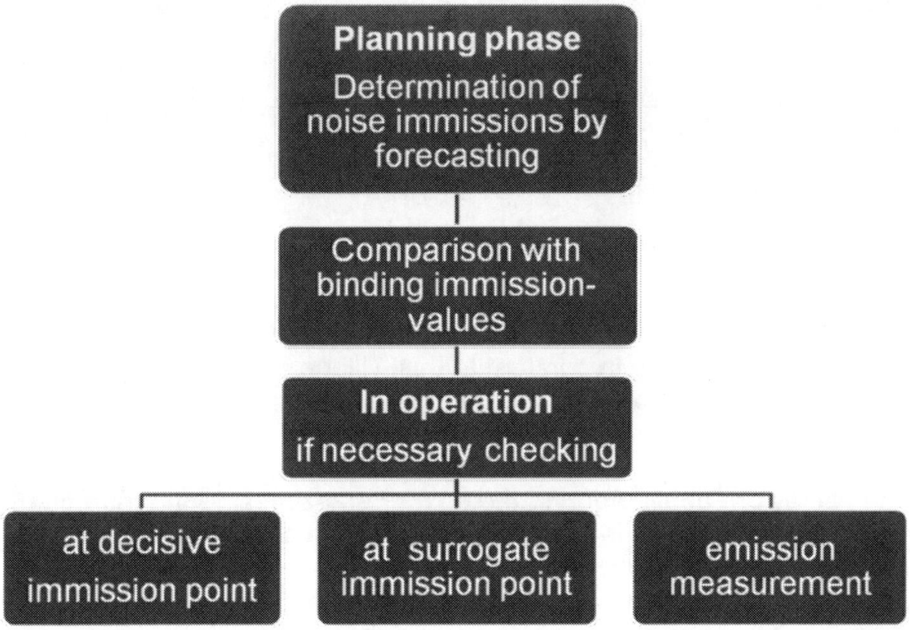

Fig.1: Noise assessment procedure according to TI Noise in Germany

The TI Noise contains criteria to identify and assess the noise immissions of installations subject to licensing, including wind turbines. The noise assessment procedure starts with the determination of noise immissions by forecasting.

3. Binding immission values according to TI Noise

For the approval of a wind turbine it is essential that the noise immission of the installation comply with binding immission values for points outside buildings which are defined in the TI Noise [3]. They depend on the type of land-use and are distinguished between daytime and nighttime. The immission values given are in table 1.

Table 1:
Binding immission values according to TI Noise in Germany

Types of areas	Day 6 a.m. - 10 p.m.	Night 10 p.m. - 6 a.m.
Industrial areas	70 dB(A)	70 dB(A)
Commercial zones	65 dB(A)	50 dB(A)
Urban areas	63 dB(A)	45 dB(A)
Core areas, village areas and mixed-use zones	60 dB(A)	45 dB(A)
General residential areas and small residential states areas	55 dB(A)	40 dB(A)
Purely residential areas	50 dB(A)	35 dB(A)
Spa areas, for hospitals and nursing homes	45 dB(A)	35 dB(A)

These types of areas correspond to the building areas, which are defined in a nationwide valid Federal Land Utilization Ordinance [5]. According to this ordinance, for example, purely residential areas are only used for living. In contrast to this, village areas may also contain agricultural and forestry holdings. Moreover, the further development of agricultural and forestry holdings is taken into consideration.

The binding immission values are applied during the day for an assessment period of 16 hours. The full hour with the highest rating level is the basis for nighttime assessment. The reason for this special regulation is that the protection against noise at night is very important because a sufficiently long period of undisturbed sleep is essential for mental and physical rest. Furthermore, the TI Noise contains regulations concerning short-term noise peaks. These peaks may exceed the values during the day by not more than 30 dB(A), and at night by not more than 20 dB(A).

The immission values displayed in table 1 must be met in order to obtain the license for the operation of a wind turbine. They refer to the forecast respectively measured rating level L_r ascertained at the immission point. In this context, it is generally assumed from the total exposure at the decisive immission point. This is the point in the area of influence of the installation at which binding immission values are most likely to be exceeded. The decisive immission points are defined as follows:

- In buildings, 0.5 m outside the middle of the open window of the room which is worst affected by noise.
- In undeveloped or built-up areas that contain no buildings with noise sensitive rooms, at the edge of the area where buildings with noise sensitive rooms may be built.

4. Examination of the sound propagation model

In the past the alternative method of DIN ISO 9613-2 [6] was used for the determination of the noise exposure caused by wind turbines. This standard is designed for sound sources with a height up to 30 m and a distance between source and immission point until 1000 m (Fig. 2).

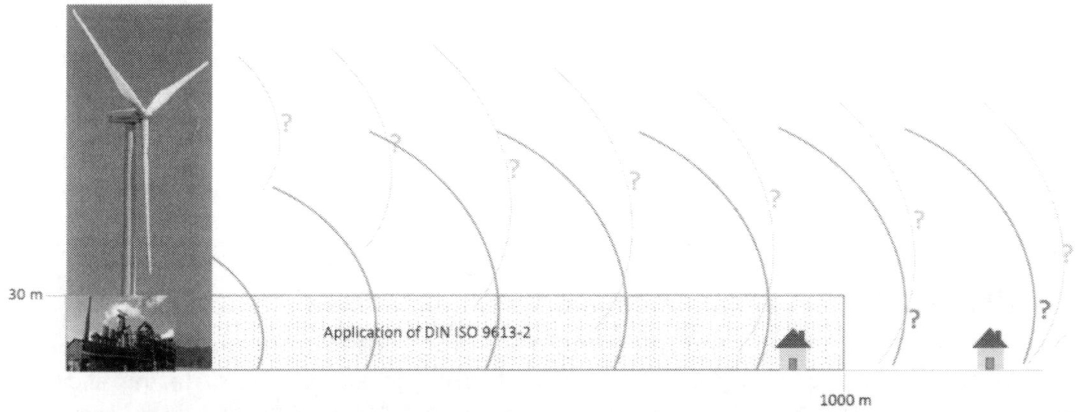

Fig. 2. Scope of DIN ISO 9613-2

Therefore, it is questionable whether DIN ISO 9613-2 is appropriate for calculating the sound propagation for modern wind turbines with a height of more than 100 m. For this reason, a research project [7] was conducted which was commissioned by the State Agency for Nature, Environment and Consumer Protection of the German Federal State of North Rhine-Westphalia. In this project, noise measurements have been carried out and compared with calculation results. This study has come to the result that up to 450 m from the wind turbine the calculation values are above the measured ones. For greater distances, the study shows contradictory results. With increasing distances from the wind turbine, the measurement results are higher than the calculation results. In view of these findings, the German organization for standardization DIN developed an interim method for the calculation of sound propagation at wind turbines [8]. This method considers the results of the above-mentioned research project and is based on a plain acoustic model. It describes the wind turbine as an omnidirectional point noise source which is frequency-dependent and located in the rotor center of the wind turbine. The input variable of the sound immission forecast is the specific octave band sound power level L_W of the wind turbine. For this purpose, the A-weighted third-octave band sound power levels $L_{WA,i}$ determined according to DIN EN 61400-11 [9] will be converted into the corresponding unweighted octave band sound power levels L_W. The equivalent continuous downwind sound power level L_{fT} is determined by the following equation [8]:

$$L_{fT}\,(DW) = L_W + D_C - A \qquad (1)$$

where

L_W	octave-band sound power level produced by a wind turbine described as a point source [dB]
D_C	directivity correction [dB]
A	attenuation [dB]

with

$$A = A_{div} + A_{atm} + A_{gr} + A_{bar} + A_{misc} \qquad (2)$$

A_{div}	attenuation owing to geometrical divergence
A_{atm}	attenuation due to atmospheric absorption
A_{gr}	ground attenuation
A_{bar}	attenuation by a barrier
A_{misc}	attenuation caused by other effects

In equation (2), the attenuation due to the ground effect is set to -3 dB. This is based on the assumption that modern wind turbines with a height of above 100 m cause only one reflection from the ground. Furthermore, the meteorological correction in the equation for the determination of the long-term average A-weighted sound pressure level $L_{AT(LT)}$ is modified [8]:

$$L_{AT(LT)} = L_{AT(DW)} - C_{met} \quad (3)$$

where

$L_{AT(DW)}$ average A-weighted sound pressure level for downwind propagation

C_{met} meteorological correction, C_{met} = 0 dB

It is envisaged to replace the interims method by a new German guideline VDI 4101 part 2 [10] in the next years.

5. Recommendations on the protection from noise of wind turbines

In addition to the TI noise there exists "recommendations on the protection from noise of wind turbines" of the German Federal States [4]. They contain further details for the determination of the noise immissions of wind turbines and are also applied by the licensing authorities. In particular, the recommendations include the interim method for the calculation of sound propagation at wind turbines. Moreover, requirements on the input data and the quality of the noise prediction are described in detail. Additionally, some practical information for the licensing authorities are also given in these regulations.

5.1 Input parameter of the noise immission forecast

The input variable of the noise immission forecast for a specific wind turbine project is the specific sound power level of a wind turbine type in the corresponding operating mode. The input parameters are divided into two groups, namely parameters for the existing and the additional exposure.

Existing exposure is defined as the noise exposure at an immission point from all installations for which the TI Noise is applied, excluding immissions emanating from the installation under assessment. Consequently, all already approved wind turbines and other installations subject to the TI Noise are to be considered. Additional exposure is the immission that is anticipated with planned installations. The sum of both is the total exposure. According to German law [2] the protection against harmful effects on the environment due to noise is ensured, if the total exposure at the decisive immission point does not exceed the binding immission value shown in table 1.

For the determination of the input data there are three different options:

1. The sound power level is given by the manufacturer for normal operation. The appropriate octave spectrum is also provided.
2. Determination of the sound power level and the corresponding octave spectrum by measurements of the wind turbine type according to the Technical Guidelines for Wind Turbines Part 1[11] in combination with the German standard DIN EN 61400-11 [9].
3. Determination of the sound power level and the corresponding octave spectrum by measurements of at least minimum three wind turbines of the same type.

5.2 Determination of the noise prediction

The determination of noise immissions by forecasting is described in detail in the TI Noise and the "recommendations on the protection from noise of wind turbines". These recommendations contain the interim method for the calculation of sound propagation at wind turbines [8]. This procedure is carried out in a frequency-selective manner for the approval of wind turbines both for the existing wind turbines as a part of the existing exposure and for newly requested installations. For air absorption calculations, the atmospheric attenuation coefficient α according Table 2 of DIN ISO 9613-2 [6] shall be considered for 70 % relative humidity and 10 °C air temperature. To ensure that the legal requirements are met a quality assurance is carried out. Within the framework of this quality assurance it must be proved that under consideration of the uncertainty of the emission data as well as the uncertainty of the propagation calculation the upper limit of confidence level of the forecasted rating level is below the binding immission value. The upper limit of confidence level is calculated from the forecasted immission value plus the total uncertainty which results from the inaccuracy of the acoustic emission measurement and the fundamental uncertainty of the prediction model.

6. Consideration of low-frequency noise

The mechanisms of the sound generation and propagation of wind turbines are complex [12]. The noise emissions of a wind turbine consist of mechanically induced and aerodynamic noise components. The aerodynamic noise is often broadband and includes low-frequency and infrasonic components. For the determination and assessment of low-frequency noise, the German standard DIN 45680 "Measurement and assessment of low frequency noise immission" [13] and in the accompanying Supplement 1 are used [14]. The supplement contains reference values, which should not be exceeded. DIN 45680 contains an auditory threshold and the draft of DIN 45680 of 2013 [15] contains a hearing and perception threshold. These thresholds are illustrated in figure 3 together with the sound pressure levels of wind turbines with 2 to 3 MW in different distances. In the usual distances between wind turbines and residential dwellings in Germany, the infrasound emission generated by wind turbines is well below the threshold of human perception. This is confirmed by extensive noise immission measurements on wind turbines in the German Federal States of Baden-Wuerttemberg and Bavaria [16]. The impact of infrasound from wind turbines can be assumed to be very low in comparison with other natural and anthropogenic sources, so that according to the current state of research this does not have negative effects on health.

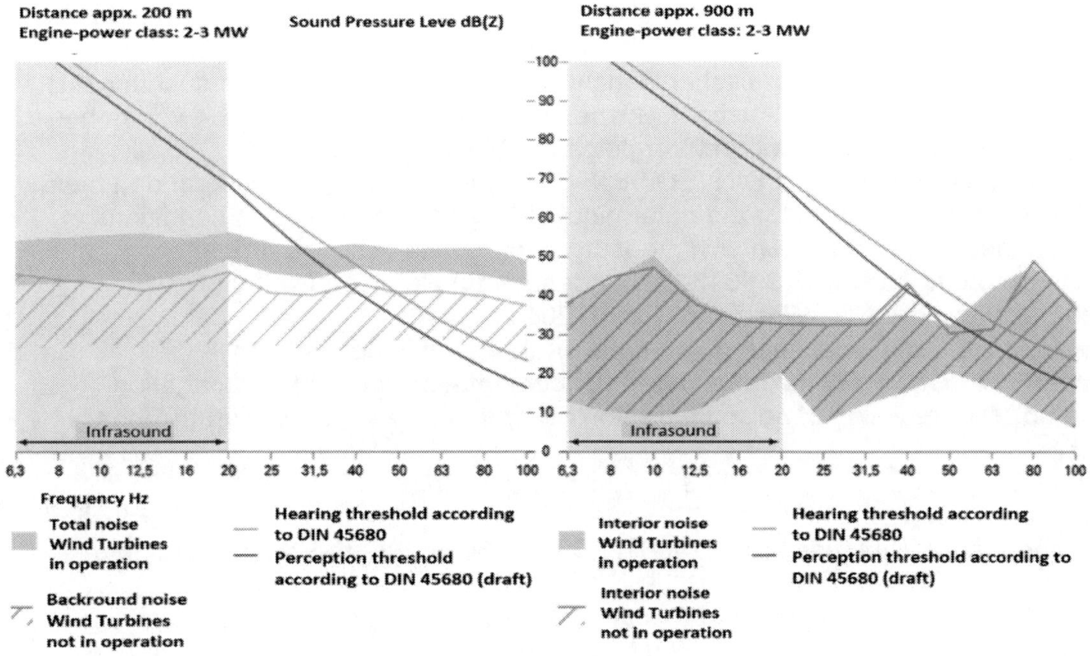

Fig. 3: Comparison of measurements results with hearing and perception thresholds. Measurements on modern wind turbines with a hub height of 140 m at wind speeds from 8 to 13 m/s, distance 200 m (left) and 900 m (right). [17] with own editing

7. Conclusions

The generation of energy by wind turbines is essential for the implementation of the energy transition. Especially in Germany, the number of these installations has continuously increased in the last years. Within this framework, the protection of the population from wind turbine noise has to be taken into consideration. Therefore, it is particularly important to have clear legal regulations for noise protection by wind turbines. The TI Noise especially contain area-related binding immission values for immission points outside buildings which may not exceeded during the operation of the wind turbine. In addition to the TI noise there exists "recommendations on the protection from noise of wind turbines" of the German Federal States.

References

1. Status of Land-based Wind Energy Development in Germany - Year 2018, Deutsche WindGuard GmbH, Varel (Germany) 2019, www.windguard.com
2. Bundes-Immissionsschutzgesetz (Federal Immission Control Act) in the version promulgated on 17 May 2013 (BGBl. I p. 1274) as last amended by Article 3 of the Act of 18 July 2017 (BGBl. I p. 2771)
3. Technische Anleitung zum Schutz gegen Lärm, TA Lärm (Technical Instructions on Noise Abatement - TI Noise), Sixth General Administrative Provision to the Federal Immission Control Act of 26 August 1998 (Joint Ministerial Gazette (GMBl.) No. 26/1998 p. 503), last amended on 1 June 2017 (BAnz AT 8 June 2017 B5)
4. Hinweise zum Schallimmissionsschutz bei Windenergieanlagen -WKA (Rekommandations on the protection from noise of wind turbines) Bund/Länder-Arbeitsgemeinschaft für Immissionsschutz (LAI), Germany 2016
5. Baunutzungsverordnung (Federal Land Utilization Ordinance) in the version promulgated on 21 November 2017 (BGBl. I p. 3786)
6. DIN ISO 9613-2: Akustik - Dämpfung des Schalls bei der Ausbreitung im Freien - Teil 2: Allgemeines Berechnungsverfahren (Acoustics - Attenuation of sound during propagation outdoors - part 2: General method of calculation), Berlin: Beuth-Verlag, October 1999
7. Engelen J., Wenzel P.: Schalltechnischer Bericht der erweiterten Hauptuntersuchung zur messtechnischen Ermittlung der Ausbreitungsbedingungen für die Geräusche von hohen Windenergieanlagen zur Nachtzeit und Vergleich der Messergebnisse mit Ausbreitungsberechnungen nach DIN ISO 9613-2 (Acoustic report on the expanded main investigation on measurements for the determination of sound propagation conditions of high wind turbines and comparsion with measurement results of sound propagation calculations according to DIN ISO 9613-2), Landesamt für Natur, Umwelt und Verbraucherschutz Nordrhein-Westfalen, Essen 2014
8. Dokumentation zur Schallausbreitung, Interimsverfahren zur Prognose der Geräuschimmissionen von Windkraftanlagen (Documentation on sound propagation, interim method for prediction of noise immission of wind turbines), Berlin: Beuth-Verlag, May 2015
9. DIN EN 61400-11; VDE 0127-11:2007-03, Windenergieanlagen - Teil 11: Schallmessverfahren (IEC 61400-11:2002 + A1:2006) (Wind turbines - part 11: Acoustic noise measurement techniques (IEC 61400-11:2012), Berlin: Beuth-Verlag, September 2013
10. VDI 4101-2: Schallausbreitung im Freien unter Berücksichtigung meteorologischer und topographischer Bedingungen, Blatt 2: Windkraftanlagen (Sound propagation outdoors

under consideration of meteorological and topographical conditions, part 2: wind turbines), ongoing project of the Association of German Engineers VDI

11. Technische Richtlinien für Windenergieanlagen - Teil 1: Bestimmung der Schallemissionswerte (Technical Guidelines for wind turbines, part 1: determination of noise emission), Fördergesellschaft Windenergie und andere Erneuerbare Energien [editor], Berlin
12. Leventhall G., Bowdler D.: Wind Turbine Noise: How it is produced, propagated, measured and received. Multi-Science Publishing, Brentwood (2011)
13. DIN 45680: Messung und Bewertung tieffrequenter Geräuschimmissionen in der Nachbarschaft (Measurement and assessment of low-frequency noise immissions in the neighborhood), Berlin: Beuth-Verlag, March 1997
14. DIN 45680, Beiblatt 1: Messung und Bewertung tieffrequenter Geräuschimmissionen in der Nachbarschaft (Measurement and assessment of low-frequency noise immissions in the neighborhood - Guidelines for the assessment for industrial plants) Berlin: Beuth-Verlag, March 1997
15. E DIN 45680 Entwurf: Messung und Bewertung tieffrequenter Geräuschimmissionen (Draft: Measurement and assessment of low-frequency noise immissions), Berlin: Beuth-Verlag, September 2013
16. Ratzel U., Bayer O., Brachat P., et al.: Tieffrequente Geräusche und Infraschall von Windkraftanlagen und anderen Quellen. Bericht über Ergebnisse des Messprojektes 2013-2015, Landesanstalt für Umwelt Baden-Württemberg (Low-frequency noise immissions including infrasound of wind turbines and other sources, report on the results of Measurement projects 2015-2015), LUBW, 2016
17. Windenergieanlagen - beeinträchtigt Infraschall die Gesundheit? (Does infrasound from wind turbines affects health?): Umwelt Wissen - Klima und Energie / Bavarian Environment Agency [editor] – 4th updated edition, Augsburg 2014, update: august 2016, Germany

INCE EUROPE

8th International Conference
on
Wind Turbine Noise
Lisbon – 12th to 14th June 2019

Applying a general noise regulation on Wind Turbines: an administrative interpretation to a specific noise source

David Parra* & Igor Valdebenito**
Environmental Noise, Light Pollution and Odors Department
Ministry of Environment, Government of Chile
San Martín 73, Santiago de Chile.
JParra@mma.gob.cl* IValdebenito@mma.gob.cl**

Camila Oyarzún
Institute of Acoustics, Austral University of Chile
General Lagos 2086, Valdivia Chile
Camila.oyarzun01@alumnos.uach.cl

Summary

In most cases, public policies management and environmental regulations development is hampered by government agendas and the interests of both, public and private sector, especially in countries that seek to foster the growth of the industry, also positioning itself in the development of non-conventional renewable energies.

In Chile, there is a general environmental noise regulation that applies for multiple noise sources, mainly productive activities, with a general procedure that can be used both for a night club or a wind farm. In this context, the possibilities of introducing a new regulation in a matter as specific as wind farms noise, are the lowest.

However, there are other ways to address the complexity of a technical problem that only seems to be solved introducing a new specific regulation. Drawing upon the legal and technical possibilities that the general noise regulation in the country presents, it was possible to introduce a path to solve in the short-term, most of the wind turbine noise technical issues that challenge the general regulation.

This paper presents an administrative interpretation of a general noise regulation to be applied on wind turbine noise through different action paths, involving the community, working with the assessment and supervising entities and establishing working groups with wind industries and other public entities, improving the regulation for wind turbine noise in Chile, without modifying the current environmental noise regulation. A comparison between the general procedure and the interpretation for wind farms noise is presented.

1. Introduction

Wind turbine noise (WTN) is a problem that has evolved around the world. It is known that currently there is no consensus among evaluation methodologies, which makes it difficult to design management strategies.

In Chile, every year more wind farms enter the System of Environmental Impact Assessment (SEIA) (2), which regulates compliance with environmental regulations in the country. Similarly, the number of people affected by WTN is increasing, as there are areas of the country with high wind potential but also have a sufficient number of communities to generate conflicts.

Given the complexity of the problem, the need arose to develop a strategy for WTN, which within its first objectives was to strengthen the regulatory framework. According to this, it was thought to elaborate a specific regulation, which led to a series of studies. However, administrative times could be very long, which would not make the regulation effective enough in relation to the large number of projects that enter year by year. This led to the definition of a dynamic strategy that would allow the establishment of WTN assessment criteria that do not avoid the technical and legal possibilities of the noise regulation in Chile. Along with an adequate follow-up to the revision of projects in the SEIA, training and dissemination of the problems associated with noise from wind farms, it has been possible to improve the assessment of WTN, through a general noise standard.

This document presents the aspects of the general noise regulation in Chile that could be interpreted to generate the specific assessment guidelines for WTN. Likewise, the key elements for the proper implementation of these criteria are mentioned, in the context of the wind farms noise strategy in Chile.

2. Noise regulation in Chile

The current noise regulation, applicable to wind farms, corresponds to the Supreme Decree N°38/11 of the Ministry of the Environment (DS 38) (4), which regulates noise from multiple sources. In this standard, wind farms are considered as elements of infrastructure, thus they must complain this regulation.

The DS 38 establishes differentiated limits, both rural and urban areas. In this case, since all the projects are located in rural areas, the maximum permissible levels are defined for each receiver, as shown in Table 1.

Table1: Noise limits for rural areas in the DS 38.

Day limit	Night limit
Background noise + 10 dB(A), maximum of 65 dB(A)	Background noise + 10 dB(A), maximum of 50 dB(A)

Limits for rural areas are directly related to the background noise of the site. Therefore, the adequate definition of background noise levels it is very important to quantify the acoustic impact on the receivers near the wind farm.

For the characterization of the emission of the source, a corrected sound pressure level is used. This value comes from the analysis of the descriptor L(A)eq 1 minute and the maximum level L(A)max recorded in the measurement period of 3 minutes. A comparison between both descriptors and the background noise level is necessary to finally define the final noise level that will characterize the noise from the source.

In relationship to the international regulation for wind farm noise, the limits of the DS 38 are located as one of the less strict, considering the night limit of the DS38. Usually the international regulation for wind farm noise is based on a standard of 45 dB(A) or less (1), with an specific measurement procedure.

2.1 Complexity for specific wind farm noise regulation

A new regulation for WTN could be part of a solution for this problem in Chile. However, the time of action is one of the most important issues to generate new regulations. Defining a specific regulation for a particular source in the Chilean regulatory framework is not easy. It will requires around 3 years of work according to the regulated procedure for making environmental regulations in Chile, which means that many projects will be unaffected by this new regulation.

On the other hand, there is an important lack of information in the public service and the private sector, regarding the noise generated by wind turbines, their causes and effects. Not knowing the phenomenon implies a low degree of depth in the noise studies presented to the SEIA. In Chile, there were no studies of noise emmission of wind farms focused on the improvement of the current noise regulation (DS 38) in this area.

It was identified that according to the massive growth of the wind industry in the country, it is necessary to generate an immediate intervention in the WTN assessment, particularly for new projects. This led to the conclusion that the definition of guidelines for the application of DS 38 is efficient in terms of time of action and technical possibilities. An adequate interpretation of DS 38 can address WTN, giving space to a more detailed analysis before the application, selecting the most representative samples according to the DS 38 requirements and possibilities.

3. Interpretation for wind turbine noise

The interpretation of the DS38 for WTN was mainly based on the following criteria that appear in the standard:

a. The measurement must be made in the most unfavorable condition for the receiver (worst case condition).
b. The background noise must be measured under the same measurement conditions of the noise source.
c. Predictions of noise levels can be made.

This led to orientate the application of DS38 for the following aspects of the regulation:

- Background noise measurement technique
- Wind farm noise prediction techniques
- Wind profile characterization
- Wind farm noise measurement technique
- Technical equipment

3.1 Background noise measurement technique

For the assessment of the background noise, which defines the maximum allowed noise level, a specific procedure has been agreed, meeting the general procedure established in the DS 38. This will require generating samples in three different wind speed ranges (6-8, 8-10 and 10-12 m/s) to assess wind farm noise and background noise at the wind farm. This also means that the assessment will require a continuous monitoring of background noise, selecting the lowest background noise average for each one of the three ranges, identifying the worst case condition. This data will be more representative of the noise immission levels but also could be stricter for the definition of maximum allowable levels. However, this will avoid that wind farms will be installed in a critical distance of the neighbourhoods.

3.2 Wind farm noise prediction techniques

To assess a new wind farm, the noise immission level must be predicted according to the following parameters:

Table 2: Specific methods and parameters for wind farm noise prediction

Method	Immission height	Humidity	Temperature	Wind direction	Ground factor	Wind speed noise map
ISO 9613-2	4 m	70 %	10 ° C	Downwind	0,5	8, 10 and 12 m/s at hub
NORD 2000	1,5 m	70 %	10 ° C	Downwind	D	8, 10 and 12 m/s at hub
CONCAWE	4 m	80 %	10 ° C	Downwind	0	8, 10 and 12 m/s at hub

These parameters have been adopted from the international regulation (3,6) and different studies in the Ministry of the Environment.

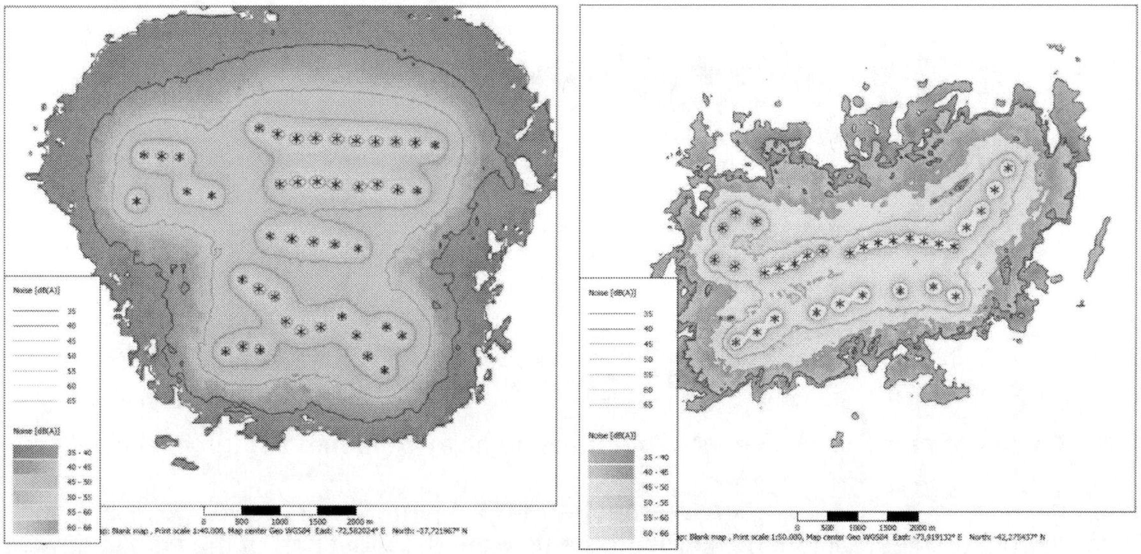

Fig. 1 – Wind farm noise maps generated with NORD 2000.

3.3 Wind profile characterization

Three methodologies for wind measurement are presented. The requirement is that all measurements must be correlated to the wind speed at hub height of wind turbines, thus identifying the worst case condition. Methodologies are the following:

1. Measuring tower with anemometer at hub height.
2. Measurement with SODAR or LIDAR system at hub height.
3. Measuring tower with anemometers or SODAR/LIDAR system.
4. SCADA monitoring system (installed wind farms).

3.4 Wind farm noise measurement technique

The noise measurement technique in the DS38 requires the definition of a receiver point, generating three noise measurements of one minute (LAeq 1 min), under the worst case condition criteria. It is clear that three arbitrary samples are not enough, due to the characteristics of the source (Fig. 2 and 3).

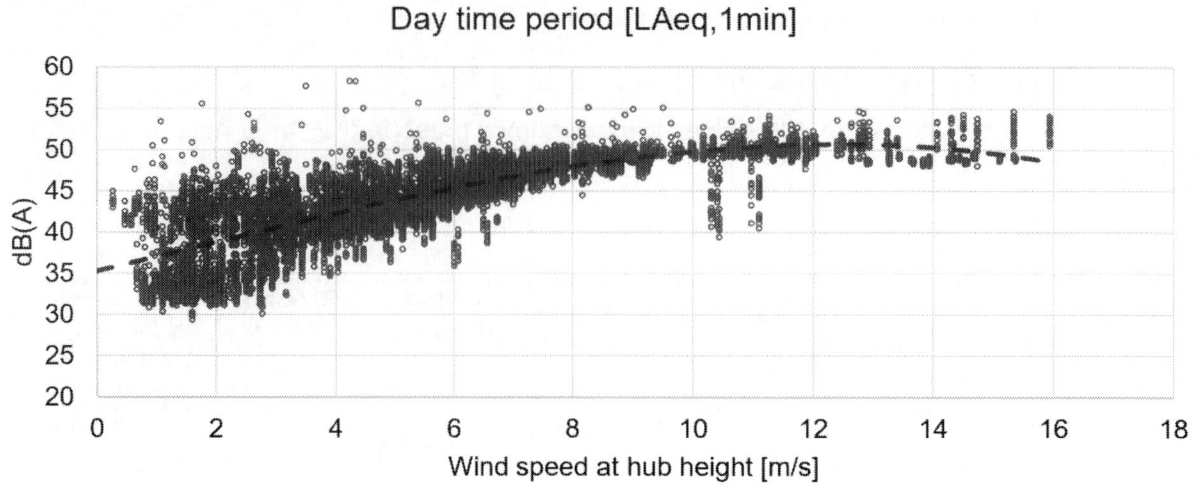

Fig. 2 – 14-day of wind farm noise monitoring 250 m away from nearest turbine (day time).

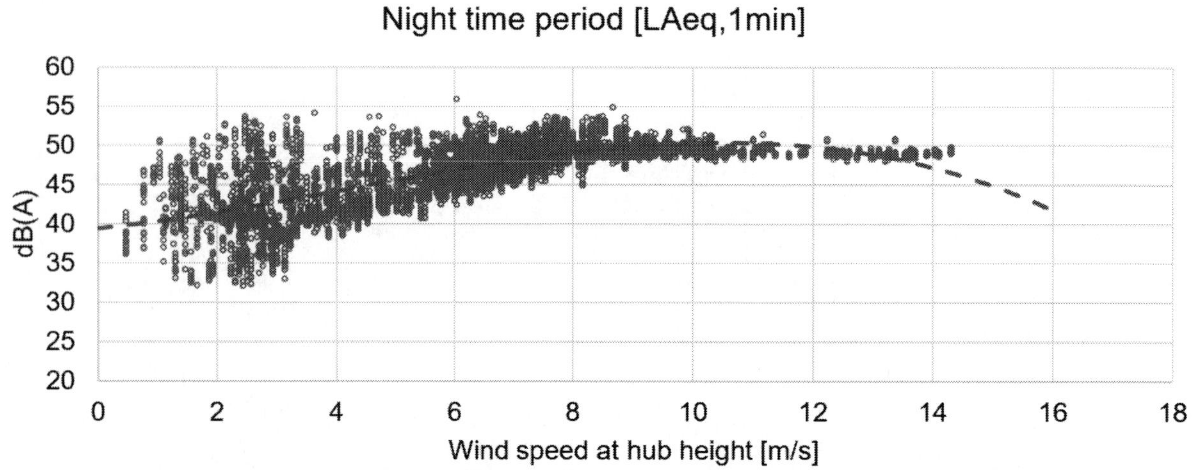

Fig. 3 – 14-day of wind farm noise monitoring 250 m away from nearest turbine (night time).

Previous figures shows a 14-day measurement at a wind farm. The receiver point was located 250 m away from the nearest turbine. Measurements were the most part of the time in downwind conditions.

Analyzing samples from 6 wind farms throughout the country, it was concluded that the maximum emission of the source is above 6 m/s, in accordance with the Danish regulation for WTN. Samples below 6 m/s show a high dispersion, not being representative. Due to the stabilization of samples over 12 m/s, the assessment will be made from 6 to 12 m/s. As the background noise, three different wind speed ranges (6-8, 8-10 and 10-12 m/s) will be studied to assess WTN.

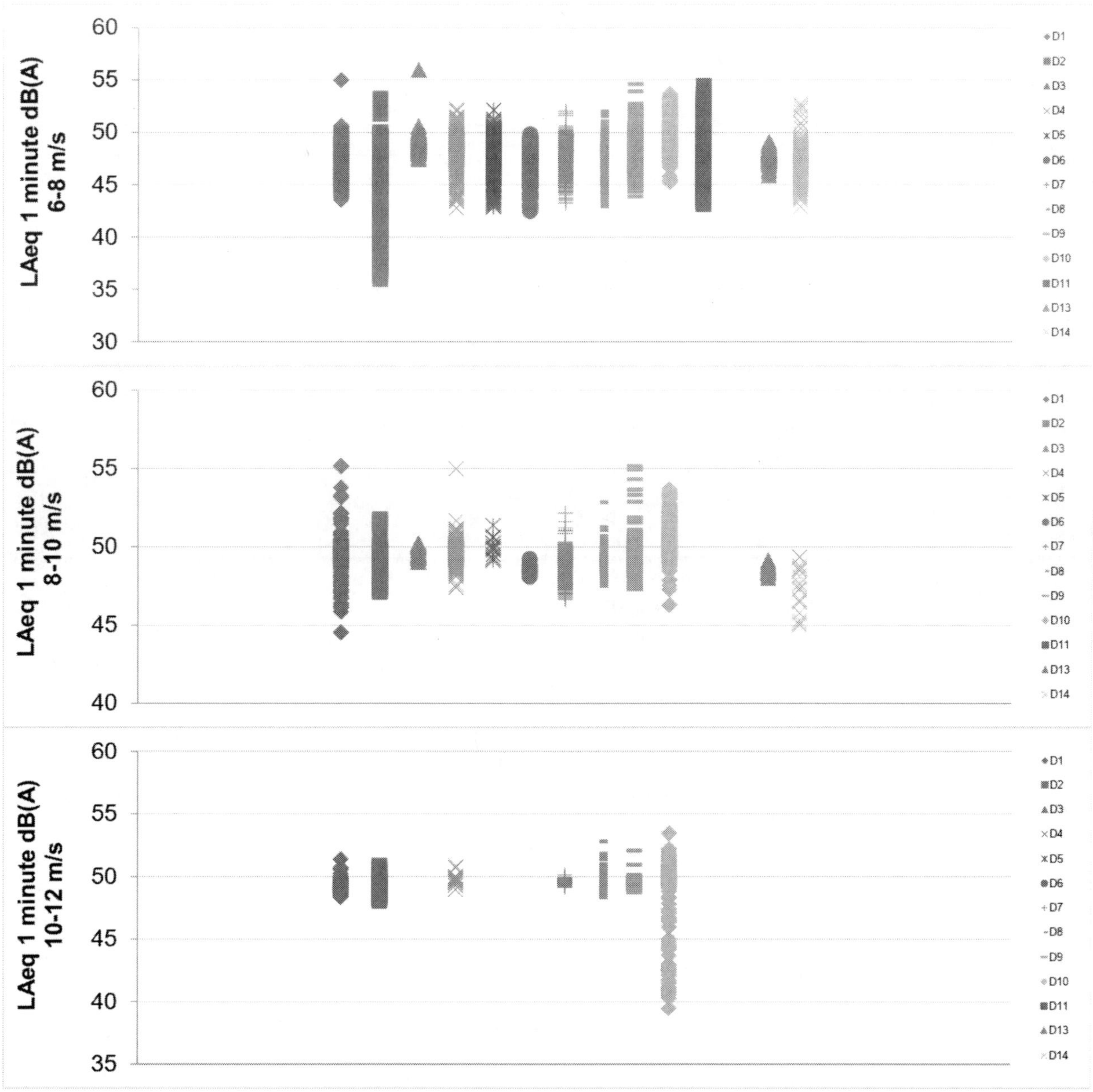

Fig. 4 – 14-day of wind farm noise monitoring 250 m away from nearest turbine (day and night time), separated by wind speed ranges.

The noise emission of the wind turbine has to be characterized with 3 samples for each wind speed range. In the figure before it is possible to notice that there are no samples for all of the days. Day 12 does not deliver any sample within the defined wind speeds, as also shown in Table. Day 9 presents all the data, for all the ranges. Through a continuous noise monitoring exercise it is possible to address these complexities, avoiding measuring on a day that is no representative for the source. This criteria is also in accordance with other international standards (3, 6).

The next step is to look for the maximum noise samples as shown in the Table, as long as they do not include occasional noises. Taking into account the lowest background noise emission for the same wind speeds, this will be representative of the worst case condition, according to the DS 38.

Table 3: Maximum samples on a 14-day of wind farm noise monitoring.

	Day time period (dB(A))			Night time period (dB(A))		
Day	**6 – 8 m/s**	**8 – 10 m/s**	**10 – 12 m/s**	**6 – 8 m/s**	**8 – 10 m/s**	**10 – 12 m/s**
1	49,9	50	**51,8**	51,1	50,4	49,8
2	49,4	-	-	50	50,1	**50**
3	-	-	-	50,2	50,2	-
4	**49,9**	**51,1**	50,8	51,3	50,6	-
5	47,9	-	-	51,3	**50,6**	-
6	45,9	-	-	49,9	49,2	-
7	**50,2**	**50,3**	-	50,6	50,3	**50,1**
8	49,9	50	**51,8**	50,5	50,4	49,6
9	**50,6**	51,1	50,1	**52,8**	**51,9**	49,8
10	-	**50,4**	**54,2**	**53,7**	**53,7**	**51,5**
11	49,9	-	-	**53,8**	-	-
12	-	-	-	-	-	-
13	-	-	-	49,1	49,1	-
14	48,6	-	-	49,9	49,4	-

Final step is to found the corrected sound pressure level. Considering the 3 maximum L(A)eq 1 minute samples for each one of the wind speed ranges, there will be a representative level for every wind speed range, for both day and night. Taking into account a minimum background noise of 35 dB(A), the maximum allowable level will be 45 dB(A).

Table 4: Selected samples for the DS38 application

Wind speed range	Day level dB(A)	Night level dB(A)	Maximum allowable level dB(A)
6-8 m/s	50	53	45
8-10 m/s	51	52	45
10-12 m/s	53	51	45

The analysis before shows that there is no compliance with the maximum allowable levels. This analysis has been implemented for new projects in the SEIA, avoiding critical distances between receivers and wind farms, therefore, reducing noise levels.

3.5 Technical equipment

Finally, the guidelines to asses WTN defines specific aspects that should be considered for the measurement of noise in relation to wind speeds. In particular, specific equipment is defined to perform measurements. With an adequate windshield, noise samples will be unaffected by wind gusts in measurement height, registering WTN with a higher degree of certainty.

Fig. 5 – Wind shield required for the application of the DS 38 in WTN.

All of these measures working together will promote an improvement of the DS38 for WTN, and the existence of an adequate distance between wind turbines and homes, avoiding the existence of any critical distance and, therefore, the acoustic impact on nearby communities will be reduced.

4. Wind turbine noise strategy

To support the presented guidelines in this work, the Wind Farms Noise Strategy in Chile integrates other action paths which are the awareness and training, information and coordination.

4.1 Awareness and training

In addition to strengthen noise regulation, the generation of training instances for environmental assessors, acoustic consultants and project developers is essential. Also, it is very important to interact with the community. It has been possible to identify that the community is very interested to be trained about WTN effects and being involved from the beginning, in the planning stage of the project. In the framework of the development of the Strategy has been possible to generate meetings with people who live near the wind farm. People wanted to be heard and involved, generating a close relationship with the wind farm owners. It has been proven that people with a constant interaction with the owners of the wind farm, does not present legal actions against the wind farm and they are willing to environmental noise management

4.2 Information of WTN

The information component is related to the studies that the state develops to assess WTN. Mainly seeks to characterize noise emission from wind farms throughout the country, perception of noise and its health effects on people. This information is the basis for interacting with the community and project developers.

4.3 Coordination

Finally, coordination with the different Government institutions it is essential to promote the development of the aforementioned components. The table 4 presents the attributions of different public and private institutions involved in the development of this Strategy.

Table 5: Involved institutions in the coordination

Action area	Institution
Environment	• Ministry of the Environment
Renewable Energies	• Ministry of Energy • Chilean Association of Renewable Energies
Projects assessment	• Service of Environmental Assessment • Project developers • Acoustic consultants
Land-use planning	• Ministry of Housing and Urban Construction • Livestock Agricultural Service
Acoustic isolation	• Ministry of Housing and Urban Construction
Perception and health effects	• Ministry of Health
Inspection	• Superintendence of the Environment
Community interaction	• City halls

5. Conclusions

The main objective of WTN noise regulation should not be the acoustic characterization of the source, but the definition of maximum allowable limits that consider reasonable exposure levels for nearby communities. In this context, it is considered that the interpretation of noise regulation in Chile for WTN is efficient.

Although, the measurement procedure and the acoustic descriptors used (L(A)eq 1 min), do not qualify in their entirety within the international standard usually seen for the analysis of WTN, through a public policy strategy the acoustic impact on nearby communities to wind projects has been reduced. This is considered a success since it has managed the system immediately, with the available tools.

From the technical point of view, it is not enough to take three noise samples arbitrarily. A proper analysis will require a noise monitoring of at least two weeks, either for the current descriptor or for a new noise descriptor. It is prudent to consider an extensive noise monitoring in order to identify the worst case condition.

The need for a specific regulation is not ruled out, defining new descriptors and more robust measurement procedures. However, this change should be gradual, keeping on mind the already defined guidelines for the assessment of WTN in Chile.

The solutions to an environmental noise problem are not always a specific regulation. A noise limit, a specific measurement procedure or the guidelines to assess an environmental noise problem responds mainly to a policy discussion. Technical decisions can influence the final decision, but they are not always the main factor that will determine that decision in a matter of public policy.

6. Acknowledgements

This work has been supported by the Ministry of Energy and the Ministry of Environment of the government of Chile, in the framework of the Wind Turbine Noise Strategy in Chile.

References

1. E. Koppen and K. Fowler, International Legislation for Wind Turbine Noise, EuroNoise 2015, Maastricht, Netherlands (2015).

2. Environmental Assessment Service, System of Environmental Impact Assessment, Government of Chile, http://www.sea.gob.cl)/ (2018).

3. New Zealand Standards (2010), "NZS 6808:2010 Acoustics – Wind farm noise", Wellington, New Zealand.

4. Republic of Chile, Ministry of the Environment (2012). Supreme Decree N ° 38/11 of the Ministry of the Environment - Norm of Emission of Noises Generated by Indicating Sources. Santiago of Chile.

5. Statutory Order on Noise from Wind Turbines, Order no. 1284. (2011). Translation of Statutory Order no. 1284 of 15 December 2011, Denmark.

6. The Working Group on Noise from Wind Turbines (1996), "ETSU-R-97 The Assessment and Rating of Noise from Wind Farms".

INCE EUROPE

8th International Conference
on
Wind Turbine Noise
Lisbon – 12th to 14th June 2019

WHO Environmental Noise Guidelines for the European Region: conditional recommendation for wind turbine noise in the context of Australian regulations

Justin Adcock, Christophe Delaire
Marshall Day Acoustics, Collingwood, Victoria, Australia
melbourne@marshallday.com

Summary

The World Health Organisation (WHO)'s publication *Environmental Noise Guidelines for the European Region* [1] published in 2018 (the 2018 WHO Guidance) provides a conditional recommendation for addressing annoyance as a result of wind turbine noise in Europe. The conditional recommendation for wind turbine noise is provided amidst guidance for health-related effects from road, rail, aircraft and leisure noise. In keeping with strategic policy of the European Union, the publication adopts the L_{den} noise metric. This paper considers the practical challenges of measuring and assessing wind farm noise in terms of the L_{den} noise metric, and compares the conditional recommendation with the typical range of noise levels observed around Australian wind farm projects.

1. Introduction

The potential health effects of environmental noise have been topical subjects for wind farms, and have attracted concern from communities neighbouring proposed and operational wind farm developments in Australia and internationally.

The potential health effects of environmental noise range from community annoyance, sleep disturbance and speech interference, through to direct physiological impacts such as hearing damage. An important aspect of this range of considerations is that some effects will be highly dependent on the listener's perception and attitude to the noise in question, such as annoyance, while other effects are primarily related to the level of sound and the direct physiological risks these may represent, such as hearing damage.

In Australia, environmental noise levels from wind farms are regulated by dedicated policies which describe measurement and assessment methodologies that are specific to the operational noise characteristics of wind farms. In common with policies for other types of noise generating development, wind farm noise policies define criteria which are chosen to prevent direct physiological risks of sound, and minimise as far as practically possible the adverse health considerations such as annoyance and sleep disturbance.

The subject of health effects related to operational wind farms has been extensively reviewed by Australian Commonwealth, national and state health authorities. The findings of these reviews [2][3][4] support that, as with any audible sound, wind farm noise can represent a potential source of annoyance or sleep disturbance for some individuals. The review did however indicate that there was no reliable evidence to support a relationship between wind farm noise and direct adverse effects on human health. These findings lend support to the suitability of the wind farm noise policies in Australia, which are intended to provide reasonable protection of health and amenity at noise sensitive locations.

The publication of guidance on wind farm noise by the WHO provides an additional valuable reference for policy makers, stakeholder groups and the wind industry.

However, the guidance is based on the L_{den} noise metric which is commonly used for transportation noise, but is not widely used for wind farm noise assessment in Australia or internationally. This introduces complications when attempting to accurately compare Australian wind farm noise policy requirements with the 2018 WHO Guidance. Further, while the L_{den} parameter is consistent with EU strategy policy requirements, there are technical challenges to its application to wind farms.

This paper presents a summary of the 2018 WHO Guidance for wind farms and a high level summary of Australian wind farm noise policies. An analysis of background sound level data from wind farm sites in terms of the L_{den} is then presented to illustrate the difficulty of measuring this parameter for a wind farm. Example wind distributions and predicted wind farm noise levels are also presented to investigate the relationship between the conditional recommendation of the 2018 WHO Guidance and the noise criteria used to assess wind farm developments in Australia.

2. Summary of 2018 WHO Guidance & Australian Wind Farm Policy

2.1 2018 WHO Guidance

The 2018 WHO Guidance conditionally recommends the following for wind turbine noise.

> *For average noise exposure, the* [2018 WHO Guidance] *conditionally recommends reducing noise levels produced by wind turbines below 45 dB L_{den}, as wind turbine noise above this level is associated with adverse health effects.*
>
> *To reduce health effects, the* [2018 WHO Guidance] *conditionally recommends that policy-makers implement suitable measures to reduce noise exposure from wind turbines in the population exposed to levels above the guideline values for average noise exposure. No evidence is available, however, to facilitate the recommendation of one particular type of intervention over another.*

The conditional recommendation was made solely on the basis of annoyance related health considerations, with the guideline being set at the noise level which corresponded to 10 % of the community being highly annoyed in the available research [5][6] for four wind farm sites. The recommendation is conditional on the basis that *"the evidence on the adverse effects of wind turbine noise was rated low quality"*. The 2018 WHO Guidance notes that no suitable studies were available in relation to cardiovascular disease, hearing impairment and learning impairment.

The recommended L_{den} metric is the day-evening-night-weighted sound pressure level as defined in ISO 1996-1:2016 and referenced in the European Noise Directive. The metric is determined as the average/equivalent noise level representing a 24 hour period, based on the aggregated noise of all day, evening and night periods in a year (times for the day, evening and night periods may vary between jurisdictions). The L_{den} includes 5 dB and 10 dB penalties for the evening and night periods respectively. The L_{den} is determined at facade locations for noise sensitive buildings but excludes the influence of facade reflections.

The use of the L_{den} metric provides the benefit of alignment with broader strategical polices for transportation noise and metrics commonly referenced in large-scale community noise exposure research. The potential benefit of this type of metric for wind farms is that it may also differentiate between locations where the highest wind farm noise levels are similar, but where the amount of time that these levels are experienced differs significantly (i.e. due to receivers being in or out of the prevailing wind direction).

However, the 2018 WHO Guidance acknowledges limitations concerning the use of the L_{den} metric for wind farm assessment. Key consideration relating to the conditional recommendation for wind farm noise are:

- No account of background noise conditions – an important consideration for wind farm developments given that background noise levels at wind farm sites are generally comparable to, and often greater than, wind farm noise levels. This is relevant in terms potential wind farm annoyance and in terms of the feasibility of measuring L_{den} wind farm noise levels at receiver locations in the presence of wind (noting that the operating range of a wind farm typically extends from 3-25 m/s or 11-90 km/h).
- No differentiation between wind farm noise containing audible characteristics such as tonality or amplitude modulation, which the 2018 WHO Guidance acknowledges as factors that are likely to give rise to increased annoyance.

- No distinction between neighbour receivers and involved receivers (i.e. receiver locations hosting or financially benefiting from a wind farm). This is particularly relevant as the conditional recommendations are based on health effects related to annoyance, and the evidence concerning annoyance is greatest for receivers who are not involved the wind farm project.

2.2 Australian Wind Farm Policy

Noise assessment requirements for wind farms in Australia vary by state. However, the underlying principles are similar and are based on:

- Assessing noise levels at receiver locations using the relationship between measured noise levels and hub height wind speeds.
- Wind speed dependent noise criteria defined as a minimum limit or the background level plus 5 dB, whichever is higher.
- Noise limits are often set in terms of equivalent noise levels at receiver locations, but L_{A90} measurements are used to estimate L_{Aeq} noise levels attributable to wind farms.
- The application of penalties for audible characteristics which are likely to increase annoyance (the types of assessable characteristics vary by state, but include tonality, amplitude modulation, impulsiveness and low frequency).

The key noise assessment publications referenced in Australia, and their applicable minimum noise limit values, are detailed in Table 1.

Table 1: Australian wind farm noise assessment – key publications and applicable minimum limits

Document	Minimum limit
South Australian publication *Wind farms: environmental noise guidelines* [7]	35 or 40 dB depending on land zoning
New Zealand Standard *NZS 6808:2010 Acoustics – Wind farm noise* [8]	35 or 40 dB depending on land zoning
NSW publication *Wind Energy: Noise Assessment Bulletin* [9]	35 dB
Queensland publication *State code 23: Wind farm development planning* [10]	Day: 37 dB / Night: 35 dB

Most regulatory approvals for new wind farm developments include mandatory requirements to prove compliance with noise criteria at receiver locations after the wind farm is operational. Compliance assessment methodologies are increasingly using noise measurement data for locations nearer to the turbines. However, compliance requirements are still predominantly based on post-construction operational noise measurements at receiver locations. This is important when considering the L_{den} noise metric in the Australian regulatory framework i.e. the feasibility of the L_{den} as a noise metric which can be reliably measured at a typical wind farm site.

3. Measured L_{den} Noise Levels at Selected Wind Farm Sites

Noise measurement data from nineteen (19) measurement locations at five (5) wind farm sites in Australia have been analysed in terms of the L_{den} metric (reference time periods for analysis: day 0700-1900 hrs, evening 1900-2200 hrs and night 2200-0700 hrs). The analysis was carried out for ambient noise level measurements at proposed wind farm sites (i.e. in the absence of turbines) and operational wind farm sites. The primary purpose of this analysis was to provide an indication of the range of measured L_{den} noise levels around wind farm sites.

The wind farm sites were located in rural areas that are remote from major transportation routes. The terrain of the sites ranged from flat to gently undulating profiles. The predominant environmental noise sources comprised wind disturbed vegetation, local fauna, occasional farming activity and wind turbines (at operational sites). The noise measurements were carried out at a combination of receiver (i.e. near to existing or proposed dwelling locations) and intermediate measurement locations (i.e. reference points between receivers and turbine locations). The receiver noise measurement locations were generally between the predicted 30 dB and 40 dB L_{Aeq} contours for the wind farms (based on downwind predicted noise levels). The intermediate measurements locations were positioned near the predicted 45 dB L_{Aeq} contour of the wind farms. All of the receiver locations where operational noise measurements were carried out had previously been assessed as compliant with local policy requirements.

The noise level measurements comprised consecutive measurements of equivalent and statistical noise levels in 10 minute intervals for periods typically spanning 3-6 weeks. All noise measurement equipment comprised 01dB CUBE or DUO Smart Noise Monitors (Class 1) fitted with enhanced wind shields designed according to the recommendations detailed in the UK IOA good practice guide. To enable screening for extraneous noise, all noise measurements included one-third octave band spectra and local measurements of rainfall in concurrent 10 minute intervals.

For consistency with the L_{den} metric, the analysis was based on measured equivalent noise levels for each period and location. However, as L_{Aeq} measurements around wind farms are highly prone to the effects of extraneous noise and variable noise influences, an alternative L_{den} parameter was calculated using L_{A90} measurements i.e. the measured L_{A90} was used as an approximation of the equivalent L_{Aeq} noise level attributable to the operation of the wind farm (consistent with common assessment practice in Australia). The same approach was adopted for the pre-construction noise measurement datasets to allow comparison with the post-construction noise measurement datasets.

The comparison shown in Figure 1 illustrates the calculated L_{den} ambient noise levels for the nineteen (19) receiver and intermediate measurement locations around proposed wind farm sites only (i.e. without the influence of turbines). The intermediate locations are identified by the designation '(i)' in the label. The results are presented with the L_{den} separately calculated using the L_{Aeq} noise levels and the L_{A90} noise levels. The results are also provided for datasets filtered for the following extraneous noise sources:

- Rainfall; and
- Prominent insect noise which is identified when the following conditions [11] are satisfied:
  - the highest A-weighted one-third octave band noise level is within 5 dB of the broadband A-weighted background noise level for that interval; and
  - the identified one-third octave band A-weighted noise level is greater than a level of 20 dB L_{A90}.

The results in Figure 1 illustrate that the ambient L_{den} noise levels calculated using the equivalent noise level metric are typically greater than the 45 dB conditional recommendation level of the 2018 WHO Guidance, even after applying filters to remove clearly identifiable extraneous noise influences. This is likely to be a result of a combination of elevated ambient equivalent noise levels and potential wind induced noise across the microphone during high local winds (noting that practical enhanced secondary windshields are validated for the measurement of statistical noise levels which exclude momentary wind-gust related noise on the microphone).

Figure 1 indicates that ambient L_{den} noise levels calculated using measured L_{A90} noise levels are significantly reduced, but the results still a high level of variation. However, even after filtering the data, the L_{den} noise levels based on the L_{A90} metric are between 40 and 45 dB for the majority of locations,

These results therefore show:

- Direct evaluation of the L_{den} wind farm noise levels from measured equivalent noise levels at receiver locations will generally not be possible. This is due to the effect of ambient noise and the limitations of practical enhanced wind shields.
- Even using L_{A90} post-construction measurements as an approximation of wind farm equivalent noise levels for the calculation of L_{den} noise levels, the background noise environment is likely to significantly influence the measured levels and, in some cases, may dominate the measured level.

The measurement difficulties noted here are not exclusive to the L_{den} metric. Standard wind farm measurement and assessment procedures based on regression analysis of statistical noise levels correlated with site wind speeds are also prone to similar ambient noise influences and complications. However, a range of supplementary measurement and analysis techniques can be used to address these difficulties with standard measurement and assessment procedures (e.g. analysis of selected data subsets for different periods or wind conditions in combination with measurement and extrapolation of intermediate location data). However, these types of supplementary techniques are either limited or not possible when the data is aggregated for all time periods and wind conditions to calculate an L_{den}.

Figure 1: Proposed rural wind farm sites – ambient L_{den} noise levels

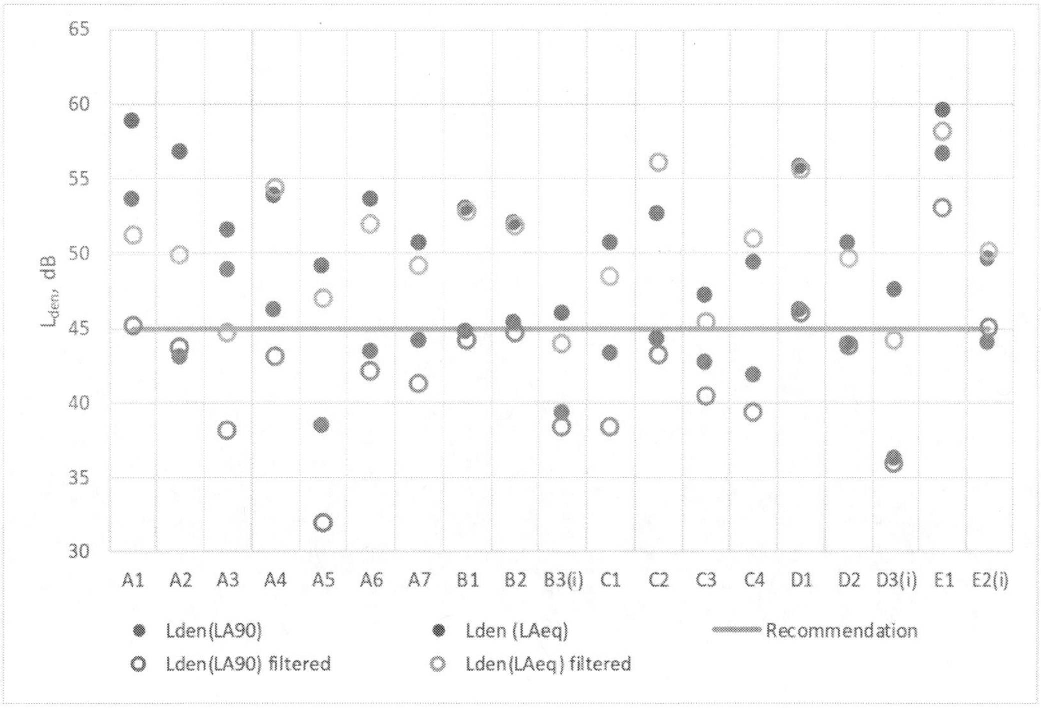

To further illustrate these complications, Figure 2 illustrates the results of pre and post construction noise level measurements at eight (8) locations around three (3) different wind farm sites. The measured levels are determined using the both the L_{A90} and L_{Aeq} noise levels. All noise levels are based on datasets which were filtered using the procedures outlined above.

The results in Figure 2 illustrate a high level of variation, particularly those relating to measurements at receiver locations. This small sample of receiver location measurements illustrates the L_{den} noise levels derived from L_{A90} measurements were lower after construction of the wind farm, suggesting background noise was a significant source of variation in the comparisons. However, despite the significant background noise influence, the results for this limited number of receiver measurements indicate the post-construction L_{den} noise levels derived from L_{A90} measurements were at, or below, the 45 dB conditional recommended level. At the two intermediate locations (identified as B3(i) and D3(i)) positioned near the predicted 45 dB L_{Aeq} contour of the wind farms, the post-construction measured noise levels show an increase relative to the pre-construction noise levels. However, despite being significantly nearer to the wind farm than the receiver locations, the post-construction L_{den} noise levels at these locations are only marginally higher than the 45 dB L_{den} conditional recommendation.

The comparison in Figure 2 further illustrates the complications of attempting to derive L_{den} noise levels from measurements at typical receiver separating distances. However, the results provide a limited indication that, for locations where Australian noise policy requirements are met, wind farm noise levels may be comparable to, or lower than, the 45 dB L_{den} conditional recommendation. This is examined further in the subsequent section using prediction-based datasets.

Figure 2: Comparison of pre and post construction L_{den} noise levels at receiver and intermediate locations

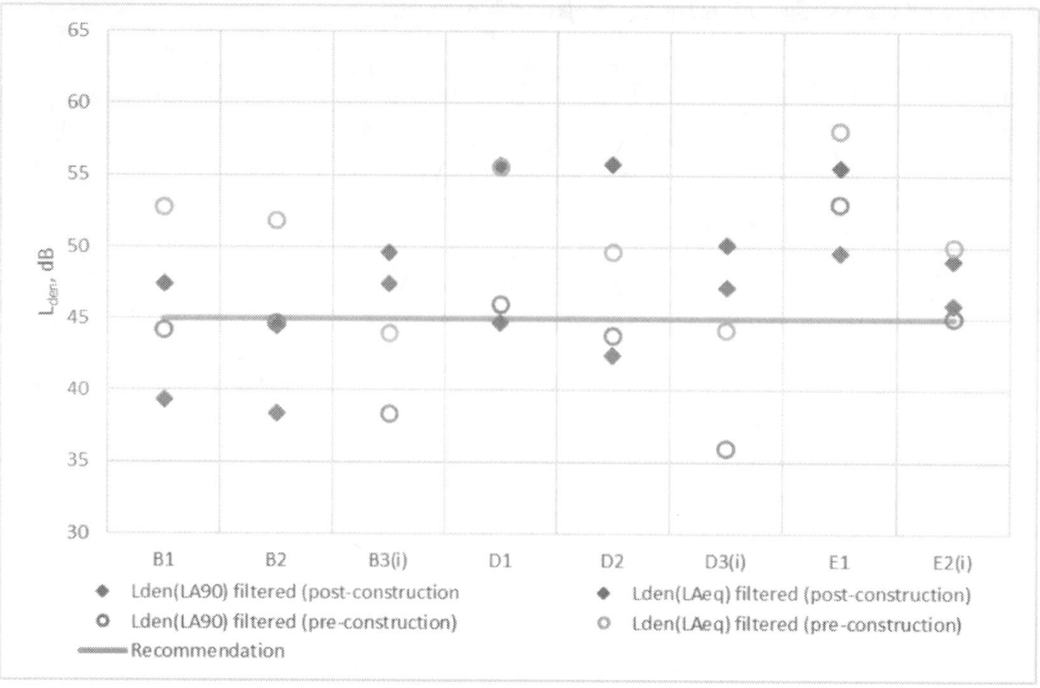

4. Predictive Comparison of the L_{den} Metric with Australian Policy Metrics

4.1 Overview

The L_{den} noise level referenced in the 2018 WHO Guidance is based on noise levels averaged over the day, evening and night period of a year. To assess the relationship between Australian wind farm noise levels and the 2018 WHO Guidance, the L_{den} has been predicted for typical receiver distances for an example wind farm layout, accounting for variations in noise levels occurring over the duration of a year.

The key sources of variation in receiver noise levels which are accounted for in this modelling are:

- the change in sound power level of the turbines for different hub height wind speeds
- the change in the noise propagation from the wind farm to the receiver due to wind direction.

The effect of these variations will vary for different sites, according to the orientation of the receivers relative to the wind farm, the turbine installed at the site and the wind characteristics of the site. The L_{den} predictions have therefore been produced for multiple receiver locations, turbine types and wind distributions.

4.2 Example wind farm layout, noise emission characteristic and wind distributions

The example wind farm layout for the modelling consists of a generic arrangement of turbines extending over a rectangular area of approximately 30 to 35 km^2, aligned along a southwest-northeast direction.

Example sound power level data for three (3) types of multi-megawatt turbines with modelled tip height of 170 m, were sourced from a combination of test and specification data. The turbines were chosen to represent a range of configurations, from turbines which are characterised by an increase to maximum sound power levels over a relatively short wind speed range, to turbines which exhibit a slower rate of increase in sound power level with increasing wind speed. Differences in the frequency spectrum of different turbines do not represent a significant source of variation for this type of analysis. As such, a single representative octave-band spectrum was used to represent all three (3) types of turbines.

To illustrate the characteristics of the selected turbines, Figure 3 presents the profile of the predicted noise levels versus hub height wind speeds, normalised to upper predicted noise levels of 35 and 40 dB for ease of comparison (the range of minimum noise limits applied to wind farms in Australia).

Figure 3: Assessed turbine characteristics – normalised to 35 dB & 40 dB upper predicted receiver levels

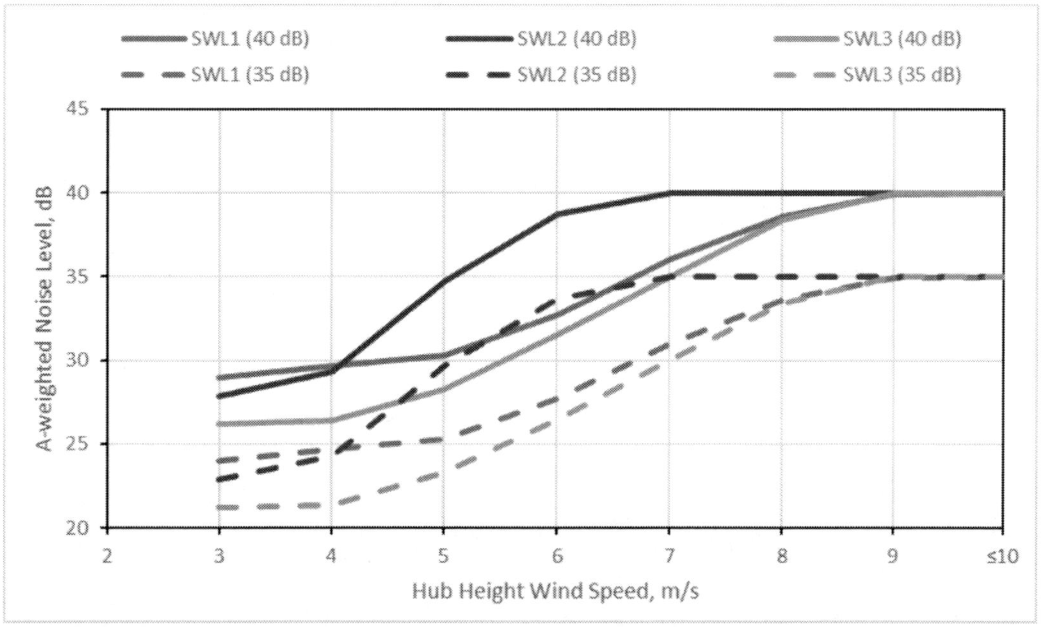

Two example wind data sets consisting of hub wind speed and direction in ten (10) minute intervals over a period of a year were sourced from other Australian wind farm sites. The wind direction and speed distributions are summarised in Figure 4.

Figure 4: Distribution of wind speed and direction for each wind data set

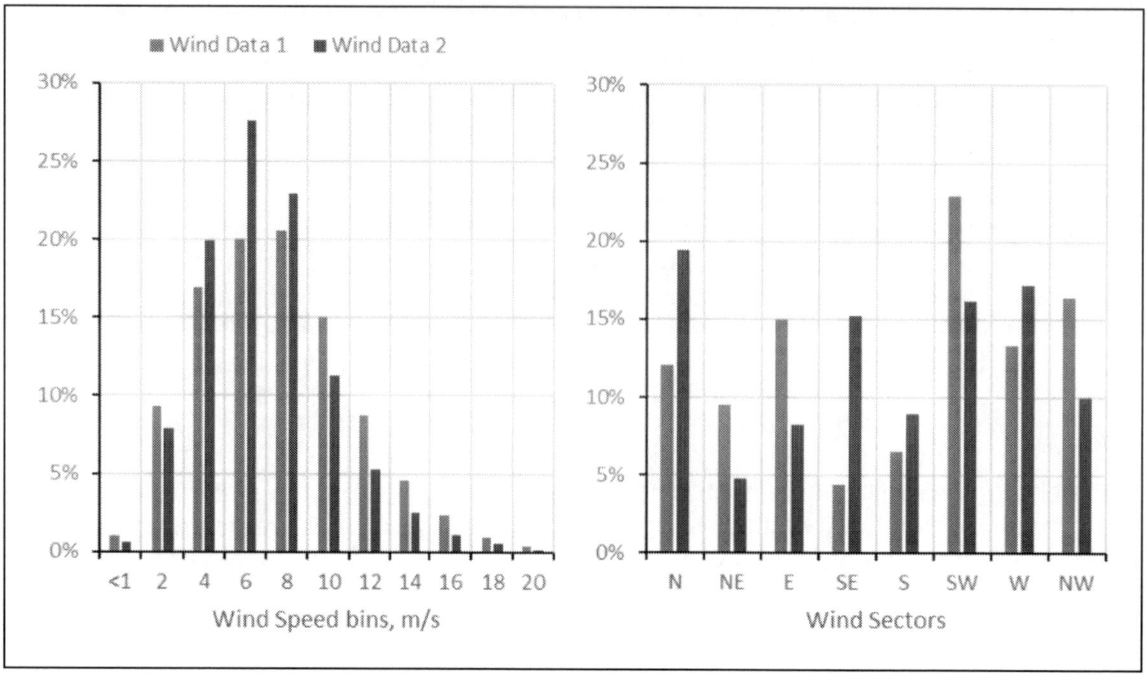

The modelling was carried out for a total of eight (8) receiver locations in various directions within the predicted 35 to 40 dB L_{Aeq} noise contours. The distance and orientation to the example layout are summarised in Table 2.

Table 2: Receiver locations

Location	Distance from nearest turbine (number of tip heights)	Direction from nearest turbine
H1	5	E
H2	8	N
H3	6	S
H4	7	E
H5	9	N
H6	14	N
H7	9	NW
H8	10	E

4.3 Prediction methodology

The noise predictions were calculated using the ISO 9613-2 [12] prediction method, based on the recommendations and adjustments detailed in the UK Institute of Acoustics guidance [13].

The ISO 9613-2 method provides predicted noise levels for conditions that are favourable to the propagation of sound, and a method of calculating a C_{met} correction to determine long term average noise levels accounting for variations in atmospheric conditions. However, the assessment of suitable C_{met} corrections for wind farm noise propagation, accounting for the wide range of wind speeds and directions that can give rise to favourable sound propagation for a wind farm, are not prescriptively defined in ISO 9613-2.

Adjustments to the predicted noise levels for wind direction have therefore been determined using the directional adjustments detailed in the UK Institute of Acoustics guidance. The applicable directional attenuation losses are illustrated in Figure 5.

Figure 5: UK Institute of Acoustics guidance – directional propagation losses vs. wind direction for flat terrain
(0 degrees being upwind propagation)

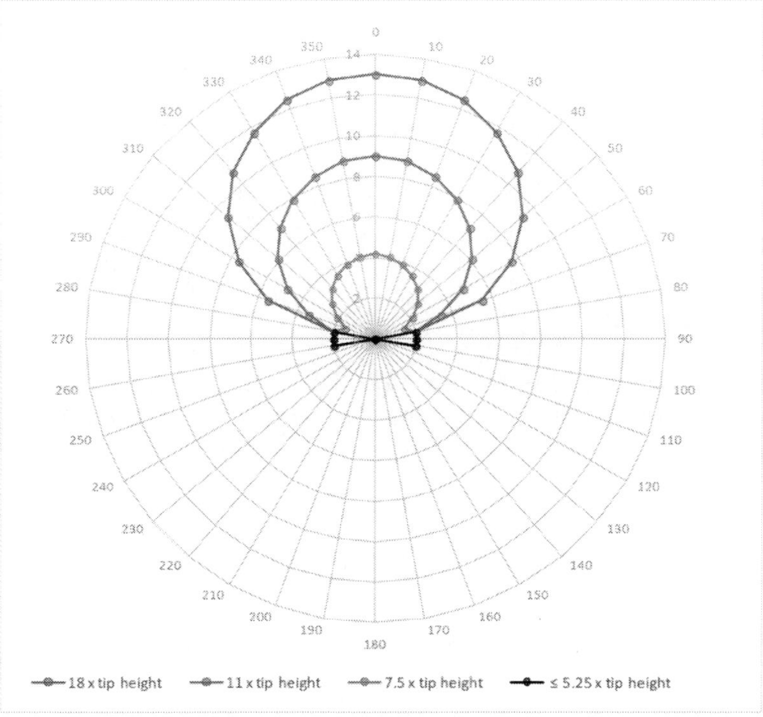

The predicted noise levels for each receiver location were determined for the highest sound power level and wind directions in 5 degree increments. The results were then used to determine the predicted noise level for each receiver location for each 10 minute period of the year, based on the corresponding site wind speed and direction from the two wind data sets. The predicted noise levels for each 10 minute period were then aggregated to calculate the L_{den} noise level for the year.

4.4 Results

The predicted directional noise levels corresponding to the highest sound power levels are presented in Figure 6 for the receiver locations adjusted to an ISO 9613-2 downwind reference level of 40 dB and 35 dB L_{Aeq}.

The directional noise predictions indicate values corresponding to the upper downwind predicted noise levels for around 30 % of wind directions or less. The reduction in noise level between favourable and least favourable noise propagation conditions varies from approximately 3 to 8 dB depending on the receiver location and their position relative to the wind farm.

Figure 6: Directional plot showing predicted noise levels for receivers adjusted to an ISO 9613-2 downwind reference level of 40 dB (left) and 35 dB (right)

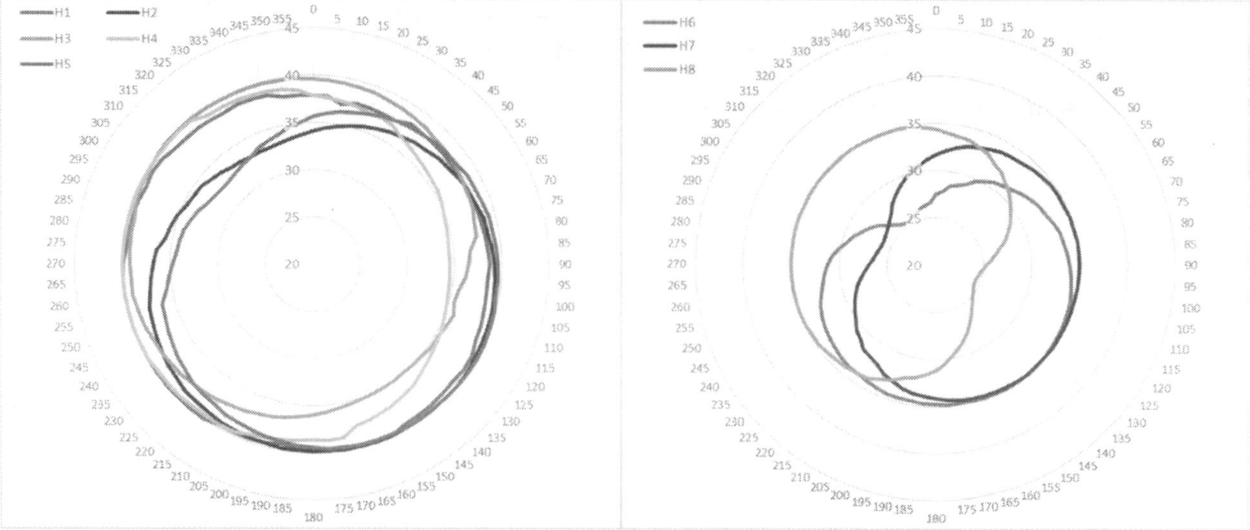

Based on the directional predictions in Figure 6 for the highest sound power levels, the variations in sound power level with wind speed for three (3) turbine types, and the site wind speed and direction distributions, the predicted L_{den} levels are presented in Table 3.

The key findings are summarised as follows:

- The predicted L_{den} levels are below the conditional recommendation level of 45 dB at all locations
- At locations where the highest predicted wind farm noise levels correspond to 40 dB L_{Aeq}, the relationship between the L_{den} and downwind predicted L_{Aeq} ranged from -0.5 to +4.3 dB (average difference of +2.1 dB), with 44.3 dB L_{den} being the highest predicted annual level
- At locations where the highest predicted wind farm noise levels correspond to 35 dB L_{Aeq}, the relationship between the L_{den} and downwind predicted L_{Aeq} ranged from -1.0 to +3.6 dB (average difference of +1.4 dB), with 38.6 dB L_{den} being the highest predicted annual level.

The results indicate that the relationship between the L_{den} and highest downwind L_{Aeq} vary considerably according to the characteristics of the turbine and orientation of the receiver relative to the wind farm site. The latter mainly relates to whether the receiver is in or out of the prevailing downwind direction range, but also relates to the reduction in noise level for unfavourable conditions being dependent on the angle of view of turbines from the receiver location. In all cases, the difference was significantly less than the maximum theoretical difference of +6.4 dB (i.e. assuming a constant noise level equal to the highest predicted noise level).

Table 3: Predicted dB L_{den} accounting for site wind speed and direction distribution

	Wind Data 1 SWL 1	Wind Data 1 SWL 2	Wind Data 1 SWL 3	Wind Data 2 SWL 1	Wind Data 2 SWL 2	Wind Data 2 SWL 3
ISO 9613-2 predicted downwind noise level of 40 dB L_{Aeq}						
H1	43.1	44.3	42.9	42.0	43.9	41.6
H2	41.5	42.8	41.3	40.1	42.3	39.6
H3	42.6	43.8	42.4	41.8	43.4	41.5
H4	42.7	43.8	42.5	41.6	43.4	41.3
H5	41.3	42.6	41.0	40.0	42.1	39.5
ISO 9613-2 predicted downwind noise level of 35 dB L_{Aeq}						
H6	36.1	37.4	35.9	34.5	36.8	34.0
H7	36.0	37.4	35.8	34.9	37.0	34.4
H8	37.5	38.6	37.3	36.7	38.2	36.4

5. Conclusion

WHO health-based guidance on wind farm noise by the is a valuable reference for policy makers, stakeholder groups and the wind industry. The use of the L_{den} metric in the 2018 WHO Guidance provides the benefit of alignment with strategic polices for transportation noise and metrics commonly referenced in large-scale community noise exposure research.

However, the use of L_{den} as an assessment tool for wind farms introduces several practical challenges. The analysis presented in this paper demonstrates that measurement of L_{den} noise levels at receiver locations is problematic because of the effect of ambient noise levels in rural environments being comparable to, or greater than, the conditional recommendation level of 45 dB L_{den}. Further, the use of L_{den} as a measurement metric for wind farm assessment potentially precludes the use of common supplementary analysis techniques that are used to separate wind farm and ambient noise.

This finding supports the commentary in the 2018 WHO Guidance which notes that:

> [...] *it may be concluded that the acoustical description of wind turbine noise by means of L_{den} or L_{night} may be a poor characterization of wind turbine noise and may limit the ability to observe associations between wind turbine noise and health outcomes.*

Consistent with the intent of the conditional recommendations in the 2018 WHO Guidance, the 45 dB L_{den} guideline for wind farms should therefore only be used to inform the development and review of noise policies for wind farms, rather than as a compliance metric for assessing individual projects.

To provide context to the policies used to assess wind farm noise in Australia, measured and predicted L_{den} noise levels for compliant wind farm scenarios were assessed and demonstrated levels below the 45 dB L_{den} conditional recommendation of the 2018 WHO Guidance.

Based on this finding, current Australian noise requirements are consistent with the WHO conditional recommendation.

References

[1] World Health Organization (2018) *Environmental Noise Guidelines for the European Region*

[2] *The* University of Adelaide (2015) *Systematic review of the human health effects of wind farms*

[3] National Health and Medical Research Council (2015) *NHMRC Statement and Information Paper - Evidence on Wind Farms and Human Health*

[4] Australian Medical Association (2014) *Position Statement - Wind Farms and Health*

[5] Janssen SA, Vos H, Eisses AR, Pedersen E (2011). *A comparison between exposure–response relationships for wind turbine annoyance and annoyance due to other noise sources*. J Acoust Soc Am. 130(6):3746–53

[6] Kuwano S, Yano T, Kageyama T, Sueoka S, Tachibanae H (2014). *Social survey on wind turbine noise in Japan*. Noise Control Eng J. 62(6):503–20.

[7] South Australian Environment Protection Authority (2009) *Wind farms environmental noise guidelines*

[8] Standards New Zealand (2010) *NZS 6808:2010 Acoustics – Wind farm noise*

[9] New South Wales Department of Planning and Environment (2016) Wind Energy: Noise Assessment Bulletin

[10] Queensland Department of Infrastructure, Local Government and Planning (20177) *State code 23: Wind farm development*

[11] Griffin, D., Delaire, C., & Pischedda, P. (2013). *Methods of identifying extraneous noise during unattended noise measurements* 20th International Congress of Sound & Vibration

[12] International Organization for Standardization (1996) *ISO 9613-2:1996 Acoustics - Attenuation of sound during propagation outdoors - Part 2: General method of calculation*

[13] UK Institute of Acoustics (2012) *A good practice guide to the application of ETSU-R-97 for the assessment and rating of wind turbine noise*

Institute of Noise Control Engineering/Europe (INCE/Europe)
Riverside House
4 Oakland Vale
New Brighton, Merseyside CH45 1LQ
United Kingdom

ISBN 978-1-5108-8970-5

CURRAN ASSOCIATES INC.
proceedings.com